Test Yourself

Take a quiz at the *Inquiry into Life* Online Learning Center to gauge your mastery of chapter concepts. Each quiz is specially constructed to test your comprehension of key concepts. Feedback on your responses explains why an answer is correct or incorrect.

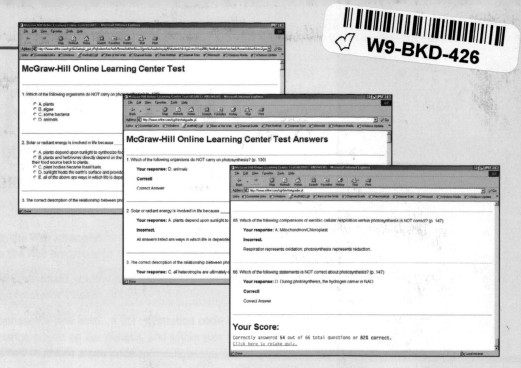

e-Learning Connection

The e-Learning Connection page from the end of each chapter in the text, which correlates relevant technology materials to each A-head, is repeated and expanded on the Online Learning Center.

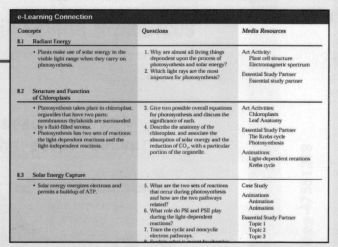

Access to Premium Learning Materials

The *Inquiry into Life* Online Learning Center is your portal to exclusive interactive study tools like McGraw-Hill's Essential Study Partner, BioCourse.com, and PowerWeb.

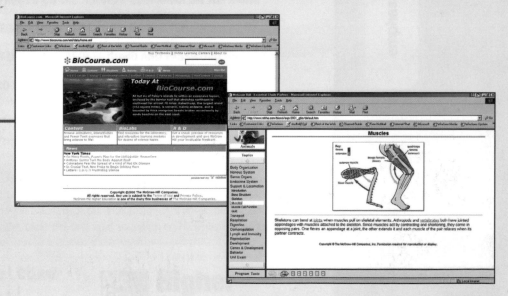

Visit www.mhhe.com/maderinquiry10 today!

Tenth Edition

Inquiry into Life

Sylvia S. Mader

For My Children

Boston   Burr Ridge, IL   Dubuque, IA   Madison, WI   New York   San Francisco   St. Louis
Bangkok   Bogotá   Caracas   Kuala Lumpur   Lisbon   London   Madrid   Mexico City
Milan   Montreal   New Delhi   Santiago   Seoul   Singapore   Sydney   Taipei   Toronto

McGraw-Hill Higher Education

A Division of The **McGraw-Hill** Companies

INQUIRY INTO LIFE, TENTH EDITION

Published by McGraw-Hill, a business unit of The McGraw-Hill Companies, Inc.,
1221 Avenue of the Americas, New York, NY 10020. Copyright © 2003, 2000, 1997 by The
McGraw-Hill Companies, Inc. All rights reserved. No part of this publication may be
reproduced or distributed in any form or by any means, or stored in a database or retrieval
system, without the prior written consent of The McGraw-Hill Companies, Inc., including,
but not limited to, in any network or other electronic storage or transmission, or broadcast
for distance learning.

Some ancillaries, including electronic and print components, may not be available to customers
outside the United States.

 This book is printed on recycled, acid-free paper containing 10% postconsumer waste.

International   1 2 3 4 5 6 7 8 9 0 VNH/VNH 0 9 8 7 6 5 4 3 2
Domestic   1 2 3 4 5 6 7 8 9 0 VNH/VNH 0 9 8 7 6 5 4 3 2

ISBN 0–07–239965–1
ISBN 0–07–115112–5 (ISE)

Publisher: *Martin J. Lange*
Senior sponsoring editor: *Patrick E. Reidy*
Developmental editor: *Margaret B. Horn*
Senior development manager: *Kristine Tibbetts*
Executive marketing manager: *Lisa Gottschalk*
Senior project manager: *Jayne Klein*
Senior production supervisor: *Sandy Ludovissy*
Designer: *K. Wayne Harms*
Cover/interior designer: *Elise Lansdon*
Cover image: *Frans Lanting/Minden Pictures*
Senior photo research coordinator: *Lori Hancock*
Photo research: *Connie Mueller*
Lead supplement producer: *Audrey A. Reiter*
Media technology producer: *Janna Martin*
Compositor: *The GTS Companies*
Typeface: *10/12 Palatino*
Printer: *Von Hoffmann Press, Inc.*

The credits section for this book begins on page C-1 and is considered an extension of the
copyright page.

Library of Congress Cataloging-in-Publication Data

Mader, Sylvia S.
    Inquiry into life / Sylvia S. Mader. — 10th ed.
        p.   cm.
    Includes bibliographical references and index.
    ISBN 0–07–239965–1
    1. Biology.   I. Title.

QH308.2   .M24   2003
570—dc21   2001044531
                                                CIP

INTERNATIONAL EDITION ISBN 0–07–115112–5
Copyright © 2003. Exclusive rights by The McGraw-Hill Companies, Inc., for manufacture
and export. This book cannot be re-exported from the country to which it is sold by McGraw-Hill.
The International Edition is not available in North America.

www.mhhe.com

Brief Contents

Contents

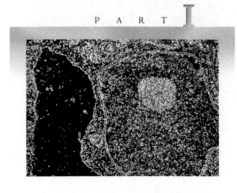

PART III

Maintenance of the Human Body 192

PART IV

Integration and Control of the Human Body 316

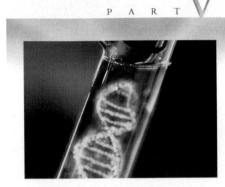
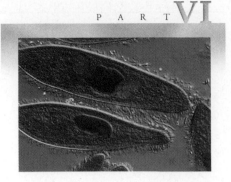

PART VII

Behavior and Ecology 660

Readings

Bioethical Focus

Ecology Focus

Health Focus

Science Focus

Preface

Inquiry into Life is written for the introductory-level student who would like to develop a working knowledge of biology. While the text covers the whole field of general biology, it emphasizes the application of this knowledge to human concerns. Along with this approach, concepts and principles are stressed, rather than detailed, high-level scientific data and terminology. Each chapter presents the topic clearly, simply, and distinctly so that the student can achieve a thorough understanding of basic biology. As with previous editions, the central themes of Inquiry into Life are understanding the workings of the human body and how humans fit into the world of living things.

Pedagogical Features

Educational theory tells us that students are most interested in knowledge of immediate practical concern. This text is consistent with and remains true to this approach. The running text and the readings stress applications to everyday life. Such topics as eating disorders, allergies, stem cell research, cloning of animals, and xenotransplantation are integral to the chapter.

As before, the chapter outline numbers the major sections of the chapter. New to this edition, instead of listing statements, students are asked questions according to the chapter outline. The questions are designed to stimulate inquisitive thinking and develop better study skills. The general nature of these questions will encourage students to concentrate on concepts and how concepts relate to one another.

The numbering system in the chapter outline is continued throughout the chapter and is repeated in the summary so that instructors can assign just certain portions of the chapter, if they like. The text is paged so that major sections begin at the top of a page and illustrations are on the same or facing page as its reference.

Each chapter begins with a vignette, a short story that applies chapter material to a real-life situation. To increase student interest, I have moved the opening vignette to the chapter opening page where it is accompanied by a photograph.

The summary at the end of the chapter also assists students in learning the concepts discussed in the chapter. New to this edition, "Testing Yourself" has been expanded to include fifteen to twenty objective style questions. "Studying the Concepts," which are essay-type questions, and "Thinking Scientifically," which are critical thinking questions, have been moved to the Online Learning Center. The last page of each chapter instructs student how best to use the Online Learning Center for that chapter.

Boxed Readings

Inquiry into Life has four types of readings. "Health Focus" readings review procedures and technology that can contribute to our well being. "Science Focus" readings describe how experimentation and observations have contributed to our knowledge about the living world. "Ecology Focus" readings show how the concepts of the chapter can be applied to ecological concerns. "Bioethical Focus" readings describe a modern situation that calls for a value judgement on the part of the reader. Students are challenged to develop a point of view by answering a series of questions that pertain to the issue.

The text contains many new Focus readings. Students will be delighted to consider "Do Animals Have Emotions?" They will be fascinated to know that the Human Genome Project will result in "New Cures on the Horizon," and intrigued by "The United States Population," a reading based on the 2000 census. New bioethical issues include the use of "Stem Cells" to treat our ills, and whether we should allow "Cloning of Humans." At adopters' requests, the bioethical focus readings now bear a title, which will facilitate their use for classroom discussions.

Revised Chapters

Inquiry into Life remains forever new and vital because it is revised from the first to the last page of every edition. These changes may be of special interest:

The introductory chapter, "The Study of Life" was rewritten to strengthen the presentation of the characteristics of life, the organization of the biosphere, and the scientific process. Feedback from many adopters allowed me to vastly improve the scientific process section so that it is useful to everyone.

In Part I, Cell Biology, Chapter 5 opens with a more lucid discussion of the cell cycle and its relevance to cancer. In Chapter 7, "Cellular Metabolism," the first illustration became an icon for the revised illustrations in the chapter.

In Part II, Plant Biology, all three chapters were rewritten. Icons are now strategically placed in Chapter 8, "Photosynthesis." Instructors will be especially pleased with the rewrite of Chapter 9, Plant Organization, which is more thorough than before. Chapter 10, now entitled "Plant Reproduction, Growth, and Development" was reorganized, and the plant hormone discussion was improved.

In Parts III and IV, the systems chapters were fine-tuned and the illustrations were improved to better present the concepts. Students should have no difficulty in following the text, understanding the concepts, and applying them to their everyday lives.

In Part V, Continuance of the Species, the genetics chapters received special attention. The new illustrations in this part will make it easier for students to study Punnett squares. New to this edition, all chromosome mutations now have an accompanying illustration.

In Part VI, Evolution and Diversity, the three-domain system of classification has replaced the five-kingdom system. The classification tables throughout these chapters have been revised to better assist students in learning the major groups of organisms. The evolution of humans was revised to reflect the latest findings.

In Part VII, Behavior and Ecology, a new chapter called "Ecosystems and Human Interferences" introduces the basics of ecology and shows how human activities have altered biogeochemical cycles to our own detriment. Another new chapter entitled "Conservation of Biodiversity" closes the text. We all need to be aware that other living things are valuable to the human species and to recognize that our activities threaten their very existence. In preserving other species, we are ultimately preserving our own species.

Online Learning Center

Students can utilize many technological resources in order to understand the content of this textbook. In addition to end of chapter questions and the printed study guide, the Online Learning Center at www.mhhe.com/maderinquiry10 contains readings, quizzes, animations, and other activities to help students master the concepts.

Each chapter in this new edition ends with an e-Learning Connection page. This page organizes the relevant online study material by major sections, helping to create a stronger association between available study activities and text material. Because this design is repeated on the Online Learning Center, the student can now easily find the appropriate learning experience.

New to this edition, the Careers pages from the previous edition, the "Studying the Concepts" questions, and the "Thinking Scientifically" questions have been moved to the Online Learning Center. Also, Further Readings now are a part of the Online Learning Center.

A complete explanation of the technology package available for students and instructors with this textbook is explained fully on pages xii through xiii of the preface.

New to This Edition

- Online study aids are organized according to the major sections of the chapter on the e-Learning Connection page found at the end of each chapter. In this way, students can easily determine the available resources that help explain difficult concepts. The same design is utilized at the Online Learning Center, and this allows students to quickly find an activity of interest.
- The chapter opening page has been revised. The chapter outline now contains questions listed according to the major sections, instead of statements. The questions are designed to start students thinking about and learning the concepts. The opening vignette now appears on this page and is accompanied by a photograph.
- Although all chapters were revised, some changes are of particular interest. Illustrations in Part I and II have new icons to assist student learning. Also, the plant chapters have been completely revised and are now more thorough. The genetics chapters have

been rewritten and are supported by new art. Two new chapters—"Ecosystems and Human Interferences" and "Conservation of Biodiversity"—strengthen the ecology section of the text.
- The classification system of the text has been modernized. The three-domain system of classification based on molecular biology replaces the five-kingdom system of classification based on structure and adaptations to the environment.
- The revised illustration program adds vitality to the art and enhances the appeal of the text. Many new micrographs provide realism. "Visual Focus" illustrations give a pictorial overview of key topics. Color coding is used for both molecular structures and for human tissues and organs.
- Relevancy of the text is increased with the inclusion or expanded treatment of topics such as eating disorders, allergies, stem cell research, hepatitis infections, xenotransplantation, human cloning, the human genome project, and gene therapy to treat cancer.

The Reviewers

Many instructors have contributed not only to this edition of *Inquiry into Life* but also to previous editions. I am extremely thankful to each one, for they have all worked diligently to remain true to our calling to provide a product that will be the most useful to our students.

In particular, it is appropriate to acknowledge the following individuals for their help on the tenth edition:

John Vincent Aliff
Georgia Perimeter College

Judith Aronow
Community College of Vermont

Tammy Atchison
Pitt Community College

Harvey Babich
Stern College for Women

Iona Baldridge
Lubbock Christian University

Laura M. Barden-Gabbei
Western Illinois University

Sarah Follis Barlow
Middle Tennessee University

James E. Bidlack
University of Central Oklahoma

Peter M. Bradley
Worcester State College

Paula Riley Burch
University of North Carolina–Greensboro

Juanita E. Cheek
Northwest Mississippi Community College

Barbara Christie-Pope
Cornell College

Pamela Anderson Cole
Shelton State Community College

Jerry L. Cook
Sam Houston State University

Tamara J. Cook
Sam Houston State University

Ronald R. Crawford
Indiana Wesleyan University

Shirley A. Crawford
SUNY–Morrisville

Forbes Davidson
Mesa State College

Clemetine A. de Angelis
Tarrant County College, South Campus

Diane M. Dixon
Southeastern Oklahoma University

Deborah A. Donovan
Western Washington University

David Emmitt
Cuyahoga Community College, East Campus

Mark Fairbrass
Georgia Military College

Dale Fogelsanger
Messiah College

Carolyn Elaine Ford
Saint Xavier University

Charlene L. Forest
Brooklyn College of SUNY

Stephanie S. Freese
Northeast Mississippi Community College

Geoffrey A. Fuller
Mount Vernon Nazarene College

Terry E. Graham
Worcester State College

Michael F. Gross
Georgian Court College

Peggy J. Guthrie
University of Central Oklahoma

James R. Hampton
Salt Lake Community College

T. M. (Mike) Hardig
University of Montevallo

Alice Long Heikens
Franklin College

William F. Hibschman
Harford Community College

Richard T. Hurley
University of St. Francis

Joseph E. Jaworski
Spring Arbor College

Beverly Knauper
University of Cincinnati

Thomas Kozel
Anderson College

Siu-Lam Lee
University of Massachusetts–Lowell

Brenda G. Leicht
University of Iowa

Kelly M. Mack
University of Maryland Eastern Shore

Kimberly Roe Maznicki
Seminole Community College

Delores McCright
Texarkana College

David E. McMillin
Clark Atlanta University

Alfred P. McQueen, Sr.
Hampton University

Carrie McVean-Waring
Mesa State College

Kimberly H. O'Donnell
Wellesley College

James O'Leary
South Suburban College

Anthony L. Palombella
Longwood College

Brenda N. Peirson
Louisiana College

Vera M. Piper
Lord Fairfax Community College

Mary E. (Betti) Pischel
Tarrant County College

William J. Pohley
Franklin College

Patricia B. Ravenell
Norfolk State University

Ramona G. Rice
Georgia Military College

Lori Ann Rose
Sam Houston State University

Brian L. Sailer
Sam Houston State University

John Richard Schrock
Emporia State University

Tracy S. Schwab
Madonna University

Mark L. Secord
Coastal Bend County College

Daniel L. Shea
Eastern Nazarene College

Patricia A. Shields
George Mason University

Gregory Sievert
Emporia State University

Thomas E. Snowden
Florida Memorial College

Amy C. Sprinkle
Jefferson Community College

James R. Sprinkle
Northeastern Illinois University

Sumesh Thomas
Baltimore City Community College

Paula L. Thompson
Florida Community College, North Campus

Ariella Olivos Vader
Messiah College

Joseph M. Wahome
Mississippi Valley State University

Kathy Webb
Bucks County Community College

Jamie D. Welling
South Suburban College

Melinda Scholl Wilder
Eastern Kentucky University

Lance R. Williams
Louisiana College

Ellen Young
University of North Carolina

TEACHING AND LEARNING SUPPLEMENTS

McGraw-Hill offers a variety of tools and technology products to support the tenth edition of *Inquiry into Life*. Students can order supplemental study materials by contacting the McGraw-Hill Customer Service Department at (800) 338-3987. Instructors can obtain teaching aids by calling the Customer Service Department or by contacting your local McGraw-Hill sales representative.

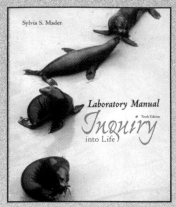

Inquiry into Life Laboratory Manual

The *Inquiry into Life Laboratory Manual*, written by Dr. Sylvia Mader, has an accompanying laboratory exercise for most chapters in the text. Every laboratory has been written to help students learn the fundamental concepts of biology and the specific content of the chapter to which the lab relates, as well as gain a better understanding of the scientific method. The *Laboratory Resource Guide* is now available on the Instructor Center of the Online Learning Center at www.mhhe.com/maderinquiry10.

Student Study Guide

Dr. Sylvia Mader has written the *Student Study Guide* that accompanies the tenth edition of *Inquiry into Life*. Each text chapter has a corresponding study guide chapter that includes a chapter review, study questions for each section of the chapter, a chapter test, and thought-provoking essay questions. Answers for all questions are provided to give students immediate feedback.

Transparencies

Every piece of line art in the textbook is included, with better visibility and contrast than ever before. Labels are large and bold for clear projection.

100 Micrograph Slides

This set contains 35mm slides of many of the photomicrographs and electron micrographs in the text.

Computerized Test Bank

Available on CD-ROM in both Mac and Windows platforms, this test bank utilizes Brownstone Diploma® testing software. This user-friendly program allows instructors to search for questions using multiple criteria, edit or add questions, and scramble questions to create customized exams.

Digital Content Manager

This multimedia collection of visual resources allows instructors to utilize artwork from the text in multiple formats to create customized classroom presentations, visually based tests and quizzes, dynamic course website content, or attractive printed support materials. The digital assets on this cross-platform CD-ROM are grouped by chapter within the following easy-to-use folders:

- **Active Art Library.** Illustrations depicting key processes have been converted to a format that allows each figure to be broken down to its core elements, thereby allowing the instructor to manipulate the art and adapt the figure to meet the needs of the lecture environment
- **Animations Library.** Harness the visual impact of key physiological processes in motion by importing these full-color animations into classroom presentations or course websites.
- **Art Libraries.** Full color digital files of all illustrations in the book, plus the same art saved in unlabeled and gray scale version, can be readily incorporated into lecture presentations, exams, or custom-made class-room materials.
- **Tables Library.** Every table that appears in the text is provided in electronic format.
- **PowerPoint Lecture Outlines.** A ready-made presentation that combines lecture notes and art is written for each chapter. They can be used as they are, or the instructor can tailor them to preferred lecture topics and sequences.
- **PowerPoint Art Slides.** Art, photographs, or tables from each chapter that have been pre-inserted into blank PowerPoint slides

Interactive Laboratories and Biological Simulations, or iLaBS, teach students real-life biomolecular applications and techniques using fun, web-based programs. These labs provide students the opportunity for repetitive practice in techniques that they would be limited to doing only once in a normal lab setting. iLaBS also allow students to virtually perform time-consuming or hazardous techniques that they would not otherwise be able to experience.

Online Learning Center

The *Inquiry into Life* Online Learning Center (OLC) at www.mhhe.com/maderinquiry10 offers access to a vast array of premium online content to fortify the learning and teaching experience.

Student Center. The Student Center of the OLC features the e-Learning Connection from the end of each chapter in the textbook. This page, which correlates online study tools such as quizzes and interactive activities to each section of the chapter, is expanded on the OLC. In addition, the following online resources are available:

- **Essential Study Partner** A collection of interactive study modules that contains hundreds of animations, learning activities, and quizzes designed to help students grasp complex concepts.
- **PowerWeb** An online supplement that offers access to current course-specific articles, news, research links, journals, and much more.

- **BioCourse.com** Accessed through the OLC, this site for students and instructors provides an exhaustive set of up-to-date resources pertaining to the life sciences.
- **Online Tutoring** A 24-hour tutorial service moderated by qualified instructors. Help with difficult concepts is only an email away!

- **BioLabs** Students can master skills vital to success in the laboratory by using these online simulations.

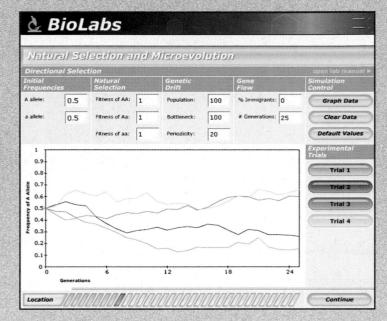

Instructor Center. The Instructor Center is an online repository for teaching aids. It houses downloadable and printable versions of traditional ancillaries plus a wealth of online content.

- **Instructor's Manual** This resource provides learning objectives, lecture outlines, lecture enrichment topics, technology resources, and essay questions with the answers.
- **Laboratory Resource Guide** A preparation guide that provides set-up instructions, sources for materials and supplies, time estimates, special requirements, and suggested answers to all questions in the laboratory manual.
- **PageOut** McGraw-Hill's exclusive tool for creating your own website for your general biology course. It requires no knowledge of coding and is hosted by McGraw-Hill.
- **Course Management System** OLC content is readily compatible with online course management software such as WebCT and Blackboard. Contact your local McGraw-Hill sales representative for details.

Life Science Animations Library 3.0 CD-ROM

This CD-ROM contains over 600 full-color animations of biological concepts and processes. Harness the visual impact of processes in motion by importing these files into classroom presentations or online course materials.

GUIDED TOUR

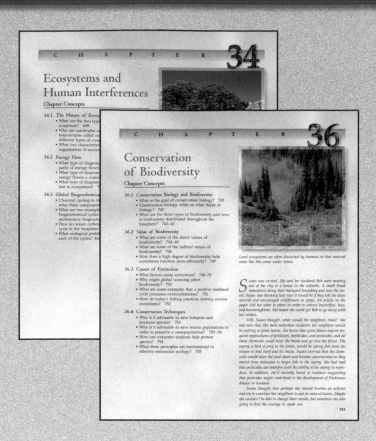

New to *Inquiry into Life,* Tenth Edition!

New Chapters in Ecology Section

Chapter 34, Ecosystems and Human Interferences, introduces the basics of ecology and shows how human activities have altered biogeochemical cycles.

Chapter 36, Conservation of Biodiversity, demonstrates that by preserving other species, we ultimately preserve our own species.

Botany Chapters Extensively Revised

New icons throughout Chapter 8 help clarify material for students. Chapter 9 has been rewritten, with new sections and new art. Chapter 10 was reorganized, and plant hormones are better represented.

Time-Proven Features That Will Enhance Your Understanding of Biology

Chapter Concepts

The chapter outline contains questions to encourage students to concentrate on concepts and how concepts relate to one another.

Opening Vignette

A short, thought-provoking vignette applies chapter material to a real-life situation.

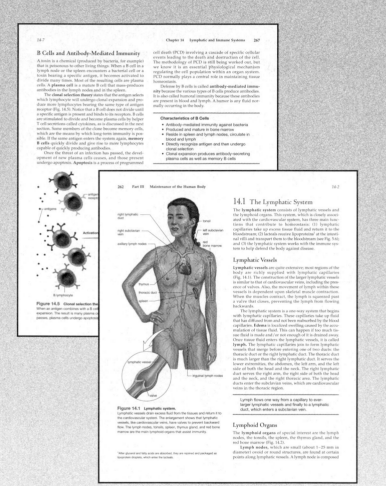

Internal Summary Statements

A summary statement appears at the end of each major section of the chapter to help students focus on the key concepts.

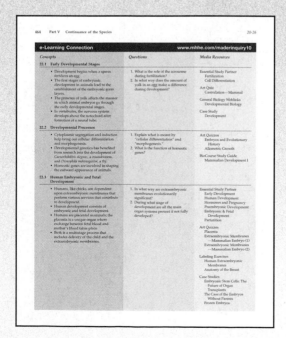

e-Learning Connection

Each chapter ends with an e-Learning Connection page, which organizes relevant online study materials by major sections. This page is repeated and expanded on the Online Learning Center at www.mhhe.com/maderinquiry10, where a click of the mouse takes you to a specific study aid.

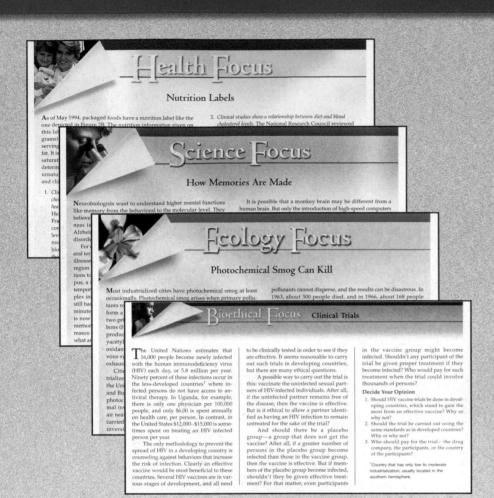

Readings

Inquiry into Life has four types of boxed readings.

- Health Focus readings review procedures and technology that can contribute to our well-being.
- Science Focus readings describe how experimentation and observations have contributed to our knowledge about the living world.
- Ecology Focus readings show how the concepts of the chapter can be applied to ecological concerns.
- Bioethical Focus readings describe modern situations that call for value judgments and challenge students to develop a point of view.

Chapter Summary

The summary is organized according to the major sections in the chapter and helps students review the important concepts and topics.

Summarizing the Concepts

14.1 The Lymphatic System

The lymphatic system consists of lymphatic vessels and lymphoid organs. The lymphatic vessels receive lipoproteins at intestinal villi and excess tissue fluid at blood capillaries, and carry these to the bloodstream.

Lymphocytes are produced and accumulate in the lymphoid organs (red bone marrow, lymph nodes, tonsils, spleen, and thymus gland). Lymph is cleansed of pathogens and/or their toxins in lymph nodes, and blood is cleansed of pathogens and/or their toxins in the spleen. T lymphocytes mature in the thymus, while B lymphocytes mature in the red bone marrow where all blood cells are produced. White blood cells are necessary for nonspecific and specific defenses.

14.2 Nonspecific Defenses

Immunity involves nonspecific and specific defenses. Nonspecific defenses include barriers to entry, the inflammatory reaction, natural killer cells, and protective proteins.

14.3 Specific Defenses

Specific defenses require B lymphocytes and T lymphocytes, also called B cells and T cells. B cells undergo clonal selection with production of plasma cells and memory B cells after their antigen receptors combine with a specific antigen. Plasma cells secrete antibodies and eventually undergo apoptosis. The IgG antibody is a Y-shaped molecule that has two binding sites for a specific antigen. Memory B cells remain in the body and produce antibodies if the same antigen enters the body at a later date.

T cells are responsible for cell-mediated immunity. The two main types of T cells are cytotoxic T cells and helper T cells. Cytotoxic T cells kill virus-infected or cancer cells on contact because they bear a nonself protein. Helper T cells produce cytokines and stimulate other immune cells. Like B cells, each T cell bears antigen receptors. However, for a T cell to recognize an antigen, the antigen must be presented by an antigen-presenting cell (APC), usually a macrophage, along with an HLA (human leukocyte-associated antigen). Thereafter, the activated T cell undergoes clonal expansion until the illness has been stemmed. Then most of the activated T cells undergo apoptosis. A few cells remain, however, as memory T cells.

14.4 Induced Immunity

Active (long-lived) immunity can be induced by vaccines when a person is well and in no immediate danger of contracting an infectious disease. Active immunity is dependent upon the presence of memory cells in the body.

Passive immunity is needed when an individual is in immediate danger of succumbing to an infectious disease. Passive immunity is short-lived because the antibodies are administered to and not made by the individual.

Cytokines, including interferon, are used in attempts to treat AIDS and to promote the body's ability to recover from cancer.

Monoclonal antibodies, which are produced by the same plasma cell, have various functions, from detecting infections to treating cancer.

14.5 Immunity Side Effects

Allergic responses occur when the immune system reacts vigorously to substances not normally recognized as foreign. Immediate allergic responses, usually consisting of coldlike symptoms, are due to the activity of antibodies. Delayed allergic responses, such as contact dermatitis, are due to the activity of T cells. Immune side effects also include blood-type reactions, tissue rejection, and autoimmune diseases.

Objective Test Questions

A full page of challenging objective questions now closes each chapter.

The Study of Life

Chapter Concepts

Biologists study life in the laboratory, on land, in the sea, and in the air. Here, a biologist swims among colorful fish while studying all the different organisms in a coral reef.

You can find a biologist almost anywhere—beneath the sea studying a coral reef or high among the trees in a tropical rain forest; among the tall grasses of a prairie or in the muck of a delta. Biologists study life wherever it occurs. They want to know how living things interact and survive, with the intent of preserving life for the health of the planet and for the benefit of human beings. Plants and animals give us our food, clothing, shelter, and medicines. They also provide us with invaluable ecological services, including taking pollutants out of the air we breathe and the water we drink.

You can also find a biologist in the laboratory among test tubes and all sorts of high-powered instruments. Due to a superhuman effort, biologists now know the exact makeup of all the genes that determine who we are—what we look like and why we have illnesses such as heart disease and cancer. This magnificent breakthrough will allow the development of new drugs to treat our ills. Gene therapy will even cure us or keep us from passing on illnesses to our children. Still, it also raises many ethical questions. Should we, as a society, permit gene therapy just to improve the looks and intelligence of our offspring?

Figure 1.1 **Living things on planet Earth.**
If aliens ever visit our corner of the universe, they will be amazed at the diversity of life on our planet. They will also note that it is the only known place where liquid water is present at the surface—and in huge quantities. This is one of the criteria for the origin of life and its diversity as we know it.

1.1 The Characteristics of Life

Life. Except for the most desolate and forbidding regions of the polar ice caps, planet Earth is teeming with life. Without life, our planet would be nothing but a barren rock hurtling through silent space. The variety of life on earth is staggering. Human beings are a part of it. So are insects, trees, fish, humans, mushrooms, and giraffes. (Fig. 1.1). The variety of living things ranges from unicellular paramecia, much too small to be seen by the naked eye, all the way up to giant sequoia trees that can reach heights of three hundred feet or more.

The variety of life seems overwhelming, yet despite its great diversity, all living things have certain characteristics in common. Taken together, these give us insight into the nature of life and help us distinguish living things from nonliving things. Living things (1) are organized, (2) acquire materials and energy, (3) reproduce, (4) respond to stimuli, (5) are homeostatic, (6) grow and develop, and (7) are adapted to their environment.

Living Things Are Organized

Living things have levels of biological organization. In trees and humans (Fig. 1.2) and all other living things, atoms join together to form molecules, such as the DNA molecules that occur within cells. The **cell** is the lowest level of biological organization to have the characteristics of life. A nerve cell is an example of the types of cells in the human body. A **tissue** is a group of similar cells that perform a particular function. Nervous tissue is composed of millions of nerve cells that transmit signals to all parts of the body. Several tissues join together to form an **organ.** The main organs in the nervous system are the brain, the spinal cord, and the nerves. Organs work together to form an **organ system.** The brain sends messages to the spinal cord, which in turn, sends them to body parts by way of the spinal nerves. Complex **organisms** such as trees and humans are a collection of organ systems. The human body contains a digestive system, a cardiovascular system, and several other systems in addition to a nervous system.

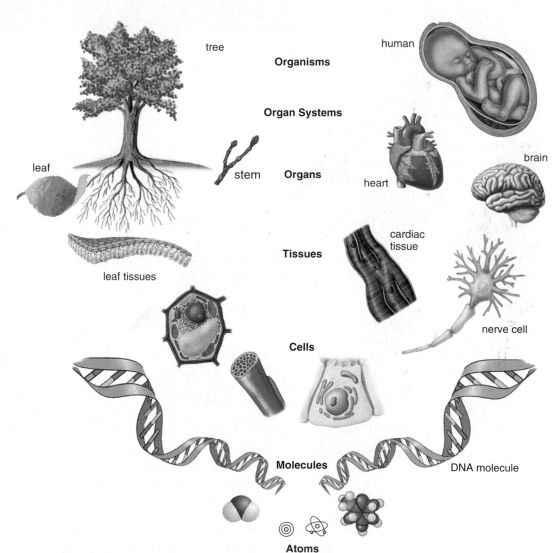

Figure 1.2 **Levels of biological organization.**

At the lowest level of organization, atoms join to form the molecules that are found in cells. Cells, the smallest units of life, differ in size, shape, and function. Tissues are groups of cells that have the same function. Organs, which are composed of different tissues, make up the organ systems of complex organisms. Each organ and organ system performs functions that keep the organism alive and well.

tree

Organisms

human

Organ Systems

Organs

brain

leaf

stem

heart

leaf tissues

Tissues

cardiac tissue

nerve cell

Cells

Molecules

DNA molecule

Atoms

Figure 1.3 Living things acquire materials and energy, and they reproduce.
A red-tailed hawk has captured a rabbit, which it is feeding to its young.

Living Things Acquire Materials and Energy

Living things cannot maintain their organization or carry on life's other activities without an outside source of materials and energy. Photosynthesizers, such as trees, use carbon dioxide, water, and solar energy to make their own food. Human beings and other animals, such as red-tailed hawks (Fig. 1.3), acquire materials and energy when they eat food.

Food provides nutrient molecules, which are used as building blocks or for energy. **Energy** is the capacity to do work, and it takes work to maintain the organization of the cell and of the organism. When nutrient molecules are used to make their parts and products, cells carry out a sequence of synthetic chemical reactions. Some nutrient molecules are broken down completely to provide the necessary energy to carry out synthetic reactions.

Most living things can convert energy into motion. Self-directed movement, as when we decide to rise from a chair, is even considered by some to be a characteristic of life.

Living Things Reproduce

Life comes only from life. The presence of **genes** in the form of DNA molecules allows cells and organisms to **reproduce**—that is, to make more of themselves (Fig. 1.3). DNA contains the hereditary information that directs the structure of the cell and its **metabolism,** all the chemical reactions in the cell. Before reproduction occurs, DNA is replicated so that exact copies of genes are produced.

Unicellular organisms reproduce asexually simply by dividing. The new cells have the same genes and structure as the single parent. Multicellular organisms usually reproduce sexually. Each parent, male and female, contributes roughly one-half the total number of genes to the offspring, which then does not resemble either parent exactly.

Living Things Respond to Stimuli

Living things respond to external stimuli, often by moving toward or away from a stimulus, such as the sight of food. Movement in animals, including humans, is dependent upon their nervous and musculoskeletal systems. Other living things use a variety of mechanisms in order to move. The leaves of plants track the passage of the sun during the day, and when a houseplant is placed near a window, its stem bends to face the sun.

The movement of an organism, whether self-directed or in response to a stimulus, constitutes a large part of its **behavior.** Behavior is largely directed toward minimizing injury, acquiring food, and reproducing (Fig. 1.3).

Living Things Are Homeostatic

Homeostasis means "staying the same." Actually, the internal environment stays *relatively* constant; for example, the human body temperature fluctuates slightly during the day. Also, the body's ability to maintain a normal internal temperature is somewhat dependent on the external temperature—we will die if the external temperature becomes overly hot or cold.

Figure 1.4   Living things grow and develop.
A small acorn gives rise to a very large oak tree by the process of growth and development.

All organ systems contribute to homeostasis. The digestive system provides nutrient molecules; the cardiovascular system transports them about the body; and the urinary system rids blood of metabolic wastes. The nervous and endocrine systems coordinate the activities of the other systems. One of the major purposes of this text is to show how all the systems of the human body help to maintain homeostasis.

Living Things Grow and Develop

Growth, recognized by an increase in the size and often the number of cells, is a part of development. In humans, **development** includes all the changes that take place between conception and death. First, the fertilized egg develops into a newborn, and then a human goes through the stages of childhood, adolescence, adulthood, and aging. Development also includes the repair that takes place following an injury.

All organisms undergo development. Figure 1.4 illustrates that an acorn progresses to a seedling before it becomes an adult oak tree.

Living Things Are Adapted

Adaptations are modifications that make an organism suited to its way of life. Consider, for example, a hawk (see Fig. 1.3), which catches and eats rabbits. A hawk can fly in part because it has hollow bones to reduce its weight and flight muscles to depress and elevate its wings. When a hawk dives, its strong feet take the first shock of the landing and its long, sharp claws reach out and hold onto the prey.

Adaptations come about through evolution. **Evolution** is the process by which a **species** (group of similarly constructed organisms that successfully interbreed) changes through time. When a new variation arises that allows certain members of the species to capture more resources, these members tend to survive and to have more offspring than the other, unchanged members. Therefore, each successive generation will include more members with the new variation. In the end, most members of a species have the same adaptation to their environment.

Evolution, which has been going on since the origin of life and will continue as long as life exists, explains both the unity and the diversity of life. All organisms share the same characteristics of life because their ancestry can be traced to the first cell or cells. Organisms are diverse because they are adapted to different ways of life.

Living things share certain characteristics. Evolution explains this unity of life and also its diversity, which is discussed next.

1.2 The Classification of Living Things

Since life is so diverse, it is helpful to have a classification system to group organisms according to their similarities. **Taxonomy** is the discipline of identifying and classifying organisms according to specific criteria. Each type of organism is placed in a species, genus, family, order, class, phylum, kingdom, and finally domain. (The category division, instead of phylum, is used for plants.) See Table 1.1 for the classification of humans.

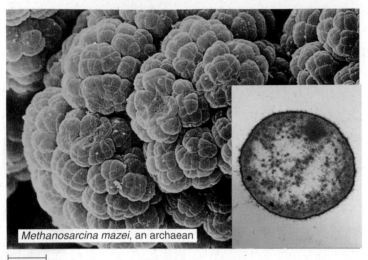

Methanosarcina mazei, an archaean

1.6 µm

a. Domain Archaea
Prokaryotes, capable of living in extreme environments.

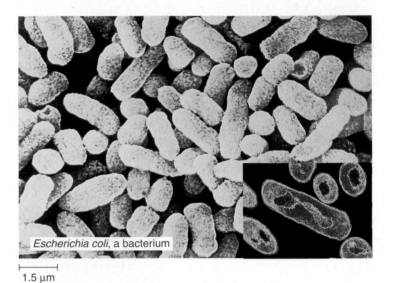

Escherichia coli, a bacterium

1.5 µm

b. Domain Bacteria
Prokaryotes, structurally similar but metabolically diverse.

Kingdom	Organization	Type of Nutrition	Representative Organisms				
Protista	Complex single cell, some multicellular	Absorb, photosynthesize, or ingest food	paramecium	euglenoid	slime mold	dinoflagellate	Protozoans, algae, water molds, and slime mold
Fungi	Some unicellular, most multicellular filamentous forms with specialized complex cells	Absorb food	black bread mold	yeast	mushroom	bracket fungus	Molds, yeast, and mushrooms
Plantae	Multicellular form with specialized complex cells	Photosynthesize food	moss	fern	pine tree	nonwoody flowering plant	Mosses, ferns, nonwoody and woody flowering plants
Animalia	Multicellular form with specialized complex cells	Ingest food	coral	earthworm	blue jay	squirrel	Invertebrates, fishes, reptiles, amphibians, birds, and mammals

c. Domain Eukarya
Eukaryotes, structurally diverse and organized into the four kingdoms depicted here.

Figure 1.5 Pictorial representation of the three domains of life.
The (**a**) archaea and (**b**) the bacteria are both prokaryotes but are so biochemically different that they are not believed to be closely related. (A transmission electron micrograph is an inset in a scanning electron micrograph.) **c.** The eukaryotes are biochemically similar but structurally dissimilar. Therefore, they have been categorized into four kingdoms. Many protists are unicellular, but the other three kingdoms are characterized by multicellular forms.

Domains

It has been common practice the past few years to recognize five kingdoms (called Monera, Protista, Fungi, Plantae, and Animalia), but biochemical evidence suggests that we should now recognize three higher categories called domains (Fig. 1.5). Both domain **Archaea** and domain **Bacteria** contain unicellular prokaryotes (*pro*, before; *karya*, nucleus), which lack the membrane-bounded nucleus found in the cells of eukaryotes (*eu*, true) in domain **Eukarya.** Genes are found in the nucleus, which thereby controls the cell.

Prokaryotes are structurally simple (Fig. 1.5*a* and *b*) but metabolically complex. The archaea live in aquatic environments that lack oxygen or are too salty, too hot, or too acidic for most other organisms. Perhaps these environments are similar to those of the primitive earth, and archaea may be representative of the first cells to have evolved. Bacteria are found almost anywhere—in the water, soil, and atmosphere, as well as on our skin and in our mouths and large intestine. Although some bacteria cause diseases, others perform many services, both environmentally and commercially. They are used for genetic research in our laboratories, help produce innumerable products in our factories, and help purify water in our sewage treatment plants, for example.

Kingdoms

Taxonomists are in the process of deciding how to categorize archaea and bacteria into kingdoms. Domain Eukarya, on the other hand, contains four kingdoms with which you may be familiar (Fig. 1.5*c*). Protists (kingdom Protista) range from unicellular to a few multicellular organisms. Some are photosynthesizers, and some must ingest their food. Among the fungi (kingdom Fungi) are the familiar molds and mushrooms that, along with bacteria, help decompose dead organisms. Plants (kingdom Plantae) are well known as multicellular photosynthesizers, while animals (kingdom Animalia) are multicellular and ingest their food.

Other Categories

Each successive classification category above species contains more different types of organisms than the one preceding. Only modern humans are in the genus *Homo*, but many different types of animals are in the animal kingdom (Table 1.1). Most genera contain several species, and these share very similar characteristics, but those that are in the same kingdom have only general characteristics in common. Thus, all species in the genus *Pisum* look pretty much the same—that is, like pea plants—while the species in the plant kingdom can be quite different, as is evident when we compare grasses to trees.

Taxonomy makes sense out of the bewildering variety of life on Earth because organisms are classified according to their presumed evolutionary relationship. Those organisms placed in the same genus are the most closely related, and those placed in separate domains are the most distantly related. Therefore, all eukarya are more closely related to one another than they are to bacteria or archaea. Similarly, all animals are more closely related to one another than they are to plants. As more is learned about evolutionary relationships between species, taxonomy changes. Taxonomists are even now making observations and performing experiments that will one day bring about changes in the classification system adopted by this text.

Scientific Names

Biologists give each living thing a binomial (*bi*, two; *nomen*, name), or two-part name. For example, the scientific name for human beings is *Homo sapiens* and for the garden pea, *Pisum sativum*. The first word is the genus, and the second word is the specific epithet of a species within a genus. (Note that both words are in italic, but only the genus is capitalized.) The genus name can be used alone to refer to a group of related species. Also, the genus can be abbreviated to a single letter if used with the specific epithet (e.g., *P. sativum*).

Scientific names are universally used by biologists to avoid confusion. Common names tend to overlap and often are in the language of a particular country. But scientific names are based on Latin, a universal language that not too long ago was well known by most scholars.

Taxonomy places species into classification categories according to their evolutionary relationships. The categories are species, genus, family, order, class, phylum, kingdom, and domain (the most inclusive).

Table 1.1	Classification of Humans
Classification Category	**Characteristics**
Domain Eukarya	Cells with nuclei
Kingdom Animalia	Multicellular, motile, ingestion of food
Phylum Chordata	Dorsal supporting rod and nerve cord
Class Mammalia	Hair, mammary glands
Order Primates	Adapted to climb trees
Family Hominidae	Adapted to walk erect
Genus *Homo*	Large brain, tool use
Species *Homo sapiens**	Body proportions of modern humans

*To specify an organism, you must use the full binomial name, such as *Homo sapiens*.

1.3 The Organization of the Biosphere

The organization of life extends beyond the individual to the **biosphere**, the zone of air, land, and water at the surface of the earth where living organisms are found. Individual organisms belong to a **population**, all the members of a species within a particular area. The populations of a **community** interact among themselves and with the physical environment (soil, atmosphere, etc.), thereby forming an **ecosystem.**

Figure 1.6 depicts a grassland inhabited by populations of rabbits, mice, snakes, hawks, and various types of grasses. These populations interchange gases with and give off heat to the atmosphere. They also take in water and give off water to the physical environment. In addition, the populations interact with each other by forming food chains in which one population feeds on another.

Mice feed on plant products, snakes feed on mice, and hawks feed on snakes, for example.

Ecosystems are characterized by chemical cycling and energy flow, both of which begin when photosynthesizers take in solar energy and inorganic nutrients to produce food (organic nutrients). The gray arrows in Figure 1.6 represent chemical cycling—chemicals move from one population to another in a food chain, until with death and decomposition, inorganic nutrients are returned to living plants once again. The yellow to red arrows represent energy flow. Energy flows from the sun through the plants and the other members of the food chain as they feed on one another. The energy gradually dissipates and returns to the atmosphere as heat. Because energy flows and does not cycle, ecosystems could not stay in existence without solar energy and the ability of photosynthesizers to absorb it.

Climate largely determines where different ecosystems are found about the globe. For example, deserts are found in areas of minimal rain, grasslands require a minimum amount of rain, and forests require much rain. The two most biologically diverse ecosystems—tropical rain forests and coral reefs occur where solar energy is most abundant. Coral reefs, which are found just offshore of the continents and islands of the southern hemisphere, are built up from the calcium carbonate skeletons of sea animals called corals. Inside the tissues of corals are tiny, one-celled protists that carry on photosynthesis and provide food to their hosts. Reefs provide a habitat for many other animals, including jellyfish, sponges, snails, crabs, lobsters, sea turtles, moray eels, and some of the world's most colorful fishes (Fig. 1.7). See the figure on page 1 also.

The Human Population

The human population tends to modify existing ecosystems for its own purposes. Humans clear forests or grasslands in order to grow crops; later, they build houses on what was once farmland; and finally, they convert

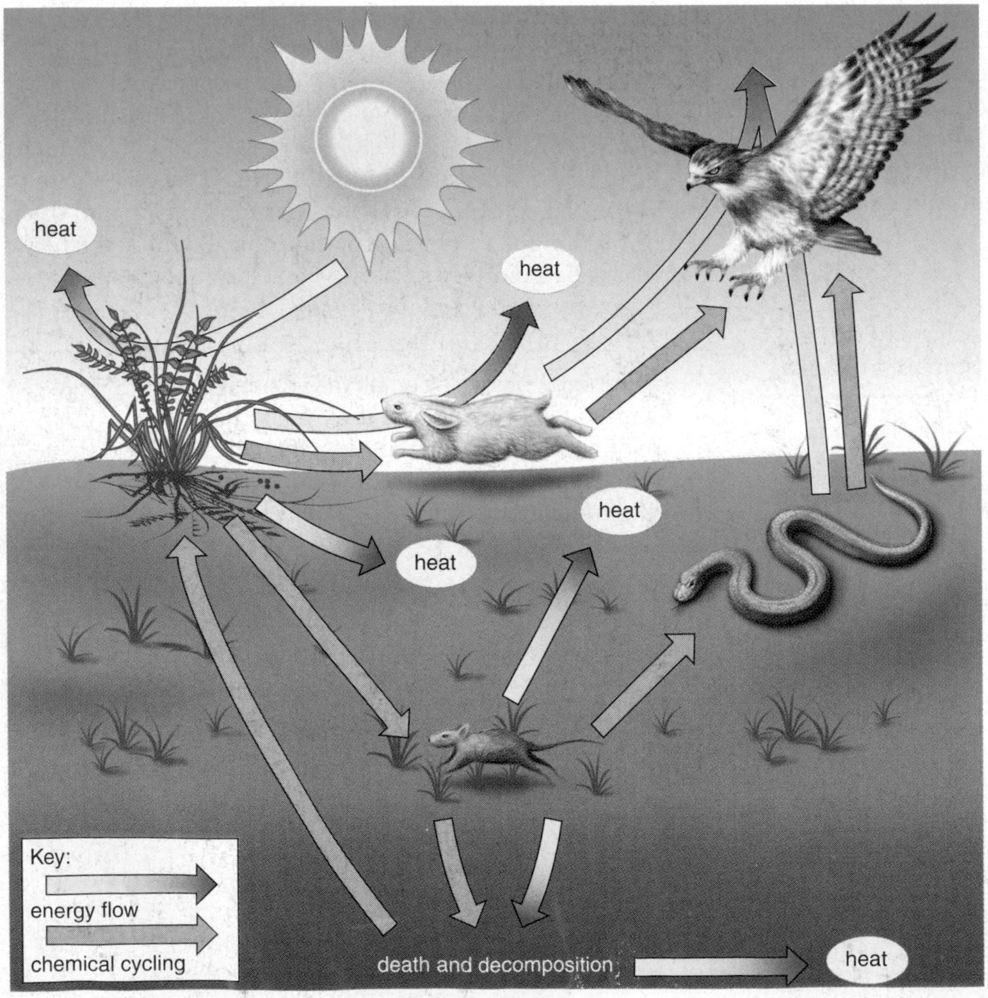

Key:

energy flow

chemical cycling

heat

heat

heat

heat

death and decomposition   heat

Figure 1.6 A grassland, a terrestrial ecosystem.
In an ecosystem, chemical cycling (gray arrows) and energy flow (yellow to red arrows) begin when plants use solar energy and inorganic nutrients to produce their own food (organic nutrients). Chemicals and energy are passed from one population to another in a food chain. Eventually, the heat energy dissipates. With death and decomposition of organisms, inorganic nutrients are returned to living plants once more.

small towns into cities. As coasts are developed, humans send sediments, sewage, and other pollutants into the sea.

Like tropical rain forests, coral reefs are severely threatened as the human population increases in size. Some reefs are 50 million years old, and yet in just a few decades, human activities have destroyed 10% of all coral reefs and seriously degraded another 30%. At this rate, nearly three-quarters could be destroyed within 50 years. Similar statistics are available for tropical rain forests.

It has long been clear that human beings depend on healthy ecosystems for food, medicines, and various raw materials. We are only now beginning to realize that we depend on them even more for the services they provide. Just as chemical cycling occurs within ecosystems, so ecosystems keep chemicals cycling throughout the entire biosphere. The workings of ecosystems ensure that the environmental conditions of the biosphere are suitable for the continued existence of humans. And many ecologists (scientists who study ecosystems) believe that ecosystems cannot function properly unless they remain biologically diverse.

Biodiversity

Biodiversity is the total number of species, the variability of their genes, and the ecosystems in which they live. The present biodiversity of our planet has been estimated to be as high as 15 million species, and so far, under 2 million have been identified and named. Extinction is the death of a species or larger taxonomic group. It is estimated that presently we are losing as many as 400 species per day due to human activities. For example, several species of fishes have all but disappeared from the coral reefs of Indonesia and along the African coast because of overfishing. Many biologists are alarmed about the present rate of extinction and believe it may eventually rival the rates of the five mass extinctions that have occurred during our planet's history. The dinosaurs became extinct during the last mass extinction, 65 million years ago.

It has been suggested that the primary bioethical issue of our time is preservation of ecosystems. Just as a native fisherman who assists in overfishing a reef is doing away with his own food source, so we as a society are contributing to the destruction of our home, the biosphere. If instead we adopt a conservation ethic that preserves the biosphere, we are helping to ensure the continued existence of our species.

Living things belong to ecosystems where populations interact among themselves in communities and with the physical environment. Preservation of ecosystems is of primary importance because they perform services that ensure our continued existence.

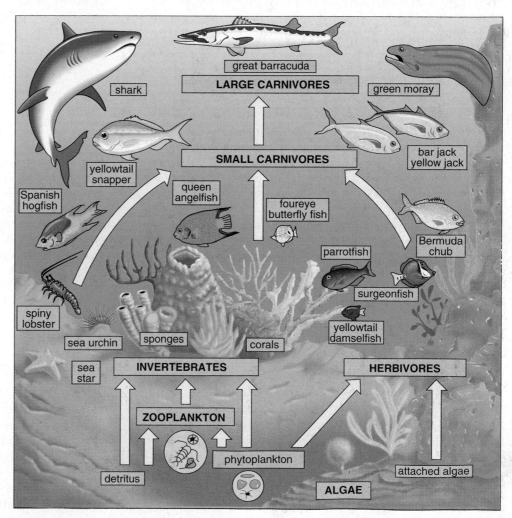

Figure 1.7   A coral reef, a marine ecosystem.
The yellow arrows represent food chains within a coral reef, a unique community of marine organisms. Food chains extend from algae and phytoplankton (unicellular, floating organisms) to the top carnivores represented here by a shark, barracuda, and moray eel. Where would humans be placed in this illustration? We are a carnivore, and feed on the shark, barracuda, and the other fishes depicted here.

1.4 The Process of Science

Biology is the scientific study of life. As discussed in the introduction, biologists are apt to be found almost anywhere studying life forms. Next, we will have an example of a study that was done in the field (outside the laboratory) and one that was done in the laboratory.

Although biologists—and all scientists—use various investigative methods depending on the topic under study, certain stages are characteristic of the general process of science (Fig. 1.8).

Observation

Scientists believe that nature is orderly and measurable; that natural laws, such as the law of gravity, do not change with time, and that a natural event, or **phenomenon,** can be understood more fully by observing it. Scientists use all their senses to make **observations.** We can observe with our noses that dinner is almost ready; observe with our fingertips that a surface is smooth and cold; and observe with our ears that a piano needs tuning. They also extend the ability of their senses by using instruments; for example, the microscope enables us to see objects that could never be seen by the naked eye. Finally, scientists may expand their understanding even further by taking advantage of the knowledge and experiences of other scientists. For instance, they may look up past studies on the Internet or at the library, or they may write or speak to others who are researching similar topics.

Chance alone can help a scientist get an idea. The most famous case pertains to penicillin. When examining a petri dish, Alexander Fleming observed an area around a mold that was free of bacteria. Upon investigating, Fleming found that the mold produced an antibacterial substance he called penicillin. This caused Fleming to think that perhaps penicillin would be useful in humans.

Hypothesis

After making observations and gathering knowledge about a phenomenon, a scientist uses inductive reasoning. **Inductive reasoning** occurs whenever a person uses creative thinking to combine isolated facts into a cohesive whole. In this case, the scientist comes up with a **hypothesis,** a possible explanation for the natural event. The scientist presents the hypothesis as an actual statement.

All of a scientist's past experiences, no matter what they might be, will most likely influence the formation of a hypothesis. But a scientist only considers hypotheses that can be tested. Moral and religious beliefs, while very important to our lives, differ between cultures, and through time and are not always testable.

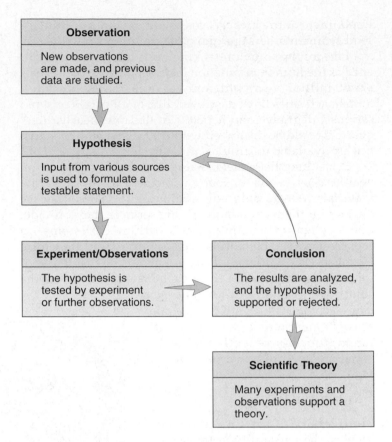

Figure 1.8 Flow diagram for the scientific method.
On the basis of new and/or previous observations, a scientist formulates a hypothesis. The hypothesis is tested by further observations and/or experiments, and new data either support or do not support the hypothesis. The return arrow indicates that a scientist often chooses to retest the same hypothesis or to test a related hypothesis. Conclusions from many different but related experiments may lead to the development of a scientific theory. For example, studies pertaining to development, anatomy, and fossil remains all support the theory of evolution.

Experiments/Further Observations

Testing a hypothesis involves either conducting an experiment or making further observations. To determine how to test a hypothesis, a scientist uses deductive reasoning. **Deductive reasoning** involves "if, then" logic. For example, a scientist might reason *if* organisms are composed of cells, *then* microscopic examination of any part of an organism should reveal cells. We can also say that the scientist has made a **prediction** that the hypothesis can be supported by doing microscopic studies. Making a prediction helps a scientist know what to do next.

The manner in which a scientist intends to conduct an experiment is called the **experimental design.** A good experimental design ensures that scientists are testing what they want to test and that their results will be meaningful. It is always best for an experiment to include a control group. A control group, or simply the **control,** goes through all the steps of an experiment but lacks the factor or is not exposed to the factor being tested.

Scientists often use a **model,** a representation of an actual object. In the experiment considered later in this section, the scientist used models of birds because it would have been impossible to get live birds to cooperate. Another type of modeling occurs when scientists use software to suggest future effects on climate due to human activities, or when they use mice instead of humans when doing, say, cancer research. Nevertheless, when a medicine is effective in mice, researchers must still test it in humans. And whenever it is impossible to study the actual phenomenon, a model remains a hypothesis in need of testing. Someday, some scientist might devise a way to test it.

Data

The results of an experiment are referred to as the **data.** Data should be observable and objective, rather than subjective. Mathematical data are often displayed in the form of a graph or table. Many studies rely on statistical data. Let's say an investigator wants to know if eating onions can prevent women from getting osteoporosis (weak bones). The scientist conducts a survey asking women about their onion-eating habits and then correlates this data with the condition of their bones. Other scientists critiquing this study would want to know: How many women were surveyed? How old were the women? What were their exercise habits? What proportion of the diet consisted of onions? And what criteria were used to determine the condition of their bones? Should the investigators conclude that eating onions does protect a woman from osteoporosis, other scientists would want to know the statistical probability of error. If the results are significant at a 0.30 level, then the probability that the correlation is incorrect is 30% or less. (This would be considered a high probability of error.) The greater the variance in the data, the greater the probability of error. And in the end, statistical data of this sort would only be suggestive until we know of some ingredient in onions that has a direct biochemical or physiological effect on bones. Therefore, you can see that scientists are skeptics who always pressure one another to keep on investigating a particular topic.

Conclusion

Scientists must analyze the data in order to reach a **conclusion** as to whether the hypothesis is supported or not. Because science progresses, the conclusion of one experiment can lead to the hypothesis for another experiment, as represented by the return arrow in Figure 1.8. In other words, results that do not support one hypothesis can often help a scientist formulate another hypothesis to be tested. Scientists report their findings in scientific journals so that their methodology and data are available to other scientists. Experiments and observations must be repeatable—that is, the reporting scientist and any scientist who repeats the experiment must get the same results, or else the data are suspect.

Scientific Theory

The ultimate goal of science is to understand the natural world in terms of **scientific theories,** which are concepts that join together well-supported and related hypotheses. In ordinary speech, the word theory refers to a speculative idea. In contrast, a scientific theory is supported by a broad range of observations, experiments, and data.

Some of the basic theories of biology are:

Name of Theory	Explanation
Cell	All organisms are composed of cells.
Biogenesis	Life comes only from life.
Evolution	All living things have a common ancestor, but each is adapted to a particular way of life.
Gene	Organisms contain coded information that dictates their form, function, and behavior.

The theory of evolution is the unifying concept of biology because it pertains to many different aspects of living things. For example, the theory of evolution enables scientists to understand the history life, the variety of living things, and the anatomy, physiology, and development of organisms—even their behavior, as we shall see in the study discussed next.

The theory of evolution has been a very fruitful theory, meaning that it has helped scientists generate new hypotheses. Because this theory has been supported by so many observations and experiments for over 100 years, some biologists refer to the **principle** of evolution, a term that may be appropriate for theories that are generally accepted by an overwhelming number of scientists. The term **law** instead of principle is preferred by some. For instance, in a subsequent chapter concerning energy relationships, we will examine the laws of thermodynamics.

Scientists carry out studies in which they test hypotheses. The conclusions of many different types of related experiments eventually enable scientists to arrive at a scientific theory that is generally accepted by all.

A Field Study

David P. Barash, while observing the mating behavior of mountain bluebirds, formulated the hypothesis that aggression of the male varies during the reproductive cycle (Fig. 1.9*a*). To test this hypothesis, he reasoned that he should evaluate male aggression at three stages: after the nest is built, after the first egg is laid, and after the eggs hatch.

> **HYPOTHESIS:** Male bluebird aggression varies during the reproductive cycle.
> **PREDICTION:** Aggression will change after the nest is built, after the first egg is laid, and after hatching.

Testing the Hypothesis

For his experiment, Barash decided to define aggression as the "number of approaches per minute" toward a rival male and his female mate. To provide a rival, Barash decided to post a male bluebird model near the nests while resident males were out foraging. The aggressive behavior of the resident male toward the male model and toward his female mate were noted during the first ten minutes of the male's return (Fig. 1.9*b*). In order to give his results validity, Barash had a control group. For his control, Barash posted a male robin model instead of a male bluebird near certain nests.

Resident males of the control group did not exhibit any aggressive behavior, but resident males of the experimental groups did exhibit aggressive behavior. Barash graphed his mathematical data (Fig. 1.9*c*). By examining the graph, you can see that the resident male was more aggressive toward the rival male model than toward his female mate, and that he was most aggressive while the nest was under construction, less aggressive after the first egg was laid, and least aggressive after the eggs hatched.

The Conclusion

The results allowed Barash to conclude that aggression in male bluebirds is related to their reproductive cycle. Therefore, his hypothesis was supported. If male bluebirds were always aggressive even toward male robin models, his hypothesis would not have been supported.

> **CONCLUSION:** The hypothesis is supported. Male bluebird aggression does vary during the reproductive cycle.

Barash reported his experiment in the *American Naturalist*.[1] In this article Barash gave an evolutionary interpretation to his results. It was adaptive, he said, for male bluebirds to be less aggressive after the first egg is laid because by then the male bird is "sure the offspring is his own." It was maladaptive for the male bird to waste energy being aggressive after hatching because his offspring are already present.

[1]Barash, D. P. 1976. The male responds to apparent female adultery in the mountain bluebird, *Sialia currucoides:* An evolutionary interpretation. *American Naturalist* 110:1097–1101.

a. Scientist makes observations, studies previous data, and formulates a hypothesis.

b. Scientist performs experiment and collects objective data.

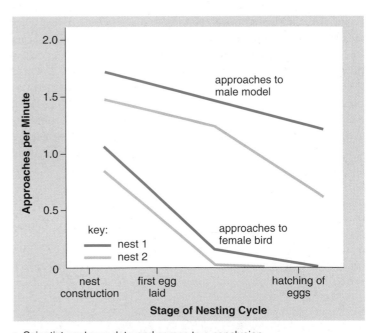

c. Scientist analyzes data and comes to a conclusion.

Figure 1.9   A field study.
a. Observation of male bluebird behavior allowed David Barash to formulate a testable hypothesis. Then he (**b**) collected data, which was (**c**) displayed in a graph. Finally, he came to a conclusion.

A Laboratory Study

When scientists are doing a study, they often perform experiments in the laboratory where conditions can be kept constant. In experiment 1 and experiment 2 discussed next, physiologists are trying to determine if sweetener S is a safe food additive.

Experiment 1

On the basis of available information, the physiologists formulate a hypothesis that sweetener S is a safe food additive even when 50% of the diet is sweetener S. If so, mice whose diet contains 50% sweetener S should suffer no ill effects.

>   **HYPOTHESIS:** Sweetener S is a safe food additive.
>   **PREDICTION:** If mice are fed a diet that is 50% sweetener S, there will be no effect on health.

In this study, the scientists decide on the following experimental design:

>   Test group: 50% of diet is sweetener S.
>   Control group: Diet contains no sweetener S.

To help ensure that conditions for the two groups are identical, the researchers place a certain number of randomly chosen inbred (genetically identical) mice into the various groups—say, 100 mice per group. If any of the mice are different from the others, it is hoped that random selection has distributed them evenly among the groups. The researchers also make sure that all conditions, such as availability of water, cage setup, and temperature of the surroundings, are constant for both groups. The food for each group is exactly the same except for the amount of sweetener S.

After several weeks, both groups of mice are examined for bladder cancer. Let's suppose that one-third of the mice in the test group are found to have bladder cancer, while none in the control group have bladder cancer (Fig. 1.10). The results of this experiment do not support the hypothesis that sweetener S is a safe food additive when 50% of the diet is sweetener S.

>   **CONCLUSION:** The hypothesis is not supported. Sweetener S is not a safe food additive when the diet is 50% sweetener S.

These results cause the scientists to believe that sweetener S may be less harmful if the diet contains a lesser amount. So, they decide to do another experiment.

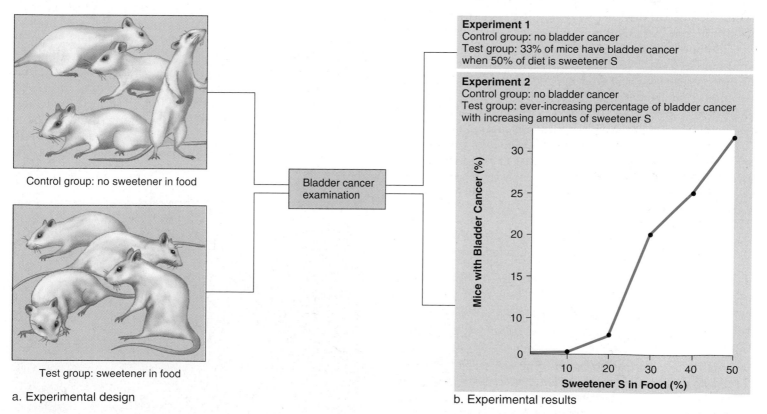

Control group: no sweetener in food

Test group: sweetener in food

a. Experimental design

Experiment 1
Control group: no bladder cancer
Test group: 33% of mice have bladder cancer when 50% of diet is sweetener S

Experiment 2
Control group: no bladder cancer
Test group: ever-increasing percentage of bladder cancer with increasing amounts of sweetener S

b. Experimental results

Figure 1.10   **A controlled experiment.**
For both experiment 1 and experiment 2, genetically similar mice are randomly divided into a control group and test group(s) that contain 100 mice each. All groups are exposed to the same conditions, such as housing, temperature, and water supply. The control group is not subjected to sweetener S in the food. At the end of the experiment, all mice are examined for bladder cancer. The results of experiment 1 and experiment 2, which are described in the text, are shown on the far right.

Experiment 2

After analyzing the results of experiment 1, the physiologists decide to hypothesize that sweetener S is safe if the diet contains a certain limited amount of sweetener S. If so, it may be possible to determine this amount by having several experimental groups, each receiving more sweetener S than the one before.

> HYPOTHESIS: Sweetener S is a safe food additive if the diet contains a limited amount.
> PREDICTION: If mice are fed a limited amount of sweetener S, there will be no effect on health.

In this second study, the researchers feed sweetener S to groups of mice at ever-greater amounts until it comprises 50% of the diet:

> Group 1: Diet contains no sweetener S (the control).
> Group 2: 5% of diet is sweetener S.
> Group 3: 10% of diet is sweetener S.
>
> ↓
>
> Group 11: 50% of diet is sweetener S.

After keeping careful records for several weeks, the researchers present their data in the form of a graph. The mice in the control group have no bladder cancer. But the incidence of bladder cancer among the experimental groups increases sharply with a diet that is 10% sweetener S or beyond (see Fig. 1.10b) . The researchers statistically analyze their data to determine if the difference in the number of cases of bladder cancer between the various experimental groups is significant and not due to simple chance. Finding that the differences are significant at a satisfactory level of probable error, the researchers conclude that they can now make a recommendation concerning the intake of sweetener S. They suggest that an intake of sweetener S up to 10% of the diet is relatively safe, but thereafter you would expect to see an ever-greater incidence of bladder cancer.

> CONCLUSION: The hypothesis is supported. Sweetener S is safe if the diet contains a limited amount.

As stated previously, experiments done in mice serve as hypotheses that need to be tested in humans. Sometimes, however, we must make use of the results in laboratory animals whenever it is not possible to design a similar experiment in humans.

Some biological studies are conducted in the field, and some are done in the laboratory where conditions can be kept constant. Regardless, scientists formulate testable hypotheses, perform experiments or make further observations, and come to conclusions after analyzing their objective results.

1.5 Science and Social Responsibility

Science improves our lives. The application of scientific knowledge for a practical purpose is called **technology.** For example, the discovery of new drugs has extended the life span of people who have AIDS. Cell biology research is helping us understand the causes of cancer. Research has also led to modern agricultural practices that are helping to feed our burgeoning world population. Most technologies have benefits but also drawbacks. The use of pesticides, as you may know, kills not only pests but also other types of organisms. The book *Silent Spring* was written to make the public aware of the harmful environmental effects of pesticide use.

Who should decide how, and even whether, a technology should be put to use? Making value judgments is not a part of science. Ethical and moral decisions must be made by all people. Therefore, the responsibility for how to use the fruits of science must reside with people from all walks of life, not with scientists alone. Scientists should provide the public with as much information as possible, but all citizens, including scientists, should make decisions about the use of technologies.

Presently, we need to decide if we want to stop producing bioengineered organisms that may be harmful to the environment. Also, through gene therapy, we are developing the ability to cure diseases and to alter the genes of our offspring. Perhaps one day we might even be able to clone ourselves (Fig. 1.11). Should we do these things? So far, as a society, we continue to believe in the sacredness of human life, and therefore we have passed laws against doing research with fetal tissues or using fetal tissues to cure human ills. Even if the procedure is perfected, we may also continue to rule against human cloning. But such judgments are subject to change, and thus we must all wrestle with the bioethical issues presented in this text and should not expect others to make these decisions for us.

Figure 1.11 Human cloning.
Your clone would be a younger version of yourself; in other words, you would be a parent to someone who has your exact genes.

Bioethical Focus   Cloning of Humans

The term cloning means making exact multiple copies of genes, a cell, or an organism. Cloning has been around for some time. Identical twins are clones of a single zygote. When a single bacterium reproduces asexually on a petri dish, a colony of cells results, and each member of the colony is a clone of the original cell. Through biotechnology, bacteria now produce cloned copies of human genes.

Now, for the first time in our history, it is possible to produce a clone of a mammal, and even one day of a human. The parent need not contribute sperm or an egg to the process. A nucleus (which contains a person's genes) from one adult cell is placed in an enucleated egg, and that egg begins developing. The developing embryo must be placed in the uterus of a surrogate mother, and when birth occurs, the clone is an exact copy, but of course, younger than the original parent. The process of cloning whole animals, and certainly humans, has not been perfected. Until it is, another type of cloning is more likely to become widespread.

Suppose it were possible to put the nucleus from a cell of a burn victim into an enucleated egg that is cajoled to become skin cells in the laboratory. These cells could be used to provide grafts of new skin that would not be rejected by the recipient. Would this be a proper use of cloning in humans?

Or suppose parents want to produce a child free of genetic disease. Scientists produce an embryo through in vitro fertilization, and then they clone the embryo to produce more of them. Genetic engineering to correct the defect doesn't work on all the embryos—only a few. They implant just those few in a uterus, where development continues to term. Would this be a proper use of cloning in humans?

What if later scientists were able to produce children with increased intelligence or athletic prowess using this same technique? Would this be an acceptable use of cloning in humans?

Decide Your Opinion

1. Presently, research on the cloning of humans is banned. Should it be? Why or why not?
2. Under what circumstances might cloning in humans be acceptable? Explain.
3. Is cloning to produce improved breeds of farm animals acceptable? Why or why

---

Summarizing the Concepts

1.1 The Characteristics of Life
Evolution accounts for both the diversity and the unity of life we see about us. All organisms share the following characteristics of life:

1. Living things are organized. The levels of biological organization extend from the cell to ecosystems: atoms and molecules → cells → tissues → organs → organ systems → organisms → populations → communities → ecosystems. In an ecosystem, populations interact with one another and with the physical environment.
2. Living things take materials and energy from the environment; they need an outside source of nutrients.
3. Living things reproduce; they produce offspring that resemble themselves.
4. Living things respond to stimuli; they react to internal and external events.
5. Living things are homeostatic; internally they stay just about the same despite changes in the external environment.
6. Living things grow and develop; during their lives, they change—most multicellular organisms undergo various stages from fertilization to death.
7. Living things are adapted; they have modifications that make them suited to a particular way of life.

1.2 The Classification of Living Things
Living things are classified according to their evolutionary relationships into these ever more inclusive categories: species, genus, family, order, class, phylum, kingdom, and domain. Species in different domains are only distantly related; species in the same genus are very closely related.

Each type of organism has a scientific name in Latin that consists of the genus and the specific epithet. Both the genus and the specific epithet are italicized; only the genus is capitalized, as in *Homo sapiens*.

1.3 The Organization of the Biosphere
All living things belong to a population, which is defined as all the members of the same species that occur in a particular locale. Populations interact with each other within a community and with the physical environment, forming an ecosystem. Ecosystems are characterized by chemical cycling and energy flow, which begin when a photosynthesizer becomes food (organic nutrients) for an animal. Food chains tell who eats whom in an ecosystem. As one population feeds on another, the energy dissipates but the nutrients do not. Eventually, inorganic nutrient molecules return to photosynthesizers, which use them and solar energy to produce more food.

Human activities have totally altered many ecosystems for their own purposes and are putting stress on most of the others. Coral reefs and tropical rain forests, for example, are quickly disappearing. We now realize that the health of the biosphere is essential to the future continuance of the human species. Therefore, we should do all we can to maintain the health of the biosphere.

1.4 The Process of Science
When studying the natural world, scientists use the scientific process. Observations, along with previous data, are used to formulate a hypothesis. New observations and/or experiments are carried out in order to test the hypothesis. A good experimental design includes a control group. The experimental and observational results are analyzed, and the scientist comes to a conclusion as to whether the results support the hypothesis or prove it false. Science is always open to change; therefore, hypotheses can be supported but not proven true.

Several conclusions in a particular area may allow scientists to arrive at a theory—such as the cell theory, the gene theory, or the theory of evolution. The theory of evolution is a unifying theory of biology.

1.5 Science and Social Responsibility
Science does not consider moral or ethical questions—these must be decided by all persons. It is up to us to decide how the various technologies that grow out of basic science should be used or regulated.

Testing Yourself

Choose the best answer for each question.
For questions 1–4, match the statements with the characteristics of life in the key.

Key:
a. Living things are organized.
b. Living things are homeostatic.
c. Living things respond to stimuli.
d. Living things reproduce.
e. Living things are adapted.

1. Genes made up of DNA are passed from parent to child.

2. Cells are made of molecules, tissues are made of cells, and organisms are made of tissues.

3. A herd of zebra will scatter when a lion approaches.

4. The long, sharp talons of a hawk can hold onto a mouse.

5. Which of these is mismatched?
   a. Domain Bacteria—mosses, ferns, pine trees
   b. Kingdom Protista—protozoans, algae, water molds
   c. Kingdom Fungi—molds and mushrooms
   d. Kingdom Plantae—woody and nonwoody flowering plants
   e. Kingdom Animalia—fish, reptiles, birds, humans

6. The level of organization that includes cells of similar structure and function would be a(n)
   a. organ.
   b. tissue.
   c. organ system.
   d. organism.

7. If an organism is unicellular, autotrophic, and eukaryotic, to which kingdom would it belong?
   a. Kingdom Protista
   b. Kingdom Fungi
   c. Kingdom Plantae
   d. Kingdom Animalia

8. The second word of a scientific name, such as *Homo sapiens*, is the
   a. genus.
   b. phylum.
   c. specific epithet.
   d. species.

9. The level of organization that includes all the populations in a given area along with the physical environment would be a(n)
   a. community.
   b. ecosystem.
   c. biosphere.
   d. tribe.

10. After performing an experiment and collecting data, the next step in the scientific method would be to
    a. propose a theory.
    b. design a model.
    c. form a hypothesis.
    d. come to a conclusion.

11. Energy is brought into ecosystems by which of the following?
    a. fungi and other decomposers
    b. cows and other animals that graze on grass
    c. meat-eating animals
    d. organisms that photosynthesize, such as plants

12. The responsibility for deciding how scientific technologies should be used is the responsibility of
    a. local government.
    b. federal government.
    c. the public.
    d. scientists.

13. Which of the following kingdoms would contain members who are photosynthetic?
    a. Animalia
    b. Protista
    c. Fungi
    d. Plantae
    e. Both b and d are correct.

14. Which of the following is a concept generally accepted because of well-supported conclusions?
    a. scientific theory.
    b. hypothesis.
    c. conclusion.
    d. law.
    e. observation.

15. An example of chemical cycling occurs when
    a. plants absorb solar energy and make their own food.
    b. energy flows through an ecosystem and becomes heat.
    c. hawks soar and nest in trees.
    d. death and decay make inorganic nutrients available to plants.
    e. we eat food and use the nutrients to grow or repair tissues.

16. Science always studies a phenomenon that
    a. has previously been published.
    b. lends itself to experimentation.
    c. is observable with the eye or with instruments.
    d. fits in with an already existing theory.
    e. Both b and c are correct.

17. After formulating a hypothesis, a scientist
    a. proves the hypothesis true or false.
    b. tests the hypothesis.
    c. decides how to best avoid having a control.
    d. makes sure environmental conditions are just right.
    e. formulates a scientific theory.

18. A scientist cannot
    a. make value judgments like everyone else.
    b. prove a hypothesis true.
    c. contribute to a long-standing scientific theory.
    d. make use of preexisting mathematical data.
    e. be as objective as possible.

19. Match the terms to these definitions. Only four of these terms are needed.

    adaptation   evolution
    biodiversity   homeostasis
    data   scientific theory
    energy   taxonomy

    a. Concept consistent with conclusions based on a large number of experiments and observations.
    b. Capacity to do work and bring about change.
    c. Suitability of an organism for its environment, enabling it to survive and produce offspring.
    d. Maintenance of the internal environment of an organism within narrow limits.

e-Learning Connection   www.mhhe.com/maderinquiry10

Concepts	Questions	Media Resources*
1.1 The Characteristics of Life		

• Although life is difficult to define, it can be recognized by certain common characteristics.	1. What are seven characteristics that distinguish living things from nonliving things?
	2. For what reason is there both great diversity and unity among living things?
		Essential Study Partner
		    Life Characteristics
		
		Labeling Exercises
		    Levels of
		        Organization—Plants
		    Levels of
		        Organization—Animals

1.2 The Classification of Living Things		

• Living things are classified into categories according to their evolutionary relationships.	1. What is the basis for classifying organisms into specific categories?
	2. Explain the difference between domains and kingdoms.
		Essential Study Partner
		    Hierarchies
		    Kingdoms
		    Three Domains
		
		Labeling Exercises
		    Kingdoms of Life
		
		General Biology Weblinks
		    Taxonomy & Phylogenetics

1.3 The Organization of the Biosphere		

• The biosphere is made up of ecosystems where living things interact with each other and the physical environment.
• Ecosystems and the biodiversity of the biosphere is being threatened by human activities.
	1. What is an ecosystem, and what determines where different ecosystems are located?
	2. Why is the preservation of biodiversity and ecosystems so important?
		Essential Study Partner
		    Biomes and Climate
		
		General Biology Weblinks
		    Biodiversity and
		        Conservation

1.4 The Process of Science		
• Biologists gather information and come to conclusions about the natural world. • Various conclusions can sometimes be used to arrive at a scientific theory, a general concept about the natural world.	1. List the steps of the scientific method. 2. What is the difference between a hypothesis and a scientific theory?	Art Quizzes Scientific Method
1.5 Science and Social Responsibility		

• All persons have the responsibility to make an ethical and moral decision about how scientific information can be used.	1. What is a "technology" and who should decide whether a technology should be put to use?	Essential Study Partner
		    Public Action
		    World Views
		
		Case Studies
		    Transplant
		    Treatment of Critically Ill
		        Newborns

*For additional Media Resources, see the Online Learning Center.

See the Online Learning Center for careers in Cell Biology.

The Molecules of Cells

2

Chapter Concepts

When grasses pollinate, windblown pollen spreads far and wide, causing allergic symptoms in many of us.

*I*t's a change of season. Henry knows what's coming. Grasses, trees, weeds, and garden plants are going to send out enormous amounts of pollen. The wind will blow and Henry will get the sniffles. Sneezing and coughing, he will reach for tissue after tissue as he endures the onslaught of allergy season. It's all a matter of chemistry, of course. Pollen bears molecules, chemical substances, that cause Henry's body to produce histamine, a substance that brings on his symptoms.

The evidence that we and all living things are composed of chemicals is overwhelming. Get sick and the medicine you take is a chemical. Henry will be given an antihistamine or maybe a steroid that will ameliorate his symptoms. As part of a physical exam, the doctor tests our blood chemistry. If the cholesterol count is too high, we will be put on a low-fat diet.

Certain types of molecules like carbohydrates, proteins, and fats make up the bulk of our bodies. The master molecule DNA carries genetic information which determines what we are like, whether an oak tree or a human who can study how oak trees produce so much pollen.

2.1 Basic Chemistry

Matter refers to anything that takes up space and has weight. It is helpful to remember that matter can exist as a solid, a liquid, or a gas. Then we can realize that not only are we humans matter, but so are the water we drink and the air we breathe.

Elements and Atoms

All matter, living and nonliving, is composed of **elements.** Considering the variety of living and nonliving things in the world, it is quite remarkable that there are only 92 naturally occurring elements—other elements have been created by special processes in the laboratory. As indicated in Figure 2.1, only six elements—carbon, hydrogen, nitrogen, oxygen, phosphorus, and sulfur—make up most (about 98%) of the body weight of organisms. The acronym CHNOPS helps us remember these six elements. Instead of writing out the names of elements, scientists use symbols to identify them. The letter C stands for carbon, and the letter N stands for nitrogen, for example. Some of the symbols used for elements are derived from Latin. For example, the symbol for sodium is Na (*natrium* in Latin means sodium).

Elements contain tiny particles called atoms. The same name is given to the element and its atoms. An **atom** is the smallest unit of matter to enter into chemical reactions. Even though an atom is extremely small, it contains even smaller subatomic particles called protons, neutrons, and electrons. Atoms have a central nucleus, where the subatomic **protons** and **neutrons** are located, and shells, which are pathways about the nucleus where **electrons** orbit. The shells represent energy levels. Figure 2.2 shows a model of an atom that has only two shells. The inner shell has the lowest energy level and can hold two electrons. The outer shell has a higher energy level and can hold eight electrons. An atom is most stable when the outer shell has eight electrons.

An atom has an atomic number; the **atomic number** is equal to the number of its protons. Notice in Table 2.1 that protons have a positive (+) electrical charge and electrons have a negative (−) charge. When an atom is electrically neutral, the number of protons equals the number of electrons. The carbon atom shown in Figure 2.3 has an atomic number of 6; therefore, it has six protons. Since it is electrically neutral, it also has six electrons. The inner shell has two electrons, and the outer shell has four electrons.

In the periodic table of the elements (Fig. 2.1), atoms are horizontally arranged in order of increasing atomic number. They are vertically arranged according to the number of electrons in the outer shell. The numeral at the top of each column indicates how many electrons are in the outer shell of the atoms in that column. An exception to this format is helium (He), which has only two electrons in the outer shell because it has only one shell. The number of electrons in the outer shell determines the chemical properties of an atom, including how readily it enters into chemical reactions.

I	II	III	IV	V	VI	VII	VIII
1 **H** Hydrogen 1						atomic number — 2 atomic symbol — **He** Helium atomic mass — 4	
3 **Li** Lithium 7	4 **Be** Beryllium 9	5 **B** Boron 11	6 **C** Carbon 12	7 **N** Nitrogen 14	8 **O** Oxygen 16	9 **F** Fluorine 19	10 **Ne** Neon 20
11 **Na** Sodium 23	12 **Mg** Magnesium 24	13 **Al** Aluminum 27	14 **Si** Silicon 28	15 **P** Phosphorus 31	16 **S** Sulfur 32	17 **Cl** Chlorine 35	18 **Ar** Argon 40
19 **K** Potassium 39	20 **Ca** Calcium 40						

Figure 2.1   Periodic Table of the Elements (shortened).
Each element has an atomic number, an atomic symbol, and an atomic mass. The elements indicated by color make up most of the body weight of organisms. See Appendix D.

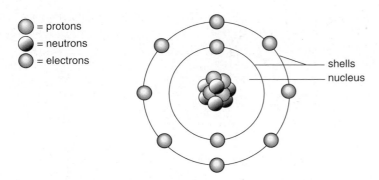

Figure 2.2 Model of an atom.
Protons and neutrons are located in the nucleus. Electrons are
located at energy levels called shells.

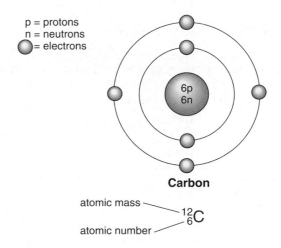

Figure 2.3 Carbon atom.
Carbon has an atomic number of 6; therefore, it has six protons and
six electrons when neutral. Carbon has a weight of 12 atomic mass
units. Therefore, it has six neutrons.

Table 2.1	Subatomic Particles	
Name	**Charge**	**Mass**
Electron	One negative unit	Almost no mass
Proton	One positive unit	One atomic mass unit
Neutron	No charge	One atomic mass unit

Isotopes

The subatomic particles are so light that their weight is indicated by special mathematical units called atomic mass units. The **atomic mass** of each atom is noted in the periodic table beneath the atomic symbol. The atomic mass of an atom equals the number of protons plus the number of neutrons. Why should that be the case? Table 2.1 shows that electrons have almost no mass, but protons and neutrons each have a weight of one atomic mass unit. Since the atomic mass of carbon is twelve and it has six protons, it is easy to calculate that carbon has six neutrons.

The atomic masses given in the periodic table are the average mass for each kind of atom. This is because the atoms of one element may differ in the number of neutrons; therefore, the mass varies. Atoms that have the same atomic number and differ only in the number of neutrons are called **isotopes.** Isotopes of carbon can be written in the following manner, where the subscript stands for the atomic number and the superscript stands for the atomic mass:

$$^{12}_{6}C \qquad ^{13}_{6}C \qquad ^{14}_{6}C*$$

$$*radioactive$$

Carbon 12 has six neutrons, carbon 13 has seven neutrons, and carbon 14 has eight neutrons. Unlike the other two isotopes, carbon 14 (which has 8 neutrons) is unstable; it breaks down into atoms with lower atomic numbers. When it decays, it emits radiation in the form of radioactive particles

or radiant energy. Therefore, carbon 14 is called a **radioactive isotope.** Carbon 14 and other types of radioactive isotopes can be used as tracers in biochemical experiments. In a now-famous experiment, carbon 14 helped scientists determine how sugar is formed during photosynthesis. And because radioactive isotopes break down at a known rate, they are also used to determine the age of fossils.

Radioactive isotopes are helpful in various forms of imaging. If a patient is injected with iodine, the thyroid gland takes it up and a scan of the thyroid shows any abnormality. Glucose labeled with a radioactive isotope is taken up by metabolically active tissues. The radiation given off can be detected by PET (positron emission tomography), which generates cross-sectional images of the body. Then researchers know which tissues are metabolically active under what circumstances.

High amounts of radiation can kill cells and compromise the immune system. When nuclear power plants employ proper safety measures, the risk of exposure to radiation is small. But accidents have occurred. On April 16, 1986, a series of explosions at the Chernobyl nuclear power plant violently threw radioactivity into the atmosphere worldwide. Over a half million people were exposed to dangerous levels of radioactivity in the immediate vicinity, and they may eventually suffer from various types of cancer and eye cataracts.

All matter is composed of atoms. Atoms have an atomic symbol, atomic number (number of protons), and atomic mass (number of protons and neutrons). Isotopes vary by the number of neutrons.

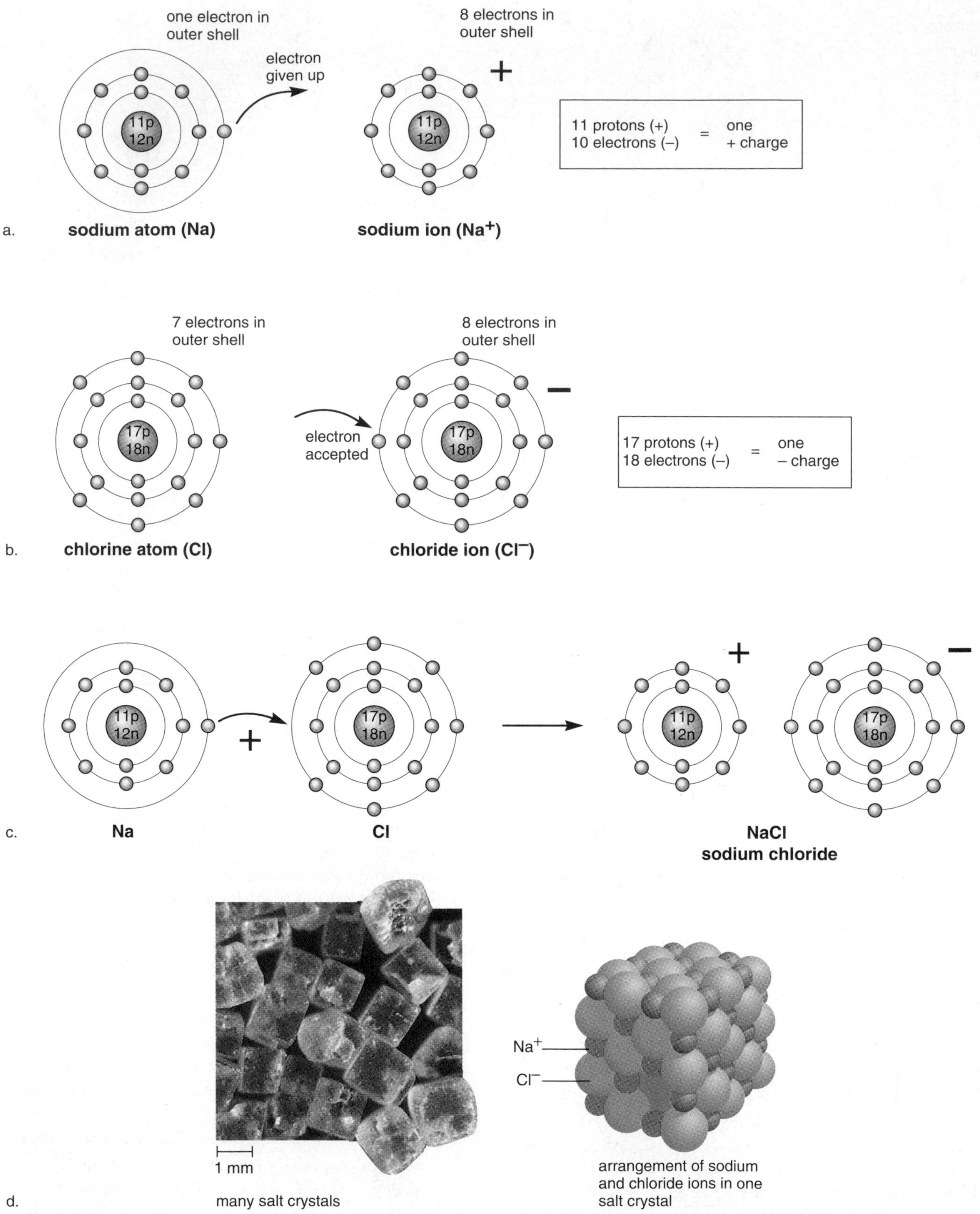

one electron in
outer shell

electron
given up

8 electrons in
outer shell

+

11p
12n

11p
12n

| 11 protons (+) | = | one |
| 10 electrons (−) | | + charge |

a.   **sodium atom (Na)**   **sodium ion (Na⁺)**

7 electrons in
outer shell

8 electrons in
outer shell

−

17p
18n

electron
accepted

17p
18n

| 17 protons (+) | = | one |
| 18 electrons (−) | | − charge |

b.   **chlorine atom (Cl)**   **chloride ion (Cl⁻)**

11p
12n

+

17p
18n

+   −

11p
12n

17p
18n

c.   **Na**   **Cl**   **NaCl**
sodium chloride

1 mm

many salt crystals

Na⁺

Cl⁻

arrangement of sodium
and chloride ions in one
salt crystal

d.

Figure 2.4   Ionic reaction.

a. When a sodium atom gives up an electron, it becomes a positive ion. **b.** When a chlorine atom gains an electron, it becomes a negative ion.
c. When sodium reacts with chlorine, the compound sodium chloride (NaCl) results. In sodium chloride, an ionic bond exists between the positive
Na⁺ and the negative Cl⁻ ions. **d.** In a sodium chloride crystal, the ionic bonding between Na⁺ and Cl⁻ causes the ions to form a three-
dimensional lattice in which each sodium ion is surrounded by six chloride ions, and each chloride ion is surrounded by six sodium ions.

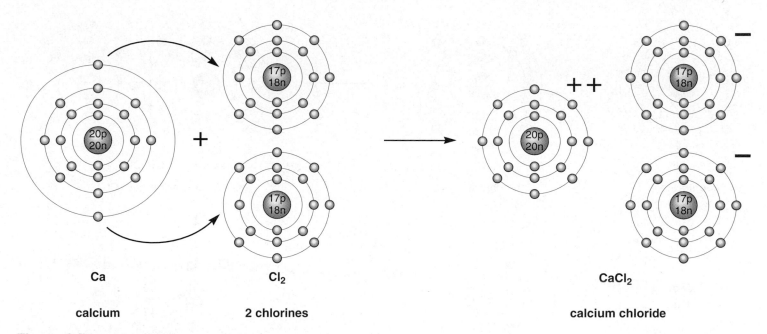

Ca   **Cl₂**   **CaCl₂**

calcium   2 chlorines   calcium chloride

Figure 2.5 Ionic reaction.
The calcium atom gives up two electrons, one to each of two chlorine atoms. In the compound calcium chloride ($CaCl_2$), the calcium ion is attracted to two chloride ions.

Molecules and Compounds

Atoms often bond with each other to form a chemical unit called a **molecule.** A molecule can contain atoms of the same kind, as when an oxygen atom joins with another oxygen atom to form oxygen gas. Or the atoms can be different, as when an oxygen atom joins with two hydrogen atoms to form water. When the atoms are different, a **compound** results.

Two types of bonds join atoms: the ionic bond and the covalent bond.

Ionic Reactions

Recall that atoms (with more than one shell) are most stable when the outer shell contains eight electrons. During an ionic reaction, atoms give up or take on an electron(s) in order to achieve a stable outer shell.

Figure 2.4 depicts a reaction between a sodium (Na) atom and a chlorine (Cl) atom in which chlorine takes an electron from sodium. **Ions** are particles that carry either a positive (+) or negative (−) charge. The sodium ion carries a positive charge because it now has one more proton than electrons, and the chloride ion carries a negative charge because it now has one fewer proton than electrons. The attraction between oppositely charged sodium ions and chloride ions forms an **ionic bond.** The resulting compound, sodium chloride, is table salt, which we use to enliven the taste of foods.

Figure 2.5 shows an ionic reaction between a calcium atom and two chlorine atoms. Notice that calcium, with two electrons in the outer shell, reacts with two chlorine atoms. Why? Because with seven electrons already, each chlorine requires only one more electron to have a stable outer shell. The resulting salt ($CaCl_2$) is called calcium chloride.

Biologically important ions in the human body are listed in Table 2.2. The balance of these ions in the body is important to our health. Too much sodium in the blood can cause high blood pressure; not enough calcium leads to rickets (a bowing of the legs) in children; too much or too little potassium results in heartbeat irregularities. Bicarbonate, hydrogen, and hydroxide ions are all involved in maintaining the acid-base balance of the body.

An ionic bond is the attraction between oppositely charged ions.

Table 2.2	Significant Ions in the Body	
Name	**Symbol**	**Special Significance**
Sodium	Na^+	Found in body fluids; important in muscle contraction and nerve conduction
Chloride	Cl^-	Found in body fluids
Potassium	K^+	Found primarily inside cells; important in muscle contraction and nerve conduction
Phosphate	PO_4^{3-}	Found in bones, teeth, and the high-energy molecule ATP
Calcium	Ca^{2+}	Found in bones and teeth; important in muscle contraction and nerve conduction
Bicarbonate	HCO_3^-	Important in acid-base balance
Hydrogen	H^+	Important in acid-base balance
Hydroxide	OH^-	Important in acid-base balance

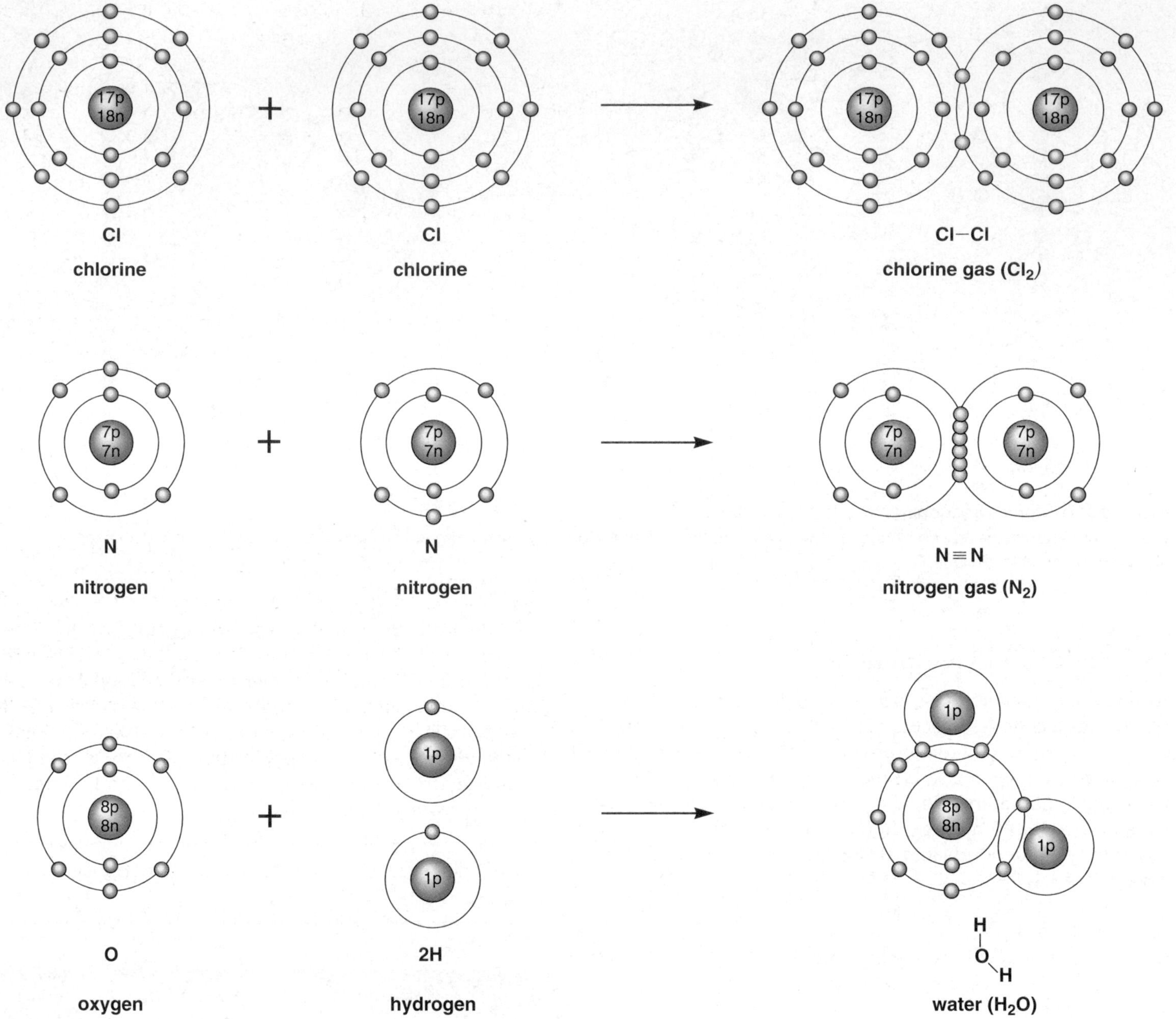

Figure 2.6 Covalent reactions.
After a covalent reaction, each atom will have filled its outer shell by sharing electrons. To determine this, it is necessary to count the shared electrons as belonging to both bonded atoms. Oxygen and nitrogen are most stable with eight electrons in the outer shell; hydrogen is most stable with two electrons in the outer shell.

Covalent Reactions

In covalent reactions, the atoms share electrons in **covalent bonds** instead of losing or gaining them. Covalent bonds can be represented in a number of ways. The overlapping outermost shells in Figure 2.6 indicate that the atoms are sharing electrons. Just as two hands participate in a handshake, each atom contributes one electron to the pair that is shared. These electrons spend part of their time in the outer shell of each atom; therefore, they are counted as belonging to both bonded atoms. When you count the electrons shared by both atoms, you can see that each atom has eight electrons in its outer shell (or two electrons in the case of hydrogen). In hydrogen, the outer shell is complete when it contains two electrons. Hydrogen can give up an electron and become H$^+$, or it can share and thereby have a completed outer shell.

Instead of drawing complex diagrams, electron-dot structures are sometimes used to depict covalent bonding between atoms. For example, in Figure 2.7, each chlorine atom can be represented by its symbol, and the electrons

Electron-Dot Formula	Structural Formula	Molecular Formula
:Ö::C::Ö: carbon dioxide	O=C=O carbon dioxide	CO_2 carbon dioxide
H ·· :N:H ·· H ammonia	H | N—H | H ammonia	NH_3 ammonia
H ·· :O:H ·· water	H | O—H water	H_2O water
H ·· H :C: H ·· H methane	H | H—C—H | H methane	CH_4 methane

Figure 2.7   Electron-dot, structural, and molecular formulas.
In the electron-dot formula, only the electrons in the outer shell are designated. In the structural formula, the lines represent a pair of electrons being shared by two atoms. The molecular formula indicates only the number of each type of atom found within a molecule.

in the outer shell can be designated by dots. The shared electrons are placed between the two sharing atoms, as shown here:

$$:\!\overset{..}{\underset{..}{Cl}}\!\cdot \;+\; \cdot\overset{..}{\underset{..}{Cl}}\!: \;\longrightarrow\; :\!\overset{..}{\underset{..}{Cl}} : \overset{..}{\underset{..}{Cl}}\!:$$

Because electron-dot structures are cumbersome, other representations are often used. Structural formulas use straight lines to show the covalent bonds between the atoms. Each line represents a pair of shared electrons. Molecular formulas indicate only the number of each type of atom making up a molecule.

Structural formula: Cl—Cl

Molecular formula: Cl_2

Additional examples of electron-dot, structural, and molecular formulas are shown in Figure 2.7. In each instance, notice that each atom has a completed outer shell.

Double and Triple Bonds   Besides a single bond, in which atoms share only a pair of electrons, a double or a triple bond can form. In a double bond, atoms share two pairs of electrons, and in a triple bond, atoms share three pairs of electrons between them. For example, in Figure 2.6, each nitrogen atom (N) requires 3 electrons to achieve a total of 8 electrons in the outermost shell. Notice that 6 electrons are placed in the outer overlapping shells in the diagram and that three straight lines are in the structural formula for nitrogen gas (N_2). Single covalent bonds between atoms are quite strong, but double and triple bonds are even stronger.

A covalent bond arises when atoms share electrons. In double covalent bonds, atoms share two pairs of electrons, and in triple covalent bonds, atoms share three pairs of electrons.

Shape of Molecules

Structural formulas make it seem as if molecules are one-dimensional, but actually molecules have a three-dimensional shape that often determines their biological function. Molecules consisting of only two atoms are always linear, but a molecule such as methane (CH_4) with five atoms has a tetrahedral shape. Why? Because carbon is sharing electrons with four hydrogen atoms. A so-called space-filling model comes closest to the actual shape of the molecule. In space-filling models, each type of atom is given a particular color; here observe that carbon is black and hydrogen is white:

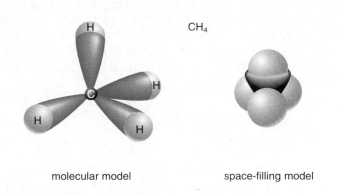

molecular model   space-filling model

The shapes of molecules are necessary to the structural and functional roles they play in living things. For example, hormones have shapes that allow them to be recognized by the cells in the body. One form of diabetes occurs when certain cell receptors fail to recognize the hormone insulin. On the other hand, AIDS occurs when certain blood cells have receptors that bind to HIV, allowing the viruses to enter, multiply, and destroy the cell.

The final shape of the molecule often determines the role it plays in cells and organisms.

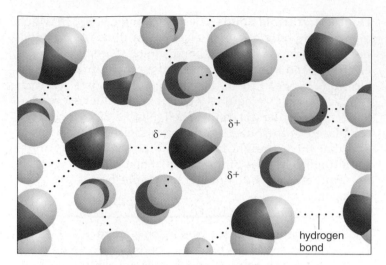

Figure 2.8   Hydrogen bonding between water molecules.
The polarity of the water molecules allows hydrogen bonds (dotted lines) to form between the molecules.

2.2 Water and Living Things

Water is the most abundant molecule in living organisms, making up about 60–70% of the total body weight of most of them. We will see that the physical and chemical properties of water make life as we know it possible.

In water, the electrons spend more time circling the larger oxygen (O) atom than the smaller hydrogen (H) atoms. When the electrons spend more time near the oxygen, they impart a slight negative charge to the oxygen and a slight positive charge to the hydrogen atoms, creating negative and positive ends of the molecule. Therefore, water is a polar molecule; the oxygen end of the molecule has a slight negative charge (δ^-), and the hydrogen end has a slight positive charge (δ^+):

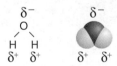

The diagram on the left shows the structural formula of water, and the one on the right shows the space-filling model of water.

Hydrogen Bonds

A **hydrogen bond** occurs whenever a covalently bonded hydrogen is positive and attracted to a negatively charged atom some distance away. A hydrogen bond is represented by a dotted line because it is relatively weak and can be broken rather easily.

In Figure 2.8, you can see that each hydrogen atom, being slightly positive, bonds to the slightly negative oxygen atom of another water molecule.

Properties of Water

Because of their polarity and hydrogen bonding, water molecules are cohesive, meaning that they cling together. Polarity and hydrogen bonding cause water to have many characteristics beneficial to life.

1. Water is a liquid at room temperature. Therefore, we are able to drink it, cook with it, and bathe in it.

Compounds with a low molecular mass are usually gases at room temperature. For example, oxygen (O_2), with a molecular mass of 32, is a gas, but water, with a molecular mass of 18, is a liquid. The hydrogen bonding between water molecules keeps water a liquid and not a gas at room temperature. Water does not boil and become a gas until 100°C, one of the reference points for the Celsius temperature scale. (See Appendix C.) Without hydrogen bonding between water molecules, our body fluids—and indeed our bodies—would be gaseous!

2. Water is the universal solvent for polar (charged) molecules and thereby facilitates chemical reactions both outside and within our bodies.

When ions and molecules disperse in water, they move about and collide, allowing reactions to occur. Therefore, water is a solvent that facilitates chemical reactions. For example, when a salt such as sodium chloride (NaCl) is put into water, the negative ends of the water molecules are attracted to the sodium ions, and the positive ends of the water molecules are attracted to the chloride ions. This causes the sodium ions and the chloride ions to separate and to dissolve in water:

The salt NaCl dissolves in water.

Ions and molecules that interact with water are said to be **hydrophilic**. Nonionized and nonpolar molecules that do not interact with water are said to be **hydrophobic**.

3. Water molecules are cohesive, and therefore liquids fill vessels, such as blood vessels.

Water molecules cling together because of hydrogen bonding, and yet, water flows freely. This property allows dissolved and suspended molecules to be evenly distributed throughout a system. Therefore, water is an excellent transport medium. Within our bodies, the blood that fills our arteries and veins is 92% water. Blood transports oxygen and nutrients to the cells and removes wastes such as carbon dioxide from cells.

a. b. c.

Figure 2.9   Characteristics of water.
a. Water boils at 100°C. If it boiled and was a gas at a lower temperature, life could not exist. **b.** It takes much body heat to vaporize sweat, which is mostly liquid water, and this helps keep bodies cool when the temperature rises. **c.** Ice is less dense than water, and it forms on top of water, making skate sailing possible.

4. The temperature of liquid water rises and falls slowly, preventing sudden or drastic changes.

The many hydrogen bonds that link water molecules cause water to absorb a great deal of heat before it boils (Fig. 2.9a). A **calorie** of heat energy raises the temperature of one gram of water 1°C. This is about twice the amount of heat required for other covalently bonded liquids. On the other hand, water holds heat, and its temperature falls slowly. Therefore, water protects us and other organisms from rapid temperature changes and helps us maintain our normal internal temperature. This property also allows great bodies of water, such as oceans, to maintain a relatively constant temperature. Water is a good temperature buffer.

5. Water has a high heat of vaporization, keeping the body from overheating.

It takes a large amount of heat to change water to steam (Fig. 2.9a). (Converting one gram of the hottest water to steam requires an input of 540 calories of heat energy.) This property of water helps moderate the earth's

temperature so that life can continue to exist. Also, in a hot environment, animals sweat and the body cools as body heat is used to evaporate sweat, which is mostly liquid water (Fig. 2.9b).

6. Frozen water is less dense than liquid water, so that ice floats on water.

As water cools, the molecules come closer together. They are densest at 4°C, but they are still moving about. At temperatures below 4°C, there is only vibrational movement, and hydrogen bonding becomes more rigid but also more open. This makes ice less dense. Bodies of water always freeze from the top down, making skate sailing possible (Fig. 2.9c). When a body of water freezes on the surface, the ice acts as an insulator to prevent the water below it from freezing. Thus, aquatic organisms are protected and have a better chance of surviving the winter.

Because of its polarity and hydrogen bonding, water has many characteristics that benefit life.

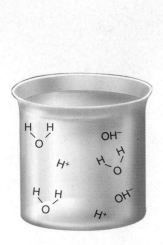

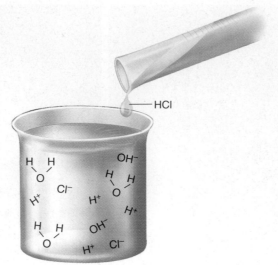

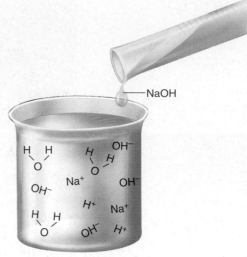

Figure 2.10   Dissociation of water molecules.
Dissociation produces an equal number of hydrogen ions (H^+) and hydroxide ions (OH^-). (These illustrations are not meant to be mathematically accurate.)

Figure 2.11   Addition of hydrochloric acid (HCl).
HCl releases hydrogen ions (H^+) as it dissociates. The addition of HCl to water results in a solution with more H^+ than OH^-.

Figure 2.12   Addition of sodium hydroxide (NaOH), a base.
NaOH releases OH^- as it dissociates. The addition of NaOH to water results in a solution with more OH^- than H^+.

Acidic and Basic Solutions

When water dissociates (breaks up), it releases an equal number of hydrogen ions (H^+) and hydroxide ions (OH^-):

$$H-O-H \rightleftharpoons H^+ + OH^-$$
$$\text{water} \qquad \text{hydrogen} \quad \text{hydroxide}$$
$$\text{ion} \qquad \text{ion}$$

Only a few water molecules at a time are dissociated (Fig. 2.10). The actual number of ions is 10^{-7} moles/liter. A mole is a unit of scientific measurement for atoms, ions, and molecules.[1]

Acidic Solutions

Lemon juice, vinegar, tomato juice, and coffee are all familiar acidic solutions. What do they have in common? Acidic solutions have a sharp or sour taste, and therefore we sometimes associate them with indigestion. To a chemist, **acids** are molecules that dissociate in water, releasing hydrogen ions (H^+). For example, an important acid in the laboratory is hydrochloric acid (HCl), which dissociates in this manner:

$$HCl \rightarrow H^+ + Cl^-$$

Dissociation is almost complete; therefore, HCl is called a strong acid. When hydrochloric acid is added to a beaker of water, the number of hydrogen ions increases (Fig. 2.11).

Basic Solutions

Milk of magnesia and ammonia are basic solutions that most people have heard of. Bases have a bitter taste and feel slippery when in water. To a chemist, **bases** are molecules that either take up hydrogen ions (H^+) or release hydroxide ions (OH^-). For example, an important inorganic base is sodium hydroxide (NaOH), which dissociates in this manner:

$$NaOH \rightarrow Na^+ + OH^-$$

Dissociation is almost complete; therefore, sodium hydroxide is called a strong base. If sodium hydroxide is added to a beaker of water, the number of hydroxide ions increases (Fig. 2.12).

It is not recommended that you taste a strong acid or base, because they are quite destructive to cells. Any container of household cleanser, such as ammonia, has a poison symbol and carries a strong warning not to ingest the product.

The Litmus Test

A simple laboratory test for acids and bases is called the litmus test. Litmus is a vegetable dye that changes color from blue to red in the presence of an acid and from red to blue in the presence of a base. The litmus test has become a common figure of speech, as when you hear a commentator say, "The litmus test for a Republican is . . ."

---

[1]A mole is the same amount of atoms, molecules, or ions as the number of atoms in exactly 12 grams of ^{12}C.

The pH Scale

The **pH**[2] **scale** is used to indicate the acidity or basicity (alkalinity) of a solution. There are normally few hydrogen ions (H^+) in a solution, and the pH scale was devised to eliminate the use of cumbersome numbers. The pH scale (Fig. 2.13) ranges from 0 to 14. A pH of 0 to 7 is an acidic solution, and a pH of 7 to 14 is a basic solution. Further, as we move down the pH scale from pH 14 to pH 0, each unit has 10 times the $[H^+]$ of the previous unit. As we move up the scale from 0 to 14, each unit has 10 times the $[OH^-]$ of the previous unit. For example, the possible hydrogen ion concentrations of a solution (in moles per liter) are on the left of this listing and the pH is on the right:

moles/liter

1×10^{-6} $[H^+]$ = pH 6
1×10^{-7} $[H^+]$ = pH 7 (neutral)
1×10^{-8} $[H^+]$ = pH 8

Pure water contains only 10^{-7} moles per liter of both hydrogen ions and hydroxide ions. Therefore, a pH of exactly 7 is a neutral pH.

To further illustrate the relationship between hydrogen ion concentration and pH, consider the following question. Which of the pH values listed above indicates a higher hydrogen ion concentration $[H^+]$ than pH 7, and therefore would be an acidic solution? A number with a smaller negative exponent indicates a greater quantity of hydrogen ions than one with a larger negative exponent. Therefore, pH 6 is an acidic solution.

The Ecology Focus on page 30 describes some detrimental environmental consequences to nonliving and living things as rain and snow have become more acidic. In humans, pH needs to be maintained within a narrow range or there are health consequences. The pH of blood is around 7.4, and blood is buffered in the manner described next to keep the pH within a normal range.

Buffers and pH

A **buffer** is a chemical or a combination of chemicals that keeps pH within normal limits. Many commercial products, such as Bufferin, shampoos, or deodorants, are buffered as an added incentive for us to buy them. Buffers resist pH changes because they can take up excess hydrogen ions (H^+) or hydroxide ions (OH^-).

The pH of our blood is usually about 7.4, in part because it contains a combination of carbonic acid and bicarbonate ions. Carbonic acid (H_2CO_3) is a weak acid that minimally dissociates.

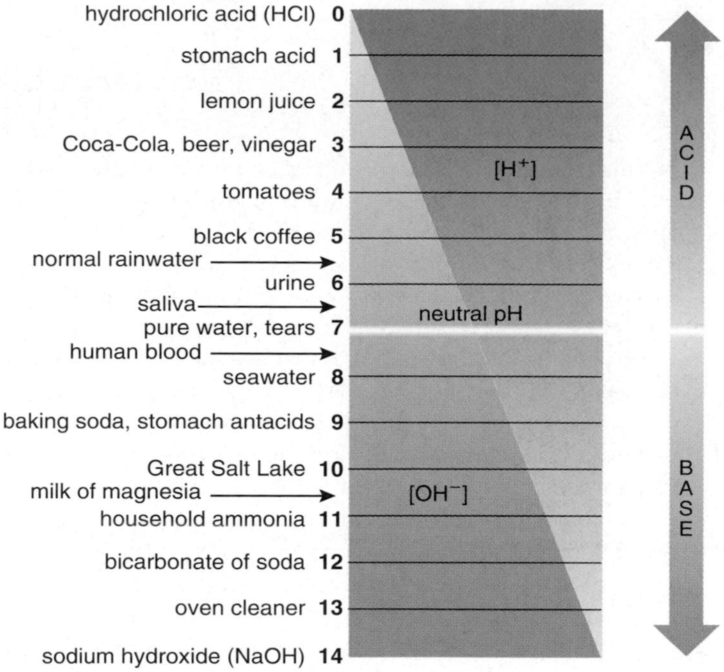

Figure 2.13 The pH scale.
The diagonal line indicates the proportionate concentration of hydrogen ions (H^+) to hydroxide ions (OH^-) at each pH value. Any pH value above 7 is basic, while any pH value below 7 is acidic.

The following reaction shows how carbonic acid dissociates and can reform:

H_2CO_3	$\underset{\text{reforms}}{\overset{\text{dissociates}}{\rightleftharpoons}}$	H^+	+	HCO_3^-
carbonic acid		hydrogen ion		bicarbonate ion

When hydrogen ions (H^+) are added to blood, the following reaction occurs:

$$H^+ + HCO_3^- \rightarrow H_2CO_3$$

When hydroxide ions (OH^-) are added to blood, this reaction occurs:

$$OH^- + H_2CO_3 \rightarrow HCO_3^- + H_2O$$

These reactions prevent any significant change in blood pH.

Acids have a pH that is less than 7, and bases have a pH that is greater than 7. Buffers, which can combine with both hydrogen ions and hydroxide ions, resist pH changes.

[2]pH is defined as the negative logarithm of the molar concentration of the hydrogen ion $[H^+]$.

The Harm Done by Acid Deposition

Normally, rainwater has a pH of about 5.6 because the carbon dioxide in the air combines with water to give a weak solution of carbonic acid. Rain falling in the northeastern United States and the southeastern Canada now has a pH of between 5.0 and 4.0. We have to remember that a pH of 4 is ten times more acidic than a pH of 5 to comprehend the increase in acidity this represents.

Very strong evidence indicates that this observed increase in rainwater acidity is a result of the burning of fossil fuels, such as coal and oil, as well as gasoline derived from oil. When fossil fuels are burned, sulfur dioxide and nitrogen oxides are produced, and they combine with water vapor in the atmosphere to form the acids sulfuric acid and nitric acid. These acids return to earth contained in rain or snow, a process properly called wet deposition, but more often called acid rain. During dry deposition, particles of sulfate and nitrate salts descend from the atmosphere.

Unfortunately, the use of tall smokestacks to reduce local air pollution only cause pollutants to be carried far from their place of origin. Acid deposition in southeastern Canada is due to the burning of fossil fuels in factories and power plants in the Midwest. Three-quarters of the acid deposition in Norway and Sweden comes from the industrialized areas of western Europe. A significant portion of Japan's acid deposition is due to emissions in China. Acid deposition adversely affects tree growth, particularly in areas where the soil is thin and lacks limestone (calcium carbonate, $CaCO_3$), whose buffering capacity prevents a lowering of pH. Acids leach aluminum from the soil, carry aluminum into the lakes, and convert mercury deposits in lake bottom sediments to soluble and toxic methyl mercury. Lakes not only become more acidic, they also show accumulation of toxic substances which can kill fish.

Pollution regulations in the United States since 1990 have reduced sulfur dioxide emissions to about 40% of what they were in 1973. But researchers say that an even greater reduction is necessary before soils and waters in the northeastern United States can recover from past acid deposition. Recovery is slow, and they predict that an overall 44% cut in emissions would still mean only a partial recovery of soil and lake health by 2050. In the meantime, the vigor of particularly red spruces and sugar maples will continue to decline. Instead of killing directly, acid deposition usually makes trees susceptible to drought and insects, which finish them off (Fig. 2A).

Trees and fish are not the only living things to suffer from acid deposition. Reduction of agricultural yields, damage to marble and limestone monuments and buildings, and even illnesses in humans have been reported. Acid deposition has been implicated in the increased incidence of lung cancer and possibly colon cancer. While acid deposition used to be primarily a problem observed on the Northeast coast, its effects are now becoming apparent in the Southeast and in the western states, as well. Tom McMillan, Canadian Minister of the Environment, says that acid rain is "destroying our lakes, killing our fish, undermining our tourism, retarding our forests, harming our agriculture, devastating our heritage, and threatening our health."

There are, of course, things that can be done. We could:

a. Use alternative energy sources, such as solar, wind, hydropower, and geothermal energy, whenever possible.
b. Use low-sulfur coal or remove the sulfur impurities from coal before it is burned.
c. Require factories and power plants to use scrubbers, which remove sulfur emissions.
d. Require people to use mass transit rather than driving their own automobiles.
e. Reduce our energy needs through other means of energy conservation.

Figure 2A Effects of acid deposition.
The burning of gasoline (left) derived from oil, a fossil fuel, causes statues (middle) to deteriorate and trees to die (right). The combustion of fossil fuels results in atmospheric acids that return to the earth in a process known as acid deposition.

2.3 Organic Molecules

Inorganic molecules constitute nonliving matter, but even so, inorganic molecules such as salts (e.g., NaCl) and water play important roles in living things. The molecules of life are organic molecules. **Organic molecules** always contain carbon (C) and hydrogen (H). The chemistry of carbon accounts for the formation of the very large variety of organic molecules found in living things. A carbon atom has four electrons in the outer shell. In order to achieve eight electrons in the outer shell, a carbon atom shares electrons covalently with as many as four other atoms, as in methane (CH_4):

$$
\begin{array}{c}
H \\
| \\
H-C-H \\
| \\
H
\end{array}
$$

A carbon atom can share with another carbon atom, and in so doing, a long hydrocarbon chain can result:

$$
H-\underset{\underset{H}{|}}{\overset{\overset{H}{|}}{C}}-\underset{\underset{H}{|}}{\overset{\overset{H}{|}}{C}}-\underset{\underset{H}{|}}{\overset{\overset{H}{|}}{C}}-\underset{\underset{H}{|}}{\overset{\overset{H}{|}}{C}}-\underset{\underset{H}{|}}{\overset{\overset{H}{|}}{C}}-\underset{\underset{H}{|}}{\overset{\overset{H}{|}}{C}}-\underset{\underset{H}{|}}{\overset{\overset{H}{|}}{C}}-\underset{\underset{H}{|}}{\overset{\overset{H}{|}}{C}}-\underset{\underset{H}{|}}{\overset{\overset{H}{|}}{C}}-H
$$

A hydrocarbon chain can also turn back on itself to form a ring compound:

So-called functional groups can be attached to carbon chains. A **functional group** is a particular cluster of atoms that always behaves in a certain way. One functional group of interest is the acidic (carboxyl) group —COOH because it can give up a hydrogen (H^+) and ionize to —COO⁻:

hydrocarbon
(hydrophobic)

acid in ionized form
(hydrophilic)

Whereas a hydrocarbon chain is *hydrophobic* (does not interact with water) because it is nonpolar, a hydrocarbon chain with an attached ionized group is *hydrophilic* (does interact with water) because it is polar.

Many molecules of life are macromolecules. Just as atoms can join to form a molecule, so molecules can join to form a macromolecule. When the same type of molecule, called a **monomer,** joins repeatedly, the macromolecule is called a **polymer.** The polymers in cells are these:

Polymer	Monomer
carbohydrate	monosaccharide
protein	amino acid
nucleic acid	nucleotide

Aside from carbohydrates, proteins, and nucleic acids, lipids are also macromolecules in cells. You are very familiar with carbohydrates, lipids, and proteins because certain foods are known to be rich in these molecules, as illustrated in Figures 2.14–2.16. The nucleic acid DNA makes up our genes, which are hereditary units that control our cells and the structure of our bodies.

Figure 2.14   Carbohydrate foods.
Breads, pasta, rice, corn, and oats all contain complex carbohydrates.

Figure 2.15   Lipid foods.
Butter and oils contain fat, the most familiar of the lipids.

Figure 2.16   Protein foods.
Meat, eggs, cheese, and beans have a high content of protein.

2.4 Carbohydrates

Carbohydrates first and foremost function for quick and short-term energy storage in all organisms, including humans. Carbohydrates play a structural role in woody plants, bacteria, and animals such as insects. In addition, carbohydrates on cell surfaces play a role in cell-to-cell recognition, as we learn in the next chapter.

Carbohydrate molecules are characterized by the presence of the atomic grouping H—C—OH, in which the ratio of hydrogen atoms (H) to oxygen atoms (O) is approximately 2:1. Since this ratio is the same as the ratio in water, the name "hydrates of carbon" seems appropriate.

Simple Carbohydrates

If the number of carbon atoms in a molecule is low (from three to seven), then the carbohydrate is a simple sugar, or **monosaccharide.** The designation **pentose** means a 5-carbon sugar, and the designation **hexose** means a 6-carbon sugar. **Glucose,** a hexose, is blood sugar (Fig. 2.17); our bodies use glucose as an immediate source of energy. Other common hexoses are fructose, found in fruits, and galactose, a constituent of milk.

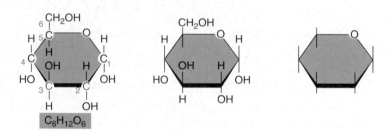

$C_6H_{12}O_6$

Figure 2.17   Three ways to represent the structure of glucose.
The *far left* structure shows the carbon atoms; $C_6H_{12}O_6$ is the molecular formula for glucose. The *far right* structure is the simplest way to represent glucose.

These three hexoses (glucose, fructose, and galactose) all occur as ring structures with the molecular formula $C_6H_{12}O_6$, but the exact shape of the ring differs, as does the arrangement of the hydrogen (—H) and hydroxyl (—OH) groups attached to the ring. A **disaccharide** (*di,* two; *saccharide,* sugar) is made by linking two monosaccharides together. Maltose is a disaccharide that contains two glucose molecules (Fig. 2.18). When glucose and fructose join, the disaccharide sucrose forms. Sucrose, which is ordinarily derived from sugarcane and sugar beets, is commonly known as table sugar.

Polysaccharides

Long polymers such as starch, glycogen, and cellulose are **polysaccharides** that contain many glucose units. Organisms have a common way of joining monomers to build polymers. **Condensation synthesis** of a larger molecule is so called because synthesis means "making of" and condensation means that water has been removed as monomers are joined. Breakdown of the larger molecule is a **hydrolysis** reaction because water is used to split bonds between monomers. Polymers are synthesized and hydrolyzed (broken down) in this manner:

$$\text{monomers} \rightleftharpoons \text{polymer} + H_2O \text{ molecules}$$

For convenience, Figure 2.18 shows how condensation synthesis results in a disaccharide called maltose and how hydrolysis of the maltose results in two glucose molecules again.

Starch and Glycogen

Starch and **glycogen** are ready storage forms of glucose in plants and animals, respectively. Some of the polymers in starch are long chains of up to 4,000 glucose units. Starch has fewer side branches, or chains of glucose that branch off from the main chain, than does glycogen, as shown in Figures 2.19 and 2.20. Flour, which we usually acquire by grinding wheat and use for baking, is high in starch, and so are potatoes.

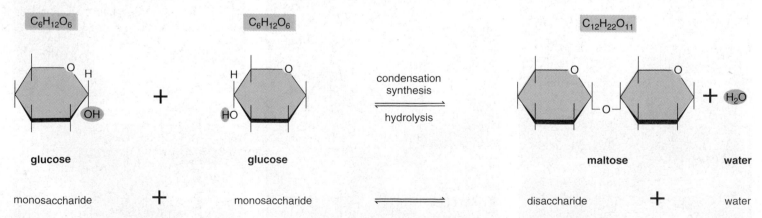

Figure 2.18   Condensation synthesis and hydrolysis of maltose, a disaccharide.
During condensation synthesis of maltose, a bond forms between the two glucose molecules, and the components of water are removed. During hydrolysis, the components of water are added, and the bond is broken.

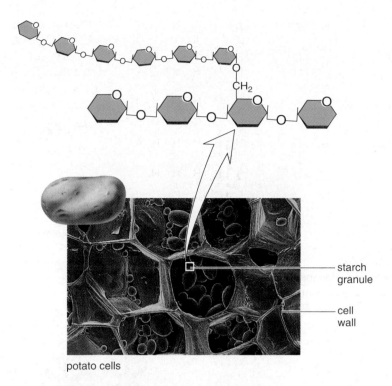

Figure 2.19   Starch structure and function.
Starch has straight chains of glucose molecules. Some chains are also branched, as indicated. The electron micrograph shows starch granules in potato cells. Starch is the storage form of glucose in plants.

After we eat starchy foods such as potatoes, bread, and cake, glucose enters the bloodstream, and the liver stores glucose as glycogen. In between eating, the liver releases glucose so that the blood glucose concentration is always about 0.1%.

Cellulose

The polysaccharide **cellulose** is found in plant cell walls, and this accounts, in part, for the strong nature of these walls. In cellulose (Fig. 2.21), the glucose units are joined by a slightly different type of linkage than that in starch or glycogen. (Observe the alternating position of the oxygen atoms in the linked glucose units.) While this might seem to be a technicality, actually it is important because we are unable to digest foods containing this type of linkage; therefore, cellulose largely passes through our digestive tract as fiber, or roughage. It is believed that fiber in the diet is necessary to good health, and some have suggested it may even help prevent colon cancer.

Cells usually use the monosaccharide glucose as an energy source. The polysaccharides starch and glycogen are storage compounds in plant and animal cells, respectively, and the polysaccharide cellulose is found in plant cell walls.

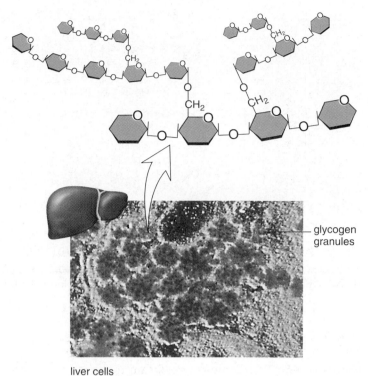

Figure 2.20   Glycogen structure and function.
Glycogen is a highly branched polymer of glucose molecules. The electron micrograph shows glycogen granules in liver cells. Glycogen is the storage form of glucose in animals.

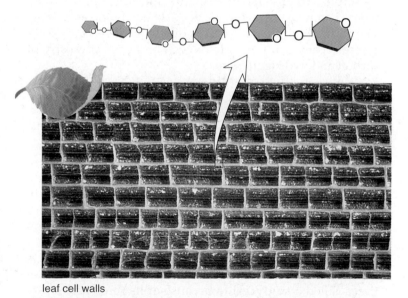

Figure 2.21   Cellulose structure and function.
Cellulose contains a slightly different type of linkage between glucose molecules than that in starch or glycogen. Plant cell walls contain cellulose, and the rigidity of the cell walls permits nonwoody plants to stand upright as long as they receive an adequate supply of water.

Figure 2.22 Condensation synthesis and hydrolysis of a fat molecule.
Fatty acids can be saturated (no double bonds between carbon atoms) or unsaturated (have double bonds, colored gray, between carbon atoms). When a fat molecule forms, three fatty acids combine with glycerol, and three water molecules are produced.

2.5 Lipids

Lipids contain more energy per gram than other biological molecules, and some function well as energy storage molecules in organisms. Others form a membrane so that the cell is separated from its environment and has inner compartments as well. The steroids are a large class of lipids that includes, among others, the sex hormones.

Lipids are diverse in structure and function, but they have a common characteristic: they do not dissolve in water. Their low solubility in water is due to an absence of polar groups. They contain little oxygen and consist mostly of carbon and hydrogen atoms.

Fats and Oils

The most familiar lipids are those found in fats and oils. **Fats,** which are usually of animal origin (e.g., lard and butter), are solid at room temperature. **Oils,** which are usually of plant origin (e.g., corn oil and soybean oil), are liquid at room temperature. Fat has several functions in the body: it is used for long-term energy storage, it insulates against heat loss, and it forms a protective cushion around major organs.

Fats and oils form when one glycerol molecule reacts with three fatty acid molecules (Fig. 2.22). A fat is sometimes called a **triglyceride** because of its three-part structure, and the term neutral fat is sometimes used because the molecule is nonpolar.

Emulsification

Emulsifiers can cause fats to mix with water. They contain molecules with a nonpolar end and a polar end. The molecules position themselves about an oil droplet so that their nonpolar ends project. Now the droplet disperses in water, which means that **emulsification** has occurred.

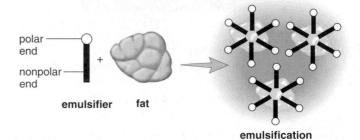

Emulsification occurs when dirty clothes are washed with soaps or detergents. Also, prior to the digestion of fatty foods, fats are emulsified by bile. A person who has had the gallbladder removed may have trouble digesting fatty foods because this organ stores bile for emulsifying fats prior to the digestive process.

Saturated and Unsaturated Fatty Acids

A **fatty acid** is a hydrocarbon chain that ends with the acidic group —COOH (Fig. 2.22). Most of the fatty acids in cells contain 16 or 18 carbon atoms per molecule, although smaller ones with fewer carbons are also known.

Fatty acids are either saturated or unsaturated. **Saturated fatty acids** have no double covalent bonds between carbon atoms. The carbon chain is saturated, so to speak, with all the hydrogens it can hold. Saturated fatty acids account for the solid nature at room temperature of fats such as lard and butter. **Unsaturated fatty acids** have double bonds between carbon atoms wherever the number of hydrogens is less than two per carbon atom. Unsaturated fatty acids account for the liquid nature of vegetable oils at room temperature. Hydrogenation of vegetable oils can convert them to margarine and products such as Crisco.

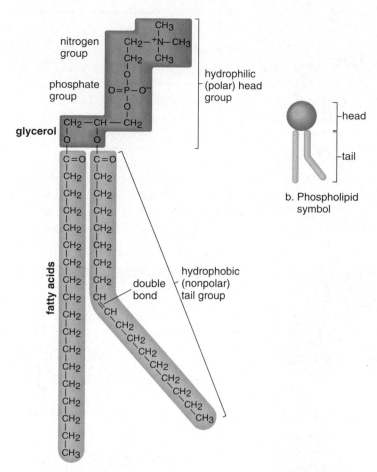

a. Lecithin, a phospholipid

b. Phospholipid symbol

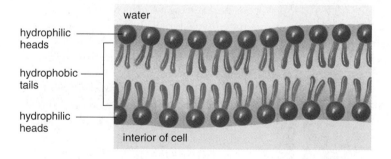

c. Phospholipid bilayer in plasma membrane

Figure 2.23   **Phospholipid structure and shape.**
a. Phospholipids are constructed like fats, except that they contain a phosphate group. This phospholipid also includes an organic group that contains nitrogen. **b.** The hydrophilic portion of the phospholipid molecule (head) is soluble in water, whereas the two hydrocarbon chains (tails) are not. **c.** This structure causes the molecules to arrange themselves within a plasma membrane as shown because the exterior and interior of a cell are mostly water.

Figure 2.24   **Steroid diversity.**
a. Cholesterol, like all steroid molecules, has four adjacent rings, but the effects of steroids on the body largely depend on the attached groups indicated in red. **b.** Testosterone is the male sex hormone.

Phospholipids

Phospholipids, as their name implies, contain a phosphate group (Fig. 2.23). Essentially, they are constructed like fats, except that in place of the third fatty acid, there is a phosphate group or a grouping that contains both phosphate and nitrogen. These molecules are not electrically neutral, as are fats, because the phosphate and nitrogen-containing groups are ionized. They form the so-called hydrophilic head of the molecule, while the rest of the molecule becomes the hydrophobic tails. Phospholipids are the backbone of cellular membranes; they spontaneously form a bilayer in which the hydrophilic heads face outward toward watery solutions and the tails form the hydrophobic interior.

Steroids

Steroids have a backbone of four fused carbon rings. Each one differs primarily by the arrangement of the atoms in the rings and the type of functional groups attached to them. Cholesterol is a component of an animal cell's plasma membrane and is the precursor of several other steroids, such as the sex hormones estrogen and testosterone (Fig. 2.24).

We know that a diet high in saturated fats and cholesterol can cause fatty material to accumulate inside the lining of blood vessels, thereby reducing blood flow. As discussed in the Science Focus on page 36, nutrition labels are now required to list the calories from fat per serving and the percent daily value from saturated fat and cholesterol.

Lipids include fats and oils (for long-term energy storage) and steroids. Phospholipids, unlike other lipids, are soluble in water because they have a hydrophilic group.

Health Focus

Nutrition Labels

As of May 1994, packaged foods have a nutrition label like the one depicted in Figure 2B. The nutrition information given on this label is based on the serving size (that is, 1¼ cup, or 57 grams) of the cereal. A Calorie[3] is a measurement of energy. One serving of the cereal provides 220 Calories, of which 20 are from fat. It is also of interest that the cereal provides no cholesterol or saturated fat. The suggestion that we study nutrition labels to determine how much cholesterol and fat (whether saturated or unsaturated) they contain is based on innumerable statistical and clinical studies of three types:

1. *Clinical trials show that an elevated blood cholesterol level is a risk factor for coronary heart disease (CHD).* The Framingham Heart Study conducted in Framingham, Massachusetts, concluded that as the blood cholesterol level in over 5,000 men and women rose, so did the risk of CHD. Elevated blood cholesterol appears to be one of the three major CHD risk factors, along with smoking and high blood pressure. Other studies have shown the same. The Multiple Risk Factor Intervention Trial followed more than 360,000 men, and also found a direct relationship between an elevated blood cholesterol level and the risk of a heart attack.

2. *Clinical trials indicate that lowering high blood cholesterol levels reduces the risk of CHD.* For example, the Coronary Primary Prevention Trial found that a 9% reduction in total blood cholesterol levels produced a 19% reduction in CHD deaths and nonfatal heart attacks. The Cholesterol Lowering Atherosclerosis Study collected X-ray evidence indicating that substantially lowering the cholesterol level produces slowed progression and even regression of plaque in coronary arteries. Plaque is a buildup of soft fatty material, including cholesterol, beneath the inner linings of arteries. Plaque can accumulate to the point that blood can no longer reach the heart and a heart attack occurs.

3. *Clinical studies show a relationship between diet and blood cholesterol levels.* The National Research Council reviewed all sorts of scientific studies before concluding that the intake of saturated fatty acids raises the blood cholesterol level, while the substitution of unsaturated fats and carbohydrates in the diet lowers the blood cholesterol level. A Los Angeles Veterans Administration study and a Finnish Mental Hospital study showed that diet alone can produce reductions of 10–15% in blood cholesterol level.

The desire of consumers to follow dietary recommendations led to the development of the type of nutrition label shown in Figure 2B, which gives the amount of total fat, saturated fat, and cholesterol in a serving of food. The carbohydrate content of a food is of interest because carbohydrates aren't usually associated with health problems. In fact, carbohydrates should compose the largest proportion of the diet. Breads and cereals containing complex carbohydrates are preferable to candy and ice cream containing simple carbohydrates because they are likely to contain dietary fiber (nondigestible plant material) as well. Insoluble fiber has a laxative effect and seems to reduce the risk of colon cancer; soluble fiber combines with the cholesterol in food and prevents the cholesterol from entering the body proper.

The amount of dietary sodium (as in table salt) is of concern because excessive sodium intake has been linked to high blood pressure in some people. It is recommended that the intake of sodium be no more than 2,400 mg per day. A serving of this cereal provides what percent of this maximum amount?

Vitamins are essential requirements needed in small amounts in the diet. Each vitamin has a recommended daily intake, and the food label tells what percent of the recommended amount is provided by a serving of this cereal.

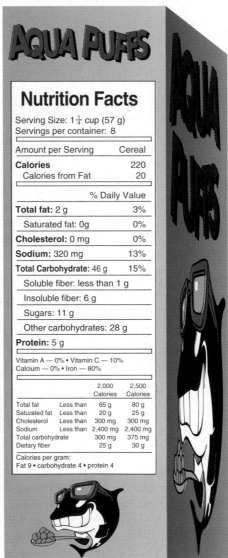

Figure 2B Nutrition label on side panel of cereal box.

---

[3]A calorie is the amount of heat required to raise the temperature of 1g of water 1°C. A Calorie (capital C), which is used to measure food energy, is equal to 1,000 calories.

Name	Structural Formula		R Group

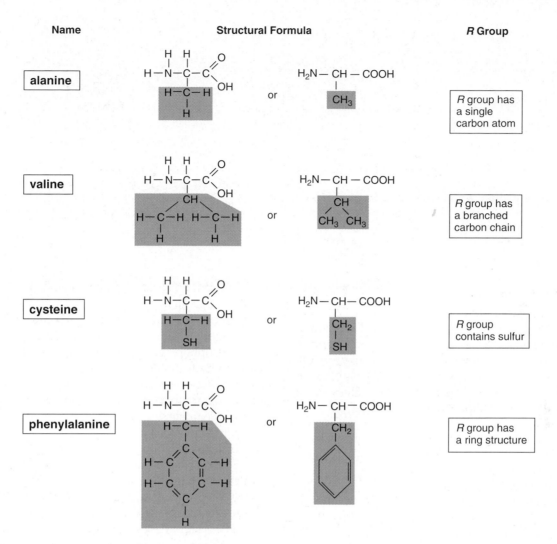

Figure 2.25   Representative amino acids.
Amino acids differ from one another by their R group; the simplest R group is a single hydrogen atom (H). The R groups (blue) that contain carbon vary as shown.

2.6 Proteins

Proteins perform many functions. Proteins such as keratin, which makes up hair and nails, and collagen, which lends support to ligaments, tendons, and skin, are structural proteins. Many hormones, which are messengers that influence cellular metabolism, are also proteins. The proteins actin and myosin account for the movement of cells and the ability of our muscles to contract. Some proteins transport molecules in the blood; hemoglobin is a complex protein in our blood that transports oxygen. Antibodies in blood and other body fluids are proteins that combine with foreign substances, preventing them from destroying cells and upsetting homeostasis.

Proteins in the plasma membrane of our cells have various functions: some form channels that allow substances to enter and exit cells; some are carriers that transport molecules into and out of the cell; and some are enzymes. Enzymes are necessary contributors to the chemical workings of the cell, and therefore of the body. **Enzymes** speed chemical reactions; they work so quickly that a reaction that normally takes several hours or days without an enzyme takes only a fraction of a second with an enzyme.

Proteins are polymers with amino acid monomers. An **amino acid** has a central carbon atom bonded to a hydrogen atom and three groups. The name of the molecule is appropriate because one of these groups is an amino group ($-NH_2$) and another is an acidic group ($-COOH$). The other group is called an R group because it is the *Remainder* of the molecule. Amino acids differ from one another by their R group; the R group varies from having a single carbon to being a complicated ring structure (Fig. 2.25).

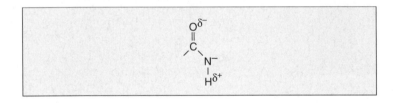

Figure 2.26   Condensation synthesis and hydrolysis of a dipeptide.
The two amino acids on the left-hand side of the equation differ by their *R* groups. As these amino acids join, a peptide bond forms, and a water molecule is produced. During hydrolysis, water is added, and the peptide bond is broken.

Peptides

Figure 2.26 shows that a condensation synthesis reaction between two amino acids results in a dipeptide and a molecule of water. A bond that joins two amino acids is called a **peptide bond.** The atoms associated with a peptide bond—oxygen (O), carbon (C), nitrogen (N), and hydrogen (H)—share electrons in such a way that the oxygen has a partial negative charge (δ^-) and the hydrogen has a partial positive charge (δ^+).

Therefore, the peptide bond is polar, and hydrogen bonding is possible between the C=O of one amino acid and the N—H of another amino acid in a polypeptide. A **polypeptide** is a single chain of amino acids.

Levels of Protein Organization

The structure of a protein has at least three levels of organization and can have four levels (Fig. 2.27). The first level, called the *primary structure*, is the linear sequence of the amino acids joined by peptide bonds. Polypeptides can be quite different from one another. If you likened a polysaccharide to a necklace that contains a single type of "bead," namely, glucose, then polypeptides make use of 20 different possible types of "beads," namely amino acids. Each particular polypeptide has its own sequence of amino acids. Therefore, each polypeptide differs by the sequence of its *R* groups.

The *secondary structure* of a protein comes about when the polypeptide takes on a certain orientation in space. A coiling of the chain results in an alpha (α) helix, or a right-handed spiral, and a folding of the chain results in a pleated sheet. Hydrogen bonding between peptide bonds holds the shape in place.

The *tertiary structure* of a protein is its final three-dimensional shape. In muscles, myosin molecules have a rod shape ending in globular (globe-shaped) heads. In enzymes, the polypeptide bends and twists in different ways. Invariably, the hydrophobic portions are packed mostly on the inside, and the hydrophilic portions are on the outside where they can make contact with water. The tertiary shape of a polypeptide is maintained by various types of bonding between the *R* groups; covalent, ionic, and hydrogen bonding all occur. One common form of covalent bonding between *R* groups is disulfide (S—S) linkages between two cysteine amino acids.

Some proteins have only one polypeptide, and others have more than one polypeptide, each with its own primary, secondary, and tertiary structures. These separate polypeptides are arranged to give some proteins a fourth level of structure, termed the *quaternary structure*. Hemoglobin is a complex protein having a quaternary structure; most enzymes also have a quaternary structure.

The final shape of a protein is very important to its function. As we will discuss in Chapter 6, for example, enzymes cannot function unless they have their usual shape. When proteins are exposed to extremes in heat and pH, they undergo an irreversible change in shape called **denaturation.** For example, we are all aware that the addition of acid to milk causes curdling and that heating causes egg white, which contains a protein called albumin, to coagulate. Denaturation occurs because the normal bonding between the *R* groups has been disturbed. Once a protein loses its normal shape, it is no longer able to perform its usual function. Researchers hypothesize that an alternation in protein organization has occurred when Alzheimer disease and Creutzfeldt-Jakob disease (the human form of "mad cow" disease) develop.

Proteins, which have levels of organization, are important in the structure and the function of cells. Some proteins are enzymes, which speed chemical reactions.

Visual Focus

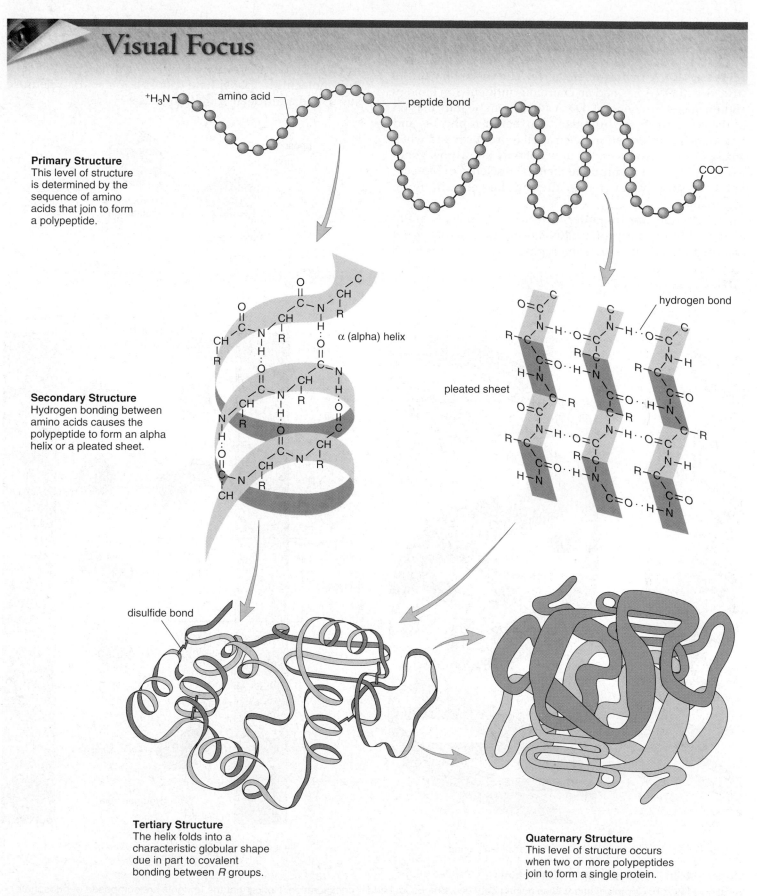

Primary Structure
This level of structure is determined by the sequence of amino acids that join to form a polypeptide.

amino acid

peptide bond

Secondary Structure
Hydrogen bonding between amino acids causes the polypeptide to form an alpha helix or a pleated sheet.

α (alpha) helix

hydrogen bond

pleated sheet

disulfide bond

Tertiary Structure
The helix folds into a characteristic globular shape due in part to covalent bonding between *R* groups.

Quaternary Structure
This level of structure occurs when two or more polypeptides join to form a single protein.

Figure 2.27   **Levels of protein organization.**

2.7 Nucleic Acids

The two types of nucleic acids are **DNA (deoxyribonucleic acid)** and **RNA (ribonucleic acid).** The discovery of the structure of DNA has had an enormous influence on biology and on society in general. DNA stores genetic information in the cell and in the organism. Further, it replicates and transmits this information when a cell reproduces and when an organism reproduces. We now not only know how genes work, but we can manipulate them. The science of biotechnology is largely devoted to altering the genes in living organisms.

DNA codes for the order in which amino acids are to be joined to form a protein. RNA is an intermediary that conveys DNA's instructions regarding the amino acid sequence in a protein.

Structure of DNA and RNA

Both DNA and RNA are polymers of nucleotides. Every **nucleotide** is a molecular complex of three types of subunit molecules—phosphate (phosphoric acid), a pentose sugar, and a nitrogen-containing base:

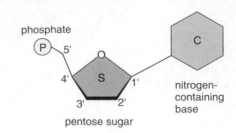

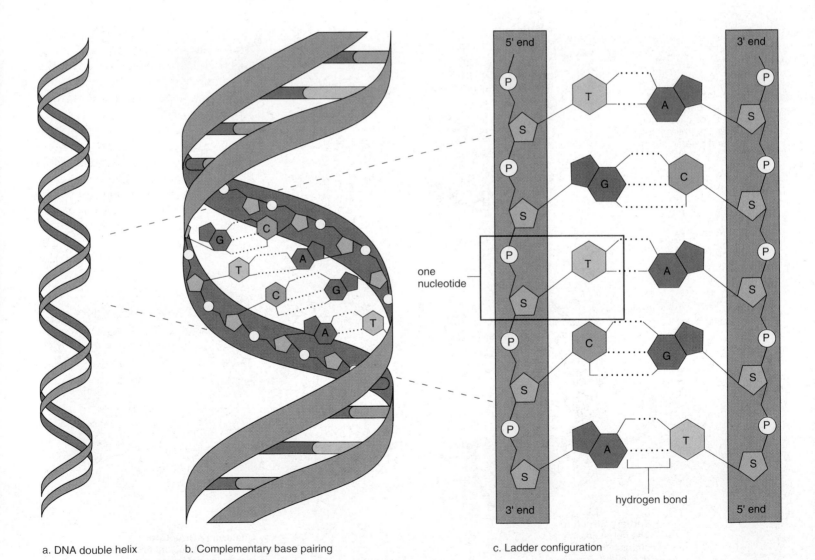

a. DNA double helix   b. Complementary base pairing

c. Ladder configuration

Figure 2.28 Overview of DNA structure.
a. Double helix. **b.** Complementary base pairing between strands. **c.** Ladder configuration. Notice that the uprights are composed of phosphate and sugar molecules and that the rungs are complementary paired bases.

The nucleotides in DNA contain the sugar deoxyribose, and the nucleotides in RNA contain the sugar ribose; this difference accounts for their respective names (Table 2.3). There are four different types of bases in DNA: A = **adenine,** T = **thymine,** G = **guanine,** and C = **cytosine.** The base can have two rings (adenine or guanine) or one ring (thymine or cytosine). These structures are called bases because their presence raises the pH of a solution. In RNA, the base **uracil** replaces the base thymine.

The nucleotides form a linear molecule called a strand, which has a backbone made up of phosphate-sugar-phosphate-sugar, with the bases projecting to one side of the backbone. Since the nucleotides occur in a definite order, so do the bases. After many years of work, researchers now know the sequence of bases along all the genes in human DNA—the human genome. This breakthrough is expected to lead to improved genetic counseling, gene therapy, and medicines to treat the cause of many human illnesses.

DNA is double stranded, with the two strands twisted about each other in the form of a double helix (Fig. 2.28). In DNA, the two strands are held together by hydrogen bonds between the bases. When unwound, DNA resembles a stepladder. The uprights (sides) of the ladder are made entirely of phosphate and sugar molecules, and the rungs of the ladder are made only of complementary paired bases. Thymine (T) always pairs with adenine (A), and guanine (G) always pairs with cytosine (C). Complementary bases have shapes that fit together.

Complementary base pairing allows DNA to replicate in a way that ensures the sequence of bases will remain the same. This sequence of the DNA bases contains a code that specifies the sequence of amino acids in the proteins of the cell. RNA is single stranded, and when it forms, complementary base pairing with one DNA strand passes this information on to RNA.

DNA has a structure like a twisted ladder: sugar and phosphate molecules make up the uprights of the ladder, and hydrogen-bonded bases make up the rungs.

Table 2.3	DNA Structure Compared to RNA Structure	
	DNA	**RNA**
Sugar	Deoxyribose	Ribose
Bases	Adenine, guanine, thymine, cytosine	Adenine, guanine, uracil, cytosine
Strands	Double stranded with base pairing	Single stranded
Helix	Yes	No

ATP (Adenosine Triphosphate)

In addition to being the monomers of nucleic acids, nucleotides have other metabolic functions in cells. When adenosine (adenine plus ribose) is modified by the addition of three phosphate groups instead of one, it becomes **ATP (adenosine triphosphate),** an energy carrier in cells. A glucose molecule contains too much energy to be used as a direct energy source in cellular reactions. Instead, the energy of glucose is converted to that of ATP molecules. ATP contains an amount of energy that makes it usable to supply energy for chemical reactions in cells.

ATP is a high-energy molecule because the last two phosphate bonds are unstable and easily broken. Usually in cells, the terminal phosphate bond is hydrolyzed, leaving the molecule **ADP (adenosine diphosphate)** and a molecule of inorganic phosphate Ⓟ (Fig. 2.29). The energy released by ATP breakdown is used by the cell to synthesize macromolecules such as carbohydrates and proteins. In muscle cells, the energy is used for muscle contraction, and in nerve cells, it is used for the conduction of nerve impulses. After ATP breaks down, it is rebuilt by the addition of Ⓟ to ADP (Fig. 2.29).

ATP is a high-energy molecule. ATP breaks down to ADP + Ⓟ, releasing energy, which is used for all metabolic work done in a cell.

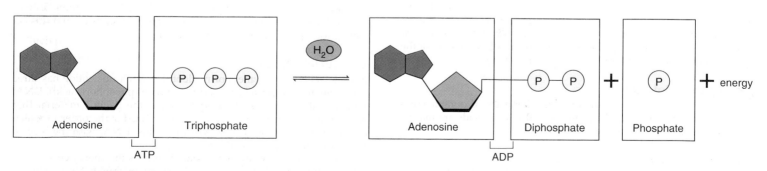

Figure 2.29 **ATP reaction.**
ATP, the universal energy currency of cells, is composed of adenosine and three phosphate groups. When cells require energy, ATP undergoes hydrolysis, producing ADP + Ⓟ, with the release of energy.

Concepts	Questions	Media Resources*

2.1   Basic Chemistry
 	
• All matter is composed of elements, each having one type of atom.
• Atoms react with one another, forming molecules and compounds.
	1. In what ways do the atoms of one element differ from those of another element?
	2. When do ionic and covalent bonds form?
		Essential Study Partner
		  Basic Chemistry
		  Atoms
		  Bonds
		
		Animation Quizzes
		  Atomic Structure
		  Covalent Bond
		  Ionic Bond

2.2   Water and Living Things
 	
• The existence of living things is dependent on the characteristics of water.
• The hydrogen ion concentration in water changes when acids or bases are added to water.
	1. List two features of water molecules that give rise to all the beneficial attributes of water.
	2. What are acids and bases, and how do living things regulate pH?
		Essential Study Partner
		  Water
		  Water Properties
		  pH
		  Acid Rain

2.3 Organic Molecules • Macromolecules are polymers that arise when their specific monomers (unit molecules) join together.	1. Describe an organic molecule. 2. What types of bonds do organic molecules contain?	Essential Study Partner Organic Chemistry

2.4   Carbohydrates
 	
• Carbohydrates function as a ready source of energy in most organisms.
• Glucose is a simple sugar; starch, glycogen, and cellulose are polymers of glucose.
• Cellulose lends structural support to plant cell walls.
	1. List several functions of carbohydrates in animal and plant cells.
	2. What monomer makes up carbohydrates?
		Essential Study Partner
		  Carbohydrates
		
		Art Quizzes
		  Disaccharides

2.5   Lipids
 	
• Lipids are varied molecules.
• Fats and oils, which function in long-term energy storage, are composed of glycerol and three fatty acids.
• Sex hormones are derived from cholesterol, a complex ring compound.
	1. How do saturated fatty acids differ from unsaturated fatty acids?
	2. How does the structure of fats differ from that of phospholipids?
		Essential Study Partner
		  Lipids
		
		Art Quizzes
		  Saturated and Unsaturated Fats

2.6   Proteins
 	
• Proteins help form structures (e.g., muscles and membranes) and function as enzymes.
• Proteins are polymers of amino acids.
	1. What are some of the many functions of proteins in cells?
	2. Why does denaturation of a protein cause it to stop functioning?
		Essential Study Partner
		  Proteins
		
		Explorations
		  How Proteins Function

2.7   Nucleic Acids
 	
• Nucleic acids are polymers of nucleotides.
• Human genes are composed of DNA (deoxyribonucleic acid). DNA specifies the correct ordering of amino acids into proteins, with RNA as intermediary.
	1. Are the building blocks of nucleic acids the same for RNA and DNA?
	2. What is the role of ATP, a modified nucleotide, in cells?
		Essential Study Partner
		  Nucleic Acids
		
		Art Quizzes
		  DNA Structure

*For additional Media Resources, see the Online Learning Center.

Cell Structure and Function

Chapter Concepts

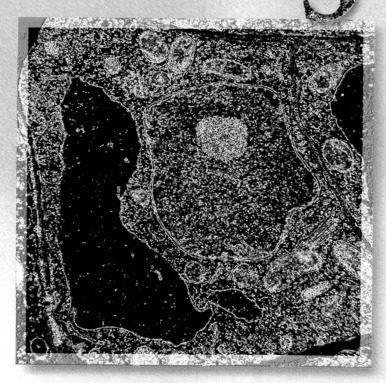

The colored micrograph of a plant cell shows you how many structures are packed inside its boundaries.

*I*f you've ever tried to stuff a week's worth of clothing, toiletries, and other necessities into a piece of carry-on luggage, you would be amazed at what cells can pack into a space smaller than the period at the end of this sentence. Like the human body, which is composed of many trillions of cells working in harmony, cells have an internal skeleton that gives them shape and controls their movement. They harbor microscopic assembly lines that manufacture a wide array of proteins. Every cell even has its own power stations that produce energy. And nestled among such structures, often in a specialized area walled off from the rest of the cell, are the all-important chromosomes, which store the instruction manual for life.

The likeness of organisms is revealed at the cellular level. Every cell, whether a plant or human cell, must carry out essentially the same processes—take in and break down nutrients, get rid of wastes, build new parts, and even divide. We will be able to study only one structure or process at a time, but it's important to remember all the activities of a cell go on at the same time. Let's begin our journey to see what cells are made of and what they do during their daily lives.

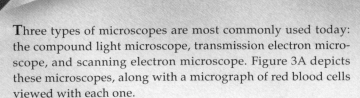

3.1 The Cellular Level of Organization

The cell marks the boundary between the nonliving and the living. The molecules that serve as food for a cell and the macromolecules that make up a cell are not alive, and yet the cell is alive. Thus, the answer to what life is must lie within the cell, because the smallest living organisms are unicellular, while larger organisms are multicellular—that is, composed of many cells. The diversity of cells is exemplified by the many types in the human body, such as muscle cells and nerve cells. But despite variety of form and function, cells contain the same components. The basic components that are common to all cells regardless of their specializations are the subject of this chapter. The Science Focus on these two pages introduces you to the microscopes most used today to study cells. Electron microscopy and biochemical analysis have revealed that the cell actually contains **organelles,** tiny specialized structures performing specific cellular functions.

Today we are accustomed to thinking of living things as being constructed of cells. But the word cell didn't enter biology until the seventeenth century. Antonie van Leeuwenhoek of Holland is now famous for making his own microscopes and observing all sorts of tiny things that no one had seen before. Robert Hooke, an Englishman, confirmed Leeuwenhoek's observations and was the first to use the term **cell.** The tiny chambers he observed in the honeycomb structure of cork reminded him of the rooms, or cells, in a monastery.

A hundred years later—in the 1830s—the German microscopist Matthias Schleiden said that plants are composed of cells; his counterpart, Theodor Schwann, said that animals are also made up of living units called cells. This was quite a feat, because aside from their own exhausting work, both had to take into consideration the studies of many other microscopists. Rudolf Virchow, another German microscopist, later came to the conclusion that cells don't suddenly appear; rather, they come from preexisting cells.

Today, the **cell theory,** which states that all organisms are made up of basic living units called cells and that cells come only from preexisting cells, is a basic theory of biology.

The cell theory states the following:

- All organisms are composed of one or more cells.
- Cells are the basic living unit of structure and function in organisms.
- All cells come only from other cells.

Three types of microscopes are most commonly used today: the compound light microscope, transmission electron microscope, and scanning electron microscope. Figure 3A depicts these microscopes, along with a micrograph of red blood cells viewed with each one.

In a compound light microscope, light rays passing through a specimen are brought to a focus by a set of glass lenses, and the resulting image is then viewed by the human eye. In the transmission electron microscope, electrons passing through a specimen are brought to a focus by a set of magnetic lenses, and the resulting image is projected onto a fluorescent screen or photographic film.

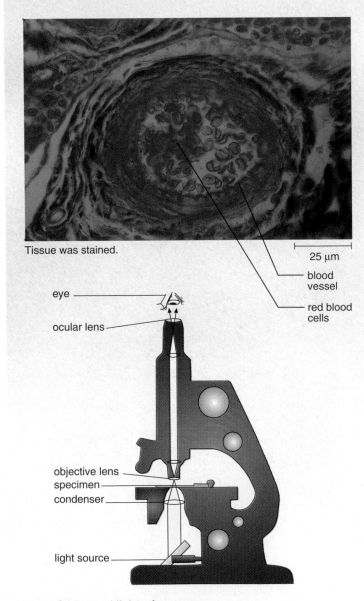

Tissue was stained.

25 μm

blood vessel

red blood cells

eye

ocular lens

objective lens

specimen

condenser

light source

Compound light microscope

Figure 3A Blood vessels and red blood cells viewed with three different types of microscopes.

Science Focus

Microscopy Today

The magnification produced by an electron microscope is much higher than that of a light microscope (50,0003 compared to 1,0003). Also, the ability of the electron microscope to make out detail is much greater. The distance needed to distinguish two points as separate is much less for an electron microscope than for a light microscope (10 nm compared to 200 nm[1]). The greater resolving power of the electron microscope is due to the fact that electrons travel at a much shorter wavelength than do light rays. However, because electrons only travel in a vacuum, the object is always dried out before viewing, whereas even living objects can be observed with a light microscope.

[1]nm = nanometer. See Appendix C, Metric System.

A scanning electron microscope provides a three-dimensional view of the surface of an object. A narrow beam of electrons is scanned over the surface of the specimen, which has been coated with a thin layer of metal. The metal gives off secondary electrons, which are collected to produce a television-type picture of the specimen's surface on a screen.

A picture obtained using a light microscope is sometimes called a photomicrograph, and a picture resulting from the use of an electron microscope is called a transmission electron micrograph (TEM) or a scanning electron micrograph (SEM), depending on the type of microscope used.

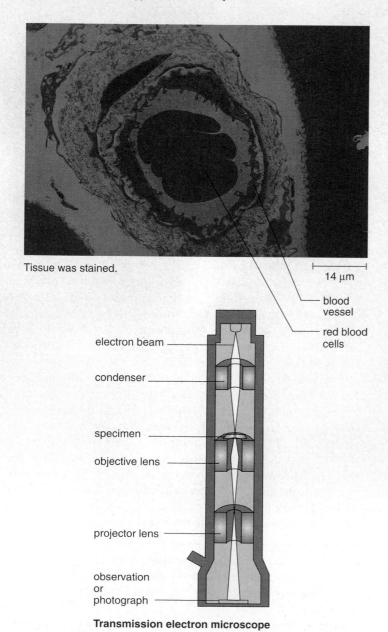

Tissue was stained.

14 μm

blood vessel

red blood cells

electron beam

condenser

specimen

objective lens

projector lens

observation or photograph

Transmission electron microscope

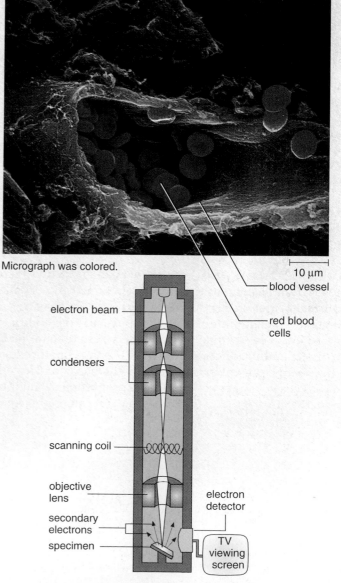

Micrograph was colored.

10 μm

blood vessel

red blood cells

electron beam

condensers

scanning coil

objective lens

electron detector

secondary electrons

specimen

TV viewing screen

Scanning electron microscope

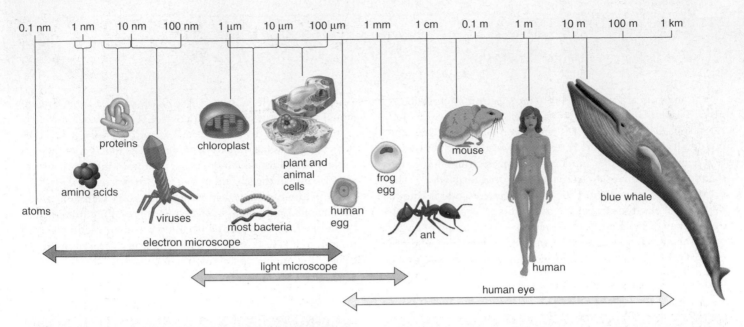

Figure 3.1   The sizes of living things and their components.
It takes a microscope to see most cells and lower levels of biological organization. Cells are visible with the light microscope, but not in much detail. An electron microscope is needed to see organelles in detail and to make out viruses and molecules. Notice that in this illustration each higher unit is 10 times greater than the lower unit. (In the metric system, 1 meter $= 10^2$ cm $= 10^3$ mm $= 10^6$ μm $= 10^9$ nm—see Appendix C.)

Cell Size

Figure 3.1 outlines the visual ranges of the eye, light microscope, and electron microscope. Cells are quite small. A frog's egg, at about one millimeter (mm) in diameter, is large enough to be seen by the human eye. But most cells are far smaller than one millimeter; some are even as small as one micrometer (μm)—one-thousandth of a millimeter. Cell inclusions and macromolecules are even smaller than a micrometer and are measured in terms of nanometers (nm).

To understand why cells are so small and why we are multicellular, consider the surface/volume ratio of cells. Nutrients enter a cell and wastes exit a cell at its surface; therefore, the amount of surface represents the ability to get material in and out of the cell. A large cell requires more nutrients and produces more wastes than a small cell. In other words, the volume represents the needs of the cell. Yet, as cells get larger in volume, the proportionate amount of surface area actually decreases, as you can see by comparing these two cells:

A small cube that is 1 mm tall has a surface area of 6 mm^2 because each side has a surface area of 1 mm^2, and 6 $\times$ 1 mm^2 is 6 mm^2. Notice that the ratio of surface area to volume is 6:1 because the surface area is 6 mm^2 and the volume is 1 mm^3. Contrast this with a larger cube that is 2 mm tall. The surface area increases to 24 mm^2 because the surface area of each side is 4 mm^2, and 6 $\times$ 4 is 24 mm^2. The volume of this larger cube is 8 mm^3 because height $\times$ width $\times$ depth is 8 mm^3. The ratio of surface area to volume of the larger cube is 3:1 because the surface area is 24 mm^2 and the volume is 8 mm^3. We can conclude then that a small cell has a greater surface area to volume ratio than does a larger cell.

Therefore, small cells, not large cells, are likely to have an adequate surface area for exchanging wastes for nutrients. We would expect, then, a size limitation for an actively metabolizing cell. A chicken's egg is several centimeters in diameter, but the egg is not actively metabolizing. Once the egg is incubated and metabolic activity begins, the egg divides repeatedly without growth. Cell division restores the amount of surface area needed for adequate exchange of materials. Further, cells that specialize in absorption have modifications that greatly increase the surface area per volume of the cell. For example, the columnar cells along the surface of the intestinal wall have surface foldings called microvilli (sing., microvillus), which increase their surface area.

small cell—
more surface area
per volume

large cell—
less surface area
per volume

A cell needs a surface area that can adequately exchange materials with the environment. Surface-area-to-volume considerations require that cells stay small.

3.2 Eukaryotic Cells

Eukaryotic cells have a nucleus, a large structure that controls the workings of the cell because it contains the genes.

Outer Boundaries of Animal and Plant Cells

All cells, including plant and animal cells, are surrounded by a **plasma membrane,** a phospholipid bilayer in which protein molecules are embedded.

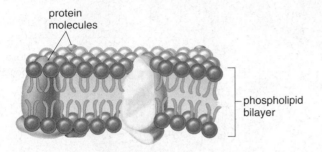

protein molecules

phospholipid bilayer

The plasma membrane is a living boundary that separates the living contents of the cell from the nonliving surrounding environment. Inside the cell, the nucleus is surrounded by the **cytoplasm,** a semifluid medium that contains organelles. The plasma membrane regulates the entrance and exit of molecules into and out of the cytoplasm.

Plant cells (but not animal cells) have a permeable but protective **cell wall** in addition to a plasma membrane. Many plant cells have both a primary and secondary cell wall. A main constituent of a primary cell wall is cellulose molecules. Cellulose molecules form fibrils that lie at right angles to one another for added strength. A cell wall sometimes forms inside the primary cell wall. Such secondary cell walls contain lignin, a substance that makes them even stronger than primary cell walls.

Organelles of Animal and Plant Cells

Animal and plant cells contain **organelles,** small bodies that have a specific structure and function. Originally the term organelle referred to only membranous structures, but we will use it to include any well-defined internal subcellular structure (Table 3.1). Still, membranes compartmentalize the cell so that its various functions are kept separate from one another. Just as all the assembly lines of a factory are in operation at the same time, so all the organelles of a cell function simultaneously. Raw materials enter a factory and then are turned into various products by different departments. In the same way, chemicals are taken up by the cell and then processed by the organelles. The cell is a beehive of activity the entire twenty-four hours of every day.

Both animal cells (Fig. 3.2) and plant cells (Fig. 3.3) contain mitochondria, while only plant cells have chloroplasts. Only animal cells have centrioles. Note that the color chosen to represent each structure in the plant and animal cell is used for that structure throughout the chapters of this part.

Table 3.1	Eukaryotic Structures in Animal Cells and Plant Cells	
Name	**Composition**	**Function**
Cell wall*	Contains cellulose fibrils	Support and protection
Plasma membrane	Phospholipid bilayer with embedded proteins	Defines cell boundary; regulation of molecule passage into and out of cells
Nucleus	Nuclear envelope nucleoplasm, chromatin, and nucleoli	Storage of genetic information; synthesis of DNA and RNA
Nucleolus	Concentrated area of chromatin, RNA, and proteins	Ribosomal subunit formation
Ribosome	Protein and RNA in two subunits	Protein synthesis
Endoplasmic reticulum (ER)	Membranous flattened channels and tubular canals	Synthesis and/or modification of proteins and other substances, and transport by vesicle formation
Rough ER	Studded with ribosomes	Protein synthesis
Smooth ER	Having no ribosomes	Various; lipid synthesis in some cells
Golgi apparatus	Stack of membranous saccules	Processing, packaging, and distribution of proteins and lipids
Vacuole and vesicle	Membranous sacs	Storage of substances
Lysosome	Membranous vesicle containing digestive enzymes	Intracellular digestion
Peroxisome	Membranous vesicle containing specific enzymes	Various metabolic tasks
Mitochondrion	Inner membrane (cristae) bounded by an outer membrane	Cellular respiration
Chloroplast*	Membranous grana bounded by two membranes	Photosynthesis
Cytoskeleton	Microtubules, intermediate filaments, actin filaments	Shape of cell and movement of its parts
Cilia and flagella	9 + 2 pattern of microtubules	Movement of cell
Centriole**	9 + 0 pattern of microtubules	Formation of basal bodies

*Plant cells only
**Animal cells only

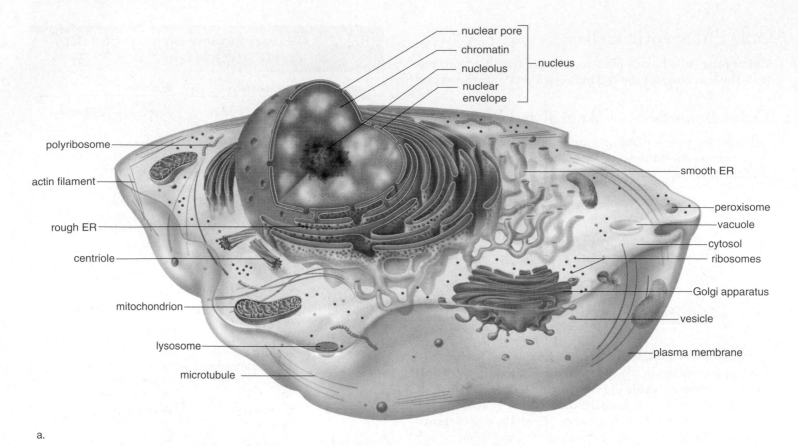

nuclear pore
chromatin
nucleolus
nuclear envelope
nucleus

polyribosome

actin filament

rough ER

centriole

mitochondrion

lysosome

microtubule

smooth ER

peroxisome

vacuole

cytosol

ribosomes

Golgi apparatus

vesicle

plasma membrane

a.

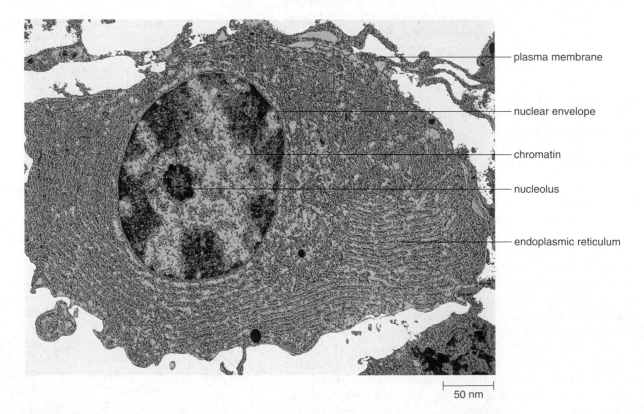

plasma membrane

nuclear envelope

chromatin

nucleolus

endoplasmic reticulum

50 nm

b.

Figure 3.2   Animal cell anatomy.
a. Generalized drawing. **b.** Transmission electron micrograph. See Table 3.1 for a description of these structures, along with a listing of their functions.

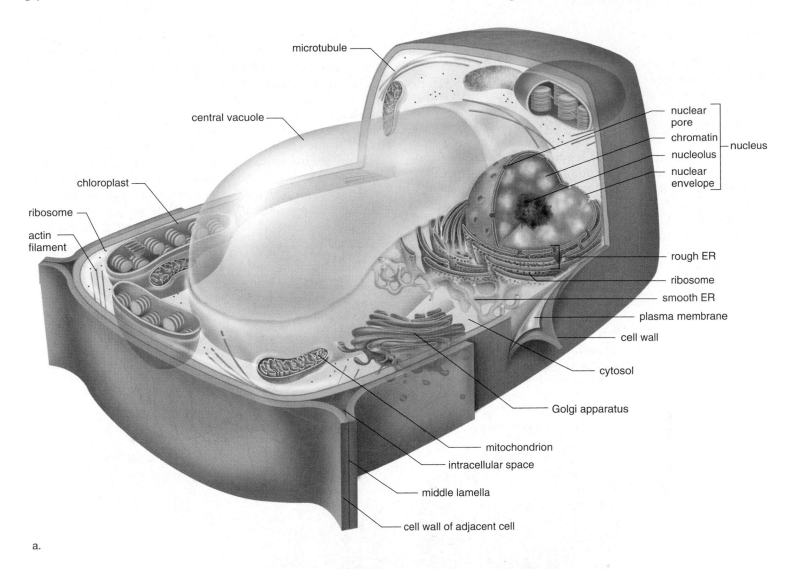

a.

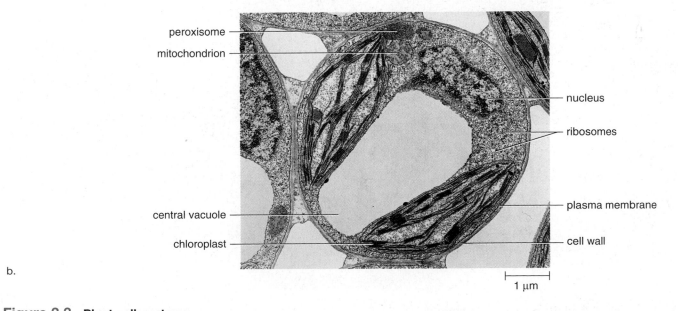

b.

Figure 3.3 Plant cell anatomy.

a. Generalized drawing. **b.** Transmission electron micrograph of a young leaf cell. See Table 3.1 for a description of these structures, along with a listing of their functions.

The Nucleus

The **nucleus,** which has a diameter of about 5 μm, is a prominent structure in the eukaryotic cell. The nucleus is of primary importance because it stores genetic information that determines the characteristics of the body's cells and their metabolic functioning. Every cell in the same individual contains the identical genetic information, but each cell type has certain genes, or segments of DNA, turned on, and others turned off. Activated DNA, with RNA acting as an intermediary, specifies the sequence of amino acids when a protein is synthesized. The proteins of a cell determine its structure and the functions it can perform.

When you look at the nucleus, even in an electron micrograph, you cannot see DNA molecules. You can see chromatin, which consists of DNA and associated proteins (Fig. 3.4). **Chromatin** looks grainy, but actually it is a threadlike material that undergoes coiling to form rodlike structures, called **chromosomes,** just before the cell divides. Chromatin is immersed in a semifluid medium called the **nucleoplasm.** A difference in pH between the nucleoplasm and the cytoplasm suggests that the nucleoplasm has a different composition.

Most likely, too, when you look at an electron micrograph of a nucleus, you will see one or more regions that look darker than the rest of the chromatin. These are nucleoli (sing., **nucleolus**) where another type of RNA, called ribosomal RNA (rRNA), is produced and where rRNA joins with proteins to form the subunits of ribosomes. (Ribosomes are small bodies in the cytoplasm that contain rRNA and proteins.)

The nucleus is separated from the cytoplasm by a double membrane known as the **nuclear envelope,** which is continuous with the endoplasmic reticulum discussed on the next page. The nuclear envelope has **nuclear pores** of sufficient size (100 nm) to permit the passage of proteins into the nucleus and ribosomal subunits out of the nucleus.

The structural features of the nucleus include the following.

Chromatin:	DNA and proteins
Nucleolus:	Chromatin and ribosomal subunits
Nuclear envelope:	Double membrane with pores

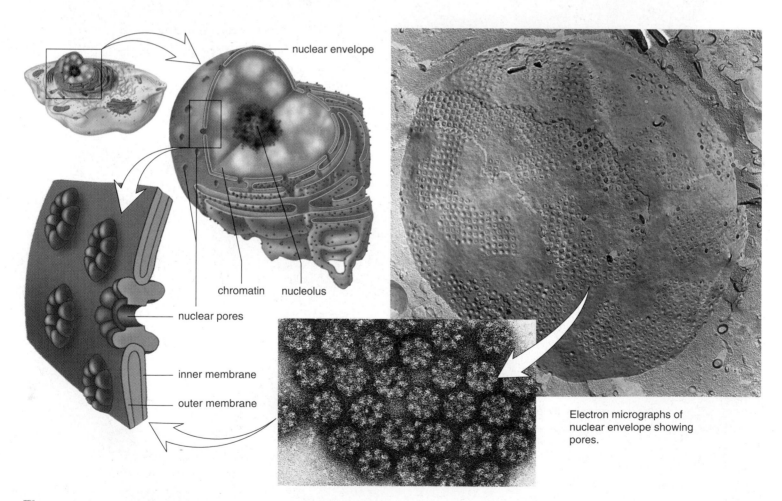

nuclear envelope

chromatin   nucleolus

nuclear pores

inner membrane

outer membrane

Electron micrographs of nuclear envelope showing pores.

Figure 3.4   The nucleus and the nuclear envelope.
The nucleoplasm contains chromatin. Chromatin has a special region called the nucleolus, where rRNA is produced and ribosomal subunits are assembled. The nuclear envelope, consisting of two membranes separated by a narrow space, contains pores. The electron micrographs show that the pores cover the surface of the envelope.

Ribosomes

Ribosomes are composed of two subunits, one large and one small. Each subunit has its own mix of proteins and rRNA. Protein synthesis occurs at the ribosomes. Ribosomes can be found free within the cytoplasm, either singly or in groups called **polyribosomes.** Ribosomes can also be found attached to the endoplasmic reticulum, a membranous system of saccules and channels discussed in the next section. Proteins synthesized at cytoplasmic ribosomes are used in the cell, such as in the mitochondria and chloroplasts. Those proteins produced at ribosomes attached to endoplasmic reticulum are eventually secreted from the cell or become a part of its external surface.

Ribosomes are small organelles where protein synthesis occurs. Ribosomes occur in the cytoplasm, both singly and in groups (i.e., polyribosomes). Numerous ribosomes are also attached to the endoplasmic reticulum.

The Endomembrane System

The endomembrane system consists of the nuclear envelope, the endoplasmic reticulum, the Golgi apparatus, and several **vesicles** (tiny membranous sacs). This system compartmentalizes the cell so that particular enzymatic reactions are restricted to specific regions. Organelles that make up the endomembrane system are connected either directly or by transport vesicles.

The Endoplasmic Reticulum

The **endoplasmic reticulum** (ER), a complicated system of membranous channels and saccules (flattened vesicles), is physically continuous with the outer membrane of the nuclear envelope. Rough ER is studded with ribosomes on the side of the membrane that faces the cytoplasm (Fig. 3.5). Here proteins are synthesized and enter the ER interior where processing and modification begin. Most of them are modified by the addition of a sugar chain, which makes them a **glycoprotein.**

Smooth ER, which is continuous with rough ER, does not have attached ribosomes. Smooth ER synthesizes the phospholipids that occur in membranes and has various other functions depending on the particular cell. In the testes, it produces testosterone, and in the liver it helps detoxify drugs. Regardless of any specialized function, smooth ER also forms vesicles in which proteins are transported to the Golgi apparatus.

ER is involved in protein synthesis (rough ER) and various other processes such as lipid synthesis (smooth ER). Vesicles transport proteins from the ER to the Golgi apparatus.

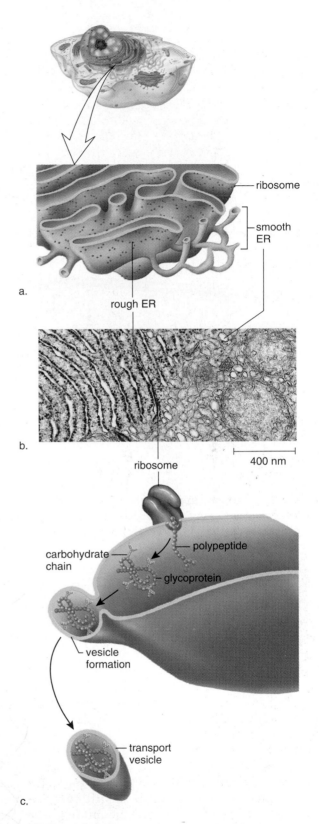

a.

b.   400 nm

c.

Figure 3.5   The endoplasmic reticulum (ER).
a. Rough ER has attached ribosomes, but smooth ER does not.
b. Rough ER appears to be flattened saccules, while smooth ER is a network of interconnected tubules. **c.** A protein made at a ribosome moves into the lumen of the system, is modified, and is eventually packaged in a transport vesicle for distribution to the Golgi apparatus.

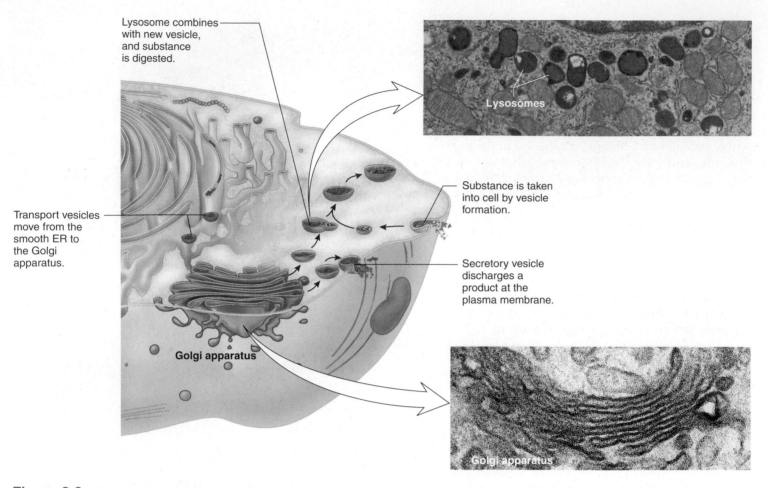

Lysosome combines with new vesicle, and substance is digested.

Lysosomes

Transport vesicles move from the smooth ER to the Golgi apparatus.

Substance is taken into cell by vesicle formation.

Secretory vesicle discharges a product at the plasma membrane.

Golgi apparatus

Golgi apparatus

Figure 3.6   The Golgi apparatus.
The Golgi apparatus receives transport vesicles containing proteins from smooth ER. After modifying the proteins, it repackages them in either secretory vesicles or lysosomes. When lysosomes combine with newly formed vesicles, their contents are digested. Lysosomes also break down cellular components.

The Golgi Apparatus

The **Golgi apparatus** is named for Camillo Golgi, who discovered its presence in cells in 1898. The Golgi apparatus consists of a stack of three to twenty slightly curved saccules whose appearance can be compared to a stack of pancakes (Fig. 3.6). In animal cells, one side of the stack (the inner face) is directed toward the ER, and the other side of the stack (the outer face) is directed toward the plasma membrane. Vesicles can frequently be seen at the edges of the saccules.

The Golgi apparatus receives protein and also lipid-filled vesicles that bud from the smooth ER. These molecules then move through the Golgi from the inner face to the outer face. How this occurs is still being debated. According to the maturation saccule model, the vesicles fuse to form an inner face saccule, which matures as it gradually becomes a saccule at the outer face. According to the stationary saccule model, the molecules move through stable saccules from the inner face to the outer face by shuttle vesicles. It is likely that both models apply, depending on the organism and the type of cell.

During their passage through the Golgi apparatus, glycoproteins have their sugar chains modified before they are repackaged in secretory vesicles. Secretory vesicles proceed to the plasma membrane, where they discharge their contents. Because this is **secretion,** the Golgi apparatus is said to be involved in processing, packaging, and secretion.

The Golgi apparatus is also involved in the formation of lysosomes, vesicles that contain proteins and remain within the cell. How does the Golgi apparatus direct traffic—in other words, what makes it direct the flow of proteins to different destinations? It now seems that proteins made at the rough ER have specific molecular tags that serve as "zip codes" to tell the Golgi apparatus whether they belong in a lysosome or secretary vesicle. The final sugar chain serves as a tag that directs proteins to their final destination.

The Golgi apparatus processes, packages, and distributes molecules about or from the cell. It is also said to be involved in secretion.

Lysosomes

Lysosomes are membrane-bounded vesicles produced by the Golgi apparatus. Lysosomes contain hydrolytic digestive enzymes.

Sometimes macromolecules are brought into a cell by vesicle formation at the plasma membrane (Fig. 3.6). When a lysosome fuses with such a vesicle, its contents are digested by lysosomal enzymes into simpler subunits that then enter the cytoplasm. For example, some white blood cells defend the body by engulfing pathogens that are then enclosed within vesicles. When lysosomes fuse with these vesicles, the bacteria are digested. It should come as no surprise then, that even parts of a cell are digested by its own lysosomes (called autodigestion). Normal cell rejuvenation takes place in this manner.

Lysosomes contain many enzymes for digesting all sorts of molecules. The absence or malfunction of one of these results in a so-called lysosomal storage disease. Instead of being degraded, the molecule accumulates inside lysosomes, and illness develops when they swell and crowd the other organelles. Occasionally, a child inherits the inability to make a lysosomal enzyme, and therefore has a lysosomal storage disease. In Tay Sachs disease, the cells that surround nerve cells cannot break down a particular lipid, and the nervous system is affected. At about six months, the infant can no longer see, and then gradually also loses hearing and even the ability to move. Death follows at about three years of age.

Lysosomes are produced by a Golgi apparatus, and their hydrolytic enzymes digest macromolecules from various sources.

Vacuoles

A **vacuole** is a large membranous sac. A vesicle is smaller than a vacuole. Animal cells have vacuoles, but they are much more prominent in plant cells. Typically, plant cells have a large central vacuole so filled with a watery fluid that it gives added support to the cell (see Fig. 3.3).

Vacuoles store substances. Plant vacuoles contain not only water, sugars, and salts but also pigments and toxic molecules. The pigments are responsible for many of the red, blue, or purple colors of flowers and some leaves. The toxic substances help protect a plant from herbivorous animals. The vacuoles present in unicellular protozoans are quite specialized, and they include contractile vacuoles for ridding the cell of excess water and digestive vacuoles for breaking down nutrients.

The endomembrane system contains the endoplasmic reticulum (rough and smooth), Golgi apparatus, lysosomes, and vacuoles.

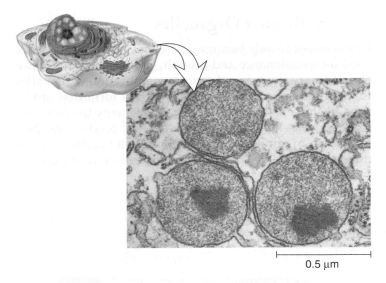

0.5 µm

Figure 3.7 **Peroxisomes.**
Peroxisomes are vesicles that oxidize organic substances with a resulting buildup of hydrogen peroxide. Peroxisomes contain the enzyme catalase, which breaks down hydrogen peroxide (H_2O_2) to water and oxygen.

Peroxisomes

Peroxisomes, similar to lysosomes, are membrane-bounded vesicles that enclose enzymes (Fig. 3.7). However, the enzymes in peroxisomes are synthesized by free ribosomes and transported into a peroxisome from the cytoplasm. All peroxisomes contain enzymes whose action results in hydrogen peroxide (H_2O_2):

$$RH_2 + O_2 \rightarrow R + H_2O_2$$

Hydrogen peroxide, a toxic molecule, is immediately broken down to water and oxygen by another peroxisomal enzyme called catalase.

The enzymes in a peroxisome depend on the function of a particular cell. However, peroxisomes are especially prevalent in cells that are synthesizing and breaking down fats. In the liver, some peroxisomes produce bile salts from cholesterol, and others break down fats. In the movie *Lorenzo's Oil*, a boy's cells lacked a carrier protein to transport a specific enzyme into peroxisomes, and he died because a type of lipid accumulated in his cells.

Plant cells also have peroxisomes. In germinating seeds, they oxidize fatty acids into molecules that can be converted to sugars needed by the growing plant. In leaves, peroxisomes can carry out a reaction that is opposite to photosynthesis—the reaction uses up oxygen and releases carbon dioxide.

The enzymes in peroxisomes produce hydrogen peroxide because they use oxygen to break down molecules.

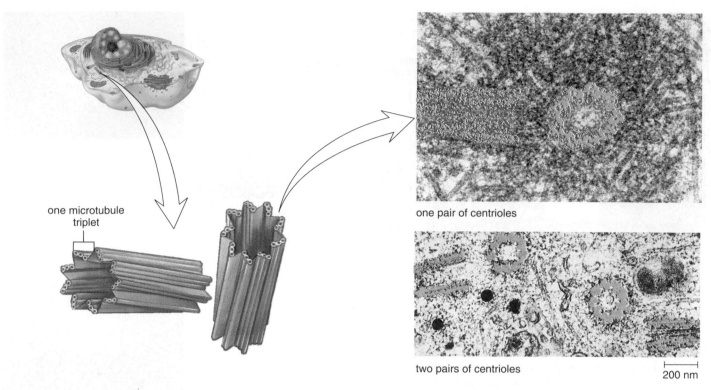

one microtubule triplet

one pair of centrioles

two pairs of centrioles

200 nm

Figure 3.11   Centrioles.

Top, right: A nondividing cell contains a pair of centrioles in a centrosome outside the nucleus. *Left and top right:* Just before a cell divides, the centrosome divides so that there are two pairs of centrioles (*Bottom, right*). During cell division, the centrosomes separate so that each new cell has one pair of centrioles.

Centrioles

Centrioles are short cylinders with a 9 + 0 pattern of microtubule triplets—that is, a ring having nine sets of triplets with none in the middle (Fig. 3.11). In animal cells, a centrosome contains two centrioles lying at right angles to each other. The centrosome is the major microtubule organizing center for the cell, and centrioles may be involved in the process of microtubule assembly and disassembly.

Before an animal cell divides, the centrioles replicate, and the members of each pair are at right angles to one another (Fig. 3.11). Then, each pair becomes part of a separate centrosome. During cell division, the centrosomes move apart and may function to organize the mitotic spindle. Plant cells have the equivalent of a centrosome, but it does not contain centrioles, suggesting that centrioles are not necessary to the assembly of cytoplasmic microtubules.

Centrioles are believed to give rise to basal bodies that direct the organization of microtubules within cilia and flagella. In other words, a basal body does for a cilium (or flagellum) what the centrosome does for the cell.

Centrioles, which are short cylinders with a 9 + 0 pattern of microtubule triplets, may be involved in microtubule formation and in the organization of cilia and flagella.

Cilia and Flagella

Cilia and **flagella** are hairlike projections that can move either in an undulating fashion, like a whip, or stiffly, like an oar. Cells that have these organelles are capable of movement. For example, unicellular paramecia move by means of cilia, whereas sperm cells move by means of flagella. The cells that line our upper respiratory tract have cilia that sweep debris trapped within mucus back up into the throat, where it can be swallowed. This action helps keep the lungs clean.

In eukaryotic cells, cilia are much shorter than flagella, but they have a similar construction. Both are membrane-bounded cylinders enclosing a matrix area. In the matrix are nine microtubule doublets arranged in a circle around two central microtubules. Therefore, they have a 9 + 2 pattern of microtubules. Cilia and flagella move when the microtubule doublets slide past one another (Fig. 3.12).

As mentioned, each cilium and flagellum has a basal body lying in the cytoplasm at its base. Basal bodies have the same circular arrangement of microtubule triplets as centrioles and are believed to be derived from them. The basal body initiates polymerization of the nine outer doublets of a cilium or flagellum

---

Cilia and flagella, which have a 9 + 2 pattern of microtubules, enable some cells to move.

Visual Focus

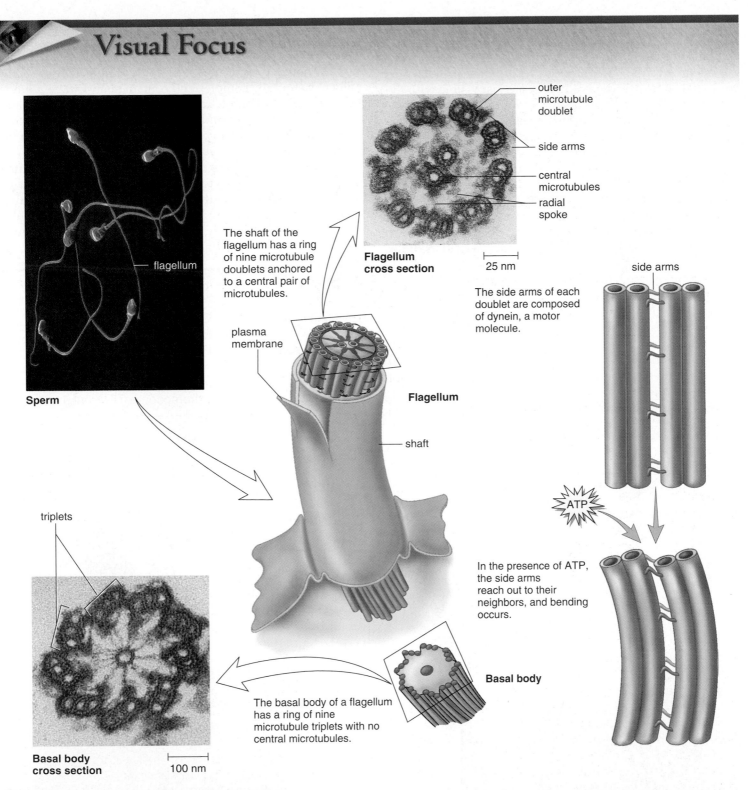

Sperm

flagellum

The shaft of the flagellum has a ring of nine microtubule doublets anchored to a central pair of microtubules.

Flagellum cross section

outer microtubule doublet

side arms

central microtubules

radial spoke

|— 25 nm

plasma membrane

Flagellum

shaft

The side arms of each doublet are composed of dynein, a motor molecule.

side arms

ATP

In the presence of ATP, the side arms reach out to their neighbors, and bending occurs.

triplets

Basal body

The basal body of a flagellum has a ring of nine microtubule triplets with no central microtubules.

Basal body cross section

100 nm

Figure 3.12 Structure of a flagellum or cilium.

A basal body derived from a centriole is at the base of a flagellum or cilium. The shaft of a flagellum (or cilium) contains microtubule doublets whose side arms are motor molecules that cause the flagellum (such as those of sperm) to move. Without the ability of sperm to move to the egg, human reproduction would not be possible.

3.3 Prokaryotic Cells

Prokaryotic cells, the archaea and bacteria, do not have the nucleus found in eukaryotic cells. Most prokaryotes are 1–10 μm in size; therefore, they are just visible with the light microscope.

Figure 3.13 illustrates the main features of bacterial anatomy. The **cell wall** contains peptidoglycan, a complex molecule with chains of a unique amino disaccharide joined by peptide chains. In some bacteria, the cell wall is further surrounded by a **capsule** and/or gelatinous sheath called a **slime layer.** Motile bacteria usually have long, very thin appendages called flagella (sing., **flagellum**) that are composed of subunits of the protein called flagellin. The flagella, which rotate like propellers, rapidly move the bacterium in a fluid medium. Some bacteria also have *fimbriae,* which are short appendages that help them attach to an appropriate surface.

Prokaryotic cells also have a **plasma membrane** which regulates the movement out of the cytoplasm. The cytoplasm consists of a semifluid medium and thousands of **ribosomes** that coordinate the synthesis of proteins. Prokaryotes have a single chromosome (loop of DNA) located within a region called the **nucleoid** because it is not bounded by membrane. Many prokaryotes also have small accessory rings of DNA called **plasmids.** In addition, the photosynthetic cyanobacteria have light-sensitive pigments, usually within the membranes of flattened disks called **thylakoids.**

Although prokaryotes seem fairly simple, they are actually metabolically complex and contain many different kinds of enzymes. Prokaryotes are adapted to living in almost any kind of environment and are diversified to the extent that almost any type of organic matter can be used as a nutrient for some particular one. Given an energy source, most prokaryotes are able to synthesize any kind of molecule they may need. Therefore, the cytoplasm is the site of thousands of chemical reactions, and they are more metabolically competent than are human beings. Indeed, the metabolic capability of bacteria is exploited by humans, who use them to produce a wide variety of chemicals and products for human use.

Bacteria are prokaryotic cells with these constant features.

Outer boundaries:	Cell wall
	Plasma membrane
Cytoplasm:	Ribosomes
	Thylakoids (cyanobacteria)
	Innumerable enzymes
Nucleoid:	Chromosome (DNA only)

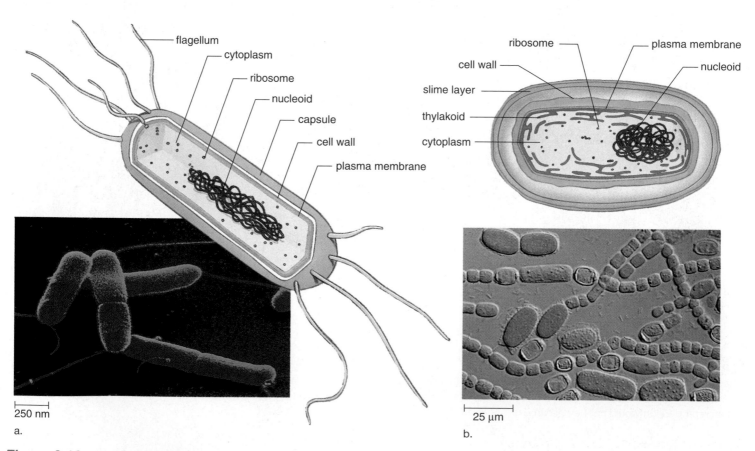

Figure 3.13   Prokaryotic cells.
a. Nonphotosynthetic bacterium. **b.** Cyanobacterium, a photosynthetic bacterium, formerly called a blue-green alga.

3.4 Evolution of the Eukaryotic Cell

How did the eukaryotic cell arise? Invagination of the plasma membrane might explain the origin of the nuclear envelope and organelles such as the endoplasmic reticulum and the Golgi apparatus. Some believe that the other organelles could also have arisen in this manner.

Another, more interesting, hypothesis has been put forth. It has been observed that in the laboratory an amoeba infected with bacteria can become dependent upon them. Some investigators believe that mitochondria and chloroplasts are derived from prokaryotes that were taken up by a much larger cell (Fig. 3.14). Perhaps mitochondria were originally aerobic heterotrophic bacteria, and chloroplasts were originally cyanobacteria. The host cell would have benefited from an ability to utilize oxygen or synthesize organic food when by chance the prokaryote was taken up and not destroyed. In other words, after these prokaryotes entered by *endocytosis*, a *symbiotic* relationship would have been established. Some of the evidence for this **endosymbiotic hypothesis** is as follows:

1. Mitochondria and chloroplasts are similar to bacteria in size and in structure.
2. Both organelles are bounded by a double membrane— the outer membrane may be derived from the engulfing vesicle, and the inner one may be derived from the plasma membrane of the original prokaryote.

3. Mitochondria and chloroplasts contain a limited amount of genetic material and divide by splitting. Their DNA (deoxyribonucleic acid) is a circular loop like that of prokaryotes.
4. Although most of the proteins within mitochondria and chloroplasts are now produced by the eukaryotic host, they do have their own ribosomes and they do produce some proteins. Their ribosomes resemble those of prokaryotes.
5. The RNA (ribonucleic acid) base sequence of the ribosomes in chloroplasts and mitochondria also suggests a prokaryotic origin of these organelles.

It is also just possible that the flagella of eukaryotes are derived from an elongated bacterium that became attached to a host cell (Fig. 3.14). However, it is important to remember that the flagella of eukaryotes are constructed differently. In any case, the acquisition of basal bodies, which could have become centrioles, may have led to the ability to form a spindle during cell division.

According to the endosymbiotic hypothesis, heterotrophic bacteria became mitochondria, and cyanobacteria became chloroplasts after being taken up by precursors to modern-day eukaryotic cells.

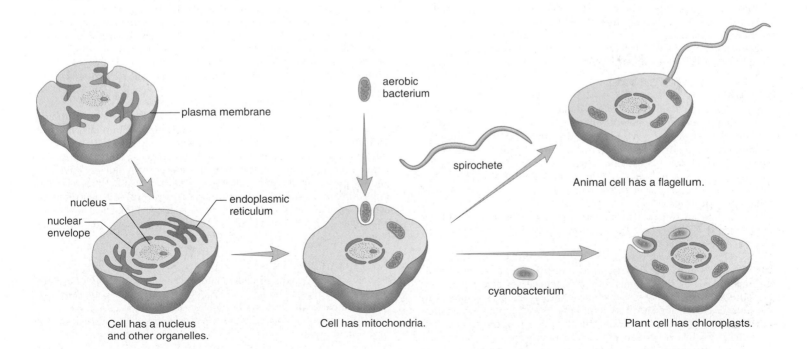

Figure 3.14 Evolution of the eukaryotic cell.
Invagination of the plasma membrane could account for the formation of the nucleus and certain other organelles. The endosymbiotic hypothesis suggests that mitochondria, chloroplasts, and flagella are derived from prokaryotes that were taken up by a much larger eukaryotic cell.

Bioethical Focus   Use of Stem Cells

Stem cells are immature cells that develop into mature, differentiated cells that make up the adult body. For example, the red bone marrow contains stem cells for all the many different types of blood cells in the bloodstream. Embryonic cells are an even more suitable source of stem cells. The early embryo is simply a ball of cells and each of these cells has the potential to become any type of cell in the body—a muscle cell, a nerve cell, or a pancreatic cell, for example.

The use of stem cells from aborted embryos or frozen embryos left over from fertility procedures is controversial. Even though quadriplegics, like Christopher Reeve and others with serious illnesses, may benefit from this research, it is difficult to get governmental approval for use of such stem cell sources. One senator said

it reminds him of the rationalization used by Nazis when they experimented on death camp inmates—"after all, they are going to be killed anyway."

Parkinson and Alzheimer are debilitating neurological disorders that people fear. It is possible that one day these disorders could be cured by supplying the patient with new nerve cells in a critical area of the brain. Suppose you had one of these disorders. Would you want to be denied a cure because the government didn't allow experimentation on human embryonic stem cells?

There are other possible sources of stem cells. It turns out that the adult body not only has blood stem cells, it also has neural stem cells in the brain. It has even been possible to coax blood stem cells and neural stem cells to become some other

types of mature cells in the body. A possible source of blood stem cells is a baby's umbilical cord and it is now possible to store umbilical blood for future use. Once researchers have the know-how, it may be possible to use any type of stem cell to cure many of the afflicting human beings.

Decide Your Opinion

1. Should researchers have access to embryonic stem cells? Any source or just certain sources? Which sources and why?
2. Should an individual have access to stem cells from just his own body? Also from a relative's body? Also from a child's umbilical cord? From embryonic cells?
3. Should differentiated cells from whatever source eventually be available for sale to patients who need them? After all, you are now able to buy artificial parts, why not living parts?

Summarizing the Concepts

3.1 The Cellular Level of Organization
All organisms are composed of cells, the smallest units of living matter. Cells are capable of self-reproduction, and new cells come only from preexisting cells. Cells are so small they are measured in micrometers. Cells must remain small in order to have an adequate amount of surface area per cell volume for exchange of molecules with the environment.

3.2 Eukaryotic Cells
The nucleus of eukaryotic cells, which include animal and plant cells, is bounded by a nuclear envelope containing pores. These pores serve as passageways between the cytoplasm and the nucleoplasm. Within the nucleus, the chromatin is a complex of DNA and protein. In dividing cells, the DNA is found in discrete structures called chromosomes. The nucleolus is a special region of the chromatin where rRNA is produced and where proteins from the cytoplasm gather to form ribosomal subunits. These subunits are joined in the cytoplasm.

Ribosomes are organelles that function in protein synthesis. They can be bound to ER or can exist within the cytoplasm singly or in groups called polyribosomes.

The endomembrane system includes the ER (both rough and smooth), the Golgi apparatus, the lysosomes, and other types of vesicles and vacuoles. The endomembrane system serves to compartmentalize the cell and keep the various biochemical reactions separate from one another. Newly produced proteins enter the ER lumen, where they may be modified before proceeding to the interior of the smooth ER. The smooth ER has various metabolic functions depending on the cell type, but it also forms vesicles that carry proteins and lipids to the Golgi apparatus. The Golgi apparatus processes proteins and repackages them into lysosomes, which carry out intracellular digestion, or into vesicles that fuse with the plasma membrane. Following fusion, secretion occurs. Vacuoles are large storage sacs, and vesicles are

smaller ones. The large single plant cell vacuole not only stores substances but also lends support to the plant cell.

Peroxisomes contain enzymes that were produced by free ribosomes in the cytoplasm. These enzymes oxidize molecules by producing hydrogen peroxide that is subsequently broken down.

Cells require a constant input of energy to maintain their structure. Chloroplasts capture the energy of the sun and carry on photosynthesis, which produces carbohydrates. Carbohydrate-derived products are broken down in mitochondria at the same time as ATP is produced. This is an oxygen-requiring process called cellular respiration.

The cytoskeleton contains actin filaments, intermediate filaments, and microtubules. These maintain cell shape and allow the cell and its organelles to move. Actin filaments, the thinnest filaments, interact with the motor molecule myosin in muscle cells to bring about contraction; in other cells, they pinch off daughter cells and have other dynamic functions. Intermediate filaments support the nuclear envelope and the plasma membrane and probably participate in cell-to-cell junctions. Microtubules radiate out from the centrosome and are present in centrioles, cilia, and flagella. They serve as tracks along which vesicles and other organelles move due to the action of specific motor molecules.

3.3 Prokaryotic Cells
The two major groups of cells are prokaryotic and eukaryotic. Both types have a plasma membrane and cytoplasm. Eukaryotic cells also have a nucleus and various organelles. Prokaryotic cells have a nucleoid that is not bounded by a nuclear envelope. They also lack most of the other organelles that compartmentalize eukaryotic cells.

3.4 Evolution of the Eukaryotic Cell
The nuclear envelope most likely evolved through invagination of the plasma membrane, but mitochondria and chloroplasts may have arisen through endosymbiotic events.

Testing Yourself

Choose the best answer for each question.

1. The cell theory states:
   a. Cells form as organelles and molecules become grouped together in an organized manner.
   b. The normal functioning of an organism does not depend on its individual cells.
   c. The cell is the basic unit of life.
   d. Only eukaryotic organisms are made of cells.

2. The small size of cells is best correlated with
   a. the fact that they are self-reproducing.
   b. their prokaryotic versus eukaryotic nature.
   c. an adequate surface area for exchange of materials.
   d. their vast versatility.
   e. All of these are correct.

3. Vesicles carrying proteins for secretion move between smooth ER and the
   a. rough ER.
   b. lysosomes.
   c. Golgi apparatus.
   d. plant cell vacuole only.
   e. cell walls of adjoining cells.

4. Lysosomes function in
   a. protein synthesis.
   b. processing and packaging.
   c. intracellular digestion.
   d. lipid synthesis.
   e. All of these are correct.

5. Mitochondria
   a. are involved in cellular respiration.
   b. break down ATP to release energy for cells.
   c. contain grana and cristae.
   d. have a convoluted outer membrane.
   e. All of these are correct.

6. Which of these is broken down during cellular respiration?
   a. carbon dioxide
   b. water
   c. carbohydrate
   d. oxygen
   e. Both c and d are correct.

7. Which of the following is NOT one of the three components of the cytoskeleton?
   a. flagella
   b. actin filaments
   c. microtubules
   d. intermediate filaments

8. Which of these is NOT true?
   a. Actin filaments are found in muscle cells.
   b. Microtubules radiate out from the ER.
   c. Intermediate filaments sometimes contain keratin.
   d. Motor molecules use microtubules as tracks.
   e. Cilia and flagella are constructed similarly.

9. Cilia and flagella
   a. bend when microtubules try to slide past one another.
   b. contain myosin that pulls on actin filaments.
   c. are organized by basal bodies derived from centrioles.
   d. are of the same length.
   e. Both a and c are correct.

10. Which of the following structures would be found in BOTH plant and animal cells?
    a. centrioles
    b. chloroplasts
    c. cell wall
    d. mitochondria
    e. All of these are found in both types of cells.

11. Which of the following organelles contains enzymes?
    a. peroxisomes
    b. lysosomes
    c. chloroplasts
    d. mitochondria
    e. All of these are correct.

12. Which of the following would NOT be in or a part of a prokaryotic cell?
    a. flagella
    b. ribosomes
    c. DNA
    d. endoplasmic reticulum
    e. enzymes

13. Which of the following organelles contains its (their) own DNA, which is evidence that supports the endosymbiotic hypothesis?
    a. Golgi apparatus
    b. mitochondria
    c. chloroplasts
    d. ribosomes
    e. Both b and c are correct.

14. Which organelle would NOT have originated by endosymbiosis?
    a. mitochondria
    b. flagella
    c. nucleus
    d. chloroplasts
    e. All of these are correct.

15. List structures found in a prokaryotic cell.
    a. cell wall, ribosomes, thylakoids, chromosome
    b. cell wall, plasma membrane, nucleus, flagellum
    c. nucleoid region, ribosomes, chloroplasts, capsule
    d. plasmid, ribosomes, enzymes, DNA, mitochondria
    e. chlorophyll, enzymes, Golgi apparatus, plasmids

16. Label these parts of the cell which are involved in protein synthesis and modification. Give a function for each structure.

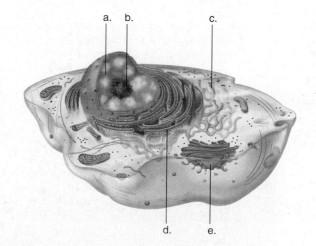

e-Learning Connection   www.mhhe.com/maderinquiry10

Concepts	Questions	Media Resources*
3.1 The Cellular Level of Organization		

• All organisms are composed of cells, which arise from preexisting cells.
• A microscope is usually needed to see a cell because most cells are quite small.
	1. List three main ideas of the cell theory.
	2. Why are cells so small?
		Essential Study Partner
		  Surface to Volume
		
		Explorations
		  Cell Size
		
		General Biology Weblinks
		  Microscopy
		
		Art Quizzes
		  Cell Size

3.2 Eukaryotic Cells		

• All cells have a plasma membrane that regulates the entrance and exit of molecules into and out of the cell. Some cells also have a protective cell wall.
• Eukaryotic cells have a number of membranous organelles that carry out specific functions.
• The nucleus controls the metabolic functions and the structural characteristics of the cell.
• A system of membranous canals and vacuoles work together to produce, modify, transport, store, secrete, and/or digest macromolecules.
• Mitochondria and chloroplasts transform one form of energy into another. Mitochondria produce ATP and chloroplasts produce carbohydrates.
• The cell has a cytoskeleton composed of microtubules, actin filaments, and intermediate filaments. The cytoskeleton gives the cell shape and allows it and its organelles to move.
• Centrioles are related to cilia and flagella, which enable the cell to move.
	1. What are the components of the endomembrane system and what is its function?
	2. What are the three cytoskeletal components of a eukaryotic cell?
		Essential Study Partner
		  Endomembrane
		  Organelles
		  Energy Organelles
		  Cytoskeleton
		
		General Biology Weblinks
		  Cell Biology
		
		BioCourse Study Guide
		  Cell Structure and Function
		
		Labeling Exercises
		  Animal Cell (1)
		  Animal Cell (2)
		  Plant Cell (1)
		  Plant Cell (2)
		  Anatomy of the Nucleus
		  Golgi Apparatus
		  Mitochondrion Structure
		  Chloroplast Structure
		  The Cytoskeleton

3.3 Prokaryotic Cells		

• In contrast to the eukaryotic cell, the prokaryotic cell lacks a well-defined nucleus.	1. Explain the statement "Although bacteria seem simple, they are metabolically diverse."
	2. What are plasmids?
		Essential Study Partner
		  Prokaryotes
		
		Labeling Exercises
		  Nonphotosynthetic
		  Bacterium
		  Cyanobacterium

3.4 Evolution of the Eukaryotic Cell		
• The endosymbiotic hypothesis suggests that certain organelles of eukaryotic cells were once prokaryotic cells.	1. How did prokaryotic cells give rise to eukaryotic cells according to the endosymbiotic hypothesis?	Essential Study Partner Eukaryotes

*For additional Media Resources, see the Online Learning Center.

Membrane Structure and Function

Chapter Concepts

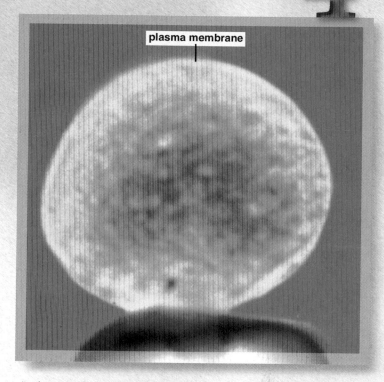

plasma membrane

A plasma membrane is the outer boundary of the cell. It regulates the contents of the cell, and affects how the cell will function in the body.

Laura liked to hang out with her friends and do all the things they did—wear nail polish and makeup, listen to rock and roll, and go to parties. But something was wrong. Her parents took her to the doctor because she didn't get a monthly period like her friends did. The news was startling. Laura's cells bore X and Y chromosomes instead of two X chromosomes like females usually do. And she didn't have the internal organs of a female. Her condition was diagnosed as androgen insensitivity. There was plenty of the male sex hormone testosterone in her blood, but her cells were unable to respond to it. So, she had the external features of a female instead of a male.

Many medical conditions that we attribute to faulty organs are actually disorders of the cell, and specifically the plasma membrane. The plasma membrane controls what gets into and out of a cell, and Laura's receptor for testosterone was faulty—the hormone couldn't get into her cells. One form of diabetes is also a disorder of the plasma membrane; in this instance, the cells do not respond to the hormone insulin. And cystic fibrosis develops when NaCl can't get out of a cell.

A plasma membrane encloses every cell, and its proper functioning is important to the health of the cell, and therefore the organism.

4.1 Plasma Membrane Structure and Function

The plasma membrane separates the internal environment of the cell from the external environment. It regulates the entrance and exit of molecules into and out of the cell. In this way, it helps the cell and the organism maintain a steady internal environment. The plasma membrane is a phospholipid bilayer in which protein molecules are either partially or wholly embedded (Fig. 4.1). The phospholipid bilayer has a *fluid* consistency, comparable to that of light oil. The proteins are scattered either just outside or within the membrane; therefore, they form a *mosaic* pattern. This description of the plasma membrane is called the **fluid-mosaic model** of membrane structure.

The hydrophilic (water-loving) polar heads of the phospholipid molecules face the outside and inside of the cell where water is found, and the hydrophobic (water-fearing) nonpolar tails face each other (Fig. 4.1). Cholesterol is another lipid found in animal plasma membranes; related steroids are found in the plasma membranes of plants. Cholesterol stiffens and strengthens the membrane, thereby helping to regulate its fluidity.

The proteins in a membrane may be peripheral proteins or integral proteins. The peripheral proteins on the inside surface of the membrane are often held in place by cytoskeletal filaments. Integral proteins are embedded in the membrane, but they can move laterally back and forth. Some integral proteins protrude from only one surface of the bilayer. Most span the membrane, with a hydrophobic region within the membrane and hydrophilic regions that protrude from both surfaces of the bilayer.

Both phospholipids and proteins can have attached carbohydrate (sugar) chains. If so, these molecules are called **glycolipids** and **glycoproteins** respectively. Since

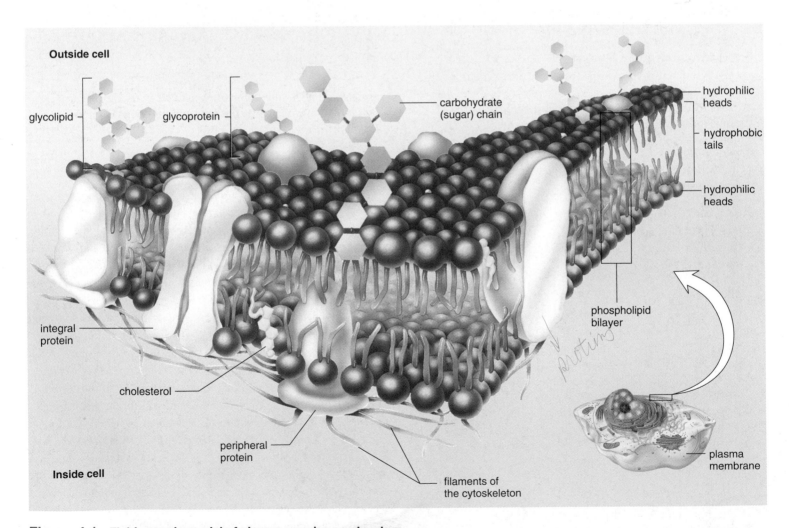

Figure 4.1   Fluid-mosaic model of plasma membrane structure.
The membrane is composed of a phospholipid bilayer in which proteins are embedded. The hydrophilic heads of phospholipids are part of both the outside surface and the inside surface of the membrane. The hydrophobic tails make up the interior of the membrane. Note the plasma membrane's asymmetry—carbohydrate chains are attached to the outside surface, and cytoskeleton filaments are attached to the inside surface.

the carbohydrate chains occur only on the outside surface and peripheral proteins occur asymmetrically on one surface or the other, the two halves of the membrane are not identical.

Functions of the Proteins

The plasma membranes of various cells and the membranes of various organelles each have their own unique collections of proteins. The integral proteins largely determine a membrane's specific functions. The plasma membrane of a red blood cell contains over 50 different types of proteins, and each has a specific function.

As we will discuss in more detail, certain plasma membrane proteins are involved in the passage of molecules through the membrane. Some of these are **channel proteins** through which a substance can simply move across the membrane; others are **carrier proteins** that combine with a substance and help it move across the membrane. Still others are receptors; each type of **receptor protein** has a shape that allows a specific molecule to bind to it. The binding of a molecule, such as a hormone (or other signal molecule), can cause the protein to change its shape and bring about a cellular response. Some plasma membrane proteins are **enzymatic proteins** that carry out metabolic reactions directly. The peripheral proteins associated with the membrane often have a structural role in that they help stabilize and shape the plasma membrane.

Figure 4.2 depicts the various functions of membrane proteins.

The Carbohydrate Chains

In animal cells, the carbohydrate chains of **cell recognition proteins** give the cell a "sugar coat," more properly called the glycocalyx. The glycocalyx protects the cell and has various other functions. For example, it facilitates adhesion between cells, reception of signal molecules, and cell-to-cell recognition.

The possible diversity of the carbohydrate (sugar) chains is enormous. The chains can vary by the number (15 is usual, but there can be several hundred) and sequence of sugars and by whether the chain is branched. Each cell within the individual has its own particular "fingerprint" because of these chains. As you probably know, transplanted tissues are often rejected by the recipient. This is because the immune system is able to recognize that the foreign tissue's cells do not have the appropriate carbohydrate chains. In humans, carbohydrate chains are also the basis for the A, B, and O blood groups.

The plasma membrane consists of a fluid phospholipid bilayer in which embedded proteins form a mosaic pattern. Carbohydrate chains project outward from the membrane.

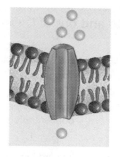

Channel Protein
Allows a particular molecule or ion to cross the plasma membrane freely. Cystic fibrosis, an inherited disorder, is caused by a faulty chloride (Cl^-) channel; a thick mucus collects in airways and in pancreatic and liver ducts.

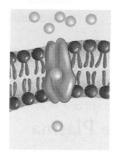

Carrier Protein
Selectively interacts with a specific molecule or ion so that it can cross the plasma membrane. A faulty carrier for glucose may be the cause of diabetes mellitus in some persons. The cells starve in the midst of plenty, and glucose spills over into the urine.

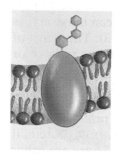

Cell Recognition Protein
The MHC (major histocompatibility complex) glycoproteins are different for each person, so organ transplants are difficult to achieve. Cells with foreign MHC glycoproteins are attacked by blood cells responsible for immunity.

Receptor Protein
Is shaped in such a way that a specific molecule can bind to it. Pygmies are short, not because they do not produce enough growth hormone, but because their plasma membrane growth hormone receptors are faulty and cannot interact with growth hormone.

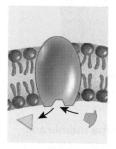

Enzymatic Protein
Catalyzes a specific reaction. Cholera bacteria release a toxin that interferes with the functioning of an enzyme that helps regulate the sodium content of cells. Sodium ions and water leave intestinal cells, and the individual may die from severe diarrhea.

Figure 4.2   **Membrane protein diversity.**
These are some of the functions performed by proteins found in the plasma membrane.

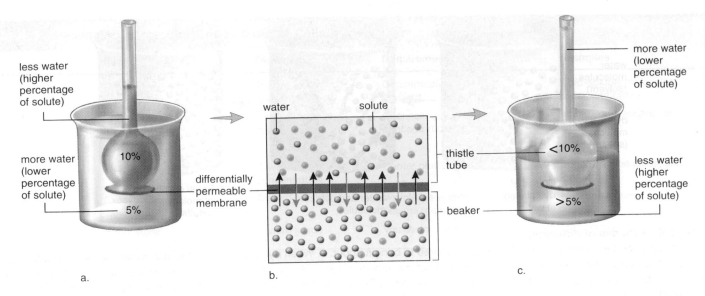

less water (higher percentage of solute)

more water (lower percentage of solute)

10%

5%

water   solute

differentially permeable membrane

thistle tube

beaker

more water (lower percentage of solute)

<10%

>5%

less water (higher percentage of solute)

a.

b.

c.

Figure 4.6   Osmosis demonstration.

a. A thistle tube, covered at the broad end by a differentially permeable membrane, contains a 10% sugar solution. The beaker contains a 5% sugar solution. **b.** The solute (gold circles) is unable to pass through the membrane, but the water (blue circles) passes through in both directions. There is a net movement of water toward the inside of the thistle tube, where there is a lower percentage of water molecules. **c.** Due to the incoming water molecules, the level of the solution rises in the thistle tube.

Osmosis

The diffusion of water across a selectively permeable membrane due to concentration differences is called **osmosis.** To illustrate osmosis, a thistle tube containing a 10% solute solution[1] is covered at one end by a differentially permeable membrane and then placed in a beaker containing a 5% sugar solution (Fig. 4.6). The beaker has a higher concentration of water molecules (lower percentage of solute), and the thistle tube has a lower concentration of water molecules (higher percentage of solute). Diffusion always occurs from higher to lower concentration. Therefore, a net movement of water takes place across the membrane from the beaker to the inside of the thistle tube.

The solute does not diffuse out of the thistle tube. Why not? Because the membrane is not permeable to the solute. As water enters and the solute does not exit, the level of the solution within the thistle tube rises (Fig. 4.6c). In the end, the concentration of solute in the thistle tube is less than 10%. Why? Because there is now less solute per unit volume. And the concentration of solute in the beaker is greater than 5%. Why? Because there is now more solute per unit volume.

Water enters the thistle tube due to the osmotic pressure of the solution within the thistle tube. **Osmotic pressure** is the pressure that develops in a system due to osmosis.[2] In other words, the greater the possible osmotic pressure, the more likely it is that water will diffuse in that direction. Due to osmotic pressure, water is absorbed by the kidneys and taken up by capillaries from tissue fluid.

Osmosis in Cells

Osmosis also occurs across the plasma membrane, as we shall now see (Fig. 4.7).

Isotonic Solution   In the laboratory, cells are normally placed in **isotonic solutions**—that is, the solute concentration and the water concentration both inside and outside the cell are equal, and therefore there is no net gain or loss of water. The prefix *iso* means the same as, and the term tonicity refers to the strength of the solution. A 0.9% solution of the salt sodium chloride (NaCl) is known to be isotonic to red blood cells. Therefore, intravenous solutions medically administered usually have this tonicity.

Hypotonic Solution   Solutions that cause cells to swell, or even to burst, due to an intake of water are said to be **hypotonic solutions.** The prefix *hypo* means less than, and refers to a solution with a lower concentration of solute (higher concentration of water) than inside the cell. If a cell is placed in a hypotonic solution, water enters the cell; the net movement of water is from the outside to the inside of the cell.

Any concentration of a salt solution lower than 0.9% is hypotonic to red blood cells. Animal cells placed in such a solution expand and sometimes burst due to the buildup of pressure. The term *lysis* is used to refer to disrupted cells; hemolysis, then, is disrupted red blood cells.

The swelling of a plant cell in a hypotonic solution creates **turgor pressure.** When a plant cell is placed in a hypotonic solution, we observe expansion of the cytoplasm

[1]Percent solutions are grams of solute per 100 ml of solvent. Therefore, a 10% solution is 10 g of solute with water added to make up 100 ml of solution.

[2]Osmotic pressure is measured by placing a solution in an osmometer and then immersing the osmometer in pure water. The pressure that develops is the osmotic pressure of a solution.

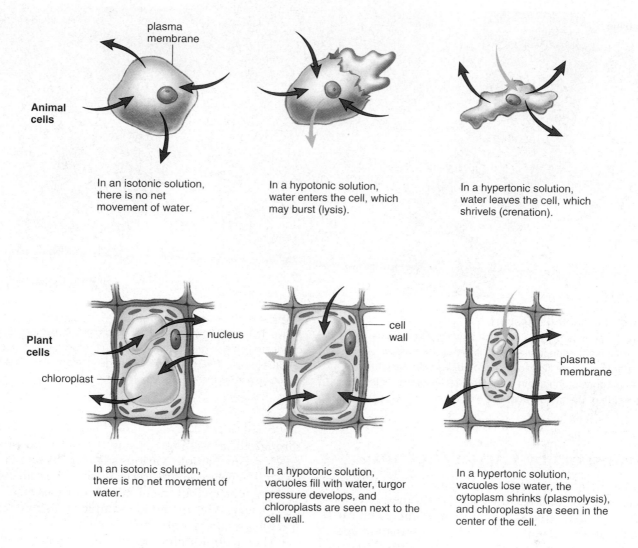

Figure 4.7   Osmosis in animal and plant cells.
The arrows indicate the movement of water molecules. To determine the net movement of water, compare the number of arrows that are taking water molecules into the cell versus the number that are taking water out of the cell. In an isotonic solution, a cell neither gains nor loses water; in a hypotonic solution, a cell gains water; and in a hypertonic solution, a cell loses water.

because the large central vacuole gains water and the plasma membrane pushes against the rigid cell wall. The plant cell does not burst because the cell wall does not give way. Turgor pressure in plant cells is extremely important to the maintenance of the plant's erect position. If you forget to water your plants, they wilt due to decreased turgor pressure.

Hypertonic Solution   Solutions that cause cells to shrink or shrivel due to loss of water are said to be **hypertonic solutions.** The prefix *hyper* means more than, and refers to a solution with a higher percentage of solute (lower concentration of water) than the cell. If a cell is placed in a hypertonic solution, water leaves the cell; the net movement of water is from the inside to the outside of the cell.

Any solution with a concentration higher than 0.9% sodium chloride is hypertonic to red blood cells. If animal cells are placed in this solution, they shrink. The term

crenation refers to red blood cells in this condition. Meats are sometimes preserved by salting them. The bacteria are not killed by the salt but by the lack of water in the meat.

When a plant cell is placed in a hypertonic solution, the plasma membrane pulls away from the cell wall as the large central vacuole loses water. This is an example of **plasmolysis,** a shrinking of the cytoplasm due to osmosis. The dead plants you may see along a salted roadside died because they were exposed to a hypertonic solution during the winter.

In an isotonic solution, a cell neither gains nor loses water. In a hypotonic solution, a cell gains water. In a hypertonic solution, a cell loses water and the cytoplasm shrinks.

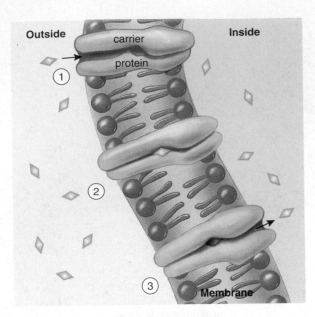

Figure 4.8   Facilitated transport.
A carrier protein speeds the rate at which a molecule crosses a membrane from higher concentration to lower concentration. (1) Molecule enters carrier. (2) Molecule combines with the carrier. (3) Carrier undergoes a change in shape that releases the molecule on the other side of the membrane.

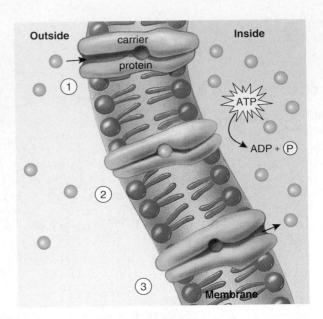

Figure 4.9   Active transport.
Active transport allows a molecule to cross the membrane from lower concentration to higher concentration. (1) Molecule enters carrier. (2) Breakdown of ATP induces a change in shape that (3) drives the molecule across the membrane.

4.4 Transport by Carrier Proteins

The plasma membrane impedes the passage of all but a few substances. Yet, biologically useful molecules do enter and exit the cell at a rapid rate because of channel and carrier proteins in the membrane. **Carrier proteins** are specific; each can combine with only a certain type of molecule, which is then transported through the membrane. How carrier proteins function is not completely understood, but after a carrier combines with a molecule, the carrier is believed to undergo a change in shape that moves the molecule across the membrane. Carrier proteins are required for facilitated transport and active transport (see Table 4.1).

Some of the proteins in the plasma membrane are carriers; they transport biologically useful molecules into and out of the cell.

Facilitated Transport

Facilitated transport explains the passage of such molecules as glucose and amino acids across the plasma membrane, even though they are not lipid soluble. The passage of glucose and amino acids is facilitated by their reversible combination with carrier proteins, which transport them through the plasma membrane. These carrier proteins are specific. For example, various sugar molecules of identical size might be present inside or outside the cell, but glucose can cross the membrane hundreds of times faster than the other sugars. This is a good example of the differential permeability of the membrane.

The carrier for glucose has been isolated, and a model has been developed to explain how it works (Fig. 4.8). It seems likely that the carrier has two conformations and that it switches back and forth between the two states. After glucose binds to the open end of a carrier, it closes behind the glucose molecule. As glucose moves along, the constricted end of the carrier opens in front of the molecule. After glucose is released into the cytoplasm of the cell, the carrier returns to its former conformation so that it can bind with glucose again. This process can occur as often as 100 times per second. Apparently, the cell has a pool of extra glucose carriers. When the hormone insulin binds to a plasma membrane receptor, more glucose carriers ordinarily appear in the plasma membrane. Some forms of diabetes are caused by insulin insensitivity; that is, the binding of insulin does not result in extra glucose carriers in the membrane.

The model shows that after a carrier has assisted the movement of a molecule to the other side of the membrane, it is free to assist the passage of other similar molecules. Neither diffusion, explained previously, nor facilitated transport requires an expenditure of chemical energy because the molecules are moving down their concentration gradient in the same direction they tend to move anyway.

Active Transport

During **active transport,** ions or molecules move through the plasma membrane, accumulating either inside or outside the cell. For example, iodine collects in the cells of the thyroid gland; nutrients are completely absorbed from the gut by the cells lining the digestive tract; and sodium ions (Na^+) can be almost completely withdrawn from urine by cells lining the kidney tubules. In these instances, substances have moved to the region of higher concentration, exactly opposite to the process of diffusion. It has been estimated that up to 40% of a cell's energy supply may be used for active transport of solute across its membrane.

Both carrier proteins and an expenditure of energy are needed to transport molecules against their concentration gradient (Fig. 4.9). In this case, energy (ATP molecules) is required for the carrier to combine with the substance to be transported. Therefore, it is not surprising that cells involved primarily in active transport, such as kidney cells, have a large number of mitochondria near the membrane through which active transport is occurring.

Proteins involved in active transport are often called *pumps,* because just as a water pump uses energy to move water against the force of gravity, proteins use energy to move a substance against its concentration gradient. One type of pump that is active in all animal cells, but is especially associated with nerve and muscle cells, moves sodium ions (Na^+) to the outside of the cell and potassium ions (K^+) to the inside of the cell. These two events are linked, and the carrier protein is called a **sodium-potassium pump.** A change in carrier shape after the attachment, and again after the detachment, of a phosphate group allows the carrier to combine alternately with sodium ions and potassium ions (Fig. 4.10). The phosphate group is donated by ATP, which is broken down enzymatically by the carrier.

The passage of salt (NaCl) across a plasma membrane is of primary importance in cells. The chloride ion (Cl^-) usually crosses the plasma membrane because it is attracted by positively charged sodium ions (Na^+). First, sodium ions are pumped across a membrane, and then chloride ions simply diffuse through channels that allow their passage. As noted in Figure 4.2, the chloride ion channels malfunction in persons with cystic fibrosis, and this leads to the symptoms of this inherited (genetic) disorder.

During facilitated transport, substances follow their concentration gradient. During active transport, substances are moved against their concentration gradient.

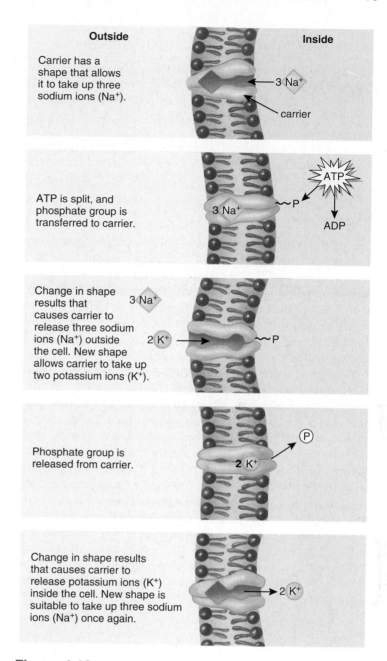

Figure 4.10 The sodium-potassium pump.
A carrier protein actively moves three sodium ions (Na^+) to the outside of the cell for every two potassium ions (K^+) pumped to the inside of the cell. Note that chemical energy of ATP is required.

4.5 Exocytosis and Endocytosis

What about macromolecules such as polypeptides, polysaccharides, or polynucleotides, which are too large to be transported by carrier proteins? They are transported into or out of the cell by vesicle formation, thereby keeping the macromolecules contained so that they do not mix with those in the cytoplasm. Vesicle formation is an energy requiring process and therefore exocytosis and endocytosis are listed as forms of active transport in Table 4.1.

Exocytosis

During **exocytosis,** vesicles fuse with the plasma membrane as secretion occurs (Fig. 4.11). Often these vesicles have been produced by the Golgi apparatus and contain proteins. Notice that during exocytosis, the membrane of the vesicle becomes a part of the plasma membrane, which is thereby enlarged. For this reason, exocytosis occurs automatically during cell growth. The proteins released from the vesicle adhere to the cell surface or become incorporated in an extracellular matrix. Some diffuse into tissue fluid where they nourish or signal other cells.

Some cells are specialized to produce and release particular molecules. In humans, molecules transported out of the cell by exocytosis include digestive enzymes, such as those produced by the pancreatic cells, and hormones, such as growth hormone produced by anterior pituitary cells. In these cells, secretory vesicles accumulate near the plasma membrane. These vesicles release their contents only when the cell is stimulated by a signal received at the plasma membrane. A rise in blood sugar, for example, signals pancreatic cells to release the hormone insulin. This is called regulated secretion, because vesicles fuse with the plasma membrane only when it is appropriate to the needs of the body.

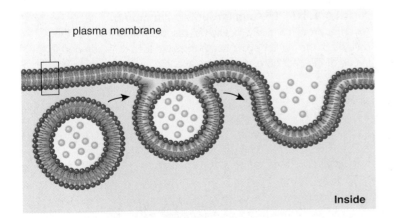

Figure 4.11 Exocytosis.
Exocytosis deposits substances on the outside of the cell and allows secretion to occur.

Endocytosis

During **endocytosis,** cells take in substances by vesicle formation. A portion of the plasma membrane invaginates to envelop the substance, and then the membrane pinches off to form an intracellular vesicle. Endocytosis occurs in one of three ways, as illustrated in Figure 4.12.

Phagocytosis

When the material taken in by endocytosis is large, such as a food particle or another cell, the process is called **phagocytosis.** Phagocytosis is common in unicellular organisms such as amoebas (Fig. 4.12a). It also occurs in humans. Certain types of human white blood cells are amoeboid—they are mobile like an amoeba, and are able to engulf debris such as worn-out red blood cells or bacteria. When an endocytic vesicle fuses with a lysosome, digestion occurs. We will see that this process is a necessary and preliminary step toward the development of immunity for bacterial diseases.

Pinocytosis

Pinocytosis occurs when vesicles form around a liquid or around very small particles. Blood cells, cells that line the kidney tubules or the intestinal wall, and plant root cells all use pinocytosis to ingest substances.

Whereas phagocytosis can be seen with the light microscope, the electron microscope must be used to observe pinocytic vesicles, which are no larger than 0.1–0.2 µm. Still, pinocytosis involves a significant amount of the plasma membrane because it occurs continuously. The loss of plasma membrane due to pinocytosis is balanced by the occurrence of exocytosis, however.

Receptor-Mediated Endocytosis

Receptor-mediated endocytosis is a form of pinocytosis that is quite specific because it uses a receptor protein shaped in such a way that a specific molecule such as a vitamin, peptide hormone, or lipoprotein can bind to it. The receptors for these substances are found at one location in the plasma membrane. This location is called a coated pit because there is a layer of protein on the cytoplasmic side of the pit. Once formed, the vesicle is uncoated and may fuse with a lysosome. If a vesicle fuses with the plasma membrane, the receptors return to their former location.

Receptor-mediated endocytosis is selective and much more efficient than ordinary pinocytosis. It is involved in uptake and also in the transfer and exchange of substances between cells. Such exchanges take place when substances move from maternal blood into fetal blood at the placenta, for example.

The importance of receptor-mediated endocytosis is demonstrated by a genetic disorder called familial hypercholesterolemia. Cholesterol is transported in blood by a complex of lipids and proteins called low-density lipopro-

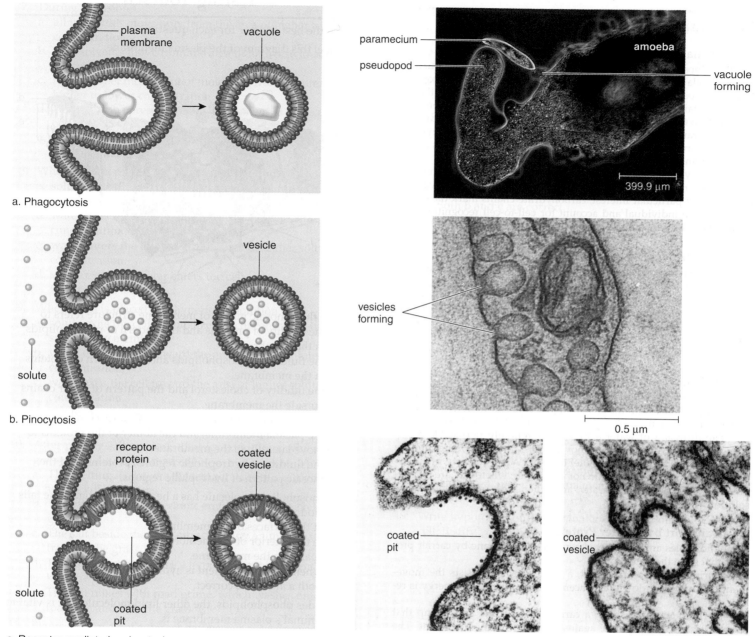

Figure 4.12 Three methods of endocytosis.
a. Phagocytosis occurs when the substance to be transported into the cell is large; amoebas ingest by phagocytosis. Digestion occurs when the resulting vacuole fuses with a lysosome. **b.** Pinocytosis occurs when a macromolecule such as a polypeptide is to be transported into the cell. The result is a vesicle (small vacuole). **c.** Receptor-mediated endocytosis is a form of pinocytosis. Molecules of substance to be taken in first bind to specific receptor proteins, which migrate to or are already in a coated pit. The vesicle that forms contains the molecules and their receptors.

tein (LDL). Ordinarily, body cells take up LDL when LDL receptors gather in a coated pit. In these individuals, the LDL receptor is unable to properly bind to the coated pit, and their cells are unable to take up cholesterol. Instead, cholesterol accumulates in the walls of arterial blood vessels, leading to high blood pressure, occluded (blocked) arteries, and heart attacks.

Substances are secreted from a cell by exocytosis. Substances enter a cell by endocytosis. Receptor-mediated endocytosis allows cells to take up specific kinds of molecules and then process them within the cell.

e-Learning Connection

Concepts	Questions	Media Resources*

4.1   Plasma Membrane Structure and Function

• The plasma membrane regulates the passage of molecules into and out of the cell.
• The membrane contains lipids and protein. Each protein has a specific function.
	1. Why is the term "fluid-mosaic" used to describe the plasma membrane?	Essential Study Partner
		  Membrane Structure
		  Cell Interactions
		
		Labeling Exercises
		  Fluid-Mosaic Model of
		    Membrane Structure (1)
		  Fluid-Mosaic Model of
		    Membrane Structure (2)
		  Membrane Protein Diversity

4.2   The Permeability of the Plasma Membrane

• Some substances, particularly small, noncharged molecules, pass freely across the plasma membrane.
• The plasma membrane is differentially permeable. Therefore, some ions and polar molecules need assistance to cross the membrane.
	1. What is meant by "differentially permeable?"
	2. How does passive transport differ from active transport?
		Essential Study Partner
		  Movement Through Cell
		  Membranes
		
		Explorations
		  Cystic Fibrosis

4.3   Diffusion and Osmosis

• Molecules spontaneously diffuse (move from an area of higher concentration to an area of lower concentration), and some can diffuse across a plasma membrane.
• Water diffuses across the plasma membrane, and this can affect cell size and shape.
	1. Why does oxygen diffuse from the lungs into the blood and not in the opposite direction?
	2. If a cell is placed into a hypotonic solution, in what direction will there be movement of water?
		Essential Study Partner
		  Diffusion
		  Osmosis
		
		Art Quizzes
		  Diffusion
		  Osmotic Pressure

4.4   Transport by Carrier Proteins

• Carrier proteins assist the transport of some ions and molecules unable to diffuse across the plasma membrane.	1. What does it mean when membrane carrier proteins are said to be "specific?"
	2. What is the sodium-potassium pump?
		Essential Study Partner
		  Facilitated Diffusion
		  Active Transport
		
		Art Quizzes
		  Functions of Plasma
		    Membrane
		  Proteins
		  Facilitated Diffusion
		  Sodium-Potassium Pump
		
		BioCourse Study Guide
		  Movement Across
		    Membranes:
		    Active Processes

4.5   Exocytosis and Endocytosis

• Vesicle formation takes other substances into the cell, and vesicle fusion with the plasma membrane discharges substances from the cell.	1. Pinocytosis differs from phagocytosis in what way?	Essential Study Partner
		  Exocytosis/Endocytosis
		
		Animation Quiz
		  Exocytosis/Endocytosis
		
		BioCourse Study Guide
		  Endocytosis and Exocytosis

*For additional Media Resources, see the Online Learning Center.

Cell Division

Chapter Concepts

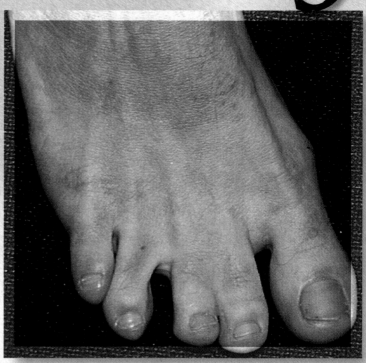

The second and third toes of this college student's feet failed to separate completely during fetal development.

*A*poptosis (ap-uh-TOH-sis, -ahp-). What does it have to do with you? Well, just ask Joe about his toes. On both feet, the second and third toes are incompletely separated. Why? Because when he was a fetus, apoptosis (programmed cell death) failed to occur to a sufficient degree. During development, hands and feet look like paddles at first, and then later, apoptosis fashions fingers and toes.

Joe is actually fortunate that the lack of apoptosis during one stage of development left him with only a small anomaly. Just as cells must divide in order for a child to grow or a cut to heal, so apoptosis must occur for parts to take shape and for cancer cells to die. From the fertilization of the egg to the death of the individual, cell division and apoptosis are in balance for the person to remain healthy.

Signals received at the plasma membrane tell the cell whether cell division or apoptosis are required. During both, specific enzymes are activated—one set makes more cells and the second set destroys cells. The genes too are involved. They respond to the signals and direct which enzymes will be present—those that build or those that tear down.

5.1 Cell Increase and Decrease

Opposing events often keep the body in balance and maintain homeostasis. As you continue your study of biology, you will come across many examples to support this statement. For now, consider that some carrier proteins transport molecules into the cell, and others transport molecules out of the cell. Some hormones increase the level of blood glucose, and others decrease the level. Similarly, two opposing processes keep the number of cells in the body at an appropriate level. Cell division increases the number of **somatic** (body) **cells.** Cell division consists of **mitosis,** which is division of the nucleus, and **cytokinesis,** which is division of the cytoplasm. **Apoptosis,** or cell death, decreases the number of cells. Both mitosis and apoptosis are normal parts of growth and development. An organism begins as a single cell that repeatedly divides to produce many cells, but eventually some cells must die for the organism to take shape. For example, when a tadpole becomes a frog, the tail disappears as apoptosis occurs. As mentioned in the introduction, the fingers and toes of a human embryo are at first webbed, but then they are usually freed from one another as a result of apoptosis.

Cell division occurs during your entire life. Even now, your body is producing thousands of new red blood cells, skin cells, and cells that line your respiratory and digestive tracts. Also, if you suffer a cut, cell division will repair the injury. Apoptosis occurs all the time too, particularly if an abnormal cell that could become cancerous appears. Death through apoptosis prevents a tumor from developing.

The Cell Cycle

Cell division is a part of the cell cycle. The **cell cycle** is an orderly set of stages that take place between the time a cell divides and the time the resulting cells also divide.

The Stages of Interphase

As Figure 5.1*a* shows, most of the cell cycle is spent in **interphase.** This is the time when a cell carries on its usual functions which are dependent on its location in the body. It also gets ready to divide: it grows larger, the number of organelles doubles, and the amount of DNA doubles. For mammalian cells, interphase lasts for about 20 hours, which is 90% of the cell cycle.

The event of DNA synthesis permits interphase to be divided into three stages: the G_1 stage occurs before DNA synthesis, the S stage includes DNA synthesis, and the G_2 stage occurs after DNA synthesis. Originally G stood for "gap" because early microscopists couldn't see anything going on during G_1 and G_2. But now that we know growth occurs during these stages, the G can be thought of as standing for growth. Let us see what specifically happens during each of these stages.

During the G_1 stage, a cell doubles its organelles (such as mitochondria and ribosomes) and it accumulates the materials needed for DNA synthesis.

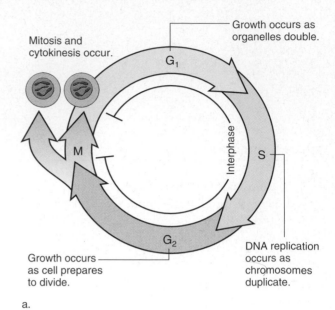

a.

Figure 5.1   The cell cycle.
a. Cells go through a cycle that consists of four stages: G_1, S, G_2, and M. The major activity for each stage is given.

During the S stage, DNA replication occurs. At the beginning of the S stage, each chromosome is composed of one DNA strand, which is called a chromatid. At the end of this stage, each chromosome consists of two identical DNA strands, and therefore is composed of two sister chromatids. Another way of expressing these events is to say that DNA replication has resulted in duplicated chromosomes.

During the G_2 stage, the cell synthesizes the proteins needed for cell division, such as the protein found in microtubules. The role of microtubules in cell division is described in a later section.

The amount of time the cell takes for interphase varies widely. Some cells, such as nerve and muscle cells, typically do not complete the cell cycle and are permanently arrested in G_1. These cells are said to have entered a G_o stage. Embryonic cells spend very little time in G_1 and complete the cell cycle in a few hours.

The Mitotic Stage

Following interphase, the cell enters the M (for mitotic) stage. This stage not only includes mitosis, it also includes cytokinesis, which usually begins before mitosis is completed. During mitosis, as we shall see, the sister chromatids of each chromosome separate, becoming daughter chromosomes that are distributed to two daughter nuclei. When cytokinesis is complete, two daughter cells are now present. Mammalian cells usually require only about four hours to complete the mitotic stage.

Control of the Cell Cycle

The cell cycle is controlled by internal and external signals. These signals ensure that the stages follow one another in

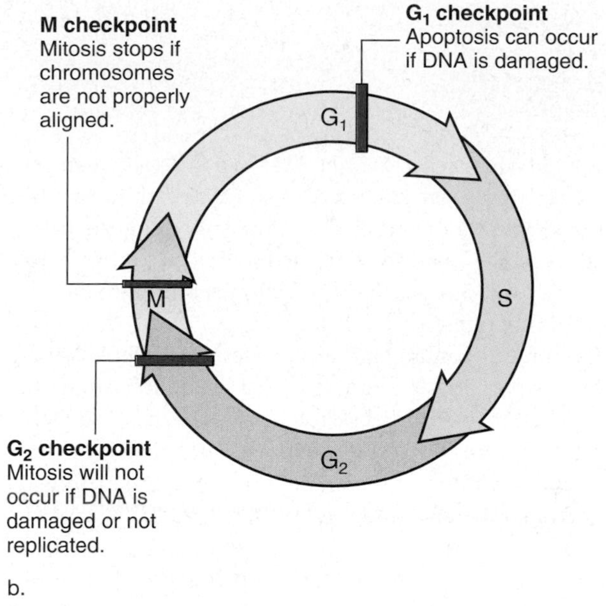

M checkpoint
Mitosis stops if chromosomes are not properly aligned.

G₁ checkpoint
Apoptosis can occur if DNA is damaged.

G₂ checkpoint
Mitosis will not occur if DNA is damaged or not replicated.

b.

Figure 5.1   (continued)
b. The cell cycle can stop at these checkpoints for the reasons noted.

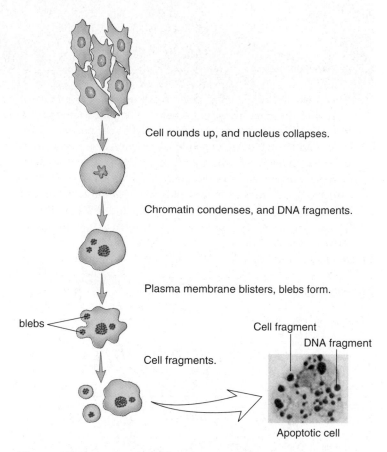

Cell rounds up, and nucleus collapses.

Chromatin condenses, and DNA fragments.

Plasma membrane blisters, blebs form.

blebs

Cell fragments.

Cell fragment

DNA fragment

Apoptotic cell

Figure 5.2   Apoptosis.
Apoptosis is a sequence of events that results in a fragmented cell. The fragments are phagocytized by white blood cells and neighboring tissue cells.

the normal sequence and that each stage is properly completed before the next stage begins. Growth factors are external signals received at the plasma membrane. Even cells that are arrested in G₀ will finish the cell cycle if stimulated to do so by growth factors. When blood platelets release a growth factor, skin fibroblasts in the vicinity are stimulated to finish the cell cycle so an injury can be repaired.

The red barriers in Figure 5.1b represent three checkpoints when the cell cycle either stops or continues on, depending on the internal signal it receives. Researchers have identified a signal called **cyclin** that increases and decreases as the cell cycle continues. Cyclin has to be present for the cell to proceed from the G₂ stage to the M stage and for the cell to proceed from the G₁ stage to the S stage.

The cell cycle stops at the G₂ checkpoint if DNA has not finished replicating. This prevents the initiation of the M stage before completion of the S stage. Also, if DNA is damaged, such as from solar radiation or X ray, stopping the cell cycle at this checkpoint allows time for the damage to be repaired so that it is not passed on to daughter cells.

Another cell cycle checkpoint occurs during the mitotic stage. The cycle stops if the chromosomes are not going to be distributed accurately to the daughter cells.

DNA damage can also stop the cell cycle at the G₁ checkpoint. In mammalian cells, the protein p53 stops the cycle at the G₁ checkpoint when DNA is damaged. First, p53 attempts to initiate DNA repair, but if that is not possible, it brings about apoptosis. We now know that many forms of tumors contain cells that lack an active *p53* gene. In other words, when the *p53* gene is faulty and unable to promote apoptosis, cancer develops.

Apoptosis

Apoptosis is often defined as programmed cell death because the cell progresses through a usual series of events that bring about its destruction (Fig. 5.2). The cell rounds up and loses contact with its neighbors. The nucleus fragments, and the plasma membrane develops blisters. Finally, the cell fragments, and its bits and pieces are engulfed by white blood cells and/or neighboring cells. A remarkable finding of the past few years is that cells routinely harbor the enzymes, now called caspases, that bring about apoptosis. The enzymes are ordinarily held in check by inhibitors, but they can be unleashed either by internal or external signals. There are two sets of caspases. The first set are the "initiators" that receive the signal to activate the "executioners," which then activate the enzymes that dismantle the cell. For example, executioners turn on enzymes that tear apart the cytoskeleton and enzymes that chop up DNA.

Cell division and apoptosis are two opposing processes that keep the number of healthy cells in balance.

Science Focus

What's in a Chromosome?

When early investigators decided that the genes are contained in the chromosomes, they had no idea of chromosome composition. By the mid-1900s, it was known that chromosomes are made up of both DNA and protein. Only in recent years, however, have investigators been able to produce models suggesting how chromosomes are organized.

A eukaryotic chromosome is more than 50% protein. Some of these proteins are concerned with DNA and RNA synthesis, but a large proportion, termed histones, seem to play primarily a structural role. The five primary types of histone molecules are designated H1, H2A, H2B, H3, and H4. Remarkably, the amino acid sequences of H3 and H4 vary little between organisms. For example, the H4 of peas is only two amino acids different from the H4 of cattle. This similarity suggests that few mutations in the histone proteins have occurred during the course of evolution and that the histones therefore have very important functions.

A human cell contains at least 2 meters of DNA. Yet all of this DNA is packed into a nucleus that is about 5 μm in diameter. The histones are responsible for packaging the DNA so that it can fit into such a small space. First the DNA double helix is wound at intervals around a core of eight histone molecules (two copies each of H2A, H2B, H3, and H4), giving the appearance of a string of beads (Fig. 5A*a* and *b*). Each bead is called a nucleosome, and the nucleosomes are said to be joined by "linker" DNA. This string is coiled tightly into a fiber that has six nucleosomes per turn (Fig. 5A*c*). The H1 histone appears to mediate this coiling process. The fiber loops back and forth (Fig. 5A*d* and *e*) and can condense to produce a highly compacted form (Fig. 5A*f*) characteristic of metaphase chromosomes. No doubt, compact chromosomes are easier to move about than extended chromatin. During interphase, extended chromatin makes DNA available for RNA synthesis and subsequent protein synthesis. Many of these proteins are enzymes, and this accounts for on-going metabolic activity during interphase.

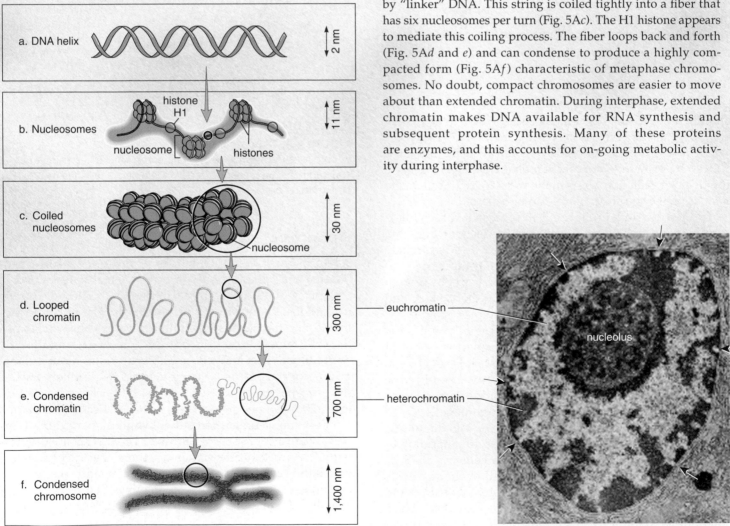

Figure 5A Levels of chromosome structure.

Figure 5B Eukaryotic nucleus.
Arrows indicate nuclear pores.

5.2 Maintaining the Chromosome Number

When a eukaryotic cell is not undergoing division, the DNA (and associated proteins) within a nucleus is a tangled mass of thin threads called **chromatin.** At the time of cell division, chromatin coils, loops, and condenses to form highly compacted structures called **chromosomes.** The Science Focus on the opposite page describes the transition from chromatin to chromosomes in greater detail.

Overview of Mitosis

When the chromosomes are visible, it is possible to photograph and count them. Each species has a characteristic chromosome number; for instance, human cells contain 46 chromosomes, corn has 20 chromosomes, and the crayfish has 200! This is the full or **diploid (2n) number** of chromosomes found in the somatic (non-sex) cells of the body. The diploid number includes two chromosomes of each kind. Half the diploid number, called the **haploid (n) number** of chromosomes, contains only one of each kind of chromosome. In the life cycle of many animals, only sperm and eggs have the haploid number of chromosomes.

The nuclei of somatic cells undergo **mitosis,** nuclear division in which the chromosome number stays constant. A 2n nucleus divides to produce daughter nuclei that are also 2n (Fig. 5.3). Before nuclear division takes place, DNA replication occurs, duplicating the chromosomes. A **duplicated chromosome** is composed of two sister chromatids held together in a region called the **centromere. Sister chromatids** are genetically identical—they contain the same genes. At the completion of mitosis, each chromosome consists of a single chromatid.

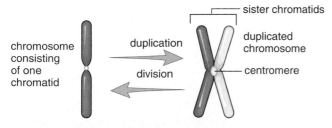

Figure 5.3 gives an overview of mitosis; for simplicity, only four chromosomes are depicted. (In determining the number of chromosomes, it is necessary to count only the number of independent centromeres.) During mitosis, the centromeres divide and then the sister chromatids separate, becoming **daughter chromosomes.** Therefore, each daughter nucleus gets a complete set of chromosomes and has the same number of chromosomes as the parental cell. Therefore, the daughter cells are genetically identical to each other and to the parental cell.

Following mitosis, a 2n parental cell gives rise to two 2n daughter cells. It would be possible to diagram this as 2n → 2n. The cells of some organisms, such as algae and

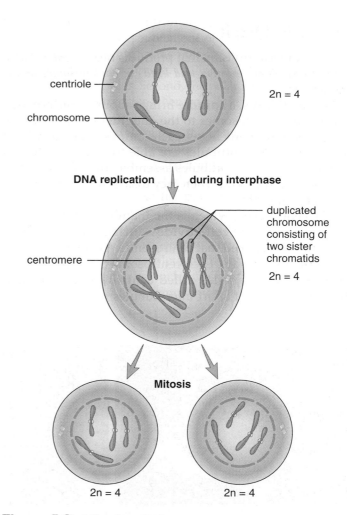

Figure 5.3   Mitosis overview.
Following DNA replication during interphase, each chromosome in the parental nucleus is duplicated and consists of two sister chromatids. During mitosis, the centromeres divide and the sister chromatids separate, becoming daughter chromosomes that move into the daughter nuclei. Therefore, daughter cells have the same number and kinds of chromosomes as the parental cell. (The blue chromosomes were inherited from one parent, and the red chromosomes were inherited from the other.)

fungi, are haploid as adults. The n parental cell gives rise to two daughter cells that are also haploid. In these organisms, n → n.

As stated, mitosis is the type of nuclear division that occurs when tissues grow or when repair occurs. Following fertilization, the zygote begins to divide mitotically, and mitosis continues during development and the life span of the individual.

After the chromosomes duplicate prior to mitosis, they consist of two sister chromatids. The sister chromatids separate during mitosis so that each of two daughter cells has the same number and kinds of chromosomes as the parental cell.

5.3 Reducing the Chromosome Number

Meiosis occurs in any life cycle that involves sexual reproduction. **Meiosis** reduces the chromosome number in such a way that the daughter nuclei receive only one of each kind of chromosome. The process of meiosis ensures that the next generation of individuals will have the diploid number of chromosomes and a combination of traits different from that of either parent.

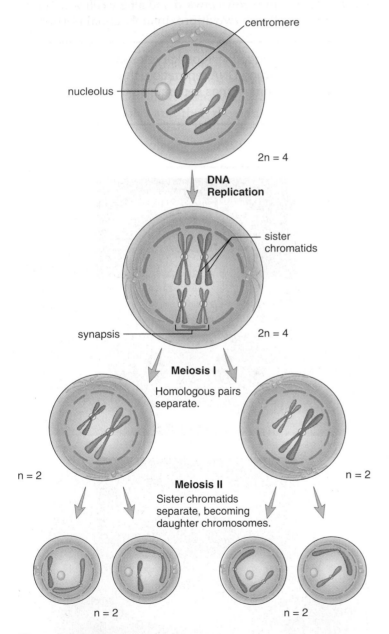

Figure 5.9   Overview of meiosis.
Following DNA replication, each chromosome is duplicated. During meiosis I, the homologous chromosomes pair during synapsis and then separate. During meiosis II, the centromeres divide and the sister chromatids separate, becoming daughter chromosomes that move into the daughter nuclei.

Overview of Meiosis

Meiosis requires two nuclear divisions and produces four haploid nuclei, each having one of each kind of chromosome and therefore half the total number of chromosomes present in the parental nucleus. The parental cell has the diploid number of chromosomes, while the daughter cells will have the haploid number of chromosomes.

When a cell is 2n, or diploid, the chromosomes occur in pairs. For example, the 46 chromosomes of humans occur in 23 pairs of chromosomes. The members of each pair are called **homologous chromosomes,** or **homologues.**

Figure 5.9 presents an overview of meiosis, indicating the two cell divisions, meiosis I and meiosis II. Prior to meiosis I, DNA replication has occurred and the chromosomes are duplicated. Each chromosome consists of two chromatids held together at a centromere. During meiosis I, the homologous chromosomes come together and line up side by side. This so-called **synapsis** results in an association of four chromatids that stay in close proximity during the first two phases of meiosis I. Synapsis is quite significant because its occurrence leads to a reduction of the chromosome number.

Because of synapsis, there are pairs of homologous chromosomes at the metaphase plate during meiosis I. Notice that only during meiosis I is it possible to observe paired chromosomes at the metaphase plate. When the members of these pairs separate, each daughter nucleus receives one member of each pair. Therefore, each daughter cell has the haploid number of chromosomes, as you can verify by counting its centromeres. Each chromosome, however, is still duplicated and no replication of DNA occurs between meiosis I and meiosis II.

No replication of DNA is needed between meiosis I and meiosis II because the chromosomes are already duplicated: they already have two sister chromatids. During meiosis II, the centromeres divide and the sister chromatids separate, becoming daughter chromosomes that are distributed to daughter nuclei. In the end, each of four daughter cells has the haploid number of chromosomes, and each chromosome consists of one chromatid.

In some life cycles, such as that of humans (see Fig. 5.15), the daughter cells mature into **gametes** (sex cells—sperm and egg) that fuse during fertilization. **Fertilization** restores the diploid number of chromosomes in a cell that will develop into a new individual. If the gametes carried the diploid instead of the haploid number of chromosomes, the chromosome number would double with each fertilization.

During meiosis I, homologous chromosomes pair up and then separate. Each daughter nucleus receives one copy of each kind of chromosome. Following meiosis II, there are four haploid daughter cells, and each chromosome consists of one chromatid.

Genetic Recombination

Meiosis helps ensure that genetic recombination occurs through two key events: crossing-over and independent assortment of homologous chromosomes. In order to appreciate the significance of these events, it is necessary to realize that the members of a homologous pair can carry slightly different instructions for the same genetic trait. For example, one homologue may carry instructions for brown eyes, while the corresponding homologue may carry instructions for blue eyes.

Crossing-Over of Nonsister Chromatids

It is often said that we inherit half our chromosomes from our mother and half from our father, but this is not strictly correct because of crossing-over. During synapsis, the homologous chromosomes come together and line up side by side. Now an exchange of genetic material may occur between the nonsister chromatids of the homologous pair (Fig. 5.10). **Crossing-over** means that the chromatids held together by a centromere are no longer identical. When the chromatids separate during meiosis I, the daughter cells may receive chromosomes with recombined genetic material.

Independent Assortment of Homologous Chromosomes

Independent assortment means that the homologous chromosomes separate independently or in a random manner. When homologues align at the metaphase plate, the maternal or paternal homologue may be orientated toward either pole. Figure 5.11 shows four possible orientations for a cell that contains only three pairs of chromosomes. Each orientation results in gametes that have a different combination of maternal and paternal chromosomes. Once all possible orientations are considered, the result will be 2^3 or eight possible combinations of maternal and paternal chromo-

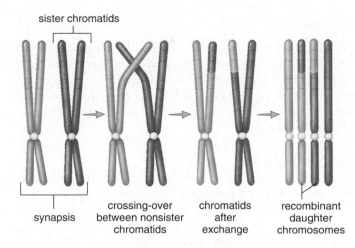

Figure 5.10   Synapsis and crossing-over.
During meiosis I, from left to right, duplicated homologous chromosomes undergo synapsis; nonsister chromatids break and then rejoin, so that two of the resulting daughter chromosomes will have a different combination of genes.

somes in the resulting gametes from this cell. In humans, where there are 23 pairs of chromosomes, the number of possible chromosomal combinations in the gametes is a staggering 2^{23}, or 8,388,608. And this does not even consider the genetic variations that are introduced due to crossing-over.

During meiosis, crossing-over mixes the genetic information of maternal and paternal chromosomes, and independent assortment leads to different combinations of these chromosomes in the gametes and offspring.

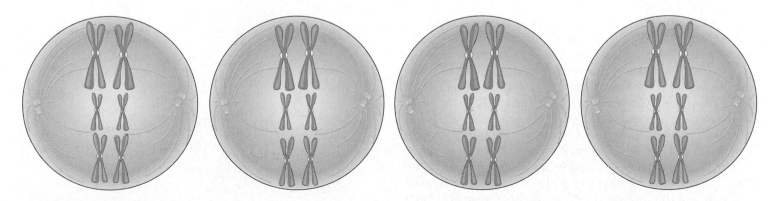

Figure 5.11   Independent assortment.
Four possible orientations of homologous pairs at the metaphase plate are shown. Each of these will result in daughter nuclei with a different combination of parental chromosomes. When a cell has three pairs of homologous chromosomes, there are 2^3 possible combinations of parental chromosomes in the daughter nuclei.

Meiosis in Detail

Meiosis which requires two nuclear divisions results *in four daughter nuclei, each having one of each kind of chromosome and therefore half the number of chromosomes as the parental cell.* The same four phases seen in mitosis—prophase, metaphase, anaphase, and telophase—occur during both meiosis I and meiosis II.

The First Division

Phases of meiosis I for an animal cell are diagrammed in Figure 5.12. During prophase I, the spindle appears while the nuclear envelope fragments and the nucleolus disappears. Due to chromosome duplication, during interphase, the homologous chromosomes each have two sister chromatids. During synapsis, crossing-over—represented by an exchange of color in Figure 5.12—can occur. If so, the sister chromatids of a duplicated chromosome are no longer identical.

During metaphase I, homologous pairs are aligned at the metaphase plate. The maternal homologue (e.g., red) may be orientated toward either pole, and the paternal (e.g., blue) homologue may be aligned toward either pole. This means that all possible combinations of chromosomes can occur in the daughter nuclei. During anaphase I, homologous chromosomes separate and move to opposite poles of the spindle. Each chromosome still consists of two chromatids.

In some species, a telophase I phase occurs at the end of meiosis I. If so, the nuclear envelopes re-form and nucleoli appear. This phase may or may not be accompanied by cytokinesis, which is separation of the cytoplasm.

Interkinesis

The period of time between meiosis I and meiosis II is called **interkinesis.** No replication of DNA occurs during interkinesis. Why is this appropriate? Because the chromosomes are already duplicated.

The Second Division

Phases of meiosis II for an animal cell are diagrammed in Figure 5.13. At the beginning of prophase II, a spindle appears while the nuclear envelope disassembles and the nucleolus disappears. Each duplicated chromosome attaches to the spindle and then they align at the metaphase plate during metaphase II. During anaphase II, sister chromatids separate, becoming daughter chromosomes that move into the daughter nuclei. In telophase II, the spindle disappears as nuclear envelopes re-form.

During the cytokinesis which can follow meiosis II, the plasma membrane furrows to give two complete cells, each of which has the haploid number of chromosomes. Each chromosome consists of one chromatid. Because each cell from meiosis I undergoes meiosis II, there are four daughter cells altogether.

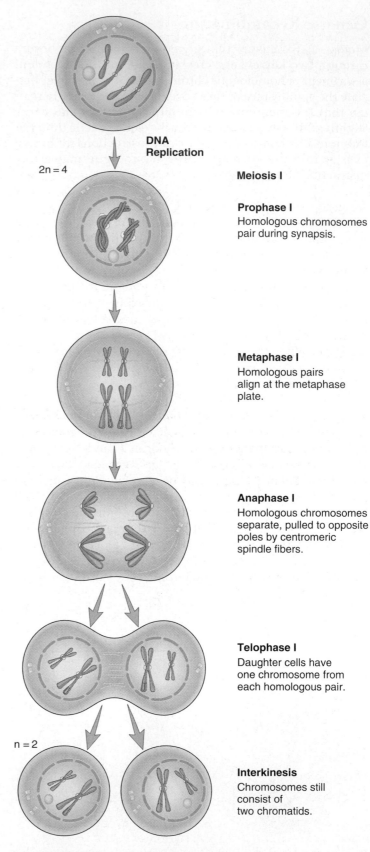

DNA Replication

2n = 4

Meiosis I

Prophase I
Homologous chromosomes pair during synapsis.

Metaphase I
Homologous pairs align at the metaphase plate.

Anaphase I
Homologous chromosomes separate, pulled to opposite poles by centromeric spindle fibers.

Telophase I
Daughter cells have one chromosome from each homologous pair.

n = 2

Interkinesis
Chromosomes still consist of two chromatids.

Figure 5.12   Meiosis I in an animal cell.
The exchange of color between nonsister chromatids represents crossing-over.

Figure 5.13   Meiosis II in an animal cell.

During meiosis II, sister chromatids separate, becoming daughter chromosomes that are distributed to the daughter nuclei. Following meiosis II, there are four haploid daughter cells. Comparing the number of centromeres in each daughter cell with the number in the parental cell at the start of meiosis I verifies that each daughter cell is haploid.

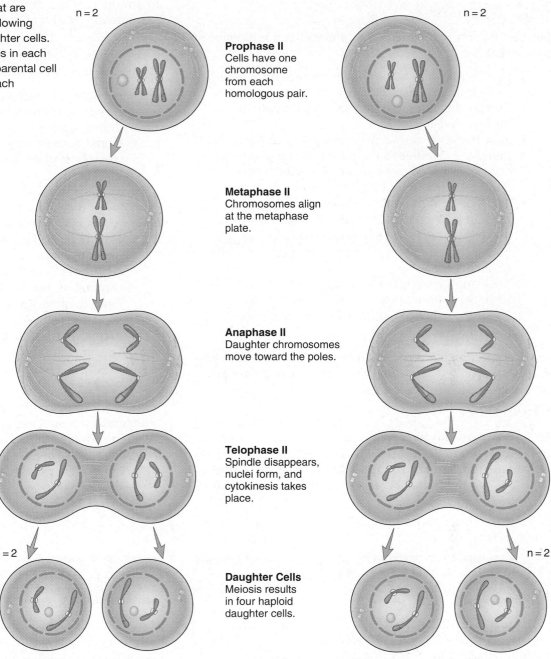

Meiosis II

n = 2

n = 2

Prophase II
Cells have one chromosome from each homologous pair.

Metaphase II
Chromosomes align at the metaphase plate.

Anaphase II
Daughter chromosomes move toward the poles.

Telophase II
Spindle disappears, nuclei form, and cytokinesis takes place.

n = 2

n = 2

Daughter Cells
Meiosis results in four haploid daughter cells.

Genetic Recombination

What are the different ways that meiosis leads to genetic recombination during sexual reproduction?

Independent alignment of paired chromosomes at the metaphase plate during meiosis I means that gametes will have different combinations of chromosomes.

Crossing-over during prophase of meiosis I means that the chromosomes in one gamete will have a different combination of genes than chromosomes in another gamete.

Upon fertilization, the combining of chromosomes from genetically different gametes occurs, and means the offspring will have a different combination of genes than either parent.

5.4 Comparison of Meiosis with Mitosis

Figure 5.14 compares meiosis to mitosis. You will want to notice that:

- DNA replication takes place only once prior to either meiosis and mitosis. However, meiosis requires two nuclear divisions, but mitosis requires only one nuclear division.
- Four daughter nuclei are produced by meiosis, and following cytokinesis there are four daughter cells. Mitosis followed by cytokinesis results in two daughter cells.
- The four daughter cells following meiosis are haploid and have half the chromosome number as the parental cell. The daughter cells following mitosis have the same chromosome number as the parental cell.
- The daughter cells from meiosis are not genetically identical to each other or to the parental cell. The daughter cells from mitosis are genetically identical to each other and to the parental cell.

The specific differences between these nuclear divisions can be categorized according to occurrence and process.

Occurrence

Meiosis occurs only at certain times in the life cycle of sexually reproducing organisms. In humans, meiosis occurs only in the reproductive organs and produces the gametes. Mitosis is more common because it occurs in all tissues during growth and repair.

Process

To summarize the process, Tables 5.1 and 5.2 separately compare meiosis I and meiosis II to mitosis.

Comparison of Meiosis I to Mitosis

Notice that these events distinguish meiosis I from mitosis:

- Homologous chromosomes pair and undergo crossing-over during prophase I of meiosis, but not during mitosis.
- Paired homologous chromosomes align at the metaphase plate during metaphase I in meiosis. These paired chromosomes have four chromatids altogether. Individual chromosomes align at the metaphase plate during metaphase in mitosis. They each have two chromatids.
- Homologous chromosomes (with centromeres intact) separate and move to opposite poles during anaphase I in meiosis. Centromeres split and sister chromatids, now called daughter chromosomes, move to opposite poles during anaphase in mitosis.

Comparison of Meiosis II to Mitosis

The events of meiosis II are just like those of mitosis except in meiosis II, the nuclei contain the haploid number of chromosomes.

Meiosis is a specialized process that reduces the chromosome number and occurs only during the production of gametes. Mitosis is a process that occurs during growth and repair of all tissues.

Table 5.1	Comparison of Meiosis I with Mitosis
Meiosis I	**Mitosis**
Prophase I	*Prophase*
Pairing of homologous chromosomes	No pairing of chromosomes
Metaphase I	*Metaphase*
Homologous duplicated chromosomes at metaphase plate	Duplicated chromosomes at metaphase plate
Anaphase I	*Anaphase*
Homologous chromosomes separate.	Sister chromatids separate, becoming daughter chromosomes that move to the poles.
Telophase I	*Telophase*
Two haploid daughter cells.	Two daughter cells, identical to the parental cell.

Table 5.2	Comparison of Meiosis II with Mitosis
Meiosis II	**Mitosis**
Prophase II	*Prophase*
No pairing of chromosomes	No pairing of chromosomes
Metaphase II	*Metaphase*
Haploid number of duplicated chromosomes at metaphase plate	Duplicated chromosomes at metaphase plate
Anaphase I	*Anaphase*
Sister chromatids separate, becoming daughter chromosomes that move to the poles.	Sister chromatids separate, becoming daughter chromosomes that move to the poles.
Telophase II	*Telophase*
Four haploid daughter cells.	Two daughter cells, identical to the parental cell.

Meiosis

Mitosis

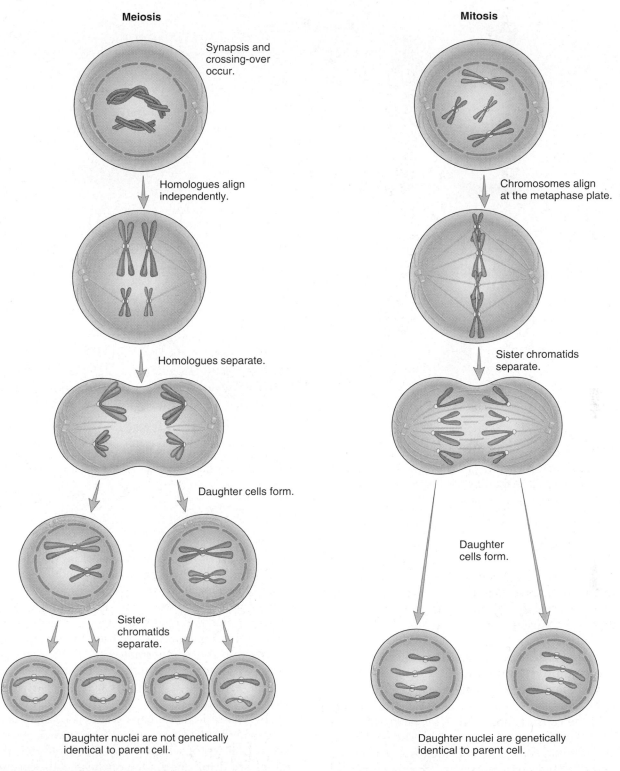

Synapsis and crossing-over occur.

Homologues align independently.

Homologues separate.

Daughter cells form.

Sister chromatids separate.

Daughter nuclei are not genetically identical to parent cell.

Chromosomes align at the metaphase plate.

Sister chromatids separate.

Daughter cells form.

Daughter nuclei are genetically identical to parent cell.

Figure 5.14   Meiosis compared to mitosis.
Why does meiosis produce daughter cells with half the number while mitosis produces daughter cells with the same number of chromosomes as the parental cell? Compare metaphase I of meiosis to metaphase of mitosis. Only in metaphase I are the homologous chromosomes paired at the metaphase plate. Members of the homologous chromosomes separate during anaphase I, and therefore the daughter cells are haploid. The blue chromosomes were inherited from one parent, and the red chromosomes were inherited from the other parent. The exchange of color between nonsister chromatids represents crossing-over during meiosis I.

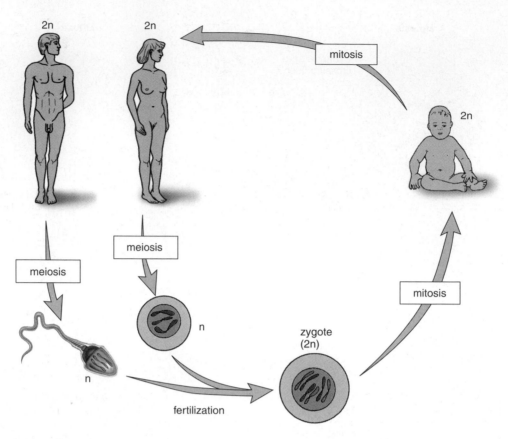

Figure 5.15   Life cycle of humans.
Meiosis in human males is a part of sperm production, and meiosis in human females is a part of egg production. When a haploid sperm fertilizes a haploid egg, the zygote is diploid. The zygote undergoes mitosis as it develops into a newborn child. Mitosis continues after birth until the individual reaches maturity; then the life cycle begins again.

5.5 The Human Life Cycle

The human life cycle requires both meiosis and mitosis (Fig. 5.15). In human males, meiosis is a part of spermatogenesis, which occurs in the testes and produces sperm. In human females, meiosis is a part of oogenesis, which occurs in the ovaries and produces eggs (Fig. 5.16). A haploid sperm and a haploid egg join at fertilization, and the resulting **zygote** has the full or diploid number of chromosomes. During development of the fetus, which is the stage of development before birth, mitosis keeps the chromosome number constant in all the cells of the body. After birth, mitosis is involved in the continued growth of the child and repair of tissues at any time. As a result of mitosis, each somatic cell in the body has the same number of chromosomes.

Spermatogenesis and Oogenesis in Humans

Spermatogenesis is the production of sperm in males, and **oogenesis** is the production of eggs in females (Fig. 5.16). In the testes of human males, primary spermatocytes, which are diploid (2n), divide during the first meiotic division to form two secondary spermatocytes, which are haploid (n).

Secondary spermatocytes divide during the second meiotic division to produce four spermatids, which are also haploid (n). What's the difference between the chromosomes in haploid secondary spermatocytes and those in haploid spermatids? The chromosomes in secondary spermatocytes are duplicated and consist of two chromatids, while those in spermatids consist of only one chromatid. Spermatids mature into sperm (spermatozoa). In human males, sperm have 23 chromosomes, which is the haploid number. The process of meiosis in males always results in four cells that become sperm.

In the ovaries of human females, a primary oocyte, which is diploid (2n), divides during the first meiotic division into two cells, each of which is haploid. Note that the chromosomes are duplicated. One of these cells, termed the **secondary oocyte,** receives almost all the cytoplasm. The other is the first polar body. A **polar body** is a nonfunctioning cell that occurs during oogenesis. The first polar body contains duplicated chromosomes and complete meiosis II occasionally. The secondary oocyte begins the second meiotic division but stops at metaphase II. The secondary oocyte leaves the ovary and enters an oviduct where it may be approached by a sperm. If a sperm does enter the oocyte, the oocyte is

Spermatogenesis

Oogenesis

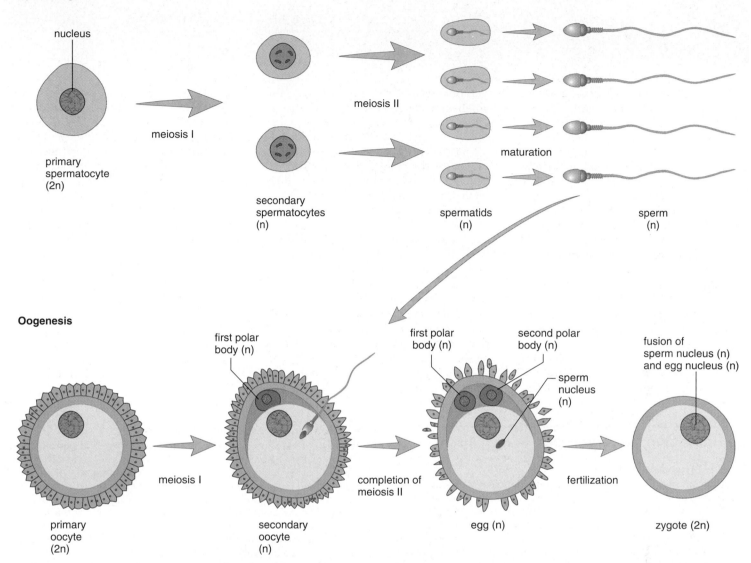

Figure 5.16 **Spermatogenesis and oogenesis.**
Spermatogenesis produces four viable sperm, whereas oogenesis produces one egg and two polar bodies. In humans, both sperm and egg have 23 chromosomes each; therefore, following fertilization, the zygote has 46 chromosomes.

activated to complete the second meiotic division. Following meiosis II, there is one egg and two or possibly three polar bodies. The mature egg has 23 chromosomes, each consisting of one chromatid. The polar bodies that disintegrate are a way to discard unnecessary chromosomes while retaining much of the cytoplasm in the egg. The cytoplasm serves as a source of nutrients for the developing embryo.

Genetic Recombination in Humans

Notice that fertilization is another means by which chromosomes are recombined in the next generation. Because each child receives both paternal and maternal chromosomes, no child is exactly like either parent. Altogether, there

are three ways in which meiosis ensures that a child has a different combination of genes from that of either parent:

1. Independent assortment of chromosomes means that all possible combinations of chromosomes occur in the gametes.
2. Crossing-over recombines genetic material so that sister chromatids are genetically dissimilar.
3. Upon fertilization, recombination of chromosomes occurs.

Sexual reproduction ensures that each generation has the same number of chromosomes, and that each individual has a different genetic makeup from that of either parent.

Cancer is more likely to develop in tissues whose cells frequently divide, such as the blood-forming cells in the bone marrow. A new cancer drug called Gleevec is particularly helpful in a deadly form of blood cancer called myeloid leukemia. Gleevec is a pill taken by mouth and it doesn't cause hair loss! For some patients the results have been dramatic. Whereas before they were bedridden and about to die, within weeks they can return to work, do volunteer work, and enjoy themselves as before.

Gleevec is expensive. It costs something like $2,400 a month, or nearly $30,000 a year and treatment may have to continue for life. How should the cost of treatment be met?

Drug companies claim that it costs them between $500 million and $1 billion to bring a single new medicine to market.

This cost may seem overblown especially when you consider that the National Cancer Institute funds basic research into cancer biology. But the drug companies tell us that they need one successful drug to pay for the many drugs they tried to develop but did not pay off. Still, it does seem as if successful drug companies try to keep lower-cost competitors out of the market.

Now that Gleevec has been taken off the experimental list, insurance companies will probably pick up the cost but, of course, it may cause the cost of insurance to rise for everyone.

Cancer most often strikes the elderly. In the future it may be that Medicare will pay for cancer drugs as well as all drugs needed by the elderly. In that case, the cost of cancer treatment will be borne by everyone who pays taxes.

The question of how much drug companies can charge for drugs and who should pay for them is a thorny problem. If drug companies don't show a profit they may go out of business and there will be no new drugs. The same is true for insurance companies if they can't raise the cost of insurance to pay for expensive drugs. If the government buys drugs for Medicare patients, taxes may go up dramatically.

Decide Your Opinion

1. How would it be possible to determine how much a drug company should charge for a particular drug?
2. Who do you think should pay for cancer treatment: the patient, the insurance company or taxpayers? Why?
3. Should an insurance company or the government be able to place a limit on payment of drugs per patient?

Summarizing the Concepts

5.1 Cell Increase and Decrease
Cell division increases the number of cells in the body, and apoptosis reduces this number when appropriate. Cells go through a cell cycle that includes (1) interphase and (2) cell division consisting of mitosis and cytokinesis. Interphase, in turn, includes G_1 (growth as certain organelles double), S (DNA synthesis), and G_2 (growth as cell prepares to divide). Cell division occurs during the mitotic stage (M) when daughter cells receive a full complement of chromosomes.

The cell cycle is controlled, and there are three checkpoints—one in G_1 prior to the S stage, one in G_2 prior to the M stage, and another near the end of mitosis. DNA damage is a reason the cell cycle stops; the *p53* gene is active at the G_1 checkpoint, and if DNA is damaged and can't be repaired, this gene initiates apoptosis. During apoptosis, the enzymes called caspases bring about destruction of the nucleus and the rest of the cell.

5.2 Maintaining the Chromosome Number
Each species has a characteristic number of chromosomes. The total number is the diploid number, and half this number is the haploid number. Among eukaryotes, cell division involves nuclear division and division of the cytoplasm (cytokinesis).

Replication of DNA precedes cell division. The duplicated chromosome is composed of two sister chromatids held together at a centromere. During mitosis, the centromeres divide, and daughter chromosomes go into each new nucleus.

Mitosis has the following phases: prophase—in early prophase, chromosomes have no particular arrangement, and in late prophase the chromosomes are attached to spindle fibers; metaphase, when the chromosomes are aligned at the metaphase plate; anaphase, when the chromatids separate, becoming daughter chromosomes that move toward the poles; and telophase, when new nuclear envelopes form around the daughter chromosomes and cytokinesis begins.

5.3 Reducing the Chromosome Number
Meiosis is found in any life cycle that involves sexual reproduction. During meiosis I, homologues separate, and this leads to daughter cells with half or the haploid number of homologous chromosomes. Crossing-over and independent assortment of chromosomes during meiosis I ensure genetic recombination in daughter cells. During meiosis II, chromatids separate, becoming daughter chromosomes that are distributed to daughter nuclei. In some life cycles, the daughter cells become gametes, and upon fertilization, the offspring have the diploid number of chromosomes, the same as their parents.

Meiosis utilizes two nuclear divisions. During meiosis I, homologous chromosomes undergo synapsis, and crossing-over between nonsister chromatids occurs. When the homologous chromosomes separate during meiosis I, each daughter nucleus receives one member from each pair of chromosomes. Therefore, the daughter cells are haploid. Distribution of daughter chromosomes derived from sister chromatids during meiosis II then leads to a total of four new cells, each with the haploid number of chromosomes.

5.4 Comparison of Meiosis with Mitosis
Figure 5.14 contrasts the phases of mitosis with the phases of meiosis.

5.5 The Human Life Cycle
The human life cycle involves both mitosis and meiosis. Mitosis ensures that each somatic cell has the diploid number of chromosomes.

Meiosis is a part of spermatogenesis and oogenesis. Spermatogenesis in males produces four viable sperm, while oogenesis in females produces one egg and two polar bodies. Oogenesis does not go on to completion unless a sperm fertilizes the developing egg.

Among sexually reproducing organisms, such as humans, meiosis results in genetic recombination due to independent assortment of homologous chromosomes and crossing-over. Fertilization also contributes to genetic recombination.

Testing Yourself

Choose the best answer for each question.

1. The cell cycle ensures that
   a. the cell grows prior to cell division.
   b. DNA replicates prior to cell division.
   c. the chromatids separate, becoming the daughter chromosomes.
   d. the cytoplasm divides.
   e. All of these are correct.

2. In human beings, mitosis is necessary to
   a. growth and repair of tissues.
   b. formation of the gametes.
   c. maintaining the chromosome number in all body cells.
   d. the death of unnecessary cells.
   e. Both a and c are correct.

For questions 3–6, match the descriptions that follow to the terms in the key. Answers may be used more than once.
   Key:
   a. centriole
   b. cell body
   c. chromosome
   d. centromere

3. Point of attachment for sister chromatids

4. Found at a pole in the center of an aster

5. Coiled and condensed chromatin

6. Consists of one or two chromatids

7. If a parental cell has fourteen chromosomes prior to mitosis, how many chromosomes will the daughter cells have?
   a. twenty-eight
   b. fourteen
   c. seven
   d. any number between seven and twenty-eight

8. In which phase of mitosis are chromosomes moving toward the poles?
   a. prophase
   b. metaphase
   c. anaphase
   d. telophase
   e. Both b and c are correct.

9. If a parental cell has twelve chromosomes, the daughter cells following meiosis II will have
   a. twelve chromosomes.
   b. twenty-four chromosomes.
   c. six chromosomes.
   d. Any one of these could be correct.

10. If a diploid cell contains 8 chromosomes, after meiosis II the two daughter cells will contain how many chromosomes?
    a. 8
    b. 4
    c. 16
    d. 2

11. At the metaphase plate during metaphase I of meiosis, there are
    a. single chromosomes.
    b. unpaired duplicated chromosomes.
    c. homologous pairs.
    d. always twenty-three chromosomes.

12. Crossing-over occurs between
    a. sister chromatids of the same chromosomes.
    b. chromatids of nonhomologous chromosomes
    c. nonsister chromatids of a homologous pair.
    d. two daughter nuclei.
    e. Both b and c are correct.

13. During which meiotic division do cells become haploid?
    a. anaphase I of meiosis I
    b. anaphase II of meiosis II
    c. anaphase I of mitosis I
    d. anaphase II of mitosis II
    e. Both a and b are correct.

14. Plant cells
    a. do not have centrioles.
    b. have a cleavage furrow like animal cells do.
    c. form a cell plate by fusion of several vesicles.
    d. do not have the four phases of mitosis: prophase, metaphase, anaphase, and telophase.
    e. Both a and c are correct.

15. Which of these is NOT a difference between spermatogenesis and oogenesis in humans?

Spermatogenesis	Oogenesis
a. Occurs in males	Occurs in females
b. Produces four sperm per meiosis	Produces one egg per meiosis
c. Produces haploid eggs	Produces diploid cells
d. Typically goes to completion	Does not typically go to completion

   e. Both c and d are not differences between spermatogenesis and oogenesis in humans

16. Label this diagram of a cell in early prophase of mitosis.

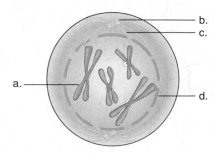

17. Which of these drawings represents metaphase I of meiosis? How do you know?

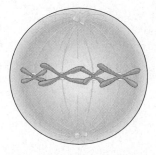
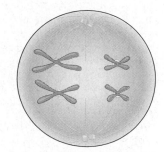

e-Learning Connection

Concepts	Questions	Media Resources*
5.1 Cell Increase and Decrease		

• During the lifetime of an individual, mitosis causes somatic cells to increase in number and apoptosis causes them to decrease in number.
• Mitosis is part of the cell cycle. Specific events must take place for mitosis to occur.
	1. What is apoptosis and what is its function?
	2. What three checkpoints help control the cell cycle?
		Essential Study Partner
		    Cell Division Introduction
		    Chromosomes
		
		Art Quizzes
		    Cell Cycle
		    Mechanisms of Cell Cycle
		        Control
		    Nucleosomes

5.2 Maintaining the Chromosome Number		

• Each species has a characteristic number of chromosomes.
• Mitosis (a type of nuclear division) maintains the chromosome number in cells.
• Mitosis is necessary to the growth and repair of body cells.
	1. What is the overall purpose of mitosis?
	2. What is cytokinesis?
		Essential Study Partner
		    Mitosis/Cell Cycle
		
		Animation Quiz
		    Mitosis
		
		BioCourse Study Guide
		    Mitosis and Cytokinesis
		
		Labeling Exercises
		    Plant Cell Mitosis

5.3 Reducing the Chromosome Number		

• Meiosis (another type of nuclear division) reduces the chromosome number in life cycles involving sexual reproduction.
• Meiosis produces cells that have half the total number of chromosomes as the parental cell had.
• Meiosis produces cells that have different combinations of chromosomes.
	1. Name two ways that genetic recombination can occur during the process of meiosis.
	2. In general, how does meiosis I differ from meiosis II?
		Essential Study Partner
		    Meiosis
		    Recombination
		
		Animation Quiz
		    Meiosis
		
		BioCourse Study Guide
		    Meiosis

5.4 Comparison of Meiosis with Mitosis		

• Meiosis differs from mitosis both in occurrence and in process.	1. Where in the human body are meiosis and mitosis likely to occur?	Essential Study Partner
		    Review of Cell Division
		
		Art Quiz
		    Alternation of Generations

5.5 The Human Life Cycle		
• The human life cycle includes both mitosis and meiosis. • In humans, and many other animals, meiosis is a part of the production of sperm in males and eggs in females. • When the sperm fertilizes the egg, the full number of chromosomes is restored in offspring.	1. What occurs during spermatogenesis and oogenesis? 2. What is the function of polar bodies during oogenesis?	Labeling Exercise Sperm and Egg Anatomy

*For additional Media Resources, see the Online Learning Center.

Metabolism: Energy and Enzymes

Chapter Concepts

A sea otter, surrounded by algae, feeds on a sea urchin. All three are solar powered because the algae uses solar energy to carry on photosynthesis and produce the nutrient molecules that sustain sea urchins, and therefore sea otters also.

*I*n the waters off the coast of California, you will find a lot more algae than sea urchins, and many more sea urchins than sea otters. In the plains of Africa, plentiful grass feeds the gazelles, which are greater in number than the lions that feed on them. Why are photosynthesizers so prevalent and why do prey always outnumber their predators?

Metabolism—all the reactions that occur in a cell—requires energy, and much of what a sea urchin or any animal eats is used to drive the reactions that maintain the organism. Cells use energy-rich nutrients to make ATP (adenosine triphosphate), and ATP fuels the metabolic reactions that produce cell parts and products. When ATP is made and when it is broken down, some energy is lost as heat. In the end only about 10% of the energy in food becomes the organism.

If energy is continually lost as heat, how is it possible to sustain a community of organisms at all? Photosynthesizers capture solar energy and they produce the nutrient molecules that become food for the biosphere. We are just as dependent on photosynthesizers as are sea urchins and sea otters. The sun is the ultimate source of energy for all organisms, whether the organism is an alga, a sea urchin, a sea otter, a gazelle, a lion, or a human.

6.1 Cells and the Flow of Energy

Energy is the ability to do work or bring about a change. Living things are constantly changing—they develop, grow, and reproduce. The need to acquire energy is one of the characteristics of life. Cells use acquired energy to maintain their organization and carry out reactions that allow organisms to develop, grow and reproduce.

Forms of Energy

Energy occurs in two basic forms: kinetic and potential energy. **Kinetic energy** is the energy of motion, as when a ball rolls, the rays of the sun transverse the atmosphere, or a moose lifts its leg (Fig. 6.1). A ball rolling and a muscle contracting are two types of mechanical energy, while the rays of the sun are referred to as solar energy. **Potential**

energy is stored energy—its capacity to do work is not being used at the moment. The food we eat has potential energy because it can be converted into various types of kinetic energy. Food is specifically called **chemical energy** because food is composed of organic molecules such as carbohydrates, proteins, and fat. When the moose lifts its leg, it has converted chemical energy into mechanical energy.

Two Laws of Thermodynamics

Figure 6.1 also demonstrates the flow of energy in ecosystems. In terrestrial ecosystems, plants capture only a small portion of solar energy and much of it dissipates as heat. When plants photosynthesize and use the food they produce, more heat results. Still, there is enough remaining to sustain moose and other animals in the ecosystem. As animals metabolize nutrient molecules, all the captured solar energy eventually dissipates as heat. Therefore, energy flows and does not cycle. Two laws of thermodynamics explain why energy flows in ecosystems and also in cells. These laws were formulated by early researchers who studied energy relationships and exchanges.

The first law of thermodynamics—the law of conservation of energy—states the following:
Energy cannot be created or destroyed, but it can be changed from one form to another.

One way to demonstrate photosynthesis is to show that leaf cells use solar energy to form carbohydrate molecules from carbon dioxide and water. (Carbohydrates are energy-rich molecules, while carbon dioxide and water are energy-poor molecules.) Not all of the captured solar energy becomes carbohydrates; some becomes heat:

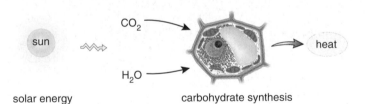

solar energy   carbohydrate synthesis

Obviously, the plant did not create the energy it used to produce carbohydrate molecules; that energy came from the sun. Was any energy destroyed? No, because heat is also a form of energy. Similarly, the muscle cells of the moose use the energy derived from carbohydrates to power its muscles. And as muscles convert chemical energy to mechanical energy, heat is given off.

carbohydrate   muscle contraction

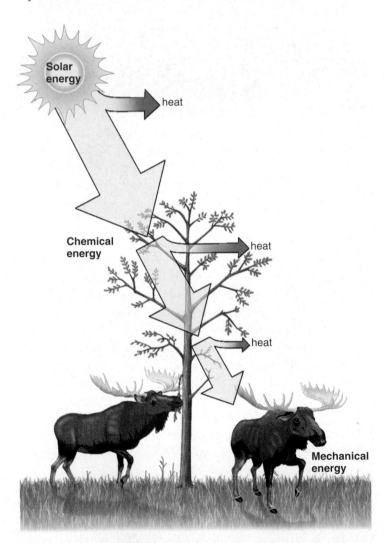

Figure 6.1   Flow of energy.
The plant converts solar energy to the chemical energy of nutrient molecules. The moose converts a portion of this chemical energy to the mechanical energy of motion. Eventually, all solar energy absorbed by the plant dissipates as heat.

This brings us to the second law of thermodynamics.

The second law of thermodynamics states the following:
Energy cannot be changed from one form to another without a loss of usable energy.

In our example, this law is upheld because some of the solar energy taken in by the plant and some of the chemical energy within the nutrient molecules taken in by a moose becomes heat. When heat dissipates into the environment, it is no longer usable—that is, it is not available to do work. With transformation upon transformation, eventually all usable forms of energy become heat that is lost to the environment. Heat cannot be converted to one of the other forms of energy.

Cells and Entropy

The second law of thermodynamics can be stated another way: Every energy transformation makes the universe less organized and more disordered. The term **entropy** is used to indicate the relative amount of disorganization. Since the processes that occur in cells are energy transformations, the second law means that every process that occurs in cells always does so in a way that increases the total entropy of the universe. Then, too, any one of these processes makes less energy available to do useful work in the future.

Figure 6.2 shows two processes that occur in cells. The second law of thermodynamics tells us that glucose tends to break apart into carbon dioxide and water. Why? Because glucose is more organized, and therefore less stable, than its breakdown products. Also, ions on one side of a membrane tend to move to the other side unless they are prevented from doing so. Why? Because when they are distributed randomly, entropy has increased. As an analogy, you know from experience that a neat room is more organized but less stable than a messy room, which is disorganized but more stable. How do you know a neat room is less stable than a messy room? Consider that a neat room always tends to become more messy.

On the other hand, you know that some cells can make glucose out of carbon dioxide and water and all cells can actively move ions to one side of the membrane. How do they do it? These cellular processes obviously require an input of energy from an outside source. This energy ultimately comes from the sun. Cells, living things, and ecosystems depend on a constant supply of energy from the sun because the ultimate fate of all solar energy in the biosphere is to become randomized in the universe as heat. A living cell is a temporary repository of order purchased at the cost of a constant flow of energy.

Energy exits in several different forms. When energy transformations occur, energy is neither created nor destroyed. However, there is always a loss of usable energy. For this reason, living things are dependent on an outside source of energy that ultimately comes from the sun.

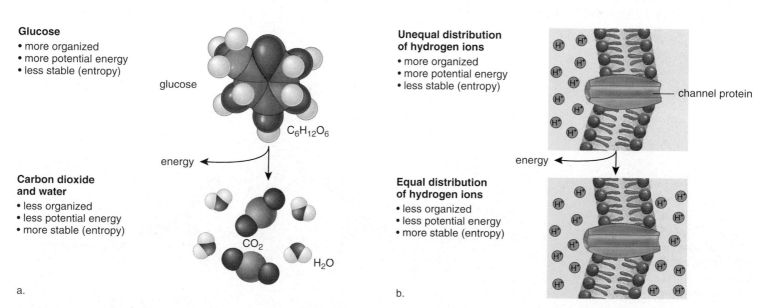

Glucose
• more organized
• more potential energy
• less stable (entropy)

glucose

$C_6H_{12}O_6$

energy

Carbon dioxide and water
• less organized
• less potential energy
• more stable (entropy)

CO_2

H_2O

a.

Unequal distribution of hydrogen ions
• more organized
• more potential energy
• less stable (entropy)

channel protein

energy

Equal distribution of hydrogen ions
• less organized
• less potential energy
• more stable (entropy)

b.

Figure 6.2   Cells and entropy.
The second law of thermodynamics tells us that (**a**) glucose, which is more organized, tends to break down to carbon dioxide and water, which are less organized. **b.** Similarly, hydrogen ions (H^+) on one side of a membrane tend to move to the other side so that the ions are randomly distributed. Both processes result in a loss of potential energy and an increase in entropy.

6.2 Metabolic Reactions and Energy Transformations

Metabolism is the sum of all the chemical reactions that occur in a cell. **Reactants** are substances that participate in a reaction, while **products** are substances that form as a result of a reaction. In the reaction A + B → C + D, A and B are the reactants while C and D are the products. How would you know that this reaction will occur spontaneously—that is, without an input of energy? Using the concept of entropy, it is possible to state that a reaction will occur spontaneously if it increases the entropy of the universe. But in cell biology, we don't usually wish to consider the entire universe. We simply want to consider this reaction. In such instances, cell biologists use the concept of free energy instead of entropy. **Free energy** is the amount of energy available—that is, energy that is still "free" to do work after a chemical reaction has occurred. Free energy is denoted by the symbol G after Josiah Gibbs, who first developed the concept. The value of ΔG is calculated by subtracting the free energy content of the reactants from that of the products. A negative ΔG (change in free energy) means that the products have less free energy than the reactants and the reaction will occur spontaneously. In our reaction, if C and D have less free energy than A and B, then the reaction will "go."

Exergonic reactions are ones in which ΔG is negative and energy is released, while **endergonic reactions** are ones in which ΔG is positive and the products have more free energy than the reactants. Endergonic reactions can only occur if there is an input of energy. In the body many reactions such as protein synthesis, nerve conduction, or muscle contraction are endergonic and they occur because the energy released by exergonic reactions is used to drive endergonic reactions. ATP is a carrier of energy between exergonic and endergonic reactions.

ATP: Energy for Cells

ATP (adenosine triphosphate) is the common energy currency of cells; when cells require energy, they "spend" ATP. You may think that this causes our bodies to produce a lot of ATP, and it does. However, the amount on hand at any one moment is minimal because ATP is constantly being generated from **ADP (adenosine diphosphate)** and ⓟ (Fig. 6.3).

The use of ATP as a carrier of energy has some advantages:

1. It provides a common energy currency that can be used in many different types of reactions.
2. When ATP becomes ADP + ⓟ, the amount of energy released is sufficient for the biological purpose, and so little energy is wasted.
3. ATP breakdown is coupled to endergonic reactions in such a way that it minimizes energy loss.

Structure of ATP

ATP is a modified nucleotide composed of the nitrogen-containing base adenine and the 5-carbon sugar ribose (together called adenosine) and three phosphate groups. ATP is called a "high-energy" compound because a phosphate group is easily removed. Under cellular conditions, the amount of energy released when ATP is hydrolyzed to ADP + ⓟ is about 7.3 kcal per mole.[1]

[1]A mole is the number of molecules present in the molecular weight of a substance (in grams).

Figure 6.3   The ATP cycle.
In cells, ATP carries energy between exergonic reactions and endergonic reactions. When a phosphate group is removed by hydrolysis, ATP releases the appropriate amount of energy for most metabolic reactions.

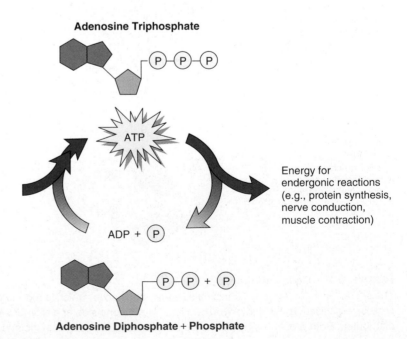

Adenosine Triphosphate

Energy from exergonic reactions (e.g., cellular respiration)

ATP

Energy for endergonic reactions (e.g., protein synthesis, nerve conduction, muscle contraction)

ADP + ⓟ

Adenosine Diphosphate + Phosphate

Coupled Reactions

In **coupled reactions,** the energy released by an exergonic reaction is used to drive an endergonic reaction. ATP breakdown is often coupled to cellular reactions that require an input of energy. Coupling, which requires that the exergonic reaction and the endergonic reaction be closely tied, can be symbolized like this:

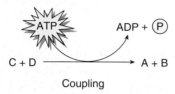

Coupling

Notice that the word energy does not appear following ATP breakdown. Why not? Because this energy was used to drive forward the coupled reaction. Figure 6.4 tells us that ATP breakdown provides the energy necessary for muscular contraction to occur. The energy released when ATP becomes ADP + Ⓟ is used to drive muscle contraction. How is a cell assured of a supply of ATP? Recall that glucose breakdown during cellular respiration provides the energy for the buildup of ATP in mitochondria. Only 39% of the free energy of glucose is transformed to ATP; the rest is lost as heat. When ATP breaks down to drive the reactions mentioned, some energy is lost as heat, and the overall reaction becomes exergonic.

Function of ATP

Recall that at various times we have mentioned at least three uses for ATP.

Chemical work. ATP supplies the energy needed to synthesize macromolecules that make up the cell, and therefore the organism.

Transport work. ATP supplies the energy needed to pump substances across the plasma membrane.

Mechanical work. ATP supplies the energy needed to permit muscles to contract, cilia and flagella to beat, chromosomes to move, and so forth.

In most cases, ATP is the immediate source of energy for these processes.

ATP is a carrier of energy in cells. It is the common energy currency because it supplies energy for many different types of reactions.

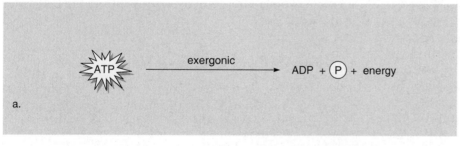

a.

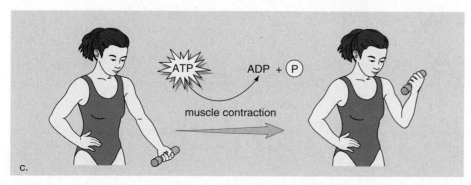

b.

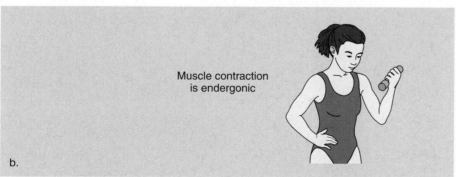

c.

Figure 6.4 Coupled reactions.
a. The breakdown of ATP is exergonic. **b.** Muscle contraction is endergonic and therefore cannot occur without an input of energy. **c.** When muscle contraction is coupled to ATP breakdown, the overall process becomes exergonic. Now muscle contraction can occur.

6.3 Metabolic Pathways and Enzymes

Reactions do not occur haphazardly in cells; they are usually part of a **metabolic pathway,** a series of linked reactions. Metabolic pathways begin with a particular reactant and terminate with an end product. While it is possible to write an overall equation for a pathway as if the beginning reactant went to the end product in one step, actually many specific steps occur in between. In the pathway, one reaction leads to the next reaction, which leads to the next reaction, and so forth in an organized, highly structured manner. This arrangement makes it possible for one pathway to lead to several others, because various pathways have several molecules in common. Also, metabolic energy is captured and utilized more easily if it is released in small increments rather than all at once.

A metabolic pathway can be represented by the following diagram:

$$E_1 \quad E_2 \quad E_3 \quad E_4 \quad E_5 \quad E_6$$
$$A \rightarrow B \rightarrow C \rightarrow D \rightarrow E \rightarrow F \rightarrow G$$

In this diagram, the letters A–F are reactants and the letters B–G are products in the various reactions. In other words, the products from the previous reaction become the reactants of the next reaction. The letters E_1–E_6 are enzymes.

An **enzyme** is a protein molecule that functions as an organic catalyst to speed a chemical reaction. In a crowded ballroom, a mutual friend can cause particular people to interact. In the cell, an enzyme brings together particular molecules and causes them to react with one another.

The reactants in an enzymatic reaction are called the **substrates** for that enzyme. In the first reaction, A is the substrate for E_1, and B is the product. Now B becomes the substrate for E_2, and C is the product. This process continues until the final product "G" forms.

Any one of the molecules (A–G) in this linear pathway could also be a substrate for an enzyme in another pathway. A diagram showing all the possibilities would be highly branched.

Energy of Activation

Molecules frequently do not react with one another unless they are activated in some way. In the lab, for example, in the absence of an enzyme, activation is very often achieved by heating the reaction flask to increase the number of effective collisions between molecules. The energy that must be added to cause molecules to react with one another is called the **energy of activation** (E_a). Figure 6.5 compares E_a when an enzyme is not present to when an enzyme is present, illustrating that enzymes lower the amount of energy required for activation to occur. Nevertheless, the addition of the enzyme does not change ΔG of the reaction.

In baseball, a home-run hitter must not only hit the ball to the fence, but over the fence. When enzymes lower the energy of activation, it is like removing the fence; then it is possible to get a home run by simply hitting the ball as far as the fence was.

Enzyme-Substrate Complexes

The following equation, which is pictorially shown in Figure 6.6, is often used to indicate that an enzyme forms a complex with its substrate:

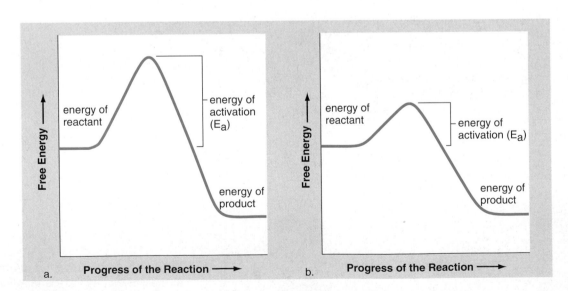

Figure 6.5   Energy of activation (E_a).
Enzymes speed the rate of chemical reactions because they lower the amount of energy required to activate the reactants. **a.** Energy of activation when an enzyme is not present. **b.** Energy of activation when an enzyme is present. Even spontaneous reactions like this one speed up when an enzyme is present.

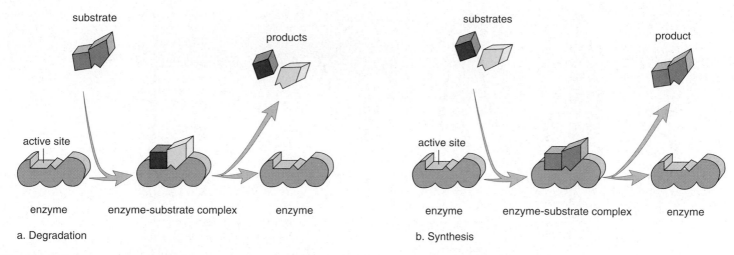

Figure 6.6   Enzymatic action.
An enzyme has an active site, where the substrates and enzyme fit together in such a way that the substrates are oriented to react. Following the reaction, the products are released and the enzyme is free to act again. **a.** Some enzymes carry out degradation; the substrate is broken down to smaller products. **b.** Other enzymes carry out synthesis; the substrates are combined to produce a larger product.

$$E + S \rightarrow ES \rightarrow E + P$$

enzyme   substrate   enzyme-substrate   product
                          complex

In most instances, only one small part of the enzyme, called the **active site,** complexes with the substrate(s). It is here that the enzyme and substrate fit together, seemingly like a key fits a lock; however, it is now known that the active site undergoes a slight change in shape in order to accommodate the substrate(s). This is called the **induced fit model** because the enzyme is induced to undergo a slight alteration to achieve optimum fit (Fig. 6.7).

The change in shape of the active site facilitates the reaction that now occurs. After the reaction has been completed, the product(s) is released, and the active site returns to its original state, ready to bind to another substrate molecule. Only a small amount of enzyme is actually needed in a cell because enzymes are not used up by the reaction.

Some enzymes do more than simply complex with their substrate(s); they participate in the reaction. Trypsin digests protein by breaking peptide bonds. The active site of trypsin contains three amino acids with R groups that actually interact with members of the peptide bond—first to break the bond and then to introduce the components of water. This illustrates that the formation of the enzyme-substrate complex is very important in speeding up the reaction.

Sometimes it is possible for a particular reactant(s) to produce more than one type of product(s). The presence or absence of an enzyme determines which reaction takes place. If a substance can react to form more than one product, then the enzyme that is present and active determines which product is produced.

Every reaction in a cell requires its specific enzyme. Because enzymes only complex with their substrates, they are named for their substrates, as in the following examples:

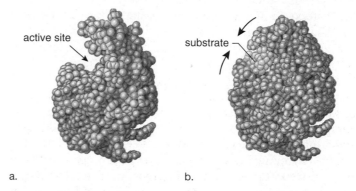

Figure 6.7   Induced fit model.
These computer-generated images show an enzyme called lysozyme that hydrolyzes its substrate, a polysaccharide that makes up bacterial cell walls. **a.** Configuration of enzyme when no substrate is bound to it. **b.** After the substrate binds, the configuration of the enzyme changes so that hydrolysis can better proceed.

Substrate	Enzyme
Lipid	Lipase
Urea	Urease
Maltose	Maltase
Ribonucleic acid	Ribonuclease
Lactose	Lactase

Enzymes are protein molecules that speed chemical reactions by lowering the energy of activation. They do this by forming an enzyme-substrate complex.

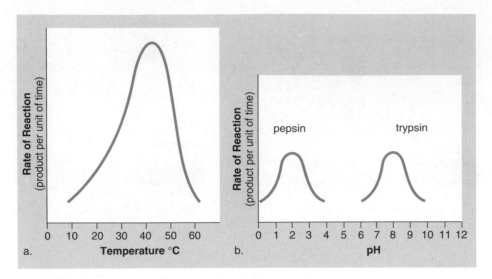

Figure 6.8   **Rate of an enzymatic reaction as a function of temperature and pH.**
a. At first, as with most chemical reactions, the rate of an enzymatic reaction doubles with every 10°C rise in temperature. In this graph, the rate of reaction is maximum at about 40°C; then it decreases until the reaction stops altogether, because the enzyme has become denatured. **b.** Pepsin, an enzyme found in the stomach, acts best at a pH of about 2, while trypsin, an enzyme found in the small intestine, performs optimally at a pH of about 8. The shape that enables these proteins to bind with their substrates is not properly maintained at other pHs.

Factors Affecting Enzymatic Speed

Enzymatic reactions proceed quite rapidly. Consider, for example, the breakdown of hydrogen peroxide (H_2O_2) as catalyzed by the enzyme catalase: $2\,H_2O_2 \rightarrow 2\,H_2O + O_2$. The breakdown of hydrogen peroxide can occur 600,000 times a second when catalase is present. To achieve maximum product per unit time, there should be enough substrate to fill active sites most of the time. Temperature and optimal pH also increase the rate of an enzymatic reaction.

Substrate Concentration
Generally, enzyme activity increases as substrate concentration increases because there are more collisions between substrate molecules and the enzyme. As more substrate molecules fill active sites, more product results per unit time. But when the enzyme's active sites are filled almost continuously with substrate, the enzyme's rate of activity cannot increase any more. Maximum rate has been reached.

Temperature and pH
As the temperature rises, enzyme activity increases (Fig. 6.8a). This occurs because higher temperatures cause more effective collisions between enzyme and substrate. However, if the temperature rises beyond a certain point, enzyme activity eventually levels out and then declines rapidly because the enzyme is **denatured.** An enzyme's shape changes during denaturation, and then it can no longer bind its substrate(s) efficiently.

Each enzyme also has an optimal pH at which the rate of the reaction is highest. Figure 6.8b shows the optimal pH for the enzymes pepsin and trypsin. At this pH value, these enzymes have their normal configurations. The globular structure of an enzyme is dependent on interactions, such as hydrogen bonding, between R groups. A change in pH can alter the ionization of these side chains and disrupt normal interactions, and under extreme conditions of pH, denaturation eventually occurs. Again, the enzyme has an altered shape and is then unable to combine efficiently with its substrate.

Enzyme Concentration
Since enzymes are specific, a cell regulates which enzymes are present and/or active at any one time. Otherwise, enzymes may be present that are not needed, or one pathway may negate the work of another pathway.

Genes must be turned on to increase the concentration of an enzyme and turned off to decrease the concentration of an enzyme.

Another way to control enzyme activity is to activate or deactivate the enzyme. Phosphorylation is one way to activate an enzyme. Molecules received by membrane receptors often turn on kinases, which then activate enzymes by phosphorylating them:

Enzyme Inhibition

Enzyme inhibition occurs when an active enzyme is prevented from combining with its substrate. The activity of almost every enzyme in a cell is regulated by feedback inhibition. In the simplest case, when a product is in abundance, it binds competitively with its enzyme's active site. As the product is used up, inhibition is reduced, and more product can be produced. In this way, the concentration of the product is always kept within a certain range.

Most metabolic pathways are regulated by a more complicated type of feedback inhibition (Fig. 6.9). In these instances, the end product of the pathway binds to an allosteric site, which is a site other than the active site of an enzyme. The binding shuts down the pathway, and no more product is produced.

Poisons are often enzyme inhibitors. Cyanide is an inhibitor for an essential enzyme (cytochrome *c* oxidase) in all cells, which accounts for its lethal effect on humans. Penicillin blocks the active site of an enzyme unique to bacteria. Therefore, penicillin is a poison for bacteria. When penicillin is administered, bacteria die, but humans are unaffected.

Enzyme Cofactors

Many enzymes require an inorganic ion or organic but nonprotein molecule to function properly; these necessary ions or molecules are called **cofactors.** The inorganic ions are metals such as copper, zinc, or iron. The organic, nonprotein molecules are called **coenzymes.** These cofactors assist the enzyme and may even accept or contribute atoms to the reactions.

It is interesting that vitamins are often components of coenzymes. **Vitamins** are relatively small organic molecules that are required in trace amounts in our diet and in the diets of other animals for synthesis of coenzymes that affect health and physical fitness. The vitamin becomes a part of the coenzyme's molecular structure. For example, the vitamin niacin is part of the coenzyme NAD, and B_{12} is part of the coenzyme FAD.

A deficiency of any one of these vitamins results in a lack of the coenzyme listed and therefore a lack of certain enzymatic actions. In humans, this eventually results in vitamin-deficiency symptoms: niacin deficiency results in a skin disease called pellagra, and riboflavin deficiency results in cracks at the corners of the mouth.

Various factors affect enzymatic speed, including substrate concentration, temperature, pH, enzyme concentration, inhibition, or necessary cofactors.

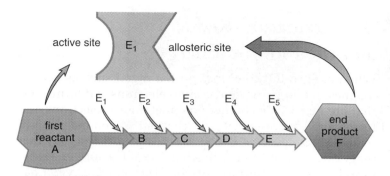

Pathway. E_1 has two sites: the active site where reactant A binds and an allosteric site where end product F binds.

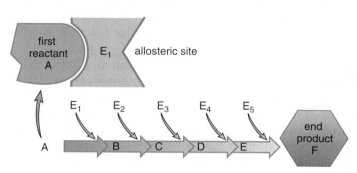

Active Pathway. Reactant A binds to the active site of E_1; therefore, the pathway is active and the end product is produced.

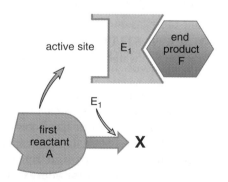

Inhibited Pathway. When there is sufficient end product F, some binds to the allosteric site of E_1. Now a change of shape prevents reactant A from binding to the active site of E_1, and the end product is no longer produced.

Figure 6.9   Feedback inhibition.
Feedback inhibition occurs when the end product of a metabolic pathway binds to the first enzyme of the pathway, preventing the reactant from binding and the reaction from occurring.

7.1 Overview of Cellular Respiration

Cellular respiration is the step-wise release of energy from carbohydrates and other molecules, accompanied by the use of this energy to synthesize ATP molecules. Cellular respiration, as implied by its name, is an **aerobic** cellular process that requires oxygen (O_2) and gives off carbon dioxide (CO_2). It usually involves the complete breakdown of glucose to carbon dioxide and water, as illustrated in Figure 7.1.

This overall equation for cellular respiration shows that it is an oxidation-reduction reaction, or redox reaction for short:

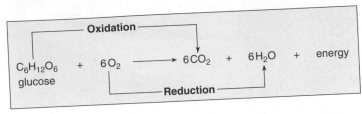

As expected in redox reactions, this equation shows changes in regard to hydrogen atom (H) distribution. A hydrogen atom consists of a hydrogen ion plus an electron ($H^+ + e^-$). When hydrogen atoms are removed from glucose, so are electrons. Since *oxidation* is the loss of electrons, and *reduction* is the gain of electrons, glucose is oxidized and O_2 is reduced.

The oxidation of glucose is an exergonic reaction (releases energy) which drives ATP synthesis which is an endergonic reaction (requires energy). The pathways of cellular respiration allow the energy within a glucose molecule to be released slowly so that ATP can be produced gradually. Cells would lose a tremendous amount of energy if glucose breakdown occurred all at once—much energy would become nonusable heat. The overall equation for cellular respiration (Fig. 7.1) shows the coupling of glucose breakdown to ATP buildup. As glucose is broken down, ATP is built up, and this is the reason the ATP reaction is drawn using a curved arrow above the glucose reaction arrow. The breakdown of one glucose molecule results in a maximum of 36 or 38 ATP molecules. This represents about 40% of the potential energy within a glucose molecule; the rest of the energy is lost as heat. This conversion is more efficient than many others; for example, only about 25% of the energy within gasoline is converted to the motion of a car.

NAD^+ and FAD

Cellular respiration involves many individual metabolic reactions, each one catalyzed by its own enzyme. Enzymes of particular significance are the redox enzymes that utilize the coenzyme **NAD^+ (nicotinamide adenine dinucleotide).** When a metabolite is oxidized, NAD^+ accepts two electrons plus a hydrogen ion (H^+) and NADH results. The electrons received by NAD^+ are high-energy electrons that are usually carried to an electron transport system. Figure 7.2 illustrates how NAD^+ carries electrons.

NAD^+ is called a coenzyme of oxidation-reduction because it can oxidize a metabolite by accepting electrons and can reduce a metabolite by giving up electrons. Only a small amount of NAD^+ need be present in a cell, because each NAD^+ molecule is used over and over again. **FAD (flavin adenine dinucleotide)** is another coenzyme of oxidation-reduction which is sometimes used instead of NAD^+. FAD accepts two electrons and two hydrogen ions (H^+) to become $FADH_2$.

NAD^+ and FAD are two coenzymes of oxidation-reduction that are active during cellular respiration.

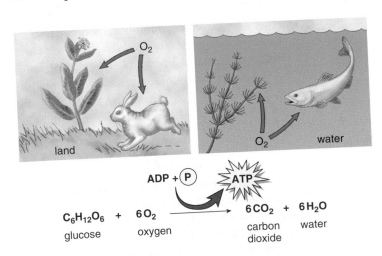

Figure 7.1   Cellular respiration.
Almost all organisms, whether they reside on land or in the water, carry on cellular respiration, which is most often glucose breakdown coupled to ATP synthesis.

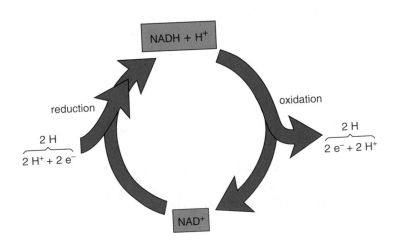

Figure 7.2   The NAD^+ cycle.
The coenzyme NAD^+ accepts two electrons (e^-) plus a hydrogen ion (H^+), and NADH + H^+ result. When NADH passes the electrons to another substrate or carrier, NAD^+ is formed.

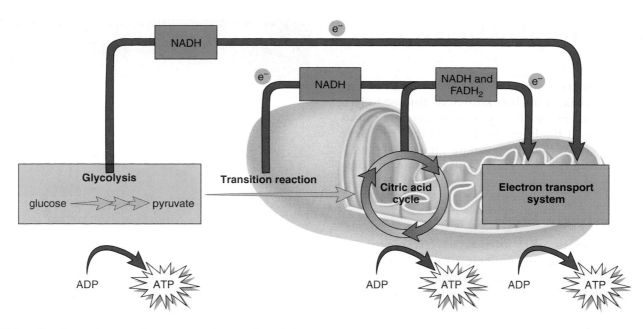

Figure 7.3   The four phases of complete glucose breakdown.
The complete breakdown of glucose consists of four phases. Glycolysis in the cytoplasm produces pyruvate, which enters mitochondria if oxygen is available. The transition reaction and the citric acid cycle that follow occur inside the mitochondria. Also inside the mitochondria, the electron transport system receives the electrons that were removed from glucose breakdown products. The net result of glucose breakdown is 36 to 38 ATP, depending on the particular cell.

Phases of Complete Glucose Breakdown

The oxidation of glucose by removal of hydrogen atoms involves four phases (Fig. 7.3). Glycolysis takes place outside the mitochondria and does not utilize oxygen. The other phases of cellular respiration take place inside the mitochondria, where oxygen is the final acceptor of electrons.

- **Glycolysis** is the breakdown of glucose to two molecules of pyruvate. Oxidation by removal of hydrogen atoms provides enough energy for the immediate buildup of two ATP.
- During the **transition reaction,** pyruvate is oxidized to a 2-carbon acetyl group carried by CoA, and CO_2 is removed. Since glycolysis ends with two molecules of pyruvate, the transition reaction occurs twice per glucose molecule.
- The **citric acid cycle** is a cyclical series of oxidation reactions that give off CO_2 and produce one ATP. The citric acid cycle used to be called the Krebs cycle in honor of the man who worked out most of the steps. The citric acid cycle turns twice because two acetyl-CoA molecules enter the cycle per glucose molecule. Altogether, the citric acid cycle accounts for two immediate ATP molecules per glucose molecule.

- The **electron transport system** is a series of carriers that accept the electrons removed from glucose and pass them along from one carrier to the next until they are finally received by O_2, which then combines with hydrogen ions and becomes water. As the electrons pass from a higher-energy to a lower-energy state, energy is released and used for ATP synthesis. The electrons from one glucose result in 32 or 34 ATP, depending on certain conditions.

Pyruvate is a pivotal metabolite in cellular respiration. If oxygen is not available to the cell, **fermentation** occurs in the cytoplasm. During fermentation, pyruvate is reduced to lactate or to carbon dioxide and alcohol, depending on the organism. As we shall see on page 125, fermentation results in a net gain of only two ATP per glucose molecule.

Cellular respiration involves the oxidation of glucose to carbon dioxide and water. As glucose breaks down, energy is made available for ATP synthesis. A total of 36 or 38 ATP molecules are produced per glucose molecule in cellular respiration (2 from glycolysis, 2 from the citric acid cycle, and 32–34 from the electron transport system.)

Energy Yield from Glucose Metabolism

Figure 7.9 calculates the ATP yield for the complete breakdown of glucose to CO_2 and H_2O. Notice that the diagram includes the number of ATP produced directly by glycolysis and the citric acid cycle, as well as the number produced as a result of electrons passing down the electron transport system.

Per glucose molecule, there is a net gain of two ATP from glycolysis, which takes place in the cytoplasm. The citric acid cycle, which occurs in the matrix of mitochondria, accounts for two ATP per glucose molecule. This means that a total of four ATP are formed by substrate-level phosphorylation outside the electron transport system.

Most ATP is produced by the electron transport system and chemiosmosis. Per glucose molecule, ten NADH and two $FADH_2$ take electrons to the electron transport system. For each NADH formed *inside* the mitochondria by the citric acid cycle, three ATP result, but for each $FADH_2$, only two ATP are produced. Figure 7.7 explains the reason for this difference: $FADH_2$ delivers its electrons to the transport system after NADH, and therefore these electrons cannot account for as much ATP production.

What about the ATP yield of NADH generated *outside* the mitochondria by the glycolytic pathway? NADH cannot cross mitochondrial membranes, but a "shuttle" mechanism allows its electrons to be delivered to the electron transport system inside the mitochondria. The shuttle consists of an organic molecule, which can cross the outer membrane, accept the electrons, and in most but not all cells, deliver them to a FAD molecule in the inner membrane. If FAD is used, only two ATP result because the electrons have not entered at the start of the electron transport system.

Efficiency of Cellular Respiration

It is interesting to calculate how much of the energy in a glucose molecule eventually becomes available to the cell. The difference in energy content between the reactants (glucose and O_2) and the products (CO_2 and H_2O) is 686 kcal. An ATP phosphate bond has an energy content of 7.3 kcal, and 36 of these are usually produced during glucose breakdown; 36 phosphates are equivalent to a total of 263 kcal. Therefore, 263/686, or 39%, of the available energy is usually transferred from glucose to ATP. The rest of the energy is lost in the form of heat.

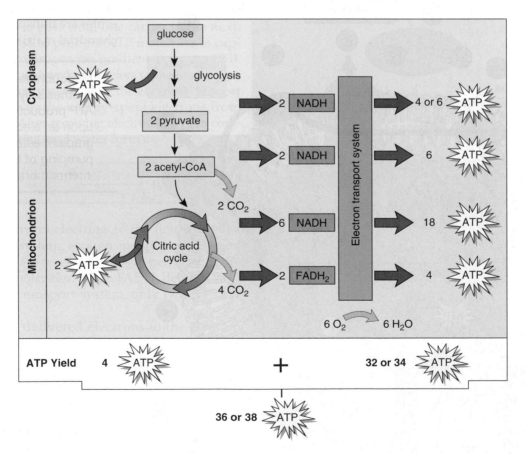

Figure 7.9 Accounting of energy yield per glucose molecule breakdown.
Substrate-level phosphorylation during glycolysis and the citric acid cycle accounts for four ATP. Oxidative phosphorylation accounts for 32 or 34 ATP, and the grand total of ATP is therefore 36 or 38 ATP. Cells differ as to the delivery of the electrons from NADH generated outside the mitochondria. If they are delivered by a shuttle mechanism to the start of the electron transport system, 6 ATP result; otherwise, 4 ATP result.

7.4 Fermentation

Most organisms prefer to carry on cellular respiration when oxygen is available. When oxygen is not available, cells turn to **fermentation.** During fermentation, the pyruvate formed by glycolysis is reduced to alcohol with the release of CO_2 or to one of several organic acids, such as lactate (Fig. 7.10). Certain anaerobic bacteria, such as lactic acid bacteria that help us manufacture cheese, consistently produce lactate in this manner. Other bacteria anaerobically produce chemicals of industrial importance: isopropanol, butyric acid, propionic acid, and acetic acid. Yeasts are good examples of organisms that generate alcohol and CO_2. Yeast is used to leaven bread; the CO_2 produced makes bread rise. Yeast is also used to ferment wine; in that case, it is the ethyl alcohol that is desired. Eventually, yeasts are killed by the very alcohol they produce.

Animal cells, including those of humans, are similar to lactic acid bacteria in that pyruvate, when produced faster than it can be oxidized by the transition reaction and the citric acid cycle, is reduced to lactate. Why is it beneficial for pyruvate to be reduced to lactate when oxygen is not available? The reaction uses NADH and regenerates NAD^+, which are "free" to return and pick up more electrons during the earlier reactions of glycolysis; this keeps glycolysis going.

Advantages and Disadvantages of Fermentation

Despite its low yield of only two ATP made by substrate-level phosphorylation during glycolysis, fermentation is essential to humans. It can provide a rapid burst of ATP—muscle cells more than other cells are apt to carry on fermentation. When our muscles are working vigorously over a short period of time, as when we run, fermentation is a way to produce ATP even though oxygen is temporarily in limited supply.

Lactate, however, is toxic to cells. At first, blood carries away all the lactate formed in muscles. Eventually, however, lactate begins to build up, changing the pH and causing the muscles to "burn" and finally to fatigue so that they no longer contract. When we stop running, our bodies are in **oxygen debt,** as signified by the fact that we continue to breathe very heavily for a time. Recovery is complete when the lactate is transported to the liver, where it is reconverted to pyruvate. Some of the pyruvate is respired completely, and the rest is converted back to glucose.

Efficiency of Fermentation

The two ATP produced per glucose molecule during fermentation are equivalent to 14.6 kcal. Complete glucose molecule breakdown to CO_2 and H_2O represents a possible

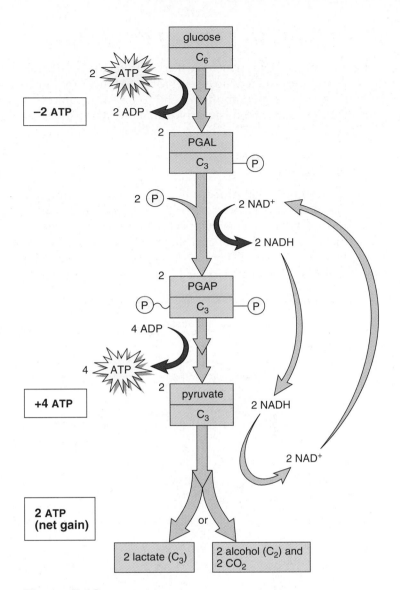

Figure 7.10   Fermentation.
Fermentation consists of glycolysis followed by a reduction of pyruvate. This "frees" NAD^+, and it returns to the glycolytic pathway to pick up more electrons.

energy yield of 686 kcal per molecule. Therefore, the efficiency for fermentation is only 14.6/686, or 2.1%. This is much less efficient than the complete breakdown of glucose. The inputs and outputs of fermentation are as follows:

Fermentation

inputs	outputs
glucose	2 lactate or 2 alcohol and 2 CO_2
2 ATP	2 ADP
4 ADP + 2 (P)	2 ATP (net)

7.5 Metabolic Pool and Biosynthesis

Degradative reactions, which participate in **catabolism,** break down molecules and tend to be exergonic. Synthetic reactions, which participate in **anabolism,** tend to be endergonic. It is correct to say that catabolism drives anabolism because catabolism results in an ATP buildup that is used by anabolism.

Catabolism

We already know that glucose is broken down during cellular respiration. However, other molecules can also undergo catabolism. When a fat is used as an energy source, it breaks down to glycerol and three fatty acids. As Figure 7.11 indicates, glycerol is converted to PGAL, a metabolite in glycolysis. The fatty acids are converted to acetyl-CoA,

which enters the citric acid cycle. An 18-carbon fatty acid results in nine acetyl-CoA molecules. Calculation shows that respiration of these can produce a total of 108 kcal ATP molecules. For this reason, fats are an efficient form of stored energy—there are three long fatty acid chains per fat molecule.

The carbon skeleton of amino acids can also be broken down. The carbon skeleton is produced in the liver when an amino acid undergoes **deamination,** or the removal of the amino group. The amino group becomes ammonia (NH_3), which enters the urea cycle and becomes part of urea, the primary excretory product of humans. Just where the carbon skeleton begins degradation depends on the length of the R group, since this determines the number of carbons left after deamination.

Anabolism

We have already mentioned that the ATP produced during catabolism drives anabolism. But catabolism is also related to anabolism in another way. The substrates making up the pathways in Figure 7.11 can be used as starting materials for synthetic reactions. In other words, compounds that enter the pathways are oxidized to substrates that can be used for biosynthesis. This is the cell's **metabolic pool,** in which one type of molecule can be converted to another. In this way, carbohydrate intake can result in the formation of fat. PGAL can be converted to glycerol, and acetyl groups can be joined to form fatty acids. Fat synthesis follows. This explains why you gain weight from eating too much candy, ice cream, or cake.

Some substrates of the citric acid cycle can be converted to amino acids through transamination, the transfer of an amino group to an organic acid, forming a different amino acid. Plants are able to synthesize all of the amino acids they need. Animals, however, lack some of the enzymes necessary for synthesis of all amino acids. Adult humans, for example, can synthesize 11 of the common amino acids, but they cannot synthesize the other 9. The amino acids that cannot be synthesized must be supplied by the diet; they are called the essential amino acids. (The amino acids that can be synthesized are called nonessential). It is quite possible for animals to suffer from protein deficiency if their diets do not contain adequate quantities of all the essential amino acids.

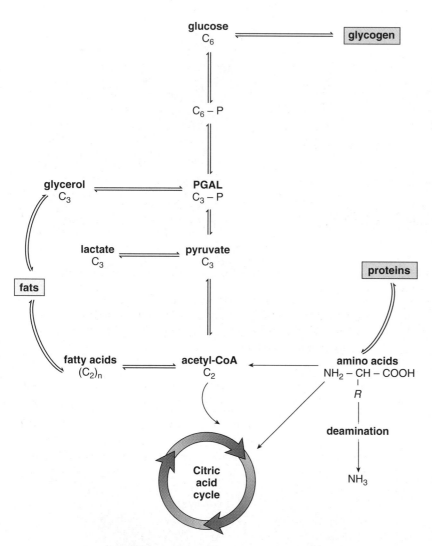

Figure 7.11 **The metabolic pool concept.**
Carbohydrates, fats, and proteins can be used as energy sources, and they enter degradative pathways at specific points. Catabolism produces molecules that can also be used for anabolism of other compounds.

All the reactions involved in cellular respiration are a part of a metabolic pool, and their substrates can be used for catabolism or for anabolism.

Bioethical Focus   Alternative Medicines

Feeling tired and run down? Want to jump start your mitochondria? If you have iron deficiency anemia, you could take iron tablets to encourage your body to build more hemoglobin, the substance that carries oxygen in your blood. If you are diabetic, medications are available to make sure glucose is entering your cells. However, if you seem to have no specific ailment, you might be tempted to turn to what is now called alternative medicine. Alternative medicine includes such non-conventional therapies as herbal supplements, acupuncture, chiropractic therapy, homeopathy, osteopathy, and therapeutic touch (e.g., the laying on of hands).

Advocates of alternative medicine have made some headway in having alternative medicine practices accepted by most anyone. In 1992, congress established what is now called the National Center for Complementary and Alternative Medicine (NCCAM), whose budget has grown from $2 million to its present $50 million for 1999. Then in 1994, the Dietary Supplement Health and Education Act allowed the marketing of vitamins, minerals, and herbs without the requirement that they be approved by the Food and Drug Administration (FDA).

Many scientists feel that this is a mistake and that alternative medicines should be subjected to the same rigorous clinical testing as traditional medicine before they can be marketed. By now plentiful studies have been done, but their quality varies tremendously. We still don't know which alternative medications are safe and/or whether or not they work. Still, sales are booming.

Decide Your Opinion

1. Do you believe that alternative medical practices should be subjected to clinical testing, or do you believe the public should simply rely on "word of mouth" recommendations?
2. What changes do you anticipate if alternative medical practices were to become a regular part of traditional medicine?

Summarizing the Concepts

7.1 Overview of Cellular Respiration
During cellular respiration, glucose is oxidized to CO_2 and H_2O. This exergonic reaction drives ATP buildup, an endergonic reaction. Four phases are required: glycolysis, the transition reaction, the citric acid cycle, and the electron transport system. Oxidation occurs by the removal of hydrogen atoms ($e^- + H^+$) from substrate molecules.

7.2 Outside the Mitochondria: Glycolysis
Glycolysis, the breakdown of glucose to two pyruvates, is a series of enzymatic reactions that occur in the cytoplasm. Oxidation by NAD^+ releases enough energy immediately to give a net gain of two ATP by substrate-level phosphorylation. Two NADH are formed.

7.3 Inside the Mitochondria
Pyruvate from glycolysis enters a mitochondrion, where the transition reaction takes place. During this reaction, oxidation occurs as CO_2 is removed. NAD^+ is reduced, and CoA receives the C_2 acetyl group that remains. Since the reaction must take place twice per glucose, two NADH result.

The acetyl group enters the citric acid cycle, a cyclical series of reactions located in the mitochondrial matrix. Complete oxidation follows, as two CO_2, three NADH, and one $FADH_2$ are formed. The cycle also produces one ATP. The entire cycle must turn twice per glucose molecule.

The final stage of glucose breakdown involves the electron transport system located in the cristae of the mitochondria. The electrons received from NADH and $FADH_2$ are passed down a chain of carriers until they are finally received by O_2, which combines with H^+ to produce H_2O. As the electrons pass down the chain, ATP is produced. The term oxidative phosphorylation is sometimes used for ATP production by the electron transport system.

The carriers of the electron transport system are located in protein complexes on the cristae of the mitochondria. Each protein complex receives electrons and pumps H^+ into the intermembrane space, setting up an electrochemical gradient. When H^+ flows down this gradient through the ATP synthase complex, energy is released and used to form ATP molecules from ADP and $\circledP$. This is ATP synthesis by chemiosmosis.

To calculate the total number of ATP per glucose breakdown, consider that for each NADH formed inside the mitochondrion, three ATP are produced. In most cells, each NADH formed in the cytoplasm results in only two ATP. This is because the carrier that shuttles the hydrogen atoms across the mitochondrial outer membrane usually passes them to a FAD. Each molecule of $FADH_2$ results in the formation of only two ATP because the electrons enter the electron transport system at a lower energy level than NADH. Of the 36 or 38 ATP formed by cellular respiration, four occur outside the electron transport system: two are formed directly by glycolysis and two are formed directly by the citric acid cycle. The rest are produced by the electron transport system.

7.4 Fermentation
Fermentation involves glycolysis, followed by the reduction of pyruvate by NADH to either lactate or alcohol and CO_2. The reduction process "frees" NAD^+ so that it can accept more electrons during glycolysis.

Although fermentation results in only two ATP, it still serves a purpose: in humans, it provides a quick burst of ATP energy for short-term, strenuous muscular activity. The accumulation of lactate puts the individual in oxygen debt because oxygen is needed when lactate is completely metabolized to CO_2 and H_2O.

7.5 Metabolic Pool and Biosynthesis
Carbohydrate, protein, and fat can be broken down by entering the degradative pathways at different locations. These pathways also provide molecules needed for the synthesis of various important substances. Both catabolism and anabolism, therefore, utilize the same metabolic pool of reactants.

Testing Yourself

Choose the best answer for each question.

1. Which of the following is needed for glycolysis to occur?
   a. pyruvate
   b. glucose
   c. NAD^+
   d. ATP
   e. All of the above are needed except a.

2. Which of the following is NOT a product, or end result, of the citric acid cycle?
   a. carbon dioxide   d. ATP
   b. pyruvate   e. $FADH_2$
   c. NADH

3. How many ATP molecules are produced from the reduction and oxidation of one molecule of NAD^+?
   a. 1   c. 36
   b. 3   d. 10

4. How many NADH molecules are produced during the complete breakdown of one molecule of glucose?
   a. 5   c. 10
   b. 30   d. 6

5. What is the name of the process that adds the third phosphate to an ADP molecule using the flow of hydrogen ions?
   a. substrate-level phosphorylation
   b. fermentation
   c. reduction
   d. chemiosmosis

6. Which are possible products of fermentation?
   a. lactic acid   c. CO_2
   b. alcohol   d. All of the above

7. The metabolic process that produces the most ATP molecules is
   a. glycolysis.   c. electron transport system.
   b. citric acid cycle.   d. fermentation.

8. The oxygen required by cellular respiration is reduced and becomes part of which molecule?
   a. ATP
   b. H_2O
   c. pyruvate
   d. CO_2

For questions 9–11, identify the pathway involved by matching them to the terms in the key.
   Key:
   a. glycolysis
   b. citric acid cycle
   c. electron transport system

9. Carbon dioxide (CO_2) given off

10. PGAL

11. Cytochrome carriers

12. The greatest contributor of electrons to the electron transport system is
   a. oxygen.
   b. glycolysis.
   c. the citric acid cycle.
   d. the transition reaction.
   e. fermentation.

13. Substrate-level phosphorylation takes place in
   a. glycolysis and the citric acid cycle.
   b. the electron transport system and the transition reaction.
   c. glycolysis and the electron transport system.
   d. the citric acid cycle and the transition reaction.

14. Fatty acids are broken down to
   a. pyruvate molecules, which take electrons to the electron transport system.
   b. acetyl groups, which enter the citric acid cycle.
   c. glycerol, which is found in fats.
   d. amino acids, which excrete ammonia.
   e. All of these are correct.

15. Which of the following is not true of fermentation? Fermentation
   a. has a net gain of only two ATP.
   b. occurs in cytoplasm.
   c. donates electrons to electron transport system.
   d. begins with glucose.
   e. is carried on by yeast.

For questions 16–18, match each term to its described location in the key.
   Key:
   a. matrix of the mitochondrion
   b. cristae of the mitochondrion
   c. intermembrane space of the mitochondrion
   d. in the cytoplasm
   e. None of these are correct.

16. Electron transport system

17. Glycolysis

18. Accumulation of hydrogen ions (H^+)

19. Match the terms to these definitions. Only four of these terms are needed.

anaerobic	oxygen debt
citric acid cycle	pyruvate
fermentation	transition reaction

   a. Occurs in mitochondria and produces CO2, ATP, NADH, and FADH2.
   b. Growing or metabolizing in the absence of oxygen.
   c. End product of glycolysis.
   d. Anaerobic breakdown of glucose that results in a gain of 2 ATP and end products such as alcohol and lactate.

20. Label this diagram of a mitochondrion.

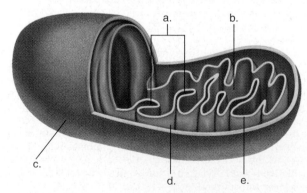

e-Learning Connection   www.mhhe.com/maderinquiry10

Concepts	*Questions*	*Media Resources**
7.1 Overview of Cellular Respiration		

• During cellular respiration, the breakdown of glucose drives the synthesis of ATP molecules.
• NAD^+ and FAD are coenzymes which carry electrons in redox reactions.
• The breakdown of glucose requires four phases: three metabolic pathways and one individual reaction.
	1. How efficient is the conversion of energy in glucose to energy in ATP as a result of cellular respiration?
	2. What is the role of NAD^+ and FAD in cellular respiration?
		Essential Study Partner
		  Introduction
		  Respiration

7.2 Outside the Mitochondria: Glycolysis		

• Glycolysis is a metabolic pathway that partially breaks down glucose outside the mitochondria.
• During glycolysis, glucose is oxidized to molecules of pyruvic acid.
	1. Why must there be an energy investment at the start of glycolysis?
	2. How is ATP formed during glycolysis?
		Essential Study Partner
		  Glycolysis
		
		Art Quizzes
		  Substrate-Level
		    Phosphorylation
		
		BioCourse Study Guides
		  Glycolysis

7.3 Inside the Mitochondria		

• The transition reaction and the citric acid cycle, which occur inside the mitochondria, continue the oxidation of glucose products until carbon dioxide is given off.
• The electron transport system, which receives electrons from NAD^+ and FAD, produces most of the ATP during cellular respiration.
• Oxygen, the final acceptor of electrons from the electron transport system, is reduced to water.
	1. Summarize the events of the citric acid cycle.
	2. In the electron transport system, how is ATP formed differently than it is during glycolysis and during the citric acid cycle?
		Essential Study Partner
		  Transition
		  Krebs Cycle (citric acid cycle)
		  Electron Transport
		
		Art Quizzes
		  Overview of ATP Synthesis
		  ATP Theoretical Yield
		
		BioCourse Study Guides
		  The Krebs Cycle (citric acid cycle)
		  The Electron Transport System
		
		Explorations
		  Oxidative Respiration

7.4 Fermentation		
• Fermentation is a metabolic pathway that partially breaks down glucose under anaerobic conditions.	1. Why do cells sometimes turn to fermentation after glycolysis? 2. What is oxygen debt?	Essential Study Partner Fermentation
7.5 Metabolic Pool and Biosynthesis		

• A number of molecules in addition to glucose can be broken down to drive ATP synthesis.	1. Explain the metabolic pool concept.	Essential Study Partner
		  Other Nutrients
		
		Art Quiz
		  Catabolism of Proteins and Fats

*For additional Media Resources, see the Online Learning Center.

8.1 Radiant Energy

Photosynthetic organisms include plants, algae, and cyanobacteria (Fig. 8.1). Cyanobacteria and algae live in bodies of water, while plants live on land. **Photosynthesis** transforms solar energy into the chemical energy of a carbohydrate, as illustrated by this overall reaction:

solar energy + carbon dioxide + water ⟶ carbohydrate + oxygen

Photosynthesizers worldwide produce an enormous amount of carbohydrate. So much, that if the carbohydrate were instantly converted to coal and the coal were loaded into standard railroad cars (each car holding about 50 tons), the photosynthesizers of the biosphere would fill more than 100 cars per second with coal.

No wonder photosynthetic organisms are able to sustain themselves and, with a few exceptions,[1] the other living things on earth. To realize this, consider that it is possible to trace any food chain back to plants. Another way to express this idea is to say that **producers,** which have the ability to synthesize organic molecules from inorganic raw materials, not only feed themselves but also **consumers,** which must take in preformed organic molecules. All organisms use the organic molecules produced by photosynthesizers as a source of building blocks for growth and repair and as a source of chemical energy for cellular work.

Photosynthesis releases oxygen into the atmosphere. As we saw in Chapter 7, oxygen is needed for cellular respiration, the process that transforms the chemical energy in glucose to the chemical energy in ATP, the energy currency molecule.

Our analogy about photosynthetic products becoming coal is reasonable because the bodies of plants did become the coal we burn in large part to produce electricity. This happened hundreds of thousands of years ago, and that's why coal is called a fossil fuel. The wood of trees is also commonly used as fuel. Then, too, the fermentation of plant materials produces alcohol, which can be used directly to fuel automobiles or as a gasoline additive.

Photosynthesis is critically important because:

- Photosynthetic organisms are able to use solar energy to produce organic nutrients.
- Almost all organisms depend either directly or indirectly on these organic nutrients to sustain themselves.
- The bodies of plants became the coal or other fossil fuels used today.

a. Prayer plant

Figure 8.1 Photosynthetic organisms.
Photosynthetic organisms include (**a**) plants, which typically live on land; (**b**) algae, which typically live in water and can range in size from microscopic to macroscopic; and (**c**) cyanobacteria, which are a type of bacterium.

[1]A few types of bacteria are chemosynthetic organisms, which obtain the necessary energy to produce their own organic nutrients by oxidizing inorganic compounds.

b. Kelp

c. *Nostoc*

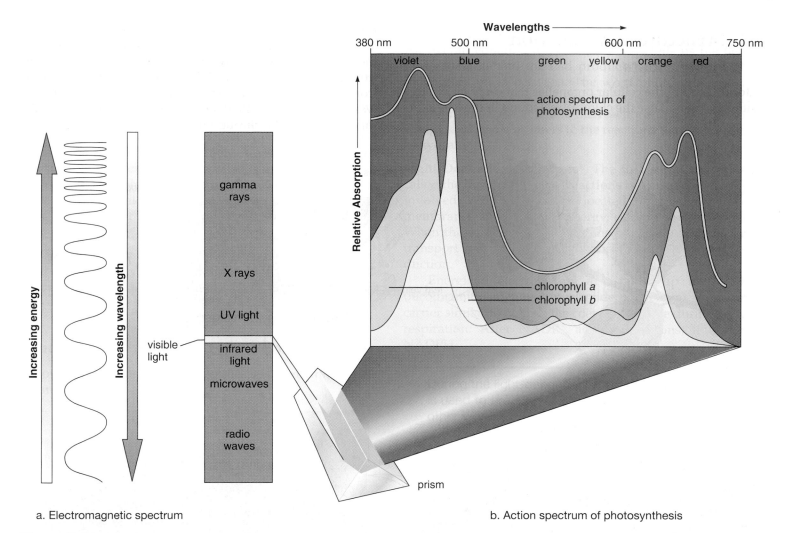

a. Electromagnetic spectrum

b. Action spectrum of photosynthesis

Figure 8.2   The electromagnetic spectrum and chlorophylls *a* and *b*.
a. The electromagnetic spectrum contains forms of energy that differ according to wavelength. Visible light is only a small portion of the electromagnetic spectrum. **b.** Chlorophylls *a* and *b* absorb certain wavelengths within visible light, and this largely accounts for the action spectrum of photosynthesis.

Visible Light

Radiant energy from the sun (solar energy) can be described in terms of its wavelength and its energy content. Figure 8.2*a* lists the different types of radiant energy, from the shortest wavelength, gamma rays, to the longest, radio waves. White or *visible light* is only a small portion of this spectrum. Visible light itself contains various wavelengths of light, as can be proven by passing it through a prism; then we see all the different colors that make up visible light. (Actually, of course, it is our eyes that interpret these wavelengths as colors.) The colors in visible light range from violet (the shortest wavelength) to blue, green, yellow, orange, and red (the longest wavelength). The energy content is highest for violet light and lowest for red light.

Only about 42% of the solar radiation that hits the earth's atmosphere ever reaches the surface, and most of this radiation is within the visible-light range. Higher-energy wavelengths are screened out by the ozone layer in the atmosphere, and lower-energy wavelengths are screened out by water vapor and carbon dioxide (CO_2) before they reach the earth's surface. The conclusion is that both the organic molecules within organisms and certain life processes, such as vision and photosynthesis, are adapted to the radiation that is most prevalent in the environment.

The pigments found within most types of photosynthesizing cells, the **chlorophylls** and **carotenoids,** are capable of absorbing various portions of visible light. The absorption spectrum for chlorophyll *a* and chlorophyll *b* is shown in Figure 8.2*b*. Both chlorophyll *a* and chlorophyll *b* absorb violet, blue, and red light better than the light of other colors. Because green light is reflected and only minimally absorbed, leaves appear green to us. Accessory pigments such as the carotenoids are yellow or orange and are able to absorb light in the violet-blue-green range. These pigments and others become noticeable in the fall when chlorophyll breaks down and the other pigments are uncovered.

Stages of the Calvin Cycle

The previous page presented a simplified overview of the Calvin cycle, but the following discussion is provided for those who want to examine the cycle in more depth. For the sake of our discussion, the cycle can be divided into: (1) fixation of CO_2; (2) reduction of CO_2; and (3) regeneration of RuBP.

Fixation of Carbon Dioxide

The first event of the Calvin cycle is fixation of carbon dioxide. **Carbon dioxide fixation** occurs when CO_2 is attached to an organic compound (Fig. 8.8). During the Calvin cycle, RuBP, a 5-carbon molecule, combines with CO_2 to form a 6-carbon molecule. The enzyme that speeds up this reaction, called RuBP carboxylase, is present in many more copies than you would expect. RuBP carboxylase makes up about 20–50% of the protein content in chloroplasts! The reason for its abundance may be that it is unusually slow (it processes only about three molecules of substrate per second, compared to about one thousand per second for a typical enzyme). Thus, a lot of RuBP carboxylase is needed to keep the Calvin cycle going.

Fixation of carbon dioxide occurs when CO_2 combines with RuBP.

Metabolites of the Calvin Cycle	
RuBP	ribulose bisphosphate
PGA	3-phosphoglycerate
PGAP	1,3-bisphosphoglycerate
PGAL	glyceraldehyde-3-phosphate

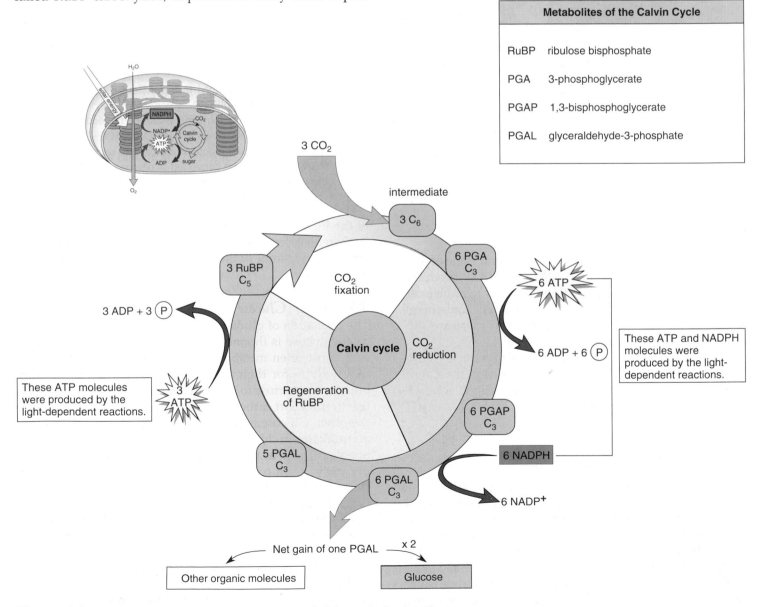

Figure 8.8   The light-independent reactions: the Calvin cycle (in detail).
The Calvin cycle is divided into three portions: CO_2 fixation, CO_2 reduction, and regeneration of RuBP. Because five PGAL are needed to re-form three RuBP, it takes three turns of the cycle to have a net gain of one PGAL. Two PGAL molecules are needed to form glucose.

Reduction of Carbon Dioxide

The 6-carbon molecule resulting from fixation of CO_2 immediately breaks down to form two PGA, which are C_3 molecules. Each of the two PGA molecules undergoes reduction to PGAL in two steps:

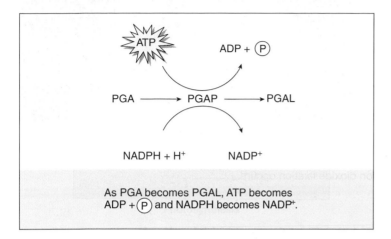

As PGA becomes PGAL, ATP becomes ADP + (P) and NADPH becomes NADP⁺.

This is a redox reaction in which CO_2 is reduced to a carbohydrate (CH_2O) and NADPH is oxidized to $NADP^+$. The energy required for this synthesis reaction is contributed by ATP, which becomes ADP + (P).

This is the reaction that utilizes NADPH and ATP from the light-dependent reactions. The $NADP^+$ and the ADP + (P) that result from this reaction return to the light-dependent reactions where they are energized once again. For every three turns of the Calvin cycle, one PGAL leaves the pathway. It takes two 3-carbon (C_3) PGAL molecules to make one 6-carbon (C_6) glucose molecule.

> The net gain from the incorporation of three CO_2 in the Calvin cycle is one PGAL molecule. Therefore, it takes six turns to form glucose.

Regeneration of RuBP

The Calvin cycle would not continue unless RuBP were regenerated. For every three turns of the Calvin cycle, five molecules of PGAL are used to re-form three molecules of RuBP, and the cycle continues:

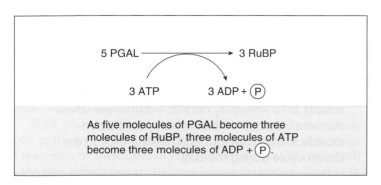

As five molecules of PGAL become three molecules of RuBP, three molecules of ATP become three molecules of ADP + (P).

Notice that this reaction also utilizes some of the ATP produced by the light-dependent reactions.

> In addition to producing PGAL, the Calvin cycle regenerates RuBP, and this allows the cycle to continue.

Let's look again at the overall reaction for photosynthesis and relate the reactants and products to the pathways studied:

$$\text{solar energy} + CO_2 + H_2O \xrightarrow{\quad\quad} (CH_2O) + O_2$$

(with **Reduction** spanning from CO_2 to (CH_2O), and **Oxidation** spanning from H_2O to O_2)

Solar energy, water, and oxygen should be associated with the light-dependent reactions. During these reactions, (1) solar energy energizes electrons, and these are used through chemiosmosis to form ATP molecules and NADPH molecules. Therefore, during the light-dependent reactions, solar energy is transformed into the chemical energy of ATP and NADPH. (2) water is oxidized and releases oxygen gas. The release of oxygen by a plant is often used as evidence that photosynthesis is occurring.

Carbon dioxide and carbohydrate (CH_2O) should be associated with the light-independent reactions. During these reactions, (1) carbon dioxide is reduced to a carbohydrate. Reduction of carbon dioxide utilizes ATP and NADPH from the light-dependent reactions. (2) ATP becomes ADP + (P), and NADPH becomes $NADP^+$; then these molecules return to the light-dependent reactions to be energized once again. Note that during the light-independent reactions, the chemical energy of ATP and NADPH is transformed into the chemical energy of a carbohydrate, notably glucose.

How efficient is photosynthesis? That is, how much of the solar energy absorbed actually ends up in carbohydrate? You know from your study of the laws of thermodynamics, that energy transformations always result in a loss of usable energy. Under ideal laboratory conditions, plants have been known to transform 25% of the solar energy they absorb into carbohydrate. Under natural conditions, the efficiency of photosynthesis is much lower, ranging from less than 1% to a maximum of 8%.

The complications of the photosynthetic process shouldn't cause us to lose sight of the importance of the process to ourselves and the biosphere. Without photosynthesis, humans probably would never have evolved and certainly wouldn't continue to exist. Why? Because only photosynthetic organisms are able to transform solar energy into the chemical energy that we and all living things can use to sustain ourselves. To release this energy, most organisms, including photosynthesizers, carry on cellular respiration, which requires the oxygen given off by photosynthesis.

9.1 Plant Organs

This chapter is about the organs and tissues of **angiosperms,** the flowering plants. In later chapters, we will also study other types of plants, such as mosses, ferns, and gymnosperms. Flowering plants are extremely diverse because they are adapted to living in varied environments. There are even flowering plants that live in water! Despite their great diversity in size and shape, flowering plants usually have three vegetative organs. An **organ** is a structure that contains different types of tissues and performs one or more specific functions. The vegetative organs of a flowering plant—the root, the stem, and the leaf—allow a plant to live and grow. The body of a plant is composed of the root system and the shoot system (Fig. 9.1).

Roots

Although we are accustomed to speaking of the root, it is more appropriate to refer to the root system. The **root system** of a plant, such as a tomato, has a main root, or taproot, and many lateral, or branch, roots (Fig. 9.2a). As a rule of thumb, the root system is at least equivalent in size and extent to the **shoot system** (the part of the plant above ground). An apple tree, then, has a much larger root system than, say, a corn plant. A single corn plant may have roots as deep as 2.5 meters and spread out over 1.5 meters, but a mesquite tree that lives in the desert may have roots that penetrate to a depth of 20 meters. The extensive root system of a plant anchors it in the soil and gives it support.

The root system absorbs water and minerals from the soil for the entire plant. The cylindrical shape of a root allows it to penetrate the soil as it grows and permits water to be absorbed from all sides. The absorptive capacity of a root is also increased by its many root hairs located in a special zone near the root tip. Root hairs, which are projections from root-hair cells, are especially responsible for the absorption of water and minerals. Root hairs are so numerous that they increase the absorptive surface of a root tremendously. It has been estimated that a single rye plant has about 14 billion hair cells, and if placed end to end, the root hairs would stretch 10,626 kilometers. Root-hair cells are constantly being replaced. So this same rye plant most likely forms about 100 million new root-hair cells every day. You are probably familiar with the fact that a plant yanked out of the soil will not fare well when transplanted; this is because the root hairs have been torn off. Transplantation is more apt to be successful if you take a part of the surrounding soil along with the plant.

Roots have still other functions. Roots produce hormones that stimulate the growth of stems and coordinate their size with the size of the root. It is more efficient for a plant to have root and stem sizes that are proportionate to one another. Also, **perennial** plants, which die back and then regrow the next season, store the products of photosynthesis in their roots. Carrots and sweet potatoes come from the roots of such plants, for example.

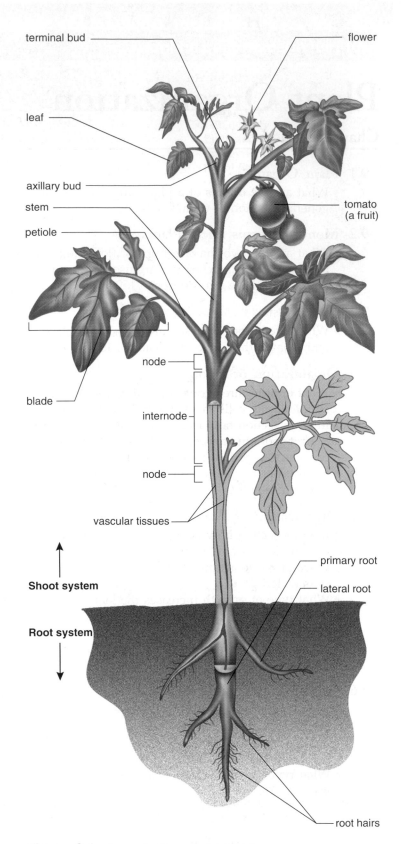

Figure 9.1   Organization of plant body.

The body of a plant consists of a root system and a shoot system. Roots are the only type of plant organ in the root system. The shoot system contains the stem and leaves, two other types of plant organs. The root system is connected to the shoot system by vascular tissue (brown) that extends from the roots to the leaves.

main (tap) root

a. Root system

b. Shoot system

leaflet

c. Leaves

Figure 9.2   Vegetative organs of the tomato.
a. The root system anchors the plant and absorbs water and minerals. **b.** The shoot system contains the stem and its branches, which support the leaves and transport water and organic nutrients. **c.** Leaves, such as these from a tomato plant, are often broad and thin. They carry on photosynthesis.

Stems

The shoot system of a plant contains both stems and leaves. A **stem** is the main axis of a plant along with its lateral branches (Fig. 9.2b). The stem of a flowering plant terminates in tissue that allows the stem to elongate and produce leaves. If vertical, as most are, stems support leaves in such a way that each leaf is exposed to as much sunlight as possible. The place where a leaf attaches to a stem is called a **node,** and an **internode** is the region between the nodes. The presence of nodes and internodes is used to identify a stem even if it happens to be an underground stem. In some plants the nodes of horizontal stems asexually produce new plants.

Aside from supporting the leaves, a stem has vascular tissue that transports water and minerals from the roots to the leaves and also transports the products of photosynthesis, usually in the opposite direction. Nonliving cells form a continuous pipeline for water and mineral transport, while living cells join end to end for organic nutrient transport. A cylindrical stem can expand in girth as well as length. As trees grow taller each year, they accumulate nonfunctional woody tissue that adds to the strength of their stems.

Some stems have functions other than transport. Some are specialized for storage. The stem stores water in cactuses and in other plants. Tubers are horizontal stems that store nutrients, as in the potato plant.

Leaves

A **leaf** is the organ of a plant that usually carries on photosynthesis, a process that requires solar energy, carbon dioxide, and water. Leaves absorb solar energy and they take up carbon dioxide from the air. Water comes to leaves by way of the stem.

In contrast to the shape of stems, leaves are often broad and thin. This shape maximizes their surface area for the absorption of carbon dioxide and the collection of solar energy. Also unlike stems, leaves are almost never woody. All their cells are living, and most of them are able to carry on photosynthesis.

The wide portion of a leaf is called the **blade.** A tomato plant has a compound blade with several leaflets (Fig. 9.2c). The **petiole** is a stalk that attaches the blade to the stem. The upper and acute angle between the petiole and the stem is designated the leaf axil, and this is where an **axillary** (lateral) **bud,** which may become a branch or a flower, originates. Not all leaves make up foliage. Some leaves are specialized to protect buds, attach to objects (tendrils), store food (bulbs), or even capture insects.

A flowering plant has three vegetative organs: the root absorbs water and minerals, the stem supports and services leaves, and the leaf carries on photosynthesis.

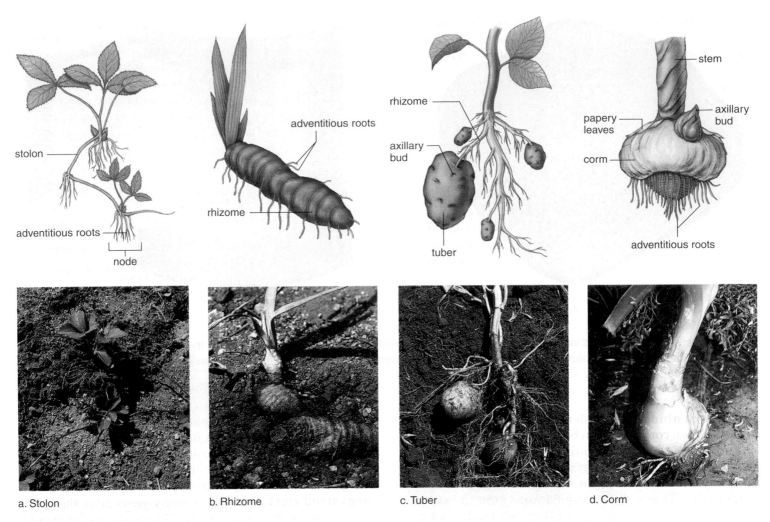

Figure 9.17   Stem diversity.
a. A strawberry plant has aboveground, horizontal stems called stolons. Every other node produces a new shoot system. **b.** The underground horizontal stem of an iris is a fleshy rhizome. **c.** The rhizomes of a potato plant have enlargements called tubers. We call the tubers potatoes. **d.** The corm of a gladiolus is a stem covered by papery leaves.

Stem Diversity

Stem diversity is illustrated in Figure 9.17. Aboveground horizontal stems, called **stolons** or runners, produce new plants where nodes touch the ground. The strawberry plant is a common example of this type of stem, which functions in vegetative reproduction.

Aboveground vertical stems can also be modified. For example, cacti have succulent stems specialized for water storage, and the tendrils of grape plants (which are stem branches) allow them to climb. Morning glory and its relatives have stems that twine around support structures. Tendrils and twining shoots help plants expose their leaves to the sun.

Underground horizontal stems, called **rhizomes,** may be long and thin, as in sod-forming grasses, or thick and fleshy, as in the iris. Rhizomes survive the winter and contribute to asexual reproduction because each node bears a bud. Some rhizomes have enlarged portions called **tubers,** which function in food storage. Potatoes are tubers, in which the eyes are buds that mark the nodes.

Corms are bulbous underground stems that lie dormant during the winter, just as rhizomes do. They also produce new plants the next growing season. Gladiolus corms are referred to as bulbs by laypersons, but the botanist reserves the term bulb for a structure composed of modified leaves attached to a short vertical stem. An onion is a bulb.

Humans use stems in many ways. The stem of the sugarcane plant is a primary source of table sugar. The spice cinnamon and the drug quinine are derived from the bark of different plants. And wood is necessary for the production of paper as discussed in the Ecology Focus on the next page.

Plants use diverse stems for such functions as reproduction, climbing, survival, and food storage. Modified stems aid adaptation to different environments.

Paper: Can We Cut Back?

The word *paper* takes its origin from papyrus, the plant Egyptians used to make the first form of paper some 5,500 years ago. The Egyptians manually made sheets from the treated stems of papyrus grass and then strung them together into scrolls. From that beginning, the production of paper has become a worldwide industry of major importance (Fig. 9B). The process is fairly simple. Plant material is ground up mechanically, then chemically treated to form a pulp that contains "fibers," which biologists know are the tracheids and vessel elements of a plant. The fibers automatically form a sheet when they are screened from the pulp. Today, a revolving wire-screen belt delivers a continuous, wet sheet of paper to heavy rollers and heated cylinders, which remove most of the remaining moisture and press the paper flat.

Each person in the United States consumes about 318 kilograms (699 pounds) of paper products per year, compared to only 2.3 kilograms (5 pounds) of paper per person in India. We are all aware that products in the United States are overpackaged and that the overabundance of our junk mail ends up in the trash almost immediately. Schools and businesses use far too much writing paper—even using two sides of a sheet instead of one side would be helpful!

Although eucalyptus plants from South America, bamboo from India, and even cotton are used to make paper, most paper is made from trees, with severe ecological consequences. Trees are sometimes clear-cut from natural forests, and if so, it will be many years before the forests will regrow. In the meantime, the community of organisms that depends on the forest ecosystem must relocate or die off. Natural ecosystems the world over have been replaced by giant tree plantations containing stands of uniform trees to serve as a source of wood. In Canada, there are temperate hardwood tree plantations of birch, beech, chestnut, poplar, and particularly aspen trees. Tropical hardwoods, which usually come from Southeast Asia and South America, are also sometimes used to make paper. In the United States, several species of pine trees have been genetically improved to have a higher wood density and to be harvestable five years earlier than ordinary pines. Southern Africa, Chile, New Zealand, and Australia also devote thousands of acres to growing pines for paper pulp production.

The making of paper uses energy and causes both air and water pollution. Caustic chemicals such as sodium hydroxide, sulfurous acid, and bleaches are used during the manufacturing process. These are released into the air and, along with paper mill wastes, also add significantly to the pollution of rivers and streams. Underground water supplies are poisoned by the ink left in the ground after paper biodegrades in landfills.

It is clear that we should all cut back on our use of paper so that less of it is made in the first place. In addition, we should recycle the paper that has already been made. When newspaper and office paper (including photocopies) are soaked in water, the fibers are released, and they can be used to make recycled paper and/or cardboard. Manufacturing recycled paper uses less energy and causes less pollution than making paper anew. Moreover, trees are conserved. It is estimated that recycling of Sunday newspapers alone would save an estimated 500,000 trees each week.

Figure 9B **Paper production.**
Machine No. 35 at Champion International's Courtland, Alabama, mill produces a 29-foot-wide roll of office paper every 60 minutes.

9.6 Organization of Leaves

Leaves are the organs of photosynthesis in vascular plants. Figure 9.18 shows a cross section of a typical dicot leaf of a temperate-zone plant. At the top and bottom is a layer of epidermal tissue that often bears protective hairs and/or glands that produce irritating substances. These features may prevent the leaf from being eaten by insects. The epidermis characteristically has an outer, waxy cuticle that keeps the leaf from drying out. The cuticle also prevents gas exchange because it is not gas permeable. However, the epidermis, particularly the lower epidermis, contains stomata that allow gases to move into and out of the leaf. Each stoma has two guard cells that regulate its opening and closing.

The body of a leaf is composed of **mesophyll** tissue, which has two distinct regions: **palisade mesophyll,** containing elongated cells, and **spongy mesophyll,** containing irregular cells bounded by air spaces. The parenchyma cells of these layers have many chloroplasts and carry on most of the photosynthesis for the plant. The loosely packed arrangement of the cells in the spongy layer increases the amount of surface area for gas exchange.

As mentioned earlier, a leaf often consists of a flattened blade and a petiole connecting the blade to the stem. The blade may be single or composed of several leaflets (Fig. 9.19). Externally, it is possible to see the pattern of the leaf veins, which contain vascular tissue. Leaf veins have a net pattern in dicot leaves and a parallel pattern in monocot leaves (see Fig. 9.3).

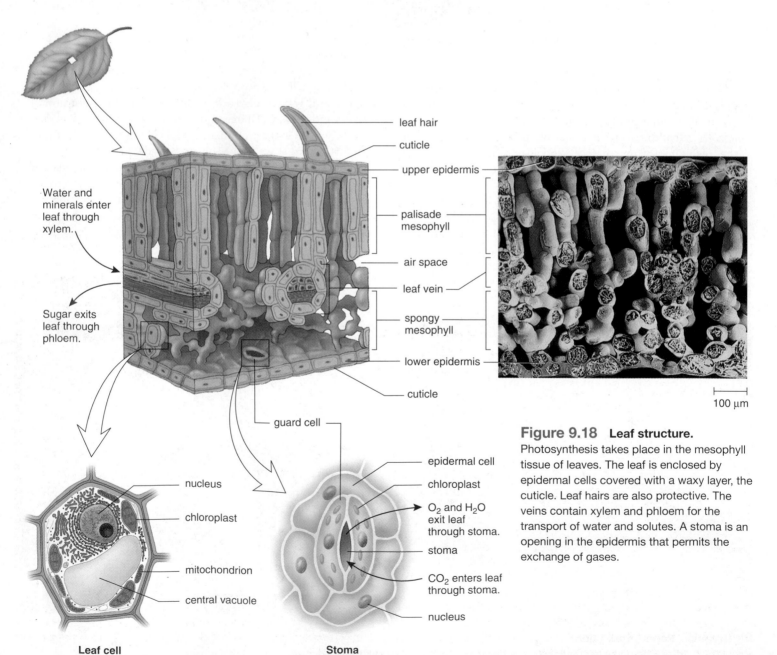

leaf hair
cuticle
upper epidermis
palisade mesophyll
air space
leaf vein
spongy mesophyll
lower epidermis
cuticle
guard cell

Water and minerals enter leaf through xylem.

Sugar exits leaf through phloem.

100 µm

epidermal cell
chloroplast
O₂ and H₂O exit leaf through stoma.
stoma
CO₂ enters leaf through stoma.
nucleus

nucleus
chloroplast
mitochondrion
central vacuole

Leaf cell

Stoma

Figure 9.18   Leaf structure.
Photosynthesis takes place in the mesophyll tissue of leaves. The leaf is enclosed by epidermal cells covered with a waxy layer, the cuticle. Leaf hairs are also protective. The veins contain xylem and phloem for the transport of water and solutes. A stoma is an opening in the epidermis that permits the exchange of gases.

Leaves are adapted to environmental conditions. Shade plants tend to have broad, wide leaves, and desert plants tend to have reduced leaves with sunken stomata. The leaves of a cactus are the spines attached to the succulent stem (Fig. 9.20*a*). Other succulents have leaves adapted to hold moisture.

An onion bulb is made up of leaves surrounding a short stem. In a head of cabbage, large leaves overlap one another. The petiole of a leaf can be thick and fleshy, as in celery and rhubarb. Climbing leaves, such as those of peas and cucumbers, are modified into tendrils that can attach to nearby objects (Fig. 9.20*b*). The leaves of a few plants are specialized for catching insects. For example, the leaves of a sundew have sticky epidermal hairs that trap insects and then secrete digestive enzymes. The Venus's-flytrap has hinged leaves that snap shut and interlock when an insect triggers sensitive hairs (Fig. 9.20*c*). The leaves of a pitcher plant resemble a pitcher and have downward-pointing hairs that lead insects into a pool of digestive enzymes. Insectivorous plants commonly grow in marshy regions, where the supply of soil nitrogen is severely limited. The digested insects provide the plants with a source of organic nitrogen.

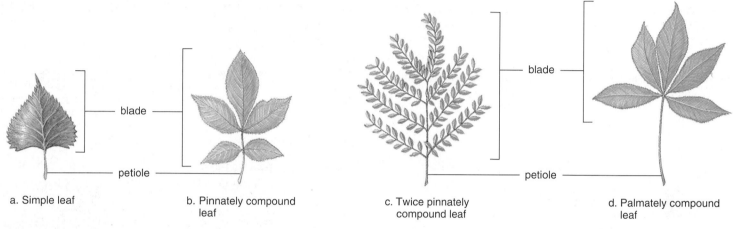

a. Simple leaf   b. Pinnately compound leaf   c. Twice pinnately compound leaf   d. Palmately compound leaf

Figure 9.19 Classification of leaves.
a. The cottonwood tree has a simple leaf. **b.** The shagbark hickory has a pinnately compound leaf. **c.** The honey locust has a twice pinnately compound leaf. **d.** The buckeye has a palmately compound leaf.

a. Cactus   b. Cucumber   c. Venus's-flytrap

Figure 9.20 Leaf diversity.
a. The spines of a cactus plant are leaves modified to protect the fleshy stem from animal consumption and limit the loss of water. **b.** The tendrils of a cucumber are leaves modified to attach the plant to a physical support. **c.** The leaves of the Venus's-flytrap are modified to serve as a trap for insect prey. When triggered by an insect, the leaf snaps shut. Once shut, the leaf secretes digestive juices, which break down the soft parts of the prey's body, allowing nutrients such as nitrogen to be absorbed by the plant body.

Summarizing the Concepts

9.1 Plant Organs

A flowering plant has three vegetative organs. A root anchors a plant, absorbs water and minerals, and stores the products of photosynthesis. Stems support leaves, conduct materials to and from roots and leaves, and help store plant products. Leaves carry on photosynthesis.

9.2 Monocot Versus Dicot Plants

Flowering plants are divided into the dicots and monocots according to the number of cotyledons in the seed; the arrangement of vascular tissue in roots, stems, and leaves; and the number of flower parts.

9.3 Plant Tissues

Three types of meristem continually divide and produce specialized tissues. Protoderm produces epidermal tissue, ground meristem produces ground tissue, and procambium becomes vascular cambium which produces vascular tissue (Table 9.1).

Epidermal tissue is composed of only epidermal cells. Ground tissue contains parenchyma cells, which are thin-walled and capable of photosynthesis when they contain chloroplasts. Collenchyma cells have thicker walls for flexible support. Sclerenchyma cells are hollow, nonliving support cells with secondary walls. Vascular tissue consists of xylem and phloem. Xylem contains vessels composed of vessel elements and tracheids. Xylem transports water and minerals. Phloem contains sieve tubes composed of sieve-tube elements, each of which has a companion cell. Phloem transports organic nutrients.

9.4 Organization of Roots

A root tip shows three zones: the zone of cell division (containing root apical meristem) protected by the root cap, the zone of elongation, and the zone of maturation. A cross section of a herbaceous dicot root reveals the epidermis (for protection), the cortex (food storage), the endodermis (regulation of the movement of minerals), and the vascular cylinder (transport). In the vascular cylinder of a dicot, the xylem appears star-shaped, and the phloem is found in separate regions, between the arms of the xylem. In contrast, a monocot root has a ring of vascular tissue with alternating bundles of xylem and phloem surrounding pith.

Roots are diversified. Taproots are specialized to store the products of photosynthesis. A fibrous root system is composed of adventitious roots, and so are prop roots, which are specialized to provide increased anchorage.

9.5 Organization of Stems

Primary growth of a stem is due to the activity of the shoot apical meristem, which is protected within a terminal bud. A terminal bud contains leaf primordia at nodes; the area between the nodes is called the internode. When stems grow, the internodes lengthen.

In cross section, a nonwoody dicot has epidermis, cortex tissue, vascular bundles in a ring, and an inner pith. Monocot stems have scattered vascular bundles, and the cortex and pith are not well defined. Secondary growth of a woody stem is due to vascular cambium, which produces new xylem and phloem every year, and cork cambium, which produces new cork cells when needed. Cork replaces epidermis in woody plants. A cross section of a woody stem shows bark, wood, and pith. The bark contains cork and phloem. Wood contains annual rings of xylem.

Stems are diverse. Besides vertical stems, there are horizontal aboveground and underground stems. Corms and some tendrils are also modified stems.

9.6 Organization of Leaves

A cross section of a leaf reveals the upper and lower epidermis, with stomata mostly in the lower epidermis. Mesophyll tissue forms the bulk of a leaf. Vascular tissue is present within leaf veins.

Leaves are diverse. The spines of a cactus are leaves. Other succulents have fleshy leaves. An onion is a bulb with fleshy leaves, and the tendrils of peas are leaves. A few plants, including the Venus's-flytrap have leaves that trap and digest insects.

Table 9.1	Vegetative Organs and Major Tissues		
	Root	**Stem**	**Leaf**
Function	Absorbs water and minerals	Transports water and nutrients	Carries on photosynthesis
	Anchors plant	Supports leaves	
	Stores materials	Helps store materials	
Tissue			
Epidermis*	Protects inner tissues	Protects inner tissues	Protects inner tissues
	Root hairs absorb water and minerals		Stomata carry on gas exchange
Cortex†	Stores water and products of photosynthesis	Carries on photosynthesis, if green Some storage of products of photosynthesis	Not present
Endodermis†	Regulates passage of water and minerals in vascular tissue	Not present	Not present
Vascular‡	Transports water and nutrients	Transports water and nutrients	Transports water and nutrients
Pith†	Stores products of photosynthesis and water (monocot only)	Stores products of photosynthesis	Not present
Mesophyll†	Not present	Not present	Primary site of photosynthesis

Note: Plant tissues belong to one of three tissue systems:

*Epidermal tissue system

†Ground tissue system

‡Vascular tissue system

Testing Yourself

Choose the best answer for each question.

1. Which of these is an incorrect contrast between monocots (stated first) and dicots (stated second)?
   a. one cotyledon—two cotyledons
   b. leaf veins parallel—net veined
   c. vascular bundles in a ring—vascular bundles scattered
   d. flower parts in threes—flower parts in fours or fives
   e. All of these are correct contrasts.

2. Which of these types of cells is most likely to divide?
   a. parenchyma
   b. meristem
   c. epidermis
   d. xylem
   e. sclerenchyma

3. Which of these cells in a plant is apt to be nonliving?
   a. parenchyma
   b. collenchyma
   c. sclerenchyma
   d. epidermal
   e. guard cells

4. Root hairs are found in the zone of
   a. cell division.
   b. elongation.
   c. maturation.
   d. apical meristem.
   e. All of these are correct.

5. Cortex is found in
   a. roots, stems, and leaves.
   b. roots and stems.
   c. roots and leaves.
   d. stems and leaves.
   e. roots only.

6. Between the bark and the wood in a woody stem, there is a layer of meristem called
   a. cork cambium.
   b. vascular cambium.
   c. apical meristem.
   d. the zone of cell division.
   e. procambium preceding bark.

7. Which part of a leaf carries on most of the photosynthesis of a plant?
   a. epidermis
   b. mesophyll
   c. epidermal layer
   d. guard cells
   e. Both a and b are correct.

8. Annual rings are the number of
   a. internodes in a stem.
   b. rings of vascular bundles in a monocot stem.
   c. layers of xylem in a stem.
   d. bark layers in a woody stem.
   e. Both b and c are correct.

9. The Casparian strip is found
   a. between all epidermal cells.
   b. between xylem and phloem cells.
   c. on four sides of endodermal cells.
   d. within the secondary wall of parenchyma cells.
   e. in both endodermis and pericycle.

10. The spines of a cactus, the tendrils of climbing vines, and the part of an onion that you eat are examples of
    a. modified leaves
    b. modified stems
    c. modified roots
    d. leaf petioles

11. Guard cells, cork, and root hairs are found in what type of plant tissue?
    a. epidermal
    b. ground
    c. vascular
    d. meristem

12. In a cross section of which of the following structures would you find scattered vascular bundles?
    a. monocot root
    b. monocot stem
    c. dicot root
    d. dicot stem

13. Secondary growth would result in which of the following?
    a. new leaves
    b. new shoots
    c. new root
    d. cork, bark, and wood
    e. All but d are correct.

14. Bark is made of which of the following?
    a. phloem
    b. cork
    c. cortex
    d. cork cambium
    e. All of these are correct.

15. Sclerenchyma, parenchyma, and collenchyma are cells found in what type of plant tissue?
    a. epidermal
    b. ground
    c. vascular
    d. meristem
    e. All but d are correct.

16. Label this root using the terms endodermis, phloem, xylem, cortex, and epidermis:

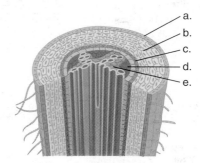

17. Label this leaf using the terms leaf vein, lower epidermis, palisade mesophyll, spongy mesophyll, and upper epidermis:

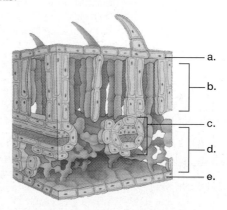

e-Learning Connection

Concepts	Questions	Media Resources*
9.1 Plant Organs		

• Flowering plants have a root system that contains the roots and a shoot system that contains the stems and the leaves.	1. List several functions of roots.
	2. How are underground stems distinguished from roots?
		Essential Study Partner
		  Plant Organs—Introduction
		
		Labeling Exercise
		  Organization of Plant Body

9.2 Monocot Versus Dicot Plants		
• Flowering plants are classified into two groups, the monocots and the dicots.	1. What are cotyledons? 2. What is the significance of the division between monocots and dicots?	
9.3 Plant Tissues		

• Plant cells can be organized into three types of tissues: epidermal tissue, ground tissue, and vascular tissue.	1. List three types of ground tissue cells and state their functions.
	2. Where is vascular tissue located in the various plant organs?
		Essential Study Partner
		  Ground Tissue
		  Dermal Tissue
		  Vascular Tissue
		
		Animation Quiz
		  Vascular System of Plants

9.4 Organization of Roots		

• In longitudinal section, a dicot root tip has a zone where new cells are produced, another where they elongate, and another where they differentiate and mature.
• In cross section, dicot and monocot roots differ in the organization of their vascular tissue.
• Some plants have a taproot, others a fibrous root, and others have adventitious roots.
	1. From what region does a root elongate, and from where can it send out lateral roots?
	2. What is the significance of having different types of roots in different plants?
		Essential Study Partner
		  Roots
		  Meristems
		
		Labeling Exercises
		  Dicot Root Tip
		  Primary Meristems

9.5 Organization of Stems		

• Dicot and monocot herbaceous stems differ in the organization of their vascular tissue.
• All stems grow in length but some plants are woody and grow in girth also.
• Stems are diverse and some plants have horizontal aboveground or underground stems.
	1. How do herbaceous stems differ from woody stems?
	2. What type of tissue is the bark of a tree?
		Essential Study Partner
		  Stems
		
		Animation Quiz
		  Girth Increase in Woody dicots
		
		Labeling Exercises
		  Stem Tip
		  Herbaceous Dicot Stem
		    Anatomy
		  Secondary Growth in a Stem

9.6 Organization of Leaves		

• The bulk of a leaf is composed of cells that perform gas exchange and carry on photosynthesis.
• Leaves are diverse; some conserve water, some help a plant climb and some help a plant capture food.
	1. Why are parenchyma cells within the spongy mesophyll loosely arranged instead of tightly packed?	Essential Study Partner
		  Leaves
		
		Labeling Exercises
		  Leaf Structure

*For additional Media Resources, see the Online Learning Center.

10

Plant Reproduction, Growth, and Development

Chapter Concepts

Fruits and vegetables provide humans with the many foods they eat.

A Japanese farmer raises a bowl of steaming rice to his lips as a Mexican mother serves freshly prepared tortillas and an American teenager reaches for another slice of white bread. Around the globe, plant products such as fruits and vegetables are the staples of peoples' diet. Italians eat pasta, Irish and Germans consume potatoes, and the bread we eat is a source of carbohydrates, the organic molecules that supply most of our energy needs.

Favorite combinations of plants, such as rice and soybeans in Asia and beans and corn in South America, supply all the amino acids and vitamins the body needs. Meats can supply all the amino acids but not all the vitamins. Plants also supply us with fats—the various cooking oils are fats. Plants concentrate and give us our minerals. Where does a cow get the calcium that goes into milk? From eating grass.

We can't digest the cellulose within cell walls of plants, and it becomes the roughage that also improves our health. "Eat your vegetables" is not the nagging expression it seems—it is a recognition that plants provide us with all the essential nutrients that allow our cells to continue metabolizing. The dietary importance of plants gives ample reason for us to know something about their anatomy and physiology.

10.1 Sexual Reproduction in Flowering Plants

Sexual reproduction is the rule in flowering plants. This may come as a surprise to those who think reproduction requires a male and female. However, sexual reproduction is defined properly as reproduction requiring gametes, often an egg and a sperm. In a flowering plant, the structures that produce the egg and sperm are located within the flower.

Structure of Flowers

Figure 10.1 (top) shows the parts of a typical **flower.** The **sepals,** which enclose the unopened flower bud, form a whorl about the **petals** once the bud has opened. The size, shape, color, and odor of a flower attract a specific pollinator. Flowers pollinated by the wind often have no petals at all.

The **stamens** make up the next whorl of flower parts. Each stamen has two parts: the **anther,** a saclike container, and the **filament,** a slender stalk. In the center of the flower is a small, vaselike structure, the **pistil,** which usually has three parts: the **stigma,** an enlarged sticky knob; the **style,** a slender stalk; and the **ovary,** an enlarged base. The ovary contains one or more **ovules,** which play a significant role in reproduction.

Not all flowers have sepals, petals, stamens, and a pistil. Those that do are said to be perfect (bisexual), and those that do not are said to be imperfect (unisexual). Imperfect flowers with only stamens are called staminate flowers, and those with only pistils are called pistillate flowers. If staminate flowers and pistillate flowers are on one plant, as in corn, the plant is monoecious. If staminate and pistillate flowers are on separate plants, the plant is dioecious. Holly trees are dioecious, and if red berries are desired, it is necessary to acquire a plant with staminate flowers and another with pistillate flowers, for reasons we will now discuss.

Alternation of Generations

Flowering plants have a life cycle called **alternation of generations** because two generations are involved: the **sporophyte** and the **gametophyte.** The sporophyte is a diploid (2n) generation that produces haploid (n) **spores** by meiosis. A flower produces two types of spores: **microspores** and **megaspores.** Microspores are produced in the anthers of stamens, and megaspores are produced within ovules (Fig. 10.1). A microspore becomes a pollen grain, which upon maturity is a sperm-containing *microgametophyte,* also called the male gametophyte. A megaspore becomes an egg-containing embryo sac, which is a *megagametophyte,* also called the female gametophyte. Following fertilization, the zygote develops into an embryo located within a seed. When the seed germinates, the new sporophyte plant begins to grow.

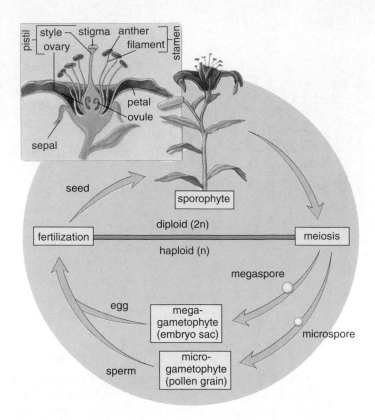

Figure 10.1   Alternation of generations in a flowering plant.

Flowering plants are heterosporous; they produce microspores and megaspores. A microspore becomes a pollen grain, which is a microgametophyte that produces sperm. A megaspore becomes an embryo sac, or megagametophyte, that produces an egg. When a sperm joins with an egg, the resulting zygote develops into an embryo that is retained within a seed.

Figure 10.2 shows these same steps in greater detail. Within an ovule, a megaspore (*mega,* large) parent cell undergoes meiosis to produce four haploid megaspores. Three of these megaspores disintegrate, leaving one functional megaspore, which divides mitotically. The result is the megagametophyte, or **embryo sac,** which typically consists of eight haploid nuclei embedded in a mass of cytoplasm. The cytoplasm differentiates into cells, one of which is an egg and another of which is the central cell with two nuclei (called the **polar nuclei**).

The anther has **pollen sacs,** which contain numerous microspore (*micro,* small) parent cells. Each parent cell undergoes meiosis to produce four haploid cells called microspores. The microspores usually separate, and each one becomes a **pollen grain.** At this point, the young pollen grain contains two cells, the **generative cell** and the **tube cell.**

Pollination occurs when pollen is windblown or carried by insects, birds, or bats to the stigma of the same type of plant. Only then does a pollen grain germinate and develop a long pollen tube. This pollen tube grows within the style until it reaches an ovule in the ovary. At this time,

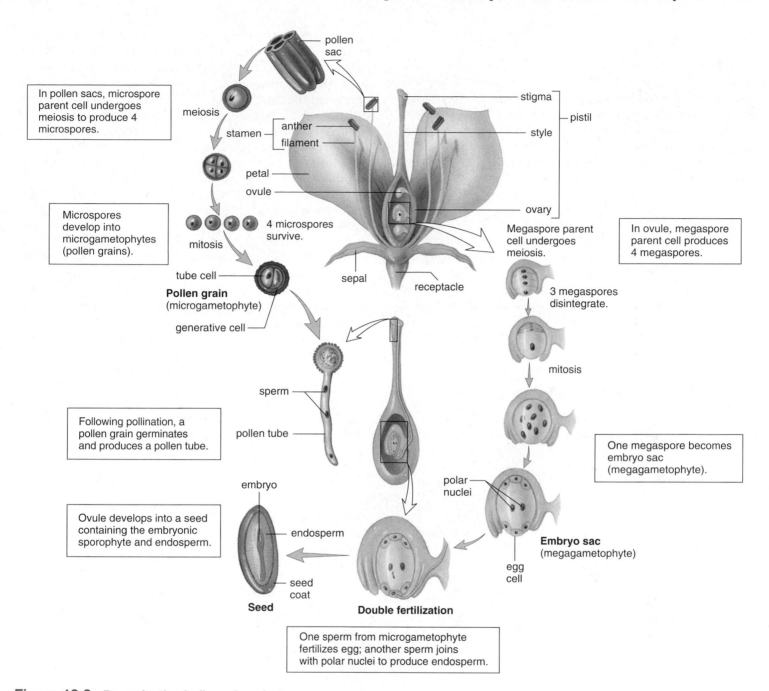

Figure 10.2   **Reproduction in flowering plants.**

Reproduction of a flowering plant involves production of sperm by a series of steps (left) and production of an egg by a series of steps (right) followed by fertilization and development of an embryo-containing seed.

too, the generative cell divides mitotically, producing two sperm, which have no flagella. Upon reaching the ovule, the pollen tube discharges the sperm. One of the two sperm migrates to and fertilizes the **egg,** forming a **zygote,** and the other sperm migrates to and unites with the two polar nuclei, producing a 3n (triploid) endosperm nucleus. These two fusions are known as **double fertilization.**

The endosperm nucleus divides to form **endosperm,** food for the developing plant. The zygote develops into a multicellular **embryo,** and the ovule wall hardens and becomes the **seed coat.** A **seed** is a structure formed by

maturation of an ovule; it consists of a sporophyte embryo, stored food, and a seed coat. The ovary, and sometimes other floral parts, develops into a **fruit.**

Flowering plants produce an egg within each ovule of an ovary, and sperm within pollen grains. They have double fertilization; one fertilization produces the zygote and the other produces endosperm. The ovule becomes a seed, and the ovary becomes a fruit.

Growth and Development in Plants

Plant growth and development involve cell division, cellular elongation, and differentiation of cells into tissues and then organs. **Development** is a programmed series of stages from a simpler to a more complex form. **Cellular differentiation,** or specialization of structure and function, occurs as development proceeds. In Chapter 9, we saw that meristem cells undergo elongation and differentiation to produce the tissues of roots and shoots. Now we will consider the stages in the development of a dicot embryo (Fig. 10.3).

Development of the Dicot Embryo

In dicot plants, after double fertilization has taken place, the single-celled zygote lies beneath the endosperm nucleus. The endosperm nucleus divides mitotically to produce a mass of endosperm tissue surrounding the embryo. The zygote also divides, forming two parts: the upper part is the embryo, and the lower part is the suspensor, which anchors the embryo and transfers nutrients to it from the sporophyte plant. Soon the **cotyledons,** or seed leaves, can be seen. At this point, the dicot embryo is heart-shaped. Later, when it becomes torpedo-shaped, it is possible to distinguish the shoot apex and the root apex. These contain apical meristems, the tissues that bring about primary growth in a plant; the shoot apical meristem is responsible for aboveground growth, and the root apical meristem is responsible for underground growth.

Monocots, unlike dicots, have only one cotyledon. Another important difference between monocots and dicots is the manner in which nutrient molecules are stored in the seed. In a monocot, the cotyledon rarely stores food; rather, it absorbs food molecules from the endosperm and passes them to the embryo. During the development of a dicot embryo, the cotyledons usually store the nutrient molecules that the embryo uses. Therefore, in Figure 10.3 we can see that the endosperm seemingly disappears. Actually, it has been taken up by the two cotyledons. In a plant embryo, the **epicotyl** is above the cotyledon and contributes to shoot development; the **hypocotyl** is that portion below the cotyledon that contributes to stem development; and the **radicle** contributes to root development. The embryo plus stored food is now contained within a seed.

> The plant embryo (which has gone through a programmed series of stages), plus stored food, is contained within a seed.

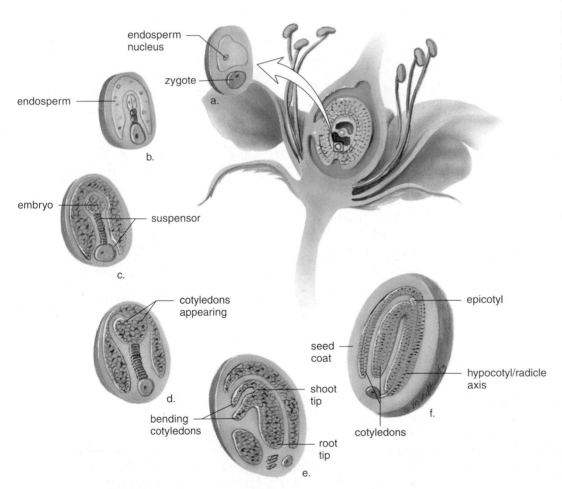

Figure 10.3   Development of a dicot embryo.
a. The unicellular zygote lies beneath the endosperm nucleus. **b, c.** The endosperm is a mass of tissue surrounding the embryo. The embryo is located above the suspensor. **d.** The embryo becomes heart shaped as the cotyledons begin to appear. **e.** There is progressively less endosperm as the embryo differentiates and enlarges. As the cotyledons bend, the embryo takes on a torpedo shape. **f.** The embryo consists of the epicotyl (represented here by the shoot apex), the hypocotyl, and the radicle, which includes the root apex.

a. Almond flower and fruit

one portion
of ovary

b. Tomato plant bearing fruit

Figure 10.4 Fruit diversity.

a. The almond fruit is fleshy, with a single seed enclosed by a hard covering. **b.** The tomato is derived from a compound ovary. **c.** Each blackberry is from a flower with many pistils and therefore ovaries.

Development of Seeds and Fruits

In flowering plants, seeds are enclosed within a fruit, which develops from the ovary and at times, from other accessory parts. Although peas, beans, tomatoes, and cucumbers are commonly called vegetables, botanists categorize them as fruits. A **fruit** is a mature ovary that usually contains seeds.

As a fruit develops from an ovary, the ovary wall thickens to become the pericarp. In fleshy fruits, the pericarp is at least somewhat fleshy; peaches and plums are good examples of fleshy fruits. In almonds, the fleshy part of the pericarp is a husk removed before marketing. We crack the remaining portion of the pericarp to obtain the seed (Fig. 10.4*a*). An apple develops from a compound ovary, but much of the flesh comes from the receptacle, which grows around the ovary. It is more obvious that a tomato comes from a compound ovary, because in cross section, you can see several seed-filled cavities (Fig. 10.4*b*).

Dry fruits have dry pericarps. Legumes, such as peas and beans, produce fruits that split open along two sides, or seams. Not all dry fruits split at maturity. The pericarp of a grain is tightly fused to the seed and cannot be separated from it. A corn kernel is a grain, as are the fruits of wheat, rice, and barley plants.

Some fruits develop from several individual ovaries. A blackberry is an aggregate fruit in which each cluster is derived from the ovaries of a single flower (Fig. 10.4*c*). The strawberry is also an aggregate fruit, but each ovary

many pistils

c. Blackberry flower and fruit

becomes a one-seeded fruit called an achene. The flesh of a strawberry is from the receptacle. In contrast, a pineapple comes from the fruit of many individual flowers attached to the same fleshy stalk. As the ovaries mature, they fuse to form a large, multiple fruit.

In flowering plants, the seed develops from the ovule and possibly other structures.

Dispersal of Seeds

For plants to be widely distributed, their seeds have to be dispersed—that is, distributed preferably long distances from the parent plant.

Plants have various means to ensure that dispersal takes place. The hooks and spines of clover, bur, and cocklebur attach to the fur of animals and the clothing of humans. Birds and mammals sometimes eat fruits, including the seeds, which are then defecated (passed out of the digestive tract with the feces) some distance from the parent plant. Squirrels and other animals gather seeds and fruits, which they bury some distance away.

The fruit of the coconut palm, which can be dispersed by ocean currents, may land many hundreds of kilometers away from the parent plant. Some plants have fruits with trapped air or seeds with inflated sacs that help them float in water. Many seeds are dispersed by wind. Woolly hairs, plumes, and wings are all adaptations for this type of dispersal. The seeds of an orchid are so small and light that they need no special adaptation to carry them far away. The somewhat heavier dandelion fruit uses a tiny "parachute"

for dispersal. The winged fruit of a maple tree, which contains two seeds, has been known to travel up to 10 kilometers from its parent. A touch-me-not plant has seed pods that swell as they mature. When the pods finally burst, the ripe seeds are hurled out.

Animals, water, and wind help plants disperse their seeds.

Germination of Seeds

Following dispersal, the seeds **germinate**: they begin to grow so that a seedling appears. Some seeds do not germinate until they have been dormant for a period of time. For seeds, **dormancy** is the time during which no growth occurs, even though conditions may be favorable for growth. In the temperate zone, seeds often have to be exposed to a period of cold weather before dormancy is broken. In deserts, germination does not occur until there is adequate moisture. This requirement helps ensure that seeds do not germinate until the most favorable growing

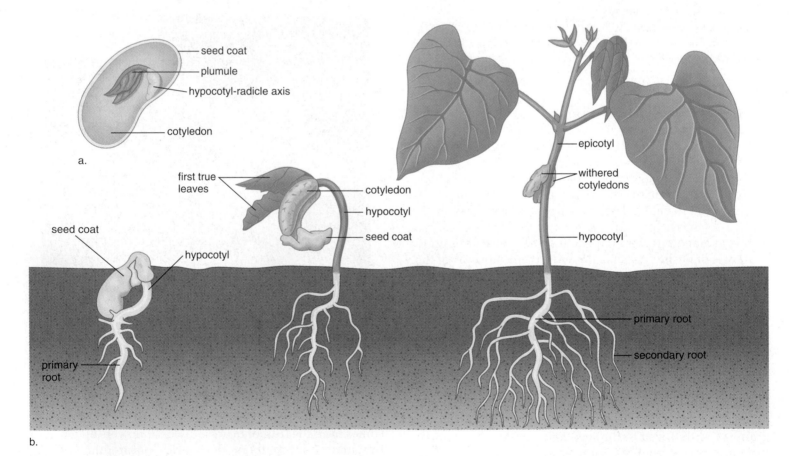

Figure 10.5   Common garden bean, a dicot.

a. Seed structure. **b.** Germination and development of the seedling. Notice that there are two cotyledons and the leaves are net veined. Also, note that the mature plant has withered cotyledons.

season has arrived. Germination takes place if there is sufficient water, warmth, and oxygen to sustain growth. Germination requires regulation, and both inhibitors and stimulators are known to exist. Fleshy fruits (e.g., apples, pears, oranges, and tomatoes) contain inhibitors so that germination does not occur until the seeds are removed and washed. In contrast, stimulators are present in the seeds of some temperate-zone woody plants. Mechanical action may also be required. Water, bacterial action, and even fire can act on the seed coat, allowing it to become permeable to water. The uptake of water causes the seed coat to burst.

Dicot Versus Monocot Development

As mentioned, the embryo of a dicot has two seed leaves, called cotyledons. The cotyledons, which have absorbed the endosperm, supply nutrients to the embryo and seedling, and eventually shrivel and disappear. If the two cotyledons of a bean seed are parted, you can see a rudimentary plant (Fig. 10.5a). The epicotyl bears young leaves and is called a **plumule.** As the dicot seedling emerges from the soil, the shoot is hook-shaped to protect the delicate plumule. The hypocotyl becomes part of the stem, and the radicle develops into the roots. When a seed germinates in darkness, it etiolates—the stem is elongated, the roots and leaves are small, and the plant lacks color and appears spindly. Phytochrome, a pigment that is sensitive to red and far-red light (see page 181), regulates this response and induces normal growth once proper lighting is available.

A corn plant is a monocot that contains a single cotyledon. In monocots, the endosperm is the food-storage tissue, and the cotyledon does not have a storage role. Corn kernels are actually fruits, and therefore the outer covering is the pericarp (Fig. 10.6). The plumule and radicle are enclosed in protective sheaths called the coleoptile and the coleorhiza, respectively. The plumule and the radicle burst through these coverings when germination occurs.

Germination is a complex event regulated by many factors. The embryo breaks out of the seed coat and becomes a seedling with leaves, stem, and roots.

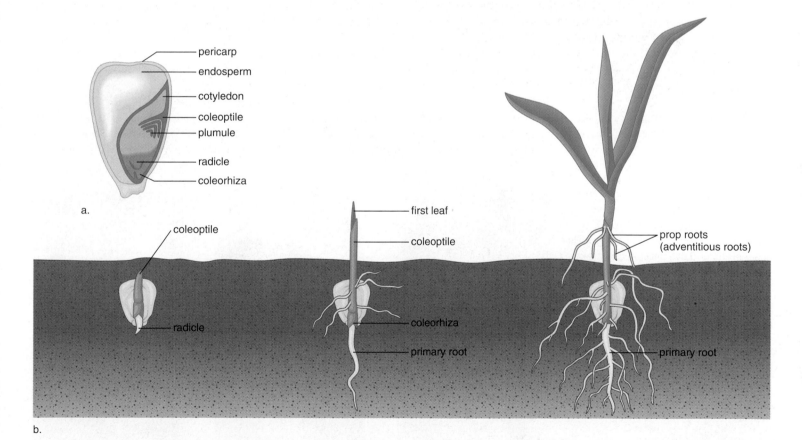

Figure 10.6   Corn, a monocot.
a. Grain structure. **b.** Germination and development of the seedling. Notice that there is one cotyledon and the leaves are parallel veined. Also, note that the mature plant has prop roots.

10.2 Asexual Reproduction in Flowering Plants

Because plants contain meristem tissue and cells that can revert to meristematic activity, they routinely reproduce asexually by vegetative propagation. In asexual reproduction, there is only one parent, instead of two as in sexual reproduction. Therefore, the parent and offspring are genetically the same. Uniform quality of flowers, fruits, and vegetables is one reason plant growers—commercial and amateur alike—appreciate vegetative propagation. Flowers (e.g., hydrangeas, roses, and geraniums), shrubs (e.g., hollies), and trees (e.g., apple and orange trees) are routinely produced by cuttings from a preferred plant.

Various wild plants provide examples of how stems and also roots produce new plants asexually. When the stolon of a wild strawberry touches the soil, new roots appear at the point of contact (see Fig. 9.17a). Similarly, irises multiply when their underground stem, called a rhizome, sends up new shoots (see Fig. 9.17b). The tubers of a white potato plant bear a number of small, whitish indentations on the surface, which are often referred to as "eyes" (see Fig. 9.17c). If you cut up the potato so that every piece has an eye, which is actually a bud, it will normally grow roots and shoots. Sweet potatoes can be propagated by planting sections of their root. Also, you may have noticed that the roots of some fruit trees, such as cherry and apple trees, produce "suckers," small plants that can be used to grow new trees.

In horticulture, a clone is all the identical individual plants descended by vegetative propagation from a single ancestor. When you grow a cutting from an African violet or a "spider" from a spider plant, you are in effect cloning it. A clone can include millions of individual plants and be in existence for some time. Certain grapevines cultivated in Europe are members of clones more than 2,000 years old.

Propagation of Plants in Tissue Culture

Vegetative propagation of stem cuttings illustrates that plant tissues can routinely grow into a complete plant. Indeed, plant cells are totipotent, meaning that each one has the full genetic potential to become a mature plant under the proper circumstances. **Tissue culture,** the growth of a *tissue* in an artificial liquid *culture* medium, can allow a plant cell to exhibit its full genetic potential.

Figure 10.7 shows the progressive steps by which plant cells become complete plants in tissue culture. Enzymes are used to digest the cell walls of a small piece of tissue, and the result is naked cells, without walls, called protoplasts. The protoplasts regenerate a new cell wall and begin to divide. These clumps of cells, called a callus, can be manipulated to produce somatic (asexually produced) embryos. Somatic embryos that are encapsulated in a protective hydrated gel (and sometimes called artificial seeds) can be shipped anywhere. It is possible to produce millions of somatic embryos at once in large tanks called bioreactors. This is done for vegetables such as tomato, celery, and asparagus and for ornamental plants such as lilies, begonias, and African violets.

Tissue culture techniques have made possible micropropagation, a commercial method of producing thousands, even millions, of identical seedlings in a limited amount of space. One favorite micropropagation method is

Figure 10.7 Tissue culture trees.
a. When plant cell walls are removed by digestive enzyme action, the result is naked cells, or protoplasts. **b.** Cell walls regenerate, and cell division begins. **c.** Cell division produces an aggregate of cells. **d.** An undifferentiated mass, called a callus. **e.** Somatic cell embryos such as this one appear. **f.** The embryos develop into plantlets that can be transferred to soil for growth into adult plants.

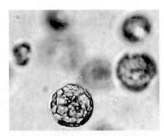
a. Protoplasts (naked cells)

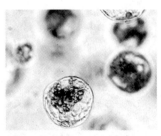
b. Cell division

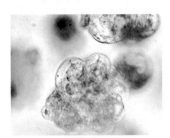
c. Aggregates of cells

d. Callus (undifferentiated mass)

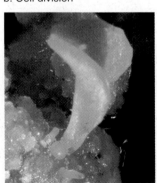
e. Somatic embryo

f. Plantlet

meristem culture. An advantage to meristem culture is that meristem, unlike other portions of a plant, is virus-free; therefore, the plants produced are also virus-free. Virus-free potato plants have been produced in this manner. (The presence of plant viruses weakens plants and makes them less productive.)

The culturing of leaf, stem, or root tissues has led to a technique called cell suspension culture. Rapidly growing calluses are cut into small pieces and shaken in a liquid nutrient medium so that single cells or small clumps of cells break off and form a suspension. These cells will produce the same chemicals as the entire plant. Scientists envision that it will be possible to maintain cell suspension cultures in bioreactors for the purpose of producing commercially valuable chemicals, cosmetics, or drugs. For example, cell suspension cultures of *Cinchona ledgeriana* produce quinine, and those of *Digitalis lanata* produce digitoxin. One day it may be possible to produce taxol, the drug of choice for ovarian cancer, in this way. Then it will not be necessary to destroy whole plants to extract the chemical from their bark.

Genetic Engineering of Plants

As will also be discussed in the Science Focus on page 188, genetically engineered plants have new and different traits because a foreign gene is present in their cells. Most often the foreign gene is introduced into protoplasts maintained in tissue culture. Then the protoplast develops into a plant that exhibits the trait governed by the foreign gene.

One aim of genetic engineering is to produce crops that have improved agricultural or food quality traits. Genetically engineered herbicide- and pest-resistant plants are now available. Researchers are working to produce crops that can grow in salty soil, with limited rain, or in cold weather. Engineered plants with improved oil, starch, and amino acid content would have a higher nutritional quality.

Recently, researchers were able to use a virus to introduce a human gene into adult tobacco plants in the field. (Note that this technology bypasses the need for tissue culture completely.) Tens of grams of an enzyme that can be used to treat a human lysosomal storage disease were later harvested per acre of tobacco plants. And it only took thirty days to get tobacco plants to produce antigens for the treatment of non-Hodgkin lymphoma after being sprayed with a differently engineered virus.

Plants reproduce asexually, and this ability facilitates tissue culture propagation of whole plants. Production of plants by tissue culture has commercial value, and it facilitates genetic engineering of plants.

10.3 Control of Plant Growth and Development

If each cell of an embryo is totipotent, what makes one become a conducting cell and another become an epidermal cell? It depends on which genes are turned on and which are turned off in that particular cell. Plant hormones play a role in determining cellular differentiation, and therefore determining which genes are expressed in that cell.

Plant Hormones

Plant hormones are small organic molecules produced by the plant that serve as chemical signals between cells and tissues. Currently, the five commonly recognized groups of plant hormones are auxins, cytokinins, gibberellins, abscisic acid, and ethylene. Each naturally occurring hormone has a specific chemical structure. Other chemicals, some of which differ only slightly from the natural hormones, also affect the growth of plants. These and the naturally occurring hormones are sometimes grouped together and called plant growth regulators.

Plant hormones bring about a physiological response in target cells after binding to a specific receptor protein in the plasma membrane. Figure 10.8 shows how any one auxin brings about elongation of a cell, a necessary step toward differentiation and maturation of a plant cell.

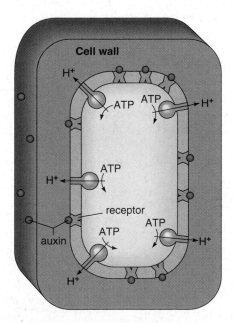

Figure 10.8   Mode of action of auxin, a plant hormone.
After auxin binds to a receptor, the combination stimulates an H^+ pump so that hydrogen ions (H^+) are transported out of the cytoplasm. The resulting acidity causes the cell wall to weaken, and the electrochemical gradient causes solutes to enter the cell. Water follows by osmosis, and the cell elongates.

Stimulatory Hormones

Table 10.1 lists the five common types of plant hormones and their major functions. Some of these are stimulators of growth, and some are inhibitors of growth. Despite the custom of assigning particular functions to particular hormones, we should realize that plant hormones often interact to control differentiation of plant cells and responses of plants to particular stimuli.

Auxins   Auxins are plant hormones whose effects have been studied for a long time. The principal naturally occurring auxin is indoleacetic acid (IAA). It is produced in shoot apical meristem and is found in young leaves and in flowers and fruits. Therefore, you would expect IAA to affect many aspects of plant growth and development. Applying auxin to a callus in tissue culture induces differentiation of procambium into vascular tissue at the site of application. It also appears that, as an embryo matures, auxin induces differentiation of vascular tissue throughout the plant.

Apically produced auxin prevents the growth of axillary buds, a phenomenon called apical dominance. When a terminal bud is removed deliberately or accidentally, the nearest axillary buds begin to grow, and the plant branches. To achieve a fuller look, one generally prunes the top (shoot apical meristem) of the plant. This removes apical dominance and causes more branching of the main body of the plant.

The application of a weak solution of auxin to a cutting causes roots to develop (Fig. 10.9a). Both commercial and home gardeners use this action of auxin to start plants from cuttings. Auxin production by seeds also promotes the growth of fruit. As long as auxin is concentrated in leaves or fruits rather than in the stem, leaves and fruits do not fall off. Therefore, trees can be sprayed with auxin to keep mature fruit from falling to the ground.

Auxin's ability to cause cell elongation (see Figs. 10.10 and 10.11) is involved in a plant's response to light and gravity, as will be discussed later.

Cytokinins   The cytokinins are a class of plant hormones that promote cell division and are involved in regulating the plant cell cycle. These substances are derivatives of adenine, one of the purine bases in DNA and RNA. A naturally occurring cytokinin called zeatin has been extracted from corn kernels. Kinetin, a synthetic cytokinin, also promotes cell division.

Tissue culture experiments reveal that auxin and cytokinin interact to affect differentiation during development. In a tissue culture that has the usual amounts of these two hormones, tobacco strips develop into a callus of undifferentiated tissue. If the ratio of auxin to cytokinin is appropriate, the callus produces roots. Change the ratio, and vegetative shoots and leaves are produced. Yet another ratio causes floral shoots.

Lateral buds begin to grow despite apical dominance caused by auxin when cytokinin is applied to them. Also, in mature plants, cytokinin prevents **senescence,** the death of a plant part, such as leaves, due to their aging. Therefore, cytokinins can be sprayed on vegetables to keep them fresh during shipping and storage.

Gibberellins   Gibberellins are growth promoters that bring about considerable elongation of the internodes of various plants (Fig. 10.9b). The largest responses usually occur in dwarfed plants, which grow tall after gibberellins are applied. Plants that are already tall may show no response to gibberellins, possibly because the appropriate genes are already fully turned on in these plants.

There are many gibberellins that differ only slightly from one another. The most common of these is GA_3 (the subscript designates the particular gibberellins). The dormancy of seeds can be broken by applying GA_3, and research with barley seeds has shown a possible mechanism of action. Barley seeds have a large, starchy endosperm, which must be broken down into sugars to provide energy for growth. After the embryo produces gibberellins, amylase, an enzyme that breaks down starch, appears in cells just inside the seed coat. It is hypothesized

Table 10.1	Plant Hormones	
Type	**Primary Example**	**Major Functions**
Stimulatory		
Auxins	Indoleacetic acid (IAA)	Promote cell elongation in stems; phototropism, gravitropism, apical dominance; formation of roots, development of fruit
Gibberellins	Gibberellic acid (GA_3)	Promote stem elongation; release some buds and seeds from dormancy
Cytokinins	Zeatin	Promote cell division and embryo development; prevent leaf senescence; promote bud activation
Inhibitory		
Abscisic acid	Abscisic acid (ABA)	Resistance to stress conditions; causes stomatal closure; maintains dormancy
Ethylene	Ethylene	Promotes fruit ripening; promotes abscission and fruit drop; inhibits growth

that after GA$_3$ is applied, it activates the gene that codes for amylase. Amylase then acts on starch to release sugars as a source of energy for the growing embryo.

Plant hormones are of enormous commercial importance. As an example, gibberellins are used to increase the size of sugarcane. Applying as little as 60 ml per acre increases the cane yield by more than 5 metric tons.

Inhibitory Hormones

The plant hormones abscisic acid and ethylene are the inhibitory hormones that oppose the action of stimulatory hormones. For example, several synthetic inhibitors are routinely used to cause leaves and fruit to drop at a time convenient to the farmer.

Abscisic Acid   Abscisic acid **(ABA)** is sometimes called the stress hormone because it initiates and maintains seed and bud dormancy and brings about the closure of stomata. In the fall, abscisic acid converts vegetative buds to winter buds. A winter bud is covered by thick and hardened scales. A reduction in the level of abscisic acid and an increase in the level of gibberellins are believed to break seed and bud dormancy. Then seeds germinate and buds send forth leaves.

Abscisic acid brings about the closing of stomata when a plant is under water stress. In some unknown way, abscisic acid causes potassium ions (K$^+$) to leave guard cells. Thereafter, the guard cells lose water, and the stomata close. It was once believed that abscisic acid functioned in **abscission,** the dropping of leaves, fruits, and flowers from a plant. Now the hormone ethylene is believed to bring about abscission (Fig. 10.9c). Abscisic acid is produced by any "green tissue" with chloroplasts, monocot endosperm, and roots.

Ethylene   The hormone **ethylene** ripens fruit by increasing the activity of enzymes that soften fruits. For example, it stimulates the production of cellulase, an enzyme that hydrolyzes cellulose in plant cell walls. Because it is a gas, ethylene moves freely through the air—a barrel of ripening apples can induce ripening of a bunch of bananas even some distance away. Ethylene is injected into airtight storage rooms to ripen bananas, honeydew melons, and tomatoes. It can also be used to degreen the rind of fruits.

The presence of ethylene in the air inhibits the growth of plants in general; homeowners who use natural gas to heat their homes sometimes report difficulties in growing houseplants. Ethylene is also present in automobile exhaust, and it's possible that plant growth is inhibited by the exhaust that enters the atmosphere. It only takes one part of ethylene per 10 million parts of air to inhibit plant growth.

Ethylene is also involved in abscission. Low levels of auxin and perhaps gibberellin in a leaf compared to the stem probably initiate abscission. But once the process of abscission has begun, ethylene stimulates an enzyme such as cellulase, which causes leaf, fruit, or flower drop (Fig. 10.9c).

Both stimulatory and inhibitory hormones help control certain growth patterns of plants. Auxins, gibberellins, and cytokinins stimulate growth. Abscisic acid and ethylene inhibit growth.

b. Untreated plant   Treated with GA$_3$

Figure 10.9
Effects of plant hormones.
a. Auxin causes cuttings to develop adventitious roots. **b.** Gibberellin (GA$_3$) promotes stem elongation. **c.** When an ethylene-producing ripe apple is placed under a glass with a holly twig, abscission (dropping of leaves) occurs.

a. Untreated plant   Treated with auxin

c. Untreated plant   Treated with ethylene

Figure 10.10 **Positive phototropism.**
The stem of a plant curves toward the light. This response is due to
the accumulation of auxin on the shady side of the stem.

Plant Responses to Environmental Stimuli

Plant growth and development are strongly influenced by
environmental stimuli (e.g., light, day length, gravity, and
touch). Indeed, environmental signals determine the sea-
sonality of growth, reproduction, and dormancy. The ability of a plant to respond to environmental signals fosters
the survival of the plant and the species in a particular environment.

Plant responses to environmental signals can be rapid,
as when stomata open in the presence of light, or they can
take some time, as when a plant flowers in season. Despite
their variety, most plant responses to environmental signals are due to growth and sometimes differentiation,
brought about at least in part by particular hormones.

Plant Tropisms

Plant growth toward or away from a directional stimulus is
called a **tropism**. Tropisms are due to differential growth—
one side of an organ elongates faster than the other, and
the result is a curving toward or away from the stimulus.
The following three well-known tropisms were each named
for the stimulus that causes the response:

phototropism	movement in response to a light stimulus
gravitropism	movement in response to gravity
thigmotropism	movement in response to touch

Figure 10.11 **Negative gravitropism.**
The stem of a plant curves away from the direction of gravity
24 hours after the plant was placed on its side. This response is due
to the accumulation of auxin on the lower side of the stem.

Growth toward a stimulus is called a positive tropism, and
growth away from a stimulus is called a negative tropism.
Figure 10.10 illustrates positive phototropism—stems curve
toward the light. Figure 10.11 illustrates negative gravitropism—stems curve away from the direction of gravity.
Roots, of course, exhibit positive gravitropism.

The role of auxin in the positive **phototropism** of stems
has been studied for quite some time. Because blue light in
particular causes phototropism to occur, it is believed that
a yellow pigment related to the vitamin riboflavin acts as a
photoreceptor for light. Following reception, auxin migrates
from the bright side to the shady side of a stem. The cells on
that side elongate faster than those on the bright side, causing the stem to curve toward the light. Positive gravitropism of stems occurs because auxin moves to the lower part
of a stem when a plant is placed on its side. Figure 10.8
explains how auxin brings about elongation.

Flowering

Flowering is a striking response in angiosperms to environmental seasonal changes. In some plants, flowering
occurs according to the **photoperiod,** which is the ratio of
the length of day to the length of night over a 24-hour
period. Plants can be divided into three groups:

1. **Short-day plants** flower when the day length is shorter
   than a critical length. (Examples are cocklebur,
   poinsettia, and chrysanthemum.)
2. **Long-day plants** flower when the day length is longer
   than a critical length. (Examples are wheat, barley,
   clover, and spinach.)
3. **Day-neutral plants** do not depend on day length for
   flowering. (Examples are tomato and cucumber.)

Further, we should note that both a long-day plant and a short-day plant can have the same critical length. Figure 10.12 illustrates that the cocklebur and clover have the same critical length, but only the cocklebur flowers when the day is shorter than 8.5 hours, and clover flowers when the day is longer than 8.5 hours.

Experiments have shown that the length of continuous darkness, not light, controls flowering. For example, the cocklebur will not flower if a suitable length of darkness is interrupted by a flash of light. On the other hand, clover will flower when an unsuitable length of darkness is interrupted by a flash of light. (Interrupting the light period with darkness has no effect on flowering.)

> In order to flower, short-day plants require a period of darkness that is longer than a critical length, and long-day plants require a period of darkness that is shorter than a critical length.

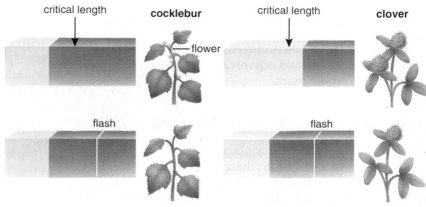

Figure 10.12   **Effect of day/night length on flowering.**
a. The cocklebur flowers when days are short and nights are long (*top*). If a long night is interrupted by a flash of light, the cocklebur will not flower (*bottom*).
b. Clover flowers when days are long and nights are short (*top*). If a long night is interrupted by a flash of light, clover will still flower (*bottom*). Therefore, we can conclude that the length of continuous darkness controls flowering.

Phytochrome and Plant Flowering

If flowering is dependent on day and night length, plants must have some way to detect these periods. This appears to be the role of **phytochrome**, a blue-green leaf pigment that alternately exists in two forms—P_r and P_{fr}:

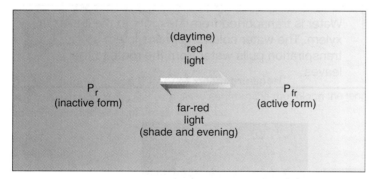

Direct sunlight contains more red light than far-red light; therefore, P_{fr} is apt to be present in plant leaves during the day. In the shade and at sunset, there is more far-red light than red light; therefore, P_{fr} is converted to P_r as night approaches. There is also a slow metabolic replacement of P_{fr} by P_r during the night.

It is possible that phytochrome conversion is the first step in a signaling pathway that results in flowering. A flowering hormone has never been discovered.

> Phytochrome alternates between two forms (P_{fr} during the day and P_r during the night), and this conversion allows a plant to detect photoperiod changes.

Other Functions of Phytochrome

The $P_r \rightarrow P_{fr}$ conversion cycle has other functions. The presence of P_{fr} indicates to some seeds that sunlight is present and conditions are favorable for germination. This is why they must be only partly covered with soil when planted. Following germination, the presence of P_r indicates that stem elongation may be needed to reach sunlight. Seedlings that are grown in the dark etiolate; that is, the stem increases in length and the leaves remain small (Fig. 10.13). Once the seedling is exposed to sunlight and P_r is converted to P_{fr}, the seedling begins to grow normally—the leaves expand and the stem branches. It appears that P_{fr} binds to regulatory proteins in the cytoplasm and the complex migrates to the nucleus, where it binds to particular genes.

a. Etiolation   b. Normal growth

Figure 10.13   **Phytochrome control of a growth pattern.**
a. If far-red light is prevalent, as it is in the shade, etiolation occurs.
b. If red light is prevalent, as it is in bright sunlight, normal growth occurs.

10.

Once

transp

in ph

Wat

Wate
throu
erals,
the va
called
eleme
one o
from
the ve
the o
ends,
move
pits,
form

V
called
It is e
woul
some
invol

Cob

The
how
(Fig.
p. 24

Adaptations of Roots for Mineral Uptake

Two symbiotic relationships assist roots in taking up mineral nutrients. Some plants, such as legumes, soybeans, and alfalfa, have roots infected by *Rhizobium* bacteria, which can fix atmospheric nitrogen (N_2). They break the N ≡ N bond and reduce nitrogen to NH_4^+ for incorporation into organic compounds. The bacteria live in **root nodules** and are supplied with carbohydrates by the host plant (Fig. 10.16). The bacteria, in turn, furnish their host with nitrogen compounds.

The second type of symbiotic relationship, called a mycorrhizal association, involves fungi and almost all plant roots (Fig. 10.17). Only a small minority of plants do not have **mycorrhizae,** sometimes called fungus roots. Ectomycorrhizae form a mantle that is exterior to the root, and they grow between cell walls. Endomycorrhizae can penetrate cell walls. In any case, the fungus increases the surface area available for mineral and water uptake and breaks down organic matter, releasing nutrients that the plant can use. In return, the root furnishes the fungus with sugars and amino acids. Plants are extremely dependent on mycorrhizae. Orchid seeds, which are quite small and contain limited nutrients, do not germinate until a mycorrhizal fungus has invaded their cells. Nonphotosynthetic plants, such as Indian pipe, use their mycorrhizae to extract nutrients from nearby trees. Plants without mycorrhizae are most often limited as to the environment in which they can grow.

Some plants have poorly developed roots or no roots at all because minerals and water are supplied by other mechanisms. **Epiphytes** are "air plants;" they do not grow in soil but on larger plants, which give them support, but no nutrients. Some epiphytes have roots that absorb moisture from the atmosphere, and many catch rain and minerals in special pockets at the base of their leaves. Parasitic plants such as dodders, broomrapes, and pinedrops send out root-like projections called haustoria that tap into the xylem and phloem of the host stem (see Fig. 9.11*d*). Carnivorous plants such as the Venus's-flytrap and the sundew obtain some nitrogen and minerals when their leaves capture and digest insects (see Fig. 9.20*c*).

Mineral uptake follows the same pathway as water uptake; however, mineral uptake requires an expenditure of energy. Most plants are assisted in acquiring minerals by a symbiotic relationship with microorganisms and/or fungi.

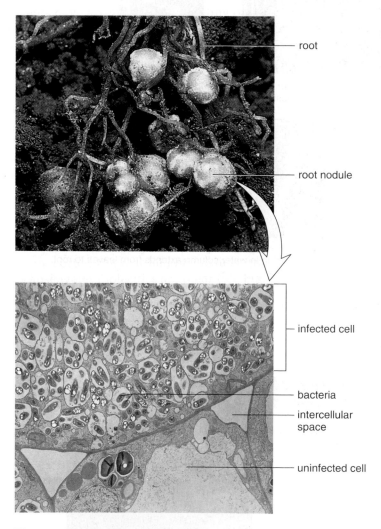

— root

— root nodule

— infected cell

— bacteria
— intercellular space

— uninfected cell

Figure 10.16   Root nodules.
Nitrogen-fixing bacteria live in nodules on the roots of plants, particularly legumes. The nodules contain bacteria that fix atmospheric nitrogen and make reduced nitrogen available to the plant. The plant passes carbohydrates to the bacteria. In infected cells, bacteria fix atmospheric nitrogen and make reduced nitrogen available to a plant. The plant passes carbohydrates to the bacteria.

a. Plants with no mycorrhizae   b. Plants with mycorrhizae

Figure 10.17   Mycorrhizae.
a. Plants without mycorrhizae do not grow as well as (**b**) plants with mycorrhizae.

Figu

The

into

vess

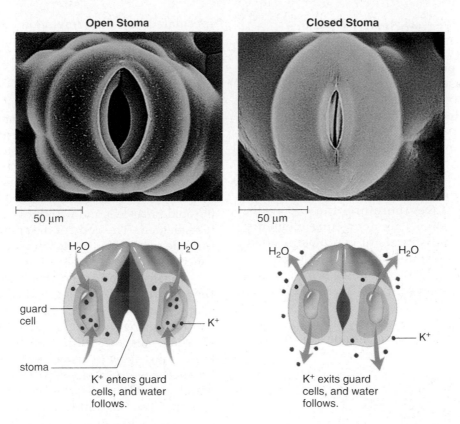

Figure 10.18 **Opening and closing of stomata.**
Stomata open when water enters guard cells and turgor pressure increases. Stomata close when water exits guard cells and turgor pressure is lost.

Opening and Closing of Stomata

For a plant to photosynthesize, stomata (sing., **stoma**), which are small openings in the epidermis of leaves, must be open so that CO_2 will enter the leaves. But when stomata are open, water is also exiting the leaves because transpiration is occurring. When a plant is water stressed, the stomata close to conserve water, and then photosynthesis ceases.

The two **guard cells** on either side of a stoma regulate whether the stoma is open or closed. When water enters guard cells, their central vacuoles fill and the stoma opens. When water exits guard cells, the stoma closes. Notice in Figure 10.18 that the guard cells are attached to each other at their ends and that the inner walls are thicker than the outer walls. When water enters guard cells, cellulose microfibrils in the walls prevent radial expansion. However, lengthwise expansion can occur, causing guard cells to buckle out from the region of their attachment so that the stoma opens.

When a plant is photosynthesizing, an ATP-driven pump actively transports H^+ out of the cell. Now potassium ions (K^+) enter the guard cells, and when water follows by osmosis, the stoma opens. When the pump is not working, K^+ moves into surrounding cells, the guard cells lose water, and the stoma closes.

What turns the H^+ pump on or off? It appears that the blue-light component of sunlight is the environmental signal for stomata to open. Evidence suggests a pigment that absorbs blue light sets in motion the cytoplasmic response that leads to activation of the H^+ pump. Similarly, there could be a receptor in the plasma membrane of guard cells that brings about inactivation of the pump when CO_2 concentration rises, as might happen when photosynthesis ceases. As mentioned on page 179, abscisic acid (ABA), which is produced by cells in wilting leaves, can also cause stomata to close. Although photosynthesis cannot occur, water is conserved.

If plants are kept in the dark, the stomata open and close on a 24-hour basis just as if they were responding to the presence of sunlight in the daytime and the absence of sunlight at night. This means that some sort of internal *biological clock* must be keeping time. Biological clocks and circadian rhythms (behaviors that occur every 24 hours) are areas of intense investigation at this time.

When stomata open, first K^+ and then water enters guard cells. Stomata open and close in response to environmental signals; the exact mechanism is being investigated.

Organic Nutrient Transport in Phloem

The green parts of the plant, commonly the leaves, carry on photosynthesis and produce the sugar sucrose, which is transported in the vascular tissue called **phloem** to all parts of a plant. The movement of organic substances in phloem is termed **translocation.** Translocation makes sugars available to those parts of a plant that are actively metabolizing, storing, or growing. The conducting cells in phloem are *sieve-tube elements,* each of which typically has a *companion cell* (Fig. 10.19). Sieve-tube elements contain cytoplasm but have no nucleus. Their end walls have pores and resemble a sieve; therefore, the end walls are said to be sieve plates. The sieve-tube elements are aligned vertically, and strands of cytoplasm called plasmodesmata (sing., **plasmodesma**) extend from one cell to the other through the sieve plates. Therefore, a continuous pathway transports organic nutrients throughout the plant.

The smaller companion cell, which does have a nucleus, is a more generalized cell than the sieve-tube element. It is speculated that the companion cell nucleus controls and maintains the lives of both itself and the sieve-tube element, and may help a sieve-tube element perform its translocating function.

Pressure-Flow Model of Phloem Transport

Chemical analysis of phloem sap shows that it is composed chiefly of sugar and that the concentration of organic nutrients is 10–13% by volume. Samples for chemical analysis most often are obtained by using aphids, small insects that are phloem feeders. The aphid drives its stylet, a short mouthpart functioning like a hypodermic needle, between the epidermal cells and withdraws phloem sap from a sieve-tube element. The body of the aphid can be cut away carefully, leaving the stylet, which exudes phloem sap for collection and analysis.

Translocation in phloem is explained by the **pressure-flow model.** During the growing season, the leaves are a **source** of sugar, meaning that they are photosynthesizing and producing more sugar than they need. This sugar is actively transported into sieve-tube elements, and water follows passively by osmosis. Active transport is possible because sieve-tube elements have a living plasma membrane, and the necessary energy is provided by the companion cells. The buildup of water within the sieve-tube elements creates pressure, which starts a flow of phloem sap. The roots (and other growth or storage areas) are a **sink** for sugar—that is, the roots remove sugar. It is actively transported out of the sieve-tube elements at this location. Water then leaves the phloem passively by osmosis (Fig. 10.20).

The pressure-flow model is supported by an experiment in which two bulbs are connected by a glass tube. The first bulb contains solute at a higher concentration than the second bulb. Each bulb is bounded by a membrane which is permeable that is permeable to water but not sugar. The entire apparatus is submerged in distilled water:

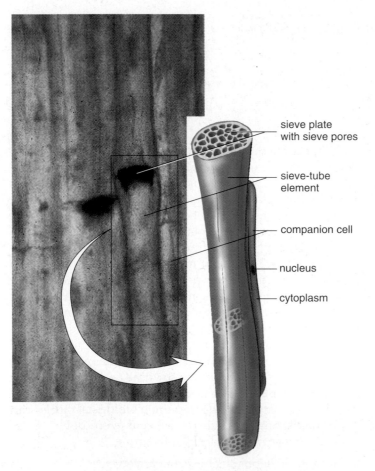

sieve plate
with sieve pores

sieve-tube
element

companion cell

nucleus

cytoplasm

Figure 10.19   Sieve-tube elements of phloem.
Sieve-tube elements contain cytoplasm, and their flat end walls have a sieve plate with large sieve pores. Plasmodesmata connect the cells one with the other. Each sieve-tube element has a companion cell that has both cytoplasm and a nucleus.

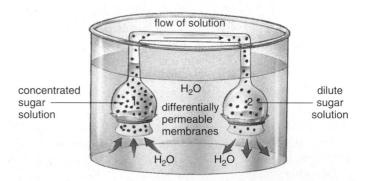

flow of solution

concentrated
sugar
solution

H_2O

differentially
permeable
membranes

dilute
sugar
solution

H_2O H_2O

Distilled water flows into the first bulb because it has the higher solute concentration. In this way, a pressure difference is created that causes water to flow from the first bulb to the second and even to exit from the second bulb. As the water flows, it carries solute with it from the first to the second bulb.

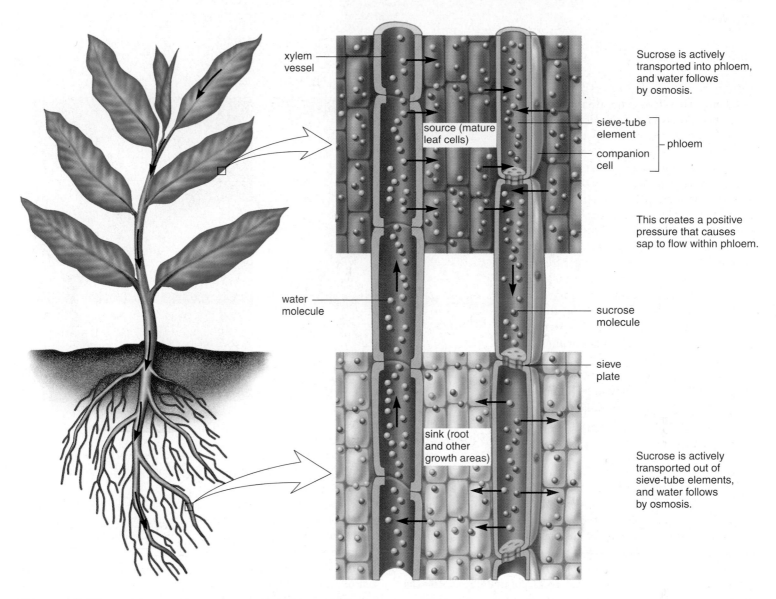

xylem
vessel

Sucrose is actively
transported into phloem,
and water follows
by osmosis.

source (mature
leaf cells)

sieve-tube
element

companion
cell

phloem

This creates a positive
pressure that causes
sap to flow within phloem.

water
molecule

sucrose
molecule

sieve
plate

sink (root
and other
growth areas)

Sucrose is actively
transported out of
sieve-tube elements,
and water follows
by osmosis.

Figure 10.20   Pressure-flow theory of phloem transport.
Sugar and water enter sieve-tube elements at a source. This creates a positive pressure, which causes phloem contents to flow. Sieve-tube elements form a continuous pipeline from a source to a sink, where sugar and water exit sieve-tube elements.

The pressure-flow model of phloem transport can account for any direction of flow in sieve tubes if we consider that the direction of flow is always *from source to sink*. In young seedlings, the cotyledons containing reserved food are a major source of sucrose, and roots are a sink. Therefore, sucrose flow is from the cotyledons to the roots. In older plants, the most recently formed leaves can be a sink and they will receive sucrose from other leaves until they begin to maximally photosynthesize. When a plant is forming fruit, phloem flow is monopolized by the fruits, little sucrose goes to the rest of the plant, and vegetative growth is slow.

Because phloem sap flows from source to a sink, a bidirectional flow sometimes arises within phloem—not just at different times in the life cycle but even at the same time! Bidirectional flow occurs because different sieve tubes can be conducting phloem sap in opposite directions.

Phloem transports organic nutrients in a plant. Typically, sugar and then water enter sieve-tube elements in the leaves. This creates pressure, which causes water to flow to the roots, carrying sugar with it.

The Valuable Weed

Plants are quite different from animals in many aspects of their development, cell biology, biochemistry, and environmental responses. Plants grow their entire lives, and contain meristem tissue that allows them to continuously produce new cells that differentiate into specific tissues. Adult cells are totipotent, and each one can easily give rise to a complete plant. Plant biochemistry involves metabolic pathways not seen in animals. Several pathways are needed for various types of photosynthesis and the synthesis of a wide range of chemicals; many of these are used defensively against herbivores and bacterial and fungal pathogens. Plant hormones, unlike those of animals, are produced by tissues and not by organs. They allow plants to respond by growing toward or away from environmental stimuli such as light and gravity. Because plants are unique, a classical and molecular study of their specific genetics is needed.

Until recently, the study of plant genetics was hampered by the lack of a suitable experimental material. Enter *Arabidopsis thaliana*, a weed of no food or economic value, even though it is a member of the mustard family, as are cabbages and radishes. Unlike crop plants used formerly, *Arabidopsis* has the characteristics needed to promote the study of both plant classical and molecular genetics. *Arabidopsis* has a short generation time; its entire life cycle takes only about four to six weeks, and each adult plant produces 10,000 seeds (potential offspring). Dozens of *Arabidopsis* plants can be grown in a single pot because of its small size (Fig. 10A), and thousands can be grown on a lab bench under fluorescent, rather than sun, light. A total of 50,000 seeds will fit in a standard 1.5 ml tube. In contrast, crop plants such as corn have generation times of at least several months, and they require a great deal of field space for a large number to grow. In addition to thriving in soil, *Arabidopsis* will grow in a liquid media, whose content can be biochemically controlled.

There are many natural mutants of *Arabidopsis*, and much can be determined by studying them. In one instance, a dwarf plant was found to be deficient in the amount of gibberellin-producing enzymes. Since the plant showed a lack of internodal elongation but showed normal development of leaves and flowers, it was known that gibberellins were affecting only internodal elongation. Researchers, working with natural mutagens, have discovered that there are three classes of genes essential to normal floral pattern formation. Triple mutants that lack all three types of genetic activities have flowers that consist entirely of leaves arranged in whorls.

Artificial mutagenesis can also be easily accomplished in *Arabidopsis* when the seeds are exposed to chemical mutagens

a.

b.

Figure 10A *Arabidopsis thaliana.*
Many investigators have turned to this weed as an experimental material to study the actions of genes, including those that control growth and development. **a.** *A. thaliana* is quite small. **b.** Enlargement of flower.

or radiation. Or the plant can be engineered to carry a foreign gene. Many artificially produced mutants have been studied, but we will consider just one example. Following mutagenesis, plants able to grow in high levels of CO_2 were isolated. This classical genetics study became a molecular genetics study when it was discovered that this particular mutation was caused by a gene which codes for a protein regulating the activity of RuBP carboxylase.

A study of the *Arabidopsis* genome will undoubtedly promote plant molecular genetics in general. *Arabidopsis* has just five small chromosomes containing only 70,000 nucleotide base pairs. Working with *Arabidopsis*, experimenters can easily find coding genes and determine what these genes do. Remarkably, it has been discovered that nearly all angiosperm plants contain approximately the same coding genes. Therefore, knowledge of the *Arabidopsis* genome can help scientists understand the genes of all plants, and also can be used to locate specific genes in the genomes of other plants.

Knowledge about *Arabidopsis* genetics is applicable to other plants. For example, one of its mutant genes that alters the development of flowers has been introduced into tobacco plants where, as expected, it caused sepals and stamens to appear where petals would ordinarily be. The investigators comment that the knowledge about the development of flowers in *Arabidopsis* can have far-ranging applications. It will undoubtedly lead someday to more productive crops.

Bioethical Focus — Tragedy of the Commons

We are in the midst of a pollination crisis due to a decline in the population of honeybees and many other insects, birds, and small mammals that transfer pollen from stamen to stigma. Pollinator populations have been decimated by pollution, pesticide use, and destruction or fragmentation of natural areas. Belatedly, we have come to realize that various types of bees are responsible for pollinating such cash crops as blueberries, cranberries, and squash, and are partly responsible for pollinating apple, almond, and cherry trees.

Why are we so short-sighted when it comes to protecting the environment and living creatures like pollinators? Because pollinators are a resource held in common. The term "commons" originally meant a piece of land where all members of a village were allowed to graze their cattle. The farmer who thought only of himself and grazed more cattle than his neighbor was better off. The difficulty is, of course, that eventually the resource is depleted and everyone loses.

So, when a farmer or property owner uses pesticides he is only thinking of his field or his lawn, and not the good of the whole. The commons can only be protected if citizens have the foresight to enact rules and regulations by which all abide. DDT was outlawed in this country in part because it led to the decline of birds of prey. Similarly, it seems we need legislation to protect pollinators from those factors that kill them off. Legislation to protect pollinators would protect the food supply for all of us.

Education would also help. Many people are afraid of bees and other stinging insects. How might it be possible to educate them to their value and the need to preserve them for the benefit of plants and the food supply?

Decide Your Opinion

1. Are you willing to stop using pesticides that kill pollinators if it means your lawn will suffer? Why or why not?
2. Are you willing to pressure your representatives for legislation to protect pollinators? Why or why not?
3. Are you willing to turn your lawn and garden into a haven for pollinators? Why or why not?

Summarizing the Concepts

10.1 Sexual Reproduction in Flowering Plants

Flowering plants have an alternation of generations life cycle, which includes separate microgametophytes and megagametophytes. The pollen grain, the microgametophyte, is produced within the stamens of a flower. The megagametophyte within the ovule of a flower produces an egg. Following pollination and double fertilization, the ovule matures to become the seed, and the ovary becomes the fruit. Prior to seed formation, the zygote undergoes growth and development to become an embryo. The seeds enclosed by a fruit contain the embryo (hypocotyl, epicotyl, plumule, radicle) and stored food (endosperm and/or cotyledons). Following dispersal a seed germinates. The root appears below, and the shoot appears above.

10.2 Asexual Reproduction in Flowering Plants

Many flowering plants reproduce asexually, as when buds located on stems (either aboveground or underground) give rise to entire plants, or when roots produce new shoots. The ability of plants to reproduce asexually means that they can be propagated in tissue culture. Micropropagation, the production of clonal plants utilizing tissue culture, is now a commercial venture. In cell suspension cultures, plant cells produce chemicals of medical importance. The practice of plant tissue culture facilitates genetic engineering of plants. One aim of genetic engineering is to produce plants that have improved agricultural or food quality traits. Plants can also be engineered to produce chemicals of use to humans.

10.3 Control of Plant Growth and Development

Plant hormones are chemical signals produced by plants that are involved in growth and development. Plant hormones bind to receptor proteins, and this leads to physiological changes within the cell when certain genes are turned on. Auxins, cytokinins, and gibberellins are the stimulatory hormones that often interact to bring about growth and development. Abscisic acid and ethylene are the inhibitory hormones that oppose growth and development.

Environmental signals play a significant role in plant growth and development. Tropisms are growth responses toward or away from unidirectional stimuli. When a plant is exposed to light, auxin moves laterally from the bright to the shady side of a stem. Thereafter, the cells on the shady side elongate, and the stem bends toward the light. Similarly, auxin is responsible for the negative gravitropism exhibited by stems. Stems grow upward opposite the direction of gravity. Thigmotropism occurs when, for example, tendrils coil about a pole.

Flowering is a striking response to environmental seasonal changes. Short-day plants flower when the days are shorter (nights are longer) than a critical length, and long-day plants flower when the days are longer (nights are shorter) than a critical length. Some plants are day-neutral. Phytochrome, a plant pigment that responds to daylight, is believed to be a part of a biological clock system that in some unknown way brings about flowering. Phytochrome has various other functions in plant cells such as seed germination, leaf expansion and stem branching. Most likely phytochrome is involved in turning particular genes on.

10.4 Transport in the Mature Plant

Water transport in plants occurs within xylem. The cohesion-tension model of xylem transport states that transpiration (evaporation of water at stomata) creates tension, which pulls water upward in xylem. This method works only because water molecules are cohesive.

Plants often expend energy to concentrate minerals, but sometimes they are assisted in mineral uptake by specific adaptations—formation of root nodules or a mycorrhizal association.

Transpiration and carbon dioxide uptake occur when stomata are open. Stomata open when guard cells take up potassium (K^+) ions and water follows by osmosis. Stomata open because the entrance of water causes the guard cells to buckle out.

Transport of organic nutrients in plants occurs within phloem. The pressure-flow model of phloem transport states that sugar is actively transported into phloem at a source, and water follows by osmosis. The resulting increase in pressure creates a flow, which moves water and sucrose to a sink.

Testing Yourself

Choose the best answer for each question.

1. The function of the flower is to _____, and the function of fruit is to _____.
   a. produce fruit; provide food for humans
   b. aid in seed dispersal; attract pollinators
   c. attract pollinators; assist in seed dispersal
   d. produce the ovule; produce the ovary

2. How is the megaspore in the plant life cycle similar to the microspore? Both
   a. have the diploid number of chromosomes.
   b. become an embryo sac.
   c. become a gametophyte that produces a gamete.
   d. are necessary to seed production.
   e. Both c and d are correct.

3. In flowering plants, microgametophytes produce _____, megagametophytes produce _____, and their union produces _____, which become plant embryos.
   a. eggs; pollen grains; seeds
   b. pollen grains; eggs; zygotes
   c. eggs; pollen grains; zygotes
   d. pollen grains; polar nuclei; eggs
   e. seeds; ovules; zygotes

4. Which of the following plant hormones is responsible for fruit ripening?
   a. auxin   d. abscisic acid
   b. gibberellins   e. ethylene
   c. cytokinins

5. Which of the following plant hormones is responsible for a plant losing its leaves?
   a. auxin   d. abscisic acid
   b. gibberellins   e. ethylene
   c. cytokinins

6. Which of the following plant hormones causes plants to grow in an upright position?
   a. auxin   d. abscisic acid
   b. gibberellins   e. ethylene
   c. cytokinins

7. After an agar block is placed on one side of an oat seedling, it bends only if
   a. unidirectional light is present.
   b. the agar block contains auxin.
   c. unidirectional light is present, and the agar block contains auxin.
   d. the agar block is acidic.
   e. the agar block is acidic, unidirectional light is present, and the agar block contains auxin.

8. Which of these is a correct statement?
   a. Both stems and roots show positive gravitropism.
   b. Both stems and roots show negative gravitropism.
   c. Only stems show positive gravitropism.
   d. Only roots show positive gravitropism.

9. Short-day plants
   a. are the same as long-day plants.
   b. are apt to flower in the fall.
   c. do not have a critical photoperiod.
   d. will not flower if a short day is interrupted by bright light.
   e. All of these are correct.

10. A plant requiring a dark period of at least 14 hours will
    a. flower if a 14-hour night is interrupted by a flash of light.
    b. not flower if a 14-hour night is interrupted by a flash of light.
    c. not flower if the days are 14 hours long.
    d. Both b and c are correct.

11. Which of the following contributes to the ability of water to move upward in a large tree?
    a. polarity of water molecules
    b. lack of evaporation on dry days
    c. pressure flow between source and sink
    d. root pressure at night

12. Which of the following accounts for the ability of plants to concentrate minerals?
    a. flow of water   d. transpiration
    b. osmosis   e. diffusion
    c. active transport

13. Stomata will be open when all but which of the following conditions is met?
    a. Guard cells are filled with water.
    b. Potassium ions move into the guard cells.
    c. Hydrogen pump is working.
    d. All of these are correct.

14. The pressure-flow model of phloem transport states that
    a. phloem sap always flows from the leaves to the root.
    b. phloem sap always flows from the root to the leaves.
    c. water flow brings sucrose from a source to a sink.
    d. Both a and c are correct.

15. Root hairs do not play a role in
    a. nitrate uptake.
    b. mineral uptake.
    c. water uptake.
    d. carbon dioxide uptake.

16. Label the arrows and dots in this diagram. The elements represented by the dots plays what role in the opening of stomata?

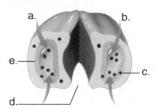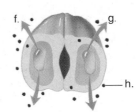

17. Label the following diagram of alternation of generations in flowering plants.

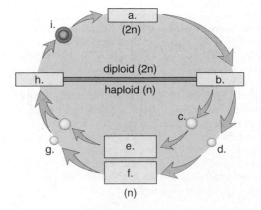

e-Learning Connection   www.mhhe.com/maderinquiry10

Concepts	Questions	Media Resources*
10.1 Sexual Reproduction in Flowering Plants		

• In the plant life cycle, the pollen grain carries sperm to the egg located within a flower part.
 • Following fertilization, growth and development are required for the zygote to become an embryo enclosed within a seed.
 • Seeds must be dispersed and must germinate to complete the plant life cycle.
	1. What is meant by alternation of generations in the flowering plant life cycle?
	 2. What is double fertilization?
		Essential Study Partner
		    Angiosperms
		    Gamete Formation
		    Fertilization
		    Embryos and Seeds
		    Fruits
		    Germination
		
		 BioCourse Study Guide
		    Plant Reproduction
		
		 Labeling Exercises
		    Garden Bean Seed Structure
		    Corn Grain Structure

10.2 Asexual Reproduction in Flowering Plants		
• Plants also reproduce asexually. This ability can be commercially utilized to mass produce identical plants.	1. List several ways in which flowering plants can reproduce asexually. 2. How might plant tissue culture be beneficial to plants as well as humans?	Essential Study Partner Asexual Reproduction
10.3 Control of Plant Growth and Development		

• Plant growth and development are controlled by many internal and external signals.
 • Various hormones help regulate plant growth and development.
 • Flowering is influenced by genes and environmental signals.
	1. What are three stimulatory plant hormones and what are their functions?
	 2. List two inhibitory plant hormones and discuss their functions.
		Essential Study Partner
		    Plant Response—
		      Introduction
		    Plant Movement
		    Hormones
		    Photoperiod
		
		 Art Quizzes
		    Plant Hormones
		    Flowering Responses to Day
		      Length
		    Leaf Abscission Zone

10.4 Transport in the Mature Plant		

• Transpiration (evaporation of water) pulls water and minerals from the roots to the leaves in xylem.
 • Stomata must be open for evaporation to occur.
 • Osmotic pressure assists the flow of organic nutrients in phloem, from where these nutrients are made to where they are used or stored.
	1. How does transpiration help water rise in xylem to the top of the plant?
	 2. What are two symbiotic relationships that help plants take up nutrients from the soil?
		Essential Study Partner
		    Uptake by Roots
		    Water Movement
		    Nutrients
		
		 Art Quizzes
		    Mass-Flow Hypothesis
		    Mineral Transport in Roots
		    Stoma

*For additional Media Resources, see the Online Learning Center.

Maintenance of the Human Body

See the Online Learning Center for careers in Human Anatomy & Physiology.

Human Organization

Chapter Concepts

Liv Arnesen and Ann Bancroft were the first women to hike to the South Pole. Despite outside temperatures averaging −34°C (−29°F), their body temperature stayed within normal limits.

*E*ven if you hike to the South Pole, swim the English Channel, climb Mt. Washington, or cross the Sahara desert by camel, your body temperature will stay at just about 37°C (98.6°F). Why is that? Because the body has mechanisms that maintain homeostasis, the relative constancy of the internal environment.

Liv Arnesen and Ann Bancroft needed all their body's systems to function properly when they decided to hike to the South Pole. Antarctica is a very inhospitable place. It's the coldest place on earth, the windiest, and strange to say, the driest. There are no plants and no animals to eat. It's desolate, dangerous, and nothing but ice. Ice bridges just large enough for a sled stretch across deep crevasses, and if you fall into the water below you freeze in a few minutes. You can get frostbitten if the subfreezing air sticks to the skin. The mechanisms for maintaining homeostasis can be overwhelmed, so it's best to take protective measures to assist the body.

You can be sure that Liv and Ann wore the latest in protective clothing, drank plenty of water, and ate regularly to keep their muscles supplied with glucose. Sleep was a good idea because while all the body's systems contribute, the brain coordinates their functioning in order to maintain homeostasis.

11.1 Types of Tissues

Recall the biological levels of organization. Cells are composed of molecules; a tissue has like cells; an organ contains several types of tissues and several organs are found in an organ system. In this chapter we consider the tissue, organ, and organ system levels of organization.

A **tissue** is composed of similarly specialized cells that perform a common function in the body. The tissues of the human body can be categorized into four major types:

Epithelial tissue covers body surfaces and lines body cavities.

Connective tissue binds and supports body parts.

Muscular tissue moves the body and its parts.

Nervous tissue receives stimuli and conducts nerve impulses.

Cancers are classified according to the type of tissue from which they arise. **Carcinomas,** the most common type, are cancers of epithelial tissue; sarcomas are cancers arising in muscle or connective tissue (especially bone or cartilage); leukemias are cancers of the blood; and lymphomas are cancers of lymphoid tissue. The chance of developing cancer in a particular tissue shows a positive correlation to the rate of cell division; epithelial cells reproduce at a high rate, and 2,500,000 new blood cells appear each second. Carcinomas and leukemias are common types of cancers.

Epithelial Tissue

Epithelial tissue, also called epithelium, consists of tightly packed cells that form a continuous layer. Epithelial tissue covers surfaces and lines body cavities. Epithelial tissue has numerous functions in the body. Usually it has a protective function but it can also be modified to carry out secretion, absorption, excretion and filtration.

On the external surface, epithelial tissue protects the body from injury, drying out, and possible **pathogen** (virus and bacterium) invasion. On internal surfaces, modifications help epithelial tissue carry out both its protective and specific functions. Epithelial tissue secretes mucus along the digestive tract and sweeps up impurities from the lungs by means of cilia (sing., **cilium**). It efficiently absorbs molecules from kidney tubules and from the intestine because of minute cellular extensions called **microvilli.**

Various types of epithelial tissue are shown in Figure 11.1. Notice that epithelial tissue can be classified according to cell type. **Squamous epithelium** is composed of flattened cells and is found lining the lungs and blood vessels. **Cuboidal epithelium** contains cube-shaped cells and is found lining the kidney tubules. **Columnar epithelium** has cells resembling rectangular pillars or columns, with nuclei usually located near the bottom of each cell. This epithelium is found lining the digestive tract. Ciliated columnar epithelium is found lining the oviducts, where it propels the egg toward the uterus, or womb.

Epithelial tissue is also classified accoding to the number of layers in the tissue. An epithelium can be simple or stratified. Simple means the tissue has a single layer of cells, and stratified means the tissue has layers of cells piled one on top of the other. The walls of the smallest blood vessels, called **capillaries,** are composed of simple squamous epithelium. The permeability of capillaries allows exchange of substances between the blood and tissue cells. Simple cuboidal epithelium lines kidney tubules and the cavity of many internal organs. The nose, mouth, esophagus, anal canal, and vagina are all lined with stratified squamous epithelium. As we shall see, the outer layer of skin is also stratified squamous epithelium, but the cells have been reinforced by keratin, a protein that provides strength.

When an epithelium is pseudostratified, it appears to be layered, but true layers do not exist because each cell touches the baseline. The lining of the windpipe, or trachea, is pseudostratified ciliated columnar epithelium. A secreted covering of mucus traps foreign particles, and the upward motion of the cilia carries the mucus to the back of the throat, where it may either be swallowed or expectorated. Smoking can cause a change in mucus secretion and inhibit ciliary action, and the result is a chronic inflammatory condition called bronchitis.

A so-called **basement membrane** often joins an epithelium to underlying connective tissue. We now know that the basement membrane consists of glycoprotein secreted by epithelial cells and collagen fibers that belong to the connective tissue.

When an epithelium secretes a product, it is said to be glandular. A **gland** can be a single epithelial cell, as in the case of mucus-secreting goblet cells found within the columnar epithelium lining the digestive tract, or a gland can contain many cells. Glands that secrete their product into ducts are called exocrine glands, and those that secrete their product directly into the bloodstream are called endocrine glands. The pancreas is both an exocrine gland, because it secretes digestive juices into the small intestine via ducts, and an endocrine gland, because it secretes insulin into the bloodstream.

Epithelial tissue is named according to the shape of the cell. These tightly packed protective cells can occur in more than one layer, and the cells lining a cavity can be ciliated and/or glandular.

Visual Focus

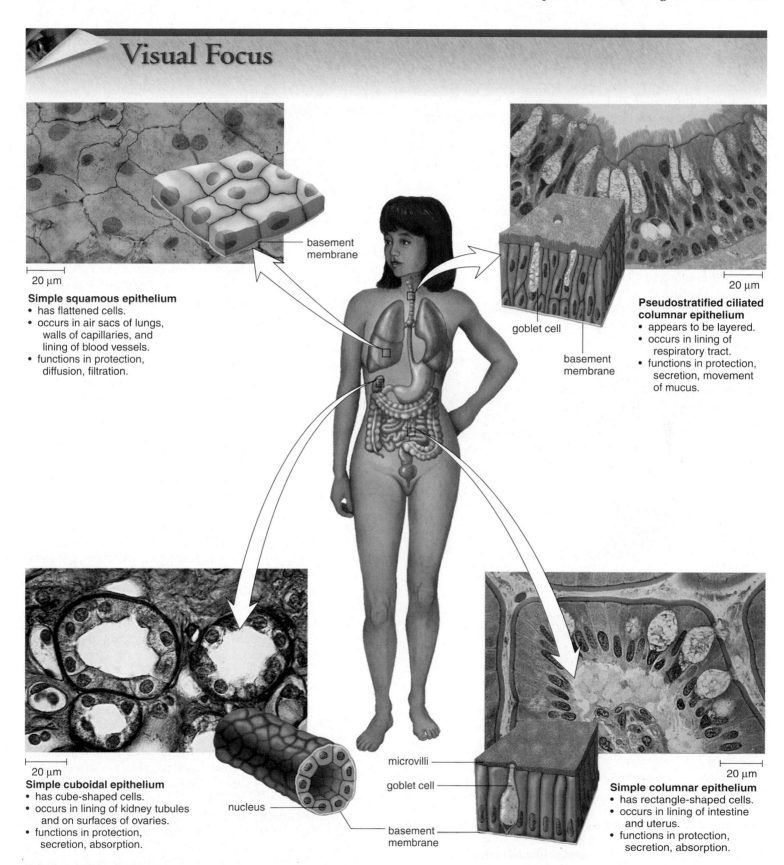

Simple squamous epithelium
- has flattened cells.
- occurs in air sacs of lungs, walls of capillaries, and lining of blood vessels.
- functions in protection, diffusion, filtration.

20 μm

basement membrane

Pseudostratified ciliated columnar epithelium
- appears to be layered.
- occurs in lining of respiratory tract.
- functions in protection, secretion, movement of mucus.

20 μm

goblet cell

basement membrane

Simple cuboidal epithelium
- has cube-shaped cells.
- occurs in lining of kidney tubules and on surfaces of ovaries.
- functions in protection, secretion, absorption.

20 μm

nucleus

microvilli

goblet cell

basement membrane

Simple columnar epithelium
- has rectangle-shaped cells.
- occurs in lining of intestine and uterus.
- functions in protection, secretion, absorption.

20 μm

Figure 11.1   Epithelial tissue.
Certain types of epithelial tissue—squamous, cuboidal, and columnar—are named for the shapes of their cells. They all have a protective function, as well as the other functions noted.

Junctions Between Epithelial Cells

The cells of a tissue can function in a coordinated manner when the plasma membranes of adjoining cells interact. The junctions that occur between cells help cells function as a tissue. A **tight junction** forms an impermeable barrier because adjacent plasma membrane proteins actually join, producing a zipperlike fastening (Fig. 11.2). In the intestine, the gastric juices stay out of the body, and in the kidneys, the urine stays within kidney tubules because epithelial cells are joined by tight junctions.

A **gap junction** forms when two adjacent plasma membrane channels join. This lends strength, but it also allows ions, sugars, and small molecules to pass between the two cells. Gap junctions in heart and smooth muscle ensure synchronized contraction. In an **adhesion junction** (desmosome), the adjacent plasma membranes do not touch but are held together by intercellular filaments firmly attached to buttonlike thickenings.

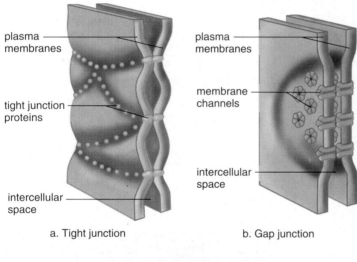

a. Tight junction   b. Gap junction

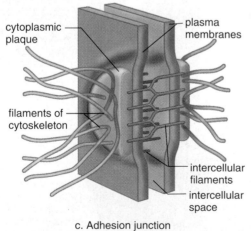

c. Adhesion junction

Figure 11.2   Junctions between epithelial cells.
Epithelial tissue cells are held tightly together by **(a)** tight junctions that hold cells together; **(b)** gap junctions that allow materials to pass from cell to cell; and **(c)** adhesion junctions that allow tissues to stretch.

Connective Tissue

Connective tissue binds organs together, provides support and protection, fills spaces, produces blood cells, and stores fat. As a rule, connective tissue cells are widely separated by a **matrix,** consisting of a noncellular material that varies in consistency from solid to semifluid to fluid. The matrix may have fibers of three possible types. White **collagen fibers** contain collagen, a protein that gives them flexibility and strength. **Reticular fibers** are very thin collagen fibers that are highly branched and form delicate supporting networks. Yellow **elastic fibers** contain elastin, a protein that is not as strong as collagen but is more elastic.

Loose Fibrous and Dense Fibrous Tissues

Both loose fibrous and dense fibrous connective tissues have cells called **fibroblasts** that are located some distance from one another and are separated by a jellylike matrix containing white collagen fibers and yellow elastic fibers.

Loose fibrous connective tissue supports epithelium and also many internal organs (Fig. 11.3a). Its presence in lungs, arteries, and the urinary bladder allows these organs to expand. It forms a protective covering enclosing many internal organs, such as muscles, blood vessels, and nerves.

Dense fibrous connective tissue contains many collagen fibers that are packed together. This type of tissue has more specific functions than does loose connective tissue. For example, dense fibrous connective tissue is found in **tendons,** which connect muscles to bones, and in **ligaments,** which connect bones to other bones at joints.

Adipose Tissue and Reticular Connective Tissue

In **adipose tissue** (Fig. 11.3b), the fibroblasts enlarge and store fat. The body uses this stored fat for energy, insulation, and organ protection. Adipose tissue is found beneath the skin, around the kidneys, and on the surface of the heart. **Reticular connective tissue** forms the supporting meshwork of lymphoid tissue present in lymph nodes, the spleen, the thymus, and the bone marrow. All types of blood cells are produced in red bone marrow, but a certain type of lymphocyte (T lymphocyte) completes its development in the thymus. The lymph nodes store lymphocytes.

Cartilage

In **cartilage,** the cells lie in small chambers called lacunae (sing., **lacuna**), separated by a matrix that is solid yet flexible. Unfortunately, because this tissue lacks a direct blood supply, it heals very slowly. There are three types of cartilage, distinguished by the type of fiber in the matrix.

Hyaline cartilage (Fig. 11.3c), the most common type of cartilage, contains only very fine collagen fibers. The matrix has a white, translucent appearance. Hyaline cartilage is found in the nose and at the ends of the long bones and the ribs, and it forms rings in the walls of respiratory passages. The fetal skeleton also is made of this type of cartilage. Later, the cartilaginous fetal skeleton is replaced by bone.

Elastic cartilage has more elastic fibers than hyaline cartilage. For this reason, it is more flexible and is found, for example, in the framework of the outer ear.

Fibrocartilage has a matrix containing strong collagen fibers. Fibrocartilage is found in structures that withstand tension and pressure, such as the pads between the vertebrae in the backbone and the wedges in the knee joint.

Bone

Bone is the most rigid connective tissue. It consists of an extremely hard matrix of inorganic salts, notably calcium salts, deposited around protein fibers, especially collagen fibers. The inorganic salts give bone rigidity, and the protein fibers provide elasticity and strength, much as steel rods do in reinforced concrete.

Compact bone makes up the shaft of a long bone (Fig. 11.3d). It consists of cylindrical structural units called osteons (Haversian systems). The central canal of each osteon

is surrounded by rings of hard matrix. Bone cells are located in spaces called lacunae between the rings of matrix. Blood vessels in the central canal carry nutrients that allow bone to renew itself. Nutrients can reach all of the bone cells because they are connected by thin processes within canaliculi (minute canals) that also reach to the central canal.

The ends of a long bone contain spongy bone, which has an entirely different structure. **Spongy bone** contains numerous bony bars and plates, separated by irregular spaces. Although lighter than compact bone, spongy bone is still designed for strength. Just as braces are used for support in buildings, the solid portions of spongy bone follow lines of stress.

Connective tissues, which bind and support body parts, differ according to the type of matrix and the abundance of fibers in the matrix.

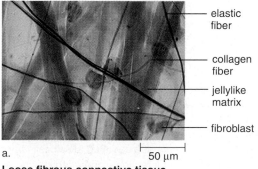

a.

├──────── 50 µm

Loose fibrous connective tissue
• has space between components.
• occurs beneath skin and most epithelial layers.
• functions in support and binds organs.

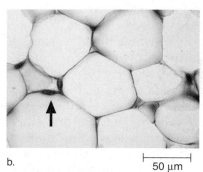

b.

├──────── 50 µm

Adipose tissue
• has cells filled with fat.
• occurs beneath skin, around organs including the heart.
• functions in insulation, stores fat.

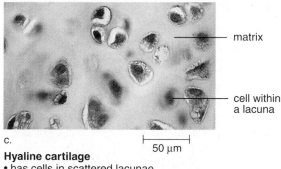

c.

├──────── 50 µm

Hyaline cartilage
• has cells in scattered lacunae.
• occurs in nose and walls of respiratory passages; at ends of bones including ribs.
• functions in support and protection.

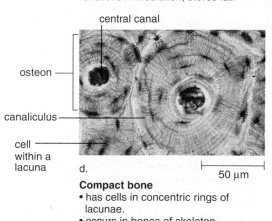

d.

├──────── 50 µm

Compact bone
• has cells in concentric rings of lacunae.
• occurs in bones of skeleton.
• functions in support and protection.

Figure 11.3   Connective tissue examples.
a. In loose fibrous connective tissue, cells called fibroblasts are separated by a jellylike matrix, which contains both collagen and elastic fibers. **b.** Adipose tissue cells have nuclei (arrow) pushed to one side because the cells are filled with fat. **c.** In hyaline cartilage, the flexible matrix has a white, translucent appearance. **d.** In compact bone, the hard matrix contains calcium salts. Concentric rings of cells in lacunae form an elongated cylinder called an osteon (Haversian system). An osteon has a central canal that contains blood vessels and nerve fibers.

Blood

Blood is unlike other types of connective tissue in that the matrix (i.e., plasma) is not made by the cells. Some people do not classify blood as connective tissue; instead, they suggest a separate tissue category called vascular tissue.

The internal environment of the body consists of blood and tissue fluid. The systems of the body help keep blood composition and chemistry within normal limits, and blood in turn creates tissue fluid. Blood transports nutrients and oxygen to tissue fluid and removes carbon dioxide and other wastes. It helps distribute heat and also plays a role in fluid, ion, and pH balance. Various components of blood, as discussed below, help protect us from disease, and blood's ability to clot prevents fluid loss.

If blood is transferred from a person's vein to a test tube and prevented from clotting, it separates into two layers (Fig. 11.4). The upper, liquid layer, called **plasma,** represents about 55% of the volume of whole blood and contains a variety of inorganic and organic substances dissolved or suspended in water (Table 11.1). The lower layer consists of red blood cells (erythrocytes), white blood cells (leukocytes), and blood platelets (thrombocytes). Collectively, these are called the formed elements and represent about 45% of the volume of whole blood. Formed elements are manufactured in the red bone marrow of the skull, ribs, vertebrae, and ends of the long bones.

The **red blood cells** are small, biconcave, disk-shaped cells without nuclei. The presence of the red pigment hemoglobin makes the cells red, and in turn, makes the blood red. Hemoglobin is composed of four units; each is composed of the protein globin and a complex iron-containing structure called heme. The iron forms a loose association with oxygen, and in this way red blood cells transport oxygen.

White blood cells may be distinguished from red blood cells by the fact that they are usually larger, have a nucleus, and without staining would appear translucent. White blood cells characteristically look bluish because they have been stained that color. White blood cells fight infection, primarily in two ways. Some white blood cells are phagocytic and engulf infectious pathogens, while other white blood cells produce antibodies, molecules that combine with foreign substances to inactivate them.

Platelets are not complete cells; rather, they are fragments of giant cells present only in bone marrow. When a blood vessel is damaged, platelets form a plug that seals the vessel, and injured tissues release molecules that help the clotting process.

Blood is a connective tissue in which the matrix is plasma.

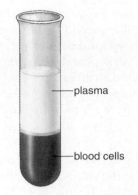

a. Blood sample

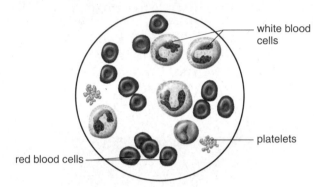

b. Blood smear

Figure 11.4 **Blood, a fluid tissue.**
a. In a test tube, a blood sample separates into its two components: blood cells and plasma. **b.** Microscopic examination of a blood smear shows red blood cells, white blood cells, and platelets. Platelets are fragments of a cell. Red blood cells transport oxygen, white blood cells fight infections, and platelets are involved in blood clotting.

Table 11.1	Components of Blood Plasma
Water (92% of Total)	
Solutes (8% of Total):	
Inorganic ions (electrolytes)	Na^+, Ca^{2+}, K^+, Mg^{2+}, Cl_2, HCO_3^-, HPO_4^{2+}, SO_4^{2+}
Gases	O_2, CO_2
Plasma proteins	Albumin, globulins, fibrinogen
Organic nutrients	Glucose, lipids, phospholipids, amino acids, etc.
Nitrogenous waste products	Urea, ammonia, uric acid
Regulatory substances	Hormones, enzymes

Muscular Tissue

Muscular (contractile) tissue is composed of cells called muscle fibers. Muscle fibers contain actin filaments and myosin filaments, whose interaction accounts for movement. Three types of vertebrate muscles are skeletal, smooth, and cardiac.

Skeletal muscle, also called voluntary muscle (Fig. 11.5a), is attached by tendons to the bones of the skeleton, and when it contracts, body parts move. Contraction of skeletal muscle is under voluntary control and occurs faster than in the other muscle types. Skeletal muscle fibers are cylindrical and quite long—sometimes they run the length of the muscle. They arise during development when several cells fuse, resulting in one fiber with multiple nuclei. The nuclei are located at the periphery of the cell, just inside the plasma membrane. The fibers have alternating light and dark bands that give them a **striated** appearance. These bands are due to the placement of actin filaments and myosin filaments in the cell.

Smooth (visceral) muscle is so named because the cells lack striations. The spindle-shaped cells form layers in which the thick middle portion of one cell is opposite the thin ends of adjacent cells. Consequently, the nuclei form an irregular pattern in the tissue (Fig. 11.5b). Smooth muscle is not under voluntary control and therefore is said to be involuntary. Smooth muscle, found in the walls of viscera (intestine, stomach, and other internal organs) and blood vessels, contracts more slowly than skeletal muscle but can remain contracted for a longer time. When the smooth muscle of the intestine contracts, food moves along its lumen (central cavity). When the smooth muscle of the blood vessels contracts, blood vessels constrict, helping to raise blood pressure.

Cardiac muscle (Fig. 11.5c) is found only in the walls of the heart. Its contraction pumps blood and accounts for the heartbeat. Cardiac muscle combines features of both smooth muscle and skeletal muscle. Like skeletal muscle, it has striations, but the contraction of the heart is involuntary for the most part. Cardiac muscle cells also differ from skeletal muscle cells in that they have a single, centrally placed nucleus. The cells are branched and seemingly fused one with the other, and the heart appears to be composed of one large interconnecting mass of muscle cells. Actually, cardiac muscle cells are separate and individual, but they are bound end to end at **intercalated disks,** areas where folded plasma membranes between two cells contain adhesion junctions and gap junctions.

All muscular tissue contains actin filaments and myosin filaments; these form a striated pattern in skeletal and cardiac muscle, but not in smooth muscle.

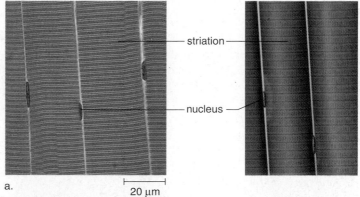

a.   ⊢——⊣ 20 μm

Skeletal muscle
- has striated cells with multiple nuclei.
- occurs in muscles attached to skeleton.
- functions in voluntary movement of body.
- is voluntary.

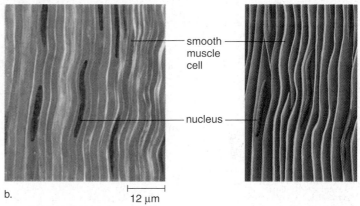

b.   ⊢——⊣ 12 μm

Smooth muscle
- has spindle-shaped cells, each with a single nucleus.
- cells have no striations.
- functions in movement of substances in lumens of body.
- is involuntary.

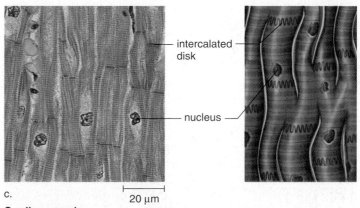

c.   ⊢——⊣ 20 μm

Cardiac muscle
- has branching striated cells, each with a single nucleus.
- occurs in the wall of the heart.
- functions in the pumping of blood.
- is involuntary.

Figure 11.5   Muscular tissue.
a. Skeletal muscle is voluntary and striated. **b.** Smooth muscle is involuntary and nonstriated. **c.** Cardiac muscle is involuntary and striated. Cardiac muscle cells branch and fit together at intercalated disks.

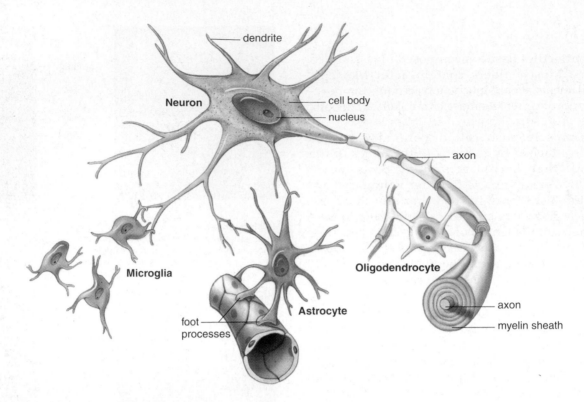

Figure 11.6   A neuron and some types of neuroglia.
Neurons conduct nerve impulses. Microglia become mobile in response to inflammation and phagocytize debris. Astrocytes lie between neurons and a capillary; therefore, substances entering neurons from the blood must first pass through astrocytes. Oligodendrocytes form the myelin sheaths around fibers in the brain and spinal cord.

Nervous Tissue

Nervous tissue, which contains nerve cells called neurons, is present in the brain and spinal cord. A **neuron** is a specialized cell that has three parts: dendrites, a cell body, and an axon (Fig. 11.6). A dendrite is a process that conducts signals toward the cell body. The cell body contains the major concentration of the cytoplasm and the nucleus of the neuron. An axon is a process that typically conducts nerve impulses away from the cell body. Long axons are covered by myelin, a white fatty substance. The term *fiber*[1] is used here to refer to an axon along with its myelin sheath if it has one. Outside the brain and spinal cord, fibers bound by connective tissue form **nerves.**

The nervous system has just three functions: sensory input, integration of data, and motor output. Nerves conduct impulses from sensory receptors to the spinal cord and the brain where integration occurs. The phenomenon called sensation occurs only in the brain, however. Nerves also conduct nerve impulses away from the spinal cord and brain to the muscles and glands, causing them to contract and secrete, respectively. In this way, a coordinated response to the stimulus is achieved.

[1]In connective tissue, a fiber is a component of the matrix; in muscle tissue, a fiber is a muscle cell; in nervous tissue, a fiber is an axon and its myelin sheath.

In addition to neurons, nervous tissue contains neuroglia.

Neuroglia

Neuroglia are cells that outnumber neurons nine to one and take up more than half the volume of the brain. Although the primary function of neuroglia is to support and nourish neurons, research is currently being conducted to determine how much they directly contribute to brain function. The three types of neuroglia found in the brain are microglia, astrocytes, and oligodendrocytes (Fig. 11.6). Microglia, in addition to supporting neurons, engulf bacterial and cellular debris. Astrocytes provide nutrients to neurons and produce a hormone known as glia-derived growth factor, which someday might be used as a cure for Parkinson disease and other diseases caused by neuron degeneration. Oligodendrocytes form myelin. Neuroglia don't have a long process, but even so, researchers are now beginning to gather evidence that they do communicate among themselves and with neurons!

Nerve cells, called neurons, have fibers (processes) called axons and dendrites. In general, neuroglia support and service neurons.

11.2 Body Cavities and Body Membranes

The human body is divided into two main categories: the ventral cavity and the dorsal cavity (Fig. 11.7). The ventral cavity, which is called a **coelom** during development, becomes divided into the thoracic and abdominal cavities. Membranes divide the thoracic cavity into the pleural cavities, containing the right and left lungs, and the pericardial cavity, containing the heart. The thoracic cavity is separated from the abdominal cavity by a horizontal muscle called the diaphragm. The stomach, liver, spleen, gallbladder, and most of the small and large intestines are in the upper portion of the abdominal cavity. The lower portion contains the rectum, the urinary bladder, the internal reproductive organs, and the rest of the large intestine. Males have an external extension of the abdominal wall, called the scrotum, containing the testes.

The dorsal cavity also has two parts: the cranial cavity within the skull contains the brain; the vertebral canal, formed by the vertebrae, contains the spinal cord.

Body Membranes

In this context, we are using the term *membrane* to refer to a thin lining or covering composed of an epithelium overlying a loose connective tissue layer. Body membranes line cavities and the internal spaces of organs and tubes that open to the outside.

Mucous membranes line the tubes of the digestive, respiratory, urinary, and reproductive systems. The epithelium of this membrane contains goblet cells that secrete mucus. This mucus ordinarily protects the body from invasion by bacteria and viruses; hence, more mucus is secreted and expelled when a person has a cold and has to blow her/his nose. In addition, mucus usually protects the walls of the stomach and small intestine from digestive juices, but this protection breaks down when a person develops an ulcer.

Serous membranes line the thoracic and abdominal cavities and the organs they contain. They secrete a watery fluid that keeps the membranes lubricated. Serous membranes support the internal organs and compartmentalize the large thoracic and abdominal cavities. This helps hinder the spread of any infection.

Serous membranes have specific names according to their location. The **pleura,** also called the pleural membranes, line the pleural cavity and cover the lungs; the pericardium lines the pericardial cavity and covers the heart; the peritoneum lines the abdominal cavity and its organs. In between these organs, there is a double layer of peritoneum called mesentery. Serous membranes are subject to infections. For example, peritonitis is a life-threatening infection of the peritoneum that may occur if an inflamed appendix bursts before it is removed.

Synovial membranes line freely movable joint cavities. They secrete synovial fluid into the joint cavity; this fluid

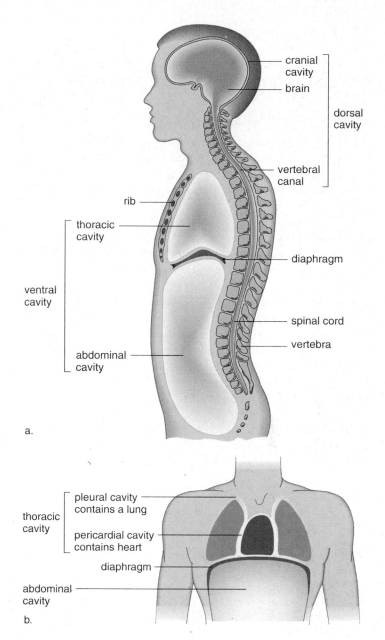

Figure 11.7 Mammalian body cavities.
a. Side view. The dorsal (toward the back) cavity contains the cranial cavity and the vertebral canal. The brain is in the cranial cavity, and the spinal cord is in the vertebral canal. The well-developed ventral (toward the front) cavity is divided by the diaphragm into the thoracic cavity and the abdominal cavity. The heart and lungs are in the thoracic cavity, and most other internal organs are in the abdominal cavity. **b.** Frontal view of the thoracic cavity.

lubricates the ends of the bones so that they can move freely. In rheumatoid arthritis, the synovial membrane becomes inflamed and grows thicker, restricting movement.

The **meninges** are membranes found within the dorsal cavity. They are composed only of connective tissue and serve as a protective covering for the brain and spinal cord. Meningitis is a life-threatening infection of the meninges.

11.3 Organ Systems

An overview of all organ systems (Fig. 11.8) precedes a discussion of the integumentary system in this section.

Overview of Organ Systems

It should be emphasized that just as organs work together in an organ system so do organ systems work together in the body. In a sense it is arbitrary to assign a particular organ to one system when it also assists the functioning of many other systems.

Integumentary System   Skin, nails, hair

The integumentary system contains skin which is made up of two main types of tissue: the epidermis is composed of stratified squamous epithelium and the dermis is a region of fibrous connective tissue.

The **integumentary system** also includes nails, located at the ends of digits, hairs, muscles that move hairs, the oil and sweat glands, blood vessels, and nerves leading to sensory receptors. It is clear that the skin has a protective function. It also synthesizes vitamin D, collects sensory data, and helps regulate body temperature.

Digestive System

The **digestive system** consists of the mouth, esophagus, stomach, small intestine, and large intestine (colon) along with these associated organs: teeth, tongue, salivary glands, liver, gallbladder, and pancreas. This system receives food and digests it into nutrient molecules, which can enter the cells of the body. The nondigested remains are eventually eliminated.

Cardiovascular System

In the **cardiovascular system** the heart pumps blood and sends it out under pressure into the blood vessels. While blood is moving throughout the body it distributes heat produced by the muscles. Blood transports nutrients and oxygen to the cells, and removes their waste molecules, including carbon dioxide. Despite the movement of molecules into and out of the blood it has a fairly constant volume and pH. Phagocytic white blood cells engulf pathogens.

Lymphatic and Immune Systems

The **lymphatic system** consists of lymphatic vessels which transport lymph, lymph nodes, and other lymphoid organs. This system protects the body from disease by purifying lymph and storing lymphocytes, the white blood cells that produce antibodies. Lymphatic vessels absorb fat from the digestive system and collect excess tissue fluid, which is returned to the cardiovascular system.

The **immune system** consists of all the cells in the body that protect us from disease. The lymphocytes in particular belong to this system.

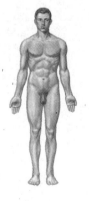 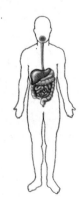 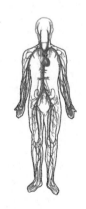 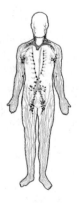 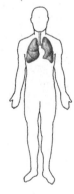 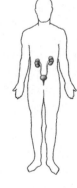

Integumentary system
- protects body.
- receives sensory input.
- helps control temperature.
- synthesizes vitamin D.

Digestive system
- ingests food.
- digests food.
- absorbs nutrients.
- eliminates waste.

Cardiovascular system
- transports blood, nutrients, gases, and wastes.
- defends against disease.
- helps control temperature, fluid, and pH balance.

Lymphatic and immune systems
- helps control fluid balance.
- absorbs fats.
- defends against disease.

Respiratory system
- maintains breathing.
- exchanges gases at lungs and tissues.
- helps control pH balance.

Urinary system
- excretion of metabolic wastes.
- helps control fluid balance.
- helps control pH balance.

Figure 11.8   Organ systems of the body.

Respiratory System *gas exchange*

The **respiratory system** consists of the lungs and the tubes that take air to and from them. The respiratory system brings oxygen into the body and removes carbon dioxide from the body at the lungs, restoring pH.

Urinary System

The **urinary system** contains the kidneys, the urinary bladder and tubes that carry urine. This system rids the body of metabolic wastes, particularly nitrogenous wastes, and helps regulate the fluid balance and pH of the blood.

Skeletal System

The bones of the **skeleton system** protect body parts. For example, the skull forms a protective encasement for the brain, as does the rib cage for the heart and lungs. The skeleton helps move the body because it serves as a place of attachment for the skeletal muscles.

The skeletal system also stores minerals, notably calcium and it produces blood cells within red bone marrow.

Muscular System

In the **muscular system,** skeletal muscle contraction maintains posture and accounts for the movement of the body and its parts. Cardiac muscle contraction results in the heart beat. The walls of internal organs contract due to the presence of smooth muscle.

Nervous System

The **nervous system** consists of the brain, spinal cord, and associated nerves. The nerves conduct nerve impulses from sensory receptors to the brain and spinal cord where integration occurs. Nerves also conduct nerve impulses from the brain and spinal cord to the muscles and glands, allowing us to respond to both external and internal stimuli.

Endocrine System *hormones*

The **endocrine system** consists of the hormonal glands, which secrete chemical messengers called hormones. Hormones have a wide range of effects, including regulation of cellular metabolism, regulation of fluid and pH balance, and helping us respond to stress. Both the nervous and endocrine systems coordinate and regulate the functioning of the body's other systems. The endocrine system also helps maintain the functioning of the male and female reproductive organs.

Reproductive System

The **reproductive system** has different organs in the male and female. The male reproductive system consists of the testes, other glands, and various ducts that conduct semen to and through the penis. The testes produce sex cells called sperm. The female reproductive system consists of the ovaries, oviducts, uterus, vagina, and external genitals. The ovaries produce sex cells called eggs. When a sperm fertilizes an egg, an offspring begins development.

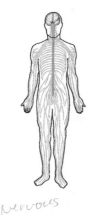
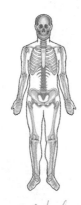
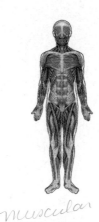
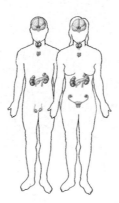
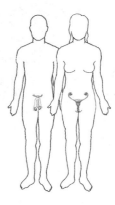

Skeletal system
- supports the body.
- protects body parts.
- helps move the body.
- stores minerals.
- produces of blood cells.

Muscular system
- maintains posture.
- moves body and internal organs.
- produces heat.

Nervous system
- receives sensory input.
- integrates and stores input.
- initiates motor output.
- helps coordinates organ systems.

Endocrine system
- produces hormones.
- helps coordinate organ systems.
- responds to stress.
- helps regulate fluid and pH balance.
- helps regulate metabolism.

Reproductive system
- produces gametes.
- transports gametes.
- produces sex hormone.
- nurtures and gives birth to offspring in females.

Figure 11.8   **Organ systems of the body—continued.**

Integumentary System

The skin and its accessory organs (hair, nails, sweat glands, and sebaceous glands) are collectively called the **integumentary system.** Skin covers the body, protecting underlying tissues from physical trauma, pathogen invasion, and water loss; it also helps regulate body temperature. Therefore, skin plays a significant role in homeostasis. The skin even synthesizes certain chemicals, such as vitamin D, that affect the rest of the body. Because skin contains sensory receptors, skin also helps us to be aware of our surroundings and to communicate through touch.

Regions of the Skin

The **skin** has two regions: the epidermis and the dermis (Fig. 11.9). A subcutaneous layer is found between the skin and any underlying structures, such as muscle or bone.

The **epidermis** is made up of stratified squamous epithelium. New cells derived from basal cells become flattened and hardened as they push to the surface. Hardening takes place because the cells produce keratin, a waterproof protein. Dandruff occurs when the rate of keratinization in the skin of the scalp is two or three times the normal rate. A thick layer of dead keratinized cells, arranged in spiral and concentric patterns, forms fingerprints and

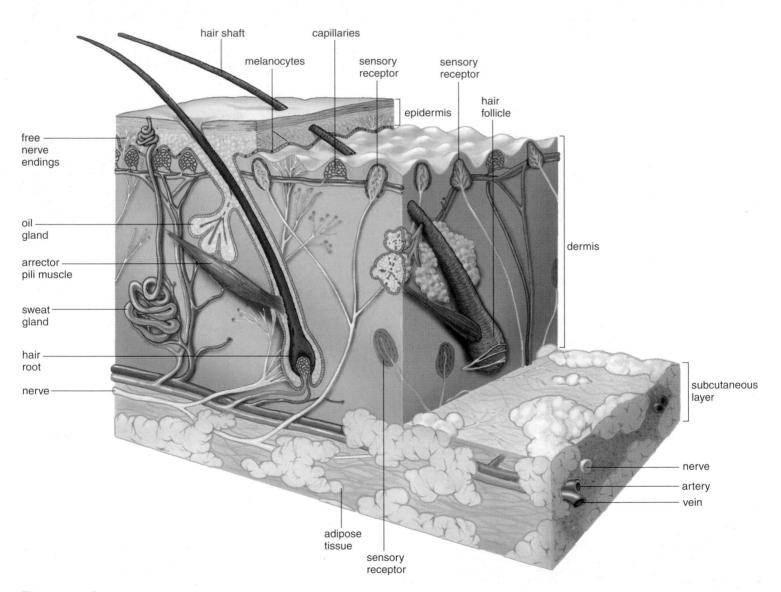

Figure 11.9   Human skin anatomy.
Skin consists of two regions, the epidermis and the dermis. A subcutaneous layer lies below the dermis.

footprints. Specialized cells in the epidermis called **melanocytes** produce melanin, the pigment responsible for skin color.

The **dermis** is a region of fibrous connective tissue beneath the epidermis. The dermis contains collagen and elastic fibers. The collagen fibers are flexible but offer great resistance to overstretching; they prevent the skin from being torn. The elastic fibers maintain normal skin tension but also stretch to allow movement of underlying muscles and joints. (The number of collagen and elastic fibers decreases with exposure to the sun, causing the skin to become less supple and more prone to wrinkling.) The dermis also contains blood vessels that nourish the skin. When blood rushes into these vessels, a person blushes, and when blood is minimal in them, a person turns "blue."

Sensory receptors are specialized nerve endings in the dermis that respond to external stimuli. There are sensory receptors for touch, pressure, pain, and temperature. The fingertips contain the most touch receptors, and these add to our ability to use our fingers for delicate tasks.

The **subcutaneous layer,** which lies below the dermis, is composed of loose connective tissue and adipose tissue, which stores fat. Fat is a stored source of energy in the body. Adipose tissue helps to thermally insulate the body from either gaining heat from the outside or losing heat from the inside. A well-developed subcutaneous layer gives the body a rounded appearance and provides protective padding against external assaults. Excessive development of the subcutaneous layer accompanies obesity.

Skin has two regions: the epidermis and the dermis. A subcutaneous layer lies beneath the dermis.

Accessory Organs of the Skin

Nails, hair, and glands are structures of epidermal origin, even though some parts of hair and glands are largely found in the dermis.

Nails are a protective covering of the distal part of fingers and toes, collectively called digits. Nails can help pry open or pick up small objects. They are also used for scratching oneself or others. Nails grow from special epithelial cells at the base of the nail in the portion called the nail root. These cells become keratinized as they grow out over the nail bed. The visible portion of the nail is called the nail body. The cuticle is a fold of skin that hides the nail root. The whitish color of the half-moon-shaped base, or lunula, results from the thick layer of cells in this area (Fig. 11.10).

Hair follicles are in the dermis and continue through the epidermis where the hair shaft extends beyond the skin. Contraction of the arrector pili muscles attached to hair follicles causes the hairs to "stand on end" and goose bumps to develop. Epidermal cells form the root of hair, and their division causes a hair to grow. The cells become keratinized and dead as they are pushed farther from the root.

Each hair follicle has one or more **oil glands,** also called sebaceous glands, which secrete sebum, an oily substance that lubricates the hair within the follicle and the skin itself. If the sebaceous glands fail to discharge, the secretions collect and form "whiteheads" or "blackheads." The color of blackheads is due to oxidized sebum. Acne is an inflammation of the sebaceous glands that most often occurs during adolescence. Hormonal changes during this time cause the sebaceous glands to become more active.

Sweat glands, also called sudoriferous glands, are quite numerous and are present in all regions of skin. A sweat gland is a coiled tubule within the dermis that straightens out near its opening. Some sweat glands open into hair follicles, but most open onto the surface of the skin. Sweat glands play a role in modifying body temperature. When body temperature starts to rise, sweat glands become active. Sweat absorbs body heat as it evaporates. Once body temperature lowers, sweat glands are no longer active.

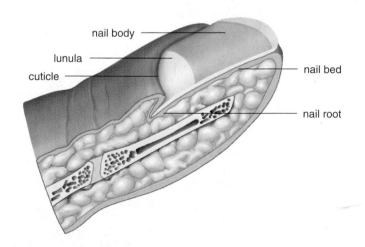

Figure 11.10   Nail anatomy.
Cells produced by the nail root become keratinized, forming the nail body.

11.4 Homeostasis

Homeostasis is the relative constancy of the body's internal environment. Because of homeostasis, even though external conditions may change dramatically, internal conditions still stay within a narrow range (Fig. 11.11). For example, regardless of how cold or hot it gets, the temperature of the body stays around 37°C (97° to 99°F). No matter how acidic your meal, the pH of your blood is usually about 7.4, and even if you eat a candy bar, the amount of sugar in your blood is just about 0.1%.

It is important to realize that internal conditions are not absolutely constant; they tend to fluctuate above and below a particular value. Therefore, the internal state of the body is often described as one of dynamic equilibrium. If internal conditions should change to any great degree, illness results. This makes the study of homeostatic mechanisms medically important.

Negative Feedback

A homeostatic mechanism in the body has three components: a sensor, a regulatory center, and an effector (Fig. 11.12*a*). The sensor detects a change in the internal environment; the regulatory center activates the effector; the effector reverses the change and brings conditions back to normal again. Now, the sensor is no longer activated.

Negative feedback is the primary homeostatic mechanism that keeps a variable, such as body temperature, close to a particular value, or set point. A home heating system illustrates how a negative feedback mechanism works (Fig. 11.12*b*). You set the thermostat at, say, 68°F. This is the set point. The thermostat contains a thermometer, a sensor that detects when the room temperature falls below the set point. The thermostat is also the regulatory center; it turns the furnace on. The furnace plays the role of the effector. The heat given off by the furnace raises the temperature of the room to 70°F. Now, the furnace turns off. Notice that a negative feedback mechanism prevents change in the same direction; the room does not get

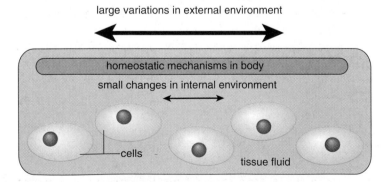

Figure 11.11   Homeostasis.
Because of homeostatic mechanisms, large external changes cause only small internal changes in such parameters as body temperature and pH of the blood.

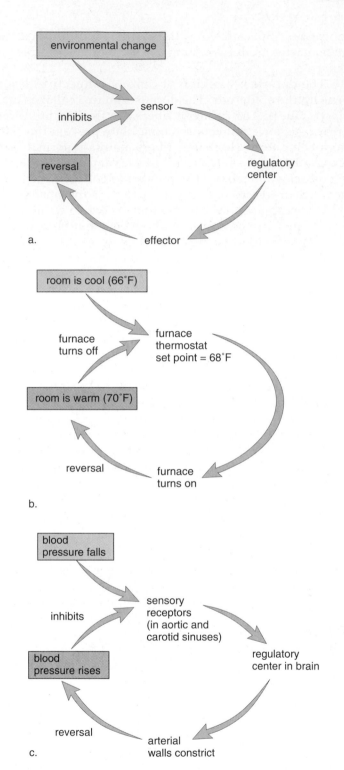

Figure 11.12   Negative feedback.
a. A sensor detects an internal environmental change and signals a regulatory center. The center activates an effector, which reverses this change. **b.** Mechanical example: When the room is cool, a thermostat that senses the room temperature signals the furnace to turn on. Once the room is warm, the furnace turns off. **c.** Biological example: When blood pressure falls, special sensory receptors in blood vessels signal a regulatory center in the brain. The brain signals the arteries to constrict, and blood pressure rises to normal. This response reverses the change.

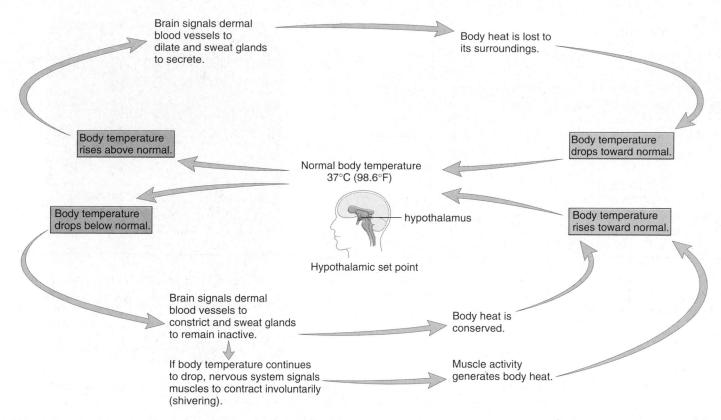

Figure 11.13   Homeostasis and body temperature regulation.
Negative feedback mechanisms control body temperature so that it remains relatively stable at 37°C. These mechanisms return the temperature to normal when it fluctuates above and below this set point.

hotter and hotter because warmth inactivates the system. Also notice that negative feedback mechanisms are activated only by a deviation from the set point, and therefore there is a fluctuation above and below this value.

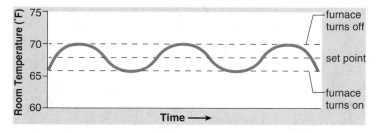

Negative feedback mechanisms in the body function similarly to this mechanical model (Fig. 11.12c). For example, when blood pressure falls, sensory receptors signal a regulatory center in the brain. This center sends out nerve impulses to the arterial walls so that they constrict. Once the blood pressure rises, the system is inactivated.

A negative feedback mechanism maintains stability by its ability to sense a change and bring about an effect that reverses that change.

Regulation of Body Temperature

The thermostat for body temperature is located in a part of the brain called the hypothalamus. When the body temperature falls below normal, the regulatory center directs (via nerve impulses) the blood vessels of the skin to constrict (Fig. 11.13). This conserves heat. If body temperature falls even lower, the regulatory center sends nerve impulses to the skeletal muscles, and shivering occurs. Shivering generates heat, and gradually body temperature rises to 37°C. When the temperature rises to normal, the regulatory center is inactivated.

When the body temperature is higher than normal, the regulatory center directs the blood vessels of the skin to dilate. This allows more blood to flow near the surface of the body, where heat can be lost to the environment. In addition, the nervous system activates the sweat glands, and the evaporation of sweat helps lower body temperature. Gradually, body temperature decreases to 37°C.

The temperature of the human body is maintained at about 37°C due to the activity of a regulatory center in the brain.

Positive Feedback

Positive feedback is a mechanism that brings about an ever greater change in the same direction. When a woman is giving birth, the head of the baby begins to press against the cervix, stimulating sensory receptors there. When nerve impulses reach the brain, the brain causes the pituitary gland to secrete the hormone oxytocin. Oxytocin travels in the blood and causes the uterus to contract. As labor continues, the cervix is ever more stimulated and uterine contractions become ever more strong until birth occurs.

A positive feedback mechanism can be harmful, as when a fever causes metabolic changes that push the fever still higher. Death occurs at a body temperature of 45°C because cellular proteins denature at this temperature and metabolism stops. Still, positive feedback loops like those involved in childbirth, blood clotting, and the stomach's digestion of protein assist the body in completing a process that has a definite cut-off point.

In contrast to negative feedback, positive feedback allows rapid change in one direction and does not achieve relative stability.

Homeostasis and Body Systems

The internal environment of the body consists of blood and tissue fluid. Tissue fluid, which bathes all the cells of the body, is refreshed when molecules such as oxygen and nutrients move into tissue fluid from the blood, and when wastes move from tissue fluid into the blood (Fig. 11.14). Tissue fluid remains constant only as long as blood composition remains constant.

All systems of the body contribute toward maintaining homeostasis and therefore a relatively constant internal environment. The cardiovascular system conducts blood to and away from capillaries, where exchange occurs. The heart pumps the blood and thereby keeps it moving toward the capillaries. The formed elements also contribute to homeostasis. Red blood cells transport oxygen and participate in the transport of carbon dioxide. White blood cells fight infection, and platelets participate in the clotting process. The lymphatic system is accessory to the cardiovascular system. Lymphatic capillaries collect excess tissue fluid, and this is returned via lymphatic veins to the cardiovascular veins. Lymph nodes help purify lymph and keep it free of pathogens.

The digestive system takes in and digests food, providing nutrient molecules that enter the blood and replace the nutrients that are constantly being used by the body cells. The respiratory system adds oxygen to and removes carbon dioxide from the blood. The chief regulators of blood composition are the liver and the kidneys. They monitor the chemical composition of plasma and alter it as required. Immediately after glucose enters the blood, it can be removed

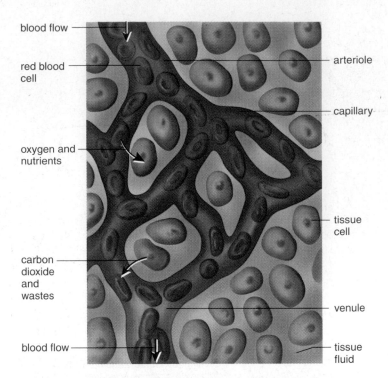

Figure 11.14 Regulation of tissue fluid composition.
Cells are surrounded by tissue fluid, which is continually refreshed because oxygen and nutrient molecules constantly exit, and carbon dioxide and waste molecules continually enter the bloodstream.

by the liver for storage as glycogen. Later, the glycogen can be broken down to replace the glucose used by the body cells; in this way, the glucose composition of blood remains constant. The hormone insulin, secreted by the pancreas, regulates glycogen storage. The liver also removes toxic chemicals, such as ingested alcohol and other drugs. The liver makes urea, a nitrogenous end product of protein metabolism. Urea and other metabolic waste molecules are excreted by the kidneys. Urine formation by the kidneys is extremely critical to the body, not only because it rids the body of unwanted substances, but also because it offers an opportunity to carefully regulate blood volume, salt balance, and pH.

The nervous system and the endocrine systems regulate the other systems of the body. They work together to control body systems so that homeostasis is maintained. We have already seen that in negative feedback mechanisms, sensory receptors send nerve impulses to regulatory centers in the brain, which then direct effectors to become active. Effectors can be muscles or glands. Muscles bring about an immediate change. Endocrine glands secrete hormones that bring about a slower, more lasting change that keeps the internal environment relatively stable.

All systems of the body contribute to homeostasis—that is, maintaining the relative constancy of the internal environment, blood, and tissue fluid.

Health Focus

Skin Cancer on the Rise

In the nineteenth century, and earlier, it was fashionable for Caucasian women who did not labor outdoors to keep their skin fair by carrying parasols when they went out. But early in this century, some fair-skinned people began to prefer the golden-brown look, and they took up sunbathing as a way to achieve a tan.

A few hours of exposure to the sun cause pain and redness due to dilation of blood vessels. Tanning occurs when melanin granules increase in keratinized cells at the surface of the skin as a way to prevent any further damage by ultraviolet (UV) rays. The sun gives off two types of UV rays: UV-A rays and UV-B rays. UV-A rays penetrate the skin deeply, affect connective tissue, and cause the skin to sag and wrinkle. UV-A rays are also believed to increase the effects of the UV-B rays, which are the cancer-causing rays. UV-B rays are more prevalent at midday.

Skin cancer is categorized as either nonmelanoma or melanoma. Nonmelanoma cancers are of two types. Basal cell carcinoma, the most common type, begins when UV radiation causes epidermal basal cells to form a tumor, while at the same time suppressing the immune system's ability to detect the tumor. The signs of a basal cell tumor are varied. They include an open sore that will not heal; a recurring reddish patch; a smooth, circular growth with a raised edge; a shiny bump; or a pale mark (Fig. 11A*a*). In about 95% of patients, the tumor can be excised surgically, but recurrence is common.

Squamous cell carcinoma begins in the epidermis proper. Squamous cell carcinoma is five times less common than basal cell carcinoma, but if the tumor is not excised promptly, it is more likely to spread to nearby organs. The death rate from squamous cell carcinoma is about 1% of cases. The signs of a tumor are the same as for basal cell carcinoma, except that a squamous cell carcinoma may also show itself as a wart that bleeds and scabs.

Melanoma that starts in pigmented cells often has the appearance of an unusual mole. Unlike a mole that is circular and confined, melanoma moles look like spilled ink spots. A variety of shades can be seen in the same mole, and they can itch, hurt, or feel numb. The skin around the mole turns gray, white, or red. Melanoma is most apt to appear in persons who have fair skin, particularly if they have suffered occasional severe sunburns as children. The chance of melanoma increases with the number of moles a person has. Most moles appear before the age of 14, and their appearance is linked to sun exposure. Any moles that become malignant are removed surgically; if the cancer has spread, chemotherapy and various other treatments are also available.

Scientists have developed a UV index to determine how powerful the solar rays are in different U.S. cities. In general, the more southern the city, the higher the UV index and the greater the risk of skin cancer. Regardless of where you live, for every 10% decrease in the ozone layer, the risk of skin cancer rises 13–20%. To prevent the occurrence of skin cancer, take the following precautions:

- Use a broad-spectrum sunscreen, which protects you from both UV-A and UV-B radiation, with an SPF (sun protection factor) of at least 15. (This means, for example, that if you usually burn after a 20-minute exposure, it will take 15 times that long before you will burn.)
- Stay out of the sun altogether between the hours of 10 A.M. and 3 P.M. This will reduce your annual exposure by as much as 60%.
- Wear protective clothing. Choose fabrics with a tight weave, and wear a wide-brimmed hat.
- Wear sunglasses that have been treated to absorb both UV-A and UV-B radiation. Otherwise, sunglasses can expose your eyes to more damage than usual because pupils dilate in the shade.
- Avoid tanning machines. Although most tanning devices use high levels of only UV-A, these rays cause the deep layers of the skin to become more vulnerable to UV-B radiation when you are later exposed to the sun.

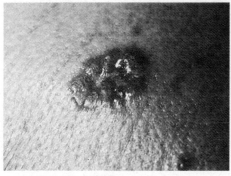

a. Basal cell carcinoma

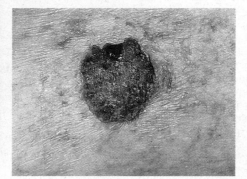

b. Squamous cell carcinoma

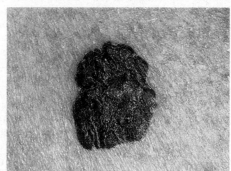

c. Melanoma

Figure 11A Skin cancer.
a. Basal cell carcinoma occurs when basal cells proliferate abnormally. **b.** Squamous cell carcinoma arises in epithelial cells derived from basal cells. **c.** Malignant melanoma is due to a proliferation of pigmented cells.

Bioethical Focus   Organ Transplants

Transplantation of the kidney, heart, liver, pancreas, lung, and other organs is now possible due to two major break-throughs. First, solutions have been developed that preserve donor organs for several hours. This made it possible for one young boy to undergo surgery for 16 hours, during which time he received five different organs. Second, rejection of transplanted organs is now prevented by immunosuppressive drugs; therefore, organs can be donated by unrelated individuals, living or dead. After death, it is possible to give the "gift of life" to someone else—over 25 organs and tissues from one cadaver can be used for transplants. The survival rate after a transplant operation is good. So many heart recipients are now

alive and healthy that they have formed basketball and softball teams, demonstrating the normalcy of their lives after surgery.

One problem persists, however, and that is the limited availability of organs for transplantation. At any one time, at least 27,000 people in the United States are waiting for a donated organ. Keen competition for organs can lead to various bioethical inequities. When the governor of Pennsylvania received a heart and lungs within a relatively short period of time, it appeared that his social status might have played a role. When Mickey Mantle received a liver transplant, people asked if it was right to give an organ to an older man who had a diseased

liver due to the consumption of alcohol. If a father gives a kidney to a child, he has to undergo a major surgical operation that leaves him vulnerable to possible serious consequences in the future. If organs are taken from those who have just died, who guarantees that the individual is indeed dead?

Decide Your Opinion

1. Is it ethical to ask a parent to donate an organ to his or her child? Why or why not?
2. Is it ethical to put a famous person at the top of the list for an organ transplant? Why or why not?
3. Is it ethical to remove organs from a newborn who is brain dead but whose organs are still functioning? Why or why not?

Summarizing the Concepts

11.1 Types of Tissues

Human tissues are categorized into four groups. Epithelial tissue covers the body and lines its cavities. The different types of epithelial tissue (squamous, cuboidal, and columnar) can be simple or stratified and have cilia or microvilli. Epithelial cells sometimes form glands that secrete either into ducts or into the blood.

Connective tissues, in which cells are separated by a matrix, often bind body parts together. Connective tissues have both white and yellow fibers and reticular fibers, and may also have fat (adipose) cells. Loose fibrous connective tissue supports epithelium and encloses organs. Dense fibrous connective tissue, such as that of tendons and ligaments, contains closely packed collagen fibers. Adipose tissue stores fat. Both cartilage and bone have cells within lacunae, but the matrix for cartilage is more flexible than that for bone, which contains calcium salts. Blood is a connective tissue in which the matrix is a liquid called plasma.

Muscular tissue is of three types. Both skeletal and cardiac muscle are striated; both cardiac and smooth muscle are involuntary. Skeletal muscle is found in muscles attached to bones, and smooth muscle is found in internal organs. Cardiac muscle makes up the heart.

Nervous tissue has one main type of conducting cell, the neuron, and several types of neuroglia. Each neuron has dendrites, a cell body, and an axon. Axons are specialized to conduct nerve impulses.

11.2 Body Cavities and Body Membranes

The internal organs occur within cavities. The thoracic cavity contains the heart and lungs; the abdominal cavity contains organs of the digestive, urinary, and reproductive systems, among others. Membranes line body cavities and the internal spaces of organs. Mucous membrane lines the tubes of the digestive system, while serous membrane lines the thoracic and abdominal cavities and covers the organs they contain.

11.3 Organ Systems

The digestive, cardiovascular, lymphatic, respiratory, and urinary systems perform processing and transporting functions that maintain the normal conditions of the body. The musculoskeletal system supports the body and permits movement. The nervous system receives sensory input from sensory receptors and directs the muscles and glands to respond to outside stimuli. The endocrine system produces hormones, some of which influence the functioning of the reproductive system, which allows humans to make more of their own kind.

Skin has two regions. The epidermis contains basal cells that produce new epithelial cells that become keratinized as they move toward the surface. The dermis, a largely fibrous connective tissue, contains epidermally derived glands and hair follicles, nerve endings, and blood vessels. Sensory receptors for touch, pressure, temperature, and pain are also present in the dermis. A subcutaneous layer, made up of loose connective tissue containing adipose cells, lies beneath the skin.

11.4 Homeostasis

Homeostasis is the relative constancy of the internal environment. Negative feedback mechanisms keep the environment relatively stable. When a sensor detects a change above or below a set point, a regulatory center activates an effector that reverses the change and brings conditions back to normal again. In contrast, a positive feedback mechanism brings about rapid change in the same direction as the stimulus. Still, positive feedback mechanisms are useful under certain conditions, such as when a child is born.

The internal environment consists of blood and tissue fluid. All organ systems contribute to the constancy of tissue fluid and blood. Special contributions are made by the liver, which keeps blood glucose constant, and the kidneys, which regulate the pH. The nervous and endocrine systems regulate the other systems.

Testing Yourself

Choose the best answer for each question.

1. Which of these is mismatched?
   a. epithelial tissue—protection and absorption
   b. muscular tissue—contraction and conduction
   c. connective tissue—binding and support
   d. nervous tissue—conduction and message sending
   e. nervous tissue—neuroglia and neurons

2. Which of these is NOT a type of epithelial tissue?
   a. simple cuboidal and stratified columnar
   b. bone and cartilage
   c. stratified squamous and simple squamous
   d. pseudostratified
   e. All of these are epithelial tissue.

3. Which tissue is more apt to line a lumen?
   a. epithelial tissue   d. muscular tissue
   b. connective tissue   e. epidermal tissue
   c. nervous tissue

4. The name of an epithelial tissue tells
   a. its location in the body.
   b. the number of cell layers.
   c. the shape of the cells.
   d. the tissue found beneath the epithelium.
   e. Both b and c are correct.

5. All glands in the body, such as salivary glands and the pancreas, are derived from which type of tissue?
   a. epithelial   c. muscle
   b. connective   d. nervous

6. A fetal skeleton is made of _____, which is later replaced by bone.
   a. elastic cartilage   c. spongy bone
   b. fibrocartilage   d. hyaline cartilage

7. Tendons and ligaments are
   a. connective tissue.   d. subject to injury.
   b. associated with the bones. e. All of these are correct.
   c. found in vertebrates.

8. Which tissue has cells in lacunae?
   a. epithelial tissue   d. bone
   b. fibrous connective   e. Both c and d are correct.
   c. cartilage

9. Cardiac muscle is
   a. striated.   d. voluntary.
   b. involuntary.   e. Both a and b are correct.
   c. smooth.

10. Which of these components of blood fights infection?
    a. red blood cells
    b. white blood cells
    c. platelets
    d. plasminogen
    e. All of these are correct.

11. Which of the following is a function of skin?
    a. temperature regulation
    b. manufacture of vitamin D
    c. collection of sensory input
    d. protection from invading pathogens
    e. All of these are correct.

12. An increase in body temperature would be followed by which of the following?
    a. constriction of dermal blood vessels
    b. activation of oil glands
    c. activation of sweat glands
    d. constriction of arrector pili muscles

13. Which of these body systems contribute to homeostasis?
    a. digestive and urinary systems
    b. respiratory and nervous systems
    c. nervous and endocrine systems
    d. All of these are correct.
    e. Body systems are not involved in homeostasis.

14. With negative feedback,
    a. the output cancels the input.
    b. there is a fluctuation above and below the average.
    c. there is self-regulation.
    d. sensory receptors communicate with a regulatory center.
    e. All of these are correct.

15. Which of the following is an example of negative feedback?
    a. Air conditioning goes off when room temperature lowers.
    b. Insulin decreases blood sugar levels after eating a meal.
    c. Heart rate increases when blood pressure drops.
    d. All of these are examples of negative feedback.

16. Which of these correctly describes a layer of the skin?
    a. The epidermis is simple squamous epithelium in which hair follicles develop and blood vessels expand when we are hot.
    b. The subcutaneous layer lies between the epidermis and the dermis. It contains adipose tissue, which keeps us warm.
    c. The dermis is a region of connective tissue that contains sensory receptors, nerve endings, and blood vessels.
    d. The skin has a special layer, still unnamed, in which there are all the accessory structures such as nails, hair, and various glands.

17. Which of the following is NOT a type of body membrane?
    a synovial membranes
    b. mucous membranes
    c. the meninges
    d. plasma membranes
    e. pleural membranes

18. The _____ separates the thoracic cavity from the abdominal cavity.
    a. liver   d. pleural membrane
    b. pancreas   e. intestines
    c. diaphragm

19. Give the name, the location, and the function for each of the tissues shown in the drawings below.
    a. Type of epithelial tissue
    b. Type of muscular tissue
    c. Type of connective tissue

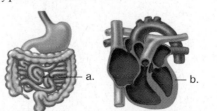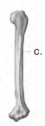

e-Learning Connection   www.mhhe.com/maderinquiry10

Concepts	Questions	Media Resources*
11.1 Types of Tissues		

• Animal tissues can be categorized into four major types: epithelial, connective, muscular, and nervous tissues.
• Epithelial tissues line body cavities and cover surfaces.
• Connective tissues protect, support, and bind other tissues.
• Muscular tissues make body parts move.
• Nervous tissues coordinate the activities of the other tissues and body parts.
	1. What is a basement membrane and what is its function?
	2. What is the difference between connective tissue fibers, muscle fibers, and nerve fibers?
		Essential Study Partner
		    Body Organization
		    Introduction to Tissues
		    Epithelial Tissue
		    Epithelial Glands
		    Connective Tissue
		    Nervous Tissue
		    Muscle Tissue
		
		General Biology Weblinks
		    Anatomy and Physiology

11.2 Body Cavities and Body Membranes		

• The internal organs occur within cavities lined by membranes that also cover the organs themselves.	1. What is the origin of the major ventral body cavities?
	2. What type of membrane lines joint cavities?
		Essential Study Partner
		    Membranes
		
		Labeling Exercise
		    Mammalian Body Cavities

11.3 Organ Systems		

• Organs are grouped into organ systems, each of which has specialized functions.
• The skin contains various tissues and has accessory organs. It is sometimes called the integumentary system.
	1. What organ systems are responsible for controlling most of the other organ systems of the body?
	2. Why is the skin considered an organ system?
		Essential Study Partner
		    Skin and Its Tissues
		    Accessory Organs of the Skin
		    Regulation of Body
		        Temperature
		    Aging
		
		Labeling Exercises
		    Human Skin Anatomy (Skin
		        Layers)
		    Human Skin Anatomy (2)
		    Human Skin Anatomy (3)

11.4 Homeostasis		
• Humans have a marked ability to maintain a relatively stable internal environment. All organ systems contribute to homeostasis.	1. How does a negative feedback system work? 2. Are there body mechanisms that function by positive feedback?	Essential Study Partner Homeostasis

*For additional Media Resources, see the Online Learning Center.

12

Digestive System and Nutrition

Chapter Concepts

Eating is one of life's pleasures. Digestion of food is essential because nutrient molecules provide us with the energy and building blocks we need to survive.

*E*njoying the summer night at an outdoor cafe, Sam washes down his last piece of pizza with a sip of wine. Even before Sam swallows his food, his mouth's saliva begins to break its carbohydrate molecules apart. The wine's alcohol is absorbed in the stomach, where the process of transforming Sam's meal into a nutrient-laden liquid begins. In the small intestine, wormlike projections from the intestinal wall absorb sugars, amino acids, and other needed molecules into Sam's bloodstream. Even the large intestine contributes by taking in needed water and salts. His body now refueled, Sam heads off for a night of dancing.

In this chapter, you will learn how the body digests food, and the importance of proper nutrition. Science is beginning to find the cellular basis for believing that fruits and vegetables, and yes, especially broccoli, can ensure a brighter and healthier future. Sam can play his part by being aware of these findings. Sugars and fats should be avoided and protein consumption should be moderate in order to maintain a normal weight and avoid certain illnesses. The contents of this chapter will be of interest to everyone.

12.1 The Digestive Tract

Digestion takes place within a tube called the digestive tract, which begins with the mouth and ends with the anus (Fig. 12.1). The functions of the digestive system are to ingest food, digest it to nutrients that can cross plasma membranes, absorb nutrients, and eliminate indigestible remains.

Digestion involves two main processes that occur simultaneously. During mechanical digestion, large pieces of food become smaller pieces, readying them for chemical digestion. Mechanical digestion begins with the chewing of the food in the mouth and continues with the churning and mixing of food that occurs in the stomach. Parts of the digestive tract produce digestive enzymes. During chemical digestion, many different enzymes break down macromolecules to small organic molecules that can be absorbed. Each enzyme has a particular job to do.

The Mouth

The mouth, which receives food, is bounded externally by the lips and cheeks. The lips extend from the base of the nose to the start of the chin. The red portion of the lips is poorly keratinized, and this allows blood to show through.

Most people enjoy eating food largely because they like its texture and taste. Sensory receptors called taste buds occur primarily on the tongue, and when these are activated by the presence of food, nerve impulses travel by way of cranial nerves to the brain. The tongue is composed of skeletal muscle whose contraction changes the shape of the tongue. Muscles exterior to the tongue cause it to move about. A fold of mucous membrane on the underside of the tongue attaches it to the floor of the oral cavity.

The roof of the mouth separates the nasal cavities from the oral cavity. The roof has two parts: an anterior (toward the front) **hard palate** and a posterior (toward the back) **soft palate** (Fig. 12.2*a*). The hard palate contains several bones, but the soft palate is composed entirely of muscle. The soft palate ends in a finger-shaped projection called the uvula. The tonsils are in the back of the mouth, on either side of the tongue and in the nasopharynx (called adenoids). The tonsils help protect the body against infections. If the tonsils become inflamed, the person has **tonsillitis.** The infection can spread to the middle ears. If tonsillitis recurs repeatedly, the tonsils may be surgically removed (called a tonsillectomy).

Three pairs of **salivary glands** send juices (saliva) by way of ducts to the mouth. One pair of salivary glands lies at the sides of the face immediately below and in front of the ears. These glands swell when a person has the mumps, a disease caused by a viral infection. Salivary glands have ducts that open on the inner surface of the cheek at the location of the second upper molar. Another

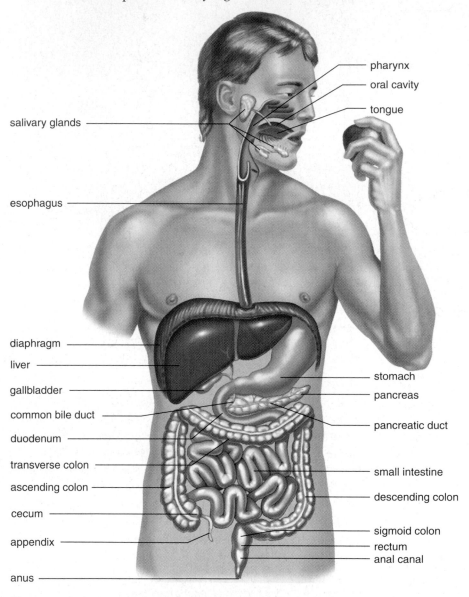

Figure 12.1 Digestive system.
Trace the path of food from the mouth to the anus. The large intestine consists of the cecum, the colon (consisting of the ascending, transverse, descending, and sigmoid colon), and the rectum and anal canal. Note also the location of the accessory organs of digestion: the pancreas, the liver, and the gallbladder.

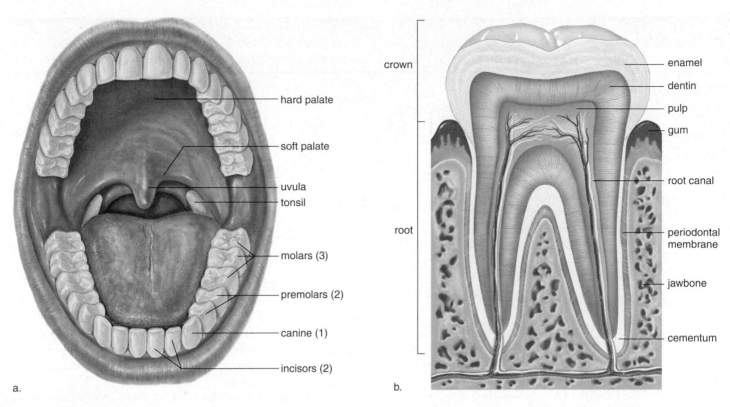

a.

b.

crown

root

hard palate

soft palate

uvula

tonsil

molars (3)

premolars (2)

canine (1)

incisors (2)

enamel

dentin

pulp

gum

root canal

periodontal membrane

jawbone

cementum

Figure 12.2   Adult mouth and teeth.
a. The chisel-shaped incisors bite; the pointed canines tear; the fairly flat premolars grind; and the flattened molars crush food. The last molar, called a wisdom tooth, may fail to erupt, or if it does, it is sometimes crooked and useless. Often dentists recommend the extraction of the wisdom teeth. **b.** Longitudinal section of a tooth. The crown is the portion that projects above the gum line and can be replaced by a dentist if damaged. When a "root canal" is done, the nerves are removed. When the periodontal membrane is inflamed, the teeth can loosen.

pair of salivary glands lies beneath the tongue, and still another pair lies beneath the floor of the oral cavity. The ducts from these salivary glands open under the tongue. You can locate the openings if you use your tongue to feel for small flaps on the inside of your cheek and under your tongue. Saliva contains an enzyme called **salivary amylase** that begins the process of digesting starch.

The Teeth
With our teeth, we chew food into pieces convenient for swallowing. During the first two years of life, the smaller 20 deciduous, or baby, teeth appear. These are eventually replaced by 32 adult teeth (Fig. 12.2a). The third pair of molars, called the wisdom teeth, sometimes fail to erupt. If they push on the other teeth and/or cause pain, they can be removed by a dentist or oral surgeon.

Each tooth has two main divisions, a crown and a root (Fig. 12.2b). The crown has a layer of enamel, an extremely hard outer covering of calcium compounds; dentin, a thick layer of bonelike material; and an inner pulp, which contains the nerves and the blood vessels. Dentin and pulp are also found in the root.

Tooth decay, called **dental caries,** or cavities, occurs when bacteria within the mouth metabolize sugar and give off acids, which erode teeth. Two measures can prevent tooth decay: eating a limited amount of sweets and daily brushing and flossing of teeth. Fluoride treatments, particularly in children, can make the enamel stronger and more resistant to decay. Gum disease is more apt to occur with aging. Inflammation of the gums (gingivitis) can spread to the periodontal membrane, which lines the tooth socket. A person then has periodontitis, characterized by a loss of bone and loosening of the teeth so that extensive dental work may be required. Stimulation of the gums in a manner advised by your dentist is helpful in controlling this condition. Medications are also available.

The tongue mixes the chewed food with saliva. It then forms this mixture into a mass called a bolus in preparation for swallowing.

The salivary glands send saliva into the mouth, where the teeth chew the food and the tongue forms it into a bolus for swallowing.

Table 12.1	Path of Food		
Organ	**Function of Organ**	**Special Feature(s)**	**Function of Special Feature(s)**
Oral cavity	Receives food; starts digestion of starch	Teeth Tongue	Chew food Forms bolus
Pharynx	Passageway	_____	_____
Esophagus	Passageway	_____	_____
Stomach	Storage of food; acidity kills bacteria; starts digestion of protein	Gastric glands	Release gastric juices
Small intestine	Digestion of all foods; absorption of nutrients	Intestinal glands Villi	Release fluids Absorb nutrients
Large intestine	Absorption of water; storage of indigestible remains	_____	_____

The Pharynx

The **pharynx** is a region that receives air from the nasal cavities and food from the mouth. The palate, which forms the roof of the mouth, consists of the hard palate anteriorly and the soft palate posteriorly. The soft palate has a projection called the uvula which people often confuse with the tonsils. The tonsils, however, are embedded in the mucous membrane of the pharynx.

Table 12.1 traces the path of food. From the oral cavity of the mouth, food passes through the pharynx and esophagus to the stomach, small intestine and large intestine. The food passage and air passage cross in the pharynx because the trachea (windpipe) is ventral to (in front of) the esophagus, a long muscular tube that takes food to the stomach. Swallowing, a process that occurs in the pharynx (Fig. 12.3), is a **reflex action** performed automatically, without conscious thought. Usually during swallowing, the soft palate moves back to close off the **nasopharynx,** and the trachea moves up under the **epiglottis** to cover the glottis. The **glottis** is the opening to the larynx (voice box) and therefore the air passage. During swallowing, food normally enters the esophagus because the air passages are blocked. We do not breathe when we swallow.

Unfortunately, we have all had the unpleasant experience of having food "go the wrong way." The wrong way may be either into the nasal cavities or into the trachea. If it is the latter, coughing will most likely force the food up out of the trachea and into the pharynx again. The up-and-down movement of the Adam's apple, the front part of the larynx, is easy to observe when a person swallows. Thus, we do not breathe when we swallow.

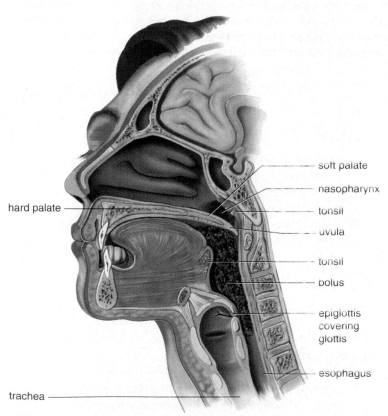

hard palate

soft palate
nasopharynx
tonsil
uvula
tonsil
bolus
epiglottis covering glottis
esophagus

trachea

Figure 12.3   Swallowing.
When food is swallowed, the soft palate closes off the nasopharynx, and the epiglottis covers the glottis, forcing the bolus to pass down the esophagus. Therefore, a person does not breathe while swallowing.

The Esophagus

The **esophagus** is a muscular tube that passes from the pharynx through the thoracic cavity and diaphragm into the abdominal cavity, where it joins the stomach. The esophagus is ordinarily collapsed, but it opens and receives the bolus when swallowing occurs.

A rhythmic contraction called **peristalsis** pushes the food along the digestive tract. Peristalsis begins in the esophagus and continues in all the organs of the digestive tract. Occasionally, peristalsis begins even though there is no food in the esophagus. This produces the sensation of a lump in the throat.

The esophagus plays no role in the chemical digestion of food. Its sole purpose is to conduct the food bolus from the mouth to the stomach. **Sphincters** are muscles that encircle tubes and act as valves; tubes close when sphincters contract, and they open when sphincters relax. The entrance of the esophagus to the stomach is marked by a constriction, often called a sphincter, although the muscle is not as developed as in a true sphincter. Relaxation of the sphincter allows the bolus to pass into the stomach, while contraction prevents the acidic contents of the stomach from backing up into the esophagus.

Heartburn, which feels like a burning pain rising up into the throat, occurs when some of the stomach contents escape into the esophagus. When vomiting occurs, a contraction of the abdominal muscles and diaphragm propels the contents of the stomach upward through the esophagus.

The air passage and food passage cross in the pharynx, which takes food to the esophagus. The esophagus conducts the bolus of food from the pharynx to the stomach. Peristalsis begins in the esophagus and occurs along the entire length of the digestive tract.

The Wall of the Digestive Tract

The wall of the esophagus in the abdominal cavity is comparable to that of the digestive tract, which has these layers (Fig. 12.4):

Mucosa (mucous membrane layer) A layer of epithelium supported by connective tissue and smooth muscle lines the lumen (central cavity) and contains glandular epithelial cells that secrete digestive enzymes and goblet cells that secrete mucus.

Submucosa (submucosal layer) A broad band of loose connective tissue that contains blood vessels lies beneath the mucosa. Lymph nodules, called Peyer's patches, are in the submucosa. Like the tonsils, they help protect us from disease.

Muscularis (smooth muscle layer) Two layers of smooth muscle make up this section. The inner, circular layer encircles the gut; the outer, longitudinal layer lies in the same direction as the gut. (The stomach also has oblique muscles.)

Serosa (serous membrane layer) Most of the digestive tract has a serosa, a very thin, outermost layer of squamous epithelium supported by connective tissue. The serosa secretes a serous fluid that keeps the outer surface of the intestines moist so that the organs of the abdominal cavity slide against one another. The esophagus has an outer layer composed only of loose connective tissue called the adventitia.

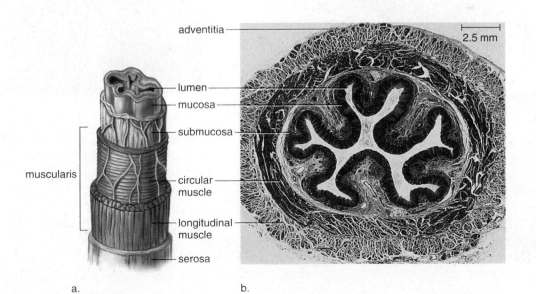

Figure 12.4   Wall of the digestive tract.
a. Several different types of tissues are found in the wall of the digestive tract. Note the placement of circular muscle inside longitudinal muscle.
b. Micrograph of the wall of the esophagus.

The Stomach

The **stomach** (Fig. 12.5) is a thick-walled, J-shaped organ that lies on the left side of the body beneath the diaphragm. The stomach is continuous with the esophagus above and the duodenum of the small intestine below. The stomach stores food and aids in digestion. The wall of the stomach has deep folds, which disappear as the stomach fills to an approximate capacity of one liter. Its muscular wall churns, mixing the food with gastric juice. The term *gastric* always refers to the stomach.

The columnar epithelial lining of the stomach (i.e., the mucosa) has millions of gastric pits, which lead into **gastric glands.** The gastric glands produce gastric juice. Gastric juice contains an enzyme called **pepsin,** which digests protein, plus hydrochloric acid (HCl) and mucus. HCl causes the stomach to have a high acidity with a pH of about 2, and this is beneficial because it kills most bacteria present in food. Although HCl does not digest food, it does break down the connective tissue of meat and activate pepsin. The wall of the stomach is protected by a thick layer of mucus secreted by goblet cells in its lining. If, by chance, HCl penetrates this mucus, the wall can begin to break down, and an ulcer results. An **ulcer** is an open sore in the wall caused by the gradual disintegration of tissue. It now appears that most ulcers are due to a bacterial (*Helicobacter pylori*) infection that impairs the ability of epithelial cells to produce protective mucus.

Alcohol is absorbed in the stomach, but food substances are not. Normally, the stomach empties in about 2–6 hours. When food leaves the stomach, it is a thick, soupy liquid called **chyme.** Chyme enters the small intestine in squirts by way of a sphincter that repeatedly opens and closes.

> The stomach can expand to accommodate large amounts of food. When food is present, the stomach churns, mixing food with acidic gastric juice.

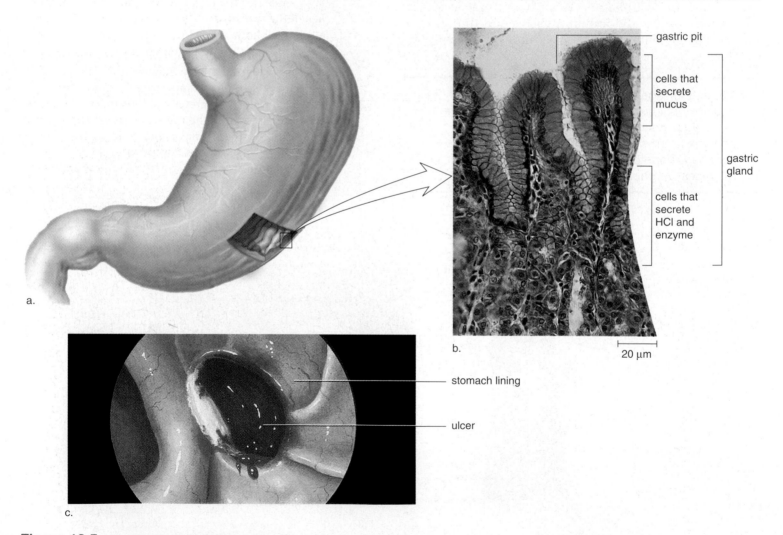

Figure 12.5   Anatomy and histology of the stomach.
a. The stomach has a thick wall with folds that allow it to expand and fill with food. **b.** The mucosa contains gastric glands, which secrete mucus and a gastric juice active in protein digestion. **c.** A bleeding ulcer viewed through an endoscope (a tubular instrument bearing a tiny lens and a light source) inserted into the abdominal cavity.

The Small Intestine

The **small intestine** is named for its small diameter (compared to that of the large intestine), but perhaps it should be called the long intestine. The small intestine averages about 6 meters (18 ft) in length, compared to the large intestine, which is about 1.5 meters (4½ ft) in length.

The first 25 cm of the small intestine is called the **duodenum.** Ducts from the liver and pancreas join to form one duct that enters the duodenum (see Fig. 12.1). The small intestine receives bile from the liver and pancreatic juice from the pancreas via this duct. **Bile** emulsifies fat—emulsification causes fat droplets to disperse in water. The intestine has a slightly basic pH because pancreatic juice contains sodium bicarbonate ($NaHCO_3$), which neutralizes chyme. The enzymes in pancreatic juice and enzymes produced by the intestinal wall complete the process of food digestion.

It has been suggested that the surface area of the small intestine is approximately that of a tennis court. What factors contribute to increasing its surface area? The wall of the small intestine contains fingerlike projections called villi (sing. **villus**), which give the intestinal wall a soft, velvety appearance (Fig. 12.6). A villus has an outer layer of columnar epithelial cells, and each of these cells has thousands of microscopic extensions called microvilli. Collectively, in electron micrographs, microvilli give the villi a fuzzy border known as a "brush border." Since the microvilli bear the intestinal enzymes, these enzymes are called brush-border enzymes. The microvilli greatly increase the surface area of the villus for the absorption of nutrients.

Nutrients are absorbed into the vessels of a villus. A villus contains blood capillaries and a small lymphatic capillary, called a **lacteal.** The lymphatic system is an adjunct to the cardiovascular system (its vessels carry a fluid called lymph to the cardiovascular veins.) Sugars and amino acids enter the blood capillaries of a villus. Glycerol and fatty acids (digested from fats) enter the epithelial cells of the villi, and within these cells are joined and packaged as lipoprotein droplets, which enter a lacteal. After nutrients are absorbed, they are eventually carried to all the cells of the body by the bloodstream.

The large surface area of the small intestine facilitates absorption of nutrients into the cardiovascular system (glucose and amino acids) and the lymphatic system (fats).

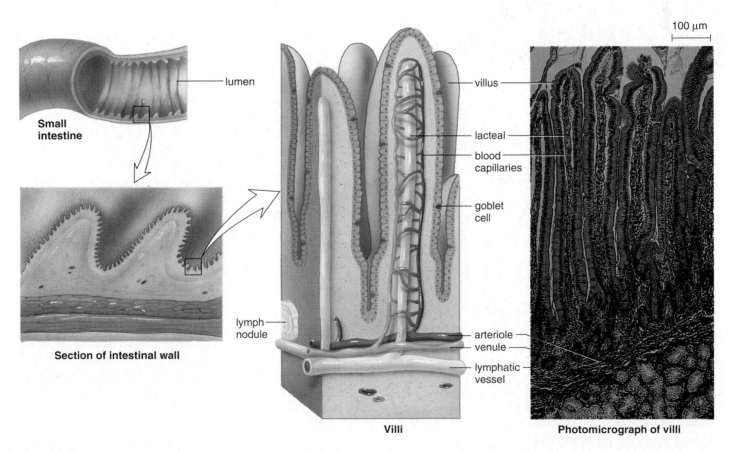

Small intestine

lumen

Section of intestinal wall

lymph nodule

villus

lacteal

blood capillaries

goblet cell

arteriole
venule

lymphatic vessel

100 µm

Villi

Photomicrograph of villi

Figure 12.6   Anatomy of the small intestine.
The wall of the small intestine has folds that bear fingerlike projections called villi. The products of digestion are absorbed into the blood capillaries and the lacteals of the villi.

Regulation of Digestive Secretions

The secretion of digestive juices is promoted by the nervous system and by hormones. A **hormone** is a substance produced by one set of cells that affects a different set of cells, the so-called target cells. Hormones are usually transported by the bloodstream. For example, when a person has eaten a meal particularly rich in protein, the stomach produces the hormone gastrin. Gastrin enters the bloodstream, and soon the stomach is churning, and the secretory activity of gastric glands is increasing. A hormone produced by the duodenal wall, GIP (gastric inhibitory peptide), works opposite to gastrin: it inhibits gastric gland secretion.

Cells of the duodenal wall produce two other hormones that are of particular interest—secretin and CCK (cholecystokinin). Acid, especially hydrochloric acid (HCl) present in chyme, stimulates the release of secretin, while partially digested protein and fat stimulate the release of CCK. Soon after these hormones enter the bloodstream, the pancreas increases its output of pancreatic juice, which helps digest food, and the liver increases its output of bile. The gallbladder contracts to release bile. Figure 12.7 summarizes the actions of gastrin, secretin, and CCK.

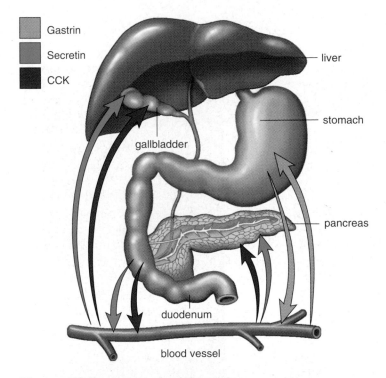

Figure 12.7   Hormonal control of digestive gland secretions.
Gastrin (blue), produced by the lower part of the stomach, enters the bloodstream and thereafter stimulates the upper part of the stomach to produce more digestive juice. Secretin (green) and CCK (purple), produced by the duodenal wall, stimulate the pancreas to secrete its digestive juice and the gallbladder to release bile.

The Large Intestine

The **large intestine,** which includes the cecum, the colon, the rectum, and the anal canal, is larger in diameter than the small intestine (6.5 cm compared to 2.5 cm), but it is shorter in length (see Fig. 12.1). The large intestine absorbs water, salts, and some vitamins. It also stores indigestible material until it is eliminated at the anus.

The **cecum,** which lies below the junction with the small intestine, is the blind end of the large intestine. The cecum has a small projection called the vermiform **appendix** (*vermiform* means wormlike) (Fig. 12.8). In humans, the appendix also may play a role in fighting infections. This organ is subject to inflammation, a condition called appendicitis. If inflamed, the appendix should be removed before the fluid content rises to the point that the appendix bursts, a situation that may cause **peritonitis,** a generalized infection of the lining of the abdominal cavity. Peritonitis can lead to death.

The **colon** includes the ascending colon, which goes up the right side of the body to the level of the liver; the transverse colon, which crosses the abdominal cavity just below the liver and the stomach; the descending colon, which passes down the left side of the body; and the sigmoid colon, which enters the **rectum,** the last 20 cm of the large intestine. The rectum opens at the **anus,** where **defecation,** the expulsion of feces, occurs. When feces are forced into the rectum by peristalsis, a defecation reflex occurs. The stretching of the rectal wall initiates nerve impulses to the spinal cord, and shortly thereafter the rectal muscles contract and the anal sphincters relax (Fig. 12.9). Ridding the body of indigestible remains is another way the digestive system helps maintain homeostasis. Feces are three-quarters water and one-quarter solids. Bacteria, **fiber** (indigestible remains), and other indigestible materials are in the solid portion. Bacterial action on indigestible materials causes the odor of feces and also accounts for the presence of gas. A breakdown product of bilirubin (see page 222) and the presence of oxidized iron causes the brown color of feces.

For many years, it was believed that facultative bacteria (bacteria that can live with or without oxygen), such as *Escherichia coli,* were the major inhabitants of the colon, but new culture methods show that over 99% of the colon bacteria are obligate anaerobes (bacteria that die in the presence of oxygen). Not only do the bacteria break down indigestible material, but they also produce some vitamins and other molecules that can be absorbed and used by our bodies. In this way, they perform a service for us.

Water is considered unsafe for swimming when the coliform (nonpathogenic intestinal) bacterial count reaches a certain number. A high count indicates that a significant amount of feces has entered the water. The more feces present, the greater the possibility that disease-causing bacteria are also present.

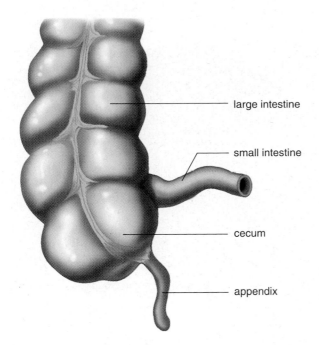

Figure 12.8   Junction of the small intestine and the large intestine.
The cecum is the blind end of the ascending colon. The appendix is attached to the cecum.

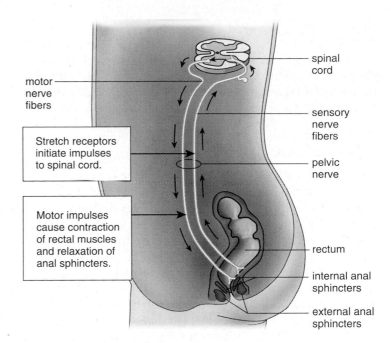

Figure 12.9   Defecation reflex.
The accumulation of feces in the rectum causes it to stretch, which initiates a reflex action resulting in rectal contraction and expulsion of the fecal material.

Polyps

The colon is subject to the development of **polyps,** small growths arising from the epithelial lining. Polyps, whether benign or cancerous, can be removed surgically. If colon cancer is detected while still confined to a polyp, the expected outcome is a complete cure. Some investigators believe that dietary fat increases the likelihood of colon cancer because dietary fat causes an increase in bile secretion. It could be that intestinal bacteria convert bile salts to substances that promote the development of cancer. On the other hand, fiber in the diet seems to inhibit the development of colon cancer. Dietary fiber absorbs water and adds bulk, thereby diluting the concentration of bile salts and facilitating the movement of substances through the intestine. Regular elimination reduces the time that the colon wall is exposed to any cancer-promoting agents in feces.

Diarrhea and Constipation

Two common everyday complaints associated with the large intestine are **diarrhea** and **constipation.** The major causes of diarrhea are infection of the lower intestinal tract and nervous stimulation. In the case of infection, such as food poisoning caused by eating contaminated food, the intestinal wall becomes irritated, and peristalsis increases. Water is not absorbed, and the diarrhea that results rids the body of the infectious organisms. In nervous diarrhea, the nervous system stimulates the intestinal wall, and diarrhea results.

Prolonged diarrhea can lead to dehydration because of water loss and to disturbances in the heart's contraction due to an imbalance of salts in the blood.

When a person is constipated, the feces are dry and hard. One reason for this condition is that socialized persons have learned to inhibit defecation to the point that the urge to defecate is ignored. Two components of the diet that can help prevent constipation are water and fiber. Water intake prevents drying out of the feces, and fiber provides the bulk needed for elimination. The frequent use of laxatives is discouraged. If, however, it is necessary to take a laxative, a bulk laxative is the most natural because, like fiber, it produces a soft mass of cellulose in the colon. Lubricants, such as mineral oil, make the colon slippery; saline laxatives, such as milk of magnesia, act osmotically—they prevent water from being absorbed and, depending on the dosage, may even cause water to enter the colon. Some laxatives are irritants, meaning that they increase peristalsis to the degree that the contents of the colon are expelled.

Chronic constipation is associated with the development of hemorrhoids, enlarged and inflamed blood vessels at the anus.

The large intestine does not produce digestive enzymes; it does absorb water, salts, and some vitamins.

12.2 Three Accessory Organs

The pancreas, liver, and gallbladder are accessory digestive organs. Figure 12.1 shows how the pancreatic duct from the pancreas and the common bile duct from the liver and gallbladder join before entering the duodenum.

The Pancreas

The **pancreas** lies deep in the abdominal cavity, resting on the posterior abdominal wall. It is an elongated and somewhat flattened organ that has both an endocrine and an exocrine function. As an endocrine gland, it secretes insulin and glucagon, hormones that help keep the blood glucose level within normal limits. In this chapter, however, we are interested in its exocrine function. Most pancreatic cells produce pancreatic juice, which contains sodium bicarbonate (NaHCO₃) and digestive enzymes for all types of food. Sodium bicarbonate neutralizes chyme; whereas pepsin acts best in an acid pH of the stomach, pancreatic enzymes require a slightly basic pH. **Pancreatic amylase** digests starch, **trypsin** digests protein, and **lipase** digests fat. In cystic fibrosis, a thick mucus blocks the pancreatic duct, and the patient must take supplemental pancreatic enzymes by mouth for proper digestion to occur.

The Liver

The **liver,** which is the largest gland in the body, lies mainly in the upper right section of the abdominal cavity, under the diaphragm (see Fig. 12.1). The liver has two main lobes, the right lobe and the smaller left lobe, which crosses the midline and lies above the stomach. The liver contains approximately 100,000 lobules that serve as its structural and functional units (Fig. 12.10). Triads consisting of these three structures are located between the lobules: a bile duct that takes bile away from the liver; a branch of the hepatic artery that brings O₂-rich blood to the liver; and a branch of the hepatic portal vein that transports nutrients from the intestines. The central veins of lobules enter a hepatic vein. In Figure 12.11, trace the path of blood from the intestines to the liver via the hepatic portal vein and from the liver to the inferior vena cava via the hepatic veins.

In some ways, the liver acts as the gatekeeper to the blood. As the blood from the hepatic portal vein passes through the liver, it removes poisonous substances and detoxifies them. The liver also removes nutrients and works to keep the contents of the blood constant. It removes and stores iron and the fat-soluble vitamins A, D, E, K, and B₁₂ per the list on the next page. The liver makes the plasma proteins from amino acids, and helps regulate the quantity of cholesterol in the blood.

The liver maintains the blood glucose level at about 100 mg/100 ml (0.1%), even though a person eats intermit-

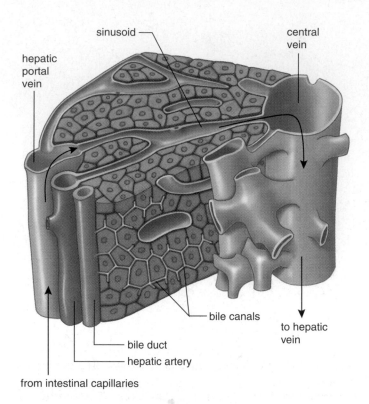

Figure 12.10 Hepatic lobules.
The liver contains over 100,000 lobules. Each lobule contains many cells that perform the various functions of the liver. They remove from and/or add materials to blood and deposit bile in bile ducts.

tently. When insulin is present, any excess glucose present in blood is removed and stored by the liver as glycogen. Between meals, glycogen is broken down to glucose, which enters the hepatic veins, and in this way, the blood glucose level remains constant.

If the supply of glycogen is depleted, the liver converts glycerol (from fats) and amino acids to glucose molecules. The conversion of amino acids to glucose necessitates deamination, the removal of amino groups. By a complex metabolic pathway, the liver then combines ammonia with carbon dioxide to form urea:

$$2\,NH_3 + CO_2 \longrightarrow H_2N - \overset{\displaystyle O}{\overset{\displaystyle \|}{C}} - NH_2$$

ammonia   carbon dioxide   urea

Urea is the usual nitrogenous waste product from amino acid breakdown in humans. After its formation in the liver, urea is excreted by the kidneys.

The liver produces bile, which is stored in the gallbladder. Bile has a yellowish-green color because it contains the bile pigment bilirubin, derived from the breakdown of hemoglobin, the red pigment of red blood

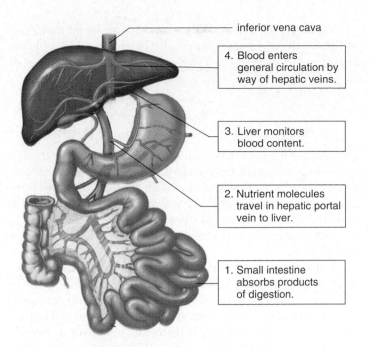

inferior vena cava

4. Blood enters general circulation by way of hepatic veins.

3. Liver monitors blood content.

2. Nutrient molecules travel in hepatic portal vein to liver.

1. Small intestine absorbs products of digestion.

Figure 12.11   Hepatic portal system.
The hepatic portal vein takes the products of digestion from the digestive system to the liver, where they are processed before entering a hepatic vein.

cells. Bile also contains bile salts. Bile salts are derived from cholesterol, and they emulsify fat in the small intestine. When fat is emulsified, it breaks up into droplets, providing a much larger surface area, which can be acted upon by a digestive enzyme from the pancreas.

Altogether, the following are significant ways in which the liver helps maintain homeostasis:

1. Detoxifies blood by removing and metabolizing poisonous substances.
2. Stores iron (Fe^{2+}) and the fat-soluble vitamins A, D, E, and K.
3. Makes plasma proteins, such as albumins and fibrinogen, from amino acids.
4. Stores glucose as glycogen after a meal, and breaks down glycogen to glucose to maintain the glucose concentration of blood between eating periods.
5. Produces urea after breaking down amino acids.
6. Removes bilirubin, a breakdown product of hemoglobin from the blood, and excretes it in bile, a liver product.
7. Helps regulate blood cholesterol level, converting some to bile salts.

Liver Disorders

Hepatitis and cirrhosis are two serious diseases that affect the entire liver and hinder its ability to repair itself. Therefore, they are life-threatening diseases. When a person has a liver

ailment, jaundice may occur. **Jaundice** is a yellowish tint to the whites of the eyes and also to the skin of light-pigmented persons. Bilirubin is deposited in the skin due to an abnormally large amount in the blood. In hemolytic jaundice, red blood cells have been broken down in abnormally large amounts; in obstructive jaundice, bile ducts are blocked or liver cells are damaged.

Jaundice can also result from **hepatitis,** inflammation of the liver. Viral hepatitis occurs in several forms. Hepatitis A is usually acquired from sewage-contaminated drinking water. Hepatitis B, which is usually spread by sexual contact, can also be spread by blood transfusions or contaminated needles. The hepatitis B virus is more contagious than the AIDS virus, which is spread in the same way. Thankfully, however, a vaccine is now available for hepatitis B. Hepatitis C, which is usually acquired by contact with infected blood and for which there is no vaccine, can lead to chronic hepatitis, liver cancer, and death.

Cirrhosis is another chronic disease of the liver. First the organ becomes fatty, and then liver tissue is replaced by inactive fibrous scar tissue. Cirrhosis of the liver is often seen in alcoholics due to malnutrition and to the excessive amounts of alcohol (a toxin) the liver is forced to break down.

The liver has amazing regenerative powers and can recover if the rate of regeneration exceeds the rate of damage. During liver failure, however, there may not be enough time to let the liver heal itself. Liver transplantation is usually the preferred treatment for liver failure, but artificial livers have been developed and tried in a few cases. One type is a cartridge that contains liver cells. The patient's blood passes through the cellulose acetate tubing of the cartridge and is serviced in the same manner as with a normal liver. In the meantime, the patient's liver has a chance to recover.

The Gallbladder

The **gallbladder** is a pear-shaped, muscular sac attached to the surface of the liver (see Fig. 12.1). About 1,000 ml of bile are produced by the liver each day, and any excess is stored in the gallbladder. Water is reabsorbed by the gallbladder so that bile becomes a thick, mucuslike material. When needed, bile leaves the gallbladder and proceeds to the duodenum via the common bile duct.

The cholesterol content of bile can come out of solution and form crystals. If the crystals grow in size, they form gallstones. The passage of the stones from the gallbladder may block the common bile duct and cause obstructive jaundice. Then the gallbladder must be removed.

The pancreas produces pancreatic juice, which contains enzymes for the digestion of food. Among the liver's many functions is the production of bile, which is stored in the gallbladder.

12.3 Digestive Enzymes

The digestive enzymes are **hydrolytic enzymes,** which break down substances by the introduction of water at specific bonds. Digestive enzymes, like other enzymes, are proteins with a particular shape that fits their substrate. They also have an optimum pH, which maintains their shape, thereby enabling them to speed up their specific reaction.

The various digestive enzymes present in the digestive juices, mentioned previously, help break down carbohydrates, proteins, nucleic acids, and fats, the major components of food. Starch is a carbohydrate, and its digestion begins in the mouth. Saliva from the salivary glands has a neutral pH and contains **salivary amylase,** the first enzyme to act on starch:

$$\text{starch} + H_2O \xrightarrow{\text{salivary amylase}} \text{maltose}$$

In this equation, salivary amylase is written above the arrow to indicate that it is neither a reactant nor a product in the reaction. It merely speeds the reaction in which its substrate, starch, is digested to many molecules of maltose, a disaccharide. Maltose molecules cannot be absorbed by the intestine; additional digestive action in the small intestine converts maltose to glucose, which can be absorbed.

Protein digestion begins in the stomach. Gastric juice secreted by gastric glands has a very low pH—about 2—because it contains hydrochloric acid (HCl). Pepsinogen, a precursor that is converted to the enzyme **pepsin** when exposed to HCl, is also present in gastric juice. Pepsin acts on protein to produce peptides:

$$\text{protein} + H_2O \xrightarrow{\text{pepsin}} \text{peptides}$$

Peptides vary in length, but they always consist of a number of linked amino acids. Peptides are usually too large to be absorbed by the intestinal lining, but later they are broken down to amino acids in the small intestine.

Starch, proteins, nucleic acids, and fats are all enzymatically broken down in the small intestine. Pancreatic juice, which enters the duodenum, has a basic pH because it contains sodium bicarbonate ($NaHCO_3$). Sodium bicarbonate neutralizes chyme, producing the slightly basic pH that is optimum for pancreatic enzymes. One pancreatic enzyme, **pancreatic amylase,** digests starch:

$$\text{starch} + H_2O \xrightarrow{\text{pancreatic amylase}} \text{maltose}$$

Another pancreatic enzyme, **trypsin,** digests protein:

$$\text{protein} + H_2O \xrightarrow{\text{trypsin}} \text{peptides}$$

Trypsin is secreted as trypsinogen, which is converted to trypsin in the duodenum.

Lipase, a third pancreatic enzyme, digests fat molecules in the fat droplets after they have been emulsified by bile salts:

$$\text{fat} \xrightarrow{\text{bile salts}} \text{fat droplets}$$

$$\text{fat droplets} + H_2O \xrightarrow{\text{lipase}} \text{glycerol} + \text{fatty acids}$$

The end products of lipase digestion, glycerol and fatty acid molecules, are small enough to cross the cells of the intestinal villi, where absorption takes place. As mentioned previously, glycerol and fatty acids enter the cells of the villi, and within these cells, they are rejoined and packaged as lipoprotein droplets before entering the lacteals (see Fig. 12.6).

Peptidases and **maltase,** enzymes produced by the small intestine, complete the digestion of protein to amino acids and starch to glucose, respectively. Amino acids and glucose are small molecules that cross into the cells of the villi. Peptides, which result from the first step in protein digestion, are digested to amino acids by peptidases:

$$\text{peptides} + H_2O \xrightarrow{\text{peptidases}} \text{amino acids}$$

Maltose, a disaccharide that results from the first step in starch digestion, is digested to glucose by maltase:

$$\text{maltose} + H_2O \xrightarrow{\text{maltase}} \text{glucose} + \text{glucose}$$

Other disaccharides, each of which has its own enzyme, are digested in the small intestine. The absence of any one of these enzymes can cause illness. For example, many people, including as many as 75% of African Americans, cannot digest lactose, the sugar found in milk, because they do not produce lactase, the enzyme that converts lactose to its components, glucose and galactose. Drinking untreated milk often gives these individuals the symptoms of **lactose intolerance** (diarrhea, gas, cramps), caused by a large quantity of nondigested lactose in the intestine. In most areas, it is possible to purchase milk made lactose-free by the addition of synthetic lactase or *Lactobacillus acidophilus* bacteria, which break down lactose.

Table 12.2 lists some of the major digestive enzymes produced by the digestive tract, salivary glands, or the pancreas. Each type of food is broken down by specific enzymes.

Digestive enzymes present in digestive juices help break down food to the nutrient molecules: glucose, amino acids, fatty acids, and glycerol. The first two are absorbed into the blood capillaries of the villi, and the last two re-form within epithelial cells before entering the lacteals as lipoprotein droplets.

Table 12.2	Major Digestive Enzymes			
Enzyme	**Produced By**	**Site of Action**	**Optimum pH**	**Digestion**
Salivary amylase	Salivary glands	Mouth	Neutral	Starch + H_2O → maltose
Pancreatic amylase	Pancreas	Small intestine	Basic	
Maltase	Small intestine	Small intestine	Basic	Maltose + H_2O → glucose + glucose
Pepsin	Gastric glands	Stomach	Acidic	Protein + H_2O → peptides
Trypsin	Pancreas	Small intestine	Basic	
Peptidases	Small intestine	Small intestine	Basic	Peptide + H_2O → amino acids
Nuclease	Pancreas	Small intestine	Basic	RNA and DNA + H_2O → nucleotides
Nucleosidases	Small intestine	Small intestine	Basic	Nucleotide + H_2O → base + sugar + phosphate
Lipase	Pancreas	Small intestine	Basic	Fat droplet + H_2O → glycerol + fatty acids

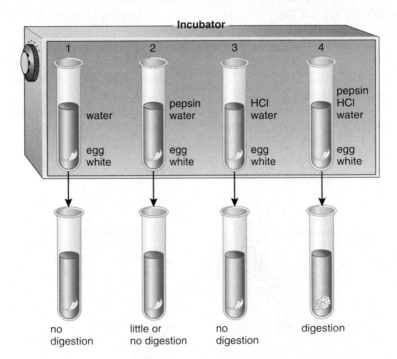

Figure 12.12   Digestion experiment.
This experiment is based on the optimum conditions for digestion by pepsin in the stomach. Knowing that the correct enzyme, optimum pH, optimum temperature, and correct substrate must be present for digestion to occur, explain the results of this experiment. Colors indicate pH of test tubes (blue, basic; red, acidic).

Conditions for Digestion

Laboratory experiments can define the necessary conditions for digestion. For example, the four test tubes shown in Figure 12.12 can be prepared and observed for the digestion of egg white, a protein digested in the stomach by the enzyme pepsin.

After all tubes are placed in an incubator at body temperature for at least one hour, the results depicted are observed. Tube 1 is a control tube; no digestion has occurred in this tube because the enzyme and HCl are missing. (If a control gives a positive result, then the experiment is invalidated.) Tube 2 shows limited or no digestion because HCl is missing, and therefore the pH is too high for pepsin to be effective. Tube 3 shows no digestion because although HCl is present, the enzyme is missing. Tube 4 shows the best digestive action because the enzyme is present and the presence of HCl has resulted in an optimum pH. This experiment supports the hypothesis that for digestion to occur, the substrate and enzyme must be present and the environmental conditions must be optimum. The optimal environmental conditions include a warm temperature and the correct pH.

12.4 Nutrition

The body requires three major classes of macronutrients in the diet: carbohydrate, protein, and fat. These supply the energy and the building blocks that are needed to synthesize cellular contents. Micronutrients—especially vitamins and minerals—are also required because they are necessary for optimum cellular metabolism.

Several modern nutritional studies suggest that certain nutrients can protect against heart disease, cancer, and other serious illnesses. These studies have analyzed the eating habits of healthy people in the United States and around the world, especially those living in areas that have lower rates of heart disease and cancer. The resulting dietary recommendations can be illustrated by a food pyramid (Fig. 12.13).

The bulk of the diet should consist of bread, cereal, rice, and pasta as energy sources. Whole grains are preferred over those that have been milled because they contain fiber, vitamins, and minerals. Vegetables and fruits are another rich source of fiber, vitamins, and minerals. Notice, then, that a largely vegetarian diet is recommended.

Animal products, especially meat, need only be minimally included in the diet; fats and sweets should be used sparingly. Dairy products and meats tend to be high in sat-urated fats, and an intake of saturated fats increases the risk of cardiovascular disease (see Lipids, p. 229). Low-fat dairy products are available, but there is no way to take much of the fat out of meat. Beef, in particular, contains a relatively high fat content. Ironically, the affluence of people in the United States contributes to a poor diet and, therefore, possible illness. Only comparatively rich people can afford fatty meats from grain-fed cattle and carbohydrates that have been highly processed to remove fiber and to add sugar and salt.

Carbohydrates

The quickest, most readily available source of energy for the body is glucose. Carbohydrates are digested to simple sugars, which are or can be converted to glucose. As mentioned earlier in this chapter, glucose is stored by the liver in the form of glycogen. Between eating periods, the blood glucose level is maintained at about 0.1% by the breakdown of glycogen or by the conversion of glycerol (from fats) or amino acids to glucose. If necessary, amino acids are taken from the muscles—even from the heart muscle. While body cells can utilize fatty acids as an energy source, brain cells require glucose. For this reason alone, it is necessary to include carbohydrates in the diet. According to Figure 12.13,

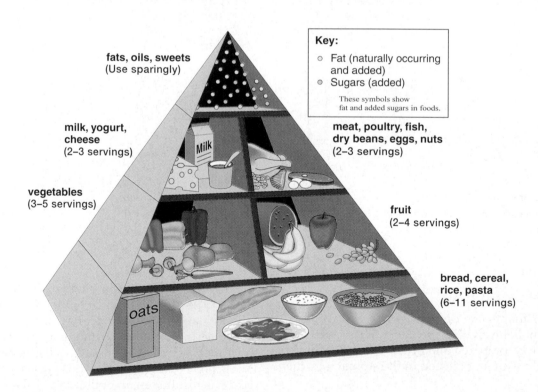

Figure 12.13   Food guide pyramid: A guide to daily food choices.
The U.S. Department of Agriculture uses a pyramid to show the ideal diet because it emphasizes the importance of eating grains, fruits, and vegetables. Meats and dairy products are needed in limited amounts; fats, oils, and sweets should be used sparingly.
Source: Data from the U.S. Department of Agriculture.

Table 12.3	Reducing Dietary Sugar

To reduce dietary sugar:

1. Eat fewer sweets, such as candy, soft drinks, ice cream, and pastry.
2. Eat fresh fruits or fruits canned without heavy syrup.
3. Use less sugar—white, brown, or raw—and less honey and syrups.
4. Avoid sweetened breakfast cereals.
5. Eat less jelly, jam, and preserves.
6. Drink pure fruit juices, not imitations.
7. When cooking, use spices, such as cinnamon, instead of sugar to flavor foods.
8. Do not put sugar in tea or coffee.

carbohydrates should make up the bulk of the diet. Further, these should be complex, not simple, carbohydrates. Complex sources of carbohydrates include preferably whole-grain pasta, rice, bread, and cereal (Fig. 12.14). Potatoes and corn, although considered vegetables, are also sources of carbohydrates.

Simple carbohydrates (e.g., sugars) are labeled "empty calories" by some dietitians because they contribute to energy needs and weight gain without supplying any other nutritional requirements. Table 12.3 gives suggestions on how to reduce dietary sugar (simple carbohydrates). In contrast to simple sugars, complex carbohydrates are likely to be accompanied by a wide range of other nutrients and by **fiber,** which is indigestible plant material.

The intake of fiber is recommended because it may decrease the risk of colon cancer, a major type of cancer, and cardiovascular disease, the number one killer in the United States. Insoluble fiber, such as that found in wheat bran, has a laxative effect and may guard against colon cancer by limiting the amount of time cancer-causing substances are in contact with the intestinal wall. Soluble fiber, such as that found in oat bran, combines with bile acids and cholesterol in the intestine and prevents them from being absorbed. The liver then removes cholesterol from the blood and changes it to bile acids, replacing the bile acids that were lost. While the diet should have an adequate amount of fiber, some evidence suggests that a diet too high in fiber can be detrimental, possibly impairing the body's ability to absorb iron, zinc, and calcium.

Complex carbohydrates, which contain fiber and a wide range of nutrients, should form the bulk of the diet.

Proteins

Foods rich in protein include red meat, fish, poultry, dairy products, legumes (i.e., peas and beans), nuts, and cereals.

Figure 12.14   Complex carbohydrates.
To meet our energy needs, dietitians recommend consuming foods rich in complex carbohydrates, such as those shown here, rather than foods consisting of simple carbohydrates, such as candy and ice cream. Simple carbohydrates provide monosaccharides but few other types of nutrients.

Following digestion of protein, amino acids enter the bloodstream and are transported to the tissues. Ordinarily, amino acids are not used as an energy source. Most are incorporated into structural proteins found in muscles, skin, hair, and nails. Others are used to synthesize such proteins as hemoglobin, plasma proteins, enzymes, and hormones.

Adequate protein formation requires 20 different types of amino acids. Of these, eight are required from the diet in adults (nine in children) because the body is unable to produce them. These are termed the **essential amino acids.** The body produces the other amino acids by simply transforming one type into another type. Some protein sources, such as meat, milk, and eggs, are complete; they provide all 20 types of amino acids. Legumes (beans and peas), other types of vegetables, seeds and nuts, and also grains supply us with amino acids, but each of these alone is an incomplete protein source because of a deficiency in at least one of the essential amino acids. Absence of one essential amino acid prevents utilization of the other 19 amino acids. Therefore, vegetarians are counseled to combine two or more incomplete types of plant products to acquire all the essential amino acids. Table 12.4 lists complementary proteins—sources of protein whose amino acid content complement each other so that all the essential amino acids are present in the diet. Soybeans and tofu, which are made from soybeans, are rich in amino acids, but even so, you have to combine tofu with a complementary protein to acquire all the essential amino acids. Table 12.4 will allow you to select various combinations of plant products in order to make sure the diet contains the essential amino acids when it does not contain meat.

Amino acids are not stored in the body, and a daily supply is needed. However, it does not take very much protein to meet the daily requirement. Two servings of meat a day (one serving is equal in size to a deck of cards) is usually enough. Some meats (e.g., hamburger) are high in protein, but also high in fat. Everything considered, it is probably a good idea to depend on protein from plant origins (e.g., whole-grain cereals, dark breads, and legumes) to a greater extent than is often the custom in the United States. A study involving native Hawaiians lends support to the belief that health improves when the diet is rich in protein from plants rather than protein from animals. Only 3% of the ancient Hawaiian diet was animal protein, whereas the modern diet is 12% animal protein (Fig. 12.15). This, in large part, accounts for why the ancient diet was only 10% fat, whereas the modern diet of Hawaiians is 42%. A statistical study showed that the rate of cardiovascular disease and cancer is higher than average among those who follow the modern diet. Diabetes is also common in persons who follow the modern diet. On the other hand, health has improved immensely among those who have switched back to the ancient diet.

Nutritionists do not recommend using protein and/or amino acid supplements. Protein supplements that athletes take to build muscle cost more than food and can be harmful. When excess protein is broken down, more urea is excreted in the urine. The water needed for excretion of urea can cause dehydration when a person is exercising and also losing water by sweating. Also, some studies suggest that protein supplements lead to calcium loss and weakened bones. Amino acid supplements can also be

Table 12.4	Complementary Proteins		
Legumes	**Seeds and Nuts**	**Grains**	**Vegetables**
Green peas	Sunflower seeds	Wheat	Leafy green (e.g., spinach)
Navy beans	Sesame seeds	Rice	Broccoli
Soybeans	Macadamia nuts	Corn	Cauliflower
Black-eyed peas	Brazil nuts	Barley	Cabbage
Pinto beans	Peanuts	Oats	Artichoke hearts
Lima beans	Cashews	Rye	
Kidney beans	Hazelnuts		
Chick peas	Almonds		
Black beans	Nut butter		

* Combine foods from any two or more columns to acquire all of the essential amino acids.

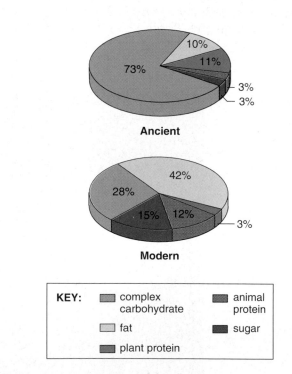

Figure 12.15 Ancient versus modern diet of native Hawaiians.
Among those Hawaiians who have switched back to the native diet, the incidence of cardiovascular disease, cancer, and diabetes has dropped.

dangerous to your health. Mistaken ideas abound. For example, contrary to popular reports, the taking of lysine does not relieve or cure herpes sores.

Lipids

Fat and cholesterol are both lipids. Fat is present not only in butter, margarine, and oils, but also in various foods high in animal protein.

The current guidelines suggest that fat should account for no more than 30% of our daily calories. The chief reason is that an intake of fat not only causes weight gain, but also increases the risk of cancer and cardiovascular disease. Dietary fat may increase the risk of colon, hepatic, and pancreatic cancers. Although recent studies suggest no link between dietary fat and breast cancer, other researchers still believe that the matter deserves further investigation.

Cardiovascular disease is often due to arteries blocked by fatty deposits, called **plaque,** that contain saturated fats and cholesterol. Cholesterol is carried in the blood by two types of lipoproteins: low-density lipoprotein (LDL) and high-density lipoprotein (HDL). LDL is thought of as "bad" because it carries cholesterol from the liver to the cells, while HDL is thought of as "good" because it carries cholesterol to the liver, which takes it up and converts it to bile salts.

Saturated fatty acids have no double bonds; monounsaturated fatty acids have one double bond; polyunsaturated fatty acids have many double bonds. Saturated fats, whether in butter or margarine, can raise LDL cholesterol levels, while monounsaturated fats and polyunsaturated fats lower LDL cholesterol levels. Olive oil and canola oil contain mostly monounsaturated fats; corn oil and safflower oil contain mostly polyunsaturated fats. These oils have a liquid consistency and come from plants. Saturated fats, which are solids at room temperature, usually have an animal origin; two well-known exceptions are palm oil and coconut oil, which contain mostly saturated fats and come from the plants mentioned.

Nutritionists stress that it is more important to consume a diet low in fat than to be overly concerned about which type of fat is in the diet. Still, polyunsaturated fats are nutritionally essential because they are the only type of fat that contains linoleic acid and linolenic acid, two fatty acids the body cannot make. The body needs these two polyunsaturated fatty acids to produce various hormones and the plasma membrane of cells. Since these fatty acids must be supplied by diet, they are called essential fatty acids. These essential fatty acids are found in small amounts in the oils of plants and cold-water fish and are readily stored in the adult body.

Table 12.5 gives suggestions on how to reduce dietary fat and cholesterol. Everyone should use diet to keep their cholesterol level within normal limits so that medications will not be needed for this purpose.

Table 12.5   Reducing Lipids
To reduce dietary fat:
1. Choose poultry, fish, or dry beans and peas as a protein source.
2. Remove skin from poultry before cooking, and place on a rack so that fat drains off.
3. Broil, boil, or bake rather than frying.
4. Limit your intake of butter, cream, hydrogenated oils, shortenings, and tropical oils (coconut and palm oils).*
5. Use herbs and spices to season vegetables instead of butter, margarine, or sauces. Use lemon juice instead of salad dressing.
6. Drink skim milk instead of whole milk, and use skim milk in cooking and baking.
7. Eat nonfat or low-fat foods.
To reduce dietary cholesterol:
1. Avoid cheese, egg yolks, liver, and certain shellfish (shrimp and lobster). Preferably, eat white fish and poultry.
2. Substitute egg whites for egg yolks in both cooking and eating.
3. Include soluble fiber in the diet. Oat bran, oatmeal, beans, corn, and fruits such as apples, citrus fruits, and cranberries are high in soluble fiber.

*Although coconut and palm oils are from plant sources, they are mostly saturated fats.

Fake Fat

Olestra is a substance made to look, taste, and act like real fat, but the digestive system is unable to digest it. It travels down the length of the digestive system without being absorbed or contributing any calories to the day's total. Therefore, it is commonly known as "fake fat." Unfortunately, the fat-soluble vitamins A, D, E, and K tend to be taken up by olestra, and thereafter they are not absorbed by the body. Similarly, people using olestra have reduced amounts of carotenoids in their blood, even as much as 20% less. Manufacturers fortify olestra-containing foods with the vitamins mentioned, but not with carotenoids.

Fake fat has other side effects. Some people who consume olestra have developed anal leakage. Others experience diarrhea, intestinal cramping, and gas. Presently, the FDA has limited the use of olestra to potato chips and other salty snacks, but the manufacturer wants approval to add it to ice cream, salad dressings, and cheese.

Dietary protein supplies the essential amino acids; proteins from plant origins generally have less accompanying fat. A diet composed of no more than 30% fat calories is recommended because fat intake, particularly saturated fats, is associated with various health problems.

Vitamins

Vitamins are organic compounds (other than carbohydrate, fat, and protein) that the body uses for metabolic purposes but is unable to produce in adequate quantity. Many vitamins are portions of coenzymes, which are enzyme helpers. For example, niacin is part of the coenzyme NAD, and riboflavin is part of another dehydrogenase, FAD. Coenzymes are needed in only small amounts because each can be used over and over again. Not all vitamins are coenzymes; vitamin A, for example, is a precursor for the visual pigment that prevents night blindness. If vitamins are lacking in the diet, various symptoms develop (Fig. 12.16). Altogether, there are 13 vitamins, which are divided into those that are fat soluble (Table 12.6) and those that are water-soluble (Table 12.7).

Antioxidants

Over the past 20 years, numerous statistical studies have been done to determine whether a diet rich in fruits and vegetables can protect against cancer. Cellular metabolism generates free radicals, unstable molecules that carry an extra electron. The most common free radicals in cells are superoxide (O_2^-) and hydroxide (OH^-). In order to stabilize themselves, free radicals donate an electron to DNA, to proteins, including enzymes, or to lipids, which are found in plasma membranes. Such donations most likely damage these cellular molecules and thereby may lead to disorders, perhaps even cancer.

Vitamins C, E, and A are believed to defend the body against free radicals, and therefore they are termed antioxidants. These vitamins are especially abundant in fruits and vegetables. The dietary guidelines shown in Figure 12.13 suggest that we eat a minimum of five servings of fruits and vegetables a day. To achieve this goal, we should include salad greens, raw or cooked vegetables, dried fruit, and fruit juice, in addition to traditional apples and oranges and such.

Dietary supplements may provide a potential safeguard against cancer and cardiovascular disease, but nutritionists do not think people should take supplements instead of improving their intake of fruits and vegetables. There are many beneficial compounds in fruits that cannot be obtained from a vitamin pill. These compounds enhance each other's absorption or action and also perform independent biological functions.

Vitamin D

Skin cells contain a precursor cholesterol molecule that is converted to vitamin D after UV exposure. Vitamin D leaves the skin and is modified first in the kidneys and then in the liver until finally it becomes calcitriol. Calcitriol promotes the absorption of calcium by the intestines. The lack of vitamin D leads to rickets in children (Fig. 12.16a). Rickets, characterized by bowing of the legs, is caused by defective mineralization of the skeleton. Most milk today is fortified with vitamin D, which helps prevent the occurrence of rickets.

> Vitamins are essential to cellular metabolism; many are protective against identifiable illnesses and conditions.

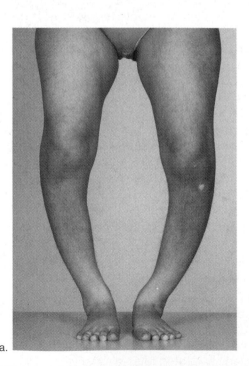

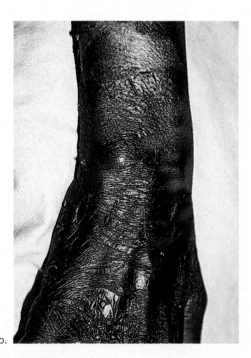

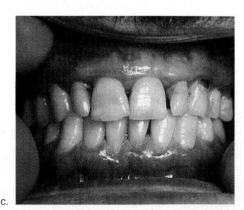

a.   b.   c.

Figure 12.16 Illnesses due to vitamin deficiency.
a. Bowing of bones (rickets) due to vitamin D deficiency. **b.** Dermatitis (pellagra) of areas exposed to light due to niacin (vitamin B_3) deficiency.
c. Bleeding of gums (scurvy) due to vitamin C deficiency.

Table 12.6   Fat-Soluble Vitamins

Vitamin	Functions	Food Sources	Conditions With	
			Too Little	**Too Much**
Vitamin A	Antioxidant synthesized from beta-carotene; needed for healthy eyes, skin, hair, and mucous membranes, and for proper bone growth	Deep yellow/orange and leafy, dark green vegetables, fruits, cheese, whole milk, butter, eggs	Night blindness, impaired growth of bones and teeth	Headache, dizziness, nausea, hair loss, abnormal development of fetus
Vitamin D	A group of steroids needed for development and maintenance of bones and teeth	Milk fortified with vitamin D, fish liver oil; also made in the skin when exposed to sunlight	Rickets, bone decalcification and weakening	Calcification of soft tissues, diarrhea, possible renal damage
Vitamin E	Antioxidant that prevents oxidation of vitamin A and polyunsaturated fatty acids	Leafy green vegetables, fruits, vegetable oils, nuts, whole-grain breads and cereals	Unknown	Diarrhea, nausea, headaches, fatigue, muscle weakness
Vitamin K	Needed for synthesis of substances active in clotting of blood	Leafy green vegetables, cabbage, cauliflower	Easy bruising and bleeding	Can interfere with anticoagulant medication

Table 12.7   Water-Soluble Vitamins

Vitamin	Functions	Food Sources	Conditions With	
			Too Little	**Too Much**
Vitamin C	Antioxidant; needed for forming collagen; helps maintain capillaries, bones, and teeth	Citrus fruits, leafy green vegetables, tomatoes, potatoes, cabbage	Scurvy, delayed wound healing, infections	Gout, kidney stones, diarrhea, decreased copper
Thiamine (vitamin B_1)	Part of coenzyme needed for cellular respiration; also promotes activity of the nervous system	Whole-grain cereals, dried beans and peas, sunflower seeds, nuts	Beriberi, muscular weakness, enlarged heart	Can interfere with absorption of other vitamins
Riboflavin (vitamin B_2)	Part of coenzymes, such as FAD; aids cellular respiration, including oxidation of protein and fat	Nuts, dairy products, whole-grain cereals, poultry, leafy green vegetables	Dermatitis, blurred vision, growth failure	Unknown
Niacin (nicotinic acid)	Part of coenzymes NAD and NADP; needed for cellular respiration, including oxidation of protein and fat	Peanuts, poultry, whole-grain cereals, leafy green vegetables, beans	Pellagra, diarrhea, mental disorders	High blood sugar and uric acid, vasodilation, etc.
Folacin (folic acid)	Coenzyme needed for production of hemoglobin and formation of DNA	Dark leafy green vegetables, nuts, beans, whole-grain cereals	Megaloblastic anemia, spina bifida	May mask B_{12} deficiency
Vitamin B_6	Coenzyme needed for synthesis of hormones and hemoglobin; CNS control	Whole-grain cereals, bananas, beans, poultry, nuts, leafy green vegetables	Rarely, convulsions, vomiting, seborrhea, muscular weakness	Insomnia, neuropathy
Pantothenic acid	Part of coenzyme A needed for oxidation of carbohydrates and fats; aids in the formation of hormones and certain neurotransmitters	Nuts, beans, dark green vegetables, poultry, fruits, milk	Rarely, loss of appetite, mental depression, numbness	Unknown
Vitamin B_{12}	Complex, cobalt-containing compound; part of the coenzyme needed for synthesis of nucleic acids and myelin	Dairy products, fish, poultry, eggs, fortified cereals	Pernicious anemia	Unknown
Biotin	Coenzyme needed for metabolism of amino acids and fatty acids	Generally in foods, especially eggs	Skin rash, nausea, fatigue	Unknown

Minerals

In addition to vitamins, various **minerals** are required by the body. Minerals are divided into major minerals and trace minerals. The body contains more than 5 grams of each major mineral and less than 5 grams of each trace mineral (Fig. 12.17). The major minerals are constituents of cells and body fluids and are structural components of tissues. For example, calcium (present as Ca^{2+}) is needed for the construction of bones and teeth and for nerve conduction and muscle contraction. Phosphorus (present as PO_4^{3-}) is stored in the bones and teeth and is a part of phospholipids, ATP, and the nucleic acids. Potassium (K^+) is the major positive ion inside cells and is important in nerve conduction and muscle contraction, as is sodium (Na^+). Sodium also plays a major role in regulating the body's water balance, as does chloride (Cl^-). Magnesium (Mg^{2+}) is critical to the functioning of hundreds of enzymes.

The trace minerals are parts of larger molecules. For example, iron is present in hemoglobin, and iodine is a part of thyroxine and triidothyronine, hormones produced by the thyroid gland. Zinc, copper, and selenium are present in enzymes that catalyze a variety of reactions. Proteins, called zinc-finger proteins because of their characteristic shapes, bind to DNA when a particular gene is to be activated. As research continues, more and more elements are added to the list of trace minerals considered essential. During the past three decades, for example, very small amounts of selenium, molybdenum, chromium, nickel, vanadium, silicon, and even arsenic have been found to be essential to good health. Table 12.8 lists the functions of various minerals and gives their food sources and signs of deficiency and toxicity.

Occasionally, individuals do not receive enough iron (especially women), calcium, magnesium, or zinc in their diets. Adult females need more iron in the diet than males (18 mg compared to 10 mg) because they lose hemoglobin each month during menstruation. Stress can bring on a magnesium deficiency, and due to its high-fiber content, a vegetarian diet may make zinc less available to the body. However, a varied and complete diet usually supplies enough of each type of mineral.

Calcium

Many people take calcium supplements to counteract **osteoporosis,** a degenerative bone disease that afflicts an estimated one-fourth of older men and one-half of older women in the United States. Osteoporosis develops because bone-eating cells called osteoclasts are more active than bone-forming cells called osteoblasts. Therefore, the bones are porous, and they break easily because they lack sufficient calcium. Due to recent studies that show consuming more calcium does slow bone loss in elderly people, the guidelines have been revised. A calcium intake of 1,000 mg a day is recommended for men and for women who are premenopausal or who use estrogen replacement therapy; 1,300 mg a day is recommended for postmenopausal women who do not use estrogen replacement therapy. To achieve this amount, supplemental calcium is most likely necessary.

Vitamin D is an essential companion to calcium in preventing osteoporosis. Other vitamins may also be helpful; for example, magnesium has been found to suppress the cycle that leads to bone loss. Estrogen replacement therapy and exercise, in addition to adequate calcium and vitamin intake, also help prevent osteoporosis. Drinking more than nine cups of caffeinated coffee per day and smoking are risk factors for osteoporosis. Medications are also available that slow bone loss while increasing skeletal mass. These are still being studied for their effectiveness and possible side effects.

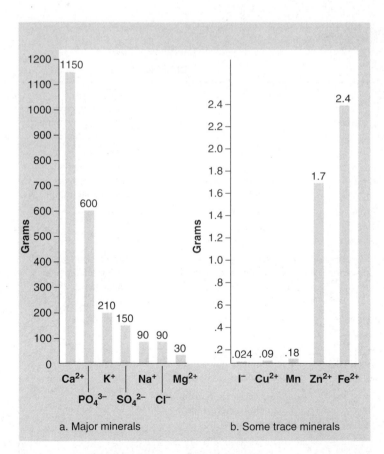

Figure 12.17   Minerals in the body.
These charts show the usual amount of certain minerals in a 60-kilogram (135 lb) person. The functions of minerals are given in Table 12.8. **a.** The major minerals are present in amounts larger than 5 grams (about a teaspoon). **b.** Trace minerals are present in lesser amounts.

Sodium

The recommended amount of sodium intake per day is 500 mg, although the average American takes in 4,000–4,700 mg every day. In recent years, this imbalance has caused

Table 12.8	Minerals			
Mineral	**Functions**	**Food Sources**	**Conditions With**	
			Too Little	**Too Much**
Macrominerals (more than 100 mg/day needed)				
Calcium (Ca^{2+})	Strong bones and teeth, nerve conduction, muscle contraction	Dairy products, leafy green vegetables	Stunted growth in children, low bone density in adults	Kidney stones; interferes with iron and zinc absorption
Phosphorus (PO_4^{3-})	Bone and soft tissue growth; part of phospholipids, ATP, and nucleic acids	Meat, dairy products, sunflower seeds, food additives	Weakness, confusion, pain in bones and joints	Low blood and bone calcium levels
Potassium (K^+)	Nerve conduction, muscle contraction	Many fruits and vegetables, bran	Paralysis, irregular heartbeat, eventual death	Vomiting, heart attack, death
Sodium (Na^+)	Nerve conduction, pH and water balance	Table salt	Lethargy, muscle cramps, loss of appetite	Edema, high blood pressure
Chloride (Cl^-)	Water balance	Table salt	Not likely	Vomiting, dehydration
Magnesium (Mg^{2+})	Part of various enzymes for nerve and muscle contraction, protein synthesis	Whole grains, leafy green vegetables	Muscle spasm, irregular heartbeat, convulsions, confusion, personality changes	Diarrhea
Microminerals (less than 20 mg/day needed)				
Zinc (Zn^{2+})	Protein synthesis, wound healing, fetal development and growth, immune function	Meats, legumes, whole grains	Delayed wound healing, night blindness, diarrhea, mental lethargy	Anemia, diarrhea, vomiting, renal failure, abnormal cholesterol levels
Iron (Fe^{2+})	Hemoglobin synthesis	Whole grains, meats, prune juice	Anemia, physical and mental sluggishness	Iron toxicity disease, organ failure, eventual death
Copper (Cu^{2+})	Hemoglobin synthesis	Meat, nuts, legumes	Anemia, stunted growth in children	Damage to internal organs if not excreted
Iodine (I^-)	Thyroid hormone synthesis	Iodized table salt, seafood	Thyroid deficiency	Depressed thyroid function, anxiety
Selenium (SeO_4^{2-})	Part of antioxidant enzyme	Seafood, meats, eggs	Vascular collapse, possible cancer development	Hair and fingernail loss, discolored skin

concern because high sodium intake has been linked to hypertension (high blood pressure) in some people. About one-third of the sodium we consume occurs naturally in foods; another one-third is added during commercial processing; and we add the last one-third either during home cooking or at the table in the form of table salt.

Clearly, it is possible for us to cut down on the amount of sodium in the diet. Table 12.9 gives recommendations for doing so.

Both macro- and microminerals play specific roles in the body. Calcium is needed for strong bones, for example. Excess sodium in the diet can lead to hypertension; therefore, excess sodium intake should be avoided.

Table 12.9	Reducing Dietary Sodium
To reduce dietary sodium:	

1. Use spices instead of salt to flavor foods.
2. Add little or no salt to foods at the table, and add only small amounts of salt when you cook.
3. Eat unsalted crackers, pretzels, potato chips, nuts, and popcorn.
4. Avoid hot dogs, ham, bacon, luncheon meats, smoked salmon, sardines, and anchovies.
5. Avoid processed cheese and canned or dehydrated soups.
6. Avoid brine-soaked foods, such as pickles or olives.
7. Read labels to avoid high-salt products.

Persons with obesity have

- weight 20% or more above appropriate weight for height.
- body fat content in excess of that consistent with optimal health, probably due to a diet rich in fats.
- low levels of exercise.

Figure 12.18   Recognizing obesity.

Eating Disorders

Authorities recognize three primary eating disorders: obesity, bulimia nervosa, and anorexia nervosa. Although they exist in a continuum as far as body weight is concerned, they all represent an inability to maintain normal body weight because of eating habits.

Obesity

As indicated in Figure 12.18, **obesity** is most often defined as a body weight 20% or more above the ideal weight for a person's height. By this standard, 28% of women and 10% of men in the United States are obese. Moderate obesity is 41–100% above ideal weight, and severe obesity is 100% or more above ideal weight.

Obesity is most likely caused by a combination of hormonal, metabolic, and social factors. It is known that obese individuals have more fat cells than normal, and when they lose weight the fat cells simply get smaller; they don't disappear. The social factors that cause obesity include the eating habits of other family members. Consistently eating fatty foods, for example, will make you gain weight. Sedentary activities, such as watching television instead of exercising, also determine how much body fat you have. The risk of heart disease is higher in obese individuals, and this alone tells us that excess body fat is not consistent with optimal health.

The treatment depends on the degree of obesity. Surgery to remove body fat may be required for those who are moderately or greatly overweight. But for most people, a knowledge of good eating habits along with behavior modification may suffice, particularly if a balanced diet is accompanied by a sensible exercise program. A lifelong commitment to a properly planned program is the best way to prevent a cycle of weight gain followed by weight loss. Such a cycle is not conducive to good health.

Bulimia Nervosa

Bulimia nervosa can coexist with either obesity or anorexia nervosa, which is discussed next. People with this condition have the habit of eating to excess (called binge eating) and then purging themselves by some artificial means, such as self-induced vomiting or use of a laxative. Bulimic individuals are overconcerned about their body shape and weight, and therefore they may be on a very restrictive diet. A restrictive diet may bring on the desire to binge, and typically the person chooses to consume sweets, such as cakes, cookies, and ice cream (Fig. 12.19). The amount of food consumed is far beyond the normal number of calories for one meal, and the person keeps on eating until every bit is gone. Then, a feeling of guilt most likely brings on the next phase, which is a purging of all the calories that have been taken in.

Bulimia can be dangerous to your health. Blood composition is altered, leading to an abnormal heart rhythm, and damage to the kidneys can even result in death. At the very least, vomiting can lead to inflammation of the pharynx and esophagus, and stomach acids can cause the teeth to erode. The esophagus and stomach may even rupture and tear due to strong contractions during vomiting.

The most important aspect of treatment is to get the patient on a sensible and consistent diet. Again, behavioral modification is helpful, and so perhaps is psychotherapy to help the patient understand the emotional causes of the behavior. Medications, including antidepressants, have sometimes helped to reduce the bulimic cycle and restore normal appetite.

Obesity and bulimia nervosa have complex causes and may be damaging to health. Therefore, they require competent medical attention.

Anorexia Nervosa

In **anorexia nervosa,** a morbid fear of gaining weight causes the person to be on a very restrictive diet. Athletes such as distance runners, wrestlers, and dancers are at risk of anorexia nervosa because they believe that being thin gives them a competitive edge. In addition to eating only low-calorie foods, the person may induce vomiting and use laxatives to bring about further weight loss. No matter how thin they have become, people with anorexia nervosa think they are overweight (Fig. 12.20). Such a distorted self-image may prevent recognition of the need for medical help.

Actually, the person is starving and has all the symptoms of starvation, including low blood pressure, irregular heartbeat, constipation, and constant chilliness. Bone density decreases and stress fractures occur. The body begins to shut down; menstruation ceases in females; the internal organs, including the brain, don't function well; and the skin dries up. Impairment of the pancreas and digestive tract means that any food consumed does not provide nourishment. Death may be imminent. If so, the only recourse may be hospitalization and force-feeding. Eventually, it is necessary to use behavior therapy and psychotherapy to enlist the cooperation of the person to eat properly. Family therapy may be necessary, because anorexia nervosa in children and teens is believed to be a way for them to gain some control over their lives.

In anorexia nervosa, the individual has a distorted body image and always feels fat. Competent medical help is often a necessity.

Persons with bulimia nervosa have

- recurrent episodes of binge eating characterized by consuming an amount of food much higher than normal for one sitting and a sense of lack of control over eating during the episode.
- an obsession about their body shape and weight, but often without exercising.
- increase in fine body hair, halitosis, and gingivitis.

Body weight is regulated by

- a restrictive diet, excessive exercise.
- purging (self-induced vomiting or misuse of laxatives).

Figure 12.19   Recognizing bulimia nervosa.

Persons with anorexia nervosa have

- a morbid fear of gaining weight; body weight no more than 85% normal.
- a distorted body image so that person feels fat even when emaciated.
- in females, an absence of a menstrual cycle for at least three months.

Body weight is kept too low by either/or

- a restrictive diet, often with excessive exercise.
- binge eating/purging (person engages in binge eating and then self-induces vomiting or misuses laxatives).

Figure 12.20   Recognizing anorexia nervosa.

Bioethical Focus   Legislation and Health

A fat-free fat, called olestra, is now being used to produce foods that are free of calories from fat. A slice of pie ordinarily contains 405 calories, but when the crust is made with olestra, it has only 252 calories. Just three chocolate chip cookies ordinarily adds 63 calories to your daily calorie count, but with olestra it's only 38 calories. And one brownie suddenly goes down from 85 to 49 calories. Olestra molecules made of six or eight fatty acids attached to a sugar molecule are much bigger than a triglyceride, and lipase can't break the molecule down. Olestra goes through the intestines without being absorbed. Procter and Gamble, who developed olestra, expects to make one billion dollars within ten years on its product.

It is generally recognized that olestra can cause intestinal cramping, flatulence (gas), and diarrhea in a small segment of the population, and can prevent the absorption of some carotenoids, a nutrient that has health benefits in all of us. Even so, the Federal Drug Administration (FDA) approved it for use on the basis that olestra was reasonably harmless. Henry Blackburn, professor of public health at the University of Minnesota, thinks this standard is too low. Instead, the standard for approval ought to be, "Does this product contribute to the nutritional health of the nation?"

David Kessler, FDA commissioner, says, "Ask the American people if government should be off the backs of business,

and you will get a resounding Yes!" However, what is the role of government in protecting public health? Do Americans want more protection than they already have from food additives? Or would they rather be free to make their own choices?

Decide Your Opinion

1. Should the government only approve food additives that will contribute to our health? Why or why not?
2. Are people responsible, themselves, for what they eat? Why or why not?
3. Is there too much emphasis in our culture on staying slim, despite what it may do to our health? Why or why not?

Summarizing the Concepts

12.1 The Digestive Tract
The digestive tract consists of the mouth, pharynx, esophagus, stomach, small intestine, and large intestine. Only these structures actually contain food, while the salivary glands, liver, and pancreas supply substances that aid in the digestion of food.

The salivary glands send saliva into the mouth, where the teeth chew the food and the tongue forms a bolus for swallowing. Saliva contains salivary amylase, an enzyme that begins the digestion of starch.

The air passage and food passage cross in the pharynx. When a person swallows, the air passage is usually blocked off, and food must enter the esophagus, where peristalsis begins.

The stomach expands and stores food. While food is in the stomach, the stomach churns, mixing food with the acidic gastric juices. Gastric juices contain pepsin, an enzyme that digests protein.

The duodenum of the small intestine receives bile from the liver and pancreatic juice from the pancreas. Bile, which is produced in the liver and stored in the gallbladder, emulsifies fat and readies it for digestion by lipase, an enzyme produced by the pancreas. The pancreas also produces enzymes that digest starch (pancreatic amylase) and protein (trypsin). The intestinal enzymes finish the process of chemical digestion.

The walls of the small intestine have fingerlike projections called villi where small nutrient molecules are absorbed. Amino acids and glucose enter the blood vessels of a villus. Glycerol and fatty acids are joined and packaged as lipoproteins before entering lymphatic vessels called lacteals in a villus.

The large intestine consists of the cecum, the colon (including the ascending, transverse, descending, and sigmoid colon), and the rectum, which ends at the anus. The large intestine does not produce digestive enzymes; it does absorb water, salts, and some vitamins. Reduced water absorption results in diarrhea. The intake of water and fiber help prevent constipation.

12.2 Three Accessory Organs
Three accessory organs of digestion—the pancreas, liver, and gallbladder—send secretions to the duodenum via ducts. The pancreas produces pancreatic juice, which contains digestive enzymes for carbohydrate, protein, and fat.

The liver produces bile, which is stored in the gallbladder. The liver receives blood from the small intestine by way of the hepatic portal vein. It has numerous important functions, and any malfunction of the liver is a matter of considerable concern.

12.3 Digestive Enzymes
Digestive enzymes are present in digestive juices and break down food into the nutrient molecules glucose, amino acids, fatty acids, and glycerol (see Table 12.2). Salivary amylase and pancreatic amylase begin the digestion of starch. Pepsin and trypsin digest protein to peptides. Lipase digests fat to glycerol and fatty acids. Intestinal enzymes finish the digestion of starch and protein.

Digestive enzymes have the usual enzymatic properties. They are specific to their substrate and speed up specific reactions at optimum body temperature and pH.

12.4 Nutrition
The nutrients released by the digestive process should provide us with an adequate amount of energy, essential amino acids and fatty acids, and all necessary vitamins and minerals.

The bulk of the diet should be carbohydrates (e.g., bread, pasta, and rice) and fruits and vegetables. These are low in saturated fatty acids and cholesterol molecules, whose intake is linked to cardiovascular disease. Aside from carbohydrates, proteins, and fats, the body requires vitamins and minerals. The vitamins A, E, and C are antioxidants that protect cell contents from damage due to free radicals. The mineral calcium is needed for strong bones.

Testing Yourself

Choose the best answer for each question.

1. The mouth, tongue, and teeth contribute to
   a. chemical digestion of proteins.
   b. mechanical breakdown of food into smaller pieces.
   c. absorption of water.
   d. mixing the food with gastric juices.

2. Tracing the path of food in the following list (a–f), which step is out of order first?
   a. mouth
   b. pharynx
   c. esophagus
   d. small intestine
   e. stomach
   f. large intestine

3. Which association is incorrect?
   a. mouth—starch digestion
   b. esophagus—protein digestion
   c. small intestine—starch, lipid, protein digestion
   d. stomach—food storage
   e. liver—production of bile

4. Why can a person not swallow food and talk at the same time?
   a. In order to swallow, the epiglottis must close off the trachea.
   b. The brain cannot control two activities at once.
   c. In order to speak, air must come through the larynx to form sounds.
   d. A swallowing reflex is only initiated when the mouth is closed.
   e. Both a and c are correct.

5. Which association is incorrect?
   a. pancreas—produces alkaline secretions and enzymes
   b. salivary glands—produce saliva and amylase
   c. gallbladder—produces digestive enzymes
   d. liver—produces bile

6. Which of these could be absorbed directly without need of digestion?
   a. glucose
   b. fat
   c. polysaccharides
   d. protein
   e. nucleic acid

7. Peristalsis occurs
   a. from the mouth to the small intestine.
   b. from the beginning of the esophagus to the anus.
   c. only in the stomach.
   d. only in the small and large intestine.
   e. only in the esophagus and stomach.

8. An organ is a structure made of two or more tissues performing a common function. Which of the four tissue types are present in the wall of the digestive tract?
   a. epithelium
   b. connective tissue
   c. nervous tissue
   d. muscle tissue
   e. All of these are correct.

9. Which association is incorrect?
   a. protein—trypsin
   b. fat—bile
   c. fat—lipase
   d. maltose—pepsin
   e. starch—amylase

10. Most of the products of digestion are absorbed across the
    a. squamous epithelium of the esophagus.
    b. striated walls of the trachea.
    c. convoluted walls of the stomach.
    d. fingerlike villi of the small intestine.
    e. smooth wall of the large intestine.

11. Bile
    a. is an important enzyme for the digestion of fats.
    b. cannot be stored.
    c. is made by the gallbladder.
    d. emulsifies fat.
    e. All of these are correct.

12. Which of these is NOT a function of the liver in adults?
    a. produces bile
    b. detoxifies alcohol
    c. stores glucose
    d. produces urea
    e. makes red blood cells

13. The large intestine
    a. digests all types of food.
    b. is the longest part of the intestinal tract.
    c. absorbs water.
    d. is connected to the stomach.
    e. is subject to hepatitis.

14. How many small servings of meat are sufficient in the daily diet?
    a. 6–11
    b. 2–4
    c. 2–3
    d. 3–4

15. The amino acids that must be consumed in the diet are called essential. Nonessential amino acids
    a. can be produced by the body.
    b. are only needed occasionally.
    c. are stored in the body until needed.
    d. can be taken in by supplements.

16. Which of the following are often organic portions of important coenzymes?
    a. minerals
    b. vitamins
    c. protein
    d. carbohydrates

17. Bulimia nervosa is NOT characterized by
    a. a restrictive diet often with excessive exercise.
    b. binge eating followed by purging.
    c. an obsession about body shape and weight.
    d. a distorted body image so person feels fat even when emaciated.
    e. a health risk due to this complex.

18. Label each organ indicated in the diagram (a–h). For the arrows (i–k), use either glucose, amino acids, lipids, or water.

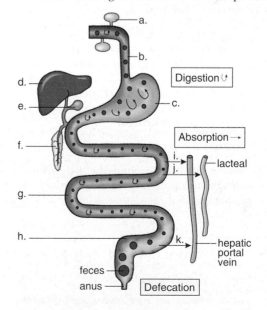

13.1 The Blood Vessels

The cardiovascular system has three types of blood vessels: the **arteries** (and arterioles), which carry blood away from the heart to the capillaries; the **capillaries,** which permit exchange of material with the tissues; and the **veins** (and venules), which return blood from the capillaries to the heart.

The blood vessels require oxygen and nutrients just as do other tissues, and therefore the larger ones have blood vessels in their own walls.

The Arteries

The largest artery in the human body, the aorta, is about 25 mm wide. An arterial wall has three layers (Fig. 13.1a). The inner layer is a simple squamous epithelium called endothelium with a connective tissue basement membrane that contains elastic fibers. The middle layer is the thickest layer and consists of smooth muscle that can contract to regulate blood flow and blood pressure. The outer layer is fibrous connective tissue near the middle layer, but it becomes loose connective tissue at its periphery.

Smaller arteries branch into a number of arterioles. **Arterioles** are small arteries just visible to the naked eye, being under 0.5 mm in diameter. The middle layer of arterioles has some elastic tissue but is composed mostly of smooth muscle whose fibers encircle the arteriole. When these muscle fibers are contracted, the vessel has a smaller diameter (is constricted); when these muscle fibers are relaxed, the vessel has a larger diameter (is dilated). Whether arterioles are constricted or dilated affects blood pressure. The greater the number of vessels dilated, the lower the blood pressure.

The Capillaries

Arterioles branch into capillaries, which are extremely narrow—about 8–10 μm wide. Capillaries have one-cell-thick walls composed only of endothelium with a basement

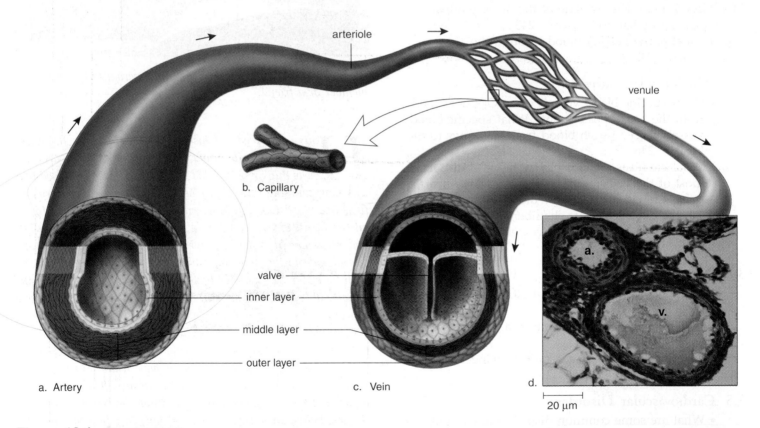

b. Capillary

arteriole

venule

valve

inner layer

middle layer

outer layer

a. Artery

c. Vein

d.

20 μm

Figure 13.1   Blood vessels.
The walls of arteries and veins have three layers. The inner layer is composed largely of endothelium, with a basement membrane that has elastic fibers; the middle layer is smooth muscle tissue; the outer layer is connective tissue (largely collagen fibers). **a.** Arteries have a thicker wall than veins because they have a larger middle layer than veins. **b.** Capillary walls are one-cell-thick endothelium. **c.** Veins are larger in diameter than arteries, so that collectively veins have a larger holding capacity than arteries. **d.** Light micrograph of an artery (a) and a vein (v).

membrane. Although each capillary is small, they form vast networks; their total surface area in humans is about 6,000 square meters. Capillary beds (networks of many capillaries) are present in all regions of the body; consequently, a cut to any body tissue draws blood. Capillaries are a very important part of the human cardiovascular system because an exchange of substances takes place across their thin walls. Oxygen and nutrients, such as glucose, diffuse out of a capillary into the tissue fluid that surrounds cells. Wastes, such as carbon dioxide, diffuse into the capillary. Some water also leaves a capillary; any excess is picked up by lymphatic vessels, as discussed later in the chapter. The relative constancy of tissue fluid is absolutely dependent upon capillary exchange.

Since capillaries serve the cells, the heart and the other vessels of the cardiovascular system can be thought of as the means by which blood is conducted to and from the capillaries. Only certain capillaries are open at any given time. For example, after eating, the capillaries that serve the digestive system are open and those that serve the muscles are closed. Shunting of blood is possible because each capillary bed has an arteriovenous shunt that allows blood to go directly from the arteriole to the venule (Fig. 13.2). Contracted sphincter muscles prevent the blood from entering the capillary vessels.

The Veins

Veins and venules take blood from the capillary beds to the heart. First, the **venules** (small veins) drain blood from the capillaries and then join to form a vein. The walls of veins (and venules) have the same three layers as arteries, but there is less smooth muscle and connective tissue (Fig. 13.1*c*). Therefore, the wall of a vein is thinner than that of an artery. Also, veins often have **valves,** which allow blood to flow only toward the heart when open and prevent the backward flow of blood when closed. Valves are found in the veins that carry blood against the force of gravity, especially the veins of the lower extremities.

Since the walls of veins are thinner, they can expand to a greater extent (Fig. 13.1*d*). At any one time, about 70% of the blood is in the veins. In this way, the veins act as a blood reservoir. If blood is lost due to hemorrhaging, nervous stimulation causes the veins to constrict, providing more blood to the rest of the body.

Arteries and arterioles carry blood away from the heart toward the capillaries; capillaries join arterioles to venules; veins and venules return blood from the capillaries to the heart.

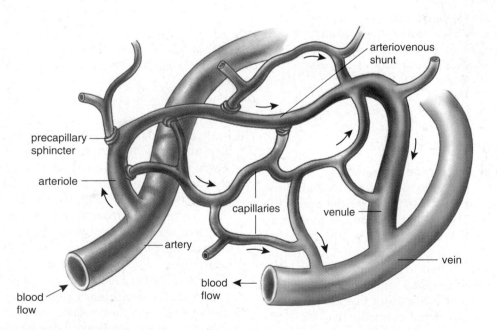

Figure 13.2   Anatomy of a capillary bed.
A capillary bed forms a maze of capillary vessels that lies between an arteriole and a venule. When sphincter muscles are relaxed, the capillary bed is open, and blood flows through the capillaries. When sphincter muscles are contracted, blood flows through a shunt that carries blood directly from an arteriole to a venule. As blood passes through a capillary in the tissues, it gives up its oxygen (O_2). Therefore, blood goes from being O_2-rich in the arteriole (red color) to being O_2-poor in the vein (blue color).

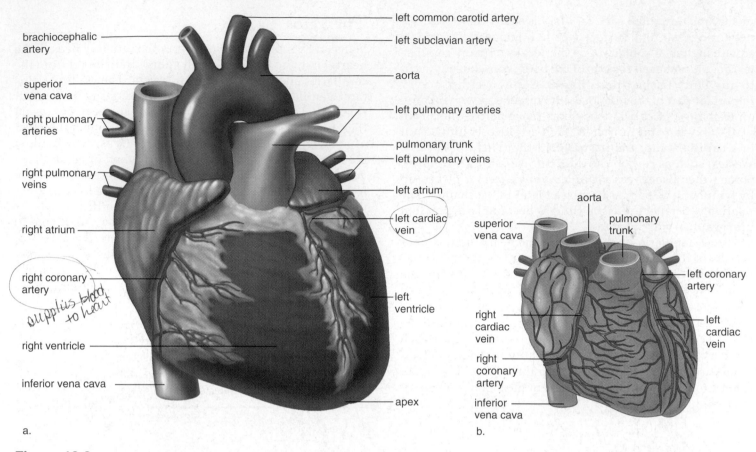

left common carotid artery
brachiocephalic artery
left subclavian artery
superior vena cava
aorta
right pulmonary arteries
left pulmonary arteries
pulmonary trunk
left pulmonary veins
right pulmonary veins
left atrium
right atrium
left cardiac vein
right coronary artery
supplies blood to heart
left ventricle
right ventricle
inferior vena cava
apex
a.

aorta
superior vena cava
pulmonary trunk
left coronary artery
right cardiac vein
left cardiac vein
right coronary artery
inferior vena cava
b.

Figure 13.3   External heart anatomy.
a. The superior vena cava and the pulmonary trunk are attached to the right side of the heart. The aorta and pulmonary veins are attached to the left side of the heart. The right ventricle forms most of the ventral surface of the heart, and the left ventricle forms most of the dorsal surface. **b.** The coronary arteries and cardiac veins pervade cardiac muscle. The coronary arteries bring oxygen and nutrients to cardiac cells, which derive no benefit from blood coursing through the heart.

13.2 The Heart

The **heart** is a cone-shaped, muscular organ about the size of a fist. It is located between the lungs directly behind the sternum (breastbone) and is tilted so that the apex (the pointed end) is oriented to the left. The major portion of the heart, called the **myocardium,** consists largely of cardiac muscle tissue. The muscle fibers of the myocardium are branched and tightly joined to one another. The heart lies within the **pericardium,** a thick, membranous sac that secretes a small quantity of lubricating liquid. The inner surface of the heart is lined with endocardium, which consists of connective tissue and endothelial tissue.

The heart has four chambers. The two upper, thin-walled atria (sing., **atrium**) have wrinkled, protruding appendages called auricles. The two lower chambers are the thick-walled **ventricles,** which pump the blood (Fig. 13.3).

Internally, a wall called the septum separates the heart into a right side and a left side (Fig. 13.4a). The heart has four valves, which direct the flow of blood and prevent its

backward movement. The two valves that lie between the atria and the ventricles are called the **atrioventricular valves.** These valves are supported by strong fibrous strings called **chordae tendineae.** The chordae, which are attached to muscular projections of the ventricular walls, support the valves and prevent them from inverting when the heart contracts. The atrioventricular valve on the right side is called the tricuspid valve because it has three flaps, or cusps. The atrioventricular valve on the left side is called the bicuspid (or mitral) valve because it has two flaps. The remaining two valves are the **semilunar valves,** whose flaps resemble half-moons, between the ventricles and their attached vessels. The pulmonary semilunar valve lies between the right ventricle and the pulmonary trunk. The aortic semilunar valve lies between the left ventricle and the aorta.

Humans have a four-chambered heart (two atria and two ventricles). A septum separates the right side from the left side.

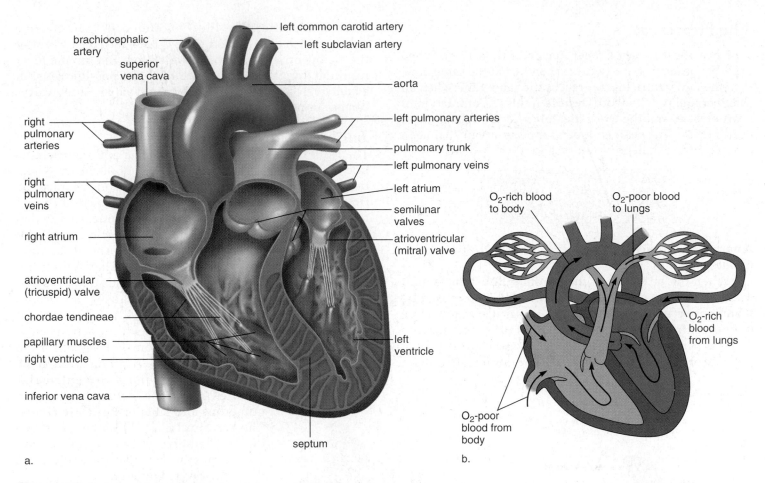

Figure 13.4   **Internal view of the heart.**
a. The heart has four valves. The atrioventricular valves allow blood to pass from the atria to the ventricles, and the semilunar valves allow blood to pass out of the heart. **b.** This diagrammatic representation of the heart allows you to trace the path of the blood through the heart.

Passage of Blood Through the Heart

We can trace the path of blood through the heart (Fig. 13.4b) in the following manner:

- The superior vena cava and the inferior vena cava, which carry O_2-poor blood, enter the right atrium.
- The right atrium sends blood through an atrioventricular valve (the tricuspid valve) to the right ventricle.
- The right ventricle sends blood through the pulmonary semilunar valve into the pulmonary trunk. The pulmonary trunk divides into two **pulmonary arteries,** which go to the lungs.
- Four **pulmonary veins,** which carry O_2-rich blood, enter the left atrium.
- The left atrium sends blood through an atrioventricular valve (the bicuspid or mitral valve) to the left ventricle.
- The left ventricle sends blood through the aortic semilunar valve into the aorta to the body proper.

From this description, you can see that O_2-poor blood never mixes with O_2-rich blood and that blood must go through the lungs in order to pass from the right side to the left side of the heart. In fact, the heart is a double pump because the right ventricle of the heart sends blood through the lungs, and the left ventricle sends blood throughout the body. Since the left ventricle has the harder job of pumping blood to the entire body, its walls are thicker than those of the right ventricle, which pumps blood a relatively short distance to the lungs.

The pumping of the heart sends blood out under pressure into the arteries. Because the left side of the heart is the stronger pump, blood pressure is greatest in the aorta. Blood pressure then decreases as the cross-sectional area of arteries and then arterioles increases.

The right side of the heart pumps blood to the lungs, and the left side of the heart pumps blood throughout the body.

The Heartbeat

Each heartbeat is called a **cardiac cycle** (Fig. 13.5). When the heart beats, first the two atria contract at the same time; then the two ventricles contract at the same time. Then all chambers relax. The word **systole** refers to contraction of heart muscle, and the word **diastole** refers to relaxation of heart muscle. The heart contracts, or beats, about 70 times a minute, and each heartbeat lasts about 0.85 second.

Time	Atria	Ventricles
0.15 sec	Systole	Diastole
0.30 sec	Diastole	Systole
0.40 sec	Diastole	Diastole

A normal adult rate at rest can vary from 60 to 80 beats per minute.

When the heart beats, the familiar "lub-dup" sound occurs. The longer and lower-pitched "lub" is caused by vibrations occurring when the atrioventricular valves close due to ventricular contraction. The shorter and sharper "dup" is heard when the semilunar valves close due to back pressure of blood in the arteries. A heart murmur, or a slight slush sound after the "lub," is often due to ineffective valves, which allow blood to pass back into the atria after the atrioventricular valves have closed. Rheumatic fever resulting from a bacterial infection is one possible cause of a faulty valve, particularly the bicuspid valve. Faulty valves can be surgically corrected.

Intrinsic Control of Heartbeat

The rhythmical contraction of the atria and ventricles is due to the intrinsic conduction system of the heart. Nodal tissue, which has both muscular and nervous characteristics, is a unique type of cardiac muscle located in two regions of the heart. The **SA (sinoatrial) node** is located in the upper dorsal wall of the right atrium; the **AV (atrioventricular) node** is located in the base of the right atrium very near the septum (Fig. 13.6a). The SA node initiates the heartbeat and automatically sends out an excitation impulse every 0.85 second; this causes the atria to contract. When impulses reach the AV node, there is a slight delay that allows the atria to finish their contraction before the ventricles begin their contraction. The signal for the ventricles to contract travels from the AV node through the two branches of the **atrioventricular bundle** (AV bundle) before reaching the numerous and smaller **Purkinje fibers.** The AV bundle, its branches, and the Purkinje fibers consist of specialized cardiac muscle fibers that efficiently cause the ventricles to contract.

The SA node is called the **pacemaker** because it usually keeps the heartbeat regular. If the SA node fails to work properly, the heart still beats due to impulses generated by the AV node. But the beat is slower (40 to 60 beats per minute). To correct this condition, it is possible to implant an artificial pacemaker, which automatically gives an electrical stimulus to the heart every 0.85 second.

The intrinsic conduction system of the heart consists of the SA node, the AV node, the atrioventricular bundle, and the Purkinje fibers.

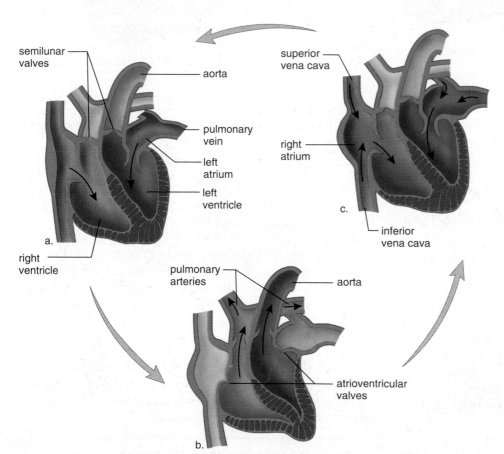

Figure 13.5   Stages in the cardiac cycle.
a. When the atria contract, the ventricles are relaxed and filling with blood. **b.** When the ventricles contract, the atrioventricular valves are closed, the semilunar valves are open, and the blood is pumped into the pulmonary trunk and aorta. **c.** When the heart is relaxed, both the atria and the ventricles are filling with blood.

Extrinsic Control of Heartbeat

The body has an extrinsic way to regulate the heartbeat. A cardiac control center in the medulla oblongata, a portion of the brain that controls internal organs, can alter the beat of the heart by way of the autonomic system, a division of the nervous system. The autonomic system has two subdivisions: the parasympathetic system, which promotes those functions we tend to associate with a resting state, and the sympathetic system, which brings about those responses we associate with increased activity and/or stress. The parasympathetic system decreases SA and AV nodal activity when we are inactive, and the sympathetic system increases SA and AV nodal activity when we are active or excited.

The hormones epinephrine and norepinephrine, which are released by the adrenal medulla, also stimulate the heart. During exercise, for example, the heart pumps faster and stronger due to sympathetic stimulation and due to the release of epinephrine and norepinephrine.

The body has an extrinsic way to regulate the heartbeat. The autonomic system and hormones can modify the heartbeat rate.

The Electrocardiogram

An **electrocardiogram (ECG)** is a recording of the electrical changes that occur in myocardium during a cardiac cycle. Body fluids contain ions that conduct electrical currents, and therefore the electrical changes in myocardium can be detected on the skin's surface. When an electrocardiogram is being taken, electrodes placed on the skin are connected by wires to an instrument that detects the myocardium's electrical changes. Thereafter, a pen rises or falls on a moving strip of paper. Figure 13.6b depicts the pen's movements during a normal cardiac cycle.

When the SA node triggers an impulse, the atrial fibers produce an electrical change called the P wave. The P wave indicates that the atria are about to contract. After that, the QRS complex signals that the ventricles are about to contract. The electrical changes that occur as the ventricular muscle fibers recover produce the T wave.

Various types of abnormalities can be detected by an electrocardiogram. One of these, called ventricular fibrillation, causes uncoordinated contraction of the ventricles (Fig. 13.6c). Ventricular fibrillation is of special interest because it can be caused by an injury or drug overdose. It is the most common cause of sudden cardiac death in a seemingly healthy person over age 35. Once the ventricles are fibrillating, they have to be defibrillated by applying a strong electrical current for a short period of time. Then the SA node may be able to reestablish a coordinated beat.

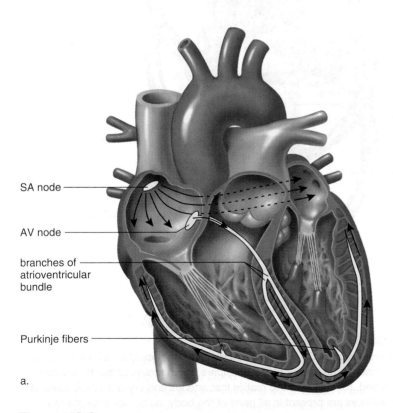

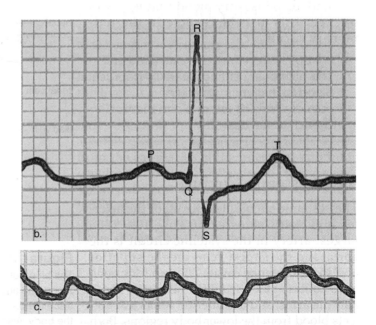

Figure 13.6 **Conduction system of the heart.**
a. The SA node sends out a stimulus, which causes the atria to contract. When this stimulus reaches the AV node, it signals the ventricles to contract. Impulses pass down the two branches of the atrioventricular bundle to the Purkinje fibers, and thereafter the ventricles contract. **b.** A normal ECG indicates that the heart is functioning properly. The P wave occurs just prior to atrial contraction; the QRS complex occurs just prior to ventricular contraction; and the T wave occurs when the ventricles are recovering from contraction. **c.** Ventricular fibrillation produces an irregular electrocardiogram due to irregular stimulation of the ventricles.

Blood Flow

The beating of the heart is necessary to homeostasis because it creates the pressure that propels blood in the arteries and the arterioles. Arterioles lead to the capillaries where exchange with tissue fluid takes place.

Blood Flow in Arteries

Blood pressure is the pressure of blood against the wall of a blood vessel. Clinicians use a sphygmomanometer to measure blood pressure, usually in the brachial artery of the arm. The highest arterial pressure, called the **systolic pressure,** is reached during ejection of blood from the heart. The lowest arterial pressure, called the **diastolic pressure,** occurs while the heart ventricles are relaxing. Blood pressure in the brachial artery is typically 120 mm mercury (Hg) over 80 mm Hg, or simply 120/80. The higher number is the systolic pressure, and the lower number is the diastolic pressure.

Although the blood pressure in the brachial artery is typically about 120/80, blood pressure actually varies throughout the body. As already stated, blood pressure is highest in the aorta and lowest in the venae cavae. Blood pressure decreases with distance from the left ventricle because there are more arterioles than arteries, which increases the total cross-sectional area of the blood vessels. The decrease in blood pressure causes the blood velocity to gradually decrease as it flows toward the capillaries (Fig. 13.9).

Blood Flow in Capillaries

Because there are many more capillaries than arterioles, blood moves even more slowly through the capillaries. This slow progress allows time for substances to be exchanged between the blood in the capillaries and the surrounding tissues.

Blood Flow in Veins

Blood pressure is minimal in venules and veins (20–0 mm Hg). Instead of blood pressure, venous return depends upon three factors: skeletal muscle contraction, the presence of valves in veins, and respiratory movements. When the skeletal muscles contract, they compress the weak walls of the veins. This causes blood to move past the next valve. Once past the valve, blood cannot flow backward. The importance of muscle contraction in moving blood in the venous vessels can be demonstrated by forcing a person to stand rigidly still for an hour or so. Frequently, the person faints because blood has collected in the limbs, depriving the brain of needed blood flow and oxygen. In this case, fainting is beneficial because the resulting horizontal position aids in getting blood to the head.

When a person inhales, the thoracic pressure falls and abdominal pressure rises as the chest expands. This also aids the flow of venous blood back to the heart because blood flows in the direction of reduced pressure. Blood velocity increases slightly in the venous vessels due to a progressive reduction in the cross-sectional area as small venules join to form veins.

The backward pressure of blood can sometimes cause the valves to become weak and ineffective. The accumulation of blood in veins is commonly referred to as **varicose veins.** Varicose veins develop when the valves of veins become weak and ineffective due to the backward pressure of blood. Abnormal and irregular dilations are particularly apparent in the superficial (near the surface) veins of the lower legs. Crossing the legs or sitting in a chair so that its edge presses against the back of the knees can contribute to the development of varicose veins. Varicose veins also occur in the rectum, where they are called piles, or more properly, hemorrhoids. **Phlebitis,** an inflammation of a vein, is a more serious condition because it can lead to blood clots. If a clot is carried to a pulmonary vessel, death can result.

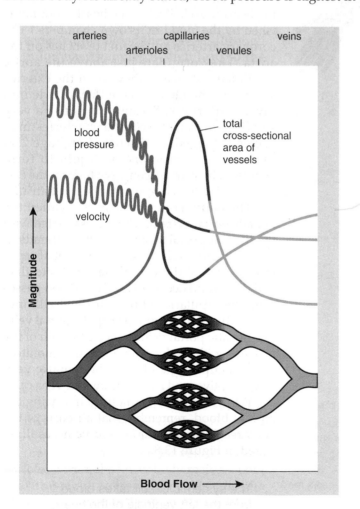

Figure 13.9   Cross-sectional area as it relates to blood pressure and blood velocity.
Blood pressure and blood velocity drop off in capillaries because capillaries have a greater cross-sectional area than arterioles.

Blood pressure accounts for the flow of blood in the arteries and the arterioles. Skeletal muscle contraction, valves in veins, and respiratory movements account for the flow of blood in the venules and the veins.

13.4 Blood

Blood is the only liquid tissue in the body. If blood is transferred from a person's vein to a test tube and is prevented from clotting, it separates into two layers (Fig. 13.10). The lower layer consists of the **formed elements** and the upper layer is **plasma.** The formed elements consist of red blood cells (erythrocytes), white blood cells (leukocytes), and

FORMED ELEMENTS	Function and Description	Source
Red Blood Cells (erythrocytes) 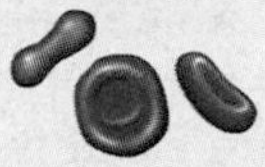 4 million–6 million per mm³ blood	Transport O₂ and help transport CO₂ 7–8 μm in diameter Bright-red to dark-purple biconcave disks without nuclei	Red bone marrow
White Blood Cells (leukocytes) 4,000–11,000 per mm³ blood	Fight infection	Red bone marrow
Granular leukocytes		
• Basophils 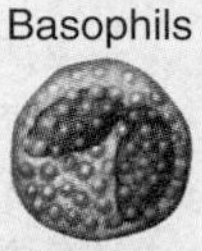 20–50 per mm³ blood	10–12 μm in diameter Spherical cells with lobed nuclei; large, irregularly shaped, deep-blue granules in cytoplasm; release histamine which promotes blood flow to injured tissues	
• Eosinophils  100–400 per mm³ blood	10–14 μm in diameter Spherical cells with bilobed nuclei; coarse, deep-red, uniformly sized granules in cytoplasm; phagocytize antigen-antibody complexes and allergens	
• Neutrophils  3,000–7,000 per mm³ blood	10–14 μm in diameter Spherical cells with multilobed nuclei; fine, pink granules in cytoplasm; phagocytize pathogens	
Agranular leukocytes		
• Lymphocytes  1,500–3,000 per mm³ blood	5–17 μm in diameter (average 9–10 μm) Spherical cells with large round nuclei; responsible for specific immunity	
• Monocytes 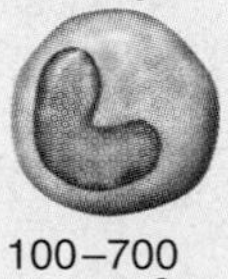 100–700 per mm³ blood	10–24 μm in diameter Large spherical cells with kidney-shaped, round, or lobed nuclei; become macrophages which phagocytize pathogens and cellular debris	
• **Platelets** (thrombocytes)  150,000–300,000 per mm³ blood	Aid clotting 2–4 μm in diameter Disk-shaped cell fragments with no nuclei; purple granules in cytoplasm	Red bone marrow

PLASMA	Function	Source
Water (90–92% of plasma)	Maintains blood volume; transports molecules	Absorbed from intestine
Plasma proteins (7–8% of plasma)	Maintain blood osmotic pressure and pH	Liver
Albumin	Maintain blood volume and pressure	
Globulins	Transport; fight infection	
Fibrinogen	Clotting	
Salts (less than 1% of plasma)	Maintain blood osmotic pressure and pH; aid metabolism	Absorbed from intestine
Gases		
Oxygen	Cellular respiration	Lungs
Carbon dioxide	End product of metabolism	Tissues
Nutrients	Food for cells	Absorbed from intestine

Lipids
  Glucose
  Amino acids		

Nitrogenous wastes	Excretion by kidneys	Liver
Urea Uric acid		
Other		
Hormones, vitamins, etc.	Aid metabolism	Varied

Plasma 55%

Formed elements 45%

• with Wright's stain

Figure 13.10 Composition of blood.
When blood is transferred to a test tube and prevented from clotting, it forms two layers. The transparent, yellow top layer is plasma, the liquid portion of blood. The formed elements are in the bottom layer. This table describes these components in detail.

blood platelets (thrombocytes). Formed elements make up about 45% of the total volume of whole blood. Plasma, which accounts for about 55% of the total volume of whole blood, contains a variety of inorganic and organic substances dissolved or suspended in water.

Plasma proteins, which make up 7–8% of plasma, assist in transporting large organic molecules in blood. For example, **albumin** transports bilirubin, a breakdown product of hemoglobin. The lipoproteins that transport cholesterol contain a type of protein called globulins. Plasma proteins also maintain blood volume because their size prevents them from readily passing through a capillary wall. Therefore, capillaries are always areas of lower water concentration compared to tissue fluid, and water automatically diffuses into capillaries. Certain plasma proteins have specific functions. As discussed later in the chapter, fibrinogen is necessary to blood clotting, and immunoglobulins are antibodies, which help fight infection.

We shall see that blood has transport functions, regulatory functions, and protective functions. Blood transports oxygen and nutrients to the capillaries and takes carbon dioxide and wastes away from the capillaries. It also transports hormones. Blood helps regulate body temperature by dispersing body heat and blood pressure because the plasma proteins contribute to the osmotic pressure of blood. It also helps regulate pH by means of the buffers it contains. It helps guard the body against invasion by **pathogens** (e.g., disease-causing viruses and bacteria), and it clots, preventing a potentially life-threatening loss of blood.

The Red Blood Cells

Red blood cells are continuously manufactured in the red bone marrow of the skull, the ribs, the vertebrae, and the ends of the long bones. Normally, there are 4 to 6 million red blood cells per mm³ of whole blood.

Red blood cells are biconcave disks (Fig. 13.11). Their shape increases their flexibility for moving through capillary beds and their surface area for the diffusion of gases. Red blood cells carry oxygen because they contain **hemoglobin,** the respiratory pigment. Since hemoglobin is a red pigment, the cells are red. A hemoglobin molecule contains four polypeptide chains that make up the protein globin. Each chain is associated with heme, a complex iron-containing group. The iron portion of hemoglobin acquires oxygen in the lungs and gives it up in the tissues. Plasma carries only about 0.3 ml of oxygen per 100 ml of blood, but whole blood carries 20 ml of oxygen per 100 ml of blood. This shows that hemoglobin increases the oxygen-carrying capacity of blood more than 60 times.

Carbon monoxide is an air pollutant that comes primarily from the incomplete combustion of natural gas and gasoline. Unfortunately, carbon monoxide combines with hemoglobin more readily than does oxygen, and it stays combined for several hours, making hemoglobin unavailable for oxygen transport. Some homes are equipped with carbon monoxide detectors because you can neither smell nor see this deadly gas.

Possibly because they lack nuclei, red blood cells live only about 120 days. They are destroyed chiefly in the

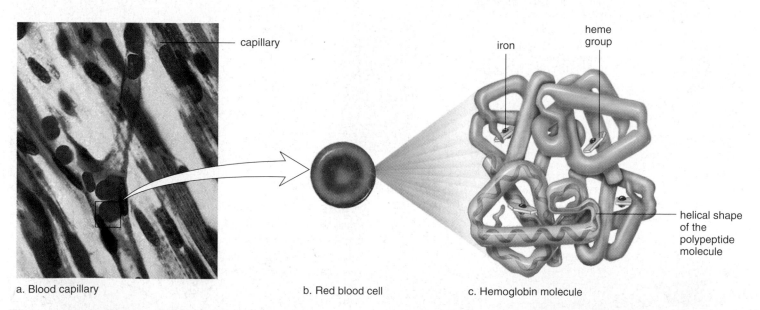

a. Blood capillary   b. Red blood cell   c. Hemoglobin molecule

Figure 13.11   Physiology of red blood cells.
a. Red blood cells move in single file through the capillaries. **b.** Each red blood cell is a biconcave disk containing many molecules of hemoglobin, the respiratory pigment. **c.** Hemoglobin contains four polypeptide chains (blue). There is an iron-containing heme group in the center of each chain. Oxygen combines loosely with iron when hemoglobin is oxygenated. Oxyhemoglobin is bright red, and deoxyhemoglobin is a dark maroon color.

liver and the spleen, where they are engulfed by large phagocytic cells. When red blood cells are broken down, the hemoglobin is released. The iron is recovered and returned to the bone marrow for reuse. The heme portion of hemoglobin undergoes chemical degradation and is excreted as bile pigments by the liver into the bile.

When the body has an insufficient number of red blood cells or the red blood cells do not contain enough hemoglobin, the individual suffers from **anemia** and has a tired, rundown feeling. In iron-deficiency anemia, the most common type, the hemoglobin level is low, probably due to a diet that does not contain enough iron. Usually, the number of red blood cells increases whenever arterial blood carries a reduced amount of oxygen, as happens when an individual first takes up residence at a high altitude. Under these circumstances, the kidneys increase their production of a hormone called **erythropoietin,** which speeds the maturation of red blood cells in the bone marrow.

The White Blood Cells

White blood cells (leukocytes) differ from red blood cells in that they are usually larger, have a nucleus, lack hemoglobin, and without staining appear translucent. White blood cells are not as numerous as red blood cells, with only 5,000–11,000 cells per mm^3. White blood cells fight infection and play a role in the development of immunity, the ability to resist disease.

On the basis of structure, it is possible to divide white blood cells into **granular leukocytes** and **agranular leukocytes.** Granular leukocytes are filled with spheres that contain various enzymes and proteins, which help white blood cells defend the body against microbes. **Neutrophils** are granular leukocytes with a multilobed nucleus joined by nuclear threads. They are the most abundant of the white blood cells and are able to phagocytize and digest bacteria. The agranular leukocytes (monocytes and lymphocytes) typically have a spherical or kidney-shaped nucleus. **Monocytes** are the largest of the white blood cells, and after they take up residence in the tissues, they differentiate into the even larger

macrophages (Fig. 13.12). Macrophages phagocytize microbes and stimulate other white blood cells to defend the body. The **lymphocytes** are of two types, B lymphocytes and T lymphocytes, and each type plays a specific role in immunity.

If the total number of white blood cells increases beyond normal, disease may be present. Sometimes an increase or decrease of only one type of white blood cell, as detected with a differential white blood cell count, is a sign of infection. A person with **infectious mononucleosis,** caused by the Epstein-Barr virus, has an excessive number of lymphocytes of the B type. A person with AIDS, caused by an HIV infection, has an abnormally low number of lymphocytes of the T type. **Leukemia** is a form of cancer characterized by uncontrolled production of abnormal white blood cells.

White blood cells live different lengths of time. Many live only a few days and are believed to die combating invading pathogens. Others live months or even years.

Red blood cells, which are more numerous and smaller than white blood cells, contain hemoglobin and carry oxygen. White blood cells, which are translucent when not stained, all fight infection but are varied as to their specific characteristics and functions.

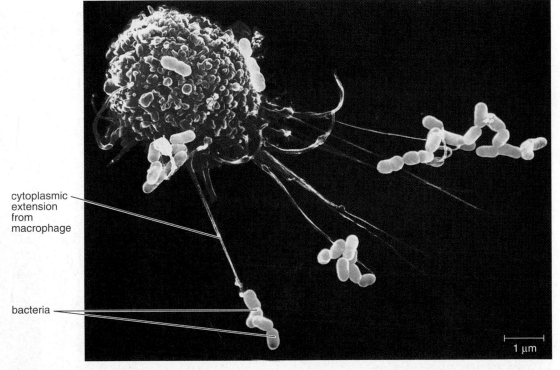

cytoplasmic extension from macrophage

bacteria

1 µm

Figure 13.12   Macrophage (red) engulfing bacteria (green).
Monocyte-derived macrophages are the body's scavengers. They engulf microbes and debris in the body's fluids and tissues, as illustrated in this colorized scanning electron micrograph.

The Platelets and Blood Clotting

Platelets (thrombocytes) result from fragmentation of certain large cells, called **megakaryocytes,** in the red bone marrow. Platelets are produced at a rate of 200 billion a day, and the blood contains 150,000–300,000 per mm^3. These formed elements are involved in the process of blood **clotting,** or coagulation.

There are at least 12 clotting factors in the blood that participate with platelets in the formation of a blood clot. We will discuss the roles played by **fibrinogen** and **prothrombin,** which are proteins manufactured and deposited in blood by the liver. Vitamin K, found in green vegetables and also formed by intestinal bacteria, is necessary for the production of prothrombin, and if by chance this vitamin is missing from the diet, hemorrhagic disorders develop.

Blood Clotting

When a blood vessel in the body is damaged, platelets clump at the site of the puncture and partially seal the leak. They and the injured tissues release a clotting factor called **prothrombin activator,** which converts prothrombin to thrombin. This reaction requires calcium ions (Ca^{2+}). **Thrombin,** in turn, acts as an enzyme that severs two short amino acid chains from each fibrinogen molecule. These activated fragments then join end to end, forming long threads of **fibrin.** Fibrin threads wind around the platelet plug in the damaged area of the blood vessel and provide the framework for the clot. Red blood cells also are trapped within the fibrin threads; these cells make a clot appear red (Fig. 13.13). A fibrin clot is present only temporarily. As soon as blood vessel repair is initiated, an enzyme called plasmin destroys the fibrin network and restores the fluidity of the plasma.

Table 13.1	Body Fluids Related to Blood
Name	**Composition**
Blood	Formed elements and plasma
Plasma	Liquid portion of blood
Serum	Plasma minus fibrinogen
Tissue fluid	Plasma minus most proteins
Lymph	Tissue fluid within lymphatic vessels

If blood is allowed to clot in a test tube, a yellowish fluid develops above the clotted material. This fluid is called **serum,** and it contains all the components of plasma except fibrinogen. Table 13.1 reviews the terms used to refer to body fluids related to blood.

Hemophilia

Hemophilia is an inherited clotting disorder caused by a deficiency in a clotting factor. The slightest bump to an affected person can cause bleeding into the joints. Cartilage degeneration in the joints and resorption of underlying bone can follow. Bleeding into muscles can lead to nerve damage and muscular atrophy. The most frequent cause of death is bleeding into the brain with accompanying neurological damage.

Platelets are cell fragments involved in blood clotting, an involved process that is necessary to keep the body in tact following an injury to a blood vessel.

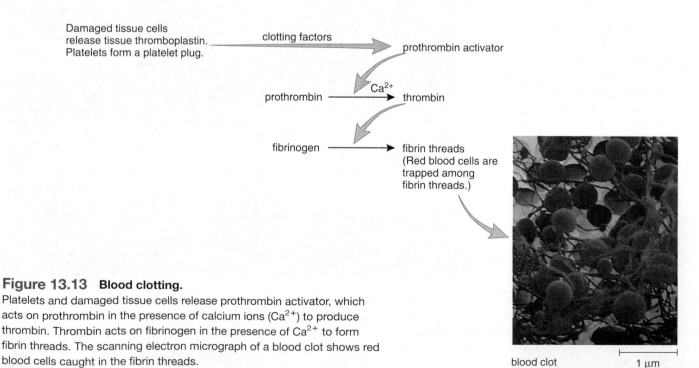

Figure 13.13   Blood clotting.
Platelets and damaged tissue cells release prothrombin activator, which acts on prothrombin in the presence of calcium ions (Ca^{2+}) to produce thrombin. Thrombin acts on fibrinogen in the presence of Ca^{2+} to form fibrin threads. The scanning electron micrograph of a blood clot shows red blood cells caught in the fibrin threads.

Bone Marrow Stem Cells

A **stem cell** is a cell that is ever capable of dividing and producing new cells that go on to differentiate into particular types of cells. It's been known for some time that the bone marrow has multipotent stem cells that give rise to other stem cells for the various formed elements (Fig. 13.14). Other tissues of the body also have stem cells. We have already mentioned the ability of the skin to rejuvenate itself. Only recently has it been discovered that several other organs, even the brain, have stem cells. Of particular interest are the so-called mesenchymal stem cells that can give rise to most any type of connective tissue, including cardiac muscle. The hope is that one day these stem cells could be used to repair the heart after a heart attack.

Many scientists want to test the use of embryonic cells to regenerate the various tissues of the adult body. The scientific benefits of using embryonic stem cells versus adult stem cells is not yet known. The use of adult stem cells does not require that an embryo be sacrificed, however.

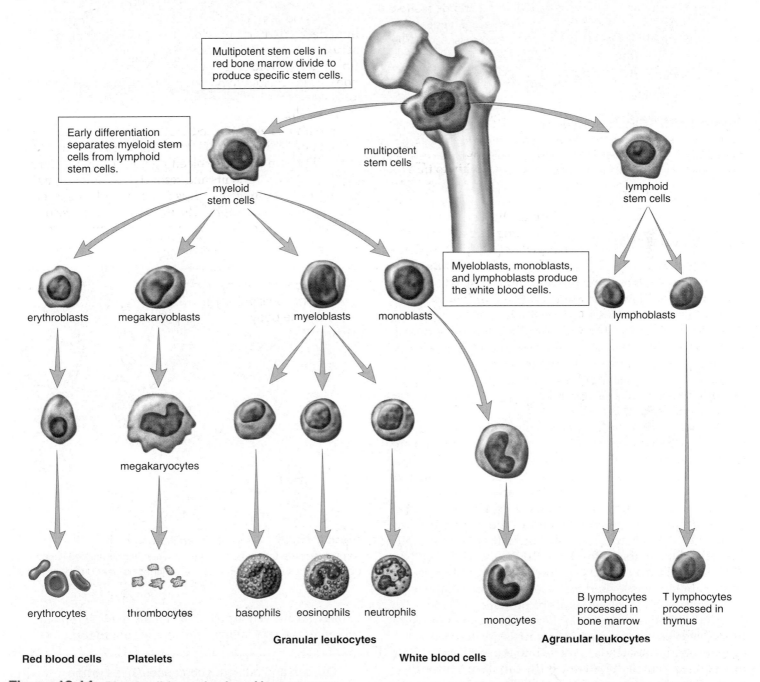

Figure 13.14   Blood cell formation in red bone marrow.
Multipotent stem cells give rise to two specialized stem cells. The myeloid stem cell gives rise to still other cells, which become red blood cells, platelets, and all the white blood cells except lymphocytes. The lymphoid stem cell gives rise to lymphoblasts, which become lymphocytes.

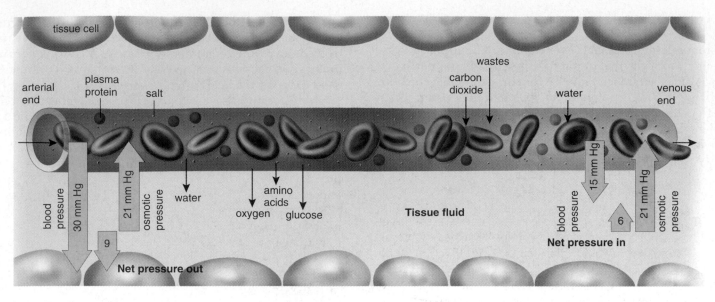

Figure 13.15   Capillary exchange.
At the arterial end of a capillary, the blood pressure is higher than the osmotic pressure; therefore, water tends to leave the bloodstream. In the midsection, the gases oxygen and carbon dioxide follow their concentration gradients, as do the nutrients glucose and amino acids. At the venous end of a capillary, the osmotic pressure is higher than the blood pressure; therefore, water tends to enter the bloodstream.

Capillary Exchange

Two forces primarily control movement of fluid through the capillary wall: osmotic pressure, which tends to cause water to move from tissue fluid to blood, and blood pressure, which tends to cause water to move in the opposite direction. At the arterial end of a capillary, blood pressure is higher than the osmotic pressure of blood (Fig. 13.15). Osmotic pressure is created by the presence of salts and the plasma proteins. Because blood pressure is higher than osmotic pressure at the arterial end of a capillary, water exits a capillary at this end.

Midway along the capillary, where blood pressure is lower, the two forces essentially cancel each other, and there is no net movement of water. Solutes now diffuse according to their concentration gradient—nutrients (glucose and oxygen) diffuse out of the capillary, and wastes (carbon dioxide) diffuse into the capillary. Red blood cells and almost all plasma proteins remain in the capillaries, but small substances leave. The substances that leave a capillary contribute to **tissue fluid,** the fluid between the body's cells. Since plasma proteins are too large to readily pass out of the capillary, tissue fluid tends to contain all components of plasma except lesser amounts of protein.

At the venous end of a capillary, where blood pressure has fallen even more, osmotic pressure is greater than blood pressure, and water tends to move into the capillary. Almost the same amount of fluid that left the capillary returns to it, although some excess tissue fluid is always collected by the lymphatic capillaries (Fig. 13.16). Tissue fluid contained within lymphatic vessels is called **lymph.** Lymph is returned

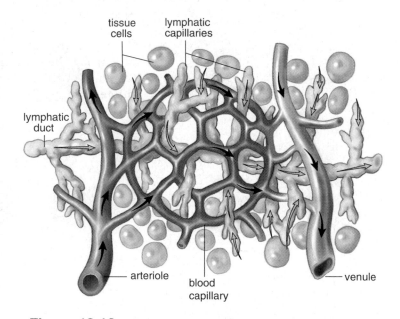

Figure 13.16   Lymphatic capillaries.
Arrows indicate that lymph is formed when lymphatic capillaries take up excess tissue fluid. Lymphatic capillaries lie near blood capillaries.

to the systemic venous blood when the major lymphatic vessels enter the subclavian veins in the shoulder region.

Oxygen and nutrient substances exit; carbon dioxide and waste molecules enter midway along a capillary.

13.5 Cardiovascular Disorders

Cardiovascular disease (CVD) is the leading cause of untimely death in the Western countries. Modern research efforts have resulted in improved diagnosis, treatment, and prevention. This section discusses the range of advances that have been made in these areas.

Atherosclerosis

Atherosclerosis is an accumulation of soft masses of fatty materials, particularly cholesterol, beneath the inner linings of arteries. Such deposits are called **plaque.** As it develops, plaque tends to protrude into the lumen of the vessel and interfere with the flow of blood (see Fig. 13A). In certain families, atherosclerosis is due to an inherited condition. The presence of the associated mutation can be detected, and this information is helpful if measures are taken to prevent the occurrence of the disease. In most instances, atherosclerosis begins in early adulthood and develops progressively through middle age, but symptoms may not appear until an individual is 50 or older. To prevent the onset and development of plaque, the American Heart Association recommends a diet low in saturated fat and cholesterol and rich in fruits and vegetables, as in the Health Focus on page 257.

Plaque can cause a clot to form on the irregular arterial wall. As long as the clot remains stationary, it is called a **thrombus,** but when or if it dislodges and moves along with the blood, it is called an **embolus. Thromboembolism,** a clot that has been carried in the bloodstream but is now stationary, must be treated, or complications can arise.

Stroke, Heart Attack, and Aneurysm

A cerebrovascular accident (CVA), also called a **stroke,** often results when a small cranial arteriole bursts or is blocked by an embolus. Lack of oxygen causes a portion of the brain to die, and paralysis or death can result. A person is sometimes forewarned of a stroke by a feeling of numbness in the hands or the face, difficulty in speaking, or temporary blindness in one eye.

A myocardial infarction (MI), also called a **heart attack,** occurs when a portion of the heart muscle dies due to a lack of oxygen. If a coronary artery becomes partially blocked, the individual may then suffer from **angina pectoris,** characterized by a squeezing sensation or a flash of burning. Nitroglycerin or related drugs dilate blood vessels and help relieve the pain. When a coronary artery is completely blocked, perhaps because of thromboembolism, a heart attack occurs.

An **aneurysm** is the ballooning of a blood vessel, most often the abdominal artery or the arteries leading to the brain. Atherosclerosis and hypertension can weaken the wall of an artery to the point that an aneurysm develops. If a major vessel such as the aorta should burst, death is likely. It is possible to replace a damaged or diseased portion of a vessel, such as an artery, with a plastic tube. Cardiovascular function is preserved, because exchange with tissue cells can still take place at the capillaries.

Coronary Bypass Operations

Each year thousands of persons have coronary bypass surgery because of an obstructed coronary artery. During this operation, a surgeon takes a segment from another blood vessel and stitches one end to the aorta and the other end to a coronary artery past the point of obstruction. Figure 13.17 shows a triple bypass in which three blood vessels have been used to allow blood to flow freely from the aorta to cardiac muscle by way of the coronary artery.

Gene therapy is now being used experimentally to grow new blood vessels that will carry blood to cardiac muscle. The surgeon need only make a small incision and inject many copies of the gene that codes for VEGF (vascular endothelial growth factor) between the ribs directly into the area of the heart that most needs improved blood flow. VEGF encourages new blood vessels to sprout out of an artery. If collateral blood vessels do form, they transport blood past clogged arteries, making bypass surgery unnecessary.

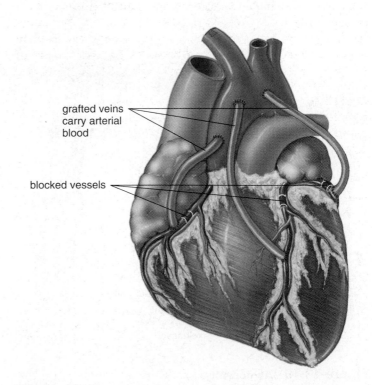

grafted veins carry arterial blood

blocked vessels

Figure 13.17 Coronary bypass operation.
During this operation, the surgeon grafts segments of another vessel, usually a small vein from the leg, between the aorta and the coronary vessels, bypassing areas of blockage. Patients who require surgery often receive two to five bypasses in a single operation.

Clearing Clogged Arteries

In **angioplasty,** a cardiologist threads a plastic tube into an artery of an arm or a leg and guides it through a major blood vessel toward the heart. When the tube reaches the region of plaque in an artery, a balloon attached to the end of the tube is inflated, forcing the vessel open (Fig. 13.18). However, the artery may not remain open because the trauma causes smooth muscle cells in the wall of the artery to proliferate and close it.

Two lines of attack are being explored. Small metal devices—either metal coils or slotted tubes, called stents—are expanded inside the artery to keep the artery open. When the stents are coated with heparin to prevent blood clotting and with chemicals to prevent arterial closing, results have been promising.

Dissolving Blood Clots

Medical treatment for thromboembolism includes the use of t-PA, a biotechnology drug. This drug converts plasminogen, a molecule found in blood, into plasmin, an enzyme that dissolves blood clots. In fact, t-PA, which stands for tissue plasminogen activator, is the body's own way of converting plasminogen to plasmin. t-PA is also being used for thrombolytic stroke patients but with limited success because some patients experience life-threatening bleeding in the brain. A better treatment might be new biotechnology drugs that act on the plasma membrane to prevent brain cells from releasing and/or receiving toxic chemicals caused by the stroke.

If a person has symptoms of angina or a stroke, aspirin may be prescribed. Aspirin reduces the stickiness of platelets and thereby lowers the probability that a clot will form. Evidence indicates that aspirin protects against first heart attacks, but there is no clear support for taking aspirin every day to prevent strokes in symptom-free people. Even so, some physicians recommend taking a low dosage of aspirin every day.

Heart Transplants and Artificial Hearts

Heart transplants are usually successful today but, unfortunately, there is a shortage of human organ donors. A functional, mechanical heart would give some persons their only hope of continuing to live. It's been nineteen years since an artificial heart, which needed a cumbersome external power supply, was implanted in a patient. In a medical breakthrough, a patient has recently received a self-contained artificial heart. A rotating centrifugal pump moves silicone hydraulic fluid between left and right sacs to force blood out of the heart into the pulmonary trunk and the aorta. An internal battery holds only a half hour charge and then has to be recharged by an external power source. The heart was considered successful because the patient lived longer than expected.

Sometimes it is possible to repair a weak heart instead of replacing it. For example, a back muscle can be wrapped around a heart to strengthen it. The muscle's nerve is stimulated with a kind of pacemaker that gives a burst of stimulation every 0.85 second. One day it may be possible to use cardiac muscle cell transplants to strengthen a heart.

Hypertension

It is estimated that about 20% of all Americans suffer from **hypertension,** which is high blood pressure. Included in this group are individuals with atherosclerosis. Hypertension is present when the systolic blood pressure is 140 or greater or the diastolic blood pressure is 90 or greater. While both the systolic and diastolic pressures are considered important, the diastolic pressure is emphasized when medical treatment is being considered.

Hypertension is sometimes called a silent killer because it may not be detected until a stroke or heart attack occurs. It has long been thought that a certain genetic makeup might account for the development of hypertension. Now researchers have discovered two genes that may be involved in some individuals. One gene codes for angiotensinogen, a plasma protein that is converted to a powerful vasoconstrictor in part by the product of the second gene.

At present, however, the best safeguard against the development of hypertension is to have regular blood pressure checks and to adopt a lifestyle that lowers the risk of hypertension.

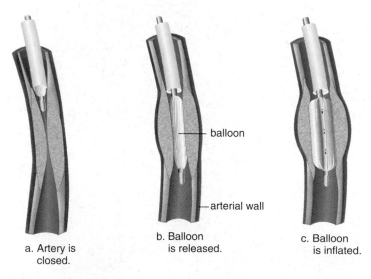

balloon

arterial wall

a. Artery is closed.

b. Balloon is released.

c. Balloon is inflated.

Figure 13.18 **Angioplasty.**
a. A plastic tube is inserted into the coronary artery until it reaches the clogged area. **b.** A metal tip with a balloon attached is pushed out the end of the plastic tube into the clogged area. **c.** When the balloon is inflated, the vessel opens. Sometimes metal coils or slotted tubes, called stents, are inserted to keep the vessel open.

Stroke, heart attack, and aneurysm are associated with both atherosclerosis and hypertension.

Prevention of Cardiovascular Disease

Certain genetic factors predispose an individual to cardiovascular disease, including family history of heart attack under age 55, male gender, and ethnicity (African Americans are at greater risk). Persons with one or more of these risk factors need not despair, however. It only means that they should pay particular attention to the following guidelines for a heart-healthy lifestyle.

The Don'ts

Smoking

When a person smokes, the drug nicotine, present in cigarette smoke, enters the bloodstream. Nicotine causes the arterioles to constrict and the blood pressure to rise. The heart has to pump harder to propel the blood through the lungs at a time when the blood's oxygen-carrying capacity is reduced by smoking.

Drug Abuse

Stimulants, such as cocaine and amphetamines, can cause an irregular heartbeat and lead to heart attacks even when using drugs for the first time. Intravenous drug use may also result in a cerebral blood clot and stroke.

Too much alcohol can destroy just about every organ in the body, the heart included. But investigators have discovered that people who take an occasional drink have a 20% lower risk of heart disease than do teetotalers. Two to four drinks a week for men, and one to three drinks for women are sufficient.

Weight Gain

Hypertension (high blood pressure) is prevalent in persons who are more than 20% above the recommended weight for their height. More tissues require servicing, and the heart sends the extra blood out under greater pressure. It may be harder to lose weight once it is gained, and therefore it is recommended that weight control be a lifelong endeavor. Even a slight decrease in weight can bring with it a reduction in hypertension.

The Do's

Healthy Diet

Diet influences the amount of cholesterol in the blood. Cholesterol is ferried by two types of plasma proteins, called LDL (low-density lipoprotein) and HDL (high-density lipoprotein). LDL (called "bad" lipoprotein) takes cholesterol from the liver to the tissues, and HDL (called "good" lipoprotein) transports cholesterol out of the tissues to the liver. When the LDL level in blood is high or the HDL level is abnormally low, plaque, which interferes with circulation, accumulates on arterial walls (Fig. 13A).

Eating foods high in saturated fat (red meat, cream, and butter) and foods containing so-called trans-fats (most margarines, commercially baked goods, and deep-fried foods) raises the LDL-cholesterol level. Replacement of these harmful fats with healthier ones, such as monounsaturated fats (olive and canola oil) and polyunsaturated fats (corn, safflower, and soybean oil), is recommended.

Evidence is mounting to suggest a role for antioxidant vitamins (A, E, and C) in preventing cardiovascular disease. Antioxidants protect the body from free radicals that oxidize cholesterol and damage the lining of an artery, leading to a blood clot that can block blood vessels. Nutritionists believe that consuming at least five servings of fruits and vegetables a day may protect against cardiovascular disease.

Cholesterol Profile

Starting at age 20, all adults are advised to have their cholesterol levels tested at least every five years. Even in healthy individuals, an LDL level above 160 mg/100 ml and an HDL level below 40 mg/100 ml are matters of concern. If a person has heart disease or is at risk for heart disease, an LDL level below 100 mg/100 ml is now recommended. Medications will most likely be prescribed for all those who do not meet these minimum guidelines.

Exercise

People who exercise are less apt to have cardiovascular disease. One study found that moderately active men who spent an average of 48 minutes a day on a leisure-time activity such as gardening, bowling, or dancing had one-third fewer heart attacks than peers who spent an average of only 16 minutes each day on those types of activities. Exercise not only helps keep weight under control, but may also help minimize stress and reduce hypertension.

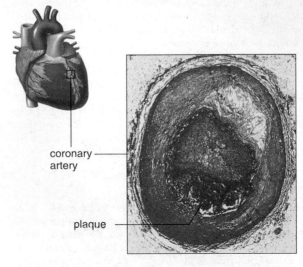

coronary artery

plaque

Figure 13A Coronary arteries and plaque.
When plaque is present in a coronary artery, a heart attack is more apt to occur because of restricted blood flow.

According to a 1993 study, about one million deaths a year in the United States could be prevented if people adopted the healthy lifestyle described in the Health Focus on page 257. Tobacco, lack of exercise, and a high-fat diet probably cost the nation about $200 billion per year in health-care costs. To what lengths should we go to prevent these deaths and reduce health-care costs?

E.A. Miller, a meat-packing entity of ConAgra in Hyrum, Utah, charges extra for medical coverage of employees who smoke. Eric Falk, Miller's director of human resources, says, "We want to teach employees to be responsible for their behavior." Anthem Blue Cross–Blue Shield of Cincinnati, Ohio, takes a more positive approach. They give insurance plan participants $240 a year in extra benefits, like additional vacation days, if they get good scores in five out of seven health-related categories. The University of Alabama, Birmingham, School of Nursing has a health-and-wellness program that councils employees about how to get into shape in order to keep their insurance coverage. Audrey Brantley, a participant in the program, has mixed feelings. She says, "It seems like they are trying to control us, but then, on the other hand, I know of folks who found out they had high blood pressure or were borderline diabetics and didn't know it."

Another question is, Does it really work? Turner Broadcasting System in Atlanta has a policy that affects all employees hired after 1986. They will be fired if caught smoking—whether at work or at home—but some admit they still manage to sneak a smoke.

Decide Your Opinion

1. Do you think employers who pay for their employees' health insurance have the right to demand, or encourage, or support a healthy lifestyle?
2. Do you think all participants in a health insurance program should qualify for the same benefits, regardless of their lifestyle?
3. What steps are ethical to encourage people to adopt a healthy lifestyle?

Summarizing the Concepts

13.1 The Blood Vessels
Blood vessels include arteries (and arterioles) that take blood away from the heart; capillaries, where exchange of substances with the tissues occurs; and veins (and venules) that take blood to the heart.

13.2 The Heart
The heart has a right and left side and four chambers. The vena cavae are connected to the right atrium; the pulmonary trunk is connected to the right ventricle. The pulmonary veins are connected to the left atria and the aorta is connected to the left ventricle. Blood from the atria passes through atrioventricular valves into the ventricles. Blood leaving the ventricles passes through semilunar valves.

The right atrium receives O_2-poor blood from the body, and the right ventricle pumps it into the pulmonary circuit. On the left side, the left atrium receives O_2-rich blood from the lungs, and the left ventricle pumps it into the systemic circuit. During the cardiac cycle, the SA node (pacemaker) initiates the heartbeat by causing the atria to contract. The AV node conveys the stimulus to the ventricles, causing them to contract. The heart sounds, "lub-dup," are due to the closing of the atrioventricular valves, followed by the closing of the semilunar valves.

13.3 The Vascular Pathways
The cardiovascular system is divided into the pulmonary circuit and the systemic circuit. In the pulmonary circuit, the pulmonary trunk from the right ventricle of the heart and the two pulmonary arteries take O_2-poor blood to the lungs, and four pulmonary veins return O_2-rich blood to the left atrium of the heart.

To trace the path of blood in the systemic circuit, start with the aorta from the left ventricle. Follow its path until it branches to an artery going to a specific organ. It can be assumed that the artery divides into arterioles and capillaries, and that the capillaries lead to venules. The vein that takes blood to the vena cava most likely has the same name as the artery that delivered blood to the organ. In the adult systemic circuit, unlike the pulmonary circuit, the arteries carry O_2-rich blood, and the veins carry O_2-poor blood.

When tracing the blood to and from the liver it is necessary to mention the hepatic portal vein which originates from intestinal capillaries and finishes in hepatic capillaries.

13.4 Blood
Blood has two main parts: plasma and the formed elements. Plasma contains mostly water (90–92%) and proteins (7–8%), but it also contains nutrients and wastes.

The formed elements are: red blood cells, white blood cells, and platelets. The red blood cells contain hemoglobin and function in oxygen transport. Defense against disease depends on the various types of white blood cells. Granular neutrophils and monocytes are phagocytic. Agranular lymphocytes are involved in the development of immunity to disease.

The platelets and a number of clotting factors are involved in blood clotting. The plasma proteins, prothrombin and fibrinogen, are well known for their contribution to the clotting process which requires a series of enzymatic reactions. When the process is over, fibrin has become fibrin threads.

When blood reaches a capillary, water moves out at the arterial end, due to blood pressure. At the venule end, water moves in, due to osmotic pressure. In between, nutrients diffuse out and wastes diffuse in.

Bone marrow stem cells forever divide and produce new blood cells and platelets. The potential of these cells to differentiate into other types of cells that could be used to cure illnesses is just now being investigated.

13.5 Cardiovascular Disorders
Hypertension and atherosclerosis are two cardiovascular disorders that lead to stroke, heart attack, and aneurysm. Medical and surgical procedures are available to control cardiovascular disease, but the best policy is prevention by following a heart-healthy diet, getting regular exercise, maintaining a proper weight, and not smoking.

Testing Yourself

Choose the best answer for each question.

1. Both the right side and the left side of the heart
   a. have semilunar valves between their chambers.
   b. consist of an atrium and ventricle.
   c. pump blood to the lungs and the body.
   d. communicate with each other.
   e. All of these are correct.

2. Systole refers to the contraction of the
   a. major arteries.   d. major veins.
   b. SA node.   e. All of these are correct.
   c. atria and ventricles.

3. During a heartbeat,
   a. the SA node initiates an impulse that passes to the AV node.
   b. first the atria contract and then the ventricles contract.
   c. the heart pumps the blood out into the attached arteries.
   d. all chambers rest for a while.
   e. All of these are correct.

4. The first heart sound, the "lub" of the "lub-dup" sound, is caused by
   a. the closing of the semilunar valves.
   b. the closing of the atrioventricular valves.
   c. the contraction of the ventricles.
   d. a heart murmur.

5. The T wave of an ECG represents
   a. atria beginning to contract.
   b. the beginning of atrial relaxation.
   c. ventricles beginning to contract.
   d. recovery following ventricular contraction.

6. Blood leaving the right ventricle will move through pulmonary arteries, then to the lungs for gas exchange. Which blood vessel was omitted?
   a. pulmonary trunk   d. superior vena cava
   b. pulmonary veins   e. coronary arteries
   c. aorta

7. Systemic arteries carry blood
   a. to the capillaries and away from the heart.
   b. away from the capillaries and toward the heart.
   c. lower in oxygen than carbon dioxide.
   d. higher in oxygen than carbon dioxide.
   e. Both a and d are correct.

8. Which of these does not correctly contrast the pulmonary circuit and the systemic circuit?

Pulmonary circuit	**Systemic circuit**
a. veins carry O_2-poor blood	veins carry O_2-rich blood
b. carries blood to and from the lungs	carries blood to and from the body
c. has a limited number of blood vessels	has a large number of blood vessels
d. goes between the right ventricle and the left atrium	goes between the left ventricle and the right atrium

9. The blood vessel which carries absorbed materials from the digestive system to the liver is the
   a. inferior vena cava.   d. hepatic portal vein.
   b. hepatic artery.   e. mesenteric artery.
   c. hepatic vein.

10. The best explanation for the slow movement of blood in capillaries is
    a. skeletal muscles press on veins, not capillaries.
    b. capillaries have much thinner walls than arteries.
    c. there are many more capillaries than arterioles.
    d. venules are not prepared to receive so much blood from the capillaries.
    e. All of these are correct.

11. Which of the following assists in the return of venous blood to the heart?
    a. valves
    b. skeletal muscle contraction
    c. respiratory movements
    d. reduction in cross-sectional area from venules to veins
    e. All of these are correct.

12. Which association is incorrect?
    a. white blood cells—infection fighting
    b. red blood cells—blood clotting
    c. plasma—water, nutrients, and wastes
    d. red blood cells—hemoglobin
    e. platelets—blood clotting

13. Blood clotting does not involve which of the following?
    a. calcium ions   d. vitamin K
    b. platelets   e. leukocytes
    c. thrombin

14. The last step in blood clotting
    a. is the only step that requires calcium ions.
    b. occurs outside the bloodstream.
    c. is the same as the first step.
    d. converts prothrombin to thrombin.
    e. converts fibrinogen to fibrin.

15. Water enters the venous side of capillaries because of
    a. active transport from tissue fluid.
    b. an osmotic pressure gradient.
    c. higher blood pressure on the venous side.
    d. higher blood pressure on the arterial side.
    e. higher red blood cell concentration on the venous side.

16. Which illness has been linked to hypertension?
    a. atherosclerosis   d. heart attack
    b. aneurysm   e. All of these are correct.
    c. stroke

17. Label this diagram of the heart.

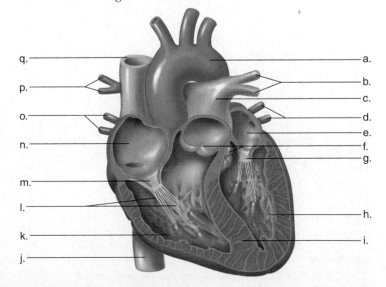

e-Learning Connection   www.mhhe.com/maderinquiry10

Concepts	Questions	Media Resources*
13.1 The Blood Vessels		

• A series of vessels delivers blood from the heart to the capillaries, where exchanges of substances takes place, and then another series of vessels delivers blood from the capillaries back to the heart.	1. How are large blood vessels nourished with oxygen and nutrients?
	2. What determines whether blood flows into one particular capillary bed or another?
		Essential Study Partner
		    Structure of Blood Vessels
		    Arteries and Arterioles
		    Capillaries
		    Capillary Exchange
		    Veins and Venules
		BioCourse Study Guide
		    Human Circulation: Blood
		        and Blood Vessels

13.2 The Heart		

• The human heart is a double pump: the right side pumps blood to the lungs, and the left side pumps blood to the rest of the body.	1. Give one example of how the heart can be used to illustrate the relationship between structure and function.
	2. Does the heart quit beating if the SA node fails?
		Essential Study Partner
		    Human Heart
		    Heart Blood Flow
		    Conduction System
		    Cardiac Cycle
		Art Quiz
		    Heart–Electrical Pathway
		Animation Quizzes
		    Cardiac Cycle Electrical
		    Cardiac Cycle Muscular
		    Cardiac Cycle Sounds
		BioCourse Study Guide
		    Human Circulation: Heart
		        Structure and Function

13.3 The Vascular Pathways		

• The pulmonary arteries transport O_2-poor blood to the lungs, and the pulmonary veins return O_2-rich blood to the heart.
• The systemic circuit transports blood from the left ventricle of the heart to the body and then returns it to the right atrium of the heart.
	1. What is a portal system; give an example of one in the human body.
	2. Is arterial blood pressure constant throughout the body?
		Essential Study Partner
		    Vessels and Pressure
		        Distribution
		    Blood Pressure
		    Systemic Arteries and
		        Arterioles
		    Systemic Veins and Venules
		Animation Exercise
		    Portal System

13.4 Blood		

• Blood is composed of cells and a fluid containing proteins and various other molecules and ions.
• Blood clotting is a series of reactions that produces a clot—fibrin threads in which red blood cells are trapped.
• Exchange of substances between blood and tissue fluid across capillary walls supplies cells with nutrients and removes wastes.
	1. What is the primary function of red blood cells and white blood cells?
	2. What portions of the blood participate in clotting?
		Essential Study Partner
		    Blood
		    Plasma
		    Erythrocytes
		    Leukocytes
		    Hemostasis
		Case Studies
		    Artificial Blood
		    Blood Doping in Cyclers

13.5 Cardiovascular Disorders		

• Although the cardiovascular system is very efficient, it is still subject to degenerative disorders.	1. What are the two primary disorders that lead to stroke, heart attack, and aneurysm?
	2. What are the recommendations for maintaining a healthy cardiovascular system?
		Essential Study Partner
		    Myocardial Infarction
		Animation Quizzes
		    Valvular Insufficiency
		    Valvular Stenosis
		    Myocardial Infarction

*For additional Media Resources, see the Online Learning Center.

Lymphatic and Immune Systems

Chapter Concepts

It's easy to spread germs from hand to face. Frequent washing of hands assists the immune system in protecting us from disease.

*K*arlin casually licks her finger, unaware she is infecting herself with a flu virus picked up when helping her small son blow his nose. The viral particles slip past the protective mucous barrier of her digestive tract, enter cells, and begin to make copies of themselves. Just before the infected cells succumb, they secrete chemicals that alert her immune system to the invaders.

The immune system reacts to antigens, any substance, usually protein or carbohydrate, that the system is capable or recognizing as being "nonself." Get a bacterial infection, and certain lymphocytes start producing antibodies. An antibody combines with the antigen, and later the complex is phagocytized by a macrophage. In Karlin's case, however, macrophages devour viruses and present their remains to other lymphocytes that gear up to kill any cell that is infected with the virus. Whereas you can take an antibiotic to help cure a bacterial infection, often there is nothing to do for a viral infection but wait for the immune system to win the battle.

Unlike the immune system of many others, Karlin's immune system mistakenly reacts to harmless antigens, and this causes her to have allergies. Otherwise, though, it does a magnificent job of keeping her well.

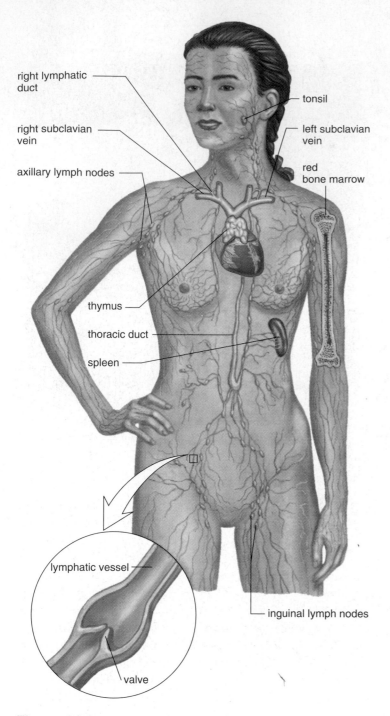

right lymphatic duct

right subclavian vein

axillary lymph nodes

tonsil

left subclavian vein

red bone marrow

thymus

thoracic duct

spleen

lymphatic vessel

valve

inguinal lymph nodes

Figure 14.1   Lymphatic system.
Lymphatic vessels drain excess fluid from the tissues and return it to the cardiovascular system. The enlargement shows that lymphatic vessels, like cardiovascular veins, have valves to prevent backward flow. The lymph nodes, tonsils, spleen, thymus gland, and red bone marrow are the main lymphoid organs that assist immunity.

[1]After glycerol and fatty acids are absorbed, they are rejoined and packaged as lipoprotein droplets, which enter the lacteals.

14.1 The Lymphatic System

The **lymphatic system** consists of lymphatic vessels and the lymphoid organs. This system, which is closely associated with the cardiovascular system, has three main functions that contribute to homeostasis: (1) lymphatic capillaries take up excess tissue fluid and return it to the bloodstream; (2) lacteals receive lipoproteins[1] at the intestinal villi and transport them to the bloodstream (see Fig. 5.6); and (3) the lymphatic system works with the immune system to help defend the body against disease.

Lymphatic Vessels

Lymphatic vessels are quite extensive; most regions of the body are richly supplied with lymphatic capillaries (Fig. 14.1). The construction of the larger lymphatic vessels is similar to that of cardiovascular veins, including the presence of valves. Also, the movement of lymph within these vessels is dependent upon skeletal muscle contraction. When the muscles contract, the lymph is squeezed past a valve that closes, preventing the lymph from flowing backwards.

The lymphatic system is a one-way system that begins with lymphatic capillaries. These capillaries take up fluid that has diffused from and not been reabsorbed by the blood capillaries. **Edema** is localized swelling caused by the accumulation of tissue fluid. This can happen if too much tissue fluid is made and/or not enough of it is drained away. Once tissue fluid enters the lymphatic vessels, it is called **lymph.** The lymphatic capillaries join to form lymphatic vessels that merge before entering one of two ducts: the thoracic duct or the right lymphatic duct. The thoracic duct is much larger than the right lymphatic duct. It serves the lower extremities, the abdomen, the left arm, and the left side of both the head and the neck. The right lymphatic duct serves the right arm, the right side of both the head and the neck, and the right thoracic area. The lymphatic ducts enter the subclavian veins, which are cardiovascular veins in the thoracic region.

Lymph flows one way from a capillary to ever-larger lymphatic vessels and finally to a lymphatic duct, which enters a subclavian vein.

Lymphoid Organs

The **lymphoid organs** of special interest are the lymph nodes, the tonsils, the spleen, the thymus gland, and the red bone marrow (Fig. 14.2).

Lymph nodes, which are small (about 1–25 mm in diameter) ovoid or round structures, are found at certain points along lymphatic vessels. A lymph node is composed

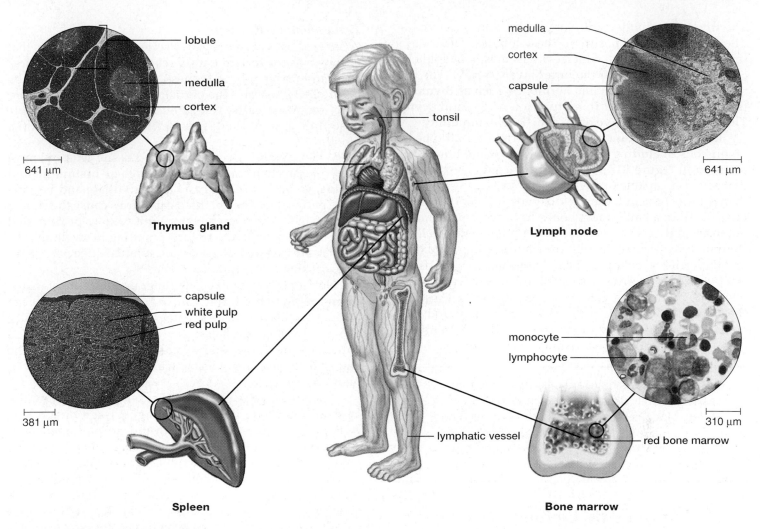

Figure 14.2 The lymphoid organs.
The lymphoid organs include the lymph nodes, the tonsils (not shown in an enlargement), the spleen, the thymus gland, and the red bone marrow, all of which contain lymphocytes.

of a capsule surrounding two distinct regions known as the cortex and medulla, which contain many lymphocytes. The cortex contains nodules where lymphocytes congregate when they are fighting off a pathogen. Macrophages, concentrated in the medulla, work to cleanse the lymph. Lymph nodes are named for their location. Inguinal nodes are in the groin, and axillary nodes are in the armpits. Physicians often feel for the presence of swollen, tender lymph nodes in the neck as evidence that the body is fighting an infection. This is a noninvasive, preliminary way to help make such a diagnosis.

The **tonsils** are patches of lymphatic tissue located in a ring about the pharynx (see Fig. 14.1). The well-known pharyngeal tonsils are also called adenoids, while the larger palatine tonsils, located on either side of the posterior oral cavity, are more apt to be infected. The tonsils perform the same functions as lymph nodes inside the body, but because of their location, they are the first to encounter

pathogens and antigens that enter the body by way of the nose and mouth.

The **spleen** is located in the upper left region of the abdominal cavity just beneath the diaphragm. It is much larger than a lymph node, about the size of a fist. Whereas the lymph nodes cleanse lymph, the spleen cleanses blood. The spleen is composed of a capsule surrounding tissue known as white pulp and red pulp. The white pulp is involved in filtering out bacteria and any debris; the red pulp is involved in filtering old worn-out red blood cells.

The spleen's outer capsule is relatively thin, and an infection or a blow can cause the spleen to burst. Although its functions are replaced by other organs, a person without a spleen is often slightly more susceptible to infections and may have to receive antibiotic therapy indefinitely.

The **thymus gland** is located along the trachea behind the sternum in the upper thoracic cavity. This gland varies in size, but it is larger in children and shrinks as we get older.

The thymus is divided into lobules by connective tissue. The T lymphocytes mature in these lobules. The interior (medulla) of the lobule, which consists mostly of epithelial cells, stains lighter than the outer layer (cortex). The thymus gland produces thymic hormones, such as thymosin, that are thought to aid in maturation of T lymphocytes. Thymosin may also have other functions in immunity.

Red bone marrow is the site of origin for all types of blood cells, including the five types of white blood cells pictured in Figure 13.10. The marrow contains stem cells that are ever capable of dividing and producing cells that then differentiate into the various types of blood cells (see Fig. 13.14). In a child, most bones have red bone marrow, but in an adult it is present only in the bones of the skull, the sternum (breastbone), the ribs, the clavicle, the pelvic bones, and the vertebral column. The red bone marrow consists of a network of connective tissue fibers, called reticular fibers, which are produced by cells called reticular cells. These and the stem cells and their progeny are packed around thin-walled sinuses filled with venous blood. Differentiated blood cells enter the bloodstream at these sinuses.

Lymphoid organs have specific functions that assist immunity. Lymph is cleansed in lymph nodes; blood is cleansed in the spleen. All blood cells are made in red bone marrow. Most white blood cells mature in the red bone marrow, but T lymphocytes mature in the thymus.

14.2 Nonspecific Defenses

The **immune system** includes the cells and tissues that are responsible for immunity. **Immunity** is the body's ability to defend itself against infectious agents, foreign cells, and even abnormal body cells, such as cancer cells. Thereby, the internal environment has a better chance of remaining stable. Immunity includes nonspecific and specific defenses. The four types of nonspecific defenses—barriers to entry, the inflammatory reaction, natural killer cells, and protective proteins—are effective against many types of infectious agents.

Barriers to Entry

Skin and the mucous membranes lining the respiratory, digestive, and urinary tracts serve as mechanical barriers to entry by pathogens. Oil gland secretions contain chemicals that weaken or kill certain bacteria on the skin. The upper respiratory tract is lined by ciliated cells that sweep mucus and trapped particles up into the throat, where they can be swallowed or expectorated (coughed out). The stomach has an acidic pH, which inhibits the growth of or kills many types of bacteria. The various bacteria that normally reside in the intestine and other areas, such as the vagina, prevent pathogens from taking up residence.

Inflammatory Reaction

Whenever tissue is damaged, a series of events occurs that is known as the **inflammatory reaction.** The inflamed area has four outward signs: redness, heat, swelling, and pain. Figure 14.3 illustrates the participants in the inflammatory reaction. **Mast cells,** which occur in tissues, resemble basophils, one of the types of white cells found in the blood.

When an injury occurs, damaged tissue cells and mast cells release chemical mediators, such as **histamine** and **kinins,** which cause the capillaries to dilate and become more permeable. The enlarged capillaries cause the skin to redden, and the increased permeability allows proteins and fluids to escape into the tissues, resulting in swelling. The swollen area, as well as the kinins, stimulate free nerve endings, causing the sensation of pain.

Neutrophils and monocytes migrate to the site of injury. They are amoeboid and can change shape to squeeze through capillary walls and enter tissue fluid. Neutrophils, and also mast cells, phagocytize pathogens. The engulfed pathogens are destroyed by hydrolytic enzymes when the endocytic vesicle combines with a lysosome, one of the cellular organelles.

As they leave the blood and enter the tissues, monocytes differentiate into **macrophages,** large phagocytic cells that are able to devour a hundred pathogens and still survive. Some tissues, particularly connective tissue, have resident macrophages, which routinely act as scavengers, devouring old blood cells, bits of dead tissue, and other debris. Macrophages can also bring about an explosive increase in the number of leukocytes by liberating colony-stimulating factors, which pass by way of blood to the red bone marrow, where they stimulate the production and the release of white blood cells, primarily neutrophils. As the infection is being overcome, some neutrophils may die. These—along with dead cells, dead bacteria, and living white blood cells—form pus, a whitish material. Pus indicates that the body is trying to overcome the infection.

When a blood vessel ruptures, the blood forms a clot to seal the break. The chemical mediators (e.g., histamine and kinins) and antigens move through the tissue fluid and lymph to the lymph nodes. Now lymphocytes are activated to react to the threat of an infection. Sometimes inflammation persists, and the result is chronic inflammation that is often treated by administering anti-inflammatory agents such as aspirin, ibuprofen, or cortisone. These medications act against the chemical mediators released by the white blood cells in the area.

The inflammatory reaction is a "call to arms"—it marshals phagocytic white blood cells to the site of bacterial invasion and stimulates the immune system to react against a possible infection.

Visual Focus

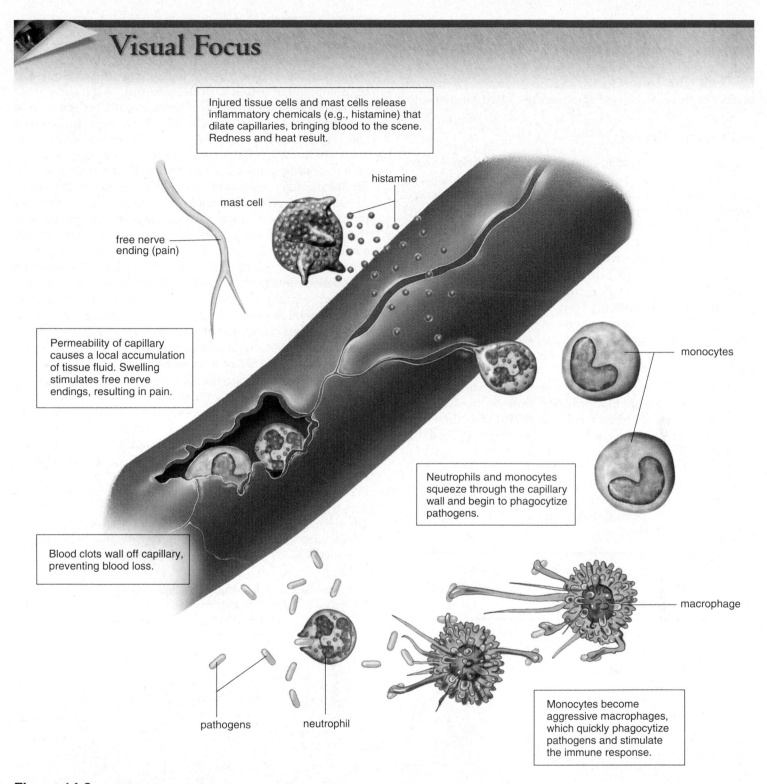

Injured tissue cells and mast cells release inflammatory chemicals (e.g., histamine) that dilate capillaries, bringing blood to the scene. Redness and heat result.

histamine

mast cell

free nerve ending (pain)

monocytes

Permeability of capillary causes a local accumulation of tissue fluid. Swelling stimulates free nerve endings, resulting in pain.

Neutrophils and monocytes squeeze through the capillary wall and begin to phagocytize pathogens.

Blood clots wall off capillary, preventing blood loss.

macrophage

pathogens   neutrophil

Monocytes become aggressive macrophages, which quickly phagocytize pathogens and stimulate the immune response.

Figure 14.3   Inflammatory reaction.
Mast cells, which are related to basophils, a type of white blood cell, are involved in the inflammatory reaction. When a blood vessel is injured, mast cells release substances like histamine. Histamine dilates blood vessels and increases permeability so that tissue fluid leaks from the vessel. Swelling in the area stimulates pain receptors (free nerve endings). Neutrophils and monocytes which become macrophages squeeze through the capillary wall. These white blood cells begin to phagocytize pathogens (e.g., disease-causing viruses and bacteria), especially those combined with antibodies. Blood clotting seals off the capillary, preventing blood loss.

Natural Killer Cells

Natural killer (NK) cells kill virus-infected cells and tumor cells by cell-to-cell contact. They are large, granular lymphocytes with no specificity and no memory. Their number is not increased by prior exposure to that kind of cell.

Protective Proteins

The **complement system,** often simply called complement, is a number of plasma proteins designated by the letter C and a subscript. A limited amount of activated complement protein is needed because a domino effect occurs: each activated protein in a series is capable of activating many other proteins.

Complement is activated when pathogens enter the body. It "complements" certain immune responses, which accounts for its name. For example, it is involved in and amplifies the inflammatory response because complement proteins attract phagocytes to the scene. Some complement proteins bind to the surface of pathogens already coated with antibodies, which ensures that the pathogens will be phagocytized by a neutrophil or macrophage.

Certain other complement proteins join to form a membrane attack complex that produces holes in the walls and plasma membranes of bacteria. Fluids and salts then enter the bacterial cell to the point that they burst (Fig. 14.4).

Interferon is a protein produced by virus-infected cells. Interferon binds to receptors of noninfected cells, causing them to prepare for possible attack by producing substances that interfere with viral replication. Interferon is specific to the species; therefore, only human interferon can be used in humans.

Immunity includes these nonspecific defenses: barriers to entry, the inflammatory reaction, natural killer cells, and protective proteins.

14.3 Specific Defenses

When nonspecific defenses have failed to prevent an infection, specific defenses come into play. An **antigen** is any foreign substance (often a protein or polysaccharide) that stimulates the immune system to react to it. Pathogens have antigens, but antigens can also be part of a foreign cell or a cancer cell. Because we do not ordinarily become immune to our own normal cells, it is said that the immune system is able to distinguish "self" from "nonself." Only in this way can the immune system aid, rather than counter, homeostasis.

Lymphocytes are capable of recognizing an antigen because they have **antigen receptors**—plasma membrane receptor proteins whose shape allows them to combine with a specific antigen. It is often said that the receptor and the antigen fit together like a lock and a key. Because we encounter a million different antigens during our lifetime, we need a great diversity of lymphocytes to protect us against them. Remarkably, diversification occurs to such an extent during the maturation process that there is a lymphocyte type for any possible antigen.

Immunity usually lasts for some time. For example, once we recover from the measles, we usually do not get the illness a second time. Immunity is primarily the result of the action of the **B lymphocytes** and the **T lymphocytes.** B lymphocytes mature in the *b*one marrow,[2] and T lymphocytes mature in the *t*hymus gland. B lymphocytes, also called B cells, give rise to plasma cells, which produce **antibodies,** proteins shaped like the antigen receptor and capable of combining with and neutralizing a specific antigen. These antibodies are secreted into the blood, lymph, and other body fluids. In contrast, T lymphocytes, also called T cells, do not produce antibodies. Instead, certain T cells directly attack cells that bear nonself proteins. Other T cells regulate the immune response.

---

[2]Historically, the B stands for bursa of Fabricius, an organ in the chicken where these cells were first identified. As it turns out, however, the B can conveniently be thought of as referring to bone marrow.

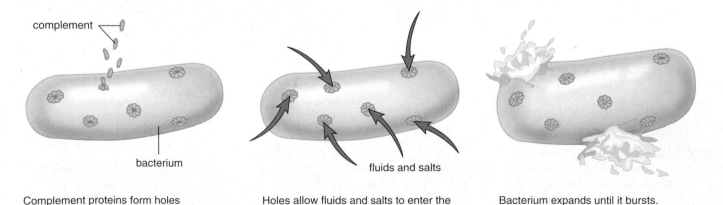

complement

bacterium

fluids and salts

Complement proteins form holes in the bacterial cell wall and membrane.

Holes allow fluids and salts to enter the bacterium.

Bacterium expands until it bursts.

Figure 14.4   Action of the complement system against a bacterium.
When complement proteins in the plasma are activated by an immune response, they form holes in bacterial cell walls and plasma membranes, allowing fluids and salts to enter until the cell eventually bursts.

B Cells and Antibody-Mediated Immunity

When a B cell in a lymph node of the spleen encounters a specific antigen, it is activated to divide many times. Most of the resulting cells are plasma cells. A **plasma cell** is a mature B cell that mass-produces antibodies against a specific antigen.

The **clonal selection theory** states that the antigen selects which lymphocyte will undergo clonal expansion and produce more lymphocytes bearing the same type of antigen receptor. Notice in Figure 14.5 that different types of antigen receptors are represented by color. The B cell with blue receptors undergoes clonal expansion because a specific antigen (red dots) is present and binds to its receptors. B cells are stimulated to divide and become plasma cells by helper T cell secretions called cytokines, as is discussed in the next section. Some members of the clone become memory cells, which are the means by which long-term immunity is possible. If the same antigen enters the system again, **memory B cells** quickly divide and give rise to more lymphocytes capable of quickly producing antibodies.

Once the threat of an infection has passed, the development of new plasma cells ceases, and those present undergo apoptosis. **Apoptosis** is a process of programmed cell death (PCD) involving a cascade of specific cellular events leading to the death and destruction of the cell. The methodology of PCD is still being worked out, but we know it is an essential physiological mechanism regulating the cell population within an organ system. PCD normally plays a central role in maintaining tissue homeostasis.

Defense by B cells is called **antibody-mediated immunity** because the various types of B cells produce antibodies. It is also called humoral immunity because these antibodies are present in blood and lymph. A humor is any fluid normally occurring in the body.

Characteristics of B Cells

- Antibody-mediated immunity against bacteria
- Produced and mature in bone marrow
- Reside in spleen and lymph nodes, circulate in blood and lymph
- Directly recognize antigen and then undergo clonal selection
- Clonal expansion produces antibody-secreting plasma cells as well as memory B cells

---

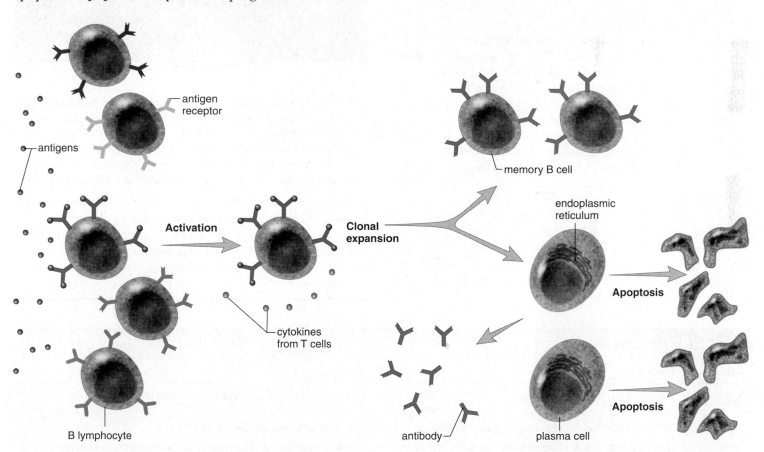

Figure 14.5   **Clonal selection theory as it applies to B cells.**
Each type B cell bears a specific antigen receptor (note different colors). When an antigen (red dots) combines with the antigen receptors in blue, that B cell is stimulated by cytokines, and it undergoes clonal expansion. The result is many plasma cells, which produce specific antibodies against this antigen and memory B cells which immediately recognize this antigen in the future. After the infection passes, plasma cells undergo apoptosis.

Structure of IgG

The most common type of antibody is IgG, a Y-shaped protein molecule with two arms. Each arm has a "heavy" (long) polypeptide chain and a "light" (short) polypeptide chain. These chains have constant regions, where the sequence of amino acids is set, and variable regions, where the sequence of amino acids varies between antibodies (Fig. 14.6). The constant regions are not identical among all the antibodies. Instead, they are almost the same within different classes of antibodies. The variable regions form an antigen-binding site, and their shape is specific to a particular antigen. The antigen combines with the antibody at the antigen-binding site in a lock-and-key manner.

The antigen-antibody reaction can take several forms, but quite often the reaction produces complexes of antigens combined with antibodies. Such antigen-antibody complexes, sometimes called immune complexes, mark the antigens for destruction. For example, an antigen-antibody complex may be engulfed by neutrophils or macrophages, or it may activate complement. Complement makes pathogens more susceptible to phagocytosis, as discussed previously.

Other Types of Antibodies

There are five different classes of circulating antibody proteins or **immunoglobulins (Igs)** (Table 14.1). IgG antibodies are the major type in blood, and lesser amounts are also found in lymph and tissue fluid. IgG antibodies bind to pathogens and their toxins. IgM antibodies are pentamers, meaning that they contain five of the Y-shaped structures shown in Figure 14.6a. These antibodies appear in blood soon after an infection begins and disappear before it is over. They are good activators of the complement system. IgA antibodies are monomers or dimers containing two Y-shaped structures. They are the main type of antibody found in bodily secretions. They bind to pathogens before they reach the bloodstream. The main function of IgD molecules seems to be to serve as antigen receptors on immature B cells. IgE antibodies, which are responsible for immediate allergic responses, are discussed on page 274 and in the Health Focus on page 275.

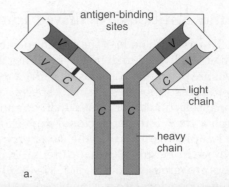

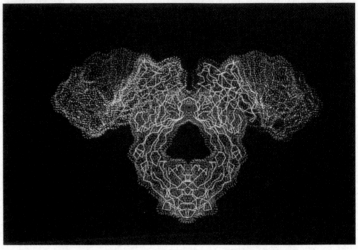

Figure 14.6 **Structure of the most common antibody (IgG).**
a. An IgG antibody contains two heavy (long) polypeptide chains and two light (short) chains arranged so there are two variable regions, where a particular antigen is capable of binding with an antibody (V = variable region, C = constant region). **b.** Computer model of an antibody molecule. The antigen combines with the two side branches.

An antigen combines with an antibody at the antigen-binding site in a lock-and-key manner. The reaction can produce antigen-antibody complexes, which contain several molecules of antibody and antigen.

Table 14.1	Antibodies	
Class	**Presence**	**Function**
IgG	Main antibody type in circulation	Binds to pathogens, activates complement, and enhances phagocytosis
IgM	Antibody type found in circulation; largest antibody	Activates complement; clumps cells
IgA	Main antibody type in secretions such as saliva and milk	Prevents pathogens from attaching to epithelial cells in digestive and respiratory tract
IgD	Antibody type found on surface of immature B cells	Presence signifies readiness of B cell
IgE	Antibody type found as antigen receptors on basophils in blood and on mast cells in tissues	Responsible for immediate allergic response and protection against certain parasitic worms

Science Focus

Antibody Diversity

In 1987, Susumu Tonegawa (Fig. 14A*a*) became the first Japanese scientist to win the Nobel Prize in Physiology or Medicine, after dedicating himself to finding the solution to an engrossing puzzle. Immunologists and geneticists knew that each B cell makes an antibody especially equipped to recognize the specific shape of a particular antigen. But they did not know how the human genome contained enough genetic information to permit the production of up to a million different antibody types needed to combat all of the pathogens we are likely to encounter during our lives.

An antibody is composed of two light and two heavy polypeptide chains, which are divided into constant and variable regions. The constant region determines the antibody class, and the variable region determines the specificity of the antibody, because this is where an antigen binds to a specific antibody (see Fig. 14.6). Each B cell must have a genetic way to code for the variable regions of both the light and heavy chains.

Tonegawa's colleagues say that he is a creative genius who intuitively knows how to design experiments to answer specific questions. In this instance, he examined the DNA sequences of lymphoblasts and compared them to mature B cells. He found that the DNA segments coding for the variable and constant regions were scattered throughout the genome in B lymphocyte stem cells and that only certain of these segments appeared in each mature antibody-secreting B cell, where they randomly came together and coded for a specific variable region. Later, the variable and constant regions are joined to give a specific antibody (Fig. 14A*b*). As an analogy, consider that each person entering a supermarket chooses various items for purchase, and that the possible combination of items in any particular grocery bag is astronomical. Tonegawa also found that mutations occur as the variable segments are undergoing rearrangements. Such mutations are another source of antibody diversity.

Invariably, some B cells with receptors that could bind to the body's own cell surface molecules arise. It is believed that these cells undergo apoptosis, or programmed cell death.

Tonegawa received his B.S. in chemistry in 1963 at Kyoto University and earned his Ph.D. in biology from the University of California at San Diego (UCSD) in 1969. After that, he worked as a research fellow at UCSD and the Salk Institute. In 1971, he moved to the Basel Institute for Immunology and began the experiments that eventually led to his Nobel Prize-winning discovery. Tonegawa also contributed to the effort to decipher the receptors of T cells. This was an even more challenging area of research than the diversity of antibodies produced by B cells. Since 1981, he has been a full professor at Massachusetts Institute of Technology (MIT), where he has a reputation for being an "aggressive, determined researcher" who often works late into the night.

a. Susumu Tonegawa in the laboratory

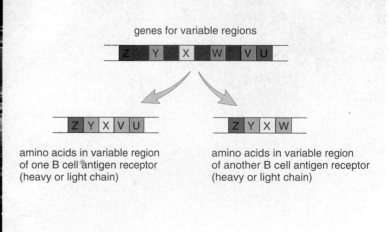

b. Antibody variable regions

Figure 14A   Antibody diversity.
a. Susumu Tonegawa received a Nobel Prize for his findings regarding antibody diversity. **b.** Different genes for the variable regions of heavy and light chains are brought together during the production of B lymphocytes so that their antigen receptors can combine with only a particular antigen.

T Cells and Cell-Mediated Immunity

When T cells leave the thymus, they have unique antigen receptors just as B cells do. Unlike B cells, however, T cells are unable to recognize an antigen present in lymph, blood, or the tissues without help. The antigen must be presented to them by an **antigen-presenting cell (APC).** When an APC presents a viral or cancer cell antigen, the antigen is first linked to a major histocompatibility complex (MHC) protein in the plasma membrane.

Human MHC proteins are called **HLA (human leukocyte-associated) antigens.** Because they mark the cell as belonging to a particular individual, HLA antigens are self proteins. The importance of self proteins in plasma membranes was first recognized when it was discovered that they contribute to the specificity of tissues and make it difficult to transplant tissue from one human to another. In other words, when the donor and the recipient are histo (tissue)-compatible (the same or nearly so), a transplant is more likely to be successful.

Figure 14.7 shows a macrophage presenting an antigen, represented by a red circle, to a particular T cell. This T cell has the type of antigen receptor that will combine with this specific antigen. In the figure, the different types of antigen receptors are represented by color. Presentation of the antigen represents activation of the T cell. An activated T cell produces cytokines and undergoes clonal expansion. **Cytokines** are signaling chemicals that stimulate various immune cells (e.g., macrophages, B cells, and other T cells) to perform their functions. Many copies of the activated T cell are produced during clonal expansion. They destroy any cell, such as a virus-infected cell or a cancer cell, that displays the antigen presented earlier.

As the illness disappears, the immune reaction wanes and fewer cytokines are produced. Now, the activated T cells become susceptible to apoptosis. As mentioned previously, apoptosis is programmed cell death that contributes to homeostasis by regulating the number of cells present in an organ, or in this case, in the immune system. When

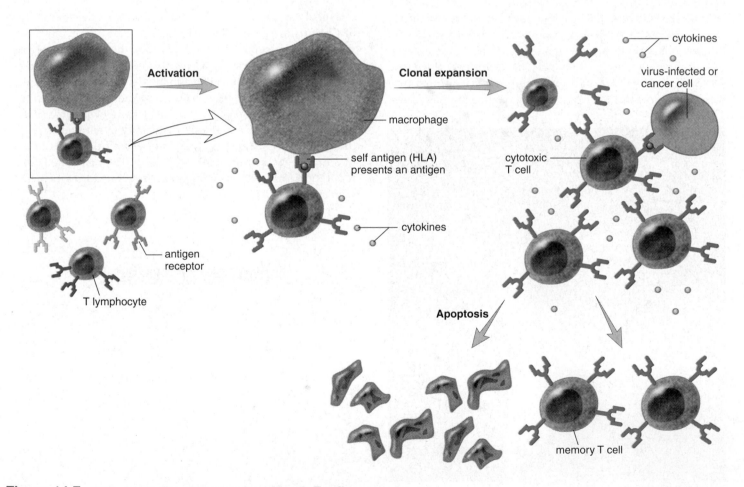

Figure 14.7　Clonal selection theory as it applies to T cells.
Each type of T cell bears a specific antigen receptor (note different colors). Activation of a T cell occurs when its antigen receptor can combine with an antigen. A macrophage presents the antigen (note red circle) in the groove of an HLA molecule. Thereafter, the T cell undergoes clonal expansion, meaning that many copies of the same type of T cell are produced. (Note that all T cells on the right have blue antigen receptors.) After the immune response has been successful, the majority of T cells undergo apoptosis, while a small number become memory T cells. Memory T cells provide protection should the same antigen enter the body again at a future time.

apoptosis does not occur as it should, T cell cancers (i.e., lymphomas and leukemias) can result.

Apoptosis also occurs in the thymus as T cells are maturing. Any T cell that has the potential to destroy the body's own cells undergoes suicide.

Types of T Cells

The two main types of T cells are cytotoxic T cells and helper T cells. **Cytotoxic T cells** can bring about the destruction of antigen-bearing cells, such as virus-infected or cancer cells. Cancer cells also have nonself proteins.

Cytotoxic T cells have storage vacuoles containing perforin molecules. **Perforin** molecules perforate a plasma membrane, forming a pore that allows water and salts to enter. The cell then swells and eventually bursts. Cytotoxic T cells are responsible for so-called **cell-mediated immunity** (Fig. 14.8).

Helper T cells regulate immunity by secreting cytokines, the chemicals that enhance the response of other immune cells. Because HIV, the virus that causes AIDS,

infects helper T cells and certain other cells of the immune system, it inactivates the immune response.

Notice in Figure 14.7 that a few of the clonally expanded T cells are labeled memory T cells. They remain in the body and can jump-start an immune reaction to an antigen previously present in the body.

Characteristics of T Cells

- Cell-mediated immunity against viruses and cancer cells
- Produced in bone marrow, mature in thymus
- Antigen must be presented in groove of an HLA molecule
- Cytotoxic T cells destroy nonself protein-bearing cells
- Helper T cells secrete cytokines that control the immune response

---

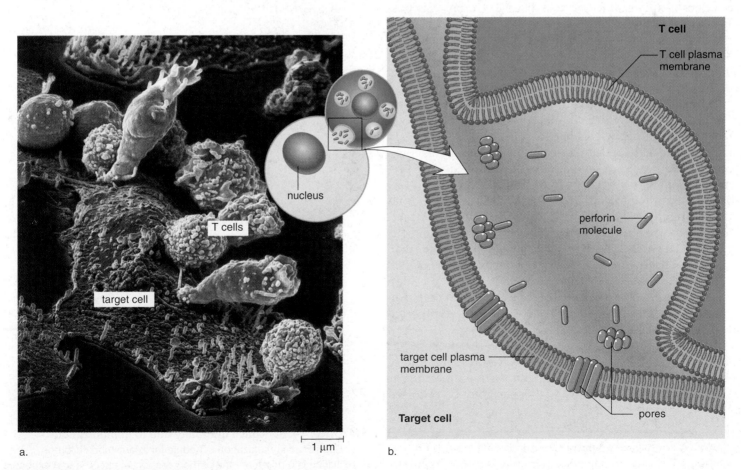

a.

b.

Figure 14.8   Cell-mediated immunity.
a. The scanning electron micrograph shows cytotoxic T cells attacking and destroying a cancer cell. Cancer cells are subject to attack because they have acquired nonself antigens. **b.** During the killing process, the vacuoles in a cytotoxic T cell release perforin molecules. These molecules combine to form pores in the target cell plasma membrane. Thereafter, fluid and salts enter so that the target cell eventually bursts.

14.4 Induced Immunity

Immunity occurs naturally through infection or is brought about artificially (induced) by medical intervention. The two types of induced immunity are active and passive. In active immunity, the individual alone produces antibodies against an antigen; in passive immunity, the individual is given prepared antibodies.

Active Immunity

Active immunity sometimes develops naturally after a person is infected with a pathogen. However, active immunity is often induced when a person is well so that future infection will not take place. To prevent infections, people can be artificially immunized against them. The United States is committed to immunizing all children against the common types of childhood diseases listed in the immunization schedule in Figure 14.9a.

Immunization involves the use of **vaccines,** substances that contain an antigen to which the immune system responds. Traditionally, vaccines are the pathogens themselves, or their products, that have been treated so they are no longer virulent (able to cause disease). Today, it is possible to genetically engineer bacteria to mass-produce a protein from pathogens, and this protein can be used as a vaccine. This method has now produced a vaccine against hepatitis B, a viral disease, and is being used to prepare a vaccine against malaria, a protozoan disease.

After a vaccine is given, it is possible to follow an immune response by determining the amount of antibody present in a sample of plasma—this is called the antibody titer. After the first exposure to a vaccine, a primary response occurs. For a period of several days, no antibodies are present; then, there is a slow rise in the titer, followed by a leveling off, and then a gradual decline as the antibodies bind to the antigen or simply break down (Fig. 14.9b). After a second exposure to the vaccine, a secondary response is expected. The titer rises rapidly to a level much greater than before, then it slowly declines. The second exposure is called a "booster" because it boosts the antibody titer to a high level. The high antibody titer now is expected to help prevent disease symptoms even if the individual is exposed to the disease-causing antigen.

Active immunity is dependent upon the presence of memory B cells and memory T cells that are capable of responding to lower doses of antigen. Active immunity is usually long-lasting, although a booster may be required every so many years.

Active (long-lasting) immunity can be induced by the use of vaccines. Active immunity is dependent upon the presence of memory B cells and memory T cells in the body.

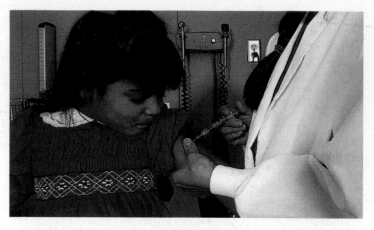

Suggested Immunization Schedule		
Vaccine	Age (months)	Age (years)
Hepatitis B	Birth–2, 1–18	11–12
Diphtheria, tetanus, pertussis	2, 4, 6, 15–18	4–6
Tetanus only		11–12, 14–16
Haemophilus influenzae, type b	2, 4, 6, 12–15	
Polio	2, 4, 6–18	4–6
Pneumococcal	2, 4, 6, 12–15	
Measles, mumps, rubella	12–15	4–6, 11–12
Varicella (chicken pox)	12–18	11–12
Hepatitis A (in selected areas)	24	4–12

a.

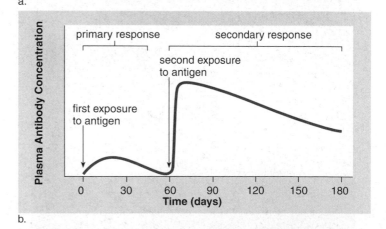

b.

Figure 14.9 Active immunity due to immunizations.
a. Suggested immunization schedule for infants and children.
b. During immunization, the primary response, after the first exposure to a vaccine, is minimal, but the secondary response, which may occur after the second exposure, shows a dramatic rise in the amount of antibody present in plasma.

Passive Immunity

Passive immunity occurs when an individual is given pre-pared antibodies (immunoglobulins) to combat a disease. Since these antibodies are not produced by the individual's plasma cells, passive immunity is temporary. For example, newborn infants are passively immune to some diseases because antibodies have crossed the placenta from the mother's blood. These antibodies soon disappear, however, so that within a few months, infants become more suscep-tible to infections. Breast-feeding prolongs the natural pas-sive immunity an infant receives from the mother because antibodies are present in the mother's milk (Fig. 14.10).

Even though passive immunity does not last, it is some-times used to prevent illness in a patient who has been unexpectedly exposed to an infectious disease. Usually, the patient receives a gamma globulin injection (serum that contains antibodies), perhaps taken from individuals who have recovered from the illness. In the past, horses were immunized, and serum was taken from them to provide the needed antibodies against such diseases as diphtheria, botulism, and tetanus. Unfortunately, a patient who received these antibodies became ill about 50% of the time, because the serum contained proteins that the individual's immune system recognized as foreign. This was called serum sickness. But problems can also occur with products produced in other ways. An immunoglobulin intravenous product called Gammagard was withdrawn from the mar-ket because of its possible implication in the transmission of hepatitis.

> Passive immunity provides protection when an individual is in immediate danger of succumbing to an infectious disease. Passive immunity is temporary because there are no memory cells.

Cytokines and Immunity

Cytokines are signaling molecules produced by T lympho-cytes, monocytes, and other cells. Because cytokines regu-late white blood cell formation and/or function, they are being investigated as possible adjunct therapy for cancer and AIDS. Both interferon and **interleukins,** which are cytokines produced by various white blood cells, have been used as immunotherapeutic drugs, particularly to enhance the ability of the individual's own T cells to fight cancer.

Interferon, discussed previously on page 266, is a sub-stance produced by leukocytes, fibroblasts, and probably most cells in response to a viral infection. Interferon is still being investigated as a possible cancer drug, but so far it has proven to be effective only in certain patients, and the exact reasons for this as yet cannot be discerned.

When and if cancer cells carry an altered protein on their cell surface, they should be attacked and destroyed by

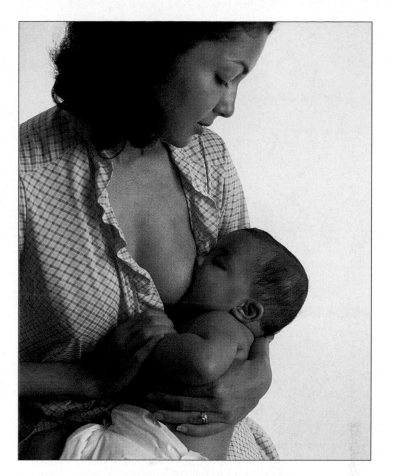

Figure 14.10   Passive immunity.
Breast-feeding is believed to prolong the passive immunity an infant receives from the mother because antibodies are present in the mother's milk.

cytotoxic T cells. Whenever cancer does develop, it is possi-ble that the cytotoxic T cells have not been activated. In that case, cytokines might awaken the immune system and lead to the destruction of the cancer. In one technique being inves-tigated, researchers first withdraw T cells from the patient, present cancer cell antigens to them, and then activate the cells by culturing them in the presence of an interleukin. The T cells are reinjected into the patient, who is given doses of interleukin to maintain the killer activity of the T cells.

Scientists who are actively engaged in interleukin research believe that interleukins soon will be used as adjuncts for vaccines, for the treatment of chronic infectious diseases, and perhaps for the treatment of cancer. Interleukin antagonists also may prove helpful in prevent-ing skin and organ rejection, autoimmune diseases, and allergies.

> The interleukins and other cytokines show some promise of enhancing the individual's own immune system.

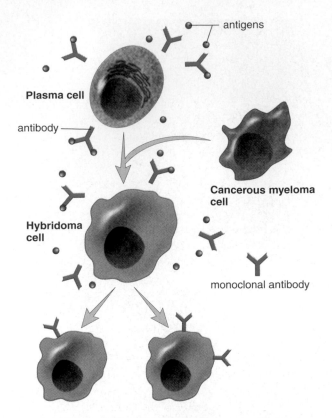

antigens

Plasma cell

antibody

Cancerous myeloma cell

Hybridoma cell

monoclonal antibody

Figure 14.11   Production of monoclonal antibodies.
Plasma cells of the same type (derived from immunized mice) are fused with myeloma (cancerous) cells, producing hybridoma cells that are "immortal." Hybridoma cells divide and continue to produce the same type of antibody, called monoclonal antibodies.

Monoclonal Antibodies

Every plasma cell derived from the same B cell secretes antibodies against a specific antigen. These are **monoclonal antibodies** because all of them are the same type and because they are produced by plasma cells derived from the same B cell. One method of producing monoclonal antibodies in vitro (outside the body in glassware) is depicted in Figure 14.11. B lymphocytes are removed from an animal (today, usually mice are used) and are exposed to a particular antigen. The resulting plasma cells are fused with myeloma cells (malignant plasma cells that live and divide indefinitely). The fused cells are called hybridomas—*hybrid*-because they result from the fusion of two different cells, and *-oma* because one of the cells is a cancer cell.

At present, monoclonal antibodies are being used for quick and certain diagnosis of various conditions. For example, a particular hormone is present in the urine of a pregnant woman. A monoclonal antibody can be used to detect this hormone; if it is present, the woman knows she is pregnant. Monoclonal antibodies are also used to identify infections. And because they can distinguish between cancer and normal tissue cells, they are used to carry radioactive isotopes or toxic drugs to tumors so that they can be selectively destroyed.

14.5 Immunity Side Effects

The immune system usually protects us from disease because it can distinguish self from nonself. Sometimes, however, it responds in a manner that harms the body, as when individuals develop allergies, receive an incompatible blood type, suffer tissue rejection, or have an auto-immune disease.

Allergies

Allergies are hypersensitivities to substances such as pollen or animal hair that ordinarily would do no harm to the body. The response to these antigens, called **allergens,** usually includes some degree of tissue damage. There are four types of allergic responses, but we will consider only two of them: immediate allergic response and delayed allergic response.

Immediate Allergic Response

An **immediate allergic response** can occur within seconds of contact with the antigen. As discussed in the Health Focus on page 275, coldlike symptoms are common. Anaphylactic shock is a severe reaction characterized by a sudden and life-threatening drop in blood pressure.

Immediate allergic responses are caused by antibodies known as IgE (see Table 14.1). IgE antibodies are attached to the plasma membrane of mast cells in the tissues and also to basophils in the blood. When an allergen attaches to the IgE antibodies on these cells, they release histamine and other substances that bring about the coldlike symptoms or, rarely, anaphylactic shock.

Allergy shots sometimes prevent the onset of an allergic response. It has been suggested that injections of the allergen may cause the body to build up high quantities of IgG antibodies, and these combine with allergens received from the environment before they have a chance to reach the IgE antibodies located in the membrane of mast cells and basophils.

Delayed Allergic Response

Delayed allergic responses are initiated by memory T cells at the site of allergen in the body. The allergic response is regulated by the cytokines secreted by both T cells and macrophages.

A classic example of a delayed allergic response is the skin test for tuberculosis (TB). When the result of the test is positive, the tissue where the antigen was injected becomes red and hardened. This shows that there was prior exposure to tubercle bacilli, the cause of TB. Contact dermatitis, which occurs when a person is allergic to poison ivy, jewelry, cosmetics, and so forth, is also an example of a delayed allergic response.

Health Focus

Immediate Allergic Responses

The runny nose and watery eyes of hay fever are often caused by an allergic reaction to the pollen of trees, grasses, and ragweed. Worse, if a person has asthma, the airways leading to the lungs constrict, resulting in difficult breathing characterized by wheezing. Windblown pollen, particularly in the spring and fall, brings on the symptoms of hay fever. Most people can inhale pollen with no ill effects. But others have developed a hypersensitivity, meaning that their immune system responds in a deleterious manner. The problem stems from a type of antibody called immunoglobulin E (IgE) that causes the release of histamine from mast cells and also basophils whenever they are exposed to an allergen. Histamine causes the mucosal membranes of the nose and eyes to release fluid as a defense against pathogen invasion. But in the case of allergies, copious fluid is released even though no real danger is present.

Most food allergies are also due to the presence of IgE antibodies, which usually bind to a protein in the food. The symptoms, such as nausea, vomiting, and diarrhea, are due to the mode of entry of the allergen. Skin symptoms may also occur, however. Adults are often allergic to shellfish, nuts, eggs, cows' milk, fish, and soybeans. Peanut allergy is a common food allergy in the United States, possibly because peanut butter is a staple in the diet. People seem to outgrow allergies to cows' milk and eggs more often than allergies to peanuts and soybeans.

Celiac disease occurs in people who are allergic to wheat, rye, barley, and sometimes oats—in short, any grain that contains gluten proteins. It is thought that the gluten proteins elicit a delayed cell-mediated immune response by T cells with the resultant production of cytokines. The symptoms of celiac disease include diarrhea, bloating, weight loss, anemia, bone pain, chronic fatigue, and weakness.

People can reduce the chances of a reaction to airborne and food allergens by avoiding the offending substances. The reaction to peanuts can be so severe that airlines are now required to have a peanut-free zone in their planes for those who are allergic. The people in Figure 14B are trying to avoid windblown allergens. Taking antihistamines can also be helpful.

If these precautions are inadequate, patients can be tested to measure their susceptibility to any number of possible allergens. A small quantity of a suspected allergen is inserted just beneath the skin, and the strength of the subsequent reaction is noted. A wheal-and-flare response at the skin prick site demonstrates that IgE antibodies attached to mast cells have reacted to an allergen. In an immunotherapy called hyposensitization, ever-increasing doses of the allergen are periodically injected subcutaneously with the hope that the body will build up a supply of IgG. IgG, in contrast to IgE, does not cause the release of histamine after it combines with the allergen. If IgG combines first upon exposure to the allergen, the allergic response does not occur. Patients know they are cured when the allergic symptoms go away. Therapy may have to continue for as long as two to three years.

Allergic-type reactions can occur without involving the immune system. Wasp and bee stings contain substances that cause swellings, even in those whose immune system is not sensitized to substances in the sting. Also, jellyfish tentacles and certain foods (e.g., fish that is not fresh and strawberries) contain histamine or closely related substances that can cause a reaction. Immunotherapy is also not possible in people who are allergic to penicillin and bee stings. High sensitivity has built up upon the first exposure, and when reexposed, anaphylactic shock can occur. Among its many effects, histamine causes increased permeability of the capillaries, the smallest blood vessels. These individuals experience a drastic decrease in blood pressure that can be fatal within a few minutes. People who know they are allergic to bee stings can obtain a syringe of epinephrine to carry with them. This medication can delay the onset of anaphylactic shock until medical help is reached.

Figure 14B Protection against allergies.
The allergic reactions that result in hay fever and asthma attacks can have many triggers, one of which is the pollen of a variety of plants. These people have found a dramatic solution to the problem.

Table 14.2   The ABO System

Blood Type	Antigen on Red Blood Cells	Antibody in Plasma	% U.S. African American	% U.S. Caucasian	% U.S. Asian	% North American Indians	% Americans of Chinese Descent
A	A	Anti-B	27	41	28	8	25
B	B	Anti-A	20	9	27	1	35
AB	A,B	None	4	3	5	0	10
O	None	Anti-A and anti-B	49	47	40	91	30

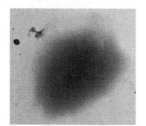
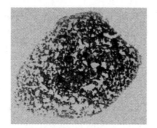

a.   No agglutination   Agglutination

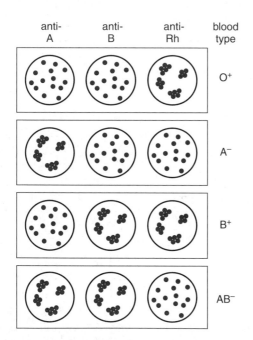

anti-A   anti-B   anti-Rh   blood type

O⁺

A⁻

B⁺

AB⁻

b.

Figure 14.12   Blood typing.
a. When typing blood, it is important to recognize no agglutination *(left)* versus agglutination *(right)*. **b.** The standard test to determine ABO and Rh blood type consists of putting a drop each of anti-A antibodies, anti-B antibodies, and anti-Rh antibodies on a slide. To each of these, a drop of the person's blood is added. If agglutination occurs, the person has this antigen on red blood cells, and therefore this type blood. Several possible results are shown.

Blood-Type Reactions

When blood transfusions were first attempted, illness and even death sometimes resulted. Eventually, it was discovered that only certain types of blood are compatible because red blood cell membranes carry proteins or carbohydrates that are antigens to blood recipients. The ABO system of typing blood is based on this principle.

ABO System

Blood typing in the ABO system uses two antigens, known as antigen A and antigen B. There are four blood types: O, A, B, and AB. Type O has neither the A antigen nor the B antigen on red blood cells; the other types of blood have antigen A, B, or both A and B, respectively (Table 14.2).

Within plasma, are naturally occurring antibodies to the antigens not present on the person's red blood cells. This is reasonable, because if the same antigen and antibody are present in blood, **agglutination,** or clumping of red blood cells, occurs. Agglutination causes the blood to stop circulating and red blood cells to burst.

Figure 14.12 shows a way to use the antibodies derived from plasma to determine the blood type. If agglutination occurs after a sample of blood is mixed with a particular antibody, the person has that type of blood.

Rh System

Another important antigen in matching blood types is the Rh factor. Persons with the Rh factor on their red blood cells are Rh positive (Rh⁺); those without it are Rh negative (Rh⁻). Rh-negative individuals normally do not have antibodies to the Rh factor, but they may make them when exposed to the Rh factor during pregnancy or blood transfusion.

If a mother is Rh negative and a father is Rh positive, a child may be Rh positive. The Rh-positive red blood cells of the child may begin leaking across the placenta into the mother's circulatory system, as placental tissues normally break down before and at birth. This sometimes causes the mother to produce anti-Rh antibodies. In this or a subsequent pregnancy with another Rh-positive child, anti-Rh antibodies may cross the placenta and destroy the child's red blood cells. This condition is called hemolytic disease of the newborn (HDN) (Fig. 14.13).

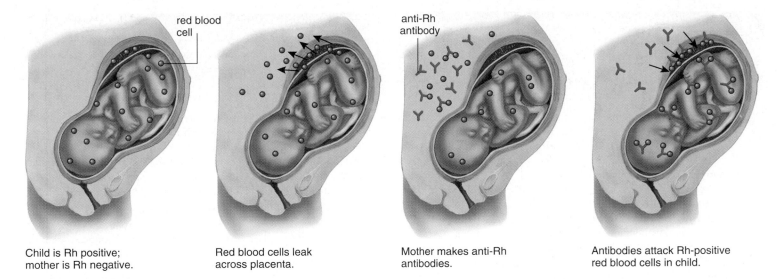

Child is Rh positive; mother is Rh negative.

Red blood cells leak across placenta.

Mother makes anti-Rh antibodies.

Antibodies attack Rh-positive red blood cells in child.

Figure 14.13   Hemolytic disease of the newborn.
Due to a pregnancy in which the child is Rh positive, an Rh negative mother can begin to produce antibodies against Rh positive red blood cells. In another pregnancy, these antibodies can cross the placenta and cause hemolysis of an Rh positive child's red blood cells.

The Rh problem has been solved by giving Rh-negative women an Rh-immunoglobulin injection (often a Rho-Gam injection) either midway through the first pregnancy or no later than 72 hours after giving birth to an Rh-positive child. This injection contains anti-Rh antibodies, which attack any of the child's red blood cells in the mother's blood before these cells can stimulate her immune system to produce her own antibodies. This injection is not beneficial if the woman has already begun to produce antibodies; therefore, the timing of the injection is most important.

> Blood is often typed according to the ABO system, combined with the Rh system. The possibility of hemolytic disease of the newborn exists when the mother is Rh negative and the father is Rh positive.

Tissue Rejection

Certain organs, such as skin, the heart, and the kidneys, could be transplanted easily from one person to another if the body did not attempt to reject them. Rejection occurs because antibodies and cytotoxic T cells bring about destruction of foreign tissues in the body. When rejection occurs, the immune system is correctly distinguishing between self and nonself.

Organ rejection can be controlled by carefully selecting the organ to be transplanted and administering immuno-suppressive drugs. It is best if the transplanted organ has the same type of HLA antigens as those of the recipient, because cytotoxic T cells recognize foreign HLA antigens. Two well known immunosuppressive drugs, cyclosporine and tacrolimus, both act by inhibiting the response of T cells to cytokines.

The hope is that tissue engineering, the production of organs that lack antigens or that can be protected in some way from the immune system, will one day do away with the problem of rejection (see page 530).

> When an organ is rejected, the immune system has recognized and destroyed cells that bear foreign HLA antigens.

Autoimmune Diseases

When cytotoxic T cells or antibodies mistakenly attack the body's own cells as if they bear foreign antigens, the resulting condition is known as an **autoimmune disease.** Exactly what causes autoimmune diseases is not known. However, sometimes they occur after an individual has recovered from an infection.

In the autoimmune disease myasthenia gravis, neuro-muscular junctions do not work properly, and muscular weakness results. In multiple sclerosis, the myelin sheath of nerve fibers breaks down, and this causes various neuromuscular disorders. A person with systemic lupus erythematosus has various symptoms prior to death due to kidney damage. In rheumatoid arthritis, the joints are affected. Researchers suggest that heart damage following rheumatic fever and type I diabetes are also autoimmune illnesses. As yet, there are no cures for autoimmune diseases, but they can be controlled with drugs.

> Autoimmune diseases occur when antibodies or cytotoxic T cells recognize and destroy the body's own cells.

Bioethical Focus   Clinical Trials

The United Nations estimates that 16,000 people become newly infected with the human immunodeficiency virus (HIV) each day, or 5.8 million per year. Ninety percent of these infections occur in the less-developed countries[1] where infected persons do not have access to antiviral therapy. In Uganda, for example, there is only one physician per 100,000 people, and only $6.00 is spent annually on health care, per person. In contrast, in the United States $12,000–$15,000 is sometimes spent on treating an HIV infected person per year.

The only methodology to prevent the spread of HIV in a developing country is counseling against behaviors that increase the risk of infection. Clearly an effective vaccine would be most beneficial to these countries. Several HIV vaccines are in various stages of development, and all need

to be clinically tested in order to see if they are effective. It seems reasonable to carry out such trials in developing countries, but there are many ethical questions.

A possible way to carry out the trial is this: vaccinate the uninfected sexual partners of HIV-infected individuals. After all, if the uninfected partner remains free of the disease, then the vaccine is effective. But is it ethical to allow a partner identified as having an HIV infection to remain untreated for the sake of the trial?

And should there be a placebo group—a group that does not get the vaccine? After all, if a greater number of persons in the placebo group become infected than those in the vaccine group, then the vaccine is effective. But if members of the placebo group become infected, shouldn't they be given effective treatment? For that matter, even participants

in the vaccine group might become infected. Shouldn't any participant of the trial be given proper treatment if they become infected? Who would pay for such treatment when the trial could involve thousands of persons?

Decide Your Opinion

1. Should HIV vaccine trials be done in developing countries, which stand to gain the most from an effective vaccine? Why or why not?
2. Should the trial be carried out using the same standards as in developed countries? Why or why not?
3. Who should pay for the trial—the drug company, the participants, or the country of the participants?

[1]Country that has only low to moderate industrialization; usually located in the southern hemisphere.

Summarizing the Concepts

14.1 The Lymphatic System
The lymphatic system consists of lymphatic vessels and lymphoid organs. The lymphatic vessels receive lipoproteins at intestinal villi and excess tissue fluid at blood capillaries, and carry these to the bloodstream.

Lymphocytes are produced and accumulate in the lymphoid organs (red bone marrow, lymph nodes, tonsils, spleen, and thymus gland). Lymph is cleansed of pathogens and/or their toxins in lymph nodes, and blood is cleansed of pathogens and/or their toxins in the spleen. T lymphocytes mature in the thymus, while B lymphocytes mature in the red bone marrow where all blood cells are produced. White blood cells are necessary for nonspecific and specific defenses.

14.2 Nonspecific Defenses
Immunity involves nonspecific and specific defenses. Nonspecific defenses include barriers to entry, the inflammatory reaction, natural killer cells, and protective proteins.

14.3 Specific Defenses
Specific defenses require B lymphocytes and T lymphocytes, also called B cells and T cells. B cells undergo clonal selection with production of plasma cells and memory B cells after their antigen receptors combine with a specific antigen. Plasma cells secrete antibodies and eventually undergo apoptosis. Plasma cells are responsible for antibody-mediated immunity. The IgG antibody is a Y-shaped molecule that has two binding sites for a specific antigen. Memory B cells remain in the body and produce antibodies if the same antigen enters the body at a later date.

T cells are responsible for cell-mediated immunity. The two main types of T cells are cytotoxic T cells and helper T cells. Cytotoxic T cells

kill virus-infected or cancer cells on contact because they bear a nonself protein. Helper T cells produce cytokines and stimulate other immune cells. Like B cells, each T cell bears antigen receptors. However, for a T cell to recognize an antigen, the antigen must be presented by an antigen-presenting cell (APC), usually a macrophage, along with an HLA (human leukocyte-associated antigen). Thereafter, the activated T cell undergoes clonal expansion until the illness has been stemmed. Then most of the activated T cells undergo apoptosis. A few cells remain, however, as memory T cells.

14.4 Induced Immunity
Active (long-lived) immunity can be induced by vaccines when a person is well and in no immediate danger of contracting an infectious disease. Active immunity is dependent upon the presence of memory cells in the body.

Passive immunity is needed when an individual is in immediate danger of succumbing to an infectious disease. Passive immunity is short-lived because the antibodies are administered to and not made by the individual.

Cytokines, including interferon, are used in attempts to treat AIDS and to promote the body's ability to recover from cancer.

Monoclonal antibodies, which are produced by the same plasma cell, have various functions, from detecting infections to treating cancer.

14.5 Immunity Side Effects
Allergic responses occur when the immune system reacts vigorously to substances not normally recognized as foreign. Immediate allergic responses, usually consisting of coldlike symptoms, are due to the activity of antibodies. Delayed allergic responses, such as contact dermatitis, are due to the activity of T cells. Immune side effects also include blood-type reactions, tissue rejection, and autoimmune diseases.

Testing Yourself

Choose the best answer for each question.

1. Which of the following is NOT a function of the lymphatic system?
   a. production of blood cells
   b. return excess fluid to the blood
   c. transport lipids absorbed from the digestive system
   d. defend the body against pathogens

2. Which of the following is a function of the spleen?
   a. produces T cells
   b. removes worn out red blood cells
   c. produces immunoglobulins
   d. produces macrophages
   e. regulates the immune system

3. What structural similarities do veins and lymphatic vessels have in common?
   a. Both have thick walls of smooth muscle.
   b. Both contain valves for one-way flow of fluids.
   c. Both empty directly into the heart.
   d. Both are fed fluids from arterioles.

4. Which of the following is a nonspecific defense of the body?
   a. immunoglobulin
   b. B cell   d. vaccine
   c. T cell   e. inflammatory response

5. Complement
   a. is a general defense mechanism.
   b. is involved in the inflammatory reaction.
   c. is a series of proteins present in the plasma.
   d. plays a role in destroying bacteria.
   e. All of these are correct.

6. Which of these pertain(s) to T cells?
   a. have specific receptors
   b. are more than one type
   c. are responsible for cell-mediated immunity
   d. stimulate antibody production by B cells
   e. All of these are correct.

7. Which one of these does NOT pertain to B cells?
   a. have passed through the thymus
   b. have specific receptors
   c. are responsible for antibody-mediated immunity
   d. synthesize and liberate antibodies

8. The clonal selection theory says that
   a. an antigen selects certain B cells and suppresses them.
   b. an antigen stimulates the multiplication of B cells that produce antibodies against it.
   c. T cells select those B cells that should produce antibodies, regardless of antigens present.
   d. T cells suppress all B cells except the ones that should multiply and divide.
   e. Both b and c are correct.

9. Plasma cells are
   a. the same as memory cells.
   b. formed from blood plasma.
   c. B cells that are actively secreting antibody.
   d. inactive T cells carried in the plasma.
   e. a type of red blood cell.

10. Antibodies combine with antigens
    a. at variable regions.
    b. at constant regions.
    c. only if macrophages are present.
    d. Both a and c are correct.

11. Which one of these is mismatched?
    a. helper T cells—help complement react
    b. cytotoxic T cells—active in tissue rejection
    c. macrophages—activate T cells
    d. memory T cells—long-living line of T cells
    e. T cells—mature in thymus

12. An antigen-presenting cell (APC)
    a. presents antigens to T cells.
    b. secretes antibodies.
    c. marks each human cell as belonging to that particular person.
    d. secretes cytokines.

13. Which of the following would NOT be a participant in cell-mediated immune responses?
    a. helper T cells   d. cytotoxic T cells
    b. macrophages   e. plasma cells
    c. cytokines

14. Vaccines are
    a. the same as monoclonal antibodies.
    b. treated bacteria or viruses, or one of their proteins.
    c. short-lived.
    d. MHC proteins.
    e. All of these are correct.

15. A person with AB⁻ type blood could receive which of the following blood types?
    a. AB⁻   d. O⁻
    b. AB⁺   e. All are correct except b.
    c. A⁻

16. During blood typing, agglutination indicates that the
    a. plasma contains certain antibodies.
    b. red blood cells carry certain antigens.
    c. plasma contains certain antigens.
    d. red blood cells carry certain antibodies.
    e. white blood cells fight infection.

17. Which of the following is NOT an example of an autoimmune disease?
    a. multiple sclerosis
    b. rheumatic fever
    c. hemolytic disease of the newborn
    d. systemic lupus erythematosus
    e. rheumatoid arthritis

18. Label a–c on this IgG molecule using these terms: antigen-binding sites, light chain, heavy chain

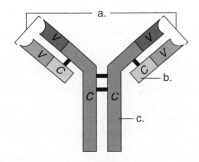

    d. What do V and C stand for in the diagram?

e-Learning Connection   www.mhhe.com/maderinquiry10

Concepts	Questions	Media Resources*
14.1 The Lymphatic System		

• The lymphatic vessels form a one-way system, which transports lymph from the tissues and fat from the lacteals to certain cardiovascular veins.
• The lymphoid organs (red bone marrow, spleen, thymus, lymph nodes, and tonsils) play critical roles in defense mechanisms.
	1. In what ways does the lymphatic system help to contribute to homeostasis?
	2. How does each of the lymphoid organs participate in defending the body against disease?
		Essential Study Partner
		   Lymph System
		   Lymph Nodes
		   Lymph Organs
		
		Animation Quiz
		   Lymphatic System
		
		Labeling Exercises
		   Bone Marrow

14.2 Nonspecific Defenses		

• Immunity consists of nonspecific and specific defenses to protect the body against disease.
• Nonspecific defenses consist of barriers to entry, the inflammatory reaction, natural killer cells, and protective proteins.
	1. How does each type barrier to entry help keep us healthy?
	2. What is the role of neutrophils and macrophages in the inflammatory reaction?
		Essential Study Partner
		   Nonspecific Immunity
		
		Animation Quizzes
		   Phagocytic Cells
		   Complement Proteins
		   Antiviral Defense
		
		BioCourse Study Guides
		   The Immune System: The
		      First Line of Defense
		   The Immune System: The
		      Second Line of Defense

14.3 Specific Defenses		

• Specific defenses require two types of lymphocytes: B lymphocytes and T lymphocytes.	1. What is an antigen?
	2. What are the targets of B lymphocytes and T lymphocytes?
		Essential Study Partner
		   Specific Immunity
		   Antibody-Mediated
		      (Humoral) Immunity
		   Cell Mediated Immunity
		
		Art Quiz
		   Activation of T Cells
		
		BioCourse Study Guide
		   The Immune System: The
		      Third Line of Defense

14.4 Induced Immunity		

• Induced immunity for medical purposes involves the use of vaccines to achieve long-lasting (active) immunity and the use of antibodies to provide temporary (passive) immunity.	1. Why does active immunity last longer than passive immunity?
	2. What are monoclonal antibodies?
		Art Quiz
		   Monoclonal Antibody
		      Production
		
		Animation Quiz
		   Vaccination
		
		Case Study
		   AIDS Vaccine

14.5 Immunity Side Effects		

• While immunity preserves life, it is also responsible for certain undesirable effects, such as allergies, blood type reactions, tissue rejection, and autoimmune diseases.	1. How do allergy shots help prevent the effects of allergic reactions?	Essential Study Partner
		   Abnormalities
		
		Art Quiz
		   Allergic Reaction
		
		Animation Quiz
		   ABO Blood Types

*For additional Media Resources, see the Online Learning Center.

Respiratory System

Chapter Concepts

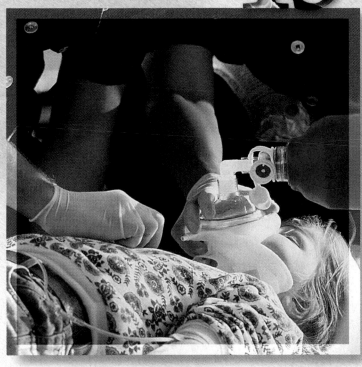

CPR (cardiopulmonary resuscitation) is an artificial means to maintain breathing and keep the blood flowing until a person has recovered the ability to do so.

Luckily the police arrived just moments after Tommy was rescued from the pond by his mother. A policeman immediately began CPR, alternately using a resusitator to give him oxygen and then pushing on his chest to have him breathe out. Eventually, Tommy could breathe on his own. All cells of the body require a constant supply of oxygen, and you have to keep breathing in order to bring oxygen into the body. Any cessation of breathing is a cause for concern, and prolonged cessation usually results in death. The heart needs oxygen to pump the blood that carries oxygen to all the cells of the body.

Cells use oxygen for cellular respiration, the process that replenishes their limited supply of ATP, without which they have no energy and cannot keep functioning. Carbon dioxide, an end product of cellular respiration, moves in the opposite direction— from the cells to the lungs, where it is expired. In this chapter, the structures and functions of the respiratory system are considered. Also, some of the medical conditions that decrease the functioning of the system will be discussed.

15.1 The Respiratory System

The organs of the respiratory system ensure that oxygen enters the body and carbon dioxide leaves the body. During **inspiration** or inhalation (breathing in), and **expiration** or exhalation (breathing out), air is conducted toward or away from the lungs by a series of cavities, tubes, and openings, illustrated in Figure 15.1.

The respiratory system also works with the cardiovascular system to accomplish respiration, which consists of:

1. Breathing: entrance and exit of air into and out of lungs.
2. External respiration: exchange of gases (oxygen and carbon dioxide) between air and blood.
3. Internal respiration: exchange of gases between blood and tissue fluid.
4. Cellular respiration: production of ATP in cells.

Cellular respiration uses the oxygen and produces the carbon dioxide that makes gas exchange with the environment necessary. Without a continuous supply of ATP, the cells cease to function. The functioning of the first three portions of respiration allow cellular respiration to continue. In this chapter, we study the first three portions of the respiratory process. Cellular respiration was discussed in chapter 7.

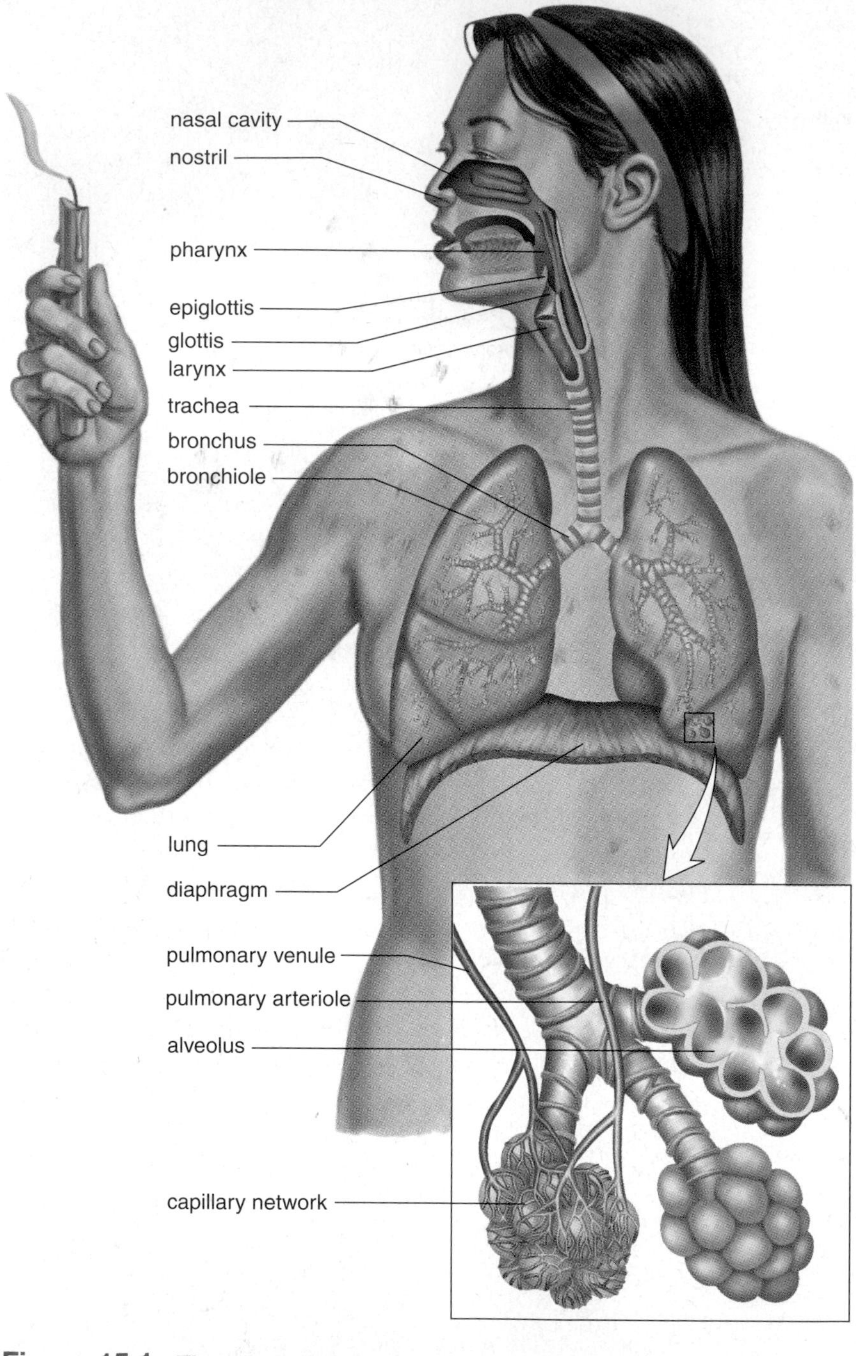

The Respiratory Tract

Table 15.1 traces the path of air from the nose to the lungs. As air moves in along the airways, it is cleansed, warmed, and moistened. Cleansing is accomplished by coarse hairs, cilia, and mucus in the region of the nostrils and by cilia alone in the rest of the nasal cavity and the other airways of the respiratory tract. In the nose, the hairs and the cilia act as a screening device. In the trachea and other airways, the cilia beat upward, carrying mucus, dust, and occasional bits of food that "went down the wrong way" into the pharynx, where the accumulation can be swallowed or expectorated. The air is warmed by heat given off by the blood vessels lying close to the surface of the lining of the airways, and it is moistened by the wet surface of these passages.

Conversely, as air moves out during expiration, it cools and loses its moisture. As the air cools, it deposits its moisture on the lining of the windpipe and the nose, and the nose may even drip as a result of this condensation. The air still retains so much moisture, however, that upon expiration on a cold day, it condenses and forms a small cloud.

The Nose

The nose contains two **nasal cavities,** which are narrow canals separated from one another by a septum composed of bone and cartilage (Fig. 15.2).

Figure 15.1   The respiratory tract.
The respiratory tract extends from the nose to the lungs, which are composed of air sacs called alveoli. Gas exchange occurs between air in the alveoli and blood within a capillary network that surrounds the alveoli. Notice that the pulmonary arteriole is colored blue—it carries O₂-poor blood away from the heart to the alveoli. Then, carbon dioxide leaves the blood and oxygen enters the blood. The pulmonary venule is colored red—it carries O₂-rich blood from alveoli toward the heart.

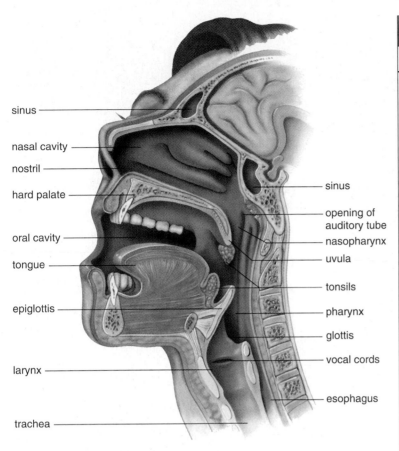

sinus

nasal cavity

nostril

hard palate

oral cavity

tongue

epiglottis

larynx

trachea

sinus

opening of
auditory tube

nasopharynx

uvula

tonsils

pharynx

glottis

vocal cords

esophagus

Figure 15.2   The path of air.
This drawing shows the path of air from the nasal cavities to the
trachea, which is a part of the lower respiratory tract. The other
organs are in the upper respiratory tract.

Table 15.1	Path of Air	
Structure	**Description**	**Function**
The Upper Respiratory Tract		
Nasal cavities	Hollow spaces in nose	Filter, warm, and moisten air
Pharynx	Chamber behind oral cavity and between nasal cavity and larynx	Connection to surrounding regions
Glottis	Opening into larynx	Passage of air into larynx
Larynx	Cartilaginous organ that contains vocal cords; voice box	Sound production
The Lower Respiratory Tract		
Trachea	Flexible tube that connects larynx with bronchi; windpipe	Passage of air to bronchi
Bronchi	Divisions of the trachea that enter lungs	Passage of air to lungs
Bronchioles	Branched tubes that lead from bronchi to alveoli	Passage of air to each alveolus
Lungs	Soft, cone-shaped organs that occupy a large portion of the thoracic cavity	Gas exchange

Special ciliated cells in the narrow upper recesses of the
nasal cavities act as receptors. Nerves lead from these cells to
the brain, where the impulses generated by the odor receptors are interpreted as smell.

The tear (lacrimal) glands drain into the nasal cavities
by way of tear ducts. For this reason, crying produces a
runny nose. The nasal cavities also communicate with the
cranial sinuses, air-filled mucosa-lined spaces in the skull.
If inflammation due to a cold or an allergic reaction blocks
the ducts leading from the sinuses, mucus may accumulate, causing a sinus headache.

The nasal cavities empty into the nasopharynx, the
upper portion of the pharynx. The auditory tubes lead from
the nasopharynx to the middle ears.

The Pharynx

The **pharynx** is a funnel-shaped passageway that connects
the nasal and oral cavities to the larynx. Therefore, the pharynx, which is commonly referred to as the "throat," has three
parts: the nasopharynx, where the nasal cavities open above
the soft palate; the oropharynx, where the oral cavity opens;

and the laryngopharynx, which opens into the larynx. The
tonsils form a protective ring at the junction of the oral cavity
and the pharynx. Being lymphoid tissue, the tonsils contain
lymphocytes that protect against invasion of foreign antigens
that are inhaled. In the tonsils, B cells and T cells are prepared to respond to antigens that may subsequently invade
internal tissues and fluids. Therefore, the respiratory tract
assists the immune system in maintaining homeostasis.

In the pharynx, the air passage and the food passage
cross because the larynx, which receives air, is ventral to
the esophagus, which receives food. The larynx lies at the
top of the trachea. The larynx and trachea are normally
open, allowing the passage of air, but the esophagus is normally closed and opens only when swallowing occurs.

The path of air starts with the nasal cavities and
ends with the lungs. Air from either the nose or the
mouth enters the pharynx, as does food. The
passage of air continues in the larynx and then in
the trachea.

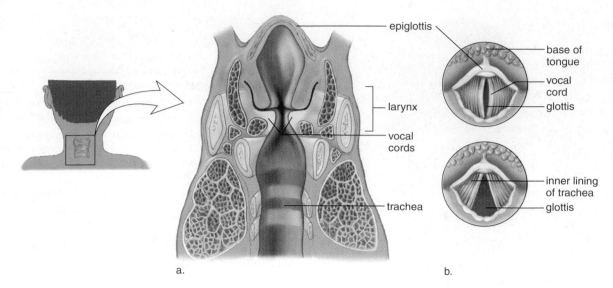

a.   b.

Figure 15.3   **Placement of the vocal cords.**

a. Frontal section of the larynx shows the location of the vocal cords. **b.** Viewed from above, it can be seen that the vocal cords are stretched across the glottis. When air passes through the glottis, the vocal cords vibrate, producing sound. The glottis is narrow when we produce a high-pitched sound *(top)*, and it widens as the pitch deepens *(bottom)*.

The Larynx

The **larynx** is a cartilaginous boxlike structure that serves as a passageway for air between the pharynx and the trachea. The larynx can be pictured as a triangular box whose apex, the Adam's apple, is located at the front of the neck. At the top of the larynx is a variable-sized opening called the **glottis.** When food is swallowed, the larynx moves upward against the **epiglottis,** a flap of tissue that prevents food from passing into the larynx. You can detect this movement by placing your hand gently on your larynx and swallowing.

The larynx is called the voice box because it houses the vocal cords. The **vocal cords** are mucosal folds supported by elastic ligaments, which are stretched across the glottis (Fig. 15.3). When air passes through the glottis, the vocal cords vibrate, producing sound. At the time of puberty, the growth of the larynx and the vocal cords is much more rapid and accentuated in the male than in the female, causing the male to have a more prominent Adam's apple and a deeper voice. The voice "breaks" in the young male due to his inability to control the longer vocal cords. These changes cause the lower pitch of the voice in males.

The high or low pitch of the voice is regulated when speaking and singing by changing the tension on the vocal cords. The greater the tension, as when the glottis becomes more narrow, the higher the pitch. When the glottis is wider, the pitch is lower (Fig. 15.3*b*). The loudness, or intensity, of the voice depends upon the amplitude of the vibrations—that is, the degree to which the vocal cords vibrate.

The Trachea

The **trachea,** commonly called the windpipe, is a tube connecting the larynx to the primary bronchi. The trachea lies ventral to the esophagus and is held open by C-shaped cartilaginous rings. The open part of the C-shaped rings faces the esophagus, and this allows the esophagus to expand when swallowing. The mucosa that lines the trachea has a layer of pseudostratified ciliated columnar epithelium. (Pseudostratified means that while the epithelium appears to be layered, actually each cell touches the basement membrane.) The cilia that project from the epithelium keep the lungs clean by sweeping mucus, produced by goblet cells, and debris toward the pharynx:

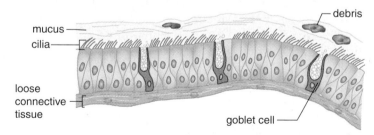

Smoking is known to destroy the cilia, and consequently the soot in cigarette smoke collects in the lungs. Smoking is discussed more fully in the Health Focus on page 297.

If the trachea is blocked because of illness or the accidental swallowing of a foreign object, it is possible to insert a breathing tube by way of an incision made in the trachea. This tube acts as an artificial air intake and exhaust duct. The operation is called a **tracheostomy.**

The Bronchial Tree

The trachea divides into right and left primary bronchi (sing., **bronchus**), which lead into the right and left lungs (see Fig. 15.1). The bronchi branch into a great number of secondary bronchi that eventually lead to **bronchioles.** The bronchi resemble the trachea in structure, but as the bronchial tubes divide and subdivide, their walls become thinner, and the small rings of cartilage are no longer present. During an asthma attack, the smooth muscle of the bronchioles contracts, causing bronchiolar constriction and characteristic wheezing. Each bronchiole terminates in an elongated space enclosed by a multitude of air pockets, or sacs, called alveoli (sing., **alveolus**). The alveoli make up the lungs.

The Lungs

The **lungs** are paired, cone-shaped organs that lie on either side of the heart within the thoracic cavity. The right lung has three lobes, and the left lung has two lobes, allowing room for the heart, which is on the left side of the body. A lobe is further divided into lobules, and each lobule has a bronchiole serving many alveoli. The base of each lung is broad and concave so that it fits the convex surface of the diaphragm. The other surfaces of the lungs follow the contours of the ribs and the diaphragm in the thoracic cavity.

The Alveoli

Each alveolar sac is made up of simple squamous epithelium surrounded by blood capillaries. Gas exchange occurs between air in the alveoli and blood in the capillaries (Fig. 15.4). Oxygen diffuses across the alveolar wall and enters the bloodstream, while carbon dioxide diffuses from the blood across the alveolar wall to enter the alveoli.

The alveoli of human lungs are lined with a surfactant, a film of lipoprotein that lowers the surface tension and prevents them from closing. The lungs collapse in some newborn babies, especially premature infants, who lack this film. The condition, called **infant respiratory distress syndrome,** is now treatable by surfactant replacement therapy.

There are altogether about 300 million alveoli, with a total cross-sectional area of 50–70 m^2. This is the surface area of a typical classroom and at least 40 times the surface area of the skin. Because of their many air spaces, the lungs are very light; normally, a piece of lung tissue dropped in a glass of water floats.

The trachea divides into the primary bronchi, which divide repeatedly to give rise to the bronchioles. The bronchioles have many branches and terminate at the alveoli, which make up the lungs.

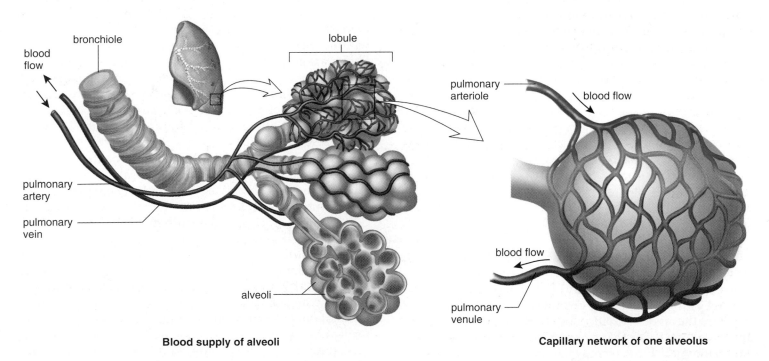

Blood supply of alveoli   **Capillary network of one alveolus**

Figure 15.4   Gas exchange in the lungs.
The lungs consist of alveoli surrounded by an extensive capillary network. Notice that the pulmonary arteriole carries O$_2$-poor blood (colored blue), and the pulmonary venule carries O$_2$-rich blood (colored red).

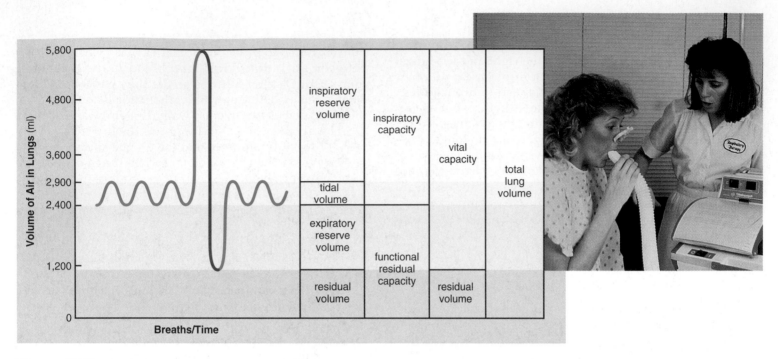

Figure 15.5   Vital capacity.
A spirometer measures the amount of air inhaled and exhaled with each breath. During inspiration, the pen moves up, and during expiration, the pen moves down. Vital capacity (red) is the maximum amount of air a person can exhale after taking the deepest inhalation possible.

15.2 Mechanism of Breathing

During breathing, air moves into the lungs from the nose or mouth (called inspiration or inhalation), and then moves out of the lungs during expiration or exhalation. A free flow of air from the nose or mouth to the lungs and from the lungs to the nose or mouth is of vital importance. Therefore, a technique has been developed that allows physicians to determine if there is a medical problem that prevents the lungs from filling with air upon inspiration and releasing it from the body upon expiration. This technique is illustrated in Figure 15.5, which shows the measurements recorded by a spirometer when a person breathes according to directions given by a technician.

Respiratory Volumes

Normally when we are relaxed, only a small amount of air moves in and out with each breath. This amount of air, called the **tidal volume,** is only about 500 ml.

It is possible to increase the amount of air inhaled, and therefore the amount exhaled, by deep breathing. The maximum volume of air that can be moved in plus the maximum amount that can be moved out during a single breath is called the **vital capacity.** It's called vital capacity because your life depends on breathing, and the more air you can move, the better off you are. There are a number of different illnesses discussed at the end of this chapter that can decrease vital capacity.

Vital capacity varies by how much we can increase inspiration and expiration over the tidal volume amount. We can increase inspiration by expanding the chest, and therefore the lungs. Forced inspiration (**inspiratory reserve volume**) usually increases by 2,900 ml, and that's quite a bit more than a tidal volume of only 500 ml! We can increase expiration by contracting the abdominal and thoracic muscles. This so-called **expiratory reserve volume** is usually about 1,400 ml of air. You can see from Figure 15.5 that vital capacity is the sum of tidal, inspiratory reserve, and expiratory reserve volumes.

It's a curious fact that some of the inhaled air never reaches the lungs; instead, it fills the nasal cavities, trachea, bronchi, and bronchioles (see Fig. 15.1). These passages are not used for gas exchange, and therefore they are said to contain dead-space air. To ensure that inhaled air reaches the lungs, it is better to breathe slowly and deeply. Also, note in Figure 15.5 that even after very deep breathing, some air (about 1,000 ml) remains in the lungs; this is called the **residual volume.** This air is no longer useful for gas exchange. In some lung diseases to be discussed later, the residual volume builds up because the individual has difficulty emptying the lungs. This means that the vital capacity is reduced because the lungs are filled with useless air.

The air used for gas exchange excludes both the air in the dead space of the respiratory tract and the residual volume in the lungs.

Ecology Focus

Photochemical Smog Can Kill

Most industrialized cities have photochemical smog at least occasionally. Photochemical smog arises when primary pollutants react with one another under the influence of sunlight to form a more deadly combination of chemicals. For example, two primary pollutants, nitrogen oxides (NO_x) and hydrocarbons (HC), react with one another in the presence of sunlight to produce nitrogen dioxide (NO_2), ozone (O_3), and PAN (peroxyacetylnitrate). Ozone and PAN are commonly referred to as oxidants. Breathing oxidants affects the respiratory and nervous systems, resulting in respiratory distress, headache, and exhaustion.

Cities with warm, sunny climates that are large and industrialized, such as Los Angeles, Denver, and Salt Lake City in the United States, Sydney in Australia, Mexico City in Mexico, and Buenos Aires in Argentina, are particularly susceptible to photochemical smog. If the city is surrounded by hills, a thermal inversion may aggravate the situation. Normally, warm air near the ground rises, so that pollutants are dispersed and carried away by air currents. But sometimes during a thermal inversion, smog gets trapped near the earth by a blanket of warm air (Fig. 15A). This may occur when a cold front brings in cold air, which settles beneath a warm layer. The trapped pollutants cannot disperse, and the results can be disastrous. In 1963, about 300 people died, and in 1966, about 168 people died, when air pollutants accumulated over New York City. Even worse were the events in London in 1957, when 700 to 800 people died, and in 1962, when 700 people died, due to the effects of air pollution.

Even though we have federal legislation to bring air pollution under control, more than half the people in the United States live in cities polluted by too much smog. In the long run, pollution prevention is usually easier and cheaper than pollution cleanup methods. Some prevention suggestions are as follows:

- Build more efficient automobiles or burn fuels that do not produce pollutants.
- Reduce the amount of waste to be incinerated by recycling materials instead.
- Reduce our energy use so that power plants need to provide less.
- Use renewable energy sources such as solar, wind, or water power.
- Require industries to meet clean air standards.

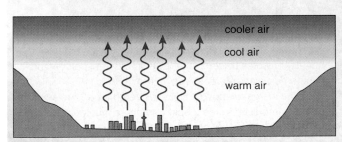

a. Normal pattern

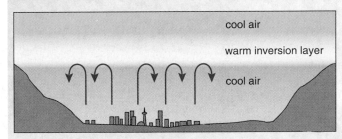

b. Thermal inversion

Figure 15A Thermal inversion.
a. Normally, pollutants escape into the atmosphere when warm air rises. **b.** During a thermal inversion, a layer of warm air (warm inversion layer) overlies and traps pollutants in cool air below. **c.** Los Angeles, a city of 8.5 million cars and thousands of factories, is particularly susceptible to thermal inversions, making this city the "air pollution capital" of the United States.

Inspiration and Expiration

To understand **ventilation,** the manner in which air enters and exits the lungs, it is necessary to remember first that normally there is a continuous column of air from the pharynx to the alveoli of the lungs.

Second, the lungs lie within the sealed-off thoracic cavity. The **rib cage** forms the top and sides of the thoracic cavity. It contains the ribs, hinged to the vertebral column at the back and to the sternum (breastbone) at the front, and the intercostal muscles that lie between the ribs. The **diaphragm,** a dome-shaped horizontal sheet of muscle and connective tissue, forms the floor of the thoracic cavity.

The lungs are enclosed by two membranes called the **pleura.** The parietal pleura adheres to the rib cage and the diaphragm, and the visceral pleura is fused to the lungs. The two pleural layers lie very close to one another, separated only by a small amount of fluid. Normally, the intrapleural pressure (pressure between the pleural membranes) is lower than atmospheric pressure by 4 mm Hg. The importance of the reduced intrapleural pressure is demonstrated when, by design or accident, air enters the intrapleural space. The affected lobules collapse. An infection of the pleura (pleural membranes) is called pleurisy.

The pleura enclose the lungs and line the thoracic cavity. Intrapleural pressure is lower than atmospheric pressure.

Inspiration

A **respiratory center** is located in the medulla oblongata of the brain. The respiratory center consists of a group of neurons that exhibit an automatic rhythmic discharge that triggers inspiration. Carbon dioxide (CO_2) and hydrogen ions (H^+) are the primary stimuli that directly cause changes in the activity of this center. This center is not affected by low oxygen (O_2) levels. Chemoreceptors in the **carotid bodies,** located in the carotid arteries, and in the **aortic bodies,** located in the aorta, are sensitive to the level of oxygen in blood. When the concentration of oxygen decreases, these bodies communicate with the respiratory center, and the rate and depth of breathing increase.

The respiratory center sends out impulses by way of nerves to the diaphragm and the external intercostal muscles of the rib cage (Fig. 15.6). In its relaxed state, the diaphragm is dome-shaped, but upon stimulation, it contracts and lowers. Also, the external intercostal muscles contract, causing the rib cage to move upward and outward. Now the thoracic cavity increases in size, and the lungs expand. As the lungs expand, air pressure within the

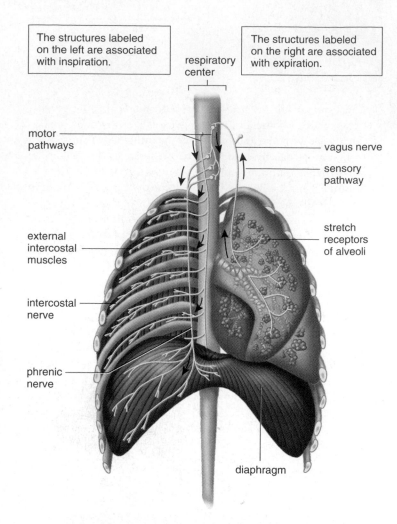

The structures labeled on the left are associated with inspiration.

respiratory center

The structures labeled on the right are associated with expiration.

motor pathways

vagus nerve

sensory pathway

external intercostal muscles

stretch receptors of alveoli

intercostal nerve

phrenic nerve

diaphragm

Figure 15.6   Nervous control of breathing.
During inspiration, the respiratory center stimulates the external intercostal (rib) muscles to contract via the intercostal nerves and stimulates the diaphragm to contract via the phrenic nerve. Should the tidal volume increase above 1.5 liters, stretch receptors send inhibitory nerve impulses to the respiratory center via the vagus nerve. In any case, expiration occurs due to lack of stimulation from the respiratory center to the diaphragm and intercostal muscles.

enlarged alveoli lowers, and air enters through the nose or the mouth.

Inspiration is the active phase of breathing (Fig. 15.7a). During this time, the diaphragm and the rib muscles contract, intrapleural pressure decreases, the lungs expand, and air comes rushing in. Note that air comes in because the lungs already have opened up; air does not force the lungs open. This is why it is sometimes said that *humans breathe by negative pressure.* The creation of a partial vacuum in the alveoli causes air to enter the lungs.

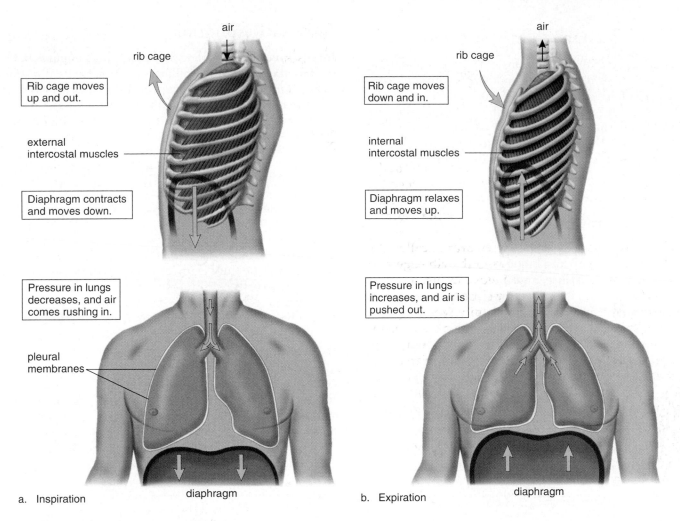

Figure 15.7   Inspiration versus expiration.
a. During inspiration, the thoracic cavity and lungs expand so that air is drawn in. **b.** During expiration, the thoracic cavity and lungs resume their original positions and pressures. Now air is forced out.

Expiration

When the respiratory center stops sending neuronal signals to the diaphragm and the rib cage, the diaphragm relaxes and resumes its dome shape. The abdominal organs press up against the diaphragm, and the rib cage moves down and inward (Fig. 15.7*b*). Now the elastic lungs recoil, and air is pushed out. What keeps the alveoli from collapsing following expiration? Recall that the presence of surfactant lowers surface tension and because of reduced intrapleural pressure, there is always some air remaining in the alveoli.

The respiratory center acts rhythmically to bring about breathing at a normal rate and volume. If by chance we inhale more deeply, the lungs are expanded and the alveoli stretch. This stimulates stretch receptors in the alveolar walls, and they initiate inhibitory nerve impulses that travel from the inflated lungs to the respiratory center. This causes the respiratory center to stop sending out nerve impulses.

While inspiration is the active phase of breathing, expiration is usually passive—that is, the diaphragm and external intercostal muscles are relaxed when expiration occurs. When breathing is deeper and/or more rapid, expiration can also be active. Contraction of internal intercostal muscles can force the rib cage to move downward and inward. Also, when the abdominal wall muscles are contracted, they push on the viscera, which push against the diaphragm, and the increased pressure in the thoracic cavity helps expel air.

During inspiration, due to nervous stimulation, the diaphragm lowers and the rib cage lifts up and out. During expiration, due to a lack of nervous stimulation, the diaphragm rises and the rib cage lowers.

15.3 Gas Exchanges in the Body

Gas exchange is critical to homeostasis. The act of breathing brings oxygen in air to the lungs and carbon dioxide from the lungs to outside the body. As mentioned previously, respiration includes not only the exchange of gases in the lungs, but also the exchange of gases in the tissues (Fig. 15.8). The principles of diffusion alone govern whether O_2 or CO_2 enters or leaves the blood in the lungs and in the tissues.

External Respiration

External respiration refers to the exchange of gases between air in the alveoli and blood in the pulmonary capillaries. Gases exert pressure, and the amount of pressure each gas exerts is called its partial pressure, symbolized as P_{O_2} and P_{CO_2}. Blood entering the pulmonary capillaries has a higher P_{CO_2} than atmospheric air. Therefore, *CO_2 diffuses out of the blood into the lungs.* Most of the CO_2 is carried as **bicarbonate ions** (HCO_3^-). As the little remaining free CO_2 begins to diffuse out, the following reaction is driven to the right:

$$H^+ + HCO_3^- \longrightarrow H_2CO_3 \longrightarrow H_2O + CO_2 \uparrow$$

hydrogen ion   bicarbonate ion   water   carbon dioxide

"Up" arrow indicates carbon dioxide is leaving the body.

The enzyme **carbonic anhydrase,** present in red blood cells, speeds up the reaction. This reaction requires that the respiratory pigment **hemoglobin,** also present in red blood cells, gives up the hydrogen ions (H^+) it has been carrying; that is, HHb becomes Hb. Hb is called deoxyhemoglobin.

The pressure pattern is the reverse for O_2. Blood entering the pulmonary capillaries is low in oxygen, and alveolar air contains a much higher partial pressure of oxygen. Therefore, *O_2 diffuses into plasma and then into red blood cells in the lungs.* Hemoglobin takes up this oxygen and becomes **oxyhemoglobin** (HbO_2):

$$Hb + \downarrow O_2 \longrightarrow HbO_2$$

deoxyhemoglobin   oxygen   oxyhemoglobin

"Down" arrow indicates that oxygen is entering the body.

Internal Respiration

Internal respiration refers to the exchange of gases between the blood in systemic capillaries and the tissue fluid. Blood that enters the systemic capillaries is a bright red color because red blood cells contain oxyhemoglobin. Oxyhemoglobin gives up O_2, which diffuses out of the blood into the tissues:

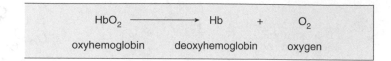

$$HbO_2 \longrightarrow Hb + O_2$$

oxyhemoglobin   deoxyhemoglobin   oxygen

Oxygen diffuses out of the blood into the tissues because the P_{O_2} of tissue fluid is lower than that of blood. The lower P_{O_2} is due to cells continuously using up oxygen in cellular respiration. *Carbon dioxide diffuses into the blood from the tissues* because the P_{CO_2} of tissue fluid is higher than that of blood. Carbon dioxide, produced continuously by cells, collects in tissue fluid.

After CO_2 diffuses into the blood, it enters the red blood cells, where a small amount is taken up by hemoglobin, forming **carbaminohemoglobin.** Most of the CO_2 combines with water, forming carbonic acid (H_2CO_3), which dissociates to hydrogen ions (H^+) and bicarbonate ions (HCO_3^-). The increased concentration of CO_2 in the blood drives the reaction to the right:

$$CO_2 + H_2O \underset{\text{carbonic anhydrase}}{\rightleftharpoons} H_2CO_3 \rightleftharpoons H^+ + HCO_3^-$$

carbon dioxide   water   carbonic acid   hydrogen ion   bicarbonate ion

The enzyme carbonic anhydrase, present in red blood cells, speeds up the reaction. Bicarbonate ions diffuse out of red blood cells and are carried in the plasma. The globin portion of hemoglobin combines with excess hydrogen ions produced by the overall reaction, and Hb becomes HHb, called **reduced hemoglobin.** In this way, the pH of blood remains fairly constant. Blood that leaves the capillaries is a dark maroon color because red blood cells contain reduced hemoglobin.

External and internal respiration are the movement of gases between blood and the alveoli and between blood and the systemic capillaries, respectively. Both processes are dependent on the process of diffusion.

Visual Focus

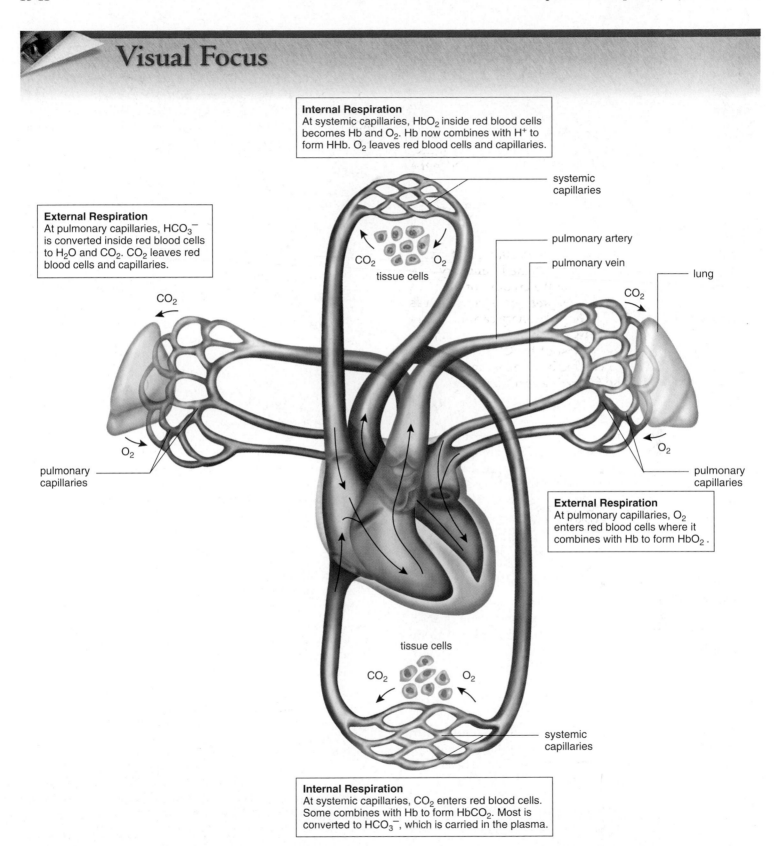

Internal Respiration
At systemic capillaries, HbO_2 inside red blood cells becomes Hb and O_2. Hb now combines with H^+ to form HHb. O_2 leaves red blood cells and capillaries.

External Respiration
At pulmonary capillaries, HCO_3^- is converted inside red blood cells to H_2O and CO_2. CO_2 leaves red blood cells and capillaries.

External Respiration
At pulmonary capillaries, O_2 enters red blood cells where it combines with Hb to form HbO_2.

Internal Respiration
At systemic capillaries, CO_2 enters red blood cells. Some combines with Hb to form $HbCO_2$. Most is converted to HCO_3^-, which is carried in the plasma.

systemic capillaries

pulmonary artery

pulmonary vein

lung

CO_2

O_2

tissue cells

CO_2

O_2

pulmonary capillaries

pulmonary capillaries

tissue cells

CO_2

O_2

systemic capillaries

Figure 15.8   External and internal respiration.
During external respiration in the lungs, CO_2 leaves the blood and O_2 enters the blood. During internal respiration in the tissues, O_2 leaves the blood and CO_2 enters the blood.

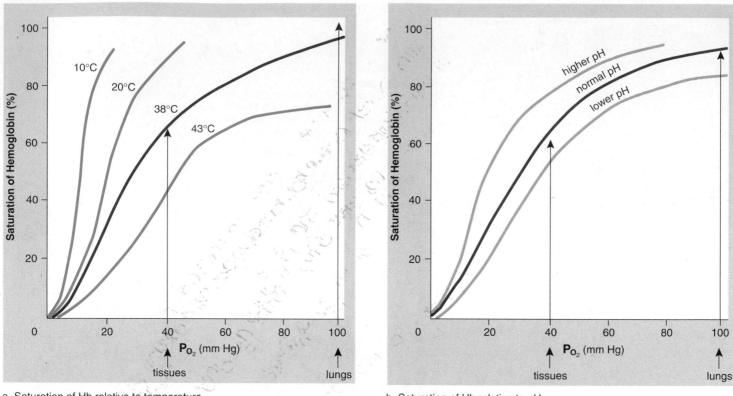

a. Saturation of Hb relative to temperature

b. Saturation of Hb relative to pH

Figure 15.9   Effect of environmental conditions on hemoglobin saturation.
The partial pressure of oxygen (P_{O_2}) in pulmonary capillaries is about 98–100 mm Hg, but in tissue capillaries it is only about 40 mm Hg. Hemoglobin is about 98% saturated in the lungs because of P_{O_2}, and also because (**a**) the temperature is cooler and (**b**) the pH is higher in the lungs. On the other hand, hemoglobin is only about 60% saturated in the tissues because of P_{O_2}, and also because (**a**) the temperature is warmer and (**b**) the pH is lower in the tissues.

Binding Capacity of Hemoglobin

The binding capacity of hemoglobin is also affected by partial pressures. The P_{O_2} of air entering the alveoli is about 100 mm Hg, and at this pressure the hemoglobin in the blood becomes saturated with O_2. This means that iron in hemoglobin molecules has combined with O_2. On the other hand, the P_{O_2} in the tissues is about 40 mm Hg, causing hemoglobin molecules to release O_2, and O_2 to diffuse into the tissues.

In addition to the partial pressure of O_2, temperature and pH also affect the amount of oxygen that hemoglobin can carry. The lungs have a lower temperature and a higher pH than the tissues:

	pH	Temperature
Lungs	7.40	37°C
Tissues	7.38	38°C

Figures 15.9*a* and *b* show that, as expected, hemoglobin is more saturated with O_2 in the lungs than in the tissues. This effect, which can be attributed to the difference in P_{O_2}

between the lungs and tissues, is enhanced by the difference in temperature and pH between the lungs and tissues. Notice in Figure 15.9*a* that the saturation curve for hemoglobin is steeper at 10°C compared to 20°C, and so forth. Also, Figure 15.9*b* shows that the saturation curve for hemoglobin is steeper at higher pH than at lower pH.

This means that the environmental conditions in the lungs are favorable for the uptake of O_2 by hemoglobin, and the environmental conditions in the tissues are favorable for the release of O_2 by hemoglobin. Hemoglobin is about 98–100% saturated in the capillaries of the lungs and about 60–70% saturated in the tissues. During exercise, hemoglobin is even less saturated in the tissues because muscle contraction leads to higher body temperature (up to 103°F in marathoners!) and lowers the pH (due to the production of lactic acid).

The difference in P_{O_2}, temperature, and pH between the lungs and tissues causes hemoglobin to take up oxygen in the lungs and release oxygen in the tissues.

15.4 Respiration and Health

The respiratory tract is constantly exposed to environmental air. The quality of this air, as discussed in the Ecology Focus on page 287, can affect our health. The presence of a disease means that homeostasis is threatened, and if the condition is not brought under control, death is a possibility.

Upper Respiratory Tract Infections

The upper respiratory tract consists of the nasal cavities, the pharynx, and the larynx. Upper respiratory infections (URI) can spread from the nasal cavities to the sinuses, middle ears, and larynx (Fig. 15.10). Viral infections sometimes lead to secondary bacterial infections. What we call "strep throat" is a primary bacterial infection caused by *Streptococcus pyogenes* that can lead to a generalized upper respiratory infection and even a systemic (affecting the body as a whole) infection. While antibiotics have no effect on viral infections, they are successfully used for most bacterial infections, including strep throat. The symptoms of strep throat are severe sore throat, high fever, and white patches on a dark red throat.

Sinusitis

Sinusitis is an infection of the cranial sinuses, cavities within the facial skeleton that drain into the nasal cavities. Only about 1–3% of upper respiratory infections are accompanied by sinusitis. Sinusitis develops when nasal congestion blocks the tiny openings leading to the sinuses. Symptoms include postnasal discharge as well as facial pain that worsens when the patient bends forward. Pain and tenderness usually occur over the lower forehead or over the cheeks. If the latter, toothache is also a complaint. Successful treatment depends on restoring proper drainage of the sinuses. Even a hot shower and sleeping upright can be helpful. Otherwise, spray decongestants are preferred over oral antihistamines, which thicken rather than liquefy the material trapped in the sinuses.

Otitis Media

Otitis media is a bacterial infection of the middle ear. The middle ear is not a part of the respiratory tract, but this infection is considered here because it is a complication often seen in children who have a nasal infection. Infection can spread by way of the **auditory tube** that leads from the nasopharynx to the middle ear. Pain is the primary symptom of a middle ear infection. A sense of fullness, hearing loss, vertigo (dizziness), and fever may also be present. Antibiotics almost always bring about a full recovery, and recurrence is probably due to a new infection. Drainage tubes (called tympanostomy tubes) are sometimes placed in the eardrums of children with multiple recurrences to help prevent the buildup of fluid in the middle ear and the

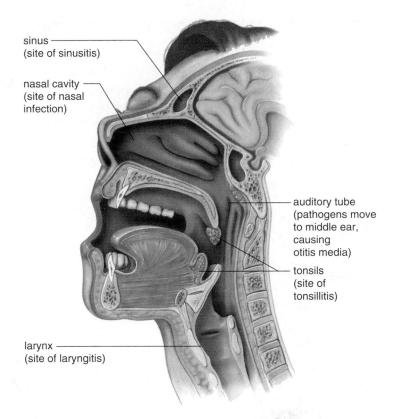

Figure 15.10 Sites of upper respiratory infections.
A nasal infection, more properly called rhinitis, is the usual symptom of a common cold due to a viral infection, but rhinitis can also be due to a bacterial infection. Secondary to a URI, the sinuses, middle ear, tonsils, and vocal cords can become infected. Allergies also cause runny nose, blocked sinuses, and laryngitis.

possibility of hearing loss. Normally, the tubes slough out with time.

Tonsillitis

Tonsillitis occurs when the **tonsils,** masses of lymphatic tissue in the pharynx, become inflamed and enlarged. The tonsils in the dorsal wall of the nasopharynx are often called adenoids. If tonsillitis occurs frequently and enlargement makes breathing difficult, the tonsils can be removed surgically in a **tonsillectomy.** Fewer tonsillectomies are performed today than in the past because it is now known that the tonsils remove many of the pathogens that enter the pharynx; therefore, they are a first line of defense against invasion of the body.

Laryngitis

Laryngitis is an infection of the larynx with accompanying hoarseness leading to the inability to talk in an audible voice. Usually laryngitis disappears with treatment of the upper respiratory infection. Persistent hoarseness without the presence of an upper respiratory infection is one of the warning signs of cancer and therefore should be looked into by a physician.

Visual Focus

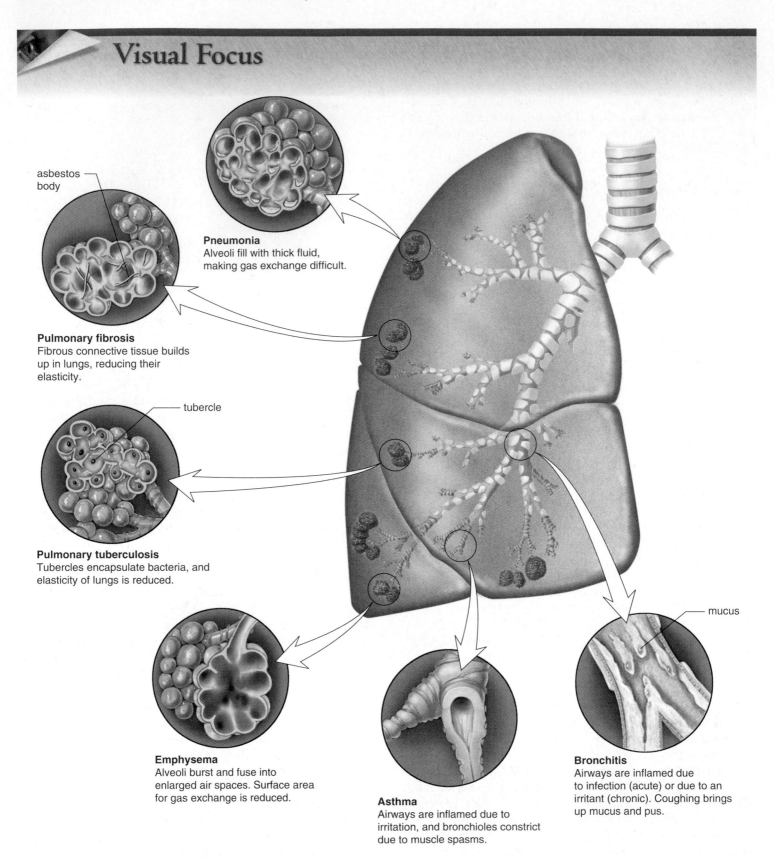

asbestos body

Pneumonia
Alveoli fill with thick fluid, making gas exchange difficult.

Pulmonary fibrosis
Fibrous connective tissue builds up in lungs, reducing their elasticity.

tubercle

Pulmonary tuberculosis
Tubercles encapsulate bacteria, and elasticity of lungs is reduced.

Emphysema
Alveoli burst and fuse into enlarged air spaces. Surface area for gas exchange is reduced.

Asthma
Airways are inflamed due to irritation, and bronchioles constrict due to muscle spasms.

mucus

Bronchitis
Airways are inflamed due to infection (acute) or due to an irritant (chronic). Coughing brings up mucus and pus.

Figure 15.11   **Lower respiratory tract disorders.**
Exposure to infectious pathogens and/or air pollutants, including cigarette and cigar smoke, can cause the diseases and disorders shown here.

Lower Respiratory Tract Disorders

Lower respiratory tract disorders, which are illustrated in Figure 15.11, include infections, restrictive pulmonary disorders, obstructive pulmonary disorders, and lung cancer.

Lower Respiratory Infections

Acute bronchitis, pneumonia, and tuberculosis are infections of the lower respiratory tract. **Acute bronchitis** is an infection of the primary and secondary bronchi. Usually it is preceded by a viral URI that has led to a secondary bacterial infection. Most likely, a nonproductive cough has become a deep cough that expectorates mucus and perhaps pus.

Pneumonia is a viral or bacterial infection of the lungs in which the bronchi and alveoli fill with thick fluid. Most often it is preceded by influenza. High fever and chills, with headache and chest pain, are symptoms of pneumonia. Rather than being a generalized lung infection, pneumonia may be localized in specific lobules of the lungs; obviously, the more lobules involved, the more serious the infection. Pneumonia can be caused by a bacterium that is usually held in check but has gained the upper hand due to stress and/or reduced immunity. AIDS patients are subject to a particularly rare form of pneumonia caused by the protozoan *Pneumocystis carinii.* Pneumonia of this type is almost never seen in individuals with a healthy immune system.

Pulmonary tuberculosis is caused by the tubercle bacillus, a type of bacterium. When tubercle bacilli invade the lung tissue, the cells build a protective capsule about the foreigners, isolating them from the rest of the body. This tiny capsule is called a tubercle. If the resistance of the body is high, the imprisoned organisms die, but if the resistance is low, the organisms eventually can be liberated. If a chest X ray detects active tubercles, the individual is put on appropriate drug therapy to ensure the localization of the disease and the eventual destruction of any live bacteria. It is possible to tell if a person has ever been exposed to tuberculosis with a test in which a highly diluted extract of the bacillus is injected into the skin of the patient. A person who has never been in contact with the tubercle bacillus shows no reaction, but one who has had or is fighting an infection shows an area of inflammation that peaks in about 48 hours.

Tuberculosis was a major killer in the United States before the middle of the twentieth century, after which antibiotic therapy brought it largely under control. In recent years, however, the incidence of tuberculosis has been on the rise, particularly among AIDS patients, the homeless, and the rural poor. Worse, the new strains are resistant to the usual antibiotic therapy. Therefore, some physicians would like to again quarantine patients in sanitariums, as was previously done.

Restrictive Pulmonary Disorders

In restrictive pulmonary disorders, vital capacity is reduced, not because air does not move freely into and out of the lungs but because the lungs have lost their elasticity. Inhaling particles such as silica (sand), coal dust, asbestos, and, now it seems, fiberglass can lead to **pulmonary fibrosis,** a condition in which fibrous connective tissue builds up in the lungs. The lungs cannot inflate properly and are always tending toward deflation. Breathing asbestos is also associated with the development of cancer. Since asbestos was formerly used widely as a fireproofing and insulating agent, unwarranted exposure has occurred. It has been projected that two million deaths caused by asbestos exposure—mostly in the workplace—will occur in the United States between 1990 and 2020.

Obstructive Pulmonary Disorders

In obstructive pulmonary disorders, air does not flow freely in the airways, and the time it takes to inhale or exhale maximally is greatly increased. Several disorders, including chronic bronchitis, emphysema, and asthma, are referred to as chronic obstructive pulmonary disorders (COPD) because they tend to recur.

In **chronic bronchitis,** the airways are inflamed and filled with mucus. A cough that brings up mucus is common. The bronchi have undergone degenerative changes, including the loss of cilia and their normal cleansing action. Under these conditions, an infection is more likely to occur. Smoking cigarettes and cigars is the most frequent cause of chronic bronchitis. Exposure to other pollutants can also cause chronic bronchitis.

Emphysema is a chronic and incurable disorder in which the alveoli are distended and their walls damaged so that the surface area available for gas exchange is reduced. Emphysema is often preceded by chronic bronchitis. Air trapped in the lungs leads to alveolar damage and a noticeable ballooning of the chest. The elastic recoil of the lungs is reduced, so not only are the airways narrowed, but the driving force behind expiration is also reduced. The victim is breathless and may have a cough. Because the surface area for gas exchange is reduced, oxygen reaching the heart and the brain is reduced. Even so, the heart works furiously to force more blood through the lungs, and an increased workload on the heart can result. Lack of oxygen to the brain can make the person feel depressed, sluggish, and irritable. Exercise, drug therapy, and supplemental oxygen, along with giving up smoking, may relieve the symptoms and possibly slow the progression of emphysema.

Asthma is a disease of the bronchi and bronchioles that is marked by wheezing, breathlessness, and sometimes a cough and expectoration of mucus. The airways are unusually sensitive to specific irritants, which can include a wide range of allergens such as pollen, animal dander, dust, cigarette smoke, and industrial fumes. Even cold air can be an irritant. When exposed to the irritant, the smooth muscle in the bronchioles undergoes spasms. It now appears that chemical mediators given off by immune cells in the bronchioles cause the spasms. Most asthma patients have some degree of bronchial inflammation that reduces the diameter of the airways and contributes to the seriousness of an attack. Asthma is not curable, but it is treatable. Special inhalers can control the inflammation and hopefully prevent an attack, while other types of inhalers can stop the muscle spasms should an attack occur.

Lung Cancer

Lung cancer used to be more prevalent in men than in women, but recently it has surpassed breast cancer as a cause of death in women. This can be linked to an increase in the number of women who smoke. Autopsies on smokers have revealed the progressive steps by which the most common form of lung cancer develops. The first event appears to be thickening and callusing of the cells lining the airways. (Callusing occurs whenever cells are exposed to irritants.) Then cilia are lost, making it impossible to prevent dust and dirt from settling in the lungs. Following this, cells with atypical nuclei appear in the callused lining. A tumor consisting of disordered cells with atypical nuclei is considered to be cancer in situ (at one location) (Fig. 15.12*b*). A final step occurs when some of these cells break loose and penetrate other tissues, a process called metastasis. Now the cancer has spread. The original tumor may grow until a bronchus is blocked, cutting off the supply of air to that lung. The entire lung then collapses, the secretions trapped in the lung spaces become infected, and pneumonia or a lung abscess (localized area of pus) results. The only treatment that offers a possibility of cure is to remove a lobe or the whole lung before metastasis has had time to occur. This operation is called **pneumonectomy.** If the cancer has spread, chemotherapy and radiation are also required.

The Health Focus on the next page lists the various illnesses, including cancer, that are apt to occur when a person smokes. Current research indicates that passive smoking can also cause lung cancer and other illnesses associated with smoking. If a person stops both voluntary and passive smoking, and if the body tissues are not already cancerous, they may return to normal over time.

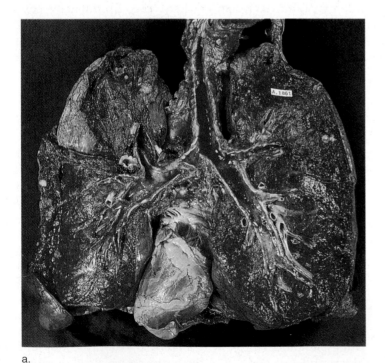

a.

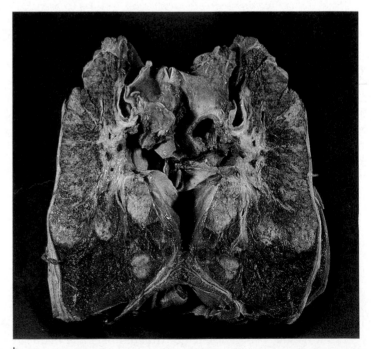

b.

Figure 15.12   Normal lung versus cancerous lung.
a. Normal lung with heart in place. Note the healthy red color. **b.** Lungs of a heavy smoker. Notice how black the lungs are except where cancerous tumors have formed.

The Most Often Asked Questions About Tobacco and Health

Is there a safe way to smoke?

No. All forms of tobacco can cause damage, and smoking even a small amount is dangerous. Tobacco is perhaps the only legal product whose advertised and intended use—that is, smoking it—will hurt the body.

Does smoking cause cancer?

Yes, and not only lung cancer. Besides lung cancer, smoking a pipe, cigarettes, or cigars is also a major cause of cancers of the mouth, larynx (voice box), and esophagus. In addition, smoking increases the risk of cancer of the bladder, kidney, pancreas, stomach, and the uterine cervix.

What are the chances of being cured of lung cancer?

Very low; the five-year survival rate is only 13%. Fortunately, lung cancer is a largely preventable disease. In other words, by not smoking, it can probably be prevented.

Does smoking cause other lung diseases?

Yes. It leads to chronic bronchitis, a disease in which the airways produce excess mucus, forcing the smoker to cough frequently. Smoking is also the major cause of emphysema, a disease that slowly destroys a person's ability to breathe. Chronic bronchitis and pulmonary emphysema are higher in smokers compared to nonsmokers.

Why do smokers have "smoker's cough"?

Normally, cilia (tiny hairlike formations that line the airways) beat outwards and "sweep" harmful material out of the lungs. Smoke, however, decreases this sweeping action, so some of the poisons in the smoke remain in the lungs.

If you smoke but don't inhale, is there any danger?

Yes. Wherever smoke touches living cells, it does harm. So, even if smokers of pipes, cigarettes, and cigars don't inhale, they are at an increased risk for lip, mouth, and tongue cancer.

Does smoking affect the heart?

Yes. Smoking increases the risk of heart disease, which is the United States' number one killer. Smoking, high blood pressure, high cholesterol, and lack of exercise are all risk factors for heart disease. Smoking alone doubles the risk of heart disease.

Is there any risk for pregnant women and their babies?

Pregnant women who smoke endanger the health and lives of their unborn babies. When a pregnant woman smokes, she really is smoking for two because the nicotine, carbon monoxide, and other dangerous chemicals in smoke enter her bloodstream and then pass into the baby's body. Smoking mothers have more stillbirths and babies of low birth weight than nonsmoking mothers.

Does smoking cause any special health problems for women?

Yes. Women who smoke and use the birth control pill have an increased risk of stroke and blood clots in the legs. In addition, women who smoke increase their chances of getting cancer of the uterine cervix.

What are some of the short-term effects of smoking cigarettes?

Almost immediately, smoking can make it hard to breathe. Within a short time, it can also worsen asthma and allergies. Only seven seconds after a smoker takes a puff, nicotine reaches the brain, where it produces a morphinelike effect.

Are there any other risks to the smoker?

Yes, there are many more risks. Smoking is a cause of stroke, which is the third leading cause of death in the United States. Smokers are more likely to have and die from stomach ulcers than nonsmokers. Smokers have a higher incidence of cancer in general. If a person smokes and is exposed to radon or asbestos, the risk for lung cancer increases dramatically.

What are the dangers of passive smoking?

Passive smoking causes lung cancer in healthy nonsmokers. Children whose parents smoke are more likely to suffer from pneumonia or bronchitis in the first two years of life than children who come from smoke-free households. Passive smokers have a 30% greater risk of developing lung cancer than nonsmokers who live in a smoke-free house.

Are chewing tobacco and snuff safe alternatives to cigarette smoking?

No, they are not. Many people who use chewing tobacco or snuff believe it can't harm them because there is no smoke. Wrong. Smokeless tobacco contains nicotine, the same addicting drug found in cigarettes and cigars. While not inhaled through the lungs, the juice from smokeless tobacco is absorbed through the lining of the mouth. There it can cause sores and white patches, which often lead to cancer of the mouth. Snuff dippers actually take in an average of over ten times more cancer-causing substances than cigarette smokers.

Bioethical Focus   Use of Antibiotics

A medical breakthrough took place in the 1940s with the introduction of the first antibiotics, chemicals able to selectively kill bacteria without harming host cells. Since that time, the number of deaths due to respiratory illnesses such as pneumonia and tuberculosis has declined dramatically. Strep throat and ear infections have also been brought under control with antibiotics.

There are problems associated with antibiotic therapy, however. Aside from a possible allergic reaction, antibiotics not only kill off disease-causing bacteria, they also reduce the number of beneficial bacteria in the intestinal tract and other locations. These beneficial bacteria hold in check the growth of other microbes; in the absence of bacteria, these microbes begin to flourish. Intestinal disorders can result, as can a vaginal yeast infection. The use of antibiotics can also prevent natural immunity from occurring, leading to the need for recurring antibiotic therapy.

Especially alarming at this time is the occurrence of resistance. Resistance takes place when vulnerable bacteria are killed off by an antibiotic, allowing resistant bacteria to become prevalent. The bacteria that cause ear, nose, and throat infections, as well as scarlet fever and pneumonia, are becoming widely resistant because we have not been using antibiotics properly. Tuberculosis is on the rise, and the new strains are resistant to the usual combined antibiotic therapy. When a disease is caused by a resistant bacterium, it cannot be cured by the administration of any presently available antibiotic.

Although drug companies now recognize the problem and have begun to develop new antibiotics that hopefully will kill bacteria resistant to today's antibiotics, every citizen needs to be aware of the present crisis. Stuart Levy, a Tufts University School of Medicine microbiologist, says that we should do what is ethical for society and ourselves. Antibiotics kill bacteria, not viruses—therefore, we shouldn't take antibiotics unless we know for sure we have a bacterial infection. And we shouldn't take them prophylactically—that is, just in case we might need one. If antibiotics are taken in low dosages and intermittently, resistant strains are bound to take over. Animal and agricultural use should be pared down, and household disinfectants should no longer be spiked with antibacterial agents. Perhaps then, Levy says, vulnerable bacteria will begin to supplant the resistant ones in the population.

Decide Your Opinion

1. With regard to antibiotics, should each person think about the needs of society as well as their own needs? Why or why not?
2. Should each person do what he or she can to help prevent the growing resistance of bacteria to disease? Why or why not?
3. Should you gracefully accept a physician's decision that an antibiotic will not help an illness you may have? Why or why not?

Summarizing the Concepts

15.1 The Respiratory System
The respiratory tract consists of the nose (nasal cavities), the nasopharynx, the pharynx, the larynx (which contains the vocal cords), the trachea, the bronchi, the bronchioles, and the lungs. The bronchi, along with the pulmonary arteries and veins, enter the lungs, which consist of the alveoli, air sacs surrounded by a capillary network.

15.2 Mechanism of Breathing
Inspiration begins when the respiratory center in the medulla oblongata sends excitatory nerve impulses to the diaphragm and the muscles of the rib cage. As they contract, the diaphragm lowers, and the rib cage moves upward and outward; the lungs expand, creating a partial vacuum, which causes air to rush in (inspiration). The respiratory center now stops sending impulses to the diaphragm and muscles of the rib cage. As the diaphragm relaxes, it resumes its dome shape, and as the rib cage retracts, air is pushed out of the lungs (expiration).

15.3 Gas Exchanges in the Body
External respiration occurs when CO_2 leaves blood via the alveoli and O_2 enters blood from the alveoli. Oxygen is transported to the tissues in combination with hemoglobin as oxyhemoglobin (HbO_2). Internal respiration occurs when O_2 leaves blood and CO_2 enters blood at the tissues. Carbon dioxide is mainly carried to the lungs within the plasma as the bicarbonate ion (HCO_3^-). Hemoglobin combines with hydrogen ions and becomes reduced (HHb).

15.4 Respiration and Health
A number of illnesses are associated with the respiratory tract. These disorders can be divided into those that affect the upper respiratory tract and those that affect the lower respiratory tract. Infections of the nasal cavities, sinuses, throat, tonsils, and larynx are all well known. In addition, infections can spread from the nasopharynx to the ears.

The lower respiratory tract is subject to infections such as acute bronchitis, pneumonia, and pulmonary tuberculosis. In restrictive pulmonary disorders, exemplified by pulmonary fibrosis, the lungs lose their elasticity. In obstructive pulmonary disorders, exemplified by chronic bronchitis, emphysema, and asthma, the bronchi (and bronchioles) do not effectively conduct air to and from the lungs. Smoking, which is associated with chronic bronchitis and emphysema, can eventually lead to lung cancer.

Testing Yourself

Choose the best answer for each question.

1. Which of these is anatomically incorrect?
   a. The nose has two nasal cavities.
   b. The pharynx connects the nasal and oral cavities to the larynx.
   c. The larynx contains the vocal cords.
   d. The trachea enters the lungs.
   e. The lungs contain many alveoli.

2. How is inhaled air modified before it reaches the lungs?
   a. It must be humidified.
   b. It must be warmed.
   c. It must be filtered.
   d. All of these are correct.

3. What is the name of the flap of tissue that prevents food from entering the trachea?
   a. glottis   c. epiglottis
   b. septum   d. Adam's apple

4. The maximum volume of air that can be moved in and out during a single breath is called the
   a. expiratory and inspiratory reserve volume.
   b. residual volume.
   c. tidal volume.
   d. vital capacity.
   e. functional residual capacity.

5. Internal respiration refers to
   a. the exchange of gases between the air and the blood in the lungs.
   b. the movement of air into the lungs.
   c. the exchange of gases between the blood and tissue fluid.
   d. cellular respiration, resulting in the production of ATP.

6. The chemical reaction that converts carbon dioxide to bicarbonate ion takes place in
   a. the blood plasma.   c. the alveolus.
   b. red blood cells.   d. the hemoglobin molecule.

7. Which of the following would affect hemoglobin's oxygen binding capacity?
   a. pH
   b. partial pressure of oxygen
   c. blood pressure
   d. temperature
   e. All of these except c are correct.

8. If air enters the intrapleural space (the space between the pleura),
   a. a lobe of the lung can collapse.
   b. the lungs could swell and burst.
   c. the diaphragm will contract.
   d. nothing will happen because air is needed in the intrapleural space.

9. The enzyme carbonic anhydrase
   a. causes the blood to be more basic in the tissues.
   b. speeds up the conversion of carbonic acid to carbon dioxide and water.
   c. actively transports carbon dioxide out of capillaries.
   d. is active only at high altitudes.
   e. All of these are correct.

10. Oxygen and carbon dioxide
    a. both exit and enter the blood in the lungs and tissues.
    b. are both carried by hemoglobin.
    c. are present only in arteries and not in veins.
    d. are both given off by mitochondria.
    e. Both a and c are correct.

11. Which of these statements is true?
    a. The P_{O_2}, temperature, and pH are higher in the lungs.
    b. The P_{O_2}, temperature, and pH are lower in the lungs.
    c. The P_{O_2} and temperature are higher and the pH is lower in the lungs.
    d. The P_{O_2} and temperature are lower and the pH is higher in the lungs.
    e. The P_{O_2} and pH are higher but the temperature is lower in the lungs.

12. Air enters the human lungs because
    a. atmospheric pressure is lower than the pressure inside the lungs.
    b. atmospheric pressure is greater than the pressure inside the lungs.
    c. although the pressures are the same inside and outside, the partial pressure of oxygen is lower within the lungs.
    d. the residual air in the lungs causes the partial pressure of oxygen to be lower than it is outside.

13. In humans, the respiratory center
    a. is stimulated by carbon dioxide.
    b. is located in the medulla oblongata.
    c. controls the rate of breathing.
    d. All of these are correct.

14. Which one of these is NOT an obstructive pulmonary disorder?
    a. pulmonary tuberculosis
    b. emphysema
    c. chronic bronchitis
    d. asthma
    e. a disorder that keeps air from flowing freely into and out of the lungs

15. Label the diagram of the human respiratory tract.

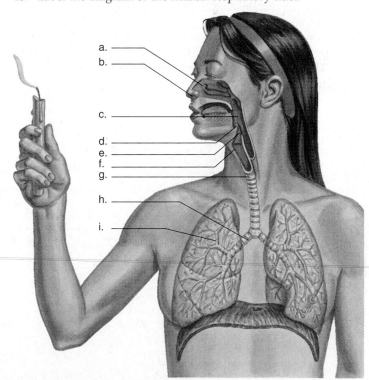

a. _____
b. _____
c. _____
d. _____
e. _____
f. _____
g. _____
h. _____
i. _____

e-Learning Connection

www.mhhe.com/maderinquiry10

Concepts	Questions	Media Resources*
15.1 The Respiratory System		

• Air passes through a series of tubes before gas exchange takes place across a very extensive moist surface.	1. Why is it better to breathe through the nose than through the mouth?
	2. Why do the alveoli of the lungs remain open, instead of collapsing like a deflated balloon, after exhalation?
		Essential Study Partner
		   Gas Exchange Systems
		   Respiratory Overview
		   Respiratory Organs
		
		Labeling Exercises
		   The Human Respiratory Tract
		   The Respiratory Tract—Alveoli
		   Section of Larynx
		   Glottis Function

15.2 Mechanism of Breathing		

• Respiration comprises breathing, external and internal respiration, and cellular respiration.
• During inspiration, the pressure in the lungs decreases, and air comes rushing in. During expiration, increased pressure in the thoracic cavity causes air to leave the lungs.
	1. Why does residual volume increase instead of decrease in individuals with certain lung diseases?
	2. Explain what is meant by the phrase "humans breathe by negative pressure."
		Essential Study Partner
		   Human Breathing
		   Mechanics of Ventilation
		   Measuring Function
		   Control of Respiration
		
		Animation Quizzes
		   Breathing
		   Respiration

15.3 Gas Exchanges in the Body		

• External respiration occurs in the lungs where oxygen diffuses into the blood and carbon dioxide diffuses out of the blood.
• Internal respiration occurs in the tissues where oxygen diffuses out of the blood and carbon dioxide diffuses into the blood.
• The respiratory pigment hemoglobin transports oxygen from the lungs to the tissues and aids in the transport of carbon dioxide from the tissues to the lungs.
	1. Differentiate between external respiration and internal respiration.
	2. What conditions lead to the higher saturation of hemoglobin with O_2 in the lungs compared to hemoglobin saturation in tissues?
		Essential Study Partner
		   Gas Exchange
		   Gas Transport
		
		Animation Quizzes
		   Hemoglobin
		   Gas Exchange
		
		Case Studies
		   Breathing Liquids: Reality or Science Fiction?

15.4 Respiration and Health		

• The respiratory tract is especially subject to disease because it is exposed to infectious agents.
• Smoking tobacco contributes to three major lung disorders—chronic bronchitis, emphysema, and cancer.
	1. Of the lower respiratory disorders discussed in this chapter, which disorders are linked with tobacco smoke?	Essential Study Partner
		   Disorders
		   Asthma
		
		Animation Quizzes
		   Bronchoscopy
		   Smoking Risks
		   Asthma
		
		Case Studies
		   Smoking Ban
		   Asthma: The New Worldwide Epidemic

*For additional Media Resources, see the Online Learning Center.

Urinary System and Excretion

Chapter Concepts

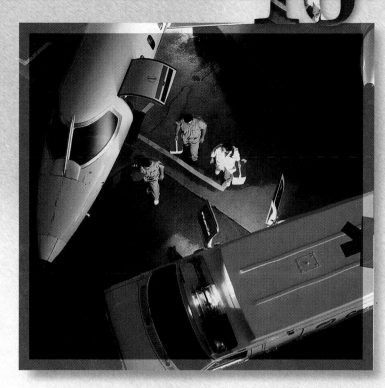

We now know how to preserve organs so they can be transported from one city to another for transplant operations.

*T*he plane arrived as expected. Paramedics were ready to rush an insulated container from the aircraft to the operating room of a nearby hospital. Brushing aside the ice in the container, a surgeon plucks out a fist-sized reddish mass—a kidney. Within hours the organ, which has replaced the diseased kidneys inside a young girl's body, is busy producing urine. The transplanted organ, if not rejected, should save the girl from a difficult life of being periodically hooked up to dialysis machines. Rejection is unlikely because it has already been determined that the tissues of the donor are very compatible with those of the recipient.

A kidney is absolutely essential for a healthy life because it helps regulate the pH and the water-salt balance of blood, and it excretes nitrogenous wastes. By regulating the amount of salt and water in the blood, a kidney helps keep blood pressure within a normal range. By excreting nitrogenous wastes, it rids the body of toxic substances. One kidney alone is all we need, and therefore the donor of a kidney will suffer no ill consequences, except the trauma of abdominal surgery. This chapter will detail exactly how a kidney performs its life-preserving functions.

16.1 Urinary System

The urinary system consists of the organs labeled in Figure 16.1. This figure also traces the path of urine. This section discusses the organs of the urinary system, urination, and the functions of the urinary system.

Urinary Organs

The **kidneys** are the primary organs of the urinary system. They are found on either side of the vertebral column, just below the diaphragm. They lie in depressions against the deep muscles of the back beneath the peritoneum, the lining of the abdominal cavity. Although they are somewhat protected by these muscles and by the lower rib cage, the kidneys can be damaged by blows to the back.

The kidneys are bean-shaped and reddish-brown in color. The fist-sized organs are covered by a tough capsule of fibrous connective tissue overlaid by adipose tissue. The concave side of a kidney has a depression called the hilum. The **renal artery** enters and the **renal vein** and ureters exit a kidney at the hilum.

The **ureters,** which extend from the kidneys to the bladder, are small muscular tubes about 25 cm long. Peristalsis moves urine within the ureters, and peristaltic contractions cause urine to enter the bladder at a rate of about five jets per minute.

The **urinary bladder,** which can hold up to 600 ml of urine, is a hollow, muscular organ that gradually expands as urine enters. A sphincter is a circular muscle that encloses a tube. Two sphincters are found in close proximity where the urethra exits the bladder. When these sphincters are closed, urination does not take place.

The **urethra,** which extends from the urinary bladder to an external opening, is a different length in females and males. In females, the urethra is only about 4 cm long. As mentioned in the Health Focus on page 304, the short length of the female urethra makes bacterial invasion easier and helps explain why females are more prone to urinary tract infections than males. In males, the urethra averages 20 cm when the penis is flaccid (limp, nonerect). As the urethra leaves the male urinary bladder, it is encircled by the prostate gland. In older men, enlargement of the prostate gland can restrict urination. A surgical procedure can usually correct the condition and restore a normal flow of urine.

In females, the reproductive and urinary systems are not connected. In males, the urethra carries urine during urination and sperm during ejaculation. This double function of the urethra in males does not alter the path of urine (Fig. 16.1).

Only the urinary system, consisting of the kidneys, the ureter, the urinary bladder, and the urethra, holds urine.

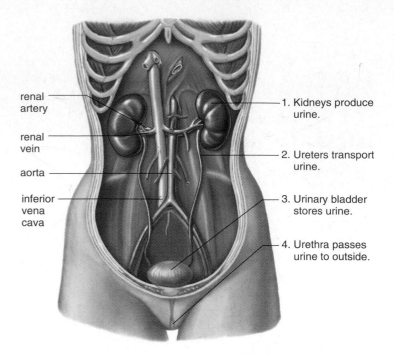

Figure 16.1   The urinary system.
Urine is found only within the kidneys, the ureters, the urinary bladder, and the urethra.

renal artery

renal vein

aorta

inferior vena cava

1. Kidneys produce urine.

2. Ureters transport urine.

3. Urinary bladder stores urine.

4. Urethra passes urine to outside.

Urination

When the urinary bladder fills to about 250 ml of urine, stretch receptors send sensory nerve impulses to the spinal cord. Subsequently, motor nerve impulses from the spinal cord cause the urinary bladder to contract and the sphincters to relax so that urination is possible (Fig. 16.2). In older children and adults, the brain controls this reflex, delaying urination until a suitable time.

Functions of the Urinary System

The function of the urinary system is to produce urine and conduct it to outside the body. The kidneys produce urine, and the other organs of the system store urine or conduct it toward the outside of the body.

Excretion is the removal of metabolic wastes from the body. People sometimes confuse the terms excretion and defecation, but they do not refer to the same process. Defecation refers to the elimination of feces from the body and is a function of the digestive system. Excretion, on the other hand, refers to the elimination of metabolic wastes, which are the products of metabolism. For example, the undigested food and bacteria that make up feces have never been a part of the functioning of the body, while the substances excreted in urine were once metabolites in the body.

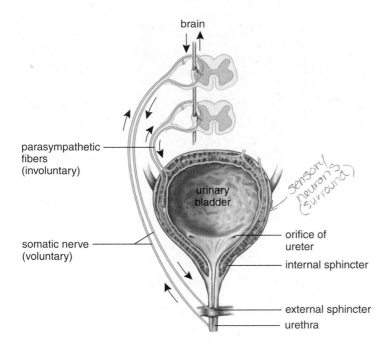

Figure 16.2   Urination.
As the bladder fills with urine, sensory impulses go to the spinal cord and then to the brain. The brain can override the urge to urinate. When urination occurs, motor nerve impulses cause the bladder to contract and an internal sphincter to open. Nerve impulses also cause an external sphincter to open.

The kidneys play a central role in homeostasis by regulating the composition of blood, and therefore tissue fluid. As urine is being produced, the kidneys (1) carry out the excretion of metabolic wastes, particularly nitrogenous wastes; (2) maintain the normal water-salt balance of the blood and, as a consequence, the normal blood volume and blood pressure; and (3) maintain the acid-base balance of blood. The kidneys also (4) have a hormonal function, as will be discussed.

Excretion of Metabolic Wastes

The kidneys excrete metabolic wastes, notably nitrogenous wastes. Urea is the primary nitrogenous end product of metabolism in human beings, but humans also excrete some ammonium, creatinine, and uric acid.

Urea is a by-product of amino acid metabolism. The breakdown of amino acids in the liver releases ammonia, which the liver combines with carbon dioxide to produce urea. Ammonia is very toxic to cells, but urea is much less toxic. Because it is less toxic, less water is required to excrete urea.

The metabolic breakdown of creatine phosphate results in **creatinine.** Creatine phosphate is a high-energy phosphate reserve molecule in muscles.

The breakdown of nucleotides, such as those containing adenine and thymine, produces **uric acid.** Uric acid is rather insoluble. If too much uric acid is present in blood, crystals form and precipitate out. Crystals of uric acid sometimes collect in the joints, producing a painful ailment called gout.

Maintenance of Water–Salt Balance

A principal function of the kidneys is to maintain the appropriate water-salt balance of the blood. As we shall see, blood volume is intimately associated with the salt balance of the body. As you know, salts, such as NaCl, have the ability to cause osmosis, the diffusion of water—in this case, into the blood. The more salts there are in the blood, the greater the blood volume and the greater the blood pressure. In this way, the kidneys are involved in regulating blood pressure.

The kidneys also maintain the appropriate level of other ions, such as potassium ions (K^+), bicarbonate ions (HCO_3^-), and calcium ions (Ca^{2+}), in the blood.

Maintenance of Acid–Base Balance

The kidneys regulate the acid-base balance of the blood. In order for us to remain healthy, the blood pH should be just about 7.4. The kidneys monitor and control blood pH, mainly by excreting hydrogen ions (H^+) and reabsorbing the bicarbonate ions (HCO_3^-) as needed to keep blood pH at 7.4. Urine usually has a pH of 6 or lower because our diet often contains acidic foods.

Secretion of Hormones

The kidneys assist the endocrine system in hormone secretion. The kidneys release renin, a substance that leads to the secretion of the hormone aldosterone from the adrenal cortex, the outer portion of the adrenal glands, which lie atop the kidneys. As described later in this chapter, aldosterone promotes the reabsorption of sodium ions (Na^+) by the kidneys.

Whenever the oxygen-carrying capacity of the blood is reduced, the kidneys secrete the hormone **erythropoietin,** which stimulates red blood cell production.

The kidneys also help activate vitamin D from the skin. Vitamin D is the precursor of the hormone calcitriol, which promotes calcium (Ca^{2+}) reabsorption from the digestive tract.

The kidneys are the primary organs of excretion, particularly of nitrogenous wastes. The kidneys are also major organs of homeostasis because they regulate the water-salt balance and the acid-base balance of the blood as well as the secretion of certain hormones.

Urinary Tract Infections Require Attention

Although males can get a urinary tract infection, the condition is 50 times more common in women. The explanation lies in a comparison of male and female anatomy (Fig. 16A). The female urethral and anal openings are closer together, and the shorter urethra makes it easier for bacteria from the bowels to enter and start an infection. Although it is possible to have no outward signs of an infection, urination is usually painful, and patients often describe a burning sensation. The urge to pass urine is frequent, but it may be difficult to start the stream. Chills with fever, nausea, and vomiting may be present.

Urinary tract infections confined to the urethra are called urethritis. If the bladder is involved, it is called cystitis. Should the infection reach the kidneys, the person has pyelonephritis. *Escherichia coli* (*E. coli*), a normal bacterial resident of the large intestine, is usually the cause of infection. Since the infection is caused by a bacterium, it is curable by antibiotic therapy. However, reinfection is possible as soon as antibiotic therapy is finished.

It makes sense to try to prevent infection in the first place. The following tips might help.

Men and women should drink from 2 to 2.5 liters of liquid a day, preferably water. Try to avoid caffeinated drinks, which may be irritating. Cranberry juice is recommended because it contains a substance that stops bacteria from sticking to the bladder wall once an infection has set in.

Since sexually transmitted diseases such as gonorrhea, chlamydia, or herpes can cause urinary tract infections, all personal behaviors should be examined carefully, and suitable adjustments made to avoid urinary tract infections.

Women may have a urinary tract infection for the first time shortly after they become sexually active. The term "honeymoon cystitis" was coined because of the common association of urinary tract infections with sexual intercourse. Washing the genitals before having sex and being careful not to introduce bacteria from the anus into the urethra are recommended. Also, urinating immediately before and after sex helps flush out any bacteria that are present. A diaphragm may press on the urethra and prevent adequate emptying of the bladder, and estrogen, such as in birth control pills, can increase the risk of cystitis. A sex partner may have an asymptomatic (no symptoms) urinary infection that causes a woman to become infected repeatedly.

Women should wipe from the front to the back after using the toilet. Perfumed toilet paper and any other perfumed products that come in contact with the genitals may be irritating. Wearing loose clothing and cotton underwear discourages the growth of bacteria, while tight clothing, such as jeans or panty hose, provides an environment for the growth of bacteria.

Personal hygiene is especially important at the time of menstruation. Hands should be washed before and after changing napkins and/or tampons. Superabsorbent tampons that are changed infrequently may encourage the growth of bacteria. Also, sexual intercourse may cause menstrual flow to enter the urethra.

In males, the prostate is a gland that surrounds the urethra just below the bladder (Fig. 16A). The prostate contributes secretions to semen whenever semen enters the urethra prior to ejaculation. An infection of the prostate, called prostatitis, is often accompanied by a urinary tract infection. Fever is present, and the prostate is tender and inflamed. The patient may have to be hospitalized and treated with a broad-spectrum antibiotic. Prostatitis, which in a young person is often preceded by a sexually transmitted disease, can lead to a chronic condition. Chronic prostatitis may be asymptomatic or, as is more typical, the person may experience irritation upon urinating and/or difficulty in urinating. The latter can lead to the need for surgery to remove the obstruction to urine flow.

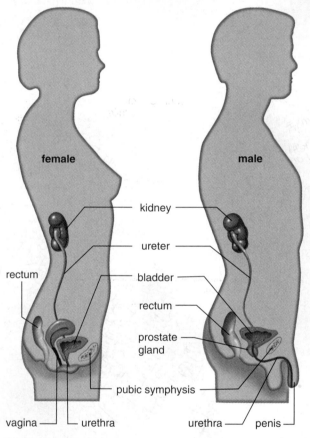

Figure 16A   Female versus male urinary tract.
Females have a short urinary tract compared to that of males. This means that it is easier for bacteria to invade the urethra and helps explain why females are 50 times more likely than males to get a urinary tract infection.

16.2 Kidney Structure

When a kidney is sliced lengthwise, it is possible to see that many branches of the renal artery and vein reach inside the kidney (Fig. 16.3a). If the blood vessels are removed, it is easier to identify the three regions of a kidney. The **renal cortex** is an outer, granulated layer that dips down in between a radially striated, or lined, inner layer called the renal medulla. The **renal medulla** consists of cone-shaped tissue masses called renal pyramids. The **renal pelvis** is a central space, or cavity, that is continuous with the ureter (Fig. 16.3b).

Microscopically, the kidney is composed of over one million **nephrons,** sometimes called renal or kidney tubules (Fig. 16.3c). The nephrons produce urine and are positioned so that the urine flows into a collecting duct. Several nephrons enter the same collecting duct; the collecting ducts enter the renal pelvis.

Macroscopically, a kidney has three regions: the renal cortex, the renal medulla, and the renal pelvis which is continuous with the ureter. Microscopically, a kidney contains over one million nephrons.

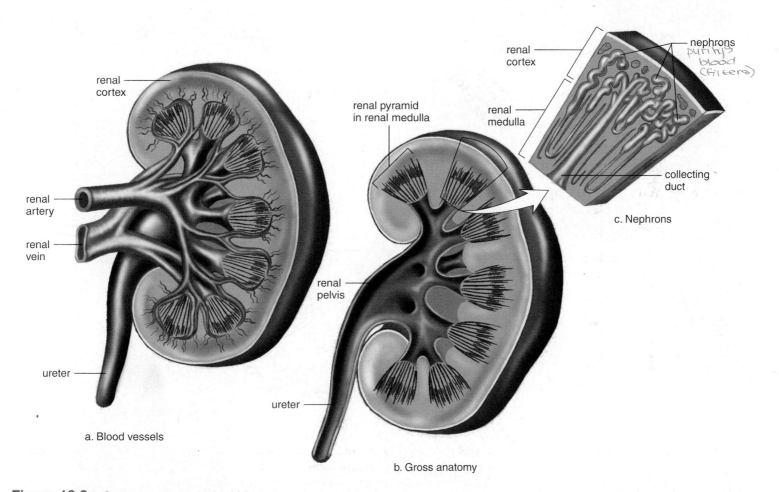

c. Nephrons

a. Blood vessels

b. Gross anatomy

Figure 16.3   Gross anatomy of the kidney.
a. A longitudinal section of the kidney showing the blood supply. Note that the renal artery divides into smaller arteries, and these divide into arterioles. Venules join to form small veins, which join to form the renal vein. **b.** The same section without the blood supply. Now it is easier to distinguish the renal cortex, the renal medulla, and the renal pelvis, which connects with the ureter. The renal medulla consists of the renal pyramids. **c.** An enlargement showing the placement of nephrons.

Anatomy of a Nephron

Each nephron has its own blood supply, including two capillary regions (Fig. 16.4). From the renal artery, an afferent arteriole leads to the **glomerulus,** a knot of capillaries inside the glomerular capsule. Blood leaving the glomerulus enters the efferent arteriole. Blood pressure is higher in the glomerulus because the efferent arteriole is narrower than the afferent arteriole. The efferent arteriole takes blood to the **peritubular capillary network,** which surrounds the rest of the nephron. From there, the blood goes into a venule that joins the renal vein.

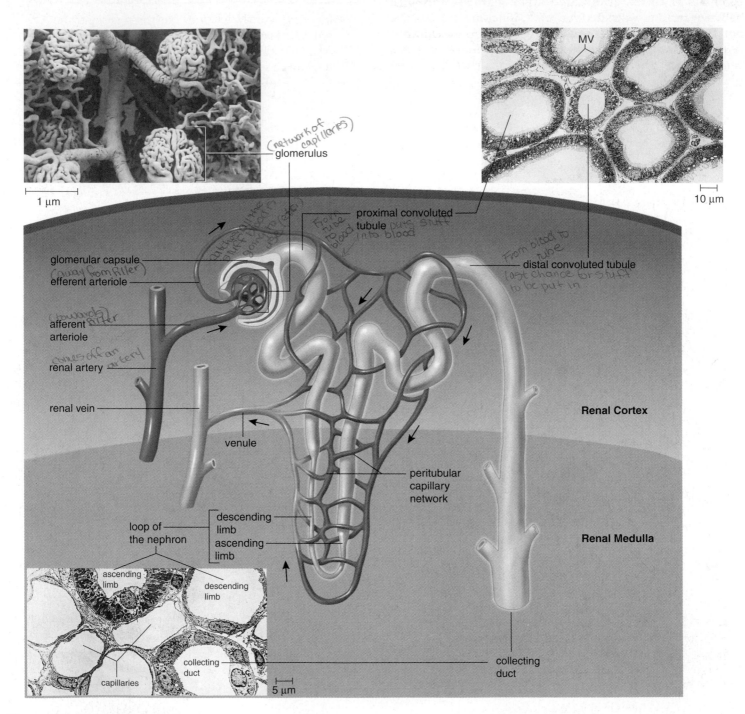

Figure 16.4   Nephron anatomy.
A nephron is made up of a glomerular capsule, the proximal convoluted tubule, the loop of the nephron, the distal convoluted tubule, and the collecting duct. The micrographs show these structures in cross section; MV = microvilli. You can trace the path of blood about the nephron by following the arrows.

Parts of a Nephron

Each nephron is made up of several parts (Fig. 16.4). The structure of each part suits its function.

First, the closed end of the nephron is pushed in on itself to form a cuplike structure called the **glomerular capsule** (Bowman's capsule). The outer layer of the glomerular capsule is composed of squamous epithelial cells; the inner layer is made up of podocytes that have long cytoplasmic processes. The podocytes cling to the capillary walls of the glomerulus and leave pores that allow easy passage of small molecules from the glomerulus to the inside of the glomerular capsule. This process, called glomerular filtration, produces a filtrate of blood.

Next, there is a **proximal convoluted tubule** (called "proximal" because it is near the glomerular capsule). The cuboidal epithelial cells lining this part of the nephron have numerous microvilli, about 1 μm in length, that are tightly packed and form a brush border (Fig. 16.5). A brush border greatly increases the surface area for the tubular reabsorption of filtrate components. Each cell also has many mitochondria, which can supply energy for active transport of molecules from the lumen to the peritubular capillary network.

Simple squamous epithelium appears as the tube narrows and makes a U-turn called the **loop of the nephron** (loop of Henle). Each loop consists of a descending limb that allows water to leave and an ascending limb that extrudes salt (NaCl). Indeed, as we shall see, this activity facilitates the reabsorption of water by the nephron and collecting duct.

The cells of the **distal convoluted tubule** have numerous mitochondria, but they lack microvilli. This is consistent with the active role they play in moving molecules from the blood into the tubule, a process called tubular secretion. The distal convoluted tubules of several nephrons enter one collecting duct. A kidney contains many collecting ducts, which carry urine to the renal pelvis.

As shown in Figure 16.4, the glomerular capsule and the convoluted tubules always lie within the renal cortex. The loop of the nephron dips down into the renal medulla; a few nephrons have a very long loop of the nephron, which penetrates deep into the renal medulla. **Collecting ducts** are also located in the renal medulla, and they give the renal pyramids their lined appearance.

Each part of a nephron is anatomically suited to its specific function in urine formation.

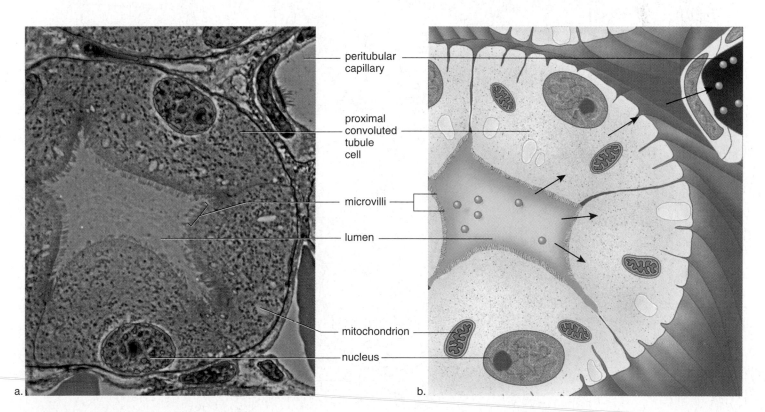

peritubular capillary

proximal convoluted tubule cell

microvilli

lumen

mitochondrion

nucleus

a.   b.

Figure 16.5 Proximal convoluted tubule.
a. This photomicrograph shows that the cells lining the proximal convoluted tubule have a brushlike border composed of microvilli, which greatly increase the surface area exposed to the lumen. The peritubular capillary network surrounds the cells. **b.** Diagrammatic representation of (**a**) shows that each cell has many mitochondria, which supply the energy needed for active transport, the process that moves molecules (green) from the lumen of the tubule to the capillary, as indicated by the arrows.

Visual Focus

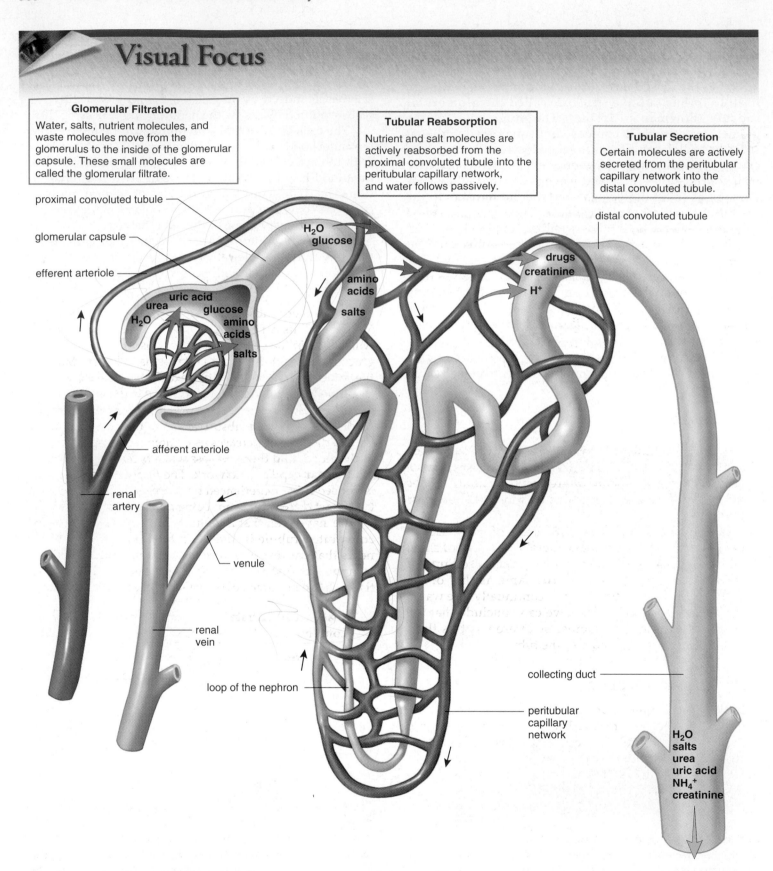

Glomerular Filtration

Water, salts, nutrient molecules, and waste molecules move from the glomerulus to the inside of the glomerular capsule. These small molecules are called the glomerular filtrate.

Tubular Reabsorption

Nutrient and salt molecules are actively reabsorbed from the proximal convoluted tubule into the peritubular capillary network, and water follows passively.

Tubular Secretion

Certain molecules are actively secreted from the peritubular capillary network into the distal convoluted tubule.

proximal convoluted tubule

glomerular capsule

efferent arteriole

H_2O
glucose

amino
acids

salts

uric acid
urea
H_2O
glucose
amino
acids
salts

drugs
creatinine

H^+

distal convoluted tubule

afferent arteriole

renal
artery

venule

renal
vein

loop of the nephron

collecting duct

peritubular
capillary
network

H_2O
salts
urea
uric acid
NH_4^+
creatinine

Figure 16.6   Steps in urine formation.

The three main steps in urine formation described in boxes at the top of this drawing are color-coded to arrows that show the movement of molecules into or out of the nephron at specific locations. In the end, urine is composed of the substances within the collecting duct (see gray arrow).

16.3 Urine Formation

Figure 16.6 gives an overview of urine formation, which is divided into these steps: glomerular filtration, tubular reabsorption, and tubular secretion.

Glomerular Filtration

Glomerular filtration occurs when whole blood enters the afferent arteriole and the glomerulus. Due to glomerular blood pressure, water and small molecules move from the glomerulus to the inside of the glomerular capsule. This is a filtration process because large molecules and formed elements are unable to pass through the capillary wall. In effect, then, blood in the glomerulus has two portions: the filterable components and the nonfilterable components:

Filterable Blood Components	Nonfilterable Blood Components
Water	Formed elements (blood cells and platelets)
Nitrogenous wastes	Proteins
Nutrients	
Salts (ions)	

The **glomerular filtrate** contains small dissolved molecules in approximately the same concentration as plasma. Small molecules that escape being filtered and the nonfilterable components leave the glomerulus by way of the efferent arteriole.

As indicated in Table 16.1, 180 liters of water are filtered per day, along with a considerable amount of small molecules (such as glucose) and ions (such as sodium). If the composition of urine were the same as that of the glomerular filtrate, the body would continually lose water, salts, and nutrients. Therefore, we can conclude that the composition of the filtrate must be altered as this fluid passes through the remainder of the tubule.

Tubular Reabsorption

Tubular reabsorption occurs as molecules and ions are both passively and actively reabsorbed from the nephron into the blood of the peritubular capillary network. The osmolarity of the blood is maintained by the presence of both plasma proteins and salt. When sodium ions (Na^+) are actively reabsorbed, chloride ions (Cl^-) follow passively. The reabsorption of salt (NaCl) increases the osmolarity of the blood compared to the filtrate, and therefore water moves passively from the tubule into the blood. About 67% of Na^+ is reabsorbed at the proximal convoluted tubule.

Nutrients such as glucose and amino acids also return to the blood at the proximal convoluted tubule. This is a selective process because only molecules recognized by carrier molecules are actively reabsorbed. Glucose is an example

Table 16.1   Reabsorption from Nephrons

Substance	Amount Filtered (per day)	Amount Excreted (per day)	Reabsorption (%)
Water, L	180	1.8	99.0
Sodium, g	630	3.2	99.5
Glucose, g	180	0.0	100.0
Urea, g	54	30.0	44.0

L = liters, g = grams

of a molecule that ordinarily is completely reabsorbed because there is a plentiful supply of carrier molecules for it. However, every substance has a maximum rate of transport, and after all its carriers are in use, any excess in the filtrate will appear in the urine. For example, as reabsorbed levels of glucose approach 1.8–2 mg/ml plasma, the rest appears in the urine. In diabetes mellitus, excess glucose occurs in the blood, and then in the filtrate, and then in the urine, because the liver and muscles fail to store glucose as glycogen, and the kidneys cannot reabsorb all of it. The presence of glucose in the filtrate increases its osmolarity compared to that of the blood, and therefore less water is reabsorbed into the peritubular capillary network. The frequent urination and increased thirst experienced by untreated diabetics are due to the fact that water is not being reabsorbed.

We have seen that the filtrate that enters the proximal convoluted tubule is divided into two portions: components that are reabsorbed from the tubule into blood, and components that are not reabsorbed and continue to pass through the nephron to be further processed into urine:

Reabsorbed Filtrate Components	Nonreabsorbed Filtrate Components
Most water	Some water
Nutrients	Much nitrogenous waste
Required salts (ions)	Excess salts (ions)

The substances that are not reabsorbed become the tubular fluid, which enters the loop of the nephron.

Tubular Secretion

Tubular secretion is a second way by which substances are removed from blood and added to the tubular fluid. Hydrogen ions, creatinine, and drugs such as penicillin are some of the substances that are moved by active transport from blood into the distal convoluted tubule. In the end, urine contains substances that have undergone glomerular filtration but have not been reabsorbed, and substances that have undergone tubular secretion.

16.4 Maintaining Water-Salt Balance

The kidneys maintain the water-salt balance of the blood within normal limits. In this way, they also maintain the blood volume and blood pressure. Most of the water and salt (NaCl) present in the filtrate is reabsorbed across the wall of the proximal convoluted tubule.

Reabsorption of Water

The excretion of a hypertonic urine (one that is more concentrated than blood) is dependent upon the reabsorption of water from the loop of the nephron and the collecting duct.

A long loop of the nephron, which typically penetrates deep into the renal medulla, is made up of a descending limb and an ascending limb. Salt (NaCl) passively diffuses out of the lower portion of the ascending limb, but the upper, thick portion of the limb actively extrudes salt out into the tissue of the outer renal medulla (Fig. 16.7). Less and less salt is available for transport as fluid moves up the thick portion of the ascending limb. Because of these circumstances, there is an osmotic gradient within the tissues of the renal medulla: the concentration of salt is greater in the direction of the inner medulla. (Note that water cannot leave the ascending limb because the limb is impermeable to water.)

The large arrow in Figure 16.7 indicates that the innermost portion of the inner medulla has the highest concentration of solutes. This cannot be due to salt because active transport of salt does not start until fluid reaches the thick portion of the ascending limb. Urea is believed to leak from the lower portion of the collecting duct, and it is this molecule that contributes to the high solute concentration of the inner medulla.

Because of the osmotic gradient within the renal medulla, water leaves the descending limb along its entire length. This is a countercurrent mechanism: as water diffuses out of the descending limb, the remaining fluid within the limb encounters an even greater osmotic concentration of solute; therefore, water continues to leave the descending limb from the top to the bottom.

Fluid enters the collecting duct from the distal convoluted tubule. This fluid is isotonic to the cells of the renal cortex. This means that to this point, the net effect of reabsorption of water and salt is the production of a fluid that has the same tonicity as blood plasma. However, the filtrate within the collecting duct also encounters the same osmotic gradient mentioned earlier (Fig. 16.7). Therefore, water diffuses out of the collecting duct into the renal medulla, and the urine within the collecting duct becomes hypertonic to blood plasma.

Antidiuretic hormone (ADH) released by the posterior lobe of the pituitary plays a role in water reabsorption at the collecting duct. In order to understand the action of this hormone, consider its name. Diuresis means increased amount of urine, and antidiuresis means decreased amount of urine. When ADH is present, more water is reabsorbed

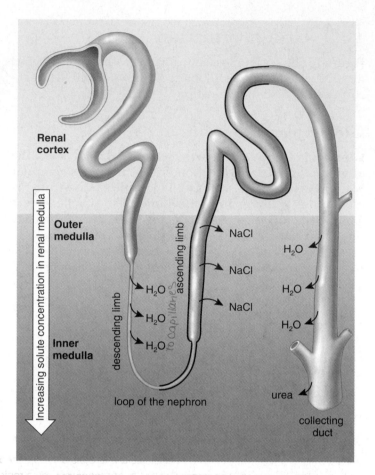

Figure 16.7   Reabsorption of water at the loop of the nephron and the collecting duct.
Salt (NaCl) diffuses and is actively transported out of the ascending limb of the loop of the nephron into the renal medulla; also, urea is believed to leak from the collecting duct and to enter the tissues of the renal medulla. This creates a hypertonic environment, which draws water out of the descending limb and the collecting duct. This water is returned to the cardiovascular system. (The thick black line means the ascending limb is impermeable to water.)

(blood volume and pressure rise), and a decreased amount of urine results. In practical terms, if an individual does not drink much water on a certain day, the posterior lobe of the pituitary releases ADH, causing more water to be reabsorbed and less urine to form. On the other hand, if an individual drinks a large amount of water and does not perspire much, ADH is not released. In that case, more water is excreted, and more urine forms.

Reabsorption of Salt

The kidneys regulate the salt balance in blood by controlling the excretion and the reabsorption of various ions. Sodium (Na^+) is an important ion in plasma that must be regulated, but the kidneys also excrete or reabsorb other ions, such as potassium ions (K^+), bicarbonate ions (HCO_3^-), and magnesium ions (Mg^{2+}), as needed.

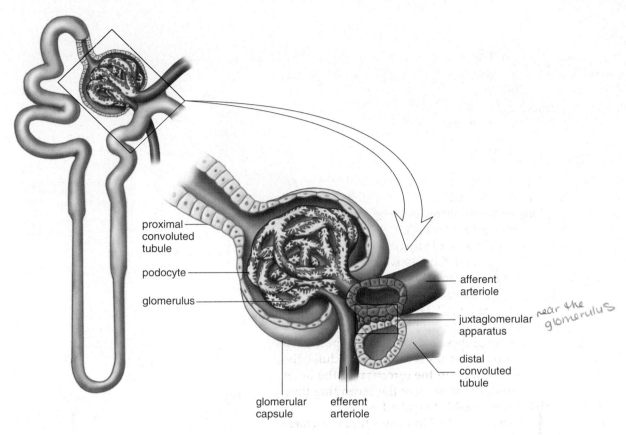

proximal convoluted tubule

podocyte

glomerulus

afferent arteriole

juxtaglomerular *near the glomerulus* apparatus

distal convoluted tubule

glomerular capsule

efferent arteriole

Figure 16.8   Juxtaglomerular apparatus.
This drawing shows that the afferent arteriole and the distal convoluted tubule usually lie next to each other. The juxtaglomerular apparatus occurs where they touch. The juxtaglomerular apparatus secretes renin, a substance that leads to the release of aldosterone by the adrenal cortex. Reabsorption of sodium ions and then water now occurs. Therefore, blood volume and blood pressure increase.

Usually, more than 99% of sodium (Na^+) filtered at the glomerulus is returned to the blood. Most sodium (67%) is reabsorbed at the proximal tubule, and a sizable amount (25%) is extruded by the ascending limb of the loop of the nephron. The rest is reabsorbed from the distal convoluted tubule and collecting duct.

Hormones regulate the reabsorption of sodium at the distal convoluted tubule. **Aldosterone** is a hormone secreted by the adrenal cortex. Aldosterone promotes the excretion of potassium ions (K^+) and the reabsorption of sodium ions (Na^+). The release of aldosterone is set in motion by the kidneys themselves. The **juxtaglomerular apparatus** is a region of contact between the afferent arteriole and the distal convoluted tubule (Fig. 16.8). When blood volume, and therefore blood pressure, is not sufficient to promote glomerular filtration, the juxtaglomerular apparatus secretes renin. **Renin** is an enzyme that changes angiotensinogen (a large plasma protein produced by the liver) into angiotensin I. Later, angiotensin I is converted to angiotensin II, a powerful vasoconstrictor that also stimulates the adrenal cortex to release aldosterone. The reabsorption of sodium ions is followed by the reab-

sorption of water. Therefore, blood volume and blood pressure increase.

Atrial natriuretic hormone (ANH) is a hormone secreted by the atria of the heart when cardiac cells are stretched due to increased blood volume. ANH inhibits the secretion of renin by the juxtaglomerular apparatus and the secretion of aldosterone by the adrenal cortex. Its effect, therefore, is to promote the excretion of Na^+, called natriuresis. When Na^+ is excreted, so is water, and therefore blood volume and blood pressure decrease.

Diuretics

Diuretics are chemicals that increase the flow of urine. Drinking alcohol causes diuresis because it inhibits the secretion of ADH. The dehydration that follows is believed to contribute to the symptoms of a hangover. Caffeine is a diuretic because it increases the glomerular filtration rate and decreases the tubular reabsorption of Na^+. Diuretic drugs developed to counteract high blood pressure inhibit active transport of Na^+ at the loop of the nephron or at the distal convoluted tubule. A decrease in water reabsorption and a decrease in blood volume follow.

Science Focus

The Artificial Kidney

After a person suffers kidney damage, perhaps due to repeated infections, waste substances accumulate in the blood. This condition is called uremia because urea is one of the substances that accumulates. Although nitrogenous wastes can cause serious damage, some believe that it is the imbalance of ions in the blood that leads to loss of consciousness and to heart failure. Patients in renal failure most often seek a kidney transplant, but in the meantime they undergo **hemodialysis** utilizing an artificial kidney.

Dialysis is the diffusion of dissolved molecules through a semipermeable membrane (an artificial membrane with pore sizes that allow only small molecules to pass through). These molecules, of course, move across a membrane from the area of greater concentration to one of lesser concentration. Substances more concentrated in blood diffuse into the dialysis solution, which is called the dialysate, and substances more concentrated in the dialysate diffuse into the blood. Accordingly, the artificial kidney is utilized either to extract substances from blood, including waste products or toxic chemicals and drugs, or to add substances to blood—for example, bicarbonate ions (HCO_3^-) if blood is acidic.

The first clinically useful artificial kidney was devised by the Dutchman William J. Kolff in 1943. Based on the work of predecessors, Kolff hypothesized that he needed a machine with these specifications: (1) only a small volume of blood is ever out of the patient's body at one time, (2) the blood should be exposed to a large surface area to lessen diffusion time, (3) the blood must be kept moving in tubes to and from the patient, and (4) the dialysate must be of a certain composition and kept fresh.

In Kolff's invention, 20 meters of cellophane tubing were wrapped around a drum. Rotation of the drum caused the blood to pass from one end of the tubing to the other. The lower half of the drum was immersed in a fluid. There was no pump, and the tube taking blood from the patient to the machine was raised or lowered to let blood in or out of the patient. Kolff used an anticoagulant to prevent the patient's blood from clotting. The only patient treated with the first artificial kidney died when all possible entry points for his blood had been utilized.

Around 1960, Dr. Belding Scribner and Mr. Wayne Quinton in Seattle developed the arteriovenous shunt. Tubes permanently placed in an artery and vein of the arm were joined to a removable Teflon shunt. When the patient needed dialysis, the shunt was disconnected and the tubes were connected to the artificial kidney. Today, it is common to surgically join an artery and vein in the arm. The vein distends, and for each dialysis, one or two needles are inserted in the distended vein. The needles are connected to tubing leading to the artificial kidney, where fresh dialysate circulates past many meters

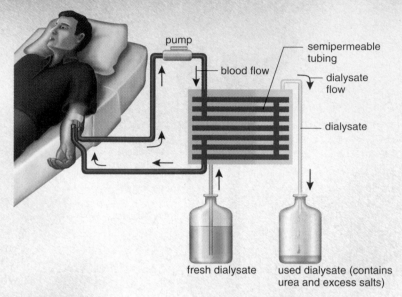

Figure 16B **An artificial kidney machine.**
As the patient's blood is pumped through dialysis tubing, it is exposed to a dialysate (dialysis solution). Wastes exit from blood into the solution because of a preestablished concentration gradient. In this way, blood is not only cleansed, but its water-salt and acid-base balance are also adjusted.

of the tubing (Fig. 16B). A pump keeps the blood moving, and there is no need for the anticoagulant that Kolff had to use. Also, today's artificial kidney uses highly permeable dialysis tubing, making it possible to shorten dialysis time. Presently, dialysis usually occurs only three times per week for three hours or less. And portable machines allow patients to dialyze at home or on trips!

Another group of Dutch investigators hypothesized that the peritoneum could serve as a dialyzing membrane. The peritoneal (abdominal) cavity has a large surface area and is richly vascularized. Therefore, an exchange does take place between a dialyzing fluid placed in the peritoneal cavity and the capillaries of the abdominal wall. A very careful mathematical analysis allowed the investigators to determine that only 1.2 liters of dialysate are needed and it should be kept in place for four hours. The old dialysate is then removed, and a new batch is instilled via a permanently implanted tube. In the meantime, the patient is free to go about his usual daily routine. This method of dialysis is appropriately called *continuous ambulatory peritoneal dialysis (CAPD)*.

Development of the artificial kidney and CAPD exemplifies that science is intimately involved with the development of suitable instruments and is a community effort that builds on the work of those who have made contributions before.

16.5 Maintaining Acid-Base Balance

The bicarbonate (HCO_3^-) buffer system and the process of breathing work together to maintain the pH of the blood. Central to the mechanism is this reaction, which you have seen before:

$$H^+ + HCO_3^- \rightleftharpoons H_2CO_3 \rightleftharpoons H_2O + CO_2$$

The excretion of carbon dioxide (CO_2) by the lungs helps keep the pH within normal limits, because when carbon dioxide is exhaled, this reaction is pushed to the right and hydrogen ions are tied up in water. Indeed, when blood pH decreases, chemoreceptors in the carotid bodies (located in the carotid arteries) and in aortic bodies (located in the aorta) stimulate the respiratory center, and the rate and depth of breathing increase. On the other hand, when blood pH begins to rise, the respiratory center is depressed, and the level of bicarbonate ions increases in the blood.

As powerful as the buffer/breathing system is, only the kidneys can rid the body of a wide range of acidic and basic substances. The kidneys are slower acting than the buffer/breathing mechanism, but they have a more powerful effect on pH. For the sake of simplicity, we can think of the kidneys as reabsorbing bicarbonate ions and excreting hydrogen ions as needed to maintain the normal pH of the blood (Fig. 16.9). If the blood is acidic, hydrogen ions are excreted and bicarbonate ions are reabsorbed. If the blood is basic, hydrogen ions are not excreted and bicarbonate ions are not reabsorbed. Since the urine is usually acidic, it fol-

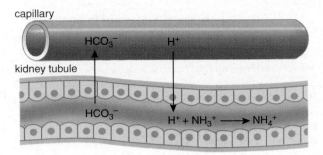

Figure 16.9   Acid-base balance.
In the kidneys, bicarbonate ions (HCO_3^-) are reabsorbed and hydrogen ions (H^+) are excreted as needed to maintain the pH of the blood. Excess hydrogen ions are buffered, for example, by ammonia (NH_3), which is produced in tubule cells by the deamination of amino acids.

lows that an excess of hydrogen ions are usually excreted. Ammonia (NH_3) provides a means for buffering these hydrogen ions in urine: ($NH_3 + H^+ \rightarrow NH_4^+$). Ammonia (whose presence is quite obvious in the diaper pail or kitty litter box) is produced in tubule cells by the deamination of amino acids. Phosphate provides another means of buffering hydrogen ions in urine.

> The acid-base balance of the blood is adjusted by the reabsorption of the bicarbonate ions (HCO_3^-) and the secretion of hydrogen ions (H^+) as appropriate.

Summarizing the Concepts

16.1 Urinary System
The kidneys produce urine, which is conducted by the ureters to the bladder where it is stored before being released by way of the urethra. The kidneys excrete nitrogenous wastes, including urea, uric acid, and creatinine. They maintain the normal water-salt balance and the acid-base balance of the blood, as well as influencing the secretion of certain hormones.

16.2 Kidney Structure
Macroscopically, the kidneys are divided into the renal cortex, renal medulla, and renal pelvis. Microscopically, they contain the nephrons.

Each nephron has its own blood supply; the afferent arteriole approaches the glomerular capsule and divides to become the glomerulus, a capillary tuft. The efferent arteriole leaves the capsule and immediately branches into the peritubular capillary network.

Each region of the nephron is anatomically suited to its task in urine formation. The spaces between the podocytes of the glomerular capsule allow small molecules to enter the capsule from the glomerulus. The cuboidal epithelial cells of the proximal convoluted tubule have many mitochondria and microvilli to carry out active transport (following passive transport) from the tubule to the blood. In contrast, the cuboidal epithelial cells of the distal convoluted tubule have

numerous mitochondria but lack microvilli. They carry out active transport from the blood to the tubule.

16.3 Urine Formation
Urine is composed primarily of nitrogenous waste products and salts in water. The steps in urine formation are glomerular filtration, tubular reabsorption, and tubular secretion.

16.4 Maintaining Water-Salt Balance
The kidneys regulate the water-salt balance of the body. Water is reabsorbed from certain parts of the tubule, and the loop of the nephron establishes an osmotic gradient that draws water from the descending loop of the nephron and also from the collecting duct. The permeability of the collecting duct is under the control of the hormone ADH.

The reabsorption of salt increases blood volume and pressure because more water is also reabsorbed. Two other hormones, aldosterone and ANH, control the kidneys' reabsorption of sodium (Na^+).

16.5 Maintaining Acid-Base Balance
The kidneys keep blood pH within normal limits. They reabsorb HCO_3^- and excrete H^+ as needed to maintain the pH at about 7.4.

Testing Yourself

Choose the best answer for each question.

1. Which of these functions of the kidneys are mismatched?
   a. excretes metabolic wastes—rids the body of urea
   b. maintains the water-salt balance—helps regulate blood pressure
   c. maintains the acid-base balance—rids the body of uric acid
   d. secretes hormones—secretes erythropoietin
   e. All of these are properly matched.

2. Which of these is out of order first?
   a. glomerular capsule
   b. proximal convoluted tubule
   c. distal convoluted tubule
   d. loop of the nephron
   e. collecting duct

3. Which of these hormones is most likely to directly cause a rise in blood pressure?
   a. aldosterone
   b. antidiuretic hormone (ADH) *decrease amt. of urine*
   c. renin
   d. atrial natriuretic hormone (ANH) *↓ blood pressure*

4. To lower blood acidity,
   a. hydrogen ions are excreted, and bicarbonate ions are reabsorbed.
   b. hydrogen ions are reabsorbed, and bicarbonate ions are excreted.
   c. hydrogen ions and bicarbonate ions are reabsorbed.
   d. hydrogen ions and bicarbonate ions are excreted.
   e. urea, uric acid, and ammonia are excreted.

5. Excretion of a *more concentrated* hypertonic urine in humans is associated best with the
   a. glomerular capsule and the tubules.
   b. proximal convoluted tubule only.
   c. loop of the nephron and collecting duct.
   d. distal convoluted tubule and peritubular capillary.

6. The presence of ADH (antidiuretic hormone) causes an individual to excrete
   a. sugars.   c. more water.
   b. less water.   d. Both a and c are correct.

7. In humans, water is
   a. found in the glomerular filtrate.
   b. reabsorbed from the nephron.
   c. in the urine.
   d. All of these are correct.

8. Filtration is associated with the
   a. glomerular capsule.
   b. distal convoluted tubule.
   c. collecting duct.
   d. All of these are correct.

9. Which of the following is a structural difference of the urinary systems of males and females?
   a. Males have a longer urethra than females.
   b. In males, the urethra passes through the prostate.
   c. In males, the urethra serves both the urinary and reproductive systems.
   d. All of these are correct.

10. The function of erythropoietin is
    a. reabsorption of sodium ions.
    b. excretion of potassium ions.
    c. reabsorption of water.
    d. to stimulate red blood cell production.
    e. to increase blood pressure.

11. The purpose of the loop of the nephron in the process of urine formation is
    a. reabsorption of water.
    b. production of filtrate.
    c. reabsorption of solutes.
    d. secretion of solutes.

12. Which of the following materials would NOT be filtered from the blood at the glomerulus?
    a. water   d. glucose
    b. urea   e. sodium ions
    c. protein

13. Which of the following materials would NOT be maximally reabsorbed from the filtrate?
    a. water   d. urea
    b. glucose   e. amino acids
    c. sodium ions

14. The renal medulla has a striped appearance due to the presence of which structures?
    a. loop of Henle
    b. collecting duct
    c. peritubular capillaries
    d. Both a and b are correct.

15. By what transport process are most molecules secreted from the blood into the distal convoluted tubule? *energy move low to high concentration*
    a. osmosis   c. active transport
    b. diffusion *high to low concentration*   d. facilitated diffusion

16. Which of the following is not correct?
    a. Uric acid is produced from the breakdown of amino acids.
    b. Creatinine is produced from breakdown reactions in the muscles. *from breakdown of nucleotides*
    c. Urea is produced from the breakdown of proteins.
    d. Ammonia results from the deamination of amino acids.

17. Label this diagram of a nephron.

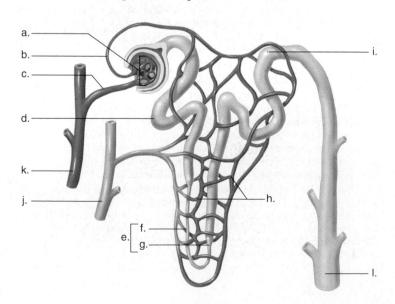

a.
b.
c.
d.
k.
j.
e. { f.
     g.
i.
h.
l.

e-Learning Connection   www.mhhe.com/maderinquiry10

Concepts	Questions	Media Resources*
16.1 Urinary System		

• The urinary system consists of organs that produce, store, and rid the body of urine.
• The kidneys excrete metabolic wastes and maintain the water-salt and the acid-base balance of the blood within normal limits.
	1. How does the reflex to urinate differ in young children compared to older children and adults?
	2. How does the urinary system help maintain homeostasis?
		Essential Study Partner
		    Nitrogenous Wastes
		    Urinary System
		    Human Excretion
		    Urine Elimination
		
		Labeling Exercises
		    Human Urinary System

16.2 Kidney Structure		

• Kidneys have a macroscopic (gross) anatomy and a microscopic anatomy.
• Urine is produced by many microscopic tubules called nephrons.
	1. What are the three regions of the kidney and which is continuous with the ureter?
	2. What adaptations of the proximal convoluted tubule maximize reabsorption?
		Essential Study Partner
		    Kidney Function
		
		Art Quizzes
		    Nephron
		
		Labeling Exercises
		    Anatomy of Kidney and Lobe

16.3 Urine Formation		

• Like many physiological processes, urine formation is a multistep process.	1. List and explain the three steps of urine formation.
	2. Why is it possible for a person to excrete glucose in urine?
		Essential Study Partner
		    Urine Formation
		    Glomerular Filtration
		    Tubular Reabsorption and Secretion
		
		Art Quizzes
		    Transport Processes in Mammalian Nephron
		
		Animation Quizzes
		    Kidney Function (1)
		    Kidney Function (2)

16.4 Maintaining Water-Salt Balance		

• The kidneys are under hormonal control as they regulate the water-salt balance of the body.	1. What portions of the nephron and kidney are involved in the reabsorption of water?
	2. If an individual drinks very little fluid on a hot day, what happens to the quantity of ADH released by the posterior pituitary?
		Essential Study Partner
		    Water Balance
		    Electrolyte Balance
		
		Art Quizzes
		    Homeostasis of Blood Volume and Pressure

16.5 Maintaining Acid-Base Balance		

• The kidneys excrete hydrogen ions and reabsorb bicarbonate ions to regulate the pH of the blood.	1. How does the buffer/breathing system compare with the ability of the kidneys to maintain acid-base balance in terms of speed and magnitude of the effect on pH?	Essential Study Partner
		    Acid-Base Balance
		
		Case Studies
		    Transplant

*For additional Media Resources, see the Online Learning Center.

Integration and Control of the Human Body

17. Nervous System 317

The nervous system coordinates and regulates all body systems, and in addition is responsible for perception, learning, and memory.

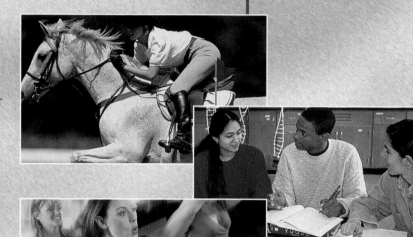

18. Senses 343

The sensory receptors are sensitive to stimuli and send sensory information to the central nervous system, which often stores it for future use.

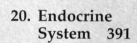

19. Musculoskel-etal System 365

The musculoskeletal system supports the body and allows it to move. This system also produces blood cells.

20. Endocrine System 391

The organs of the endocrine system secrete hormones, chemical signals that are carried in the blood and regulate all the body's tissues.

See the Online Learning Center for careers in Human Anatomy and Physiology.

17

Nervous System

Chapter Concepts

We can ride a horse and perform other athletic feats only as long as the brain can send commands to the muscles by way of the spinal cord.

*I*t was a warm spring afternoon when the crowd gathered to enjoy a horse-riding competition. Everything was fine until—suddenly—one horse stopped dead in its tracks. The horse's rider went tumbling through the air. Hitting the ground, she crushed the top two vertebrae in her neck, damaging the sensitive spinal cord underneath. In a split second, she was paralyzed. Doctors say she is lucky to be alive.

When Frances hurt her spinal cord, the brain lost its avenue of communication with the portion of her body located below the site of damage. She cannot command her arms and legs to move, nor can she receive any sensation from most of her body. But cranial nerves from her eyes and ears still allow her to see and hear, and her brain still enables her to have emotions, to remember, and to reason. Also, her internal organs still function normally, a sign that her injury was not as severe as it could have been.

In this chapter, we will examine the structure of the nervous system and how the central nervous system, consisting of the brain and spinal cord, communicates with the body by way of the peripheral nervous system, consisting of nerves.

17.1 Nervous Tissue

The nervous system has two major divisions. **The central nervous system (CNS)** consists of the brain and spinal cord which are located in the midline of the body. **The peripheral nervous system (PNS)** consists of nerves that carry sensory messages to the CNS and motor commands from the CNS to the muscles and glands (Fig. 17.1). The division between the CNS and the PNS is arbitrary; the two systems work together and are connected to one another.

The nervous system contains two types of cells: neurons and neuroglia (neuroglial cells). **Neurons** are the cells that transmit nerve impulses between parts of the nervous system; neuroglia support and nourish neurons. This section discusses the structure and function of neurons.

Neuron Structure

Although the nervous system is extremely complex, the principles of its operation are simple. There are three classes of neurons: sensory neurons, interneurons, and motor neurons (Fig. 17.2). Their functions are best described in relation to the CNS. A **sensory neuron** takes messages from a sensory receptor to the CNS. Sensory receptors are special structures that detect changes in the environment. An **interneuron** lies entirely within the CNS. Interneurons can receive input from

sensory neurons and also from other interneurons in the CNS. Thereafter, they sum up all the messages received from these neurons before they communicate with motor neurons. A **motor neuron** takes messages away from the CNS to an effector (muscle fiber or gland). Effectors carry out our responses to environmental changes.

Neurons vary in appearance, but all of them have just three parts: a cell body, dendrites, and an axon. The **cell body** contains the nucleus as well as other organelles. The **dendrites** are the many extensions from the cell body that receive signals from other neurons and send them on to the cell body. An **axon** conducts nerve impulses away from the cell body toward other neurons or target structures.

Myelin Sheath

Some axons are covered by a protective **myelin sheath.** In the PNS, this covering is formed by a type of neuroglia called **Schwann cells,** which contain the lipid substance myelin in their plasma membranes. The myelin sheath develops when Schwann cells wrap themselves around an axon many times and in this way lay down several layers of plasma membrane. Since each Schwann cell myelinates only part of an axon, the myelin sheath is interrupted. The gaps where there is no myelin sheath are called **nodes of Ranvier** (Fig. 17.3).

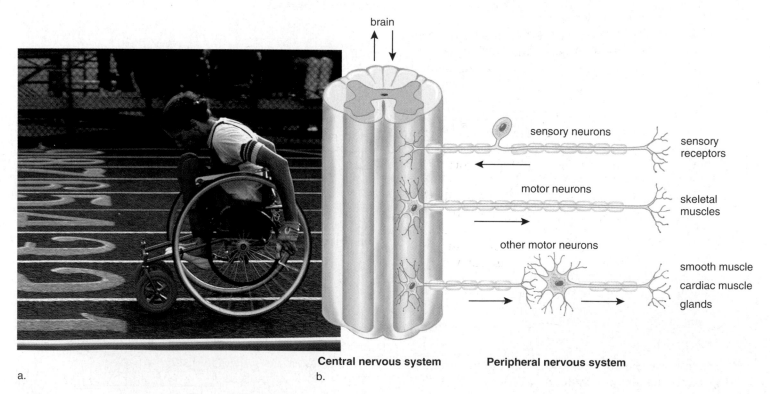

a.

b.

Central nervous system   **Peripheral nervous system**

brain

sensory neurons — sensory receptors

motor neurons — skeletal muscles

other motor neurons — smooth muscle, cardiac muscle, glands

Figure 17.1   Organization of the nervous system.

a. In paraplegics, messages no longer flow between the lower limbs and the central nervous system (the spinal cord and brain). **b.** The sensory neurons of the peripheral nervous system take nerve impulses from sensory receptors to the central nervous system (CNS), and motor neurons take nerve impulses from the CNS to the organs, muscles, and glands mentioned.

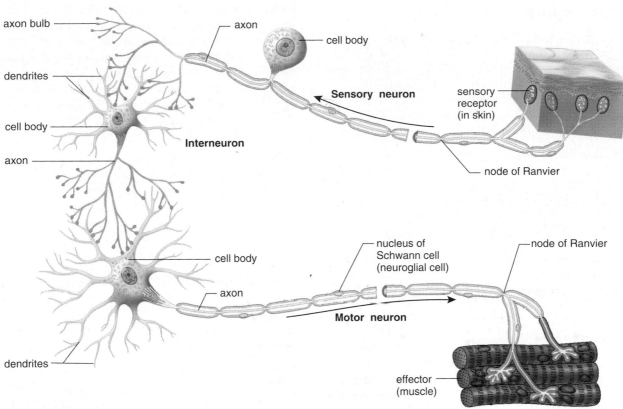

Figure 17.2   Types of neurons.
A sensory neuron, an interneuron, and a motor neuron are drawn here to show their arrangement in the body. (The breaks indicate that the fibers are much longer than shown.) How does this arrangement correlate with the function of each neuron?

Myelin gives nerve fibers their white, glistening appearance and serves as an excellent insulator. The myelin sheath also plays an important role in nerve regeneration within the PNS. If an axon is accidentally severed, the myelin sheath remains and serves as a passageway for new fiber growth. Multiple sclerosis (MS) is a disease of the myelin sheath in the CNS. Lesions develop and become hardened scars that interfere with normal conduction of nerve impulses, and the result is various neuromuscular symptoms.

Long axons tend to have myelin sheaths, while short axons do not have a myelin sheath. The gray matter of the CNS consists of neurons that have no myelin sheath. The white matter of the CNS consists of neurons that have long axons and a myelin sheath. These axons are carrying messages from one part of the nervous system to another. The surface layer of the brain is gray matter, and the white matter lies deep within the gray matter. The central part of the spinal cord consists of gray matter, and the white matter surrounds the gray matter.

All neurons have three parts: a cell body, dendrites, and an axon. Sensory neurons take messages to the CNS, and interneurons sum up sensory input before motor neurons take commands away from the CNS.

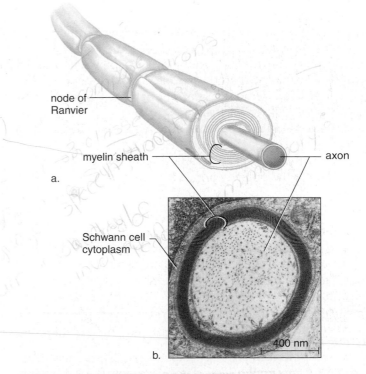

Figure 17.3   Myelin sheath.
a. In the PNS, a myelin sheath forms when Schwann cells wrap themselves around an axon. **b.** Electron micrograph of a cross section of an axon surrounded by a myelin sheath.

The Nerve Impulse

The nervous system uses the **nerve impulse** to convey information. The nature of a nerve impulse has been studied by using excised axons and a voltmeter called an **oscilloscope.** Voltage, often measured in millivolts (mV), is a measure of the electrical potential difference between two points, which in this case are the inside and the outside of the axon. Voltage is displayed on the oscilloscope screen as a trace, or pattern, over time.

Resting Potential

In the experimental setup shown in Figure 17.4a, an oscilloscope is wired to two electrodes: one electrode is placed inside an axon, and the other electrode is placed outside. The axon is essentially a membranous tube filled with axoplasm (cytoplasm of the axon). When the axon is not conducting an impulse, the oscilloscope records a potential difference across a membrane equal to about −65 mV. This reading indicates that the inside of the axon is negative compared to the outside. This is called the **resting potential** because the axon is not conducting an impulse.

The existence of this polarity (charge difference) correlates with a difference in ion distribution on either side of the axomembrane (plasma membrane of the axon). As Figure 17.4a shows, the concentration of sodium ions (Na^+) is greater outside the axon than inside, and the concentration of potassium ions (K^+) is greater inside the axon than outside. The unequal distribution of these ions is due to the action of the **sodium-potassium pump,** a membrane protein that actively transports Na^+ out of and K^+ into the axon. The work of the pump maintains the unequal distribution of Na^+ and K^+ across the membrane.

The pump is always working because the membrane is somewhat permeable to these ions, and they tend to diffuse toward their lesser concentration. Since the membrane is more permeable to K^+ than to Na^+, there are always more positive ions outside the membrane than inside. This accounts for the polarity recorded by the oscilloscope. Large, negatively charged organic ions in the axoplasm also contribute to the polarity across a resting axomembrane.

Because of the sodium-potassium pump, there is a greater concentration of Na^+ outside an axon and a greater concentration of K^+ inside an axon. An unequal distribution of ions causes the inside of an axon to be negative compared to the outside.

Action Potential

An **action potential** is a rapid change in polarity across an axomembrane as the nerve impulse occurs. An action potential is an all-or-none phenomenon. If a stimulus causes the axomembrane to depolarize to a certain level, called **threshold,** an action potential occurs. The strength of an action potential does not change, but an intense stimulus can cause an axon to fire (start an axon potential) more often in a given time interval than a weak stimulus.

The action potential requires two types of gated channel proteins in the membrane. One gated channel protein opens to allow Na^+ to pass through the membrane, and another opens to allow K^+ to pass through the membrane (Fig. 17.4b).

Sodium Gates Open   When an action potential occurs, the gates of sodium channels open first, and Na^+ flows into the axon. As Na^+ moves to inside the axon, the membrane potential changes from −65 mV to +40 mV. This is a *depolarization* because the charge inside the axon changes from negative to positive (Fig. 17.4c).

Potassium Gates Open   Second, the gates of potassium channels open, and K^+ flows to outside the axon. As K^+ moves to outside the axon, the action potential changes from +40 mV back to −65 mV. This is a *repolarization* because the inside of the axon resumes a negative charge as K^+ exits the axon.

The nerve impulse consists of an electrochemical change that occurs across an axomembrane. During depolarization, Na^+ moves to inside the axon, and during repolarization K^+ moves to outside the axon.

Propagation of an Action Potential

When an action potential travels down an axon, each successive portion of the axon undergoes a depolarization and then a repolarization. Like a domino effect, each preceding portion causes an action potential in the next portion of an axon.

As soon as an action potential has moved on, the previous portion of an axon undergoes a **refractory period** during which the sodium gates are unable to open. This ensures that the action potential cannot move backward and instead always moves down an axon toward its branches.

In myelinated axons, the gated ion channels that produce an action potential are concentrated at the nodes of Ranvier. Since ion exchange occurs only at the nodes, the action potential travels faster than in nonmyelinated axons. This is called saltatory conduction, meaning that the action potential "jumps" from node to node. Speeds of 200 meters per second (450 miles per hour) have been recorded.

An action potential travels along the length of an axon.

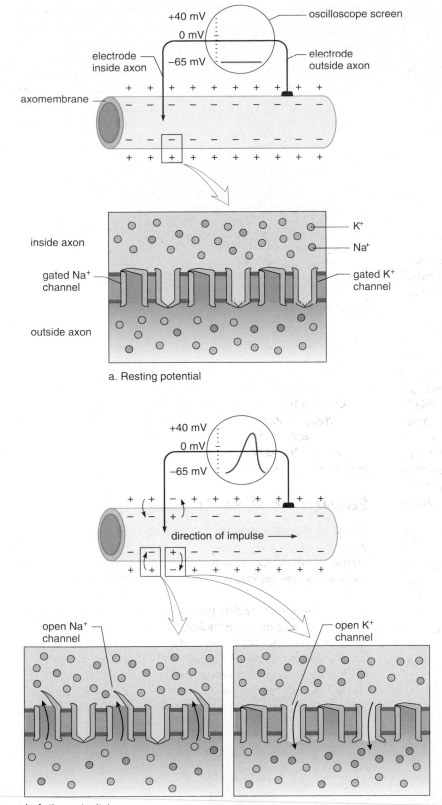

a. Resting potential

b. Action potential

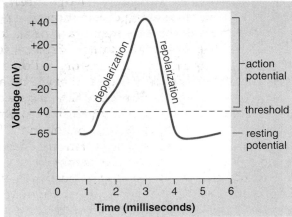

c. Enlargement of action potential

Figure 17.4   Resting and action potential.

a. Resting potential. An oscilloscope, an instrument that records voltage changes, records a resting potential of −65 mV. There is a preponderance of Na$^+$ outside the axon and a preponderance of K$^+$ inside the axon. The permeability of the membrane to K$^+$ compared to Na$^+$ causes the inside to be negative compared to the outside. **b.** Action potential. A depolarization occurs when Na$^+$ gates open and Na$^+$ moves to inside the axon; a repolarization occurs when K$^+$ gates open and K$^+$ moves to outside the axon. **c.** Enlargement of the action potential in (**b**), as seen by an experimenter using an oscilloscope.

Visual Focus

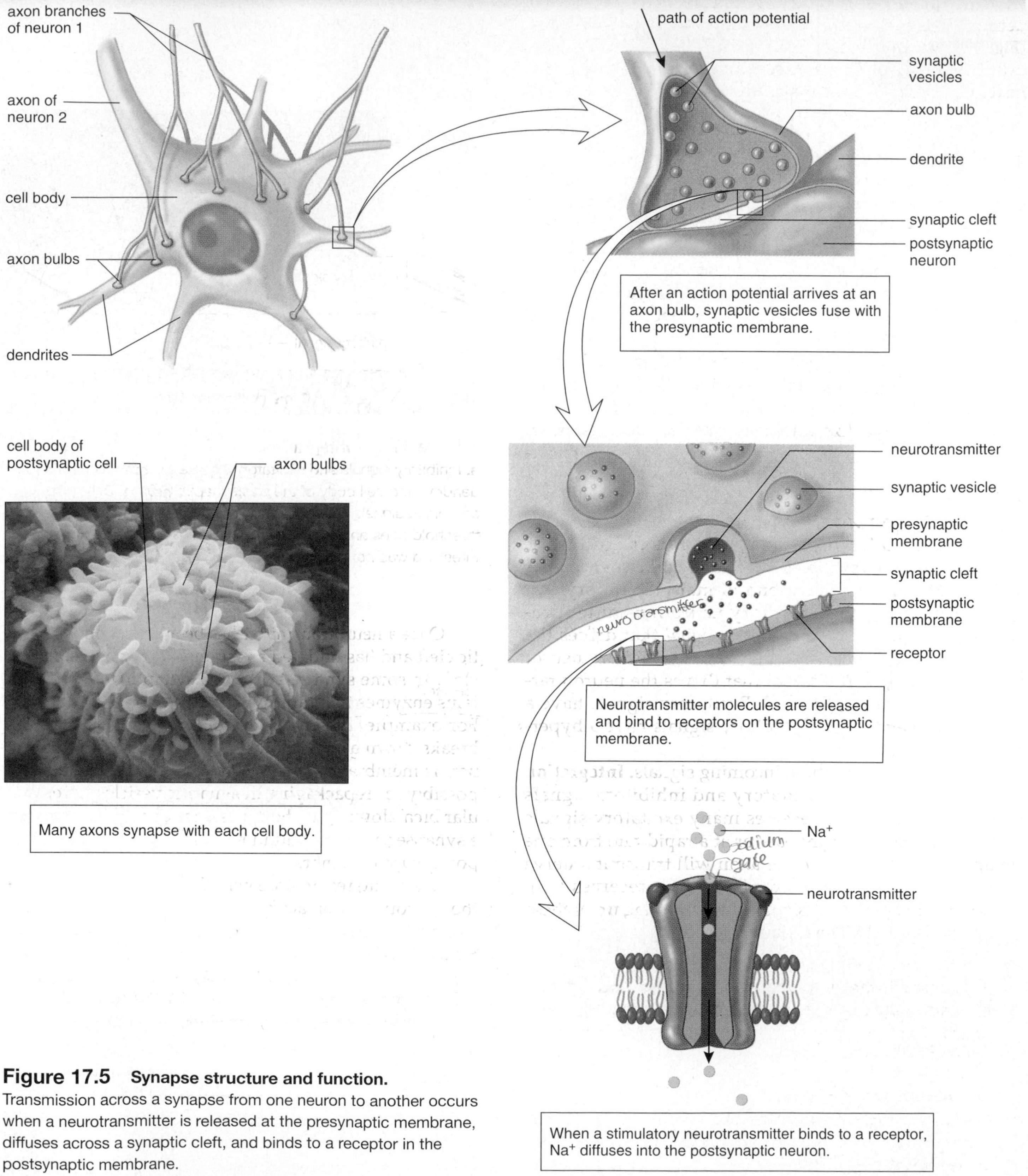

axon branches of neuron 1

axon of neuron 2

cell body

axon bulbs

dendrites

path of action potential

synaptic vesicles

axon bulb

dendrite

synaptic cleft

postsynaptic neuron

After an action potential arrives at an axon bulb, synaptic vesicles fuse with the presynaptic membrane.

cell body of postsynaptic cell

axon bulbs

Many axons synapse with each cell body.

neurotransmitter

synaptic vesicle

presynaptic membrane

synaptic cleft

postsynaptic membrane

receptor

neuro transmitter

Neurotransmitter molecules are released and bind to receptors on the postsynaptic membrane.

Na+

sodium gate

neurotransmitter

Figure 17.5 **Synapse structure and function.**
Transmission across a synapse from one neuron to another occurs when a neurotransmitter is released at the presynaptic membrane, diffuses across a synaptic cleft, and binds to a receptor in the postsynaptic membrane.

When a stimulatory neurotransmitter binds to a receptor, Na+ diffuses into the postsynaptic neuron.

Transmission Across a Synapse

Every axon branches into many fine endings, each tipped by a small swelling called an **axon bulb.** Each bulb lies very close to either the dendrite or the cell body of another neuron. This region of close proximity is called a **synapse** (Fig. 17.5). At a synapse, the membrane of the first neuron is called the *pre*synaptic membrane, and the membrane of the next neuron is called the *post*synaptic membrane. The small gap between is the **synaptic cleft.**

Transmission across a synapse is carried out by molecules called **neurotransmitters,** which are stored in synaptic vesicles in the axon bulbs. When nerve impulses traveling along an axon reach an axon bulb, gated channels for calcium ions (Ca^{2+}) open, and calcium enters the bulb. This sudden rise in Ca^{2+} stimulates synaptic vesicles to merge with the presynaptic membrane, and neurotransmitter molecules are released into the synaptic cleft. They diffuse across the cleft to the postsynaptic membrane, where they bind with specific receptor proteins.

Depending on the type of neurotransmitter and the type of receptor, the response of the postsynaptic neuron can be toward excitation or toward inhibition. Excitatory neurotransmitters that utilize gated ion channels are fast acting. Other neurotransmitters affect the metabolism of the postsynaptic cell and therefore are slower acting.

Synaptic Integration

A single neuron has many dendrites plus the cell body, and both can have synapses with many other neurons. A neuron is on the receiving end of many excitatory and inhibitory signals. An excitatory neurotransmitter produces a potential change called a *signal* that drives the neuron closer to an action potential; an inhibitory neurotransmitter produces a signal that drives the neuron farther from an action potential. Excitatory signals have a depolarizing effect, and inhibitory signals have a hyperpolarizing effect.

Neurons integrate these incoming signals. **Integration** is the summing up of excitatory and inhibitory signals (Fig. 17.6). If a neuron receives many excitatory signals (either from different synapses or at a rapid rate from one synapse), the chances are the axon will transmit a nerve impulse. On the other hand, if a neuron receives both inhibitory and excitatory signals, the summing up of these signals may prohibit the axon from firing.

> Integration is the summing up of inhibitory and excitatory signals received by a postsynaptic neuron.

Neurotransmitter Molecules

At least 25 different neurotransmitters have been identified, but two very well-known ones are **acetylcholine (ACh)** and **norepinephrine (NE).**

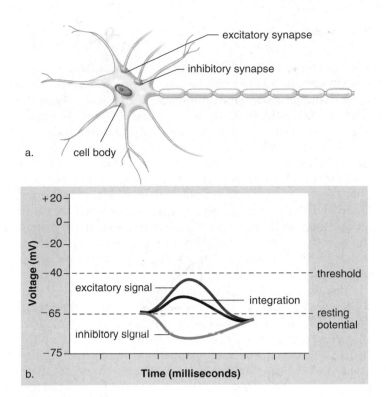

excitatory synapse

inhibitory synapse

a.   cell body

Figure 17.6   Integration.
a. Inhibitory signals and excitatory signals are summed up in the dendrite and cell body of the postsynaptic neuron. Only if the combined signals cause the membrane potential to rise above threshold does an action potential occur. **b.** In this example, threshold was not reached.

Once a neurotransmitter has been released into a synaptic cleft and has initiated a response, it is removed from the cleft. In some synapses, the postsynaptic membrane contains enzymes that rapidly inactivate the neurotransmitter. For example, the enzyme **acetylcholinesterase (AChE)** breaks down acetylcholine. In other synapses, the presynaptic membrane rapidly reabsorbs the neurotransmitter, possibly for repackaging in synaptic vesicles or for molecular breakdown. The short existence of neurotransmitters at a synapse prevents continuous stimulation (or inhibition) of postsynaptic membranes.

It is of interest to note here that many drugs that affect the nervous system act by interfering with or potentiating (enhancing) the action of neurotransmitters. As described in Figure 17.18, drugs can enhance or block the release of a neurotransmitter, mimic the action of a neurotransmitter or block the receptor, or interfere with the removal of a neurotransmitter from a synaptic cleft.

> Transmission across a synapse is dependent on the release of neurotransmitters, which diffuse across the synaptic cleft from one neuron to the next.

17.2 The Central Nervous System

The spinal cord and the brain make up the central nervous system (CNS), where sensory information is received and motor control is initiated. Figure 17.7 illustrates how the CNS relates to the PNS. Both the spinal cord and the brain are protected by bone; the spinal cord is surrounded by vertebrae, and the brain is enclosed by the skull. Also, both the spinal cord and the brain are wrapped in protective membranes known as **meninges** (sing., meninx). The spaces between the meninges are filled with **cerebrospinal fluid,** which cushions and protects the CNS. A small amount of this fluid is sometimes withdrawn from around the cord for laboratory testing when a spinal tap (lumbar puncture) is performed. Meningitis is an infection of the meninges.

Cerebrospinal fluid is also contained within the ventricles of the brain and in the central canal of the spinal cord. The brain's **ventricles** are interconnecting cavities that produce and serve as a reservoir for cerebrospinal fluid. Normally, any excess cerebrospinal fluid drains away into the cardiovascular system. However, blockages can occur. In an infant, the brain can enlarge due to cerebrospinal fluid accumulation, resulting in a condition called hydrocephalus ("water on the brain"). If cerebrospinal fluid collects in an adult, the brain cannot enlarge, and instead is pushed against the skull, possibly becoming injured.

The CNS is composed of two types of nervous tissue—gray matter and white matter. **Gray matter** is gray because it contains cell bodies and short, nonmyelinated fibers. **White matter** is white because it contains myelinated axons that run together in bundles called **tracts.**

The CNS, which lies in the midline of the body and consists of the brain and the spinal cord, receives sensory information and initiates motor control.

The Spinal Cord

The **spinal cord** extends from the base of the brain through a large opening in the skull called the foramen magnum and into the vertebral canal formed by openings in the vertebrae.

Structure of the Spinal Cord

Figure 17.8*a* shows how an individual vertebra protects the spinal cord. The spinal nerves project from the cord between the vertebrae that make up the vertebral column. Intervertebral disks separate the vertebrae, and if a disk slips a bit and presses on the spinal cord, pain results.

A cross section of the spinal cord shows a central canal, gray matter, and white matter (Fig. 17.8*b,c*). The central canal contains cerebrospinal fluid, as do the meninges that

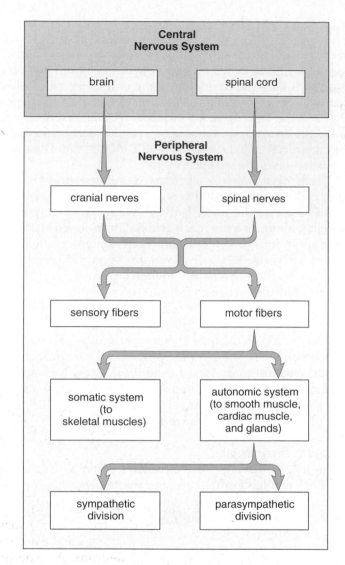

Figure 17.7   Organization of the nervous system.
The CNS, composed of the spinal cord and brain, communicates with the PNS, which contains nerves. In the somatic system, nerves conduct impulses from sensory receptors to the CNS and motor impulses from the CNS to the skeletal muscles. In the autonomic system, consisting of the sympathetic and parasympathetic divisions, motor impulses travel to smooth muscle, cardiac muscle, and the glands.

protect the spinal cord. The gray matter is centrally located and shaped like the letter H. Portions of sensory neurons and motor neurons are found there, as are interneurons that communicate with these two types of neurons. The dorsal root of a spinal nerve contains sensory fibers entering the gray matter, and the ventral root of a spinal nerve contains motor fibers exiting the gray matter. The dorsal and ventral roots join before the spinal nerve leaves the vertebral canal. Spinal nerves are a part of the PNS.

The white matter of the spinal cord occurs in areas around the gray matter. The white matter contains ascending tracts taking information to the brain (primarily located dorsally) and descending tracts taking information from the brain (primarily located ventrally). Because the tracts cross just after they enter and exit the brain, the left side of the brain controls the right side of the body, and the right side of the brain controls the left side of the body.

The spinal cord extends from the base of the brain into the vertebral canal formed by the vertebrae. A cross section shows that the spinal cord has a central canal, gray matter, and white matter.

Functions of the Spinal Cord

The spinal cord provides a means of communication between the brain and the peripheral nerves that leave the cord. When someone touches your hand, sensory receptors generate nerve impulses that pass through sensory fibers to the spinal cord and up ascending tracts to the brain. When we voluntarily move our limbs, motor impulses originating in the brain pass down descending tracts to the spinal cord and out to our muscles by way of motor fibers. Therefore, if the spinal cord is severed, we suffer a loss of sensation and a loss of voluntary control—that is, paralysis. If the cut occurs in the thoracic region, the lower body and legs are paralyzed, a condition known as paraplegia. If the injury is in the neck region, all four limbs are usually affected, a condition called quadriplegia.

We will see that the spinal cord is also the center for thousands of reflex arcs. A stimulus causes sensory receptors to generate nerve impulses that travel in sensory axons to the spinal cord. Interneurons integrate the incoming data and relay signals to motor neurons. A response to the stimulus occurs when motor axons cause skeletal muscles to contract. Each interneuron in the spinal cord has synapses with many other neurons, and therefore they send signals to several other interneurons and motor neurons.

The spinal cord plays a similar role for the internal organs. For example, when blood pressure falls, internal receptors in the carotid arteries and aorta generate nerve impulses that pass through sensory fibers to the cord and then up an ascending tract to a cardiovascular center in the brain. Thereafter, nerve impulses pass down a descending tract to the spinal cord. Motor impulses then cause blood vessels to constrict so that the blood pressure rises.

The spinal cord serves as a means of communication between the brain and much of the body. The spinal cord is also a center for reflex actions.

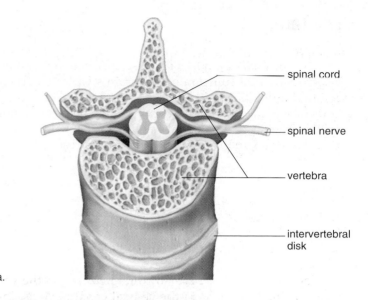

a.

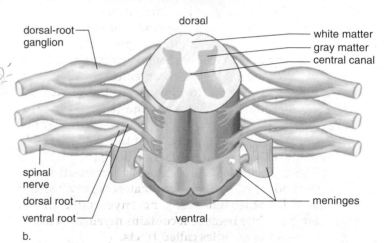

b.

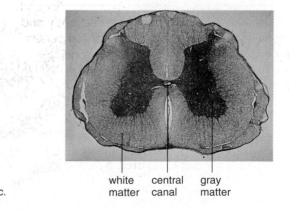

c.

Figure 17.8   Spinal cord.
a. The spinal cord passes through the vertebral canal formed by the vertebrae. **b.** The spinal cord has a central canal filled with cerebrospinal fluid, gray matter in an H-shaped configuration, and white matter around the outside. The white matter contains tracts that take nerve impulses to and from the brain. **c.** Photomicrograph of a cross section of the spinal cord.

The Brain

The human **brain** has been called the last great frontier of biology. The goal of modern neuroscience is to understand the structure and function of the brain's various parts so well that it will be possible to prevent or correct the thousands of mental disorders that rob human beings of a normal life. This section gives only a glimpse of what is known about the brain and the modern avenues of research.

We will discuss the parts of the brain with reference to the cerebrum, the diencephalon, the cerebellum, and the brain stem. The brain has four ventricles called, in turn, the two lateral ventricles, the third ventricle, and the fourth ventricle. It may be helpful to you to associate the cerebrum with the two lateral ventricles, the diencephalon with the third ventricle, and the brain stem and the cerebellum with the fourth ventricle (Fig. 17.9*a*).

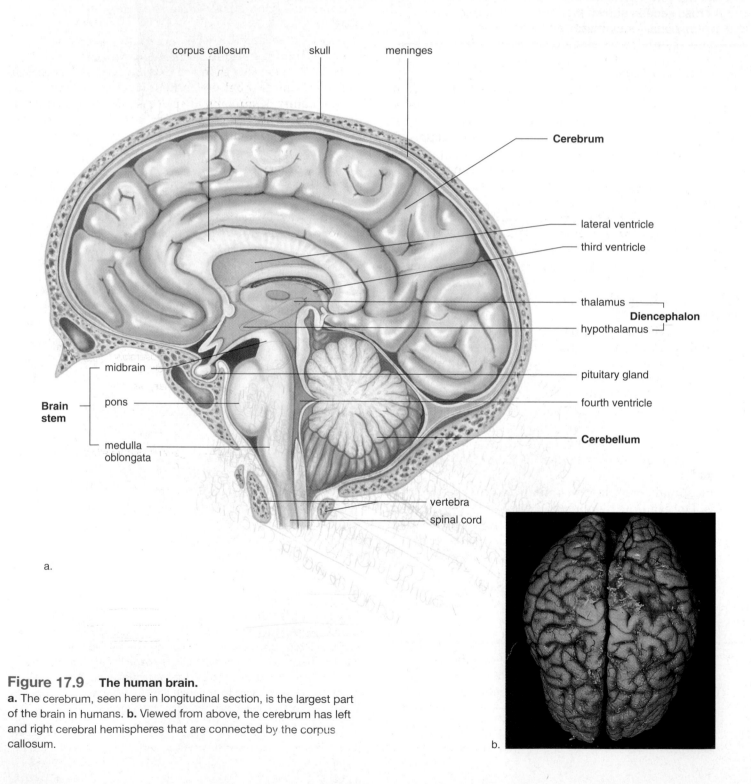

Figure 17.9   The human brain.
a. The cerebrum, seen here in longitudinal section, is the largest part of the brain in humans. **b.** Viewed from above, the cerebrum has left and right cerebral hemispheres that are connected by the corpus callosum.

The Cerebrum

The **cerebrum,** also called the telencephalon, is the largest portion of the brain in humans. The cerebrum is the last center to receive sensory input and carry out integration before commanding voluntary motor responses. It communicates with and coordinates the activities of the other parts of the brain. As we shall see, the cerebrum carries out higher thought processes required for learning and memory and for language and speech.

Just as the human body has two halves, so does the cerebrum. These halves are called the left and right **cerebral hemispheres** (see Fig. 17.9b). A deep groove called the longitudinal fissure divides the left and right cerebral hemispheres. Still, the two cerebral hemispheres are connected by a bridge of tracts within the corpus callosum.

Shallow grooves called sulci (sing., sulcus) divide each hemisphere into lobes (Fig. 17.10). The *frontal lobe* is toward the front of a cerebral hemisphere, and the *parietal lobe* is toward the back. The *occipital lobe* is dorsal to (behind) the parietal lobe, and the *temporal lobe* lies below the frontal and parietal lobes.

The Cerebral Cortex   The **cerebral cortex** is a thin but highly convoluted outer layer of gray matter that covers the cerebral hemispheres. The cerebral cortex contains over one billion cell bodies and is the region of the brain that accounts for sensation, voluntary movement, and all the thought processes we associate with consciousness.

The cerebral cortex contains motor areas and sensory areas as well as association areas. The **primary motor area** is in the frontal lobe just ventral to (before) the central sulcus. Voluntary commands to skeletal muscles begin in the primary motor area, and each part of the body is controlled by a certain section. For example, the versatile human hand takes up an especially large portion of the primary motor area. Ventral to the primary motor area is a premotor area. The *premotor area* organizes motor functions for skilled motor activities, and then the primary motor area sends signals to the cerebellum, which integrates them. The unique ability of humans to speak is partially dependent upon *Broca's area*, a motor speech area in the left frontal lobe. Signals originating here pass to the premotor area before reaching the primary motor area.

The **primary somatosensory area** is just dorsal to the central sulcus. Sensory information from the skin and skeletal muscles arrives here, where each part of the body is sequentially represented. A *primary visual area* in the occipital lobe receives information from our eyes, and a *primary*

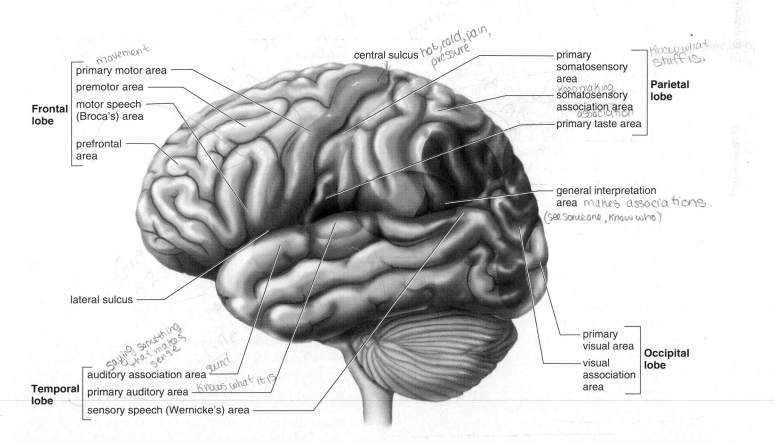

Figure 17.10   The lobes of a cerebral hemisphere.
Each cerebral hemisphere is divided into four lobes: frontal, parietal, temporal, and occipital. The cerebral cortex of a frontal lobe has motor areas and an association area called the prefrontal area. Sensory areas and association areas are in the cerebral cortex of each lobe.

auditory area in the temporal lobe receives information from our ears. A *primary taste area* in the parietal lobe accounts for taste sensations.

Association areas are places where integration occurs. For example, the *somatosensory association area*, located just dorsal to the primary somatosensory area, processes and analyzes sensory information from the skin and muscles. The *visual association area* in the occipital lobe associates new visual information with previously received visual information. It might "decide," for example, if we have seen this face or tool or whatever before. The *auditory association area* in the temporal lobe performs the same functions with regard to sounds. These association areas meet near the dorsal end of the lateral sulcus. This region is called the general interpretation area because it receives information from all the sensory association areas and allows us to quickly integrate incoming signals and send them on to the prefrontal area so that an immediate response is possible. The general interpretation area is the part of the brain that is in operation when people are able to quickly assess a situation and take actions that save themselves or others from danger. The **prefrontal area,** an association area in the frontal lobe, receives information from the other association areas and uses this information to reason and plan our actions. Integration in this area accounts for our most cherished human abilities to think critically and to formulate appropriate behaviors.

White Matter   Much of the rest of the cerebrum is composed of white matter. As you know, white matter in the CNS consists of long myelinated axons organized into tracts. Descending tracts from the primary motor area communicate with lower brain centers, and ascending tracts from lower brain centers send sensory information up to the primary somatosensory area. Because the tracts cross over in the medulla, the left side of the cerebrum controls the right side of the body and vice versa. Tracts within the cerebrum also take information between the different sensory, motor, and association areas pictured in Figure 17.10. As previously mentioned, the corpus callosum contains tracts that join the two cerebral hemispheres.

Basal Nuclei   While the bulk of the cerebrum is composed of tracts, there are masses of gray matter located deep within the white matter. These so-called **basal nuclei** (formerly termed basal ganglia) integrate motor commands, ensuring that proper muscle groups are activated or inhibited. Huntington disease and Parkinson disease, which are both characterized by uncontrollable movements, are believed to be due to malfunctioning of the basal nuclei.

The gray matter of the cerebrum consists of the cerebral cortex and the basal nuclei. The white matter consists of tracts.

The Diencephalon

The hypothalamus and the thalamus are in the **diencephalon,** a region that encircles the third ventricle. The **hypothalamus** forms the floor of the third ventricle. The hypothalamus is an integrating center that helps maintain homeostasis by regulating hunger, sleep, thirst, body temperature, and water balance. The hypothalamus controls the pituitary gland and thereby serves as a link between the nervous and endocrine systems.

The **thalamus** consists of two masses of gray matter located in the sides and roof of the third ventricle. The thalamus is on the receiving end for all sensory input except smell. Visual, auditory, and somatosensory information arrives at the thalamus via the cranial nerves and tracts from the spinal cord. The thalamus integrates this information and sends it on to the appropriate portions of the cerebrum. The thalamus is involved in arousal of the cerebrum, and it also participates in higher mental functions such as memory and emotions.

The pineal gland, which secretes the hormone melatonin, is located in the diencephalon. Presently there is much popular interest in the role of melatonin in our daily rhythms; some researchers believe it can help ameliorate jet lag or insomnia. Scientists are also interested in the possibility that the hormone may regulate the onset of puberty.

The Cerebellum

The **cerebellum** is separated from the brain stem by the fourth ventricle. The cerebellum has two portions that are joined by a narrow median portion. Each portion is primarily composed of white matter, which in longitudinal section has a treelike pattern. Overlying the white matter is a thin layer of gray matter that forms a series of complex folds.

The cerebellum receives sensory input from the eyes, ears, joints, and muscles about the present position of body parts, and it also receives motor output from the cerebral cortex about where these parts should be located. After integrating this information, the cerebellum sends motor impulses by way of the brain stem to the skeletal muscles. In this way, the cerebellum maintains posture and balance. It also ensures that all of the muscles work together to produce smooth, coordinated voluntary movements. The cerebellum assists the learning of new motor skills such as playing the piano or hitting a baseball.

The Brain Stem

The **brain stem** contains the midbrain, the pons, and the medulla oblongata (see Fig. 17.9a). The **midbrain** acts as a relay station for tracts passing between the cerebrum and the spinal cord or cerebellum. It also has reflex centers for visual, auditory, and tactile responses. The word **pons** means "bridge" in Latin, and true to its name, the pons contains bundles of axons traveling between the cerebellum and the rest of the CNS. In addition, the pons functions with the medulla oblongata to regulate breathing rate and has reflex centers concerned with head movements in response to visual and auditory stimuli.

The **medulla oblongata** contains a number of reflex centers for regulating heartbeat, breathing, and vasoconstriction (blood pressure). It also contains the reflex centers for vomiting, coughing, sneezing, hiccuping, and swallowing. The medulla oblongata lies just superior to the spinal cord and it contains tracts that ascend or descend between the spinal cord and higher brain centers.

The Reticular Formation   The **reticular formation** is a complex network of **nuclei** (masses of gray matter) and fibers that extend the length of the brain stem (Fig. 17.11). The reticular formation receives sensory signals, which it sends up to higher centers, and motor signals, which it sends to the spinal cord.

One portion of the reticular formation, called the reticular activating system (RAS), arouses the cerebrum via the thalamus and causes a person to be alert. It is believed to filter out unnecessary sensory stimuli; this may explain why you can study with the TV on. An inactive reticular formation results in sleep, and a severe injury to the RAS can cause a person to be comatose.

The other portions of the brain work with the cerebrum, and they are essential to maintaining homeostasis.

17.3 The Limbic System and Higher Mental Functions

The limbic system is intimately involved in our emotions and higher mental functions. After a short description, we will discuss the functions of the limbic system.

Limbic System

The **limbic system** is a complex network of tracts and nuclei that incorporates medial portions of the cerebral lobes, the basal nuclei, and the diencephalon (Fig. 17.12). The limbic system blends primitive emotions and higher mental functions into a united whole. It accounts for why activities like sexual behavior and eating seem pleasurable and also for why, say, mental stress can cause high blood pressure.

Two significant structures within the limbic system are the hippocampus and the amygdala, which are essential for learning and memory. The hippocampus is well situated in the brain to make the prefrontal area aware of past experiences stored in association areas. The amygdala, in particular, can cause these experiences to have emotional overtones.

The prefrontal area consults the hippocampus in order to use memories to modify our behavior. However, the inclusion of the frontal lobe in the limbic system means that reason can keep us from acting out strong feelings.

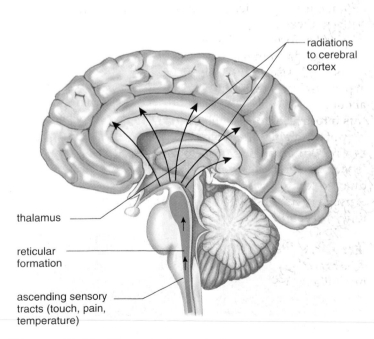

Figure 17.11   The reticular activating system.
The reticular formation receives and sends on motor and sensory information to various parts of the CNS. One portion, the reticular activating system (RAS; see arrows), arouses the cerebrum and in this way controls alertness versus sleep.

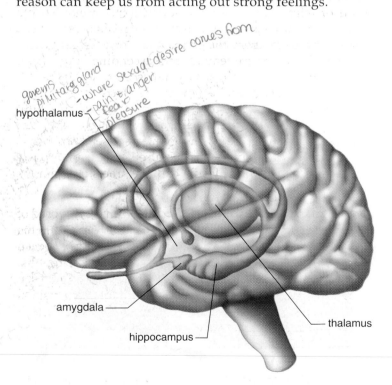

Figure 17.12   The limbic system.
Structures deep within each cerebral hemisphere and surrounding the diencephalon join higher mental functions such as reasoning with more primitive feelings such as fear and pleasure.

Figure 17.18   Drug actions at a synapse.
A drug can affect a neurotransmitter in these ways: (**a**) cause leakage out of a synaptic vesicle into the axon bulb; (**b**) prevent release of the neurotransmitter into the synaptic cleft; (**c**) promote release of the neurotransmitter into the synaptic cleft; (**d**) prevent reuptake by the presynaptic membrane; (**e**) block the enzyme that causes breakdown of the neurotransmitter; or (**f**) bind to a receptor, mimicking the action or preventing the uptake of a neurotransmitter.

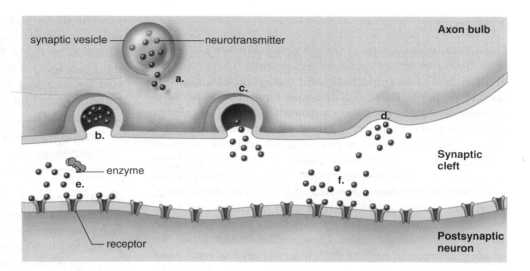

17.5 Drug Abuse

A wide variety of drugs affect the nervous system and can alter the mood and/or emotional state. Such drugs have two general effects: (1) they impact the limbic system, and (2) they either promote or decrease the action of a particular neurotransmitter (Fig. 17.18). Stimulants are drugs that increase the likelihood of neuron excitation, and depressants decrease the likelihood of excitation. Increasingly, researchers believe that dopamine, among other neurotransmitters in the brain, is responsible for mood. Cocaine is known to potentiate the effects of dopamine by interfering with its uptake from synaptic clefts. Many of the new medications developed to counter drug dependence and mental illness affect the release, reception, or breakdown of dopamine.

Drug abuse is apparent when a person takes a drug at a dose level and under circumstances that increase the potential for a harmful effect (Fig. 17.19). Drug abusers are apt to display a psychological and/or physical dependence on the drug. Dependence has developed when the person spends much time thinking about the drug or arranging to get it and often takes more of the drug than was intended. With physical dependence, formerly called "addiction," the person has become tolerant to the drug—that is, more of the drug is needed to get the same effect, and withdrawal symptoms occur when he or she stops taking the drug.

Taking drugs that affect the nervous system lead to physical dependence and withdrawal symptoms.

Alcohol

It is possible that alcohol influences the action of GABA, an inhibiting neurotransmitter, or glutamate, an excitatory neurotransmitter. Once imbibed, alcohol is primarily metabolized in the liver, where it disrupts the normal workings of this organ so that fats cannot be broken down. Fat accumulation, the first stage of liver deterioration, begins after only a single night of heavy drinking. If heavy drinking continues, fibrous scar tissue appears during a second stage of deterioration. If heavy drinking stops, the liver can still recover and become normal once again. If not, the final and irrevocable stage, cirrhosis of the liver, occurs: liver cells die, harden, and turn orange (cirrhosis means orange).

Alcohol is used by the body as an energy source, but it lacks the vitamins, minerals, essential amino acids, and fatty acids the body needs to stay healthy. For this reason, many alcoholics are undernourished and prone to illness.

The surgeon general recommends that pregnant women drink no alcohol at all. Alcohol crosses the placenta freely and causes fetal alcohol syndrome in newborns, which is characterized by mental retardation and various physical defects.

Nicotine

Nicotine, an alkaloid derived from tobacco, is a widely used neurological agent. When a person smokes a cigarette, nicotine is quickly distributed to the central and peripheral nervous systems. In the central nervous system, nicotine causes neurons to release the neurotransmitter dopamine. The excess dopamine has a reinforcing effect that leads to dependence on the drug. In the peripheral nervous system, nicotine stimulates the same postsynaptic receptors as acetylcholine and leads to increased activity of the skeletal muscles. It also increases the heart rate and blood pressure, as well as digestive tract mobility.

Many cigarette and cigar smokers find it difficult to give up the habit because nicotine induces both physiological and psychological dependence. Withdrawal symptoms include headache, stomach pain, irritability, and insomnia. Cigarette smoking in young women who are sexually active is most unfortunate because if they become pregnant, nicotine, like other psychoactive drugs, adversely affects a developing embryo and fetus.

Cocaine

Cocaine is an alkaloid derived from the shrub *Erythroxylon coca*. It is sold in powder form and as crack, a more potent extract. Cocaine prevents the synaptic uptake of dopamine, and this causes the user to experience a rush sensation. The epinephrine-like effects of dopamine account for the state of arousal that lasts for several minutes after the rush experience.

A cocaine binge can go on for days, after which the individual suffers a crash. During the binge period, the user is hyperactive and has little desire for food or sleep but has an increased sex drive. During the crash period, the user is fatigued, depressed, and irritable, has memory and concentration problems, and displays no interest in sex. Indeed, men are often impotent.

Cocaine causes extreme physical dependence. With continued cocaine use, the body begins to make less dopamine to compensate for a seemingly excess supply. The user, therefore, experiences tolerance, withdrawal symptoms, and an intense craving for cocaine. These are indications that the person is highly dependent upon the drug. Overdosing on cocaine can cause seizures and cardiac and respiratory arrest. It is possible that long-term cocaine abuse causes brain damage. Babies born to addicts suffer withdrawal symptoms and may have neurological and developmental problems.

Heroin

Heroin is derived from morphine, an alkaloid of opium. Once heroin is injected into a vein, a feeling of euphoria, along with relief of any pain, occurs within 3 to 6 minutes. Side effects can include nausea, vomiting, dysphoria, and respiratory and circulatory depression.

Heroin binds to receptors meant for the endorphins, the special neurotransmitters that kill pain and produce a feeling of tranquility. With time, the body's production of endorphins decreases. Tolerance develops so that the user needs to take more of the drug just to prevent withdrawal symptoms. The euphoria originally experienced upon injection is no longer felt.

Heroin withdrawal symptoms include perspiration, dilation of pupils, tremors, restlessness, abdominal cramps, gooseflesh, vomiting, and increase in systolic blood pressure and respiratory rate. People who are excessively dependent may experience convulsions, respiratory failure, and death. Infants born to women who are physically dependent also experience these withdrawal symptoms.

Marijuana

The dried flowering tops, leaves, and stems of the Indian hemp plant *Cannabis sativa* contain and are covered by a

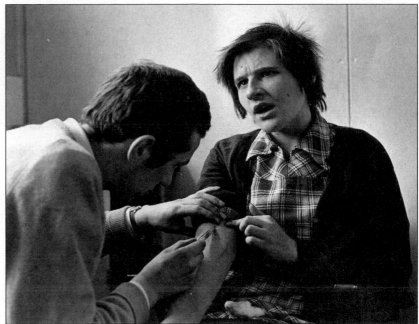

Figure 17.19   Drug use.
Blood-borne diseases such as AIDS and hepatitis B pass from one drug abuser to another when they share needles.

resin that is rich in THC (tetrahydrocannabinol). The names *cannabis* and *marijuana* apply to either the plant or THC. Usually, marijuana is smoked in a cigarette form called a "joint."

The occasional user of marijuana experiences a mild euphoria along with alterations in vision and judgment, which result in distortions of space and time. Motor incoordination, including the inability to speak coherently, takes place. Heavy use can result in hallucinations, anxiety, depression, rapid flow of ideas, body image distortions, paranoid reactions, and similar psychotic symptoms. The terms cannabis psychosis and cannabis delirium refer to such reactions. Craving and difficulty in stopping usage can occur as a result of regular use.

Recently, researchers have found that marijuana binds to a receptor for anandamide, a normal molecule in the body. Some researchers believe that long-term marijuana use leads to brain impairment. Fetal cannabis syndrome, which resembles fetal alcohol syndrome, has been reported. Some psychologists believe that marijuana use among adolescents is a way to avoid dealing with the personal problems that often develop during that stage of life.

Neurological drugs either potentiate or dampen the effect of the body's neurotransmitters.

An electroencephalogram is a record of the brain's activity picked up from an array of electrodes on the forehead and scalp. The complete and persistent absence of brain activity is often used as a clinical and legal criterion of brain death. Brain death can be present even though the body is still warm and moist. Indeed, the heart has been known to continue beating for more than a month after a declaration of brain death. Even so, should it be permissible to remove organs for transplantation as soon as the person is declared legally dead on the basis of no brain activity? Organs that have never been deprived of oxygen and nutrients are more likely to be successfully transplanted.

As a safeguard, should we have to sign a card that permits organ removal if we have been declared legally dead on the basis of brain activity? Also, should organs only be removed if family members confirm that the person is legally dead and consents to their removal. What evidence should the family require?

On the other hand, perhaps it would be best if a society simply disallowed the removal of organs unless the person is no longer breathing and the heart has stopped beating. In that case, society would be denying some of its citizens the gift of life. On the other hand, does the end justify the means—we can live with the remote possibility that organs will be removed from a living person because of the benefit to recipients?

Decide Your Opinion

1. Do you support the concept of brain death on the basis of absence of brain activity?
2. Who should decide that a person is legally brain dead?
3. Should it be permissible to remove organs for transplantation from a person who has been declared legally dead? Why or why not?

Summarizing the Concepts

17.1 Nervous Tissue

There are three types of neurons. Sensory neurons take information from sensory receptors to the CNS; interneurons occur within the CNS; and motor neurons take information from the CNS to effectors (muscles or glands). A neuron is composed of dendrites, a cell body, and an axon. Long axons are covered by a myelin sheath.

When an axon is not conducting a nerve impulse, the inside of the axon is negative (-65 mV) compared to the outside. The sodium-potassium pump actively transports Na^+ out of an axon and K^+ to the inside of an axon. The resting potential is due to the leakage of K^+ to the outside of the neuron. When an axon is conducting a nerve impulse (action potential), Na^+ first moves into the axoplasm, and then K^+ moves out of the axoplasm.

Transmission of a nerve impulse from one neuron to another takes place when a neurotransmitter molecule is released into a synaptic cleft. The binding of the neurotransmitter to receptors in the postsynaptic membrane causes excitation or inhibition. Integration is the summing of excitatory and inhibitory signals.

17.2 The Central Nervous System

The CNS consists of the spinal cord and brain, which are both protected by bone. The CNS receives and integrates sensory input and formulates motor output. The gray matter of the spinal cord contains neuron cell bodies; the white matter consists of myelinated axons that occur in bundles called tracts. The spinal cord sends sensory information to the brain, receives motor output from the brain, and carries out reflex actions.

In the brain, the cerebrum has two cerebral hemispheres connected by the corpus callosum. Sensation, reasoning, learning and memory, and language and speech take place in the cerebrum. The cerebral cortex is a thin layer of gray matter covering the cerebrum. The cerebral cortex of each cerebral hemisphere has four lobes: a frontal, parietal, occipital, and temporal lobe. The primary motor area in the frontal lobe sends out motor commands to lower brain centers, which pass them on to motor neurons. The primary somatosensory area in the parietal lobe receives sensory information from lower brain centers in communication with sensory neurons. Association areas are located in all the lobes.

The brain has a number of other regions. The hypothalamus controls homeostasis, and the thalamus specializes in sending sensory input on to the cerebrum. The cerebellum primarily coordinates skeletal muscle contractions. The medulla oblongata and the pons have centers for vital functions such as breathing and the heartbeat.

17.3 The Limbic System and Higher Mental Functions

The limbic system connects portions of the cerebral cortex with the hypothalamus, the thalamus, and basal nuclei. In the limbic system, the hippocampus acts as a conduit for sending messages to long-term memory and retrieving them once again. The amygdala adds emotional overtones to memories.

17.4 The Peripheral Nervous System

The peripheral nervous system contains only nerves and ganglia. Voluntary actions always involve the cerebrum, but reflexes are automatic, and some do not require involvement of the brain. In the somatic system, for example, a stimulus causes sensory receptors to generate nerve impulses that are then conducted by sensory fibers to interneurons in the spinal cord. Interneurons signal motor neurons, which conduct nerve impulses to a skeletal muscle that contracts, producing the response to the stimulus.

The autonomic (involuntary) system controls the smooth muscle of the internal organs and glands. The sympathetic division is associated with responses that occur during times of stress, and the parasympathetic system is associated with responses that occur during times of relaxation.

17.5 Drug Abuse

Although neurological drugs are quite varied, each type has been found to either promote or prevent the action of a particular neurotransmitter.

Testing Yourself

Choose the best answer for each question.

1. Which of these are the first and last elements in a spinal reflex?
   a. axon and dendrite
   b. sensory receptor and muscle effector
   c. ventral horn and dorsal horn
   d. brain and skeletal muscle
   e. motor neuron and sensory neuron

2. A spinal nerve takes nerve impulses
   a. to the CNS.
   b. away from the CNS.
   c. both to and away from the CNS.
   d. from the CNS to the spinal cord.

3. Which of these correctly describes the distribution of ions on either side of an axon when it is not conducting a nerve impulse?
   a. more sodium ions (Na^+) outside and more potassium ions (K^+) inside
   b. more K^+ outside and less Na^+ inside
   c. charged protein outside; Na^+ and K^+ inside
   d. Na^+ and K^+ outside and water only inside
   e. chlorine ions (Cl^-) on outside and K^+ and Na^+ on inside

4. When the action potential begins, sodium gates open, allowing Na^+ to cross the membrane. Now the polarity changes to
   a. negative outside and positive inside.
   b. positive outside and negative inside.
   c. There is no difference in charge between outside and inside.
   d. neutral outside and positive inside.
   e. Any one of these could be correct.

5. Transmission of the nerve impulse across a synapse is accomplished by the
   a. movement of Na^+ and K^+.
   b. release of a neurotransmitter by a dendrite.
   c. release of a neurotransmitter by an axon.
   d. release of a neurotransmitter by a cell body.
   e. Any one of these is correct.

6. The autonomic system has two divisions, called the
   a. CNS and PNS.
   b. somatic and skeletal systems.
   c. efferent and afferent systems.
   d. sympathetic and parasympathetic systems.

7. Synaptic vesicles are
   a. at the ends of dendrites and axons.
   b. at the ends of axons only.
   c. along the length of all long fibers.
   d. All of these are correct.

8. Which of these would be covered by a myelin sheath?
   a. short dendrites
   b. globular cell bodies
   c. long axons
   d. interneurons
   e. All of these are correct.

9. When you remove your hand from a hot stove, which system is least likely to be involved?
   a. somatic system
   b. parasympathetic system
   c. central nervous system
   d. sympathetic system
   e. peripheral nervous system

10. The spinal cord communicates with the brain via
    a. the gray matter of the cord and brain.
    b. sensory nerve fibers in a spinal nerve.
    c. the sympathetic system.
    d. tracts in the white matter.
    e. ventricles in the brain and the spinal cord.

11. Which two parts of the brain are least likely to work together?
    a. thalamus and cerebrum
    b. cerebrum and cerebellum
    c. hypothalamus and medulla oblongata
    d. cerebellum and medulla oblongata
    e. reticular formation and thalamus

12. Which of these is NOT a proper contrast between these two areas of the brain?
    primary motor—primary somatosensory
    a. ventral to the central sulcus—dorsal to the central sulcus
    b. controls skeletal muscles—receives sensory information
    c. communicates directly with association areas in the parietal lobe—communicates directly with association areas in the frontal lobe
    d. has connections with the cerebellum—has connections with the thalamus
    e. All of these are contrasts between the two areas.

13. The limbic system
    a. involves portions of the cerebral lobes, subcortical nuclei, and the diencephalon.
    b. is responsible for our deepest emotions, including pleasure, rage, and fear.
    c. is a system necessary to memory storage.
    d. is not responsible for reason and self control.
    e. All of these are correct.

14. Label this diagram.

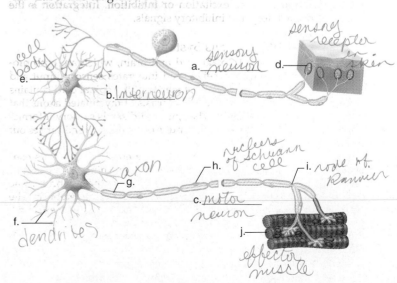

e-Learning Connection   www.mhhe.com/maderinquiry10

Concepts	Questions	Media Resources*
17.1 Nervous Tissue		

• The nervous system contains cells called neurons, which are specialized to carry nerve impulses.
• A nerve impulse is an electrochemical change that travels along the length of a neuron axon.
• Transmission of signals between neurons is dependent on neurotransmitter molecules.
	1. What are two important functions of the myelin sheath surrounding a neuron?
	2. What causes the resting potential and action potential of a neuron?
		Essential Study Partner
		   Nervous Tissue
		   Action Potential
		   Synapse
		   Human Nervous System
		
		Art Quizzes
		   Sodium-Potassium Pump
		
		BioCourse Study Guide
		   Nervous Systems

17.2 The Central Nervous System		

• The central nervous system is made up of the spinal cord and the brain.
• The spinal cord transmits messages to and from the brain and coordinates reflex responses.
• The parts of the brain are specialized for particular functions.
• The cerebral cortex contains motor areas, sensory areas, and association areas that communicate with each other.
• The reticular formation contains fibers that arouse the brain when they are active and account for sleep when they are inactive.
	1. What are the two primary functions of the spinal cord?
	2. What portion of the brain regulates many of the body's basic metabolic functions by controlling the pituitary gland?
		Essential Study Partner
		   Central Nervous System
		   Overview of the Brain
		   Spinal Cord Structure
		   Spinal Cord Functions
		   Brain Stem
		
		Art Quizzes
		   Evolution of Vertebrate Brain
		
		Labeling Exercises
		   Central Nervous System
		   The Human Brain

17.3 The Limbic System and Higher Mental Functions		
• The limbic system contains cortical and subcortical areas that are involved in higher mental functions and emotional responses. • Long-term memory depends upon association areas that are in contact with the limbic system. • Particular areas in the left hemisphere are involved in language and speech.	1. What two structures within the limbic system are essential for learning and memory? 2. How do semantic memory and episodic memory differ?	Art Quiz Limbic System
17.4 The Peripheral Nervous System		

• The peripheral nervous system contains nerves that conduct nerve impulses toward and away from the central nervous system.	1. What are the two main branches of the peripheral nervous system and what is the primary function of each?
	2. Under what circumstances are the sympathetic and parasympathetic divisions of the autonomic system active?
		Essential Study Partner
		   Peripheral Nervous System
		   ANS Structure
		   ANS Function
		
		Art Quizzes
		   Autonomic Nervous System
		   Divisions

17.5 Drug Abuse		

• The use of psychoactive drugs, such as alcohol, nicotine, cocaine, heroin, marijuana, and ecstasy is detrimental to the body.	1. List the two general physiological effects of mood-altering drugs.	Art Quiz
		   Drug Addiction and the
		   Synapse

*For additional Media Resources, see the Online Learning Center.

Senses

Chapter Concepts

Our senses help us collect data about the natural world and communicate our findings to others.

Sarah tells the group that when she had her eyes closed, she really couldn't tell whether her lab partner had placed a piece of apple or carrot on her tongue. Joe said that when he was looking through a tube with one eye, he saw a hole in his free hand with the other eye. Mary related how a letter she wrote on a piece of paper had disappeared when she moved the paper closer to her eye. This laboratory session had been a lot more fun than many of the others. The old expression "Don't believe everything you hear and only half of what you see" had more meaning after performing some of the exercises on sensory perception.

Sense organs at the periphery of the body are the "windows of the brain" because they keep the brain aware of what is going on in the external world. When stimulated, sensory receptors generate nerve impulses that travel to the central nervous system (CNS). Nerve impulses arriving at the cerebral cortex of the brain result in sensation. Stimulation of specific parts of the cerebral cortex result in seeing, hearing, tasting, and all our other senses. As the laboratory exercises mentioned above demonstrate, the brain's interpretation of data received from sensory receptors can lead to a mistaken perception of environmental circumstances.

18.1 Sensory Receptors and Sensations

Sensory receptors are specialized to detect certain types of stimuli (sing., stimulus). Exteroceptors are sensory receptors that detect stimuli from outside the body, such as those that result in taste, smell, vision, hearing, and equilibrium (Table 18.1). Interoceptors receive stimuli from inside the body. In other chapters, we have mentioned interoceptors, including pressoreceptors that respond to changes in blood pressure, osmoreceptors that detect changes in blood volume, and chemoreceptors that monitor the pH of the blood.

Interoceptors are directly involved in homeostasis and are regulated by a negative feedback mechanism. For example, when blood pressure rises, pressoreceptors signal a regulatory center in the brain, which then sends out nerve impulses to the arterial walls, causing them to relax, and the blood pressure now falls. Therefore, the pressoreceptors are no longer stimulated, and the system shuts down.

The exteroceptors such as those in the eyes and ears are not directly involved in homeostasis and continually send messages to the central nervous system regarding environmental conditions.

Types of Sensory Receptors

Sensory receptors in humans can be classified into just four categories: chemoreceptors, photoreceptors, mechanoreceptors, and thermoreceptors.

Chemoreceptors respond to chemical substances in the immediate vicinity. As Table 18.1 indicates, taste and smell are dependent on this type of sensory receptor, but certain chemoreceptors in various other organs are sensitive to internal conditions. Chemoreceptors that monitor blood pH are located in the carotid arteries and aorta. If the pH lowers, the breathing rate increases. As more carbon dioxide is expired, the blood pH rises.

Pain receptors (nociceptors) are a type of chemoreceptor. They are naked dendrites that respond to chemicals released by damaged tissues. Pain receptors are protective because they alert us to possible danger. For example, without the pain of appendicitis, we might never seek the medical help needed to avoid a ruptured appendix.

Photoreceptors respond to light energy. Our eyes contain photoreceptors that are sensitive to light rays and thereby provide us with a sense of vision. Stimulation of the photoreceptors known as rod cells results in black-and-white vision, while stimulation of the photoreceptors known as cone cells results in color vision.

Mechanoreceptors are stimulated by mechanical forces, which most often result in pressure of some sort. When we hear, air-borne sound waves are converted to fluid-borne pressure waves that can be detected by mechanoreceptors in the inner ear. Similarly, mechanoreceptors are responding to fluid-borne pressure waves when we detect changes in gravity and motion, helping us keep our balance. These receptors are in the vestibule and semicircular canals of the inner ear, respectively. The sense of touch is dependent on pressure receptors that are sensitive to either strong or slight pressures. Pressoreceptors located in certain arteries detect changes in blood pressure, and stretch receptors in the lungs detect the degree of lung inflation. Proprioceptors, which respond to the stretching of muscle fibers, tendons, joints, and ligaments, make us aware of the position of our limbs.

Thermoreceptors located in the hypothalamus and skin are stimulated by changes in temperature. Those that respond when temperatures rise are called warmth receptors, and those that respond when temperatures lower are called cold receptors.

The sensory receptors are categorized as chemoreceptors, photoreceptors, mechanoreceptors, and thermoreceptors.

Table 18.1   Exteroceptors

Sensory Receptor	Stimulus	Category	Sense	Sensory Organ
Taste cells	Chemicals	Chemoreceptor	Taste	Taste buds
Olfactory cells	Chemicals	Chemoreceptor	Smell	Olfactory epithelium
Rod cells and cone cells in retina	Light rays	Photoreceptor	Vision	Eye
Hair cells in spiral organ	Sound waves	Mechanoreceptor	Hearing	Ear
Hair cells in semi-circular canals	Motion	Mechanoreceptor	Rotational equilibrium	Ear
Hair cells in vestibule	Gravity	Mechanoreceptor	Gravitational equilibrium	Ear

How Sensation Occurs

Sensory receptors respond to environmental stimuli by generating nerve impulses. **Sensation** occurs when nerve impulses arrive at the cerebral cortex of the brain. **Perception** occurs when the cerebral cortex interprets the meaning of sensations.

As we discussed in the previous chapter, sensory receptors are the first element in a reflex arc. We are only aware of a reflex action when sensory information reaches the brain. At that time, the brain integrates this information with other information received from other sensory receptors. After all, if you burn yourself and quickly remove your hand from a hot stove, the brain receives information not only from your skin, but also from your eyes, nose, and all sorts of sensory receptors.

Some sensory receptors are free nerve endings or encapsulated nerve endings, while others are specialized cells closely associated with neurons. The plasma membrane of a sensory receptor contains receptor proteins that react to the stimulus. For example, the receptor proteins in the plasma membrane of chemoreceptors bind to certain molecules. When this happens, ion channels open, and ions flow across the plasma membrane. If the stimulus is sufficient, nerve impulses begin and are carried by a sensory nerve fiber within the PNS to the CNS (Fig. 18.1). The stronger the stimulus, the greater the frequency of nerve impulses. Nerve impulses that reach the spinal cord first are conveyed to the brain by ascending tracts. If nerve impulses finally reach the cerebral cortex, sensation and perception occur.

All sensory receptors initiate nerve impulses; the sensation that results depends on the part of the brain receiving the nerve impulses. Nerve impulses that begin in the optic nerve eventually reach the visual areas of the cerebral cortex, and thereafter, we see objects. Nerve impulses that begin in the auditory nerve eventually reach the auditory areas of the cerebral cortex, and thereafter, we hear sounds. If it were possible to switch these nerves, stimulation of the eyes would result in hearing! On the other hand, when a blow to the eye stimulates photoreceptors, we "see stars" because nerve impulses from the eyes can only result in sight.

Before sensory receptors initiate nerve impulses, they carry out **integration,** the summing up of signals. One type of integration is called **sensory adaptation,** a decrease in response to a stimulus. We have all had the experience of smelling an odor when we first enter a room and then later not being aware of it at all. Some authorities believe that when sensory adaptation occurs, sensory receptors have stopped sending impulses to the brain. Others believe that the reticular activating system (RAS) has filtered out the ongoing stimuli. You will recall that sensory information is conveyed from the brain stem through the thalamus to the cerebral cortex by the RAS. The thalamus acts as a gatekeeper and only passes on information of immediate importance. Just as we gradually become unaware of particular

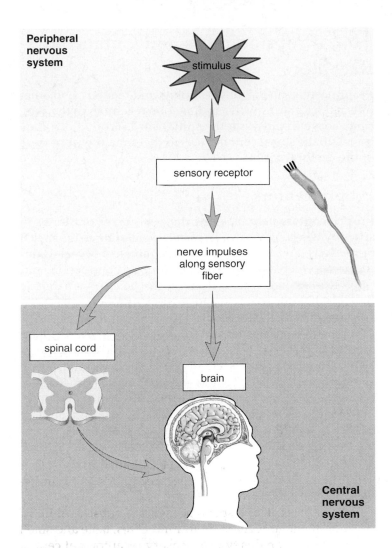

Figure 18.1   Sensation.
The stimulus is received by a sensory receptor, which generates nerve impulses (action potentials). Nerve impulses are conducted to the CNS by sensory fibers within the PNS, and only those impulses that reach the cerebral cortex result in sensation and perception.

environmental stimuli, we can suddenly become aware of stimuli that may have been present for some time. This can be attributed to the workings of the RAS, which has synapses with all the great ascending sensory tracts.

The functioning of our sensory receptors makes a significant contribution to homeostasis. Without sensory input, we would not receive information about our internal and external environment. This information leads to appropriate reflex and voluntary actions to keep the internal environment constant.

Sensation occurs when nerve impulses reach the cerebral cortex of the brain. Perception, which also occurs in the cerebral cortex, is an interpretation of the meaning of sensations.

18.2 Proprioceptors and Cutaneous Receptors

Proprioceptors in the muscles, joints and tendons, and other internal organs, as well as cutaneous receptors in the skin, send nerve impulses to the spinal cord. From there, they travel up the spinal cord in tracts to the somatosensory areas of the cerebral cortex.

Proprioceptors

Proprioceptors help us know the position of our limbs in space by detecting the degree of muscle relaxation, the stretch of tendons, and the movement of ligaments. Muscle spindles act to increase the degree of muscle contraction, and Golgi tendon organs act to decrease it. The result is a muscle that has the proper length and tension, or muscle *tone*.

Figure 18.2 illustrates the activity of a muscle spindle. In a muscle spindle, sensory nerve endings are wrapped around thin muscle cells within a connective tissue sheath. When the muscle relaxes and undue stretching of the muscle spindle occurs, nerve impulses are generated. The rapidity of the nerve impulses generated by the muscle spindle is proportional to the stretching of a muscle. A reflex action then occurs, which results in contraction of muscle fibers adjoining the muscle spindle. The knee-jerk reflex, which involves muscle spindles, offers an opportunity for physicians to test a reflex action.

The information sent by muscle spindles to the CNS is used to maintain the body's equilibrium and posture despite the force of gravity always acting upon the skeleton and muscles.

Proprioceptors are involved in reflex actions that maintain muscle tone and thereby the body's equilibrium and posture.

Cutaneous Receptors

The skin is composed of two layers: the epidermis and the dermis. In Figure 18.3, the artist has dramatically indicated these two layers by separating the epidermis from the dermis in one location. The epidermis is stratified squamous epithelium in which cells become keratinized as they rise to the surface where they are sloughed off. The dermis is a thick connective tissue layer. The dermis contains cutaneous receptors, which include sensory receptors for touch, pressure, pain, and temperature. The dermis is a mosaic of these tiny receptors, as you can determine by slowly passing a metal probe over your skin. At certain points, you will feel pressure, and at others, you will feel of hot or cold (depending on the probe's temperature).

Figure 18.2   Muscle spindle.
① When a muscle is stretched, a muscle spindle sends sensory nerve impulses to the spinal cord.
② Motor nerve impulses from the spinal cord result in muscle fiber contraction so that muscle tone is maintained.

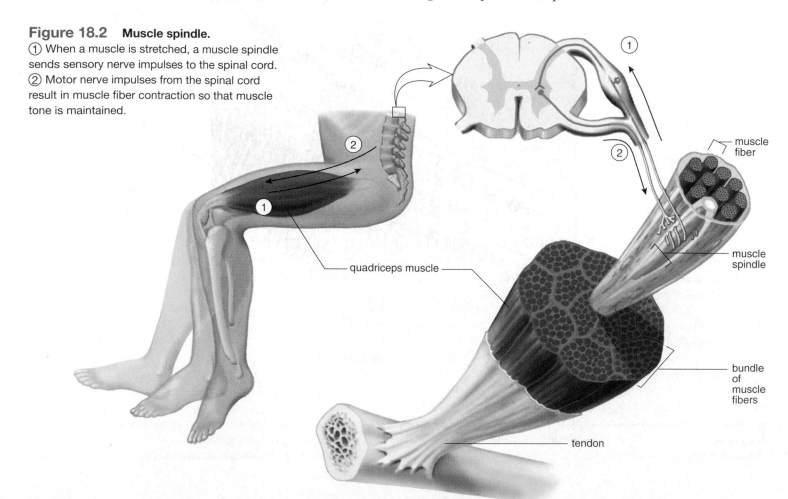

quadriceps muscle

muscle fiber

muscle spindle

bundle of muscle fibers

tendon

Three types of cutaneous receptors are sensitive to fine touch. The Meissner corpuscles are concentrated in the fingertips, the palms, the lips, the tongue, the nipples, the penis, and the clitoris. Merkel disks are found where the epidermis meets the dermis. A free nerve ending, called a root hair plexus, winds around the base of a hair follicle and fires if the hair is touched.

The three different types of pressure receptors are Pacinian corpuscles, Ruffini endings, and Krause end bulbs. Pacinian corpuscles are onion-shaped sensory receptors that lie deep inside the dermis. Ruffini endings and Krause end bulbs are encapsulated by sheaths of connective tissue and contain lacy networks of nerve fibers.

Temperature receptors are simply free nerve endings in the epidermis. Some free nerve endings are responsive to cold; others are responsive to warmth. Cold receptors are far more numerous than warmth receptors, but there are no known structural differences between the two.

Sensory receptors in the human skin are sensitive to touch, pressure, pain, and temperature (warmth and cold).

Pain Receptors

Like the skin, many internal organs have pain receptors, also called nociceptors. Pain receptors are sensitive to extremes in temperature or pressure and to chemicals released by damaged tissues. When inflammation occurs, cells release chemicals that stimulate pain receptors. Aspirin and ibuprofen reduce pain by inhibiting the synthesis of one class of these chemicals.

Sometimes, stimulation of internal pain receptors is felt as pain from the skin as well as the internal organs. This is called **referred pain.** Some internal organs have a referred pain relationship with areas located in the skin of the back, groin, and abdomen; pain from the heart is felt in the left shoulder and arm. This most likely happens when nerve impulses from the pain receptors of internal organs travel to the spinal cord and synapse with neurons also receiving impulses from the skin.

Pain receptors, also called nociceptors, are present in the skin and internal organs.

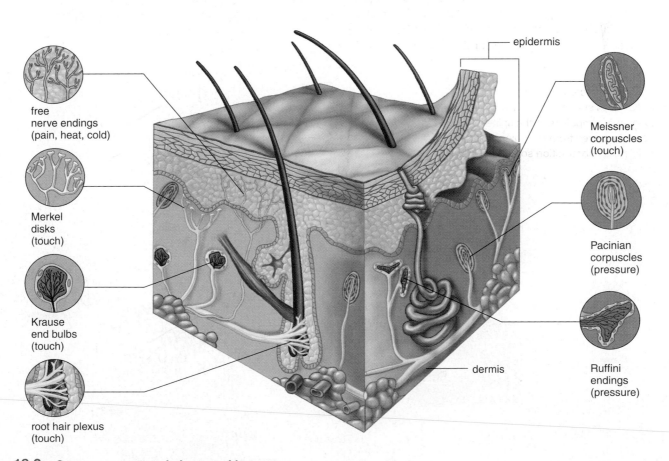

free nerve endings (pain, heat, cold)

Merkel disks (touch)

Krause end bulbs (touch)

root hair plexus (touch)

epidermis

Meissner corpuscles (touch)

Pacinian corpuscles (pressure)

dermis

Ruffini endings (pressure)

Figure 18.3   Sensory receptors in human skin.
The classical view is that each sensory receptor has the main function shown here. However, investigators report that matters are not so clear-cut. For example, microscopic examination of the skin of the ear shows only free nerve endings (pain receptors), and yet the skin of the ear is sensitive to all sensations. Therefore, it appears that the receptors of the skin are somewhat, but not completely, specialized.

18.3 Chemical Senses

Chemoreceptors in the carotid arteries and in the aorta are primarily sensitive to the pH of the blood. These bodies communicate via sensory nerve fibers with the respiratory center located in the medulla oblongata. When the pH drops, they signal this center, and immediately thereafter the breathing rate increases. The expiration of CO_2 raises the pH of the blood.

Taste and smell are called the chemical senses because their receptors are sensitive to molecules in the food we eat and the air we breathe.

Sense of Taste

The receptors for taste are found in **taste buds** located primarily on the tongue (Fig. 18.4). Many lie along the walls of the papillae, the small elevations on the tongue that are visible to the naked eye. Isolated ones are also present on the hard palate, the pharynx, and the epiglottis.

Taste buds are embedded in tongue epithelium and open at a taste pore. They have supporting cells and a number of elongated taste cells that end in microvilli. The microvilli bear receptor proteins for certain molecules. When these molecules bind to receptor proteins, nerve impulses are generated in associated sensory nerve fibers. These nerve impulses go to the brain, including the cortical areas, which interpret them as tastes.

There are four primary types of taste (sweet, sour, salty, bitter), and taste buds for each are concentrated on the tongue in particular regions (Fig. 18.4a). Sweet receptors are most plentiful near the tip of the tongue. Sour receptors occur primarily along the margins of the tongue. Salty receptors are most common on the tip and upper front portion of the tongue. Bitter receptors are located toward the back of the tongue. Actually, the response of taste buds can result in a range of sweet, sour, salty, and bitter tastes. The brain appears to survey the overall pattern of incoming sensory impulses and to take a "weighted average" of their taste messages as the perceived taste.

Taste buds contain taste cells. The microvilli of taste cells have receptor proteins for molecules that cause the brain to distinguish between sweet, sour, salty, and bitter tastes.

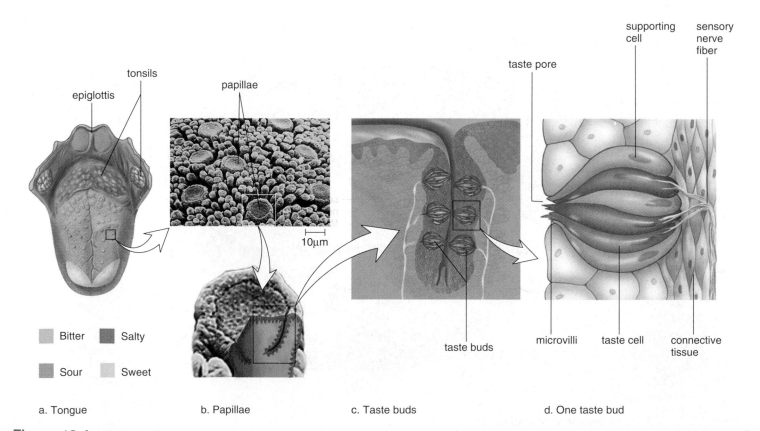

a. Tongue   b. Papillae   c. Taste buds   d. One taste bud

Figure 18.4   Taste buds.
a. Papillae on the tongue contain taste buds that are sensitive to sweet, sour, salty, and bitter tastes in the regions indicated. **b.** Enlargement of papillae. **c.** Taste buds occur along the walls of the papillae. **d.** Taste cells end in microvilli that bear receptor proteins for certain molecules. When molecules bind to the receptor proteins, nerve impulses are generated that go to the brain where the sensation of taste occurs.

Sense of Smell

Our sense of smell is dependent on **olfactory cells** located within olfactory epithelium high in the roof of the nasal cavity (Fig. 18.5). Olfactory cells are modified neurons. Each cell ends in a tuft of about five olfactory cilia, which bear receptor proteins for odor molecules. Each olfactory cell has only one out of 1,000 different types of receptor proteins. Nerve fibers from like olfactory cells lead to the same neuron in the olfactory bulb, an extension of the brain. An odor contains many odor molecules, which activate a characteristic combination of receptor proteins. A rose might stimulate olfactory cells, designated by purple and green in Figure 18.5, while a hyacinth might stimulate a different combination. An odor's signature in the olfactory bulb is determined by which neurons are stimulated. When the neurons communicate this information via the olfactory tract to the olfactory areas of the cerebral cortex, we know we have smelled a rose or a daffodil.

Have you ever noticed that a certain aroma vividly brings a certain person or place? A person's perfume may remind you of someone else, or the smell of boxwood may remind you of your grandfather's farm. The olfactory bulbs have direct connections with the limbic system and its centers for emotions and memory. One investigator showed that when subjects smelled an orange while viewing a painting, they not only remembered the painting when asked about it later, they had many deep feelings about it.

Actually, the sense of taste and the sense of smell work together to create a combined effect when interpreted by the cerebral cortex. For example, when you have a cold, you think food has lost its taste, but most likely you have lost the ability to sense its smell. This method works in reverse also. When you smell something, some of the molecules move from the nose down into the mouth region and stimulate the taste buds there. Therefore, part of what we refer to as smell may in fact be taste.

Olfactory epithelium contains olfactory cells. The cilia of olfactory cells have receptor proteins for odor molecules that cause the brain to distinguish odors.

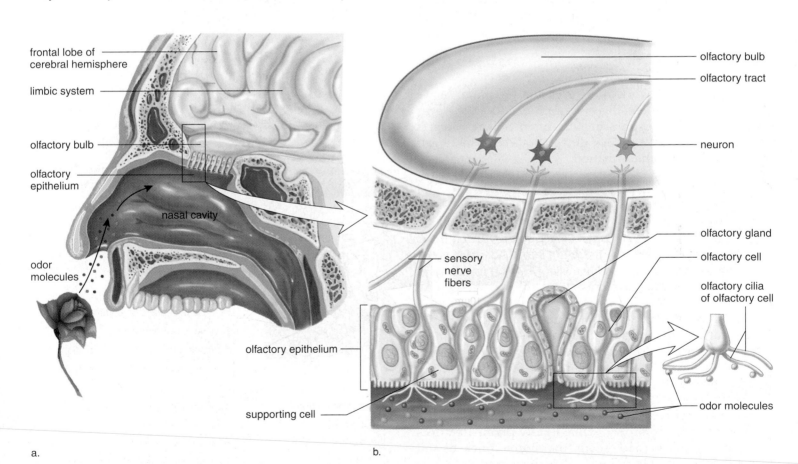

a. b.

Figure 18.5 Olfactory cell location and anatomy.

a. The olfactory epithelium in humans is located high in the nasal cavity. **b.** Olfactory cells end in cilia that bear receptor proteins for specific odor molecules. The cilia of each olfactory cell can bind to only one type of odor molecule (signified here by color). If a rose causes olfactory cells sensitive to "purple" and "green" odor molecules to be stimulated, then neurons designated by purple and green in the olfactory bulb are activated. The primary olfactory area of the cerebral cortex interprets the pattern of stimulation as the scent of a rose.

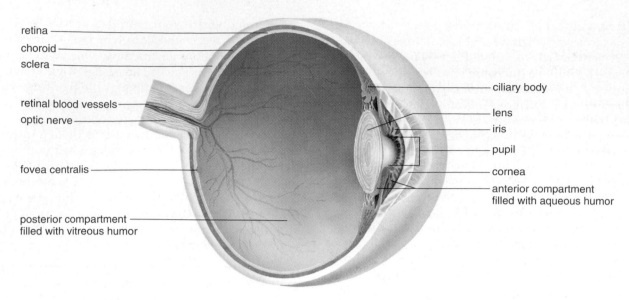

retina
choroid
sclera

retinal blood vessels
optic nerve

fovea centralis

posterior compartment
filled with vitreous humor

ciliary body
lens
iris
pupil
cornea
anterior compartment
filled with aqueous humor

Figure 18.6   Anatomy of the human eye.
Notice that the sclera, the outer layer of the eye, becomes the cornea and that the choroid, the middle layer, is continuous with the ciliary body and the iris. The retina, the inner layer, contains the photoreceptors for vision; the fovea centralis is the region where vision is most acute.

18.4 Sense of Vision

Vision requires the work of the eyes and the brain. As we shall see, much processing of stimuli occurs in the eyes before nerve impulses are sent to the brain. Still, researchers estimate that at least a third of the cerebral cortex takes part in processing visual information.

Anatomy of the Eye

The eyeball, which is an elongated sphere about 2.5 cm in diameter, has three layers, or coats: the sclera, the choroid, and the retina (Fig. 18.6 and Table 18.2). The outer layer, the **sclera,** is white and fibrous except for the transparent **cornea,** which is made of transparent collagen fibers. The cornea is the window of the eye.

The middle, thin, darkly pigmented layer, the **choroid,** is vascular and absorbs stray light rays that photoreceptors have not absorbed. Toward the front, the choroid becomes the donut-shaped **iris.** The iris regulates the size of the **pupil,** a hole in the center of the iris through which light enters the eyeball. The color of the iris (and therefore the color of your eyes) correlates with its pigmentation. Heavily pigmented eyes are brown, while lightly pigmented eyes are green or blue. Behind the iris, the choroid thickens and forms the circular ciliary body. The **ciliary body** contains the ciliary muscle, which controls the shape of the lens for near and far vision.

The **lens,** attached to the ciliary body by ligaments, divides the eye into two compartments; the one in front of the lens is the anterior compartment, and the one behind the lens is the posterior compartment. The anterior compartment is filled with a clear, watery fluid called the

Table 18.2	Functions of the Parts of the Eye
Part	**Function**
Sclera	Protects and supports eyeball
Cornea	Refracts light rays
Pupil	Admits light
Choroid	Absorbs stray light
Ciliary body	Holds lens in place, accommodation
Iris	Regulates light entrance
Retina	Contains sensory receptors for sight
Rods	Make black-and-white vision possible
Cones	Make color vision possible
Fovea centralis	Makes acute vision possible
Other	
Lens	Refracts and focuses light rays
Humors	Transmit light rays and support eyeball
Optic nerve	Transmits impulse to brain

aqueous humor. A small amount of aqueous humor is continually produced each day. Normally, it leaves the anterior compartment by way of tiny ducts. When a person has **glaucoma,** these drainage ducts are blocked, and aqueous humor builds up. If glaucoma is not treated, the resulting pressure compresses the arteries that serve the nerve fibers of the retina, where photoreceptors are located. The nerve fibers begin to die due to lack of nutrients, and the person becomes partially blind. Eventually, total blindness can result.

The third layer of the eye, the **retina,** is located in the posterior compartment, which is filled with a clear gelatinous material called the **vitreous humor.** The retina contains photoreceptors called rod cells and cone cells. The rods are very sensitive to light, but they do not see color; therefore, at night or in a darkened room we see only shades of gray. The cones, which require bright light, are sensitive to different wavelengths of light, and therefore we have the ability to distinguish colors. The retina has a very special region called the **fovea centralis** where cone cells are densely packed. Light is normally focused on the fovea when we look directly at an object. This is helpful because vision is most acute in the fovea centralis. Sensory fibers from the retina form the **optic nerve,** which takes nerve impulses to the brain.

The eye has three layers: the outer sclera, the middle choroid, and the inner retina. Only the retina contains photoreceptors for light energy.

Focusing

When we look at an object, light rays pass through the pupil and are **focused** on the retina (Fig. 18.7*a*). The image produced is much smaller than the object because light rays are bent (refracted) when they are brought into focus. Focusing starts with the cornea and continues as the rays pass through the lens and the humors. Notice that the image on the retina is inverted (upside down) and reversed from left to right.

The lens provides additional focusing power as **visual accommodation** occurs for close vision. The shape of the lens is controlled by the ciliary muscle within the ciliary body. When we view a distant object, the ciliary muscle is relaxed, causing the suspensory ligaments attached to the ciliary body to be taut; therefore, the lens remains relatively flat (Fig. 18.7*b*). When we view a near object, the ciliary muscle contracts, releasing the tension on the suspensory ligaments, and the lens rounds up due to its natural elasticity (Fig. 18.7*c*). Because close work requires contraction of the ciliary muscle, it very often causes muscle fatigue known as eyestrain.

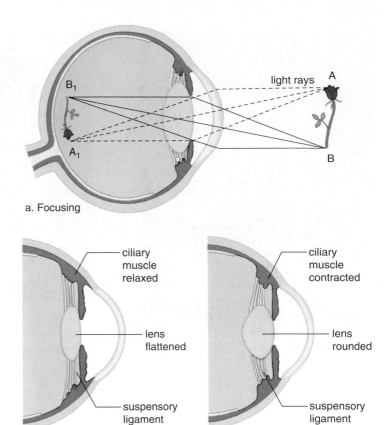

a. Focusing

b. Focusing on distant object   c. Focusing on near object

Figure 18.7   Focusing.
a. Light rays from each point on an object are bent by the cornea and the lens in such a way that an inverted and reversed image of the object forms on the retina. **b.** When focusing on a distant object, the lens is flat because the ciliary muscle is relaxed and the suspensory ligament is taut. **c.** When focusing on a near object, the lens accommodates; it becomes rounded because the ciliary muscle contracts, causing the suspensory ligament to relax.

Usually after the age of 40, the lens loses some of its elasticity and is unable to accommodate. Bifocal lenses may then be necessary for those who already have corrective lenses. Also with aging, or possibly exposure to the sun (see the Health Focus on page 356), the lens is subject to **cataracts.** The lens becomes opaque and therefore incapable of transmitting rays of light. Today, the lens is usually surgically replaced with an artificial lens. In the future, it may be possible to restore the original configuration of the proteins making up the lens.

The lens, assisted by the cornea and the humors, focuses images on the retina.

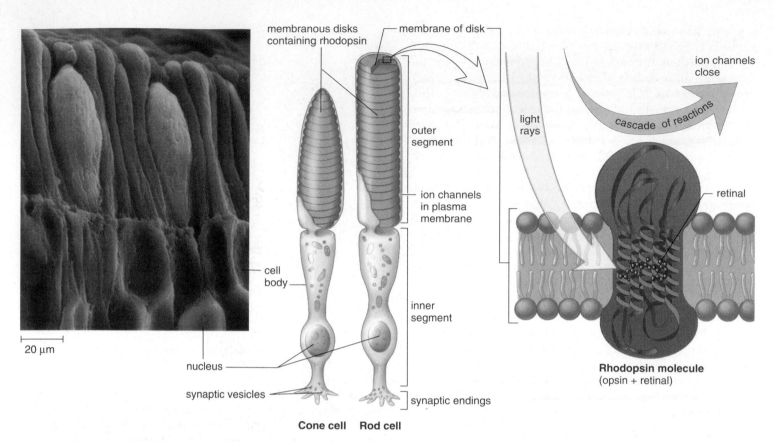

Figure 18.8   Photoreceptors in the eye.

The outer segment of rods and cones contains stacks of membranous disks, which contain visual pigments. In rods, the membrane of each disk contains rhodopsin, a complex molecule containing the protein opsin and the pigment retinal. When rhodopsin absorbs light energy, it splits, releasing opsin, which sets in motion a cascade of reactions that cause ion channels in the plasma membrane to close. Thereafter, nerve impulses go to the brain.

Photoreceptors

Vision begins once light has been focused on the photoreceptors in the retina. Figure 18.8 illustrates the structure of the photoreceptors called **rod cells** and **cone cells.** Both rods and cones have an outer segment joined to an inner segment by a stalk. Pigment molecules are embedded in the membrane of the many disks present in the outer segment. Synaptic vesicles are located at the synaptic endings of the inner segment.

The visual pigment in rods is a deep purple pigment called rhodopsin. **Rhodopsin** is a complex molecule made up of the protein opsin and a light-absorbing molecule called **retinal,** which is a derivative of vitamin A. When a rod absorbs light, rhodopsin splits into opsin and retinal, leading to a cascade of reactions and the closure of ion channels in the rod cell's plasma membrane. The release of inhibitory transmitter molecules from the rod's synaptic vesicles ceases. Thereafter, nerve impulses go to the visual areas of the cerebral cortex. Rods are very sensitive to light and therefore are suited to night vision. (Since carrots are rich in vitamin A, it is true that eating carrots can improve your night vision.) Rod cells are plentiful throughout the entire retina; therefore, they also provide us with peripheral vision and perception of motion.

The cones, on the other hand, are located primarily in the fovea and are activated by bright light. They allow us to detect the fine detail and the color of an object. **Color vision** depends on three different kinds of cones, which contain pigments called the B (blue), G (green), and R (red) pigments. Each pigment is made up of retinal and opsin, but there is a slight difference in the opsin structure of each, which accounts for their individual absorption patterns. Various combinations of cones are believed to be stimulated by in-between shades of color.

The receptors for sight are the rods and the cones. The rods permit vision in dim light at night, and the cones permit vision in the bright light needed for color vision.

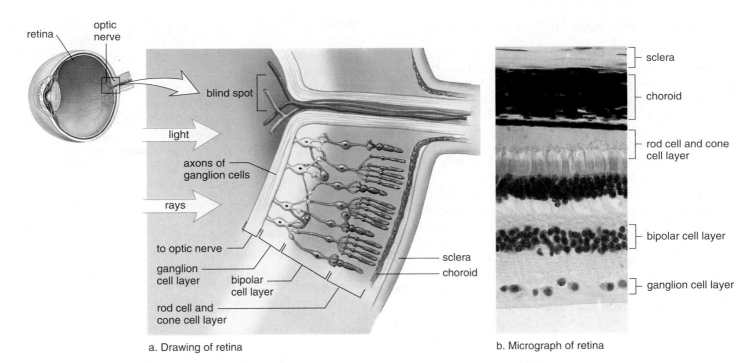

a. Drawing of retina

b. Micrograph of retina

Figure 18.9 Structure and function of the retina.
a. The retina is the inner layer of the eyeball. Rod cells and cone cells, located at the back of the retina nearest the choroid, synapse with bipolar cells, which synapse with ganglion cells. Integration of signals occurs at these synapses; therefore, much processing occurs in bipolar and ganglion cells. Further, notice that many rod cells share one bipolar cell, but cone cells do not. Certain cone cells synapse with only one ganglion cell. Cone cells, in general, distinguish more detail than do rod cells. **b.** Micrograph shows that the sclera and choroid are relatively thin compared to the retina, which has several layers of cells.

Integration of Visual Signals in the Retina

The retina has three layers of neurons (Fig. 18.9). The layer closest to the choroid contains the rod cells and cone cells; the middle layer contains bipolar cells; and the innermost layer contains ganglion cells, whose sensory fibers become the optic nerve. Only the rod cells and the cone cells are sensitive to light, and therefore light must penetrate to the back of the retina before they are stimulated.

The rod cells and the cone cells synapse with the bipolar cells, which in turn synapse with ganglion cells that initiate nerve impulses. Notice in Figure 18.9 that there are many more rod cells and cone cells than ganglion cells. In fact, the retina has as many as 150 million rod cells and 6 million cone cells but only one million ganglion cells. The sensitivity of cones versus rods is mirrored by how directly they connect to ganglion cells. As many as 100 rods may synapse with the same ganglion cell. No wonder stimulation of rods results in vision that is blurred and indistinct. In contrast, some cone cells in the fovea centralis synapse with only one ganglion cell. This explains why cones, especially in the fovea, provide us with a sharper, more detailed image of an object.

As signals pass to bipolar cells and ganglion cells, integration occurs. Each ganglion cell receives signals from rod cells covering about one square millimeter of retina (about the size of a thumbtack hole). This region is the ganglion cell's receptive field. Some time ago, scientists discovered that a ganglion cell is stimulated only by messages received from the center of its receptive field; otherwise, it is inhibited. If all the rod cells in the receptive field receive light, the ganglion cell responds in a neutral way—that is, it reacts only weakly or perhaps not at all. This supports the hypothesis that considerable processing occurs in the retina before nerve impulses are sent to the brain. Additional integration occurs in the visual areas of the cerebral cortex.

Synaptic integration and processing begin in the retina before nerve impulses are sent to the brain.

Blind Spot

Figure 18.9 provides an opportunity to point out that there are no rods and cones where the optic nerve exits the retina. Therefore, no vision is possible in this area. You can prove this to yourself by putting a dot to the right of center on a piece of paper. Use your right hand to move the paper slowly toward your right eye while you look straight ahead. The dot will disappear at one point—this is your **blind spot.**

Integration of Visual Signals in the Brain

Sensory fibers from ganglion cells assemble to form the optic nerves. At the X-shaped optic chiasma, fibers from the right half of each retina converge and continue on together in the right optic tract, and fibers from the left half of each retina converge and continue on together in the left optic tract.

Notice in Figure 18.10 that the image is split because the left optic tract carries information about the right portion of the **visual field,** and the right optic tract carries information about the left portion of the visual field.

The optic tracts sweep around the hypothalamus, and most fibers synapse with neurons in nuclei (masses of neuron cell bodies) of the thalamus. Axons from the thalamic nuclei form optic radiations that take nerve impulses to the primary visual areas of the cerebral cortex. Since each primary visual area receives information regarding only half the visual field, these areas must eventually share information to form a unified image. Also, the inverted and reversed image (see also Figure 18.7) must be righted in the brain for us to correctly perceive the visual field.

The most surprising finding has been that the brain has a further way of taking the field apart. Each primary visual area of the cerebral cortex acts like a post office, parceling out information regarding color, form, motion, and possibly other attributes to different portions of the adjoining visual association area. Therefore, the brain has taken the field apart even though we see a unified visual field. The cerebral cortex is believed to rebuild the visual field and give us an understanding of it at the same time.

The visual pathway begins in the retina and passes through the thalamus before reaching the cerebral cortex. The pathway and the visual cortex take the visual field apart, but then the cortex rebuilds it so that we correctly perceive the field.

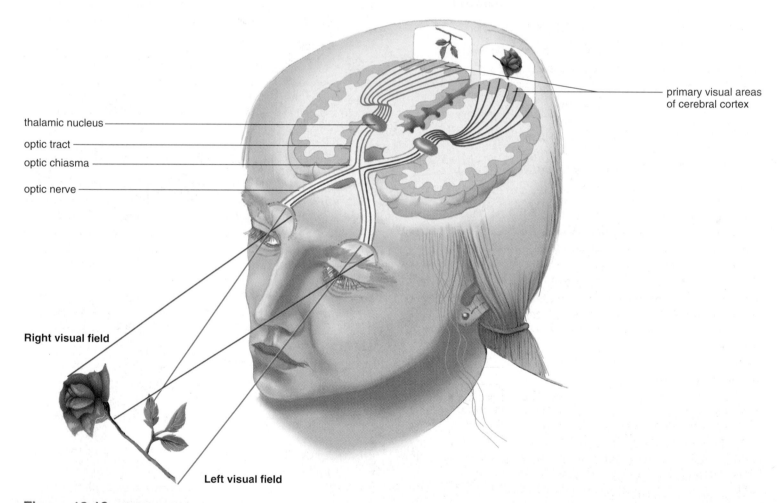

Figure 18.10   Optic chiasma.
Both eyes "see" the entire visual field. Because of the optic chiasma, data from the right half of each retina go to the right visual area of the cerebral cortex, and data from the left half of the retina go to the left visual area of the cerebral cortex. These data are then combined to allow us to see the entire visual field. Note that the visual pathway to the brain includes the thalamus, which has the ability to filter sensory stimuli.

Abnormalities of the Eye

Color blindness and misshapen eyeballs are two common abnormalities of the eye. More serious abnormalities are discussed in the Health Focus on page 356.

Color Blindness

Complete color blindness is extremely rare. In most instances, a particular type of cone is lacking or deficient in number. The most common mutation is a lack of red or green cones. This abnormality affects 5–8% of the male population. If the eye lacks red cones, the green colors are accentuated, and vice versa.

Distance Vision

The majority of people can see what is designated as a size 20 letter 20 feet away, and so are said to have 20/20 vision. Persons who can see close objects but cannot see the letters from this distance are said to be nearsighted. **Nearsighted** people can see close objects better than they can see objects at a distance. These individuals have an elongated eyeball, and when they attempt to look at a distant object, the image is brought to focus in front of the retina (Fig. 18.11*a*). They can see close objects because the lens can compensate for the long eyeball. To see distant objects, these people can wear concave lenses, which diverge the light rays so that the image focuses on the retina.

Rather than wear glasses or contact lenses, many nearsighted people are now choosing to undergo laser surgery. First, specialists determine how much the cornea needs to be flattened to achieve visual acuity. Controlled by a computer, the laser then removes this amount of the cornea. Most patients achieve at least 20/40 vision, but a few complain of glare and varying visual acuity.

Persons who can easily see the optometrist's chart but cannot see close objects well are **farsighted;** these individuals can see distant objects better than they can see close objects. They have a shortened eyeball, and when they try to see close objects, the image is focused behind the retina. When the object is distant, the lens can compensate for the short eyeball. When the object is close, these persons must wear

convex lenses to increase the bending of light rays so that the image can be focused on the retina.

When the cornea or lens is uneven, the image is fuzzy. The light rays cannot be evenly focused on the retina. This condition, called **astigmatism,** can be corrected by an unevenly ground lens to compensate for the uneven cornea (Fig. 18.11*c*).

The shape of the eyeball determines the need for corrective lenses.

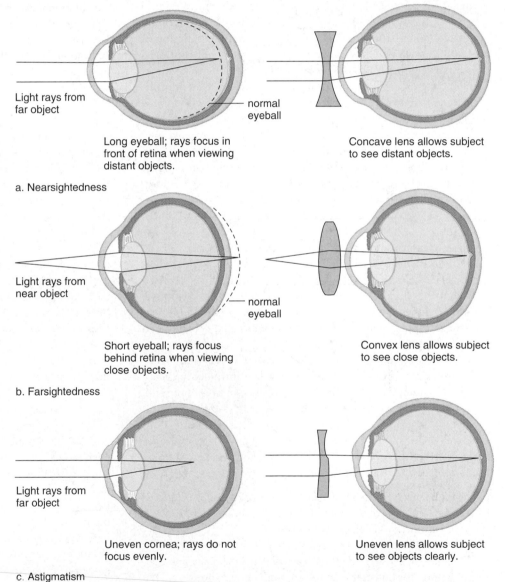

Light rays from far object — normal eyeball

Long eyeball; rays focus in front of retina when viewing distant objects.

Concave lens allows subject to see distant objects.

a. Nearsightedness

Light rays from near object — normal eyeball

Short eyeball; rays focus behind retina when viewing close objects.

Convex lens allows subject to see close objects.

b. Farsightedness

Light rays from far object

Uneven cornea; rays do not focus evenly.

Uneven lens allows subject to see objects clearly.

c. Astigmatism

Figure 18.11   Common abnormalities of the eye, with possible corrective lenses.
a. A concave lens in nearsighted persons focuses light rays on the retina. **b.** A convex lens in farsighted persons focuses light rays on the retina. **c.** An uneven lens in persons with astigmatism focuses light rays on the retina.

Health Focus

Protecting Vision and Hearing

Preventing a Loss of Vision

The eye is subject to both injuries and disorders. Although flying objects sometimes penetrate the cornea and damage the iris, lens, or retina, careless use of contact lenses is the most common cause of injuries to the eye. However, injuries cause only 4% of all cases of blindness.

The most frequent causes of blindness are retinal disorders, glaucoma, and cataracts, in that order. Retinal disorders are varied. In diabetic retinopathy, which blinds many people between the ages of 20 and 74, capillaries to the retina burst, and blood spills into the vitreous fluid. Careful regulation of blood glucose levels in these patients may be protective. In macular degeneration, the cones are destroyed because thickened choroid vessels no longer function as they should. Glaucoma occurs when the drainage system of the eyes fails, so that fluid builds up and destroys the nerve fibers responsible for peripheral vision. Eye doctors always check for glaucoma, but it is advisable to be aware of the disorder in case it comes on quickly. Those who have experienced acute glaucoma report that the eyeball feels as heavy as a stone. In cataracts, cloudy spots on the lens of the eye eventually pervade the whole lens. The milky, yellow-white lens scatters incoming light and blocks vision.

Regular visits to an eye-care specialist, especially by the elderly, are a necessity in order to catch conditions such as glaucoma early enough to allow effective treatment. It has not been proven that consuming particular foods or vitamin supplements reduces the risk of cataracts. Even so, it's possible that estrogen replacement therapy may be somewhat protective in postmenopausal women.

Accumulating evidence suggests that both macular degeneration and cataracts, which tend to occur in the elderly, are caused by long-term exposure to the ultraviolet rays of the sun. It is recommended, therefore, that everyone, especially those who live in sunny climates or work outdoors, wear sunglasses that absorb ultraviolet light. Large lenses worn close to the eyes offer further protection. The Sunglass Association of America has devised the following system for categorizing sunglasses:

- Cosmetic lenses absorb at least 70% of UV-B, 20% of UV-A, and 60% of visible light. Such lenses are worn for comfort rather than protection.
- General-purpose lenses absorb at least 95% of UV-B, 60% of UV-A, and 60–92% of visible light. They are good for outdoor activities in temperate regions.
- Special-purpose lenses block at least 99% of UV-B, 60% of UV-A, and 20–97% of visible light. They are good for bright sun combined with sand, snow, or water.

Health-care providers have found an increased incidence of cataracts in heavy cigarette smokers. The risk of cataracts doubles in men who smoke 20 cigarettes or more a day, and in women who smoke 35 cigarettes or more a day. A possible reason is that smoking reduces the delivery of blood, and therefore nutrients, to the lens.

Preventing a Loss of Hearing

Especially when we are young, the middle ear is subject to infections that can lead to hearing impairment if not treated promptly by a physician. The mobility of ossicles decreases with age, and in otosclerosis, new filamentous bone grows over the stirrup, impeding its movement. Surgical treatment is the only remedy for this type of conduction deafness. However, age-associated nerve deafness due to stereocilia damage from exposure to loud noises is preventable. Hospitals are now aware that even the ears of the newborn need to be protected from noise, and are taking steps to make sure neonatal intensive care units and nurseries are as quiet as possible.

In today's society, exposure to the types of noises listed in Table 18A is common. Noise is measured in decibels, and any noise above a level of 80 decibels could result in damage to the hair cells of the organ of Corti. Eventually, the stereocilia and then the hair cells disappear completely (Fig. 18A). If listening to city traffic for extended periods can damage hearing, it stands to reason that frequent attendance at rock concerts, constantly playing a stereo loudly, or using earphones at high volume is also damaging to hearing. The first hint of danger could be temporary hearing loss, a "full" feeling in the ears, muffled hearing, or tinnitus (e.g., ringing in the ears). If you have any of these symptoms, modify your listening habits immediately to prevent further damage. If exposure to noise is unavoidable, specially designed noise-reduction earmuffs are available, and it is also possible to purchase earplugs made from a compressible, spongelike material at the drugstore or sporting-goods store. These earplugs are not the same as those worn for swimming, and they should not be used interchangeably.

Aside from loud music, noisy indoor or outdoor equipment, such as a rug-cleaning machine or a chain saw, is also troublesome. Even motorcycles and recreational vehicles such as snowmobiles and motocross bikes can contribute to a gradual loss of hearing. Exposure to intense sounds of short duration, such as a burst of gunfire, can result in an immediate hearing loss. Hunters may have a significant hearing reduction in the ear opposite the shoulder where the gun is carried. The butt of the rifle offers some protection to the ear nearest the gun when it is shot.

Finally, people need to be aware that some medicines are ototoxic. Anticancer drugs, most notably cisplatin, and certain antibiotics (e.g., streptomycin, kanamycin, and gentamicin) make ears especially susceptible to a hearing loss. Anyone taking such medications needs to be especially careful to protect the ears from loud noises.

Table 18A Noises That Affect Hearing

Type of Noise	Sound Level (decibels)	Effect
"Boom car," jet engine, shotgun, rock concert	Over 125	Beyond threshold of pain; potential for hearing loss high
Discotheque, "boom box," thunderclap	Over 120	Hearing loss likely
Chain saw, pneumatic drill, jackhammer, symphony orchestra, snowmobile, garbage truck, cement mixer	100–200	Regular exposure of more than one minute risks permanent hearing loss
Farm tractor, newspaper press, subway, motorcycle	90–100	Fifteen minutes of unprotected exposure potentially harmful
Lawn mower, food blender	85–90	Continuous daily exposure for more than eight hours can cause hearing damage
Diesel truck, average city traffic noise	80–85	Annoying; constant exposure may cause hearing damage

Source: National Institute on Deafness and Other Communication Disorders, January 1990, National Institute of Health.

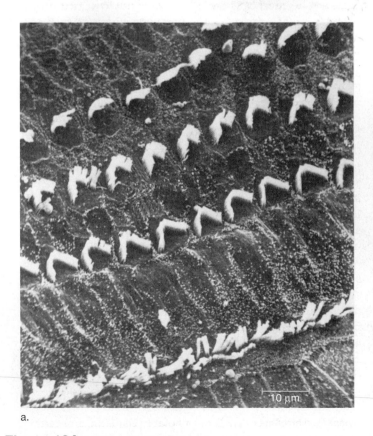

a.

b.

Figure 18A Hair cell damage.
a. Normal hair cells in the spiral organ of a guinea pig. **b.** Damaged cells. This damage occurred after 24-hour exposure to a noise level equivalent to that at a heavy-metal rock concert (see Table 18A). Hearing is permanently impaired because lost cells will not be replaced, and damaged cells may also die.

18.5 Sense of Hearing

The ear has two sensory functions: hearing and balance (equilibrium). The sensory receptors for both of these are located in the inner ear, and each consists of **hair cells** with stereocilia (long microvilli) that are sensitive to mechanical stimulation. They are mechanoreceptors.

Anatomy of the Ear

Figure 18.12 shows that the ear has three divisions: outer, middle, and inner. The **outer ear** consists of the pinna (external flap) and the auditory canal. The opening of the auditory canal is lined with fine hairs and sweat glands. Modified sweat glands are located in the upper wall of the canal; they secrete earwax, a substance that helps guard the ear against the entrance of foreign materials, such as air pollutants.

The **middle ear** begins at the **tympanic membrane** (eardrum) and ends at a bony wall containing two small openings covered by membranes. These openings are called the **oval window** and the **round window.** Three small bones are found between the tympanic membrane and the oval window. Collectively called the **ossicles,** individually they are the **malleus** (hammer), the **incus** (anvil), and the **stapes** (stirrup) because their shapes resemble these objects. The malleus adheres to the tympanic membrane, and the stapes touches the oval window. An **auditory tube** (eustachian tube), which extends from each middle ear to the nasopharynx, permits equalization of air pressure. Chewing gum, yawning, and swallowing in elevators and airplanes help move air through the auditory tubes upon ascent and descent. As this occurs, we often hear the ears "pop."

Whereas the outer ear and the middle ear contain air, the inner ear is filled with fluid. Anatomically speaking, the **inner ear** has three areas: the **semicircular canals** and the **vestibule** are both concerned with equilibrium; the **cochlea** is concerned with hearing. The cochlea resembles the shell of a snail because it spirals.

Process of Hearing

The process of hearing begins when sound waves enter the auditory canal. Just as ripples travel across the surface of a pond, sound waves travel by the successive vibrations of molecules. Ordinarily, sound waves do not carry much energy, but when a large number of waves strike the tympanic membrane, it moves back and forth (vibrates) ever so slightly. The malleus then takes the pressure from the inner surface of the tympanic membrane and passes it by means of the incus to the stapes in such a way that the pressure is multiplied about 20 times as it moves. The stapes strikes the membrane of the oval window, causing it to vibrate, and in this way, the pressure is passed to the fluid within the cochlea.

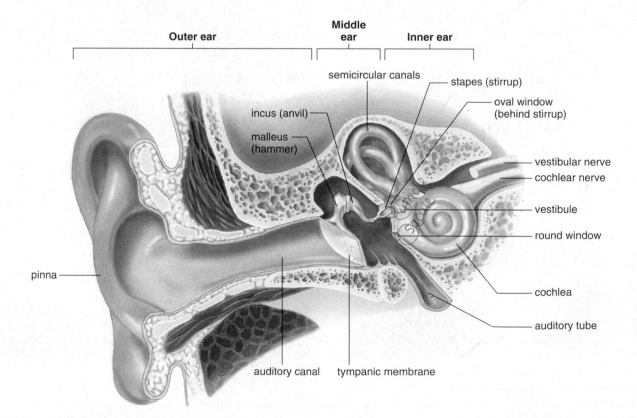

Figure 18.12   Anatomy of the human ear.

In the middle ear, the malleus (hammer), the incus (anvil), and the stapes (stirrup) amplify sound waves. In the inner ear, the mechanoreceptors for equilibrium are in the semicircular canals and the vestibule, and the mechanoreceptors for hearing are in the cochlea.

If the cochlea is unwound and examined in cross section (Fig. 18.13), you can see that it has three canals: the vestibular canal, the **cochlear canal,** and the tympanic canal. Along the length of the basilar membrane, which forms the lower wall of the cochlear canal, are little hair cells whose stereocilia are embedded within a gelatinous material called the **tectorial membrane.** The hair cells of the cochlear canal, called the spiral organ (organ of Corti), synapse with nerve fibers of the **cochlear nerve** (auditory nerve).

When the stapes strikes the membrane of the oval window, pressure waves move from the vestibular canal to the tympanic canal across the basilar membrane, and the round window membrane bulges. The basilar membrane moves up and down, and the stereocilia of the hair cells embedded in the tectorial membrane bend. Then nerve impulses begin in the cochlear nerve and travel to the brain stem. When they reach the auditory areas of the cerebral cortex, they are interpreted as a sound.

Each part of the spiral organ is sensitive to different wave frequencies, or pitch. Near the tip, the spiral organ responds to low pitches, such as a tuba, and near the base,

it responds to higher pitches, such as a bell or a whistle. The nerve fibers from each region along the length of the spiral organ lead to slightly different areas in the brain. The pitch sensation we experience depends upon which region of the basilar membrane vibrates and which area of the brain is stimulated.

Volume is a function of the amplitude of sound waves. Loud noises cause the fluid within the vestibular canal to exert more pressure and the basilar membrane to vibrate to a greater extent. The resulting increased stimulation is interpreted by the brain as volume. It is believed that the brain interprets the tone of a sound based on the distribution of the hair cells stimulated.

The mechanoreceptors for sound are hair cells on the basilar membrane (the spiral organ). When the basilar membrane vibrates, the stereocilia of the hair cells bend, and nerve impulses are transmitted to the brain.

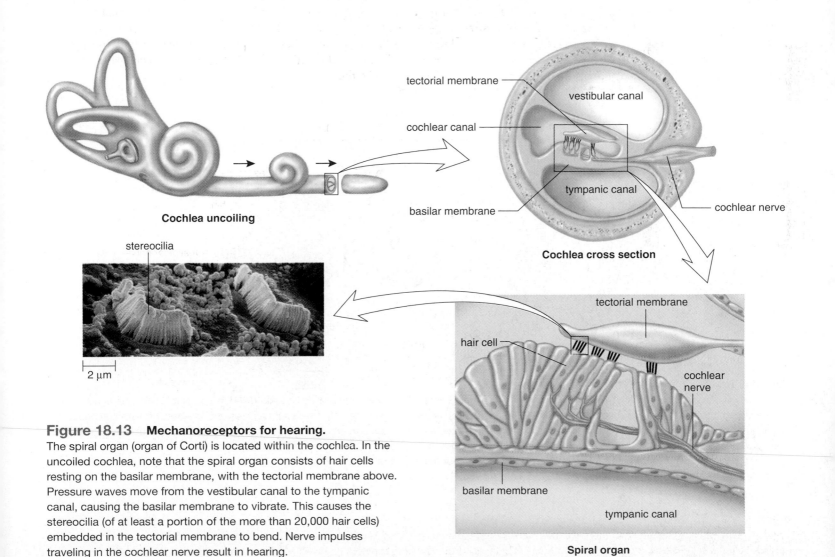

Cochlea uncoiling

Cochlea cross section

tectorial membrane

vestibular canal

cochlear canal

basilar membrane

tympanic canal

cochlear nerve

stereocilia

2 μm

tectorial membrane

hair cell

cochlear nerve

basilar membrane

tympanic canal

Spiral organ

Figure 18.13   Mechanoreceptors for hearing.
The spiral organ (organ of Corti) is located within the cochlea. In the uncoiled cochlea, note that the spiral organ consists of hair cells resting on the basilar membrane, with the tectorial membrane above. Pressure waves move from the vestibular canal to the tympanic canal, causing the basilar membrane to vibrate. This causes the stereocilia (of at least a portion of the more than 20,000 hair cells) embedded in the tectorial membrane to bend. Nerve impulses traveling in the cochlear nerve result in hearing.

Visual Focus

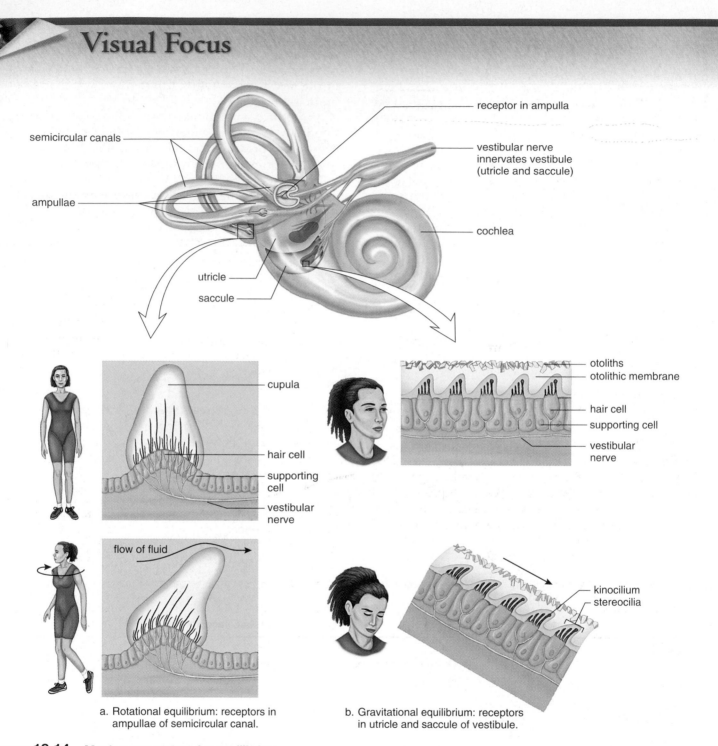

a. Rotational equilibrium: receptors in
   ampullae of semicircular canal.

b. Gravitational equilibrium: receptors
   in utricle and saccule of vestibule.

Figure 18.14   Mechanoreceptors for equilibrium.

a. Rotational equilibrium. The ampullae of the semicircular canals contain hair cells with stereocilia embedded in a cupula. When the head rotates, the cupula is displaced, bending the stereocilia. Thereafter, nerve impulses travel in the vestibular nerve to the brain. **b.** Gravitational equilibrium. The utricle and the saccule contain hair cells with stereocilia embedded in an otolithic membrane. When the head bends, otoliths are displaced, causing the membrane to sag and the stereocilia to bend. If the stereocilia bend toward the kinocilium, the longest of the stereocilia, nerve impulses increase in the vestibular nerve. If the stereocilia bend away from the kinocilium, nerve impulses decrease in the vestibular nerve. This difference tells the brain in which direction the head moved.

18.6 Sense of Equilibrium

Mechanoreceptors in the semicircular canals detect rotational and/or angular movement of the head **(rotational equilibrium)**, while mechanoreceptors in the utricle and saccule detect movement of the head in the vertical or horizontal planes **(gravitational equilibrium)** (Fig. 18.14).

These mechanoreceptors help achieve equilibrium, but other structures in the body are also involved. For example, we already mentioned that proprioceptors are necessary for maintaining our equilibrium. Vision, if available, is also extremely helpful.

Rotational Equilibrium

Rotational equilibrium involves the semicircular canals, which are arranged so that there is one in each dimension of space. The base of each of the three canals, called the **ampulla,** is slightly enlarged. Little hair cells, whose stereocilia are embedded within a gelatinous material called a cupula, are found within the ampullae. Because there are three semicircular canals, each ampulla responds to head rotation in a different plane of space. As fluid within a semicircular canal flows over and displaces a cupula, the stereocilia of the hair cells bend, and the pattern of impulses carried by the vestibular nerve to the brain changes. Continuous movement of fluid in the semicircular canals causes one form of motion sickness.

Vertigo is dizziness and a sensation of rotation. It is possible to simulate a feeling of vertigo by spinning rapidly and stopping suddenly. When the eyes are rapidly jerked back to a midline position, the person feels like the room is spinning. This shows that the eyes are also involved in our sense of equilibrium.

Gravitational Equilibrium

Gravitational equilibrium depends on the **utricle** and **saccule,** two membranous sacs located in the vestibule. Both of these sacs contain little hair cells, whose stereocilia are embedded within a gelatinous material called an otolithic membrane. Calcium carbonate ($CaCO_3$) granules, or **otoliths,** rest on this membrane. The utricle is especially sensitive to horizontal (back-forth) movements and the bending of the head, while the saccule responds best to vertical (up-down) movements.

When the body is still, the otoliths in the utricle and the saccule rest on the otolithic membrane above the hair cells. When the head bends or the body moves in the horizontal and vertical planes, the otoliths are displaced and the otolithic membrane sags, bending the stereocilia of the hair cells beneath. If the stereocilia move toward the largest stereocilium, called the kinocilium, nerve impulses increase in the vestibular nerve. If the stereocilia move away from the kinocilium, nerve impulses decrease in the vestibular nerve. If you are upside down, nerve impulses in the vestibular nerve cease. These data tell the brain the direction of the movement of the head.

Movement of a cupula within the semicircular canals contributes to the sense of rotational equilibrium. Movement of the otolithic membrane within the utricle and the saccule accounts for gravitational equilibrium.

Table 18.3 reviews the functions of the parts of the ear for easy reference.

Table 18.3	Functions of the Parts of the Ear		
Part	**Medium**	**Function**	**Mechanoreceptor**
Outer Ear	Air		
Pinna		Collects sound waves	—
Auditory canal		Filters air	—
Middle Ear	Air		
Tympanic membrane and ossicles		Amplify sound waves	—
Auditory tube		Equalizes air pressure	—
Inner Ear	Fluid		
Semicircular canals		Rotational equilibrium	Stereocilia embedded in cupula
Vestibule (contains utricle and saccule)		Gravitational equilibrium	Stereocilia embedded in otolithic membrane
Cochlea (spiral organ)		Hearing	Stereocilia embedded in tectorial membrane

Bioethical Focus   Personal Responsibility for Health

A cataract is a cloudiness of the lens that usually occurs in the elderly. A dense centrally placed cataract causes severe blurring of vision. Certain factors have been identified as contributing to the chances of having a cataract:

- Smoking 20 or more cigarettes a day doubles the risk of cataracts in men—smoking more than 30 cigarettes a day in women increases the chances of a cataract.
- Exposure to the ultraviolet radiation in sunlight can more than double the risk of a cataract.

- Also, as many as one-third of cataracts may be caused by being overweight. Diet, rather than exercise, seems to reduce cataract formation.

Should we be responsible in our actions and take all possible steps to prevent developing a cataract, such as not smoking, wearing sunglasses and a wide-brimmed hat, and watching our weight? Or should we simply rely on medical science to restore our eyesight? Most cataract operations today are performed on an out-

patient basis with minimal postoperative discomfort and with a high expectation of restoration of sight. Perhaps it's better, though, to take all possible steps to prevent the occurrence of cataracts, just in case our experience is atypical?

Decide Your Opinion

1. To what extent do you feel each person is responsible for his or her own health?
2. Should Medicare pay for cataract surgery in heavy smokers?
3. At what age should we make people aware that their behavior today can affect their health tomorrow?

Summarizing the Concepts

18.1 Sensory Receptors and Sensations
Each type of sensory receptor detects a particular kind of stimulus. When stimulation occurs, sensory receptors initiate nerve impulses that are transmitted to the spinal cord and/or brain. Sensation occurs when nerve impulses reach the cerebral cortex. Perception is an interpretation of the meaning of sensations.

18.2 Proprioceptors and Cutaneous Receptors
Proprioception is illustrated by the action of muscle spindles that are stimulated when muscle fibers stretch. A reflex action, which is illustrated by the knee reflex, causes the muscle fibers to contract. Proprioception helps maintain equilibrium and posture.

The skin contains sensory receptors, called cutaneous receptors, for touch, pressure, pain, and temperature (warmth and cold). The pain of internal organs is sometimes felt in the skin and is called referred pain.

18.3 Chemical Senses
Taste and smell are due to chemoreceptors that are stimulated by molecules in the environment. The taste buds contain taste cells that communicate with sensory fibers, while the chemoreceptors for smell are neurons.

After molecules bind to plasma membrane receptor proteins on the microvilli of taste cells and the cilia of olfactory cells, nerve impulses eventually reach the cerebral cortex, which determines the taste and odor according to the pattern of stimulation.

18.4 Sense of Vision
Vision is dependent on the eye, the optic nerves, and the visual areas of the cerebral cortex. The eye has three layers. The outer layer, the sclera, can be seen as the white of the eye; it also becomes the transparent bulge in the front of the eye called the cornea. The middle pigmented layer, called the choroid, absorbs stray light rays. The rod cells (sensory receptors for dim light) and the cone cells (sensory receptors for bright light and color) are located in the retina, the inner layer of the eyeball. The cornea, the humors, and especially the lens bring the light rays to

focus on the retina. To see a close object, accommodation occurs as the lens rounds up.

When light strikes rhodopsin within the membranous disks of rod cells, rhodopsin splits into opsin and retinal. A cascade of reactions leads to the closing of ion channels in a rod cell's plasma membrane. Inhibitory transmitter molecules are no longer released, and nerve impulses are carried in the optic nerve to the brain.

Integration occurs in the retina, which is composed of three layers of cells: the rod and cone layer, the bipolar cell layer, and the ganglion cell layer. Integration also occurs in the brain. The visual field is taken apart by the optic chiasma, and the primary visual area in the cerebral cortex, which parcels out signals for color, form, and motion to the visual association area. Then the cortex rebuilds the field.

18.5 Sense of Hearing
Hearing is dependent on the ear, the cochlear nerve, and the auditory areas of the cerebral cortex.

The ear is divided into three parts: outer, middle, and inner. The outer ear consists of the pinna and the auditory canal, which direct sound waves to the middle ear. The middle ear begins with the tympanic membrane and contains the ossicles (malleus, incus, and stapes). The malleus is attached to the tympanic membrane, and the stapes is attached to the oval window, which is covered by a membrane. The inner ear contains the cochlea and the semicircular canals, plus the utricle and the saccule.

Hearing begins when the outer ear receives and the middle ear amplifies the sound waves that then strike the oval window membrane. Its vibrations set up pressure waves across the cochlear canal, which contains the spiral organ, consisting of hair cells whose stereocilia are embedded within the tectorial membrane. When the basilar membrane vibrates, the stereocilia of the hair cells bend. Nerve impulses begin in the cochlear nerve and are carried to the brain.

18.6 Sense of Equilibrium
The ear also contains mechanoreceptors for our sense of equilibrium. Rotational equilibrium is dependent on the stimulation of hair cells within the ampullae of the semicircular canals. Gravitational equilibrium relies on the stimulation of hair cells within the utricle and the saccule.

Testing Yourself

Choose the best answer for each question.

1. A receptor
   a. is the first portion of a reflex arc.
   b. initiates nerve impulses.
   c. can be internal or external.
   d. All of these are correct.

2. Which of these is NOT a proper contrast between olfactory receptors and equilibrium receptors?

Olfactory receptors	Equilibrium receptors
a. located in nasal cavities	located in the inner ear
b. chemoreceptors	mechanoreceptors
c. respond to molecules in air	respond to movements of the body
d. communicate with brain via a tract	communicate with brain via vestibular nerve
e. All of these contrasts are correct.	

3. Which of these is NOT a proper contrast between proprioceptors and cutaneous receptors?

Proprioceptors	Cutaneous receptors
a. located in muscles and tendons	located in the skin
b. mechanoreceptors	chemoreceptors
c. respond to tension	respond to pain, hot, cold touch, pressure
d. type of somatic sense	type of special sense
e. Both b and d are not correct.	

4. Which of these gives the correct path for light rays entering the human eye?
   a. sclera, retina, choroid, lens, cornea
   b. fovea centralis, pupil, aqueous humor, lens
   c. cornea, pupil, lens, vitreous humor, retina
   d. cornea, fovea centralis, lens, choroid, rods
   e. optic nerve, sclera, choroid, retina, humors

5. Which pair gives an incorrect function for the structure?
   a. lens—focusing
   b. cones—color vision
   c. iris—regulation of amount of light
   d. choroid—location of cones
   e. sclera—protection

6. Which one of these wouldn't you mention if you were tracing the path of sound vibrations?
   a. auditory canal
   b. tympanic membrane
   c. ossicles
   d. semicircular canals
   e. cochlea

7. Which one of these correctly describes the location of the spiral organ?
   a. between the tympanic membrane and the oval window in the inner ear
   b. in the utricle and saccule within the vestibule
   c. between the tectorial membrane and the basilar membrane in the cochlear canal
   d. between the nasal cavities and the throat
   e. between the outer and inner ear within the semicircular canals

8. Which of these pairs is mismatched?
   a. semicircular canals—inner ear
   b. utricle and saccule—outer ear
   c. auditory canal—outer ear
   d. cochlea—inner ear
   e. ossicles—middle ear

9. Retinal is
   a. a derivative of vitamin A.
   b. sensitive to light energy.
   c. a part of rhodopsin.
   d. found in both rods and cones.
   e. All of these are correct.

10. Both olfactory receptors and sound receptors have cilia, and they both
    a. are chemoreceptors.
    b. are a part of the brain.
    c. are mechanoreceptors.
    d. initiate nerve impulses.
    e. All of these are correct.

11. In order to focus on objects that are close to the viewer,
    a. the suspensory ligaments must be pulled tight.
    b. the lens needs to become more rounded.
    c. the ciliary muscle will be relaxed.
    d. the image must focus on the area of the optic nerve.

12. Which abnormality of the eye is mismatched?
    a. astigmatism—either the lens or cornea is not even
    b. farsightedness—eyeball is shorter than usual
    c. nearsightedness—image focuses behind the retina
    d. color blindness—genetic disorder in which certain types of cones may be missing

13. Which of the following would allow you to know that you were upside down, even if you were in total darkness?
    a. utricle and saccule
    b. cochlea
    c. semicircular canals
    d. tectorial membrane

14. Which of the following could result in hearing loss?
    a. certain antibiotics
    b. earphone use
    c. consistent use of loud equipment such as a jackhammer
    d. use of firearms
    e. All of these are correct.

15. Label this diagram of an eye. State a function for each structure labeled.

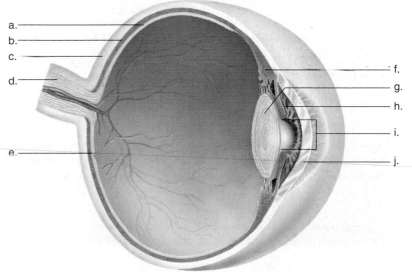

a.
b.
c.
d.
e.
f.
g.
h.
i.
j.

19.1 Anatomy and Physiology of Bones

The bones provide attachment sites for the muscles, whose contraction makes the bones move so that we can work out, play tennis, type papers, and do all manner of activities. The bones also support and protect. For example, the large heavy bones of the legs support the entire body against the pull of gravity; the skull protects our brain; and the rib cage protects the heart and lungs. The organs of the skeleton are largely composed of connective tissues. Connective tissue, including bone and cartilage, contains cells separated by a matrix that contains fibers.

Structure of Bone

Bones are strong because their matrix contains mineral salts, notably calcium phosphate. **Compact bone** is highly organized and composed of tubular units called **osteons.** Bone cells called **osteocytes** lie in lacunae, or tiny chambers, arranged in concentric circles around a central canal (Fig. 19.1). The mineralized matrix, which also contains collagen fibers, fills the spaces between the lacunae. Central canals contain blood vessels, lymphatic vessels, and nerves. Tiny canals called canaliculi run through the matrix, connecting the lacunae with each other and with the central canal. The canaliculi contain extensions of osteocytes surrounded by tissue fluid. In this way, the canaliculi bring nutrients from the blood vessel in the central canal to the cells in the lacunae.

Compared to compact bone, **spongy bone** has an unorganized appearance (Fig. 19.1). Osteocytes are found in trabeculae which are numerous thin plates separated by unequal spaces. Although the latter make spongy bone lighter than compact bone, spongy bone is still designed for strength. Just as braces are used for support in buildings, the plates of spongy bone follow lines of stress. The spaces of spongy bone are often filled with **red bone marrow,** a specialized tissue that produces all types of blood cells.

In infants, red bone marrow is found in the cavities of most bones. In adults, red bone marrow occurs in the spongy bone of the skull, ribs, sternum (breastbone), and vertebrae, and in the ends of the long bones.

Tissues Associated with Bones

Cartilage and dense fibrous connective tissue are two types of tissues associated with bones.

Cartilage

Cartilage is not as strong as bone, but it is more flexible because the matrix is gel-like and contains many collagenous and elastic fibers. The cells lie within lacunae that are irregularly grouped. Cartilage has no blood vessels, and therefore, injured cartilage is slow to heal.

All three types of cartilage—hyaline cartilage, fibrocartilage, and elastic cartilage—are associated with bones. Hyaline cartilage is firm and somewhat flexible. The matrix appears uniform and glassy (Fig. 19.1), but actually it contains a generous supply of collagen fibers. Hyaline cartilage is found at the ends of long bones and in the nose, at the ends of the ribs, and in the larynx and trachea.

Fibrocartilage is stronger than hyaline cartilage because the matrix contains wide rows of thick collagen fibers. Fibrocartilage is able to withstand both tension and pressure. This type of cartilage is found where support is of prime importance—in the disks between the vertebrae and in the cartilaginous disks between the bones of the knee.

Elastic cartilage is more flexible than hyaline cartilage because the matrix contains mostly elastin fibers. This type of cartilage is found in the ear flaps and epiglottis.

Dense Fibrous Connective Tissue

Dense fibrous connective tissue contains rows of cells called fibroblasts separated by bundles of collagen fibers. Dense fibrous connective tissue forms the flared sides of the nose. **Ligaments,** which bind bone to bone, are dense fibrous connective tissue, and so are **tendons,** which connect muscles to a bone at **joints.** Joints are also called articulations.

Structure of a Long Bone

Bones are classified by their shape; there are long, short, flat, or irregular bones. Long bones do not have to be terribly long—the bones in your fingers are classified as long. But a long bone must be longer than it is wide. The bones of the arms and legs are long bones except for the knee cap and the wrist and ankle bones.

A long bone of the arm or leg can be used to illustrate general principles of bone anatomy. A bone is enclosed by a tough fibrous, connective tissue covering called the periosteum, which is continuous with the ligaments that go across a joint. The periosteum contains blood vessels that enter the bone and give off branches that service osteocytes located in lacunae. Each expanded end of a long bone is termed an epiphysis; the portion between the epiphyses is called the diaphysis.

When a bone is split open as in Figure 19.1, the section shows that the diaphysis is not solid but has a medullary cavity containing yellow bone marrow. Yellow bone marrow contains a large amount of fat. The medullary cavity is bounded at the sides by compact bone. The epiphyses of a long bone are composed largely of spongy bone. Even in adults the epiphyses contain red bone marrow where all the types of blood cells are formed. Beyond the spongy bone, there is a thin shell of compact bone and, finally, a layer of hyaline cartilage called **articular cartilage** because it occurs at a joint.

Figure 19.1 shows how the spongy bone of an epiphysis is connected to the compact bone of the diaphysis and micrographs illustrate the structure of compact bone and

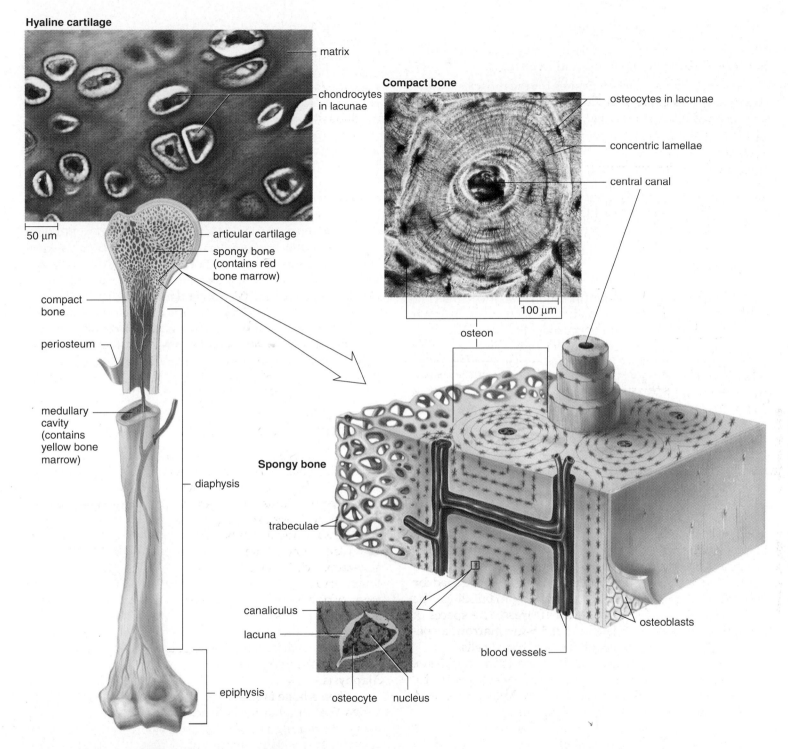

Hyaline cartilage

matrix

chondrocytes in lacunae

50 μm

articular cartilage

spongy bone (contains red bone marrow)

compact bone

periosteum

medullary cavity (contains yellow bone marrow)

diaphysis

epiphysis

Compact bone

osteocytes in lacunae

concentric lamellae

central canal

100 μm

osteon

Spongy bone

trabeculae

canaliculus

lacuna

osteocyte   nucleus

blood vessels

osteoblasts

Figure 19.1   Anatomy of a bone from the macroscopic to the microscopic level.
A long bone is encased by the periosteum, except at the ends, where it is covered by hyaline (articular) cartilage (see micrograph, *top left*).
Spongy bone located in each epiphysis may contain red bone marrow. The medullary cavity contains yellow bone marrow and is bordered by
compact bone, which is shown in the enlargement and micrograph *(right)*.

hyaline cartilage. Notice that the blood vessel coming from
outside the bone penetrates the periosteum and enters the
central canals of compact bone before continuing on into
spongy bone. The next section will illustrate that bone is a
living tissue capable of repairing itself.

Various types of connective tissue, including bone,
cartilage, and fibrous connective tissue, make up
the skeleton. A long bone contains all these tissues.

Bone Growth and Repair

Bones are composed of living tissues as exemplified by their ability to undergo remodeling and to grow.

Remodeling of Bones

In the adult, bone is continually being broken down and built up again. **Osteoclasts,** which are derived from monocytes in red bone marrow, break down bone, remove worn-out cells, and deposit calcium in the blood. After a period of about three weeks, the osteoclasts disappear, and the bone is repaired by the work of osteoblasts. As they form new bone, osteoblasts take calcium from the blood. Eventually, some of these cells get caught in the matrix they secrete and are converted to osteocytes, the cells found within the lacunae of osteons.

Because of continual remodeling, the thickness of bones can change. Physical use and hormone balance affect the thickness of bones. Strange as it may seem, adults apparently require more calcium in the diet (about 1,000 to 1,500 mg daily) than do children in order to promote the work of osteoblasts. Otherwise, osteoporosis, a condition in which weak and thin bones easily fracture, may develop. The likelihood of osteoporosis is greater in older women due to reduced estrogen levels after menopause. Although it is not known how estrogen acts on bone maintenance, it seems to play a role in calcium metabolism.

Bone Development and Growth

The bones of the human skeleton, except those of the skull, first appear during embryonic development as hyaline cartilage. The cartilaginous models are then gradually replaced by bone, a process called endochondral ossification (Fig. 19.2).

During endochondral ossification, the cartilage begins to break down in the center of a long bone, which is now covered by a periosteum. **Osteoblasts** invade the region and begin to lay down spongy bone in what is called a primary ossification center. Other osteoblasts lay down compact bone beneath the periosteum. As the compact bone thickens, the spongy bone is broken down by osteoclasts, and the cavity created becomes the medullary cavity.

The ends of developing bone continue to grow, but soon secondary ossification centers appear in these regions. Here spongy bone forms and does not break down. Also, a band of cartilage called a **growth plate** remains between the primary ossification center and each secondary center. The limbs keep increasing in length as long as the growth plates are still present. The rate of growth is controlled by hormones, such as growth hormones and sex hormones. Eventually, the growth plates become ossified, and the bone stops growing.

Bone is living tissue. It develops, grows, and remodels itself. In all these processes, osteoclasts break down bone, and osteoblasts build bone.

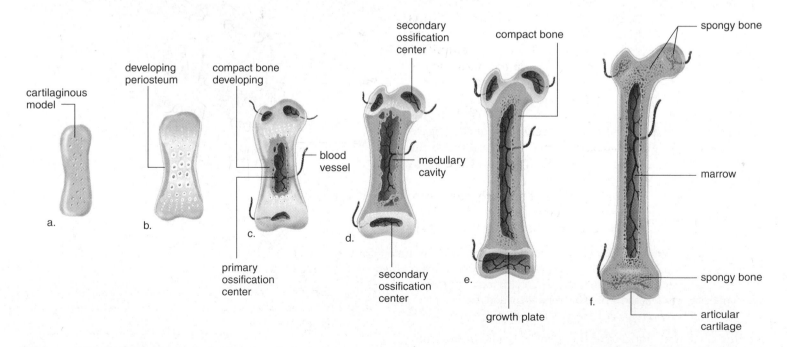

Figure 19.2   Endochondral ossification of a long bone.
a. A cartilaginous model develops during fetal development. **b.** A periosteum develops. **c.** A primary ossification center contains spongy bone surrounded by compact bone. **d.** The medullary cavity forms in the shaft, and secondary ossification centers develop at the ends of a long bone. **e.** Growth is still possible as long as cartilage remains at the growth plates. **f.** When the bone is fully formed, the growth plate disappears.

19.2 Bones of the Skeleton

The functions of the skeleton pertain to particular bones.

The skeleton supports the body. The bones of the lower limbs (the femur in particular and also the tibia) support the entire body when we are standing, and the coxal bones of the pelvic girdle support the abdominal cavity.

The skeleton protects soft body parts. The bones of the skull protect the brain; the rib cage, composed of the ribs, thoracic vertebrae, and sternum, protects the heart and lungs.

The skeleton produces blood cells. All bones in the fetus have spongy bone with red bone marrow that produces blood cells. In the adult, the flat bones of the skull, ribs, sternum, clavicles, and also the vertebrae and pelvis produce blood cells. Fat is stored in yellow bone marrow.

The skeleton stores minerals and fat. All bones have a matrix that contains calcium phosphate. When bones are remodeled, osteoclasts break down bone and return calcium ions and phosphorus ions to the bloodstream.

The skeleton, along with the muscles, permits flexible body movement. While articulations (joints) occur between all the bones, we associate body movement in particular with the bones of the lower limbs (especially the femur and tibia) and the feet (tarsals, metatarsals, and phalanges) because we use them when walking.

Classification of the Bones

The bones are classified according to their shape. Long bones, exemplified by the humerus and femur, are longer than they are wide. Short bones, such as the carpals and tarsals, are cube-shaped—that is, their lengths and widths are about equal. Flat bones, such as those of the skull, are platelike, with broad surfaces. Round bones, exemplified by the patella, are circular in shape. Irregular bones, such as the vertebrae and facial bones, have varied shapes that permit connections with other bones.

The 206 bones of the skeleton are also classified according to whether they occur in the axial skeleton or the appendicular skeleton. The axial skeleton is in the midline of the body, and the appendicular skeleton consists of the limbs along with their girdles (Fig. 19.3).

The bones of the skeleton are not smooth; they have articulating depressions and protuberances at various joints. They also have projections, often called processes, where the muscles attach, as well as openings for nerves and/or blood vessels to pass through.

The skeleton is divided into the axial and appendicular skeletons. Each has different types of bones, with protuberances at joints and processes where the muscles attach.

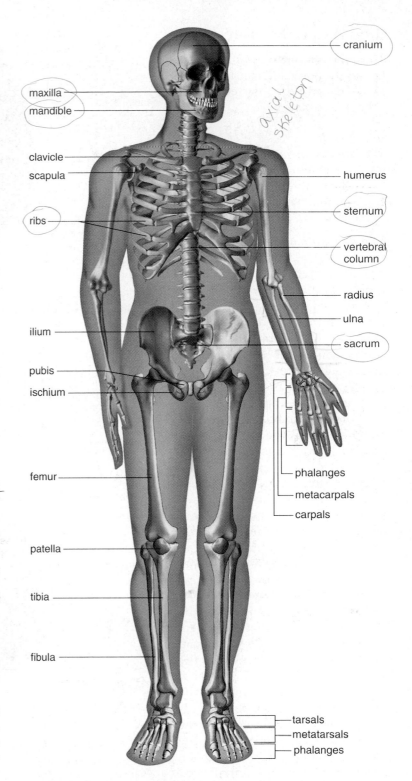

Figure 19.3   The skeleton.
The skeleton of a human adult contains bones that belong to the axial skeleton (red labels) and those that belong to the appendicular skeleton (black labels).

The Axial Skeleton

The **axial skeleton** lies in the midline of the body and consists of the skull, hyoid bone, vertebral column, and rib cage.

The Skull

The **skull** is formed by the cranium (braincase) and the facial bones. It should be noted, however, that some cranial bones also contribute to the face.

The Cranium   The cranium protects the brain. In adults, it is composed of eight flat bones fitted tightly together. In newborns, certain cranial bones are not completely formed and instead are joined by membranous regions called **fontanels.** The fontanels usually close through the age of 16 months by the process of intramembranous ossification.

Some of the bones of the cranium contain the **sinuses,** air spaces lined by mucous membrane. The sinuses reduce the weight of the skull and give a resonant sound to the voice. Two sinuses, called the mastoid sinuses, drain into the middle ear. **Mastoiditis,** a condition that can lead to deafness, is an inflammation of these sinuses.

The major bones of the cranium have the same names as the lobes of the brain: frontal, parietal, occipital, and temporal. On the top of the cranium (Fig. 19.4*a*), the **frontal bone** forms the forehead, the **parietal bones** extend to the sides, and the **occipital bone** curves to form the base of the skull. At the base, there is a large opening, the **foramen magnum** (Fig. 19.4*b*), through which the spinal cord passes and becomes the brain stem. Below the much larger parietal bones, each **temporal bone** has an opening (external auditory canal) that leads to the middle ear.

The **sphenoid bone,** which is shaped like a bat with wings outstretched, extends across the floor of the cranium from one side to the other. The sphenoid is considered the keystone bone of the cranium because all the other bones articulate with it. The sphenoid completes the sides of the skull and also contributes to forming the orbits (eye sockets).

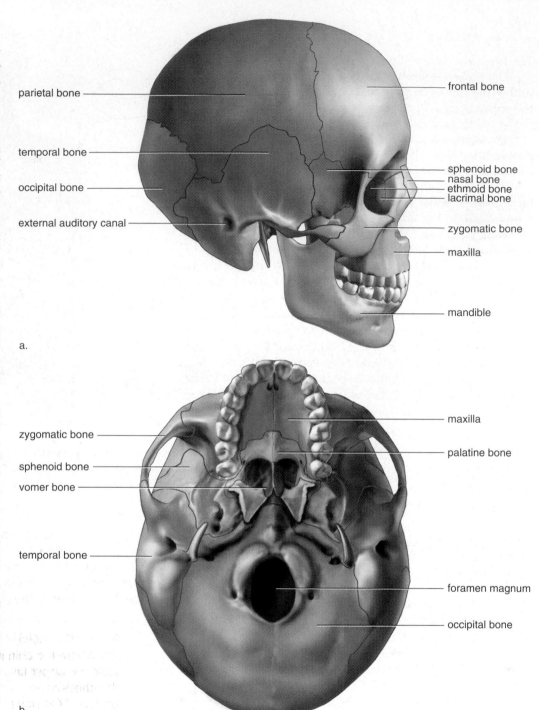

Figure 19.4   Bones of the skull.
a. Lateral view. **b.** Inferior view.

The **ethmoid bone,** which lies in front of the sphenoid, also helps form the orbits and the nasal septum. The orbits are completed by various facial bones. The eye sockets are called orbits because of our ability to rotate the eyes.

The cranium contains eight bones: the frontal, two parietal, the occipital, two temporal, the sphenoid, and the ethmoid.

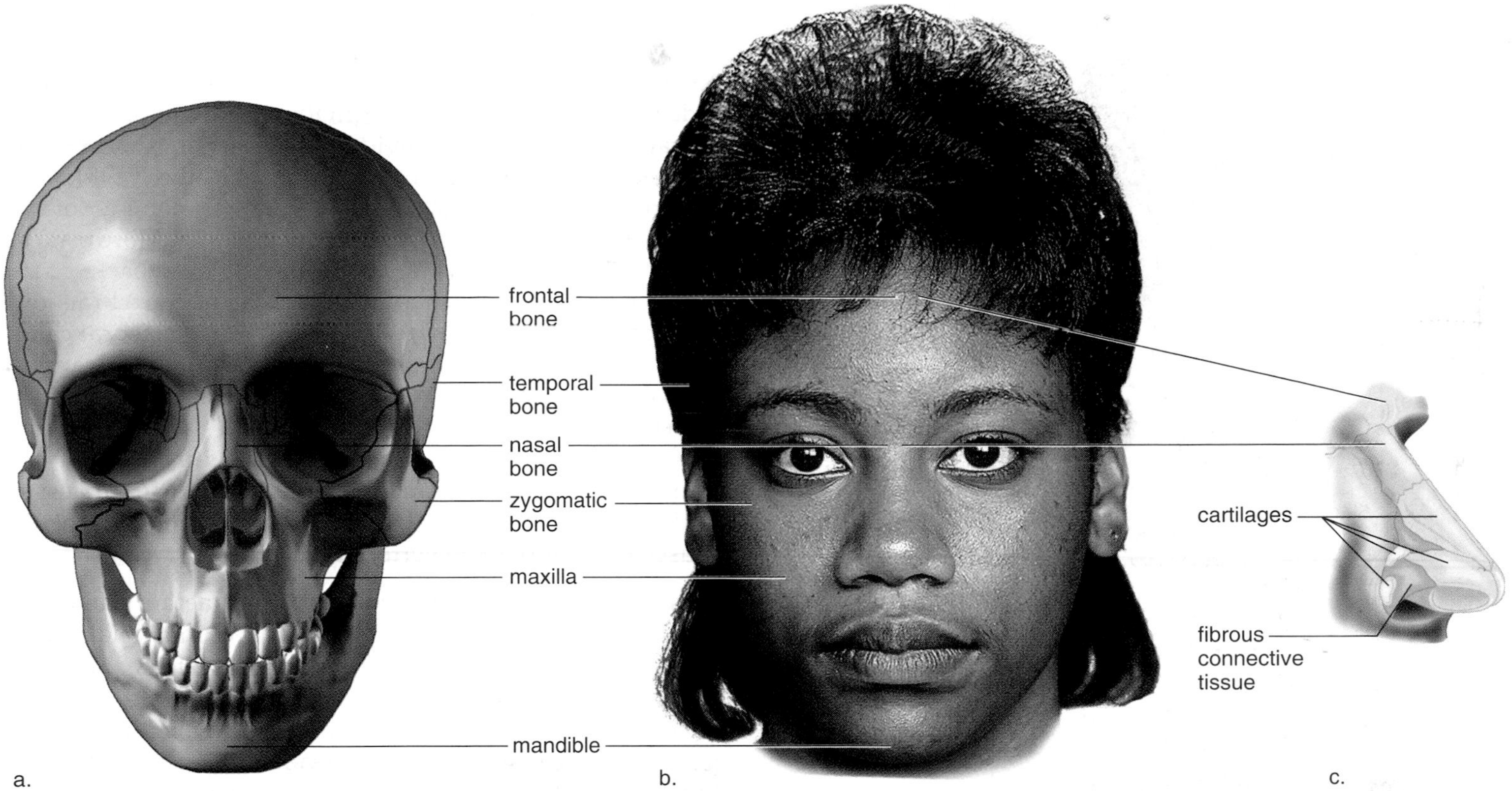

a.   b.   c.

Figure 19.5   Bones of the face, including the nose.
a. The frontal bone forms the forehead and eyebrow ridges; the zygomatic bones form the cheekbones; and the maxillae form the upper jaw. The maxillae are the most expansive of the facial bones. The mandible has a projection we call the chin. **b.** The maxillae, frontal bones, and nasal bones help form the external nose. **c.** The rest of the nose is formed by cartilages and fibrous connective tissue.

The Facial Bones   The most prominent facial bones are the mandible, the maxillae (maxillary bones), the zygomatic bones, and the nasal bones.

The **mandible,** or lower jaw, is the only movable portion of the skull (Fig. 19.5*a*), and its action permits us to chew our food. It also forms the chin. Tooth sockets are located on the mandible and on the **maxillae,** the upper jaw that also forms the anterior portion of the hard palate. The palatine bones make up the posterior portion of the hard palate and the floor of the nose (see Fig. 19.4*b*).

The lips and cheeks have a core of skeletal muscle. The **zygomatic bones** are the cheekbone prominences, and the **nasal bones** form the bridge of the nose. Other bones (e.g., ethmoid and vomer) are a part of the nasal septum, which divides the interior of the nose into two nasal cavities. The lacrimal bone (see Fig. 19.4*a*) contains the opening for the nasolacrimal canal, which brings tears from the eyes to the nose.

The temporal and frontal bones are cranial bones that contribute to the face. The temporal bones account for the flattened areas we call the temples. The frontal bone forms the forehead and has supraorbital ridges where the eyebrows are located. Glasses sit where the frontal bone joins the nasal bones (Fig. 19.5*b*).

While the ears are formed only by elastic cartilage and not by bone, the nose (Fig. 19.5*c*) is a mixture of bones, cartilages, and fibrous connective tissue. The cartilages complete the tip of the nose, and fibrous connective tissue forms the flared sides of the nose.

Among the facial bones, the mandible is the lower jaw where the chin is located, the two maxillae form the upper jaw, the two zygomatic bones are the cheekbones, and the two nasal bones form the bridge of the nose.

The Hyoid Bone

Although the **hyoid bone** is not part of the skull, it will be mentioned here because it is part of the axial skeleton. The hyoid is the only bone in the body that does not articulate with another bone. It is attached to the temporal bones by muscles and ligaments and to the larynx by a membrane. The larynx is the voice box at the top of the trachea in the neck region. The hyoid bone anchors the tongue and serves as the site of attachment for the muscles associated with swallowing.

Calcitonin

Calcium (Ca^{2+}) plays a significant role in both nervous conduction and muscle contraction. It is also necessary to blood clotting. The blood calcium level is regulated in part by **calcitonin,** a hormone secreted by the thyroid gland when the blood calcium level rises (Fig. 20.7). The primary effect of calcitonin is to bring about the deposit of calcium in the bones. It does this by temporarily reducing the activity and number of osteoclasts. When the blood calcium lowers to normal, the release of calcitonin by the thyroid is inhibited, but a low level stimulates the release of **parathyroid hormone (PTH)** by the parathyroid glands. *raise blood calcium*

Parathyroid Glands

Many years ago, the four parathyroid glands were sometimes mistakenly removed during thyroid surgery because they are so small. Parathyroid hormone (PTH), the hormone produced by the **parathyroid glands,** causes the blood phosphate (HPO_4^{2-}) level to decrease and the blood calcium level to increase.

A low blood calcium level stimulates the release of PTH. PTH promotes the activity of osteoclasts and the release of calcium from the bones. PTH also promotes the reabsorption of calcium by the kidneys, where it activates vitamin D. Vitamin D, in turn, stimulates the absorption of calcium from the intestine. These effects bring the blood calcium level back to the normal range so that the parathyroid glands no longer secrete PTH.

When insufficient parathyroid hormone production leads to a dramatic drop in the blood calcium level, tetany results. In **tetany,** the body shakes from continuous muscle contraction. This effect is brought about by increased excitability of the nerves, which initiate nerve impulses spontaneously and without rest.

The antagonistic actions of calcitonin from the thyroid gland and parathyroid hormone from the parathyroid glands maintain the blood calcium level within normal limits.

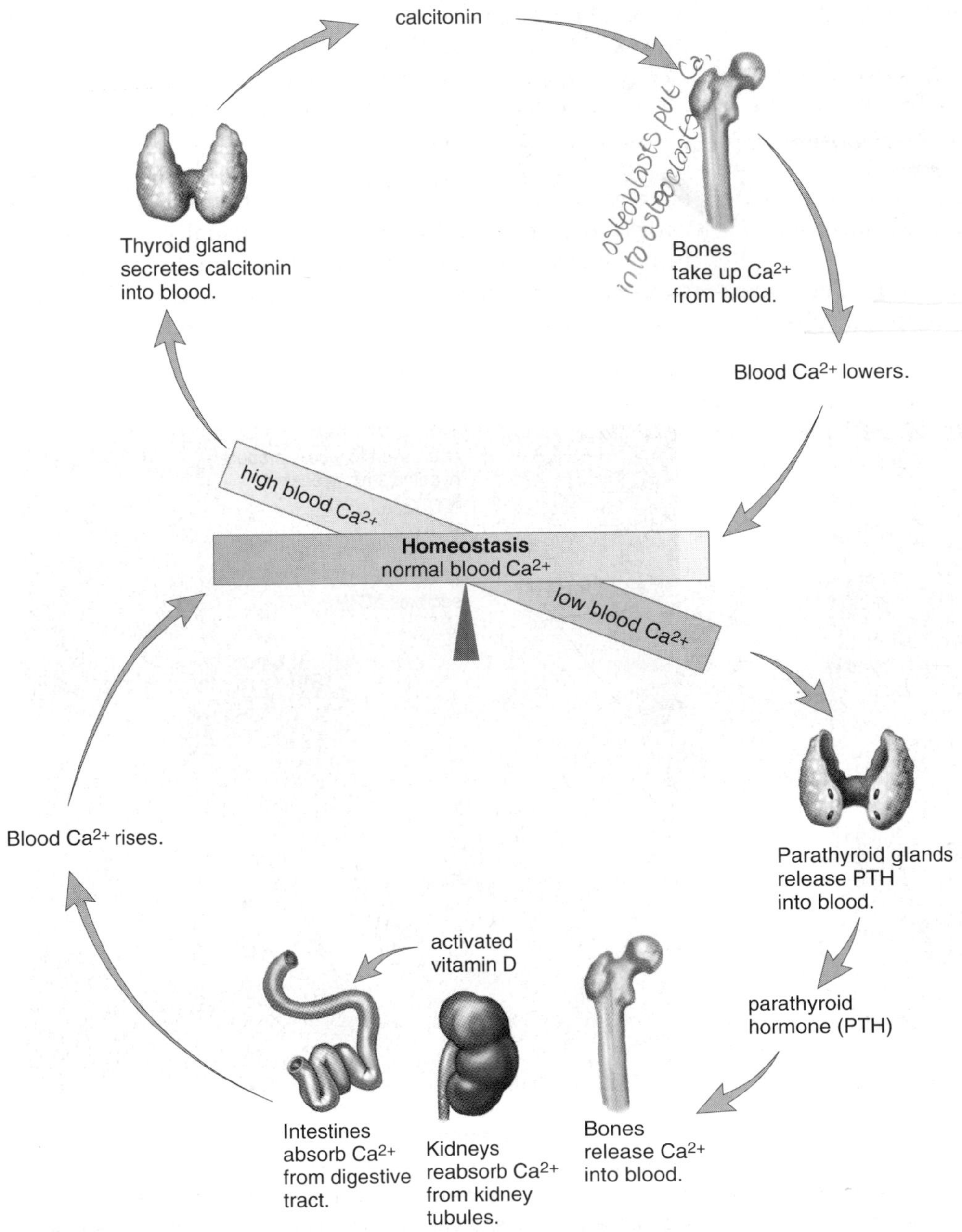

calcitonin

osteoblasts put Ca^{2+} into osteoclasts

Thyroid gland secretes calcitonin into blood.

Bones take up Ca^{2+} from blood.

Blood Ca^{2+} lowers.

high blood Ca^{2+}

Homeostasis normal blood Ca^{2+}

low blood Ca^{2+}

Blood Ca^{2+} rises.

Parathyroid glands release PTH into blood.

activated vitamin D

parathyroid hormone (PTH)

Intestines absorb Ca^{2+} from digestive tract.

Kidneys reabsorb Ca^{2+} from kidney tubules.

Bones release Ca^{2+} into blood.

Figure 20.7 **Regulation of blood calcium level.**

(Top) When the blood calcium (Ca^{2+}) level is high, the thyroid gland secretes calcitonin. Calcitonin promotes the uptake of Ca^{2+} by the bones, and therefore the blood Ca^{2+} level returns to normal. *(Bottom)* When the blood Ca^{2+} level is low, the parathyroid glands release parathyroid hormone (PTH). PTH causes the bones to release Ca^{2+}, the kidneys to reabsorb Ca^{2+} and activate vitamin D, and thereafter the intestines absorb Ca^{2+}. Therefore, the blood Ca^{2+} level returns to normal.

20.4 Adrenal Glands

Two **adrenal glands** sit atop the kidneys (see Fig. 20.1). Each adrenal gland consists of an inner portion called the **adrenal medulla** and an outer portion called the **adrenal cortex.** These portions, like the anterior pituitary and the posterior pituitary, have no physiological connection with one another.

The hypothalamus exerts control over the activity of both portions of the adrenal glands. It initiates nerve impulses that travel by way of the brain stem, spinal cord, and sympathetic nerve fibers to the adrenal medulla, which then secretes its hormones. The hypothalamus, by means of ACTH-releasing hormone, controls the anterior pituitary's secretion of ACTH, which in turn stimulates the adrenal cortex. Stress of all types, including both emotional and physical trauma, prompts the hypothalamus to stimulate the adrenal glands.

Epinephrine (adrenaline) and **norepinephrine** (noradrenaline) produced by the adrenal medulla rapidly bring about all the bodily changes that occur when an individual reacts to an emergency situation. The effects of these hormones are short-term (Fig. 20.8). In contrast, the hormones produced by the adrenal cortex provide a long-term response to stress. The two major types of hormones produced by the adrenal cortex are the mineralocorticoids and the glucocorticoids. The **mineralocorticoids** regulate salt and water balance, leading to increases in blood volume and blood pressure. The **glucocorticoids** regulate carbohydrate, protein, and fat metabolism, leading to an increase in blood glucose level. Cortisone, the medication that is often administered for inflammation of joints, is a glucocorticoid.

The adrenal cortex also secretes a small amount of male sex hormones and a small amount of female sex hormones in both sexes—that is, in the male, both male and female sex hormones are produced by the adrenal cortex, and in the female, both male and female sex hormones are also produced by the adrenal cortex.

The adrenal medulla is under nervous control, and the adrenal cortex is under the control of ACTH, an anterior pituitary hormone. The adrenal hormones help us respond to stress.

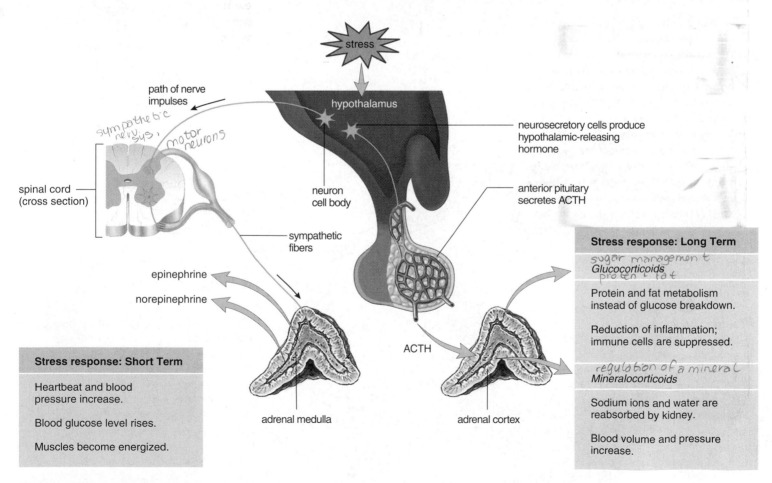

Figure 20.8   Adrenal glands.
Both the adrenal medulla and the adrenal cortex are under the control of the hypothalamus when they help us respond to stress. *(Left)* The adrenal medulla provides a rapid, but short-term, stress response. *(Right)* The adrenal cortex provides a slower, but long-term, stress response.

Glucocorticoids

Cortisol is a biologically significant glucocorticoid produced by the adrenal cortex. Cortisol raises the blood glucose level in at least two ways: (1) it promotes the breakdown of muscle proteins to amino acids, which are taken up by the liver from the bloodstream. The liver then breaks down these excess amino acids to glucose, which enters the blood. (2) Cortisol promotes the metabolism of fatty acids rather than carbohydrates, and this spares glucose.

Cortisol also counteracts the inflammatory response that leads to the pain and swelling of joints in arthritis and bursitis. The administration of cortisol aids these conditions because it reduces inflammation. Very high levels of glucocorticoids in the blood can suppress the body's defense system, including the inflammatory response that occurs at infection sites. Cortisone and other glucocorticoids can relieve swelling and pain from inflammation, but by suppressing pain and immunity, they can also make a person highly susceptible to injury and infection.

Mineralocorticoids

Aldosterone is the most important of the mineralocorticoids. The aldosterone primarily targets the kidney, where it promotes renal absorption of sodium (Na$^+$) and renal excretion of potassium (K$^+$).

The secretion of mineralocorticoids is not controlled by the anterior pituitary. When the blood sodium level and therefore blood pressure are low, the kidneys secrete **renin** (Fig. 20.9). Renin is an enzyme that converts the plasma protein angiotensinogen to angiotensin I, which is changed to angiotensin II by a converting enzyme found in lung

capillaries. Angiotensin II stimulates the adrenal cortex to release aldosterone. The effect of this system, called the renin-angiotensin-aldosterone system, is to raise blood pressure in two ways: angiotensin II constricts the arterioles, and aldosterone causes the kidneys to reabsorb sodium. When the blood sodium level rises, water is reabsorbed, in part because the hypothalamus secretes ADH (see page 394). Then blood pressure increases to normal.

There is an antagonistic hormone to aldosterone, as you might suspect. When the atria of the heart are stretched due to increased blood volume, cardiac cells release a hormone called **atrial natriuretic hormone (ANH),** which inhibits the secretion of aldosterone from the adrenal cortex. The effect of this hormone is to cause the excretion of sodium—that is, *natriuresis*. When sodium is excreted, so is water, and therefore blood pressure lowers to normal.

Figure 20.9   Regulation of blood pressure and volume.

(Bottom) When the blood sodium (Na$^+$) level is low, a low blood pressure causes the kidneys to secrete renin. Renin leads to the secretion of aldosterone from the adrenal cortex. Aldosterone causes the kidneys to reabsorb Na$^+$, and water follows, so that blood volume and pressure return to normal. *(Top)* When the blood Na$^+$ is high, a high blood volume causes the heart to secrete atrial natriuretic hormone (ANH). ANH causes the kidneys to excrete Na$^+$, and water follows. The blood volume and pressure return to normal.

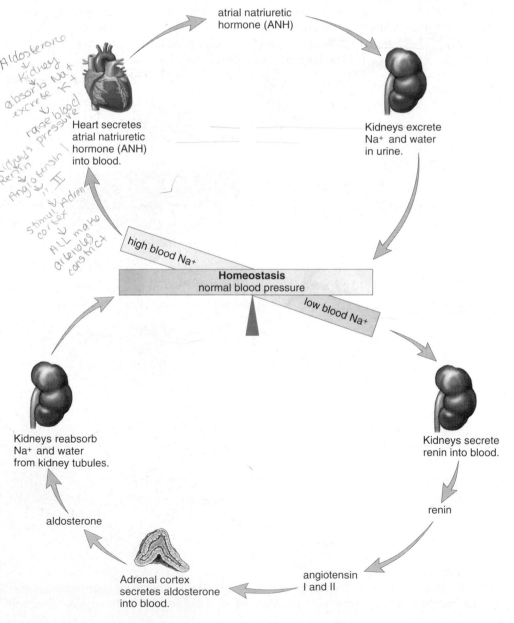

[handwritten margin notes: Addison disease ↓H adrenal cortex bronze ; ↑ H "Cushing Syndrome"]

Malfunction of the Adrenal Cortex

When the level of adrenal cortex hormones is low due to hyposecretion, a person develops **Addison disease.** The presence of excessive but ineffective ACTH causes a bronzing of the skin because ACTH, like MSH, can lead to a buildup of melanin (Fig. 20.10). Without cortisol, glucose cannot be replenished when a stressful situation arises. Even a mild infection can lead to death. The lack of aldosterone results in a loss of sodium and water, the development of low blood pressure, and possibly severe dehydration. Left untreated, Addison disease can be fatal.

When the level of adrenal cortex hormones is high due to hypersecretion, a person develops **Cushing syndrome** (Fig. 20.11). The excess cortisol results in a tendency toward diabetes mellitus as muscle protein is metabolized and subcutaneous fat is deposited in the midsection. The trunk is obese, while the arms and legs remain a normal size. An excess of aldosterone and reabsorption of sodium and water by the kidneys leads to a basic blood pH and hypertension. The face is moon-shaped due to edema. Masculinization may occur in women because of excess adrenal male sex hormones.

> The adrenal cortex hormones are essential to homeostasis. Addison disease is due to adrenal cortex hyposecretion, and Cushing syndrome is due to adrenal cortex hypersecretion.

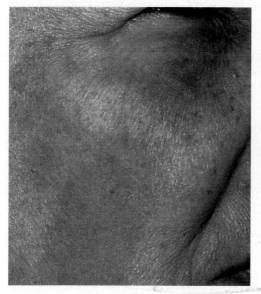
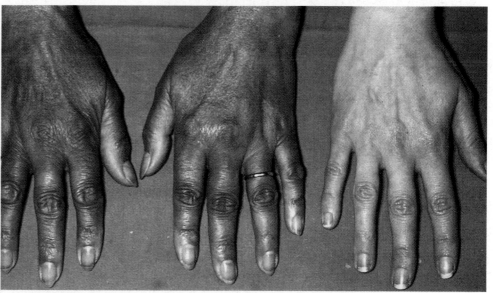

a.   b.

Figure 20.10   Addison disease.
Addison disease is characterized by a peculiar bronzing of the skin, particularly noticeable in these light-skinned individuals. Note the color of (**a**) the face and (**b**) the hands compared to the hand of an individual without the disease.

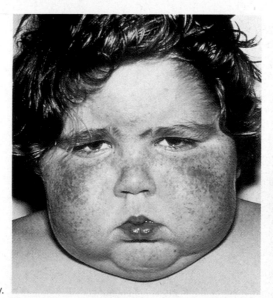
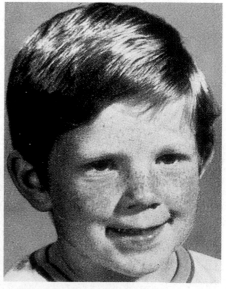

Figure 20.11   Cushing syndrome.
Cushing syndrome results from hypersecretion of hormones due to an adrenal cortex tumor. *(Left)* Patient first diagnosed with Cushing syndrome. *(Right)* Four months later, after therapy.

20.5 Pancreas

The **pancreas** is a long organ that lies transversely in the abdomen between the kidneys and near the duodenum of the small intestine. It is composed of two types of tissue. Exocrine tissue produces and secretes digestive juices that go by way of ducts to the small intestine. Endocrine tissue, called the **pancreatic islets** (islets of Langerhans), produces and secretes the hormones **insulin** and **glucagon** directly into the blood (Fig. 20.12).

Insulin is secreted when the blood glucose level is high, which usually occurs just after eating. Insulin stimulates the uptake of glucose by cells, especially liver cells, muscle cells, and adipose tissue cells. In liver and muscle cells, glucose is then stored as glycogen. In muscle cells, the breakdown of glucose supplies energy for protein metabolism, and in fat cells the breakdown of glucose supplies

glycerol for the formation of fat. In these various ways, insulin lowers the blood glucose level.

Glucagon is secreted from the pancreas, usually between meals, when the blood glucose level is low. The major target tissues of glucagon are the liver and adipose tissue. Glucagon stimulates the liver to break down glycogen to glucose and to use fat and protein in preference to glucose as energy sources. Adipose tissue cells break down fat to glycerol and fatty acids. The liver takes these up and uses them as substrates for glucose formation. In these various ways, glucagon raises the blood glucose level.

The two antagonistic hormones insulin and glucagon, both produced by the pancreas, maintain the normal level of glucose in the blood.

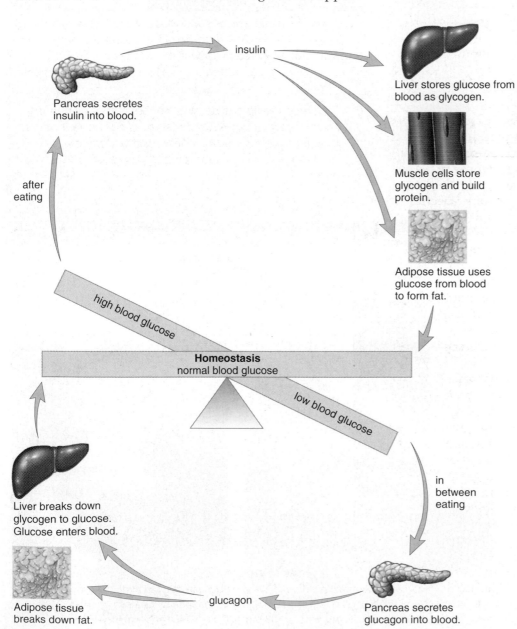

Pancreas secretes insulin into blood.

insulin

Liver stores glucose from blood as glycogen.

Muscle cells store glycogen and build protein.

Adipose tissue uses glucose from blood to form fat.

after eating

high blood glucose

Homeostasis
normal blood glucose

low blood glucose

Liver breaks down glycogen to glucose. Glucose enters blood.

Adipose tissue breaks down fat.

glucagon

in between eating

Pancreas secretes glucagon into blood.

Figure 20.12   Regulation of blood glucose level.
(Top) When the blood glucose level is high, the pancreas secretes insulin. Insulin promotes the storage of glucose as glycogen and the synthesis of proteins and fats (as opposed to their use as energy sources). Therefore, insulin lowers the blood glucose level. *(Bottom)* When the blood glucose level is low, the pancreas secretes glucagon. Glucagon acts opposite to insulin; therefore, glucagon raises the blood glucose level to normal.

Diabetes Mellitus

Diabetes mellitus is a fairly common hormonal disease in which liver cells, and indeed all body cells, are unable to take up and/or metabolize glucose. Therefore, cellular famine exists in the midst of plenty, and the person becomes extremely hungry. As the blood glucose level rises, glucose, along with water, is excreted in the urine. The loss of water in this way causes the diabetic to be extremely thirsty. Since glucose is not being metabolized, the body turns to the breakdown of protein and fat for energy. The metabolism of fat leads to the buildup of ketones in the blood and acidosis (acid blood), which can eventually cause coma and death.

The glucose tolerance test assists in the diagnosis of diabetes mellitus. After the patient is given 100 g of glucose, the blood glucose concentration is measured at intervals. In a diabetic, the blood glucose level rises greatly and remains elevated for several hours. In the meantime, glucose appears in the urine (Fig. 20.13). In a nondiabetic, the blood glucose level rises somewhat and then returns to normal in about 1 1/2 hours.

Types of Diabetes

There are two types of diabetes mellitus. In *type I (insulin-dependent) diabetes,* the pancreas is not producing insulin. This condition is believed to be brought on by exposure to an environmental agent, most likely a virus, whose presence causes cytotoxic T cells to destroy the pancreatic islets. As a result, the individual must have daily insulin injections. These injections control the diabetic symptoms but can still cause inconveniences, since either an overdose of insulin or missing a meal can bring on the symptoms of hypoglycemia (low blood sugar). These symptoms include perspiration, pale skin, shallow breathing, and anxiety. Because the brain requires a constant supply of sugar, unconsciousness can result. The cure is quite simple: immediate ingestion of a sugar cube or fruit juice can very quickly counteract hypoglycemia.

It is possible to transplant a working pancreas into patients with type I diabetes. To do away with the necessity of taking immunosuppressive drugs after the transplant, fetal pancreatic islet cells have been injected into patients. Another experimental procedure is to place pancreatic islet cells in a capsule that allows insulin to get out but prevents antibodies and T lymphocytes from getting in. This artificial organ is implanted in the abdominal cavity.

Of the 16 million people who now have diabetes in the United States, most have *type II (non-insulin-dependent) diabetes.* This type of diabetes mellitus usually occurs in people of any age who are obese and inactive. The pancreas produces insulin, but the liver and muscle cells do not respond to it in the usual manner. They may increasingly lack the receptor proteins that bind to insulin. It is possible to prevent or at least control type II diabetes by adhering to a low-fat, low-sugar diet and exercising regularly. If this fails, oral drugs that stimulate the pancreas to secrete more insulin and enhance the metabolism of glucose in the liver and muscle cells are available.

The symptoms of type I diabetes are compelling, and therefore most people seek help right away. The symptoms of type II diabetes are more likely to be overlooked. It's projected that as many as 7 million Americans may have type II diabetes without being aware of it. Yet the results of untreated type II diabetes are as serious as those for type I diabetes.

Long-term complications of both types of diabetes are blindness, kidney disease, and circulatory disorders, including atherosclerosis, heart disease, stroke, and reduced circulation. The latter can lead to gangrene in the arms and legs. Pregnancy carries an increased risk of diabetic coma, and the child of a diabetic is somewhat more likely to be stillborn or to die shortly after birth. These complications of diabetes are not expected to appear if the mother's blood glucose level is carefully regulated and kept within normal limits.

Diabetes mellitus is caused by the lack of insulin or the insensitivity of cells to insulin, a hormone that lowers the blood glucose level, particularly by causing the liver to store glucose as glycogen.

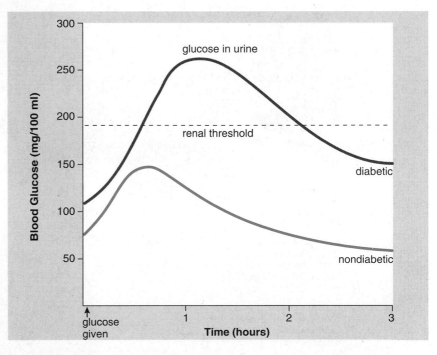

Figure 20.13   Glucose tolerance test.
Following the administration of 100 g of glucose, the blood glucose level rises dramatically in the diabetic but not in the nondiabetic. Glucose appears in the urine when its level exceeds 190 mg/100 ml (called the renal threshold).

20.6 Other Endocrine Glands

The **gonads** are the testes in males and the ovaries in females. The gonads are endocrine glands. Other lesser known glands and some tissues also produce hormones.

Testes and Ovaries

The **testes** are located in the scrotum, and the **ovaries** are located in the pelvic cavity. The testes produce **androgens** (e.g., **testosterone**), which are the male sex hormones, and the ovaries produce estrogens and progesterone, the female sex hormones. The hypothalamus and the pituitary gland control the hormonal secretions of these organs in the same manner previously described for the thyroid gland.

Greatly increased testosterone secretion at the time of puberty stimulates the growth of the penis and the testes. Testosterone also brings about and maintains the male secondary sex characteristics that develop during puberty. Testosterone causes growth of a beard, axillary (underarm) hair, and pubic hair. It prompts the larynx and the vocal cords to enlarge, causing the voice to change. It is partially responsible for the muscular strength of males, and this is the reason some athletes take supplemental amounts of **anabolic steroids,** which are either testosterone or related chemicals. The contraindications of taking anabolic steroids are listed in Figure 20.14. Testosterone also stimulates oil and sweat glands in the skin; therefore, it is largely responsible for acne and body odor. Another side effect of testosterone is baldness. Genes for baldness are probably inherited by both sexes, but baldness is seen more often in males because of the presence of testosterone.

The female sex hormones, **estrogens** and **progesterone,** have many effects on the body. In particular, estrogens secreted at the time of puberty stimulate the growth of the uterus and the vagina. Estrogen is necessary for egg maturation and is largely responsible for the secondary sex characteristics in females, including female body hair and fat distribution. In general, females have a more rounded appearance than males because of a greater accumulation of fat beneath the skin. Also, the pelvic girdle is wider in females than in males, resulting in a larger pelvic cavity. Both estrogen and progesterone are required for breast development and regulation of the uterine cycle, which includes monthly menstruation (discharge of blood and mucosal tissues from the uterus).

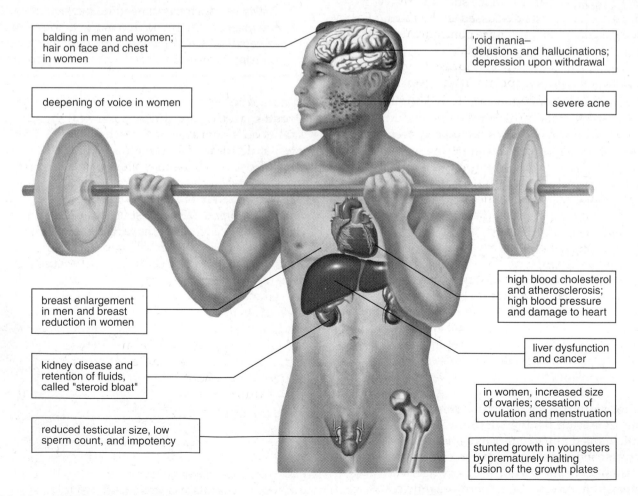

balding in men and women; hair on face and chest in women

deepening of voice in women

breast enlargement in men and breast reduction in women

kidney disease and retention of fluids, called "steroid bloat"

reduced testicular size, low sperm count, and impotency

'roid mania— delusions and hallucinations; depression upon withdrawal

severe acne

high blood cholesterol and atherosclerosis; high blood pressure and damage to heart

liver dysfunction and cancer

in women, increased size of ovaries; cessation of ovulation and menstruation

stunted growth in youngsters by prematurely halting fusion of the growth plates

Figure 20.14   **The effects of anabolic steroid use.**

Thymus Gland

The **thymus** is a lobular gland that lies just beneath the sternum (see Fig. 20.1). This organ reaches its largest size and is most active during childhood. With aging, the organ gets smaller and becomes fatty. Lymphocytes that originate in the bone marrow and then pass through the thymus are transformed into T lymphocytes. The lobules of the thymus are lined by epithelial cells that secrete hormones called thymosins. These hormones aid in the differentiation of lymphocytes packed inside the lobules. Although the hormones secreted by the thymus ordinarily work in the thymus, there is hope that these hormones could be injected into AIDS or cancer patients where they would enhance T lymphocyte function.

Pineal Gland

The **pineal gland,** which is located in the brain (see Fig. 20.1), produces the hormone **melatonin,** primarily at night. Melatonin is involved in our daily sleep-wake cycle; normally we grow sleepy at night when melatonin levels increase and awaken once daylight returns and melatonin levels are low. Daily 24-hour cycles such as this are called **circadian rhythms** and, as discussed in the Health Focus on page 406, circadian rhythms are controlled by a biological clock located in the hypothalamus.

Based on animal research, it appears that melatonin also regulates sexual development. It has been noted that children whose pineal gland has been destroyed due to a brain tumor experience early puberty.

The gonads, thymus, and pineal gland are also endocrine organs. The gonads secrete the sex hormones, the thymus secretes thymosins, and the pineal gland secretes melatonin.

Hormones from Other Tissues

Some organs that are usually not considered endocrine glands do indeed secrete hormones. We have already mentioned that the heart produces atrial natriuretic hormone. And you will recall that the stomach and the small intestine produce peptide hormones that regulate digestive secretions. A number of other types of tissues produce hormones.

Leptin

Leptin is a protein hormone produced by adipose tissue. Leptin acts on the hypothalamus, where it signals satiety—that the individual has had enough to eat. Strange to say, the blood of obese individuals may be rich in leptin. It is possible that the leptin they produce is ineffective because of a genetic mutation, or else their hypothalamic cells lack a suitable number of receptors for leptin.

Growth Factors

A number of different types of organs and cells produce peptide **growth factors,** which stimulate cell division and mitosis. They are like hormones in that they act on cell types with specific receptors to receive them. Some, like lymphokines, are released into the blood; others diffuse to nearby cells. Growth factors of particular interest are:

Granulocyte and macrophage colony-stimulating factor (GM-CSF) is secreted by many different tissues. GM-CSF causes bone marrow stem cells to form either granulocyte or macrophage cells, depending on whether the concentration is low or high.

Platelet-derived growth factor is released from platelets and from many other cell types. It helps in wound healing and causes an increase in the number of fibroblasts, smooth muscle cells, and certain cells of the nervous system.

Epidermal growth factor and nerve growth factor stimulate the cells indicated by their names, as well as many others. These growth factors are also important in wound healing.

Tumor angiogenesis factor stimulates the formation of capillary networks and is released by tumor cells. One treatment for cancer is to prevent the activity of this growth factor.

Prostaglandins

Prostaglandins are potent chemical signals produced within cells from arachidonate, a fatty acid. Prostaglandins are not distributed in the blood; instead, they act locally, quite close to where they were produced. In the uterus, prostaglandins cause muscles to contract; therefore, they are implicated in the pain and discomfort of menstruation in some women. Also, prostaglandins mediate the effects of pyrogens, chemicals that are believed to reset the temperature regulatory center in the brain. Aspirin reduces body temperature and controls pain because of its effect on prostaglandins.

Certain prostaglandins reduce gastric secretion and have been used to treat ulcers; others lower blood pressure and have been used to treat hypertension; and yet others inhibit platelet aggregation and have been used to prevent thrombosis. However, different prostaglandins have contrary effects, and it has been very difficult to successfully standardize their use. Therefore, prostaglandin therapy is still considered experimental.

Many tissues aside from the traditional endocrine glands produce hormones. Some of these enter the bloodstream, and some act only locally.

Health Focus

Melatonin

The hormone melatonin has been sold as a nutritional supplement for about five years. The popular press promotes its use in pill form for sleep, aging, cancer treatment, sexuality, and more. At best, melatonin may have some benefits in certain sleep disorders. But most physicians will not yet recommend it for that use because so little is known about its dosage requirements and possible side effects.

Melatonin is a hormone produced by the pineal gland in greatest quantity at night and smallest quantity during the day. Notice in Figure 20A that melatonin's production cycle accompanies our natural sleep-wake cycle. Rhythms with a period of about 24 hours are called circadian ("about a day") rhythms. All circadian rhythms seem to be controlled by an internal biological clock because they are free-running—that is, they have a regular cycle even in the absence of environmental cues. In scientific experiments, humans have lived in underground bunkers where they never see the light of day. In a few people, the sleep-wake cycle drifts badly, but in most, the daily activity schedule is just about 25 hours. How do we normally manage to stay on a 24-hour schedule? An individual's internal biological clock is reset each day by the environmental day-night cycle. Characteristically, biological clocks that control circadian rhythms are reset by environmental cues, or else they drift out of phase with the environment.

Recent research suggests that our biological clock lies in a cluster of neurons within the hypothalamus called the suprachiasmatic nucleus (SCN). The SCN undergoes spontaneous cyclical changes in activity, and therefore it can act as a pacemaker for circadian rhythms such as the rise and fall of body temperature and our sleep-wake cycle. Neural connections between the retina and the SCN indicate that reception of light by the eyes most likely resets the SCN and keeps our biological rhythms on a 24-hour cycle. Some people suffer from seasonal affective disorder, or SAD. As the days get darker and darker during the fall and winter, they become depressed, sometimes severely. They find it difficult to keep up because their biological clock has fallen behind without early morning light to reset it. If so, a half-hour dose of simulated daylight from a portable light box first thing in the morning makes them feel operational again. The SCN also controls the secretion of melatonin by the pineal gland, and in turn melatonin may quiet the operation of the neurons in the SCN.

Research is still going forward to see if melatonin will be effective for circadian rhythm disorders such as SAD, jet lag, sleep phase problems, recurrent insomnia in the totally blind, and some other less common disorders. So-called jet lag occurs when you travel across several time zones and your biological clock is out of phase with local time. You feel wide awake when it is time to sleep because your body is receiving internal signals that it is morning. Likewise, acute periods of sleepiness and fatigue occur during the day, because of internal signals indicating it is nighttime. Jet-lag symptoms gradually disappear as your biological clock adjusts to the environmental signals of the local time zone. Many young people have a sleep phase problem because their circadian cycle lasts 25 to 26 hours. As they lengthen their day, they get out of sync with normal times for sleep and activity. The totally blind have the same problem because they have no chance of using the environmental day-night cycle to reset their internal clock.

In clinical trials, it was found that melatonin could shift circadian rhythms and reset our biological clock. The experimenters found that melatonin given in the afternoon shifts rhythms earlier, while melatonin given in the morning shifts rhythms later. For most people, the process was gradual: the average rate of change was about an hour a day. Before you try melatonin, however, you might want to consider that melatonin is known to affect reproductive behavior in other mammals. It's a matter of deciding whether any potential side effect from melatonin use is worth the possible benefits.

Figure 20A Melatonin production.
Melatonin production is greatest at night when we are sleeping. Light suppresses melatonin production (**a**), so its duration is longer in the winter (**b**) than in the summer (**c**).

20.7 Chemical Signals

A chemical signal is any substance that affects the metabolism of cells or the behavior of the individual. Chemical signals are a means of communication between cells, between body parts, and even between individuals. Hormones are one type of chemical signal. Some hormones, such as these in Figure 20.15a, act at a distance between body parts. Insulin travels in the bloodstream from the pancreas to the liver and most of the cells in the body. Also, secretions produced by neurosecretory cells in the hypothalamus are now called hypothalamic-releasing hormones. They travel in the capillary network that runs between the hypothalamus and the pituitary gland. Some of these secretions stimulate and others prevent the pituitary from secreting its hormones.

Figure 20.15b illustrates that some chemical signals act locally between adjacent cells. Prostaglandins are a good example of a local hormone. After prostaglandins are produced, they are not carried in the bloodstream; instead they affect neighboring cells, sometimes promoting pain and inflammation. Growth factors are local hormones that promote cell division and mitosis. Some growth factors are now being used to treat various human conditions. Aside from local hormones, neurotransmitters released by neurons belong to this category of chemical signals.

Chemical signals that act between individuals are called **pheromones.** Pheromones are well exemplified in other animals but not in humans. For example, female moths release a sex attractant that is received by male moth antennae even several miles away. The receptors on the antennae of the male silkworm moth are so sensitive that only 40 out of 40,000 receptor proteins need to be activated in order for the male to respond and find the female.

Do humans release and receive pheromones? This is a question that has kept researchers busy for many years. Humans produce a rich supply of airborne chemicals from a variety of areas, including the scalp, oral cavity, arm pits, genital areas, and feet. Numerous human studies suggest that the axillary (arm pit) odor profile can identify the particular individual. Further, women seem to prefer the axillary odors of men who are a different HLA type from themselves. (HLA are plasma membrane proteins involved in immunity). Choosing a mate of a different HLA type could conceivably improve the immune response of offspring.

Several studies indicate that axillary secretions can affect the menstrual cycle. The cycle length becomes more normal when women with irregular cycles are exposed to extracts of male axillary secretions. Women who live in the same household often have menstrual cycles in synchrony. Researchers have found that a woman's axillary extract can alter another woman's cycle by a few days.

Chemical signals work at a distance between body parts, locally between adjacent cells, or even possibly between individuals.

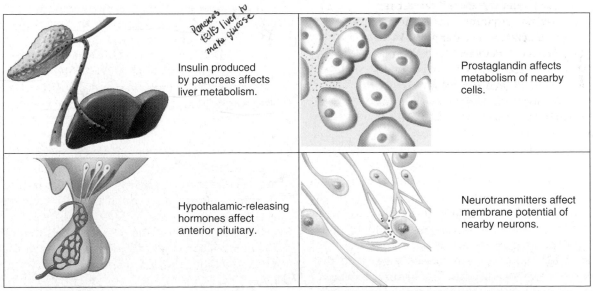

Pancreas tells liver to make glucose

Insulin produced by pancreas affects liver metabolism.

Prostaglandin affects metabolism of nearby cells.

Hypothalamic-releasing hormones affect anterior pituitary.

Neurotransmitters affect membrane potential of nearby neurons.

a. Signal acts at a distance between body parts.

b. Signal acts locally between adjacent cells.

Figure 20.15   Chemical signals.

a. Hormones are chemical signals that are usually carried in the bloodstream and act at a distance within the body of a single individual. **b.** Some hormones, such as prostaglandins, have local effects only; they pass between cells that are adjacent to one another. Neurotransmitters belong to this category of chemical signals.

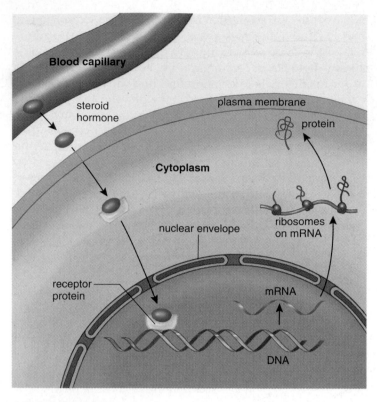

a. Action of steroid hormone

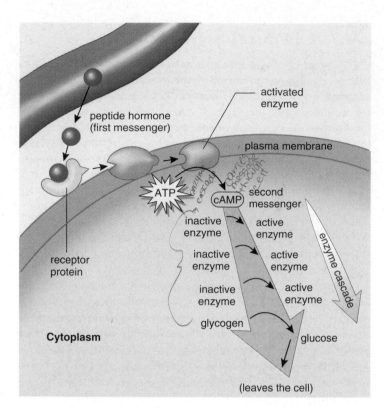

b. Action of peptide hormone

Figure 20.16   Cellular activity of hormones.
a. After passing through the plasma membrane and possibly not until it passes through the nuclear envelope, a steroid hormone binds to a receptor protein inside the nucleus. The hormone-receptor complex then binds to DNA, and this leads to activation of certain genes and protein synthesis. **b.** Peptide hormones, called first messengers, bind to a specific receptor protein in the plasma membrane. A protein relay in the plasma membrane ends when an enzyme converts ATP to cAMP, the second messenger, which activates an enzyme cascade.

The Action of Hormones

Hormones fall into two basic chemical classes: (1) in the class **steroid hormone,** the hormone always has the same complex of four carbon rings, but each one has different side chains; (2) the class **peptide hormone** includes those hormones that are peptides, proteins, glycoproteins, or modified amino acids. Because their effect is amplified through cellular mechanisms, both classes of hormones can function at extremely low concentrations.

A hormone does not seek out a particular organ; rather the organ is awaiting the arrival of the hormone. Cells that can react to a hormone have receptor proteins that combine with the hormone in a lock-and-key manner. Steroid hormones are lipids, and therefore they cross cell membranes (Fig. 20.16a). Only after they are inside the cytoplasm or the nucleus do steroid hormones, such as estrogen and progesterone, bind to receptor proteins. The hormone-receptor complex then binds to DNA, activating particular genes. Activation leads to production of a cellular enzyme in multiple quantities.

Most peptide hormones cannot pass through cell membranes, and they bind to a receptor protein in the plasma membrane (Fig. 20.16b). After a peptide hormone binds to a receptor protein, a relay system leads to the conversion of ATP to cyclic AMP (cyclic adenosine monophosphate). Cyclic AMP (cAMP) contains only one phosphate group attached to the adenosine portion of the molecule at two spots. Therefore, the molecule is cyclic. The peptide hormone is called the **first messenger** and cAMP is called the **second messenger.** Calcium is also a common second messenger, and this helps explain why calcium regulation in the body is so important.

The second messenger sets in motion an enzyme cascade. In muscle cells, the reception of epinephrine leads to the breakdown of glycogen to glucose (Fig. 20.16b). An enzyme cascade is so called because each enzyme in turn activates another. Because enzymes work over and over, every step in an enzyme cascade leads to more reactions—the binding of a single peptide hormone molecule can result even in a thousandfold response.

Hormones are chemical signals that influence the metabolism of the cell either indirectly by regulating the production of a particular protein (steroid hormone) or directly by activating an enzyme cascade (peptide hormone).

Bioethical Focus   Fertility Drugs

Higher-order multiple births (triplets or more) in the United States increased 19% between 1980 and 1994. During these years, it became customary to use fertility drugs (gonadotropic hormones) to stimulate the ovaries. The risks for premature delivery, low birth weight, and developmental abnormalities rise sharply for higher-order multiple births. And the physical and emotional burden placed on the parents is extraordinary. They face endless everyday chores and find it difficult to maintain normal social relationships, if only because they get insufficient sleep. Finances are strained in order to provide for the children's needs, including housing and child-care assistance. About one-third report that they received no help from relatives, friends, or neighbors in the first year after the birth. Trips to the hospital for accidental injury are more frequent because parents with only two arms and legs cannot keep so many children safe at one time.

Many clinicians are now urging that all possible steps be taken to ensure that the chance of higher-order multiple births be reduced. However, none of the ethical choices to bring this about are attractive. If fertility drugs are outlawed, some couples might be denied the possibility of ever having a child. A higher-order multiple pregnancy can be terminated, or selective reduction can be done. During selective reduction, one or more of the fetuses is killed by an injection of potassium chloride. Selective reduction could very well result in psychological and social complications for the mother and surviving children. The parents could opt to utilize in vitro fertilization (in which the eggs are fertilized in the lab), with the intent that only one or two zygotes will be placed in the woman's womb. But then any leftover zygotes may never have an opportunity to continue development.

Decide Your Opinion

1. Do you approve of the use of fertility drugs despite the risk of higher-order multiple births? Why or why not?
2. Not many clinics will carry out selective reduction. Should a woman be forced to carry all fetuses to term even if it increases the likelihood that some children will be born with physical abnormalities?
3. Should society decide the fate of zygotes left over from in vitro fertilization, or should that be left up to the couple?

Summarizing the Concepts

20.1 Endocrine Glands
Endocrine glands secrete hormones into the bloodstream, and from there they are distributed to target organs or tissues. The major endocrine glands and hormones are listed in Table 20.1. Negative feedback and antagonistic hormonal actions control the secretion of hormones.

20.2 Hypothalamus and Pituitary Gland
Neurosecretory cells in the hypothalamus produce antidiuretic hormone (ADH) and oxytocin, which are stored in axon endings in the posterior pituitary until they are released.

The hypothalamus produces hypothalamic-releasing and hypothalamic-inhibiting hormones, which pass to the anterior pituitary by way of a portal system. The anterior pituitary produces at least six types of hormones, and some of these stimulate other hormonal glands to secrete hormones.

20.3 Thyroid and Parathyroid Glands
The thyroid gland requires iodine to produce thyroxine and triiodothyronine, which increase the metabolic rate. If iodine is available in limited quantities, a simple goiter develops; if the thyroid is overactive, an exophthalmic goiter develops. The thyroid gland also produces calcitonin, which helps lower the blood calcium level. The parathyroid glands secrete parathyroid hormone, which raises the blood calcium and decreases the blood phosphate levels.

20.4 Adrenal Glands
The adrenal glands respond to stress: immediately, the adrenal medulla secretes epinephrine and norepinephrine, which bring about responses we associate with emergency situations. On a long-term basis, the adrenal cortex produces the glucocorticoids (e.g., cortisol) and the mineralocorticoids (e.g., aldosterone). Cortisol stimulates hydrolysis of proteins to amino acids that are converted to glucose; in this way, it raises the blood glucose level. Aldosterone causes the kidneys to reabsorb sodium ions (Na^+) and to excrete potassium ions (K^+). Addison disease develops when the adrenal cortex is underactive, and Cushing syndrome develops when the adrenal cortex is overactive.

20.5 Pancreas
The pancreatic islets secrete insulin, which lowers the blood glucose level, and glucagon, which has the opposite effect. The most common illness caused by hormonal imbalance is diabetes mellitus, which is due to the failure of the pancreas to produce insulin or the failure of the cells to take it up.

20.6 Other Endocrine Glands
The gonads produce the sex hormones; and the thymus secretes thymosins, which stimulate T lymphocyte production and maturation; the pineal gland produces melatonin, which may be involved in circadian rhythms and the development of the reproductive organs.

Tissues also produce hormones. Adipose tissue produces leptin, which acts on the hypothalamus, and various tissues produce growth factors. Prostaglandins are produced and act locally.

20.7 Chemical Signals
In the human body, some chemical signals, such as traditional endocrine hormones and secretions of neurosecretory cells, act at a distance. Others, such as prostaglandins, growth factors, and neurotransmitters, act locally. Whether humans have pheromones is under study.

Hormones are either steroids or peptides. Steroid hormones combine with a receptor in the cell, and the complex attaches to and activates DNA. Protein synthesis follows. Reception of a peptide hormone at the plasma membrane activates an enzyme cascade inside the cell.

Testing Yourself

Choose the best answer for each question. Match the hormones in questions 1–5 to the correct gland in the key.

Key:
a. pancreas
b. anterior pituitary
c. posterior pituitary
d. thyroid
e. adrenal medulla
f. adrenal cortex

1. cortisol _f_
2. growth hormone (GH) _b_
3. oxytocin storage _c_
4. insulin _a_
5. epinephrine _e_

6. The blood cortisol level controls the secretion of
   a. hypothalamic-releasing hormone from the hypothalamus.
   b. adrenocorticotropic hormone (ACTH) from the anterior pituitary.
   c. cortisol from the adrenal cortex.
   d. All of these are correct.

7. The anterior pituitary controls the secretion(s) of
   a. both the adrenal medulla and the adrenal cortex. _hypothalamus_
   b. both thyroid and adrenal cortex.
   c. both ovaries and testes.
   d. Both b and c are correct.

8. Diabetes mellitus is associated with
   a. too much insulin in the blood.
   b. too high a blood glucose level.
   c. blood that is too dilute.
   d. All of these are correct.

9. Which of these is not a pair of antagonistic hormones?
   a. insulin—glucagon
   b. calcitonin—parathyroid hormone
   c. cortisol—epinephrine
   d. aldosterone—atrial natriuretic hormone (ANH)
   e. thyroxine—growth hormone

10. Which hormone and condition is mismatched?
    a. growth hormone—acromegaly
    b. thyroxine—goiter
    c. parathyroid hormone—tetany
    d. cortisol—myxedema
    e. insulin—diabetes

11. Which of the following hormones could affect fat metabolism?
    a. growth hormone   d. glucagon
    b. thyroxine   e. All of these are correct.
    c. insulin

12. The difference between Type I and Type II diabetes is that
    a. for Type II diabetes, insulin is produced, but not utilized; Type I results from lack of insulin production.
    b. treatment for Type II involves insulin injections, while Type I can be controlled usually by diet.
    c. only Type I can result in complications such as kidney disease, reduced circulation, or stoke.
    d. Type I can be a result of lifestyle and Type II is thought to be caused by a virus or other agent.

13. Which of the following hormones is/are found in females?
    a. estrogen
    b. testosterone
    c. follicle-stimulating hormone
    d. Both a and c are correct.
    e. All of these are correct.

14. Which of the following is true of growth factors?
    a. They are peptide molecules.
    b. Certain growth factors stimulate production of macrophages and wound healing.
    c. Cancer cells may secrete growth factors.
    d. Growth factors stimulate mitosis of several different tissues and cells.
    e. All of these are correct.

15. Parathyroid hormone causes
    a. the kidneys to excrete more calcium ions.
    b. bone tissue to break down and release calcium into the bloodstream.
    c. fewer calcium ions to be absorbed by the intestines.
    d. more calcium ions to be deposited in bone tissue.

16. Hormones from all but which of the following glands can affect glucose levels in the body?
    a. pancreas   d. hypothalamus
    b. adrenal glands   e. thymus
    c. pituitary

17. Tropic hormones are hormones that affect other endocrine tissues. Which of the following would be considered a tropic hormone?
    a. calcitonin   d. melatonin
    b. oxytocin   e. follicle-stimulating
    c. glucagon   hormone

18. One of the chief differences between endocrine hormones and local hormones is
    a. the distance over which they act.
    b. that one is a chemical signal and the other is not.
    c. only endocrine hormones are made by humans.
    d. All of these are correct.

19. Peptide hormones
    a. are received by a receptor located in the plasma membrane.
    b. are received by a receptor located in the cytoplasm.
    c. bring about the transcription of DNA.
    d. Both b and c are correct.

20. Complete this diagram.

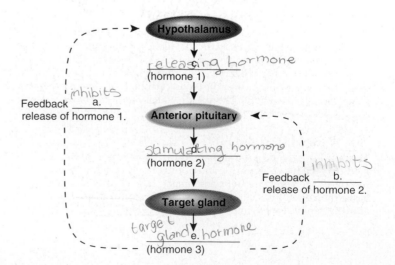

Hypothalamus

releasing hormone
(hormone 1)

Feedback ___a.___ _inhibits_
release of hormone 1.

Anterior pituitary

stimulating hormone
(hormone 2)

Feedback ___b.___ _inhibits_
release of hormone 2.

Target gland

target gland
e. hormone
(hormone 3)

Concepts	Questions	Media Resources*
20.1 Endocrine Glands		

• Endocrine glands secrete hormones into the bloodstream and are thereby distributed to target organs or tissues.	1. How are the effects of hormones regulated?	Essential Study Partner
		    Human System Overview
		    Endocrine Glands

20.2 Hypothalamus and Pituitary Gland		

• The hypothalamus, a part of the brain, controls the function of the pituitary gland, which consists of the anterior pituitary and the posterior pituitary.
• The anterior pituitary produces several hormones, some of which control other endocrine glands.
	1. By what means does the hypothalamus control the releases of the anterior pituitary and of the posterior pituitary?
	2. What anterior pituitary hormone influences skeletal and muscular growth?
		Essential Study Partner
		    The Hypothalamus
		    Pituitary Gland
		
		Art Quiz
		    Anterior Pituitary Control by Hypothalamus

20.3 Thyroid and Parathyroid Glands		

• The thyroid produces hormones that speed metabolism and another that lowers the blood calcium level. Antagonistic to this activity, parathyroid glands produce a hormone that raises the blood calcium level.	1. What is the primary effect of T_3 and T_4 on body cells?
	2. How do calcitonin and parathyroid hormone influence blood calcium levels?
		Essential Study Partner
		    Thyroid Gland
		    Parathyroid Glands
		
		Art Quiz
		    Thyroxine Action

20.4 Adrenal Glands		
• The adrenal glands produce hormones that help us respond to stress.	1. What hormones are released by the adrenal medulla? 2. How do the mineralocorticoids and glucocorticoids help maintain homeostasis?	Essential Study Partner Adrenal Glands
20.5 Pancreas		

• The pancreas secretes hormones that help regulate the blood glucose level.
• Diabetes mellitus occurs when cells do not take up glucose and it spills over into the urine.
	1. What are the effects of glucagon and insulin on blood glucose and when are these hormones released?
	2. How do the causes of type I and type II diabetes differ?
		Essential Study Partner
		    Pancreas
		
		Animation Quiz
		    Glucose Regulation

20.6 Other Endocrine Glands		

• Many other tissues, although not traditionally considered endocrine glands, secrete hormones.	1. What hormone is secreted by the pineal gland and what is its effect?
	2. What is the effect of leptin, a hormone released by adipose tissue?
		Case Studies
		    Could Your Inner Clock
		        Make You a Junk Food Junkie?

20.7 Chemical Signals		

• Hormones are chemical signals that influence the metabolism of their target cells.	1. What is the role of cAMP in the action of peptide hormones?	Essential Study Partner
		    Hormone Action
		    Hormonal Secretion
		    Types of Hormones

*For additional Media Resources, see the Online Learning Center.

Continuance of the Species

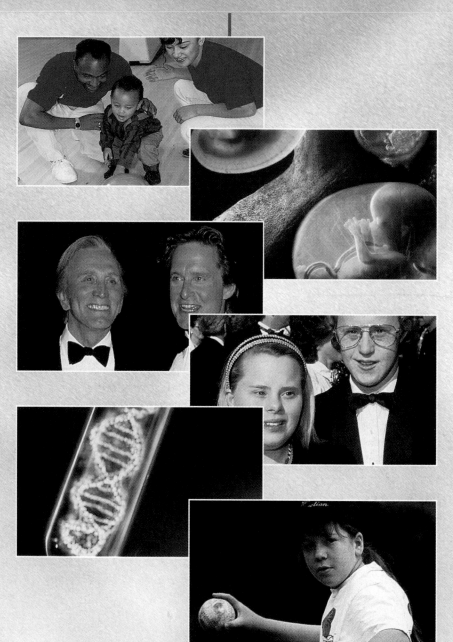

21. Reproductive System 413

The reproductive systems of male and female produce the gametes. A new individual comes into being after the sperm fertilizes the egg.

23. Patterns of Gene Inheritance 465

Genes are inherited from each parent, and they bring about our characteristics, including any genetic disorders.

25. Molecular Basis of Inheritance 501

The genes are composed of DNA, the molecule that stores genetic information from generation to generation.

22. Development and Aging 439

Development is a series of controlled events that results in a human being, who is like, and at the same time different from, his or her parents.

24. Patterns of Chromosome Inheritance 483

The genes are parts of chromosomes, and chromosome inheritance has a marked effect on our individual characteristics.

26. Biotechnology 525

We have learned how to control DNA and therefore the characteristics of cells and individuals.

See the Online Learning Center for careers in Genetics.

Reproductive System

Chapter Concepts

Reproduction, the cycle of life.

*I*t had seemed simple enough. Leigh Anne and Joe graduated from college, launched their careers, and got married. Some years later, they bought a house. Then, they decided to have a baby.

That's when things got complicated.

For some reason, Leigh Anne just didn't get pregnant. After two years of trying to conceive, the couple headed to a well-known fertility specialist. After some tests, the doctor explained a variety of fertility treatments and drugs designed to help couples conceive.

Leigh Anne and Joe weighed their options and decided to try in vitro fertilization. During a series of visits to the clinic, a doctor removed eggs from Leigh Anne and combined them with sperm from Joe. Nurtured in the lab, the combination formed fertilized eggs, which the doctor then placed in Leigh Anne's uterus.

Fortunately, the procedure worked. Leigh Anne's pregnancy was normal and healthy. Today, the couple's three-year-old races around, plays leapfrog over the family dog, and wants to be involved in everything his parents are doing. At some point, Leigh Anne and Joe might try in vitro fertilization again. For now, though, they'd just like their toddler to try a nap.

21.1 Male Reproductive System

The male reproductive system includes the organs depicted in Figure 21.1 and listed in Table 21.1. The male gonads are paired testes (sing., **testis**), which are suspended within the sacs of the **scrotum.**

Genital Tract

Sperm produced by the testes mature within the **epididymis** (pl., epididymides), which is a tightly coiled duct lying just outside each testis. Maturation seems to be required in order for sperm to swim to the egg. When sperm leave an epididymis, they enter a **vas deferens** (pl., vasa deferentia) where they may also be stored for a time. Each vas deferens passes into the abdominal cavity, where it curves around the bladder and empties into an ejaculatory duct. The ejaculatory ducts enter the **urethra.**

At the time of ejaculation, sperm leave the penis in a fluid called **semen** (seminal fluid). The seminal vesicles, the prostate gland, and the bulbourethral glands (Cowper glands) add secretions to seminal fluid. The pair of **seminal vesicles** lie at the base of the bladder, and each has a duct that joins with a vas deferens. The **prostate gland** is a single, doughnut-shaped gland that surrounds the upper portion of the urethra just below the bladder. In older men, the prostate can enlarge and squeeze off the urethra, making urination painful and difficult. The condition can be treated medically. **Bulbourethral glands** are pea-sized organs that lie posterior to the prostate on either side of the urethra.

Each component of seminal fluid seems to have a particular function. Sperm are more viable in a basic solution, and seminal fluid, which is milky in appearance, has a slightly basic pH (about 7.5). Swimming sperm require energy, and seminal fluid contains the sugar fructose, which presumably serves as an energy source. Semen also contains prostaglandins, chemicals that cause the uterus to contract. Some investigators believe that uterine contractions help propel the sperm toward the egg.

Sperm mature in the epididymis and are stored in the vas deferens before entering the urethra just prior to ejaculation. The accessory glands (seminal vesicles, prostate gland, and bulbourethral glands) produce seminal fluid.

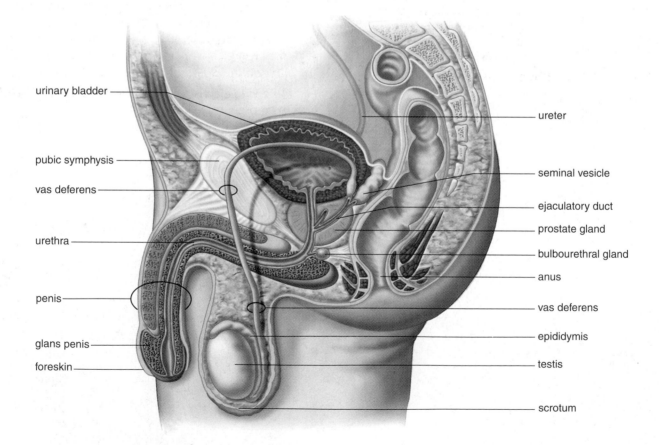

Figure 21.1   The male reproductive system.
The testes produce sperm. The seminal vesicles, the prostate gland, and the bulbourethral glands provide a fluid medium for the sperm which move from the vas deferens through the ejaculatory duct to the urethra in the penis. The foreskin (prepuce) is removed when a penis is circumcised.

Table 21.1　Male Reproductive Organs

Organ	Function
Testes	Produce sperm and sex hormones
Epididymides	Ducts where sperm mature and some sperm are stored
Vasa deferentia	Conduct and store sperm
Seminal vesicles	Contribute nutrients and fluid to semen
Prostate gland	Contributes basic fluid to semen
Urethra	Conducts sperm
Bulbourethral glands	Contribute mucoid fluid to semen
Penis	Organ of sexual intercourse

Orgasm in Males

The **penis** (Fig. 21.2) is the male organ of sexual intercourse. The penis has a long shaft and an enlarged tip called the glans penis. The glans penis is normally covered by a layer of skin called the foreskin. Circumcision, the surgical removal of the foreskin, is usually done soon after birth.

Spongy, erectile tissue containing distensible blood spaces extends through the shaft of the penis. During sexual arousal, autonomic nerve impulses lead to the production of cGMP (cyclic guanosine monophosphate) in smooth muscle cells, and the erectile tissue fills with blood. The veins that take blood away from the penis are compressed, and the penis becomes erect. **Erectile dysfunction** (formerly called impotency) exists when the erectile tissue doesn't expand enough to compress the veins. The drug Viagra inhibits an enzyme that breaks down cGMP, ensuring that a full erection will take place. However, vision problems may occur when taking Viagra because the same enzyme occurs in the retina.

As sexual stimulation intensifies, sperm enter the urethra from each vas deferens, and the glands contribute secretions to the seminal fluid. Once seminal fluid is in the urethra, rhythmic muscle contractions cause it to be expelled from the penis in spurts (ejaculation). During ejaculation, a sphincter closes off the bladder so that no urine enters the urethra. (Notice that the urethra carries either urine or semen at different times.)

The contractions that expel seminal fluid from the penis are a part of male orgasm, the physiological and psychological sensations that occur at the climax of sexual stimulation. The psychological sensation of pleasure is centered in the brain, but the physiological reactions involve the genital (reproductive) organs and associated muscles, as well as the entire body. Marked muscular tension is followed by contraction and relaxation.

Following ejaculation and/or loss of sexual arousal, the penis returns to its normal flaccid state. After ejaculation, a male typically experiences a period of time, called the refractory period, during which stimulation does not bring about an erection. The length of the refractory period increases with age.

There may be in excess of 400 million sperm in the 3.5 ml of semen expelled during ejaculation. The sperm count can be much lower than this, however, and fertilization of the egg by a sperm still can take place.

Semen is ejaculated from the penis at the time of male orgasm.

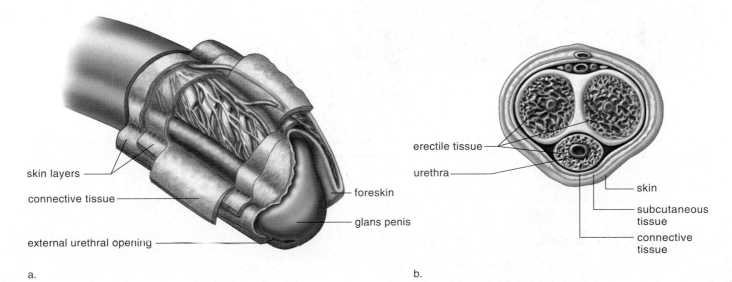

a.

erectile tissue

urethra

skin

subcutaneous tissue

connective tissue

skin layers

connective tissue

foreskin

glans penis

external urethral opening

b.

Figure 21.2　Penis anatomy.
a. Beneath the skin and the connective tissue lies the urethra, surrounded by erectile tissue. This tissue expands to form the glans penis, which in uncircumcised males is partially covered by the foreskin (prepuce). **b.** Two other columns of erectile tissue in the penis are located dorsally.

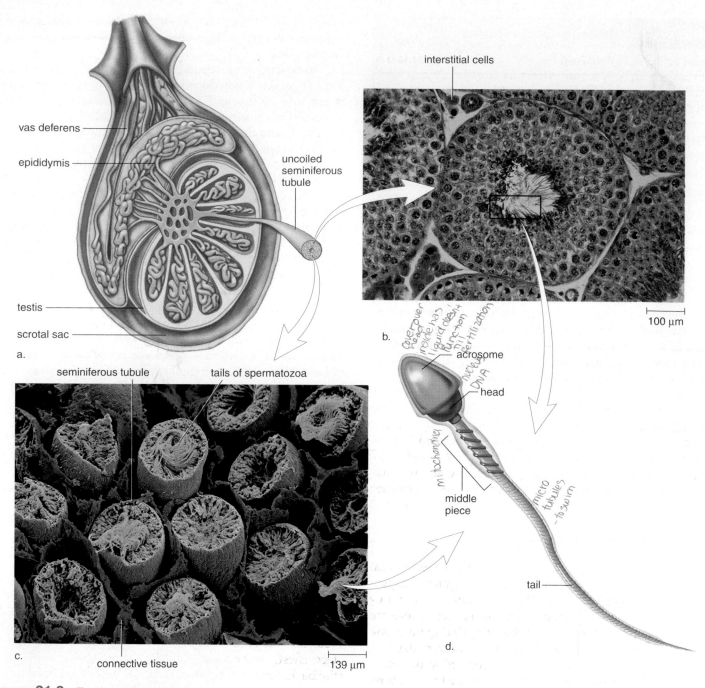

Figure 21.3   Testis and sperm.
a. The lobules of a testis contain seminiferous tubules. **b.** Light micrograph of a cross section of the seminiferous tubules shows interstitial cells occurring in clumps among the seminiferous tubules. **c.** Scanning electron micrograph of a cross section of the seminiferous tubules, where spermatogenesis occurs. **d.** A sperm has a head, a middle piece, and a tail. The nucleus is in the head, which is capped by the enzyme-containing acrosome.

Male Gonads, the Testes

The testes, which produce sperm and also the male sex hormones, lie outside the abdominal cavity of the male within the scrotum. The testes begin their development inside the abdominal cavity but descend into the scrotal sacs during the last two months of fetal development. If, by chance, the testes do not descend and the male is not treated or operated on to place the testes in the scrotum, sterility—the inability to produce offspring—usually follows. This is because the internal temperature of the body is too high to produce viable sperm. The scrotum helps regulate the temperature of the testes by holding them closer or further away from the body.

Seminiferous Tubules

A longitudinal section of a testis shows that it is composed of compartments called lobules, each of which contains one to three tightly coiled **seminiferous tubules** (Fig. 21.3*a*). Altogether, these tubules have a combined length of approximately 250 meters. A microscopic cross section of a seminiferous tubule reveals that it is packed with cells undergoing **spermatogenesis** (Fig. 21.3*b* and *c*), the production of sperm. Newly formed cells move away from the outer wall, increase in size, and undergo meiosis to become spermatids, which contain twenty-three chromosomes. Spermatids then differentiate into sperm. Also present are sustentacular cells (Sertoli cells), which support, nourish, and regulate the spermatogenic cells.

Mature **sperm,** or spermatozoa, have three distinct parts: a head, a middle piece, and a tail (Fig. 21.3*d*). Mitochondria in the middle piece provide energy for the movement of the tail, which has the structure of a flagellum. The head contains a nucleus covered by a cap called the **acrosome,** which stores enzymes needed to penetrate the egg. The ejaculated semen of a normal human male contains several hundred million sperm, but only one sperm normally enters an egg. Sperm usually do not live more than 48 hours in the female genital tract.

Interstitial Cells

The male sex hormones, the androgens, are secreted by cells that lie between the seminiferous tubules. Therefore, they are called **interstitial cells** (see Fig. 21.3*b*). The most important of the androgens is testosterone, whose functions are discussed next.

Hormonal Regulation in Males

The hypothalamus has ultimate control of the testes' sexual function because it secretes a hormone called **gonadotropin-releasing hormone,** or **GnRH,** that stimulates the anterior pituitary to secrete the gonadotropic hormones. There are two gonadotropic hormones—**follicle-stimulating hormone (FSH)** and **luteinizing hormone (LH)**—in both males and females. In males, FSH promotes the production of sperm in the seminiferous tubules, which also release the hormone inhibin.

LH in males is sometimes given the name **interstitial cell-stimulating hormone (ICSH)** because it controls the production of testosterone by the interstitial cells. All these hormones are involved in a negative feedback relationship that maintains the fairly constant production of sperm and testosterone (Fig. 21.4).

Testosterone, the main sex hormone in males, is essential for the normal development and functioning of the organs listed in Table 21.1. Testosterone also brings about and maintains the male secondary sex characteristics that develop at the time of puberty. Males are generally taller than females and have broader shoulders and longer legs relative to trunk length. The deeper voices of males compared to those of

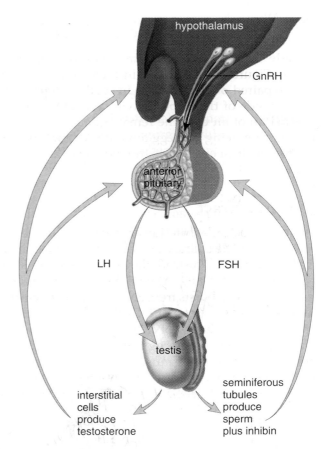

Figure 21.4 Hormonal control of testes.
GnRH (gonadotropin-releasing hormone) stimulates the anterior pituitary to secrete the gonadotropic hormones: FSH stimulates the production of sperm, and LH stimulates the production of testosterone. Testosterone and inhibin exert negative feedback control over the hypothalamus and the anterior pituitary, and this regulates the level of testosterone in the blood.

females are due to a larger larynx with longer vocal cords. Since the so-called Adam's apple is a part of the larynx, it is usually more prominent in males than in females. Testosterone causes males to develop noticeable hair on the face, chest, and occasionally other regions of the body such as the back. Testosterone also leads to the receding hairline and pattern baldness that occur in males.

Testosterone is responsible for the greater muscular development in males. Knowing this, both males and females sometimes take anabolic steroids, which are either testosterone or related steroid hormones resembling testosterone. Health problems involving the kidneys, the circulatory system, and hormonal imbalances can arise from such use. The testes shrink in size, and feminization of other male traits occurs.

The gonads in males are the testes, which produce sperm as well as testosterone, the most significant male sex hormone.

The Uterine Cycle

The female sex hormones, **estrogen** and **progesterone,** have numerous functions. These hormones affect the endometrium, causing the uterus to undergo a cyclical series of events known as the **uterine cycle** (Fig. 21.9). Twenty-eight-day cycles are divided as follows:

During *days 1–5,* a low level of female sex hormones in the body causes the endometrium to disintegrate and its blood vessels to rupture. On day one of the cycle, a flow of blood and tissues, known as the menses, passes out of the vagina during **menstruation,** also called the menstrual period.

During *days 6–13,* increased production of estrogen by a new ovarian follicle in the ovary causes the endometrium to thicken and become vascular and glandular. This is called the proliferative phase of the uterine cycle.

On *day 14* of a 28-day cycle, ovulation usually occurs.

During *days 15–28,* increased production of progesterone by the corpus luteum in the ovary causes the endometrium of the uterus to double or triple in thickness (from 1 mm to 2–3 mm) and the uterine glands to mature, producing a thick mucoid secretion. This is called the secretory phase of the uterine cycle. The endometrium is now prepared to receive the developing embryo. If this does not occur, the corpus luteum in the ovary degenerates, and the low level of sex hormones in the female body results in the endometrium breaking down.

Table 21.3 compares the stages of the uterine cycle with those of the ovarian cycle.

During the uterine cycle, the endometrium builds up and then is broken down during menstruation.

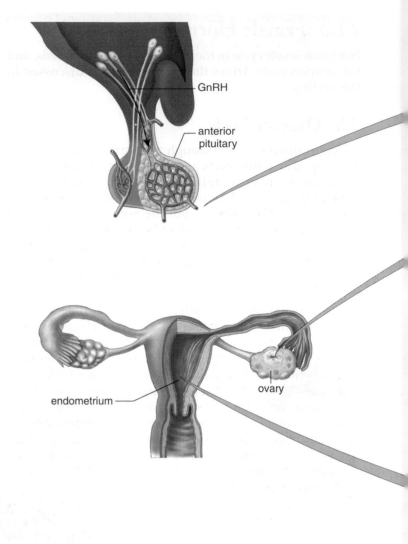

Table 21.3	Ovarian and Uterine Cycles		
Ovarian Cycle	**Events**	**Uterine Cycle**	**Events**
Follicular phase—Days 1–13	FSH secretion begins.	Menstruation—Days 1–5	Endometrium breaks down.
	Follicle maturation occurs.	Proliferative phase—Days 6–13	Endometrium rebuilds.
	Estrogen secretion is prominent.		
Ovulation—Day 14*	LH spike occurs.		
Luteal phase—Days 15–28	LH secretion continues.	Secretory phase—Days 15–28	Endometrium thickens, and glands are secretory.
	Corpus luteum forms.		
	Progesterone secretion is prominent.		

*Assuming a 28-day cycle.

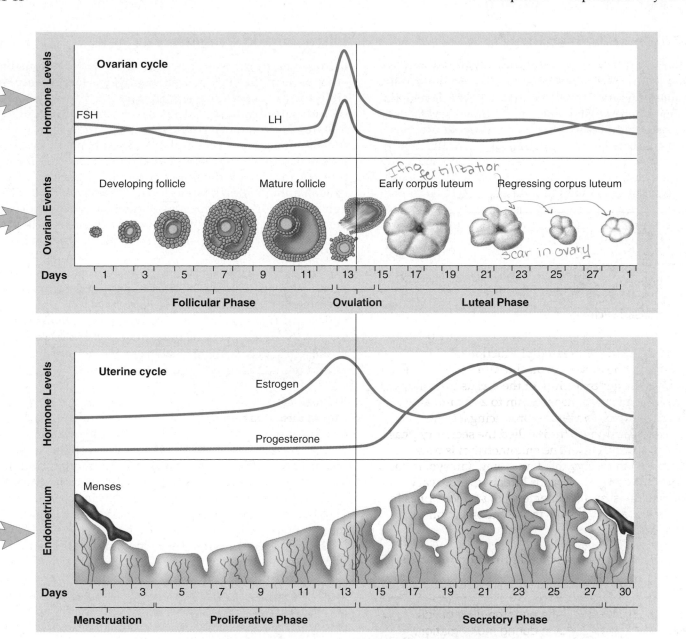

Figure 21.9   Female hormone levels.
During the follicular phase of the ovarian cycle, FSH released by the anterior pituitary promotes the maturation of a follicle in the ovary. The ovarian follicle produces increasing levels of estrogen, which causes the endometrium to thicken during the proliferative phase of the uterine cycle. After ovulation and during the luteal phase of the ovarian cycle, LH promotes the development of the corpus luteum. This structure produces increasing levels of progesterone, which causes the endometrial lining to become secretory. Menses due to the breakdown of the endometrium begins when progesterone production declines to a low level.

Fertilization and Pregnancy

If fertilization does occur, an embryo begins development even as it travels down the oviduct to the uterus. The endometrium is now prepared to receive the developing embryo, which becomes implanted in the lining several days following fertilization (Fig. 21.10). The **placenta** originates from both maternal and fetal tissues. It is the region of exchange of molecules between fetal and maternal blood, although the two rarely mix. At first, the placenta produces **human chorionic gonadotropin (HCG),** which maintains the corpus luteum in the ovary until the placenta begins its own production of progesterone and estrogen. Progesterone and estrogen have two effects: they shut down the anterior pituitary so that no new follicle in the ovaries matures, and they maintain the endometrium so that the corpus luteum in the ovary is no longer needed. Usually, there is no menstruation during pregnancy.

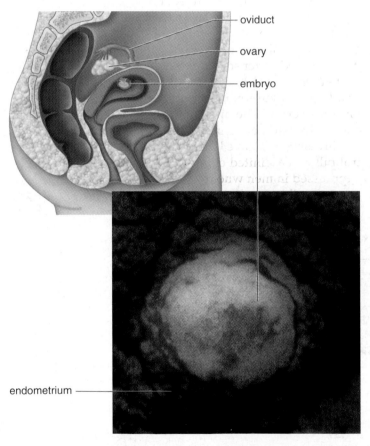

oviduct

ovary

embryo

endometrium

Figure 21.10   Implantation.
A scanning electron micrograph showing an embryo implanted in the endometrium on day 12 following fertilization.

Estrogen and Progesterone

Estrogen and progesterone affect not only the uterus but other parts of the body as well. Estrogen is largely responsible for the secondary sex characteristics in females, including body hair and fat distribution. In general, females have a more rounded appearance than males because of a greater accumulation of fat beneath the skin. Like males, females develop axillary and pubic hair during puberty. In females, the upper border of pubic hair is horizontal, but in males, it tapers toward the navel. Both estrogen and progesterone are also required for breast development. Other hormones are involved in milk production following pregnancy and milk letdown when a baby begins to nurse.

The pelvic girdle is wider and deeper in females, so the pelvic cavity usually has a larger relative size compared to that of males. This means that females have wider hips than males and that their thighs converge at a greater angle toward the knees. Because the female pelvis tilts forward, females tend to have more of a lower back curve than males, an abdominal bulge, and protruding buttocks.

Menopause

Menopause, the period in a woman's life during which the ovarian and uterine cycles cease, is likely to occur between ages 45 and 55. The ovaries are no longer responsive to the gonadotropic hormones produced by the anterior pituitary, and the ovaries no longer secrete estrogen or progesterone. At the onset of menopause, the uterine cycle becomes irregular, but as long as menstruation occurs, it is still possible for a woman to conceive. Therefore, a woman is usually not considered to have completed menopause until menstruation is absent for a year.

The hormonal changes during menopause often produce physical symptoms, such as "hot flashes" (caused by circulatory irregularities), dizziness, headaches, insomnia, sleepiness, and depression. These symptoms may be mild or even absent. If they are severe, medical attention should be sought. Women sometimes report an increased sex drive following menopause. It has been suggested that this may be due to androgen production by the adrenal cortex.

Estrogen and progesterone produced by the ovaries are the female sex hormones. They foster the development of the reproductive organs, maintain the uterine cycle, and bring about the secondary sex characteristics in females.

21.4 Control of Reproduction

Several means are available to dampen or enhance our reproductive potential. **Contraceptives** are medications and devices that reduce the chance of pregnancy.

Birth Control Methods

The most reliable method of birth control is abstinence—that is, not engaging in sexual intercourse. This form of birth control has the added advantage of preventing transmission of a sexually transmitted disease. Table 21.4 lists other means of birth control used in the United States, and rates their effectiveness. For example, with natural family planning, one of the least effective methods given in the table, we expect that within a year, 70 out of 100, or 70%, of sexually active women will not get pregnant, while 30 women will get pregnant.

Figure 21.11 features some of the most effective and commonly used means of birth control. Oral contraception **(birth control pills)** often involves taking a combination of estrogen and progesterone on a daily basis. The estrogen and progesterone in the birth control pill effectively shut down the pituitary production of both FSH and LH so that no follicle in the ovary begins to develop in the ovary; and since ovulation does not occur, pregnancy cannot take place. Because of possible side effects, women taking birth control pills should see a physician regularly.

An **intrauterine device (IUD)** is a small piece of molded plastic that is inserted into the uterus by a physician. IUDs are believed to alter the environment of the uterus and oviducts so that fertilization probably will not occur—but if fertilization should occur, implantation cannot take place. The type of IUD featured in Figure 21.11 has copper wire wrapped around the plastic.

The **diaphragm** is a soft latex cup with a flexible rim that lodges behind the pubic bone and fits over the cervix. Each woman must be properly fitted by a physician, and the diaphragm can be inserted into the vagina no more than two hours before sexual relations. Also, it must be used with spermicidal jelly or cream and should be left in place at least six hours after sexual relations. The cervical cap is a minidiaphragm.

A male **condom** is most often a latex sheath that fits over the erect penis. The ejaculate is trapped inside the sheath, and thus does not enter the vagina. When used in conjunction with a spermicide, the protection is better than with the condom alone.

Contraceptive implants utilize a synthetic progesterone to prevent ovulation by disrupting the ovarian cycle. Six match-sized, time-release capsules are surgically inserted under the skin of a woman's upper arm. The implants may remain effective for five years. **Depo-Provera injections** utilize a synthetic progesterone that must be administered every three months. The injections cause changes in the endometrium that make pregnancy less likely to occur.

There has been a revival of interest in barrier methods of birth control, including the latex male condom, because these methods offer some protection against sexually transmitted diseases. A female condom, now available, consists of a large polyurethane tube with a flexible ring that fits onto the cervix. The open end of the tube has a ring that covers the external genitals.

Investigators have long searched for a "male birth control pill." In a limited clinical trial, sperm production was suppressed in men who received a daily oral dose of progesterone. Skin implants of testosterone were required to maintain secondary sex characteristics. Larger studies are being planned.

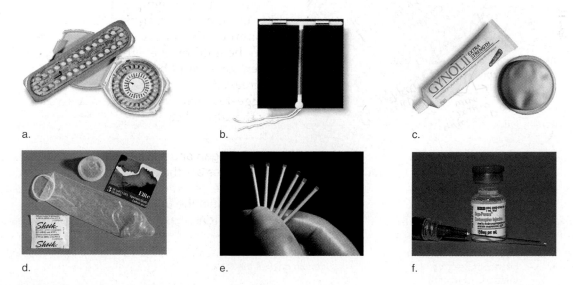

a.   b.   c.

d.   e.   f.

Figure 21.11   Various birth-control devices.
a. Oral contraception (birth control pills). **b.** Intrauterine device. **c.** Spermicidal jelly and diaphragm. **d.** Male condom. **e.** Contraceptive implants. **f.** Depo-Provera injection.

Contraceptive vaccines are now being developed. For example, a vaccine intended to immunize women against HCG, the hormone so necessary to maintaining the implantation of the embryo, was successful in a limited clinical trial. Since HCG is not normally present in the body, no untoward autoimmune reaction is expected, but the immunization does wear off with time. Others believe that it would also be possible to develop a safe antisperm vaccine that could be used in women.

Morning-After Pills

The expression "morning-after pill" refers to a medication that will prevent pregnancy after unprotected intercourse. The expression is a misnomer in that medication can begin one to several days after unprotected intercourse.

One type, a kit called Preven, contains four synthetic progesterone pills; two are taken up to 72 hours after unprotected intercourse, and two more are taken 12 hours later. The medication upsets the normal uterine cycle, making it difficult for an embryo to implant itself in the endometrium. In a recent study, it was estimated that the medication was 85% effective in preventing unintended pregnancies.

Mifepristone, better known as RU-486, is a pill that is presently used to cause the loss of an implanted embryo by blocking the progesterone receptor proteins of endometrial cells. Without functioning receptors for progesterone, the endometrium sloughs off, carrying the embryo with it. When taken in conjunction with a prostaglandin to induce uterine contractions, RU-486 is 95% effective. It is possible that some day this medication will also be a "morning-after pill," taken when menstruation is late without evidence that pregnancy has occurred.

The birth control methods and devices now available vary in effectiveness. New methods of birth control are expected to be developed.

Table 21.4	Common Birth Control Methods			
Name	Procedure	Methodology	Effectiveness	Risk
Abstinence	Refrain from sexual intercourse	No sperm in vagina	100%	None
Vasectomy	Vasa deferentia cut and tied	No sperm in seminal fluid	Almost 100%	Irreversible sterility
Tubal ligation	Oviducts cut and tied	No eggs in oviduct	Almost 100%	Irreversible sterility
Oral contraception	Hormone medication taken daily	Anterior pituitary does not release FSH and LH	Almost 100%	Thromboembolism, especially in smokers
Depo-Provera injection	Four injections of progesterone-like steroid given per year	Anterior pituitary does not release FSH and LH	About 99%	Breast cancer? Osteoporosis?
Contraceptive implants	Tubes of progestin (form of progesterone) implanted under skin	Anterior pituitary does not release FSH and LH	More than 90%	Presently none known
Intrauterine device (IUD)	Plastic coil inserted into uterus by physician	Prevents implantation	More than 90%	Infection (pelvic inflammatory disease, PID)
Diaphragm	Latex cup inserted into vagina to cover cervix before intercourse	Blocks entrance of sperm to uterus	With jelly, about 90%	Presently none known
Cervical cap	Latex cap held by suction over cervix	Delivers spermicide near cervix	Almost 85%	Cancer of cervix
Male condom	Latex sheath fitted over erect penis	Traps sperm and prevents STDs	About 85%	Presently none known
Female condom	Polyurethane liner fitted inside vagina	Blocks entrance of sperm to uterus and prevents STDs	About 85%	Presently none known
Coitus interruptus	Penis withdrawn before ejaculation	Prevents sperm from entering vagina	About 75%	Presently none known
Jellies, creams, foams	These spermicidal products inserted before intercourse	Kills a large number of sperm	About 75%	Presently none known
Natural family planning	Day of ovulation determined by record keeping; various methods of testing	Intercourse avoided on certain days of the month	About 70%	Presently none known
Douche	Vagina cleansed after intercourse	Washes out sperm	Less than 70%	Presently none known

Ecology Focus

Endocrine-Disrupting Contaminants

Do when ecology chapt comes [handwritten note]

Rachel Carson's book, *Silent Spring*, published in 1962 predicted that pesticides would have a deleterious effect on animal life. Soon, thereafter, it was found that pesticides caused the thinning of egg shells in bald eagles to the point that their eggs broke and the chicks died. Additionally, populations of terns, gulls, cormorants, and lake trout declined after they ate fish contaminated by high levels of environmental toxins. The concern was so great that the United States Environmental Protection Agency came into existence. This agency and civilian environmental groups have brought about a reduction in pollution release and a cleaning up of emissions. Even so, we are now aware of more subtle effects that pollutants can have.

Hormones influence nearly all aspects of physiology and behavior in animals, including tissue differentiation, growth, and reproduction. Therefore, when wildlife in contaminated areas began to exhibit certain types of abnormalities, researchers began to think that certain pollutants can affect the endocrine system. In England, male fish exposed to sewage developed ovarian tissue and produced a metabolite normally found only in females during egg formation. In California, western gulls displayed abnormalities in gonad structure and nesting behaviors. Hatchling alligators in Florida possessed abnormal gonads and hormone concentrations linked to nesting.

At first, such effects seemed to indicate only the involvement of the female hormone estrogen, and researchers therefore called the contaminants ecoestrogens. Many of the contaminants interact with hormone receptors, and in that way cause developmental effects. Others bind directly with sex hormones like testosterone and estradiol. Still others alter the physiology of growth hormones and neurotransmitters responsible for brain development and behavior. Therefore, the preferred term today

for these pollutants is endocrine-disrupting contaminants (EDCs).

Many EDCs are chemicals used as pesticides and herbicides in agriculture, and some are associated with the manufacture of various other organic molecules such as PCBs (polychlorinated biphenyls). Some chemicals shown to influence hormones are found in plastics, food additives, and personal hygiene products. In mice, phthalate esters, which are plastic components, affect neonatal development when present in the part-per-trillion range. It is, therefore, of great concern that EDCs have been found to exist at levels one thousand times greater than this—even in amounts comparable to functional hormone levels in the human body. Therefore, it is not surprising that EDCs are affecting the endocrine systems of a wide range of organisms (Fig. 21A).

Scientists and those representing industrial manufacturers continue to debate whether EDCs pose a health risk to humans. Some suspect that EDCs lower sperm counts, reduce male and female fertility, and increase rates of certain cancers (breast, ovary, testis, and prostate). Additionally, some studies suggest that EDCs contribute to learning deficits and behavioral problems in children. Laboratory and field research continues to identify chemicals that have the ability to influence the endocrine system. Millions of tons of potential EDCs are produced annually in the United States, and the United States Environmental Protection Agency is under pressure to certify these compounds as safe. The European Economic Community has already restricted the use of certain EDCs, and has banned the production of specific plastic components that are found in items intended for use by children, specifically in toys. Only through continued scientific research and the cooperation of industry can we identify the risks that EDCs pose to the environment, wildlife, and humans.

Figure 21A Exposure to endocrine-disrupting contaminants.
These types of wildlife and also humans are exposed to endocrine-disrupting contaminants that can seriously affect their health and reproductive abilities.

Infertility

Infertility is the failure of a couple to achieve pregnancy after one year of regular, unprotected intercourse. The American Medical Association estimates that 15% of all couples are infertile. The cause of infertility can be attributed to the male (40%), the female (40%), or both (20%).

Causes of Infertility

Common causes of infertility in females are blocked oviducts and endometriosis. Most often, pelvic inflammatory disease, as discussed on page 434, is the cause of blocked oviducts. **Endometriosis** is the presence of uterine tissue outside the uterus, particularly in the oviducts and on the abdominal organs. Endometriosis occurs when the menstrual discharge flows up into the oviducts and out into the abdominal cavity. This backward flow allows living uterine cells to establish themselves in the abdominal cavity, where they go through the usual uterine cycle, causing pain and structural abnormalities that make it more difficult for a woman to conceive.

Sometimes the causes of infertility can be corrected by medical intervention so that couples can have children. If no obstruction is apparent and body weight is normal, it is possible to give females fertility drugs, which are gonadotropic hormones that stimulate the ovaries and bring about ovulation. Such hormone treatments may cause multiple ovulations and higher-order multiple births (see the Bioethical Focus in Chapter 20).

The most frequent cause of infertility in males is low sperm count and/or a large proportion of abnormal sperm. Disease, radiation, chemical mutagens, high testes temperature, and the use of psychoactive drugs can contribute to this condition. Conforming to a healthier lifestyle can sometimes lead to an improved sperm count, but thus far no hormonal treatment has proven to be especially successful. For men who have had a vasectomy (a portion of the vasa deferentia removed), reversal surgery is available, but the pregnancy success rate is only about 40% unless the vasectomy occurred less than three years earlier.

When reproduction does not occur in the usual manner, many couples adopt a child. Others sometimes try one of the assisted reproductive technologies discussed in the following paragraphs.

Assisted Reproductive Technologies

Assisted reproductive technologies (ART) consist of techniques used to increase the chances of pregnancy. Often, sperm and/or eggs are retrieved from the testes and ovaries, and fertilization takes place in a clinical or laboratory setting.

Artificial Insemination by Donor (AID)   During artificial insemination, sperm are placed in the vagina by a physician. Sometimes a woman is artificially inseminated by her partner's sperm. This is especially helpful if the partner has a low sperm count, because the sperm can be collected over a period of time and concentrated so that the sperm count is sufficient to result in fertilization. Often, however, a woman is inseminated by sperm acquired from a donor who is a complete stranger to her. At times, a combination of partner and donor sperm is used.

A variation of AID is *intrauterine insemination (IUI)*. In IUI, fertility drugs are given to stimulate the ovaries, and then the donor's sperm is placed in the uterus rather than in the vagina.

If the prospective parents wish, sperm can be sorted into those that are believed to be X-bearing or Y-bearing to increase the chances of having a child of the desired sex.

In Vitro Fertilization (IVF)   During IVF, conception occurs in laboratory glassware. Ultrasound machines can now spot follicles in the ovaries that hold immature eggs; therefore, the latest method is to forgo the administration of fertility drugs and retrieve immature eggs by using a needle. The immature eggs are then brought to maturity in glassware before concentrated sperm are added. After about two to four days, the embryos are ready to be transferred to the uterus of the woman, who is now in the secretory phase of her uterine cycle. If desired, the embryos can be tested for a genetic disease, and only those found to be free of disease will be used. If implantation is successful, development is normal and continues to term.

Gamete Intrafallopian Transfer (GIFT)   The term **gamete** refers to a sex cell, either a sperm or an egg. Gamete intrafallopian transfer was devised to overcome the low success rate (15–20%) of in vitro fertilization. The method is exactly the same as in vitro fertilization, except the eggs and the sperm are placed in the oviducts immediately after they have been brought together. GIFT has the advantage of being a one-step procedure for the woman—the eggs are removed and reintroduced all in the same time period. A variation on this procedure is to fertilize the eggs in the laboratory and then place the zygotes in the oviducts.

Surrogate Mothers   In some instances, women are contracted and paid to have babies. These women are called surrogate mothers. The sperm and even the egg can be contributed by the contracting parents.

Intracytoplasmic Sperm Injection (ICSI)   In this highly sophisticated procedure, a single sperm is injected into an egg. It is used effectively when a man has severe infertility problems.

If all the alternative methods discussed were employed simultaneously, it would be possible for a baby to have five parents: (1) sperm donor, (2) egg donor, (3) surrogate mother, and (4) and (5) contracting mother and father.

When corrective procedures fail to reverse infertility, assisted reproductive technologies may be considered.

21.5 Sexually Transmitted Diseases

Not testing

Many diseases are transmitted by sexual contact, and are therefore called sexually transmitted diseases (STDs). Our discussion centers on those that are most prevalent. AIDS, genital herpes, and genital warts are viral diseases; therefore, they are difficult to treat because viral infections do not respond to traditional antibiotics. However, other types of drugs have been developed to treat them. And although gonorrhea and chlamydia are treatable with appropriate antibiotic therapy, they are not always promptly diagnosed. Unfortunately, the human body does not generally become immune to STDs; only one (Hepatitis B) can be prevented by vaccine therapy.

AIDS

Acquired immunodeficiency syndrome (AIDS) is caused by a group of related retroviruses known as HIV (human immunodeficiency viruses). Much has been learned about the structure and reproductive cycle of HIV-1, the most frequent cause of AIDS in the United States. HIV are *retroviruses*, meaning that their genetic material consists of RNA instead of DNA. Figure 21.12 shows HIV-1 in the process of entering a cell. Once inside the host cell, the viral enzyme called reverse transcriptase makes a DNA copy of viral RNA. Now the enzyme integrase helps integrate viral DNA into a host chromosome. Following integration, viral DNA directs the production of more viral RNA. Some of the viral RNA proceeds to the ribosomes where it brings about the synthesis of capsid proteins. The viral enzyme protease helps shape the capsids before the assembly of new viruses takes place. Then HIV-1 buds from the host cell.

The immune system usually protects us from disease. Two ways the immune system does this is by producing antibodies, proteins that attack foreign proteins called antigens, and by attacking infected cells outright. The primary hosts for HIV are helper T lymphocytes, the type of white blood cell that stimulates B lymphocytes to produce antibodies and cytotoxic T lymphocytes to attack and kill virus-infected cells.

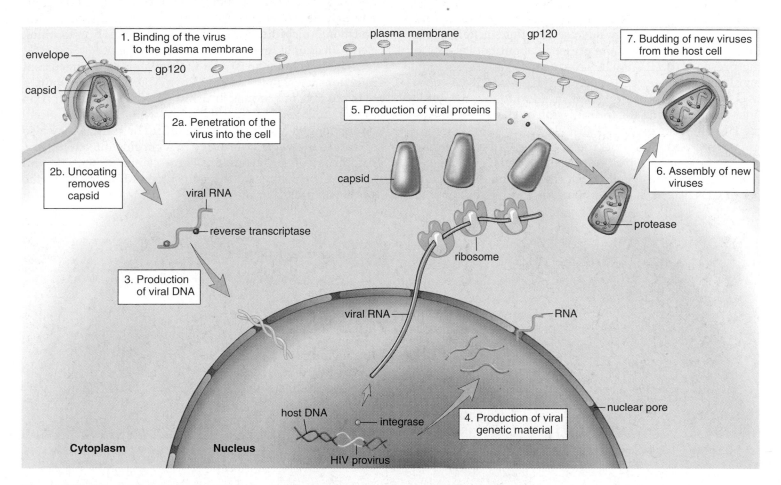

Figure 21.12   Reproduction of HIV.
HIV-1 consists of an inner core of RNA plus three viral enzymes, a protein coat called a capsid, and an outer envelope having gp120 protein spikes. Only the capsid and its contents enter the cell where the three enzymes help produce more viruses: reverse transcriptase produces viral DNA, integrase helps integrate viral DNA into the host chromosome (called a provirus), and protease helps shape more viral capsids. The parts of a virus assemble before HIV-1 buds from the cell. The virus picks up an envelope containing gp120 spikes when it buds from the cell.

Phases of an HIV Infection

HIV primarily infect helper T lymphocytes that are also called CD4 T lymphocytes, or simply CD4 T cells. These cells have molecules called CD4 on their surface and HIV-1 attaches to this molecule before it enters the cell.

An HIV-1 infection is now divided into three categories, designated by the letters A, B, and C, by the Centers for Disease Control and Prevention.

Category A: Acute Phase   During Category A stage, the T lymphocyte count is 500 per mm^3 or greater. This count is sufficient for the immune system to function normally.

Today, investigators are able not only to track the blood level of CD4 T cells, but also to monitor the viral load. The viral load is the number of HIV particles in the blood. At the start of an HIV infection, the virus is replicating ferociously, and the killing of CD4 T cells is evident because the blood level of these cells drops dramatically. For a few weeks, however, people don't usually have any symptoms at all. Then, a few (1–2%) do have mononucleosis-like symptoms that may include fever, chills, aches, swollen lymph nodes, and an itchy rash. These symptoms disappear, and no other symptoms appear for quite some time. The HIV blood test commonly used at clinics is not yet positive because it tests for the presence of antibodies, not for the presence of HIV itself. This means that the person is highly infectious, even though the HIV blood test is negative. For this reason, all persons need to follow the guidelines for preventing the transmission of HIV as outlined in the Health Focus on page 432.

After a period of time, the body responds to the infection by increased activity of immune cells, and the HIV blood test becomes positive. During this phase, the number of CD4 T cells is greater than the viral load. But some investigators believe that a great unseen battle is going on. The body is staying ahead of the hordes of viruses entering the blood by producing as many as one to two billion new helper T lymphocytes each day. This is called the "kitchen sink model" for CD4 loss. The sink's faucet (production of new CD4 T cells) and the sink's drain (destruction of CD4 T cells) are wide open. As long as the body's production of new CD4 T cells is able to keep pace with the destruction of these cells by HIV and by cytotoxic T cells, the person has a healthy immune system that can deal with the infection.

Category B: Chronic Phase   Several months to several years after infection, an untreated individual will probably progress to Category B. During this stage, the CD4 T cell count is 200 to 499 per mm^3 of blood, and most likely, symptoms will begin to appear. Symptoms of a Category B level infection include swollen lymph nodes in the neck, armpits, or groin that persist for three months or more; severe fatigue not related to exercise or drug use; unexplained persistent or recurrent fevers, often with night sweats; persistent cough not associated with smoking, a cold, or the flu; and persistent diarrhea. Also possible are signs of nervous system impairment, including loss of memory, inability to think clearly, loss of judgment, and/or depression.

The development of non-life-threatening but recurrent infections is a signal that full-blown AIDS will occur shortly. One possible infection is thrush, a *Candida albicans* infection that is identified by the presence of white spots and ulcers on the tongue and inside the mouth. The fungus may also spread to the vagina, resulting in a chronic infection there. Another frequent infection is herpes simplex, with painful and persistent sores on the skin surrounding the anus, the genital area, and/or the mouth.

a. AIDS patient, Tom Moran July 1987

b. AIDS patient Tom Moran, early January 1988

c. AIDS patient Tom Moran, late January 1988

Figure 21.13   The course of an AIDS infection.
These photos show the effect of an HIV infection in one individual who progressed through all the phases of AIDS.

Category C: AIDS When a person has AIDS, the CD4 T cell count has fallen below 200 per mm^3, and the lymph nodes have degenerated. The patient is extremely thin and weak due to persistent diarrhea and coughing, and will most likely develop one of the opportunistic infections. An **opportunistic infection** is one that has the *opportunity* to occur only because the immune system is severely weakened. Persons with AIDS die from one or more of the following diseases rather than from the HIV infection itself:

- *Pneumocystis carinii* pneumonia. The lungs become useless as they fill with fluid and debris due to an infection with a protozoan.
- *Mycobacterium tuberculosis.* This bacterial infection, usually of the lungs, is seen more often as an infection of lymph nodes and other organs in patients with AIDS.
- Toxoplasmic encephalitis. A protozoan parasite that ordinarily lives in cats and other animals, including humans, causes this disease. Many people harbor a latent infection in the brain or muscle, but in AIDS patients, the infection leads to loss of brain cells, seizures, and weakness.
- Kaposi's sarcoma. This unusual cancer of the blood vessels gives rise to reddish-purple, coin-sized spots and lesions on the skin.
- Invasive cervical cancer. This cancer of the cervix spreads to nearby tissues. This condition was added to the list when AIDS became more common in women.

Although there are newly developed drugs to deal with opportunistic diseases, most AIDS patients are repeatedly hospitalized due to weight loss, constant fatigue, and multiple infections (Fig. 21.13). Death usually follows in 2–4 years.

Treatment for HIV

Until a few years ago, an HIV infection almost invariably led to AIDS and an early death. The medical profession was able to treat the opportunistic infections that stemmed from immune failure, but had no drugs for controlling HIV itself. But since late 1995, scientists have gained a much better understanding of the structure of HIV and its life cycle. Now therapy is available that successfully controls HIV replication and keeps patients in the chronic phase of infection for a variable number of years so that the development of AIDS is postponed.

Drug Therapy There is no cure for AIDS, but a treatment called highly active antiretroviral therapy (HAART) is usually able to stop HIV replication to such an extent that the viral load becomes undetectable. Even so, investigators know that about one million viruses are still present, including those that exist only as inactive proviruses.

HAART utilizes a combination of two types of drugs: reverse transcriptase and protease inhibitors. The well-publicized drug called AZT and several others are reverse transcriptase inhibitors, which prevent viral DNA from being produced. When HIV protease is blocked, the resulting viruses lack the capacity to cause infection. It is important to realize that these drugs are very expensive, that they cause side effects such as diarrhea, neuropathy (painful or numb feet), hepatitis, and possibly diabetes, and that the regimen of pill taking throughout the day is very demanding. If the drugs are not taken as prescribed or if therapy is stopped, resistance may occur. Researchers are trying to develop new drugs that might be helpful against resistant strains. A new class of drugs called fusion inhibitors seems to block HIV's entry into cells. Even more experimental is the discovery of compounds that sabotage integrase, the enzyme that splices viral RNA into host-cell DNA.

Investigators have found that when HAART is discontinued, the virus rebounds. It may then be possible to help the immune system counter the virus by injecting the patient with a stimulatory cytokine such as interferon or interleukin-2.

A pregnant woman who is infected with HIV and takes reverse transcriptase inhibitors during her pregnancy reduces the chances of HIV transmission to her newborn by nearly 66%. If possible, drug therapy should be delayed until the 10th to 12th week of pregnancy to minimize any adverse effects of AZT on fetal development. Treatment with reverse transcriptase inhibitors during the last few weeks of a pregnancy only cuts transmission of HIV to an offspring by half.

Vaccines There is a general consensus that control of the AIDS epidemic will not occur until a vaccine that prevents an HIV infection is developed. An effective vaccine should bring about a twofold immune response: production of antibodies by B cells and stimulation of cytotoxic T cells.

Traditionally, vaccines are made by weakening a pathogen so that it will not cause disease when it is injected into the body. One group of investigators using this approach announced that they have found a way to expose hidden parts of gp120 (the viral spike shown in Fig. 21.12) so the immune system can better learn to recognize this antigen.

Some scientists have developed DNA vaccines by inserting pieces of viral DNA into plasmids, which are rings of DNA from bacterial cells. This vaccine is expected to enter cells and start producing proteins that will migrate to the surface, and thereby alert cytotoxic T cells.

Others have been working on subunit vaccines that utilize just a single HIV protein, such as gp120, as the vaccine. So far this approach has not resulted in sufficient antibodies to keep an infection at bay. After many clinical trials, none too successful, most investigators now agree that a combination of various vaccines may be the best strategy.

HIV is a retrovirus that infects immune cells carrying a CD4 receptor. The diseases progresses through two phases before the final phase AIDS occurs. Combination drug therapy has met with encouraging success against an HIV infection. Development of a vaccine is also being pursued.

Preventing Transmission of HIV

HIV is transmitted by sexual contact with an infected person or by passing virus-infected T lymphocytes in blood from one person to another. To date, as many as 40 million people worldwide may have contracted HIV, and almost 12 million have died. HIV infections are not distributed equally throughout the world. Most infected people live in Africa (66%), where the infection first began, but new infections are now occurring at the fastest rate in Southeast Asia and the Indian subcontinent. The largest proportion of people with AIDS in the United States are homosexual men, but the fastest rate of increase is now seen among homosexuals, particularly minority females. Babies born to HIV-infected women may become infected before or during birth, or through breast-feeding after birth. Transmission at birth can be prevented if the mother takes AZT and delivers by planned cesarean section.

Following the guidelines listed here can help prevent transmission of HIV.

Sexual Activities Transmit HIV

Abstain from sexual intercourse or develop a long-term monogamous (always the same partner) sexual relationship with a partner who is free of HIV.

Refrain from multiple sex partners or having relations with someone who has multiple sex partners. If you have sex with two other people and each of these has sex with two people and so forth, the number of people who are relating is quite large.

Remember that although the prevalence of AIDS is presently higher among homosexuals and bisexuals the highest rate of increase is now occurring among heterosexuals. The lining of the uterus is only one cell thick and it does allow CD4 T cells to enter (Fig. 21B).

Be aware that having relations with an intravenous drug user is risky because the behavior of this group risks AIDS. Be aware that anyone who already has another sexually transmitted disease is more susceptible to an HIV infection.

Uncircumcised males are more likely to become infected than circumcised males because vaginal secretions can remain under the foreskin for a long period of time.

Avoid anal-rectal intercourse (in which the penis is inserted into the rectum) because the lining of the rectum is thin and infected CD4 T cells can easily enter the body here.

Figure 21B **Transmission by way of the uterus.**
HIV is spread by passing virus-infected CD4 cells found in body secretions or in blood from one person to another. Arrows show lymphocytes moving from the uterine cavity through the endometrium to enter the body.

Unsafe Sexual Practices Transmit HIV

Always use a latex condom during sexual intercourse if you do not know for certain that your partner has been free of HIV for the past five years. Be sure to follow the directions supplied by the manufacturer. Use of a water-based spermicide containing nonoxynol-9 in addition to the condom can offer further protection because nonoxynol-9 immobilizes the virus and virus-infected lymphocytes.

Avoid fellatio (kissing and insertion of the penis into a partner's mouth) *and cunnilingus* (kissing and insertion of the tongue into the vagina) because they may be a means of transmission. The mouth and gums often have cuts and sores that facilitate the entrance of infected CD4 T cells.

Be cautious about the use of alcohol or any drug that may prevent you from being able to control your behavior.

Drug Use Transmits HIV

Stop, if necessary, or do not start the habit of injecting drugs into your veins. Be aware that HIV can be spread by blood-to-blood contact.

Always use a new sterile needle for injection or one that has been cleaned in bleach if you are a drug user and cannot stop your behavior.

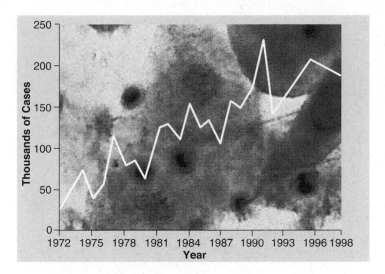

Figure 21.14 **Genital herpes.**
A graph depicting the incidence of new reported cases of genital herpes in the United States from 1972 to 1998 is superimposed on a photomicrograph of cells infected with the herpesvirus.

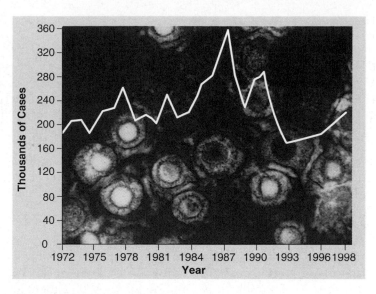

Figure 21.15 **Genital warts.**
A graph depicting the incidence of new cases of genital warts reported in the United States from 1972 to 1998 is superimposed on a photomicrograph of human papillomaviruses.

Genital Herpes

Genital herpes is caused by the herpes simplex virus, of which there are two types: type 1, which usually causes cold sores and fever blisters, and type 2, which more often causes genital herpes. Crossover infections do occur, however. That is, type 1 has been known to cause a genital infection, while type 2 has been known to cause cold sores and fever blisters.

Genital herpes is one of the more prevalent sexually transmitted diseases today (Fig. 21.14). At any one time, millions of persons could be having recurring symptoms. After infection, the first symptoms may be a tingling or itching sensation before blisters appear at the infected site, usually within 2–20 days. Once the blisters rupture, they leave painful ulcers, which may take as long as three weeks or as little as five days to heal. These symptoms may be accompanied by fever, pain upon urination, and swollen lymph nodes.

After the ulcers heal, the disease is only dormant, not gone. Blisters can recur repeatedly at variable intervals. Sunlight, sexual intercourse, menstruation, and stress seem to cause the symptoms of genital herpes to recur. While the virus is dormant, it primarily resides in the ganglia of sensory nerves associated with the affected skin. Although type 2 was once thought to cause a form of cervical cancer, this is no longer believed.

Infection of the newborn can occur if the child comes in contact with a lesion in the birth canal. In 1–3 weeks, the infant is gravely ill and can become blind, have neurological disorders, including brain damage, or die. Birth by cesarean section prevents these adverse developments.

Genital Warts

Genital warts are caused by the human papillomaviruses (HPVs), which are sexually transmitted. A newborn can become infected by passage through the birth canal.

Over a million persons become infected each year with a form of HPV, but only a portion seek medical help (Fig. 21.15). Sometimes carriers do not have any sign of warts. If warts are present, they can be flat or raised. Warts can be found on the penis and the foreskin of males and at the vaginal orifice and cervix in females. Warts on the cervix are always flat and they can be hard to detect.

HPVs, rather than genital herpes, are now associated with cancer of the cervix as well as tumors of the vulva, the vagina, the anus, and the penis. Abnormal Pap tests and cervical tissue sections often indicate the presence of an HPV infection. More than 90% of human cervical cancers have been found to harbor high risk HPVs. Teenagers who have or have had multiple sex partners seem to be particularly susceptible to HPV infections. More and more cases of cancer of the cervix are being seen in this age group.

Sometimes warts disappear on their own. Presently, there is no cure for those that persist, but the warts can be effectively treated by surgery, freezing, application of an acid, or laser burning. If the warts are removed, they may recur. Also even after treatment, the virus can be transmitted. Therefore, abstinence or use of a condom with a vaginal spermicide containing nonoxynol-9 is necessary to prevent the spread of genital warts. A suitable medication to treat genital warts before cancer occurs is being sought. Efforts are also underway to develop a vaccine.

Hepatitis Infections

There are several types of hepatitis. Hepatitis A, caused by HAV (hepatitis A virus), is usually acquired from sewage-contaminated drinking water. However, hepatitis A can also be sexually transmitted through oral/anal contact. Hepatitis B, caused by HBV (hepatitis B virus) is usually transmitted by sexual contact. Hepatitis C, caused by HCV (hepatitis C virus), is called the post-transfusion form of hepatitis. Infection can lead to chronic hepatitis, liver cancer, and death. Still other types of hepatitis are now under investigation, and how many will be found is not yet known.

Hepatitis B

HBV is more likely to be spread by sexual contact than the HIV virus. Like HIV, it can also be spread by blood transfusions or contaminated needles. Because the mode of transmission is similar, it is common for an AIDS patient to also have an HBV infection. Also, like HIV, HBV can be passed from mother to child by way of the placenta.

Only about 50% of persons infected with HBV have flu-like symptoms, including fatigue, fever, headache, nausea, vomiting, muscle aches, and dull pain in the upper right of the abdomen. Jaundice, a yellowish cast to the skin, can also be present. Some persons have an acute infection that lasts only three to four weeks. Others have a chronic form of the disease that leads to liver failure and a need for a liver transplant.

To prevent an infection, the same directions given below to stop the spread of AIDS can be followed. However, inoculation with the HBV vaccine is the best protection. The vaccine, which is safe and does not have any major side effects, is now on the list of recommended immunizations for children.

Chlamydia

Chlamydia is named for the tiny bacterium that causes it, *Chlamydia trachomatis*. New chlamydial infections are more numerous than any other sexually transmitted disease (Fig. 21.16). Some estimate that the actual incidence could be as high as 6 million new cases per year. For every reported case in men, more than five cases are detected in women. The low rates in men suggest that many of the sex partners of women with chlamydia are not diagnosed or reported.

Chlamydial infections of the lower reproductive tract usually are mild or asymptomatic. About 8–21 days after infection, men experience a mild burning sensation on urination and a mucous discharge. Women may have a vaginal discharge, along with the symptoms of a urinary tract infection. If not properly treated, the infection can eventually cause PID, sterility, or ectopic pregnancy.

Some believe that chlamydial infections increase the possibility of premature and stillborn births. If a newborn is exposed to chlamydia during delivery, inflammation of the eyes or pneumonia can result. Some believe that chlamydial

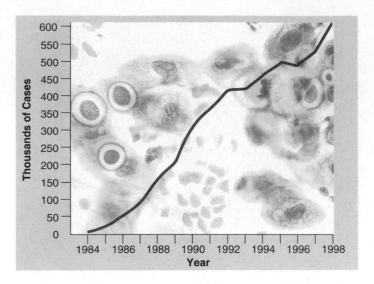

Figure 21.16   Chlamydial infection.
A graph depicting the incidence of reported cases of chlamydia in the United States from 1984 to 1998 is superimposed on a photomicrograph of a cell containing different stages of the organism.

infections increase the possibility of premature births and stillbirths.

Detection and Treatment of Chlamydia

New and faster laboratory tests are not available for detecting a chlamydial infection. Their expense sometimes prevents public clinics from testing for chlamydia. Thus, the following criteria have been suggested to help physicians decide which women should be tested: no more than 24 years old; a new sex partner within the preceding two months; cervical discharge; bleeding during parts of the vaginal exam; and use of a nonbarrier method of contraception. Some doctors, however, are routinely prescribing additional antibiotics appropriate to treating chlamydia for anyone who has gonorrhea, because 40% of females and 20% of males with gonorrhea also have chlamydia.

Gonorrhea

Gonorrhea is caused by the bacterium *Neisseria gonorrhoeae*. This bacterium is a diplococcus, meaning that two cells generally stay together.

The diagnosis of gonorrhea in the male is not difficult as long as he displays typical symptoms (as many as 20% of males may be asymptomatic). The patient complains of pain on urination and has a thick, greenish-yellow urethral discharge 3–5 days after contact with an infected partner. In the female, the bacteria may first settle within the urethra or near the cervix, from which they may spread to the oviducts, causing **pelvic inflammatory disease (PID).** As

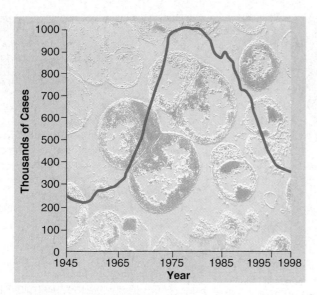

Figure 21.17   Gonorrhea.
A graph depicting the incidence of new cases of gonorrhea in the United States from 1945 to 1998 is superimposed on a photomicrograph of a urethral discharge from an infected male. Gonorrheal bacteria *(Neisseria gonorrhoeae)* occur in pairs; for this reason, they are called diplococci.

the inflamed tubes heal, they may become partially or completely blocked by scar tissue. As a result, the female is sterile, or at best, subject to ectopic pregnancy. Similarly, inflammation may occur in untreated males followed by scarring of each vas deferens. Unfortunately, 60–80% of females are asymptomatic until they develop severe PID-induced pains in the abdominal region.

Homosexual males develop gonorrhea proctitis, or infection of the anus, with symptoms that may include pain in the anus and blood or pus in the feces. Oral sex can cause infection of the throat and the tonsils. Gonorrhea can also spread to other parts of the body, causing heart damage or arthritis. If, by chance, the person touches infected genitals and then his or her eyes, a severe eye infection can result.

Eye infection leading to blindness can occur as a baby passes through the birth canal. Because of this, all newborn infants receive eyedrops containing antibacterial agents such as silver nitrate, tetracycline, or penicillin as a protective measure.

The incidence of gonorrhea has been declining since reaching an all-time high in 1978 (Fig. 21.17). However, it can be noted that gonorrhea rates among African Americans is 30 times greater than the rate among Caucasians. Also, women using the birth control pill have a greater risk of contracting gonorrhea because hormonal contraceptives cause the genital tract to be more receptive to pathogens. As with AIDS, condoms protect against gonorrheal and chlamydial infections. The concomitant use of a spermicide containing nonoxynol-9 enhances protection.

Syphilis

Syphilis is caused by a type of bacterium called *Treponema pallidum.* As with many other bacterial diseases, penicillin has been used as an effective antibiotic. Syphilis has three stages, which can be separated by intervening latent periods, during which the bacteria are resting before multiplying again. During the *primary stage,* a hard chancre (ulcerated sore with hard edges) indicates the site of infection. The chancre can go unnoticed, especially since it usually heals spontaneously, leaving little scarring. During the *secondary stage,* the victim breaks out in a rash, evidence that the bacteria have invaded and spread throughout the body. Curiously, the rash does not itch and is seen even on the palms of the hands and the soles of the feet. Hair loss and infectious gray patches on the mucous membranes, including the mouth, may also occur. These symptoms disappear of their own accord.

During the *tertiary stage,* which lasts until the patient dies, syphilis may affect the cardiovascular system; weakened arterial walls (aneurysms) are seen, particularly in the aorta. In other instances, the disease may affect the nervous system; an infected person may show psychological disturbances. In still another variety of the tertiary stage, gummas, large destructive ulcers, may develop on the skin or within the internal organs.

Congenital syphilis (present at birth) is caused by syphilitic bacteria crossing the placenta. The child is born blind and/or with numerous anatomical malformations.

Syphilis is a devastating disease. Control of syphilis depends on prompt and adequate treatment of all new cases; therefore, it is crucial for all sexual contacts to be traced so they can be treated. Diagnosis of syphilis can be made by blood tests or by microscopic examination of fluids from lesions. The cure for all stages of syphilis is some form of penicillin.

Two Other Infections

Females very often have vaginitis, or infection of the vagina, caused by either the flagellated protozoan *Trichomonas vaginalis* or the yeast *Candida albicans.* The protozoan infection causes a frothy white or yellow, foul-smelling discharge accompanied by itching, and the yeast infection causes a thick, white, curdy discharge, also accompanied by itching. **Trichomoniasis** is most often acquired through sexual intercourse, and the asymptomatic male is usually the reservoir of infection. *Candida albicans,* however, is a normal organism found in the vagina; its growth simply increases beyond normal under certain circumstances. For example, women taking birth control pills are sometimes prone to yeast infections. Also, the indiscriminate use of antibiotics can alter the normal balance of organisms in the vagina so that a yeast infection flares up.

Bioethical Focus   Assisted Reproductive Technologies

The dizzying array of assisted reproductive technologies has progressed from simple in vitro fertilization to the ability to freeze eggs or sperm or even embryos for future use. Older women who never had the opportunity to freeze their eggs can still have children if they use donated eggs—perhaps today harvested from a fetus.

Legal complications abound, ranging from which mother has first claim to the child—the surrogate mother, the woman who donated the egg, or the primary caregiver—to which partner has first claim to frozen embryos following a divorce. Legal issues about who has the right to use what techniques have rarely been discussed, much less decided upon. Some clinics will help anyone, male or female, no questions asked, as long as they have the ability to pay. And most clinics are heading toward doing any type of procedure, including guaranteeing the sex of the child and making sure the child will be free from a particular genetic disorder. It would not be surprising if, in the future, zygotes could be engineered to have any particular trait desired by the parents.

Even today eugenic (good gene) goals are evidenced by the fact that reproductive clinics advertise for egg and sperm donors, primarily in elite college newspapers. The question becomes, "Is it too late for us as a society to make ethical decisions about reproductive issues?" Should we come to a consensus about what techniques should be allowed and who should be able to use them? Should a woman investigate a couple before she has a child for them?

Decide Your Opinion

1. Should assisted reproduction be regulated when unassisted reproduction is not regulated? Why or why not?
2. Should the state be the guardian of frozen embryos and make sure they all get a chance for life? Why or why not?
3. Is it appropriate for physicians and parents to select which embryos will be implanted in the uterus? On the basis of sex? On the basis of genetic inheritance? Why or why not?

Summarizing the Concepts

21.1 Male Reproductive System

In males, spermatogenesis, occurring in seminiferous tubules of the testes, produces sperm that mature and are stored in the epididymides. Sperm may also be stored in the vasa deferentia before entering the urethra, along with secretions produced by the seminal vesicles, prostate gland, and bulbourethral glands. Sperm and these secretions are called semen, or seminal fluid.

The external genitals of males are the penis, the organ of sexual intercourse, and the scrotum, which contains the testes. Orgasm in males is a physical and emotional climax during sexual intercourse that results in ejaculation of semen from the penis.

Hormonal regulation, involving secretions from the hypothalamus, the anterior pituitary, and the testes, maintains testosterone, produced by the interstitial cells of the testes, at a fairly constant level. FSH from the anterior pituitary promotes spermatogenesis in the seminiferous tubules, and LH promotes testosterone production by the interstitial cells.

21.2 Female Reproductive System

In females, oogenesis occurring within the ovaries typically produces one mature follicle each month. This follicle balloons out of the ovary and bursts, releasing an egg, which enters an oviduct. The oviducts lead to the uterus, where implantation and development occur. The external genital area includes the vaginal opening, the clitoris, the labia minora, and the labia majora.

The vagina is the organ of sexual intercourse and the birth canal in females. The vagina and the external genitals, especially the clitoris, play an active role in orgasm, which culminates in uterine and oviduct contractions.

21.3 Female Hormone Levels

In the nonpregnant female, the ovarian cycle is under hormonal control of the hypothalamus and anterior pituitary. During the first half of the cycle, FSH from the anterior pituitary causes maturation of a follicle that secretes estrogen and some progesterone. After ovulation and during the second half of the cycle, LH from the anterior pituitary converts the follicle into the corpus luteum, which secretes progesterone and some estrogen.

Estrogen and progesterone regulate the uterine cycle. Estrogen causes the endometrium to rebuild. Ovulation usually occurs on day 14 of a 28-day cycle. Progesterone produced by the corpus luteum causes the endometrium to thicken and become secretory. Then a low level of hormones causes the endometrium to break down, as menstruation occurs.

If fertilization takes place, the embryo implants itself in the thickened endometrium. If fertilization and implantation occur, the corpus luteum in the ovary is maintained because of HCG production by the placenta, and therefore progesterone production does not cease. Menstruation usually does not occur during pregnancy.

21.4 Control of Reproduction

Numerous birth control methods and devices, such as the birth control pill, diaphragm, and condom, are available for those who wish to prevent pregnancy. Effectiveness varies, and research is being conducted to find new and possibly better methods.

A morning-after pill is now on the market, and a male pill may soon be available. Some couples are infertile, and if so, they may use assisted reproductive technologies in order to have a child. Artificial insemination and in vitro fertilization have been followed by more sophisticated techniques such as intracytoplasmic sperm injection.

21.5 Sexually Transmitted Diseases

Sexually transmitted diseases include AIDS, an epidemic disease; hepatitis, especially types A and B; genital herpes, which repeatedly flares up; genital warts, which lead to cancer of the cervix; gonorrhea and chlamydia, which cause pelvic inflammatory disease (PID); and syphilis, which has cardiovascular and neurological complications if untreated.

Testing Yourself

Choose the best answer for each question.

1. Label this diagram of the male reproductive system, and trace the path of sperm.

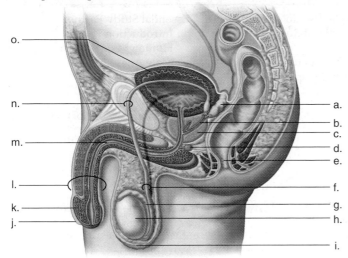

2. Which of these pairs is mismatched?
   a. interstitial cells—testosterone
   b. seminiferous tubules—sperm production
   c. vasa deferentia—seminal fluid production
   d. urethra—conducts sperm

3. Follicle-stimulating hormone (FSH)
   a. is secreted by females but not males.
   b. stimulates the seminiferous tubules to produce sperm.
   c. secretion is controlled by gonadotropic-releasing hormone (GnRH).
   d. Both b and c are correct.

4. In tracing the path of sperm, you would mention the vasa deferentia before the
   a. testes.   c. urethra.
   b. epididymides.   d. uterus.

5. An oocyte is fertilized in the
   a. vagina.   c. oviduct.
   b. uterus.   d. ovary.

6. Semen does NOT contain
   a. prostate fluid.   d. prostaglandins.
   b. urine.   e. Both b and d are correct.
   c. fructose.

7. Luteinizing hormone in males
   a. stimulates sperm development.
   b. triggers ovulation.
   c. is responsible for secondary sexual characteristics.
   d. controls testosterone production by interstitial cells.

8. The release of the oocyte from the follicle is caused by
   a. a decreasing level of estrogen.
   b. a surge in the level of follicle-stimulating hormone.
   c. a surge in the level of luteinizing hormone.
   d. progesterone released from the corpus luteum.

9. For nine months, pregnancy is maintained by the
   a. anterior pituitary.   c. corpus luteum.
   b. ovaries.   d. placenta.

Match the method of protection in questions 10–12 with these means of birth control:
   a. vasectomy   d. diaphragm
   b. oral contraception   e. male condom
   c. intrauterine device (IUD) f. coitus interruptus

10. Blocks entrance of sperm to uterus.

11. Traps sperm and also prevents STDs.

12. Prevents implantation of an embryo.

13. Home pregnancy tests are based on the presence of
    a. estrogen.
    b. progesterone.
    c. follicle-stimulating hormone.
    d. human chorionic gonadotropin.

14. During pregnancy,
    a. the ovarian and uterine cycles occur more quickly than before.
    b. GnRH is produced at a higher level than before.
    c. the ovarian and uterine cycles do not occur.
    d. the female secondary sexual characteristics are not maintained.

15. Female oral contraceptives prevent pregnancy because
    a. the pill inhibits the release of luteinizing hormone.
    b. oral contraceptives prevent the release of an egg.
    c. follicle-stimulating hormone is not released.
    d. All of these are correct.

16. Which of the following is a sexually transmitted disease caused by a bacterium?
    a. gonorrhea   d. genital herpes
    b. hepatitis B   e. HIV
    c. genital warts

17. The HIV virus has a preference for binding to
    a. B lymphocytes.   c. helper T lymphocytes.
    b. cytotoxic T lymphocytes. d. All of these are correct.

18. A complication of contracting a sexually transmitted disease is
    a. STDs are often asymptomatic.
    b. pelvic inflammatory disease.
    c. sterility.
    d. ectopic pregnancy.
    e. All of these are correct.

Match the description in questions 19–21 with the sexually transmitted disease:
    a. AIDS   e. gonorrhea
    b. hepatitis B   f. chlamydia
    c. genital herpes   g. syphilis
    d. genital warts

19. Blisters, ulcers, pain on urination, swollen lymph nodes

20. Flu-like symptoms, jaundice; eventual liver failure possible

21. Males have a thick, greenish-yellow discharge; no symptoms in female; can lead to PID

Model Organisms

Developmental genetics has benefited from research utilizing the roundworm, *Caenorhabditis elegans,* and the fruit fly, *Drosophila melanogaster.* These organisms are referred to as model organisms because the study of their development produced concepts that help us understand development in general.

The Roundworm Experiments

The roundworm is only one millimeter long, and vast numbers can be raised in the laboratory in either petri dishes or a liquid medium. Development of *C. elegans* takes only three days, and the adult worm contains only 959 cells. Investigators have been able to watch the process from beginning to end, especially since the worm is transparent. Many modern genetic studies have been done utilizing this worm. Its entire genome has been sequenced. Individual genes have been altered and cloned, and their products have been injected into cells or extracellular fluid.

What have we learned? **Fate maps** have been developed that show the destiny of each cell as it arises following successive cell divisions (Fig. 22.8). Some investigators have studied in detail the development of the vulva, a pore through which eggs are laid. A cell called the anchor cell induces the vulva to form. The cell closest to the anchor cell receives the most inducer and becomes the inner vulva. This cell in turn produces another inducer, which acts on its two neighboring cells, and they become the outer vulva. The inducers are signals that alter the metabolism of the receiving cell and activate particular genes. Work with *C. elegans* has shown that induction requires the regulation of genes in a particular sequence. This diagram shows how induction can occur sequentially:

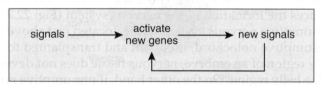

Apoptosis We have already discussed the importance of **apoptosis** (programmed cell death) in the normal day-to-day operation of the body. Apoptosis is also an important part of development in all organisms. In humans, for example, we know that apoptosis is necessary to the shaping of the hands and feet; if it does not occur, the child is born with webbing between the fingers and toes. The fate maps of *C. elegans* indicate that apoptosis occurs in 131 cells as development takes place. When a cell-death signal is received, an inhibiting protein becomes inactive, allowing a cell-death cascade to proceed, which ends in enzymes destroying the cell.

The Fruit Fly Experiments

A fruit fly may be larger than a roundworm, but a few pairs can produce hundreds of offspring in a couple of weeks, all within a small bottle kept on a laboratory bench. Work

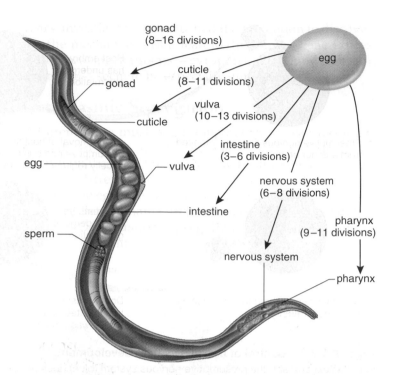

Figure 22.8   Development of *C. elegans,* a small worm.
A fate map of the worm shows that, as cells arise by cell division, they are destined to become particular structures.

with the common fruit fly has suggested how later morphogenesis comes about. Investigators have discovered certain genes, now called **morphogen genes,** that determine the pattern of the animal and its individual parts. For example, some genes control which end of the animal will be the head and which the tail, and others determine how many segments the animal will have (Fig. 22.9).

Each of these genes codes for a protein that is present in a gradient—cells at the start of the gradient contain high levels of the protein, and cells at the end of the gradient contain low levels of the protein. These gradients are called morphogen gradients because they determine the shape or form of the organism. A morphogen gradient is efficient because it has a range of effects, depending on the particular concentration in a portion of the animal. Sequential sets of master genes code for morphogen gradients that activate the next set of master genes in turn, and that's one reason development turns out to be so orderly.

Homeotic Genes **Homeotic genes** control the organization of differentiated cells into specific three-dimensional structures. They are also believed to be morphogen genes that function by bringing about protein gradients. Certain homeotic genes control whether a particular segment will bear antennae, legs, or wings. A homeotic mutation causes these appendages to be misplaced—thus, a mutant fly could have extra legs where antennae should be or two pairs of wings (Fig. 22.10).

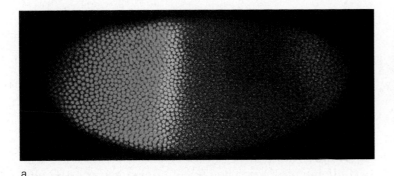

a.

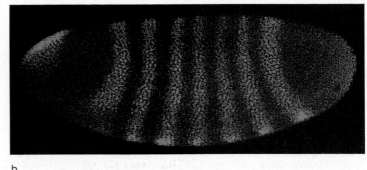

b.

Figure 22.9   Pattern formation in the fruit fly.
a. Different protein gradients (color) appear as development proceeds. One early gradient determines which end is the head and which the tail.
b. Another gradient determines the number of segments.

Homeotic genes have now been found in many other organisms, and surprisingly, they all contain the same particular sequence of nucleotides, called a **homeobox.** (Because homeotic genes contain a homeobox in mammals, they are called *Hox* genes.) The homeobox codes for a particular sequence of 60 amino acids called a homeodomain:

homeotic gene   homeobox

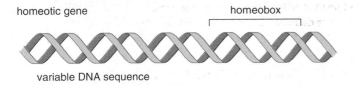

variable DNA sequence

a.

homeodomain protein   homeodomain

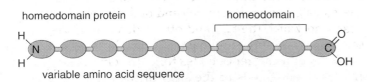

variable amino acid sequence

A homeodomain protein binds to DNA and helps determine which particular genes are turned on. Researchers envision that a homeodomain protein produced by one homeotic gene binds to and turns on the next homeotic gene, and so forth. This orderly process in the end determines the morphology of particular segments.

In *Drosophila*, homeotic genes are located on a single chromosome. In mice and also humans, the same four clusters of homeotic genes are located on four different chromosomes. In all three types of animals, homeotic genes are expressed from anterior to posterior in the same order. The first clusters determine the final development of anterior segments of the embryo, while those later in the sequence determine the final development of posterior segments of the embryo.

Since the homeotic genes of so many different organisms contain the same homeodomain, we know that this nucleotide sequence arose early in the history of life and that it has been largely conserved as evolution occurred. In general, it has been very surprising to learn that developmental genetics is

b.

Figure 22.10   Homeotic mutations.
When homeotic genes mutate, morphogenetic abnormalities occur.
a. This fly has four wings. **b.** This one has legs on its head.

similar in organisms ranging from yeasts to plants to a wide variety of animals. Certainly, the genetic mechanisms of development appear to be quite similar in all animals.

A hierarchy of gene activity causes development to be orderly. Morphogen gradients cause cells to produce yet other morphogen gradients, until finally specific structures are formed.

22.3 Human Embryonic and Fetal Development

In humans, the length of the time from conception (fertilization followed by **implantation**) to birth (parturition) is approximately nine months. It is customary to calculate the time of birth by adding 280 days to the start of the last menstrual period because this date is usually known, whereas the day of fertilization is usually unknown. Because the time of birth is influenced by so many variables, only about 5% of babies actually arrive on the forecasted date.

Human development before birth is often divided into embryonic development (months 1 and 2) and fetal development (months 3–9). The **embryonic period** consists of early formation of the major organs, and fetal development is the refinement of these structures.

Before we consider human development chronologically, we must understand the placement of **extraembryonic membranes.** Extraembryonic membranes are best understood by considering their function in reptiles and birds. In reptiles, these membranes made development on land possible. If an embryo develops in the water, the water supplies oxygen for the embryo and takes away waste products. The surrounding water prevents desiccation, or drying out, and provides a protective cushion. For an embryo that develops on land, all these functions are performed by the extraembryonic membranes.

In the chick, the extraembryonic membranes develop from extensions of the germ layers, which spread out over the yolk. Figure 22.11 shows the chick surrounded by the membranes. The **chorion** lies next to the shell and carries on gas exchange. The **amnion** contains the protective amniotic fluid, which bathes the developing embryo. The **allantois** collects nitrogenous wastes, and the **yolk sac** surrounds the remaining yolk, which provides nourishment.

Humans (and other mammals) also have these extraembryonic membranes. The chorion develops into the fetal half of the placenta; the yolk sac, which has little yolk, is the first site of blood cell formation; the allantoic blood vessels become the umbilical blood vessels; and the amnion contains fluid to cushion and protect the embryo, which develops into a fetus. Therefore, the function of the membranes in humans has been modified to suit internal development, but their very presence indicates our relationship to birds and to reptiles. It is interesting to note that all chordate animals develop in water, either in bodies of water or within amniotic fluid.

The presence of extraembryonic membranes in reptiles made development on land possible. Humans also have these membranes, but their function has been modified for internal development.

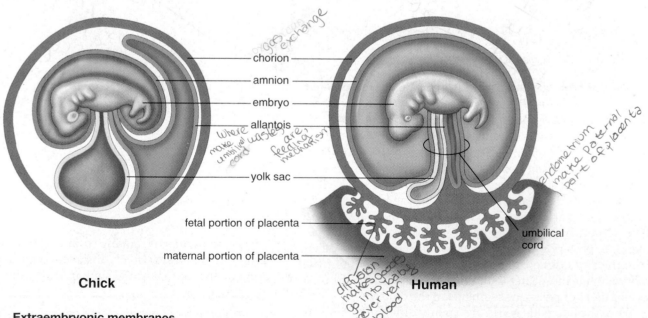

Figure 22.11   Extraembryonic membranes.
Extraembryonic membranes, which are not part of the embryo, are found during the development of chicks and humans. Each has a specific function.

Embryonic Development

Embryonic development encompasses the first two months of development following fertilization.

The First Week

Fertilization occurs in the upper third of an oviduct (Fig. 22.12), and cleavage begins even as the embryo passes down this duct to the uterus. By the time the embryo reaches the uterus on the third day, it is a morula. The morula is not much larger than the zygote because, even though multiple cell divisions have occurred, the newly formed cells do not grow. By about the fifth day, the morula is transformed into the blastocyst. The **blastocyst** has a

fluid-filled cavity, a single layer of outer cells called the **trophoblast,** and an inner cell mass. Later, the trophoblast, reinforced by a layer of mesoderm, gives rise to the chorion, one of the extraembryonic membranes (see Fig. 22.11). The inner cell mass eventually becomes the embryo, which develops into a fetus.

The Second Week

At the end of the first week, the embryo begins the process of implanting in the wall of the uterus. The trophoblast secretes enzymes to digest away some of the tissue and blood vessels of the endometrium of the uterus (Fig. 22.12). The embryo is now about the size of the period at the end of this sentence. The trophoblast begins to secrete **human**

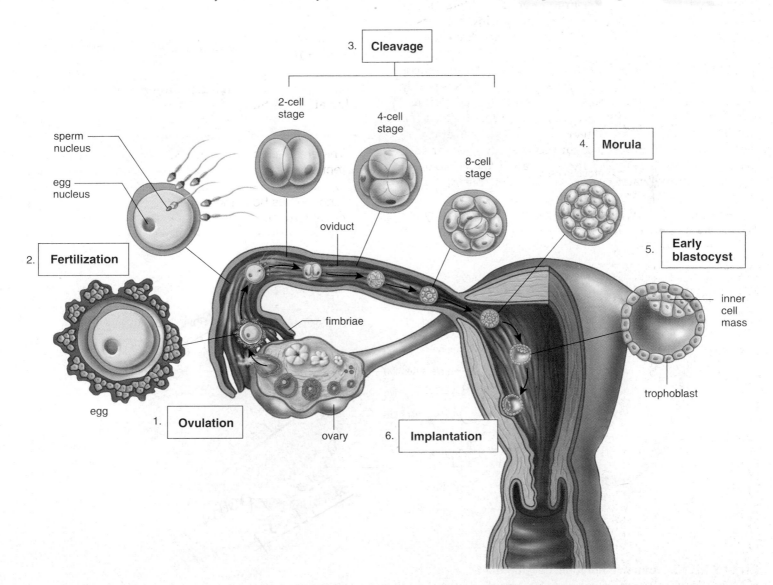

Figure 22.12 Human development before implantation.
Structures and events proceed counterclockwise. At ovulation (1), the secondary oocyte leaves the ovary. A single sperm nucleus enters the egg, and fertilization (2) occurs in the oviduct. As the zygote moves along the oviduct, it undergoes cleavage (3) to produce a morula (4). The blastocyst forms (5) and implants itself in the uterine lining (6).

chorionic gonadotropin (HCG), the hormone that is the basis for the pregnancy test and that serves to maintain the corpus luteum past the time it normally disintegrates. Because of this, the endometrium is maintained and menstruation does not occur.

As the week progresses, the inner cell mass detaches itself from the trophoblast, and two more extraembryonic membranes form (Fig. 22.13a). The yolk sac, which forms below the embryonic disk, has no nutritive function as it does in chicks, but it is the first site of blood cell formation. The amnion and its cavity are where the embryo (and then the fetus) develops. In humans, amniotic fluid acts as an insulator against cold and heat and also absorbs shock, such as that caused by the mother exercising.

Gastrulation occurs during the second week. The inner cell mass now has flattened into the **embryonic disk,** composed of two layers of cells: ectoderm above and endoderm below. Once the embryonic disk elongates to form the so-called primitive streak, the third germ layer, mesoderm, forms by invagination of cells along the streak. The trophoblast is reinforced by mesoderm and becomes the chorion (Fig. 22.13b).

It is possible to relate the development of future organs to these germ layers (see page 441).

The Third Week

Two important organ systems make their appearance during the third week. The nervous system is the first organ system to be visually evident. At first, a thickening appears

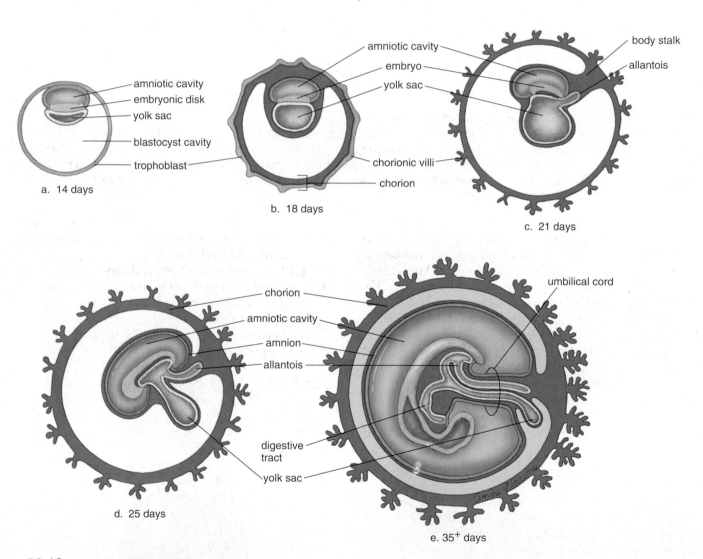

Figure 22.13 **Human embryonic development.**
a. At first, the embryo contains no organs, only tissues. The amniotic cavity is above the embryo, and the yolk sac is below. **b.** The chorion develops villi, the structures so important to exchange between mother and child. **c, d.** The allantois and yolk sac, two more extraembryonic membranes, are positioned inside the body stalk as it becomes the umbilical cord. **e.** At 35+ days, the embryo has a head region and a tail region. The umbilical cord takes blood vessels between the embryo and the chorion (placenta).

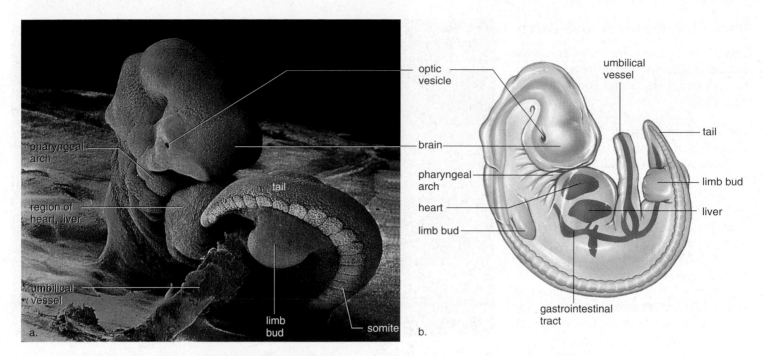

Figure 22.14 **Human embryo at beginning of fifth week.**
a. Scanning electron micrograph. **b.** The embryo is curled so that the head touches the region of the heart and liver. The organs of the gastrointestinal tract are forming, and the arms and the legs develop from the bulges called limb buds. The tail is an evolutionary remnant; its bones regress and become those of the coccyx (tailbone). The pharyngeal arches become functioning gills only in fishes and amphibian larvae; in humans, the first pair of pharyngeal pouches becomes the auditory tubes. The second pair becomes the tonsils, while the third and fourth become the thymus gland and the parathyroid glands.

along the entire dorsal length of the embryo, and then invagination occurs as neural folds appear. When the neural folds meet at the midline, the neural tube, which later develops into the brain and the nerve cord, is formed (see Fig. 22.4). After the notochord is replaced by the vertebral column, the nerve cord is called the spinal cord.

Development of the heart begins in the third week and continues into the fourth week. At first, there are right and left heart tubes; when these fuse, the heart begins pumping blood, even though the chambers of the heart are not fully formed. The veins enter posteriorly and the arteries exit anteriorly from this largely tubular heart, but later the heart twists so that all major blood vessels are located anteriorly.

The Fourth and Fifth Weeks
At four weeks, the embryo is barely larger than the height of this print. A bridge of mesoderm called the body stalk connects the caudal (tail) end of the embryo with the chorion, which has treelike projections called **chorionic villi** (Fig. 22.13c and d). The fourth extraembryonic membrane, the allantois, is contained within this stalk, and its blood vessels become the umbilical blood vessels. The head and the tail then lift up, and the body stalk moves anteriorly by constriction. Once this process is complete, the **umbilical**

cord, which connects the developing embryo to the placenta, is fully formed (Fig. 22.13e).

Little flippers called limb buds appear (Fig. 22.14); later, the arms and the legs develop from the limb buds, and even the hands and the feet become apparent. At the same time—during the fifth week—the head enlarges and the sense organs become more prominent. It is possible to make out the developing eyes, ears, and even nose.

The Sixth Through Eighth Weeks
A remarkable change in external appearance takes place during the sixth through eighth weeks of development. The embryo changes from a form that is difficult to recognize as a human to one easily recognized as human. Concurrent with brain development, the head achieves its normal relationship with the body as a neck region develops. The nervous system is developed well enough to permit reflex actions, such as a startle response to touch. At the end of this period, the embryo is about 38 mm (1.5 inches) long and weighs no more than an aspirin tablet, even though all organ systems have been established.

During the embryonic period of development, all major organs start to develop.

Fetal Development and Birth

Fetal development includes the third through ninth months of development. At this time, the fetus looks human (Fig. 22.15).

The Third and Fourth Months

At the beginning of the third month, the fetal head is still very large, the nose is flat, the eyes are far apart, and the ears are well formed. Head growth now begins to slow down as the rest of the body increases in length. Epidermal refinements, such as eyelashes, eyebrows, hair on head, fingernails, and nipples, appear.

Cartilage begins to be replaced by bone as ossification centers appear in most of the bones. Cartilage remains at the ends of the long bones, and ossification is not complete until age 18 or 20 years. The skull has six large membranous areas called **fontanels,** which permit a certain amount of flexibility as the head passes through the birth canal and allow rapid growth of the brain during infancy. Progressive fusion of the skull bones causes the fontanels to usually close by 2 years of age.

Sometime during the third month, it is possible to distinguish males from females. Researchers have discovered a series of genes on the X and Y chromosomes that cause the differentiation of gonads into testes and ovaries. Once these have differentiated, they produce the sex hormones that influence the differentiation of the genital tract.

At this time, either testes or ovaries are located within the abdominal cavity, but later, in the last trimester of fetal development, the testes descend into the scrotal sacs (scrotum). Sometimes the testes fail to descend, and in that case, an operation may be done later to place them in their proper location.

During the fourth month, the fetal heartbeat is loud enough to be heard when a physician applies a stethoscope to the mother's abdomen. By the end of this month, the fetus is about 152 mm (6 inches) in length and weighs about 171 grams (6 oz).

During the third and fourth months, it is obvious that the skeleton is becoming ossified. The sex of the individual can now be distinguished.

The Fifth Through Seventh Months

During the fifth through seventh months, the mother begins to feel movement. At first, there is only a fluttering sensation, but as the fetal legs grow and develop, kicks and jabs are felt. The fetus, though, is in the fetal position, with the head bent down and in contact with the flexed knees.

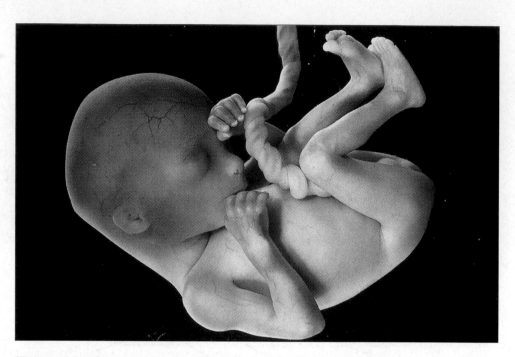

Figure 22.15   **The three- to four-month-old fetus looks human.**
Face, hands, and fingers are well defined.

The wrinkled, translucent, pink-colored skin is covered by a fine down called **lanugo.** This in turn is coated with a white, greasy, cheeselike substance called **vernix caseosa,** which probably protects the delicate skin from the amniotic fluid. The eyelids are now fully open.

At the end of this period, the length has increased to about 300 mm (12 inches), and the weight is about 1,380 grams (3 lb). It is possible that, if born now, the baby will survive.

Fetal Circulation

The fetus has circulatory features that are not present in the adult circulation (Fig. 22.16). All of these features can be related to the fact that the fetus does not use its lungs for gas exchange. For example, much of the blood entering the right atrium is shunted into the left atrium through the **oval opening** (foramen ovale) between the two atria. Also, any blood that does enter the right ventricle and is pumped into the pulmonary trunk is shunted into the aorta by way of the **arterial duct** (ductus arteriosus).

Blood within the aorta travels to the various branches, including the iliac arteries, which connect to the **umbilical arteries** leading to the placenta. Exchange of gases and nutrients between maternal blood and fetal blood takes place at the placenta. The **umbilical vein** carries blood rich in nutrients and oxygen to the fetus. The umbilical vein enters the liver and then joins the **venous duct,** which merges with the inferior vena cava, a vessel that returns blood to the heart. It is interesting to note that the umbilical arteries and vein run alongside one another in the umbilical cord, which is cut at birth, leaving only the umbilicus (navel).

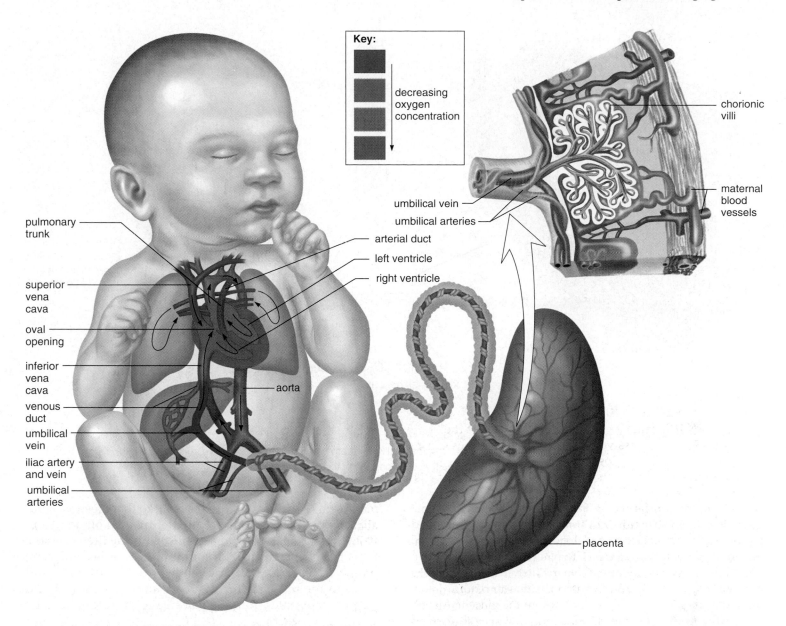

Figure 22.16 **Fetal circulation and the placenta.**
The lungs are not functional in the fetus, and the blood passes directly from the right atrium to the left atrium or from the right ventricle via the pulmonary trunk and arterial duct to the aorta. The umbilical arteries take fetal blood to the placenta where exchange of molecules between fetal and maternal blood takes place across the walls of the chorionic villi. Oxygen and nutrient molecules diffuse into the fetal blood, and carbon dioxide and urea diffuse from the fetal blood. The umbilical vein returns blood from the placenta to the fetus.

The most common of all cardiac defects in the newborn is the persistence of the oval opening. With the tying of the cord and the expansion of the lungs, blood enters the lungs in quantity. Return of this blood to the left side of the heart usually causes a flap to cover the opening. Incomplete closure occurs in nearly one out of four individuals, but even so, passage of the blood from the right atrium to the left atrium rarely occurs because either the opening is small or it closes when the atria contract. In a small number of cases, the passage of impure blood from the right side to the left side of the heart is sufficient to cause a "blue baby." Such a condition can now be corrected by open heart surgery.

The arterial duct closes because endothelial cells divide and block off the duct. Remains of the arterial duct and parts of the umbilical arteries and vein are later transformed into connective tissue.

In the fetus, blood is shunted away from the lungs, and moves to and away from the placenta through the umbilical blood vessels located in the umbilical cord. Exchange of substances between fetal blood and maternal blood takes place at the placenta, which forms from the chorion and uterine tissue.

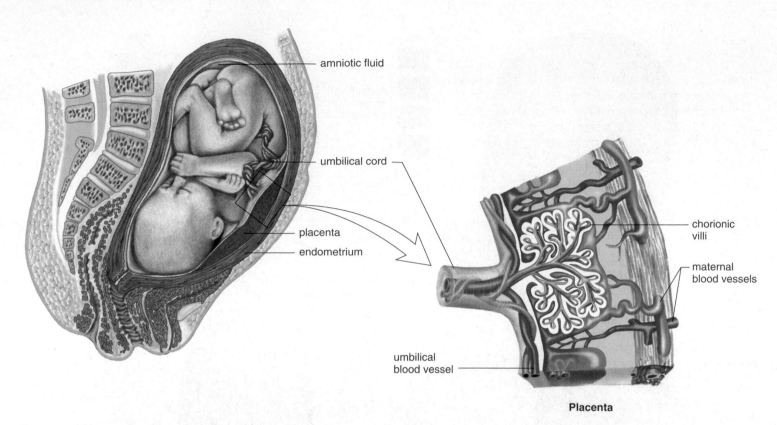

Figure 22.17   Anatomy of the placenta in a fetus at six to seven months.
The placenta is composed of both fetal and maternal tissues. Chorionic villi penetrate the uterine lining and are surrounded by maternal blood. Exchange of molecules between fetal and maternal blood takes place across the walls of the chorionic villi.

The Structure and Function of the Placenta   The **placenta** is a mammalian structure that functions in gas, nutrient, and waste exchange between the embryonic (later fetal) and maternal circulatory systems. The placenta begins formation once the embryo is fully implanted. At first, the entire chorion has chorionic villi that project into the endometrium. Later, these disappear in all areas except where the placenta develops. By the tenth week, the placenta (Fig. 22.17) is fully formed and is producing progesterone and estrogen. These hormones have two effects: (1) due to their negative feedback control of the hypothalamus and the anterior pituitary, they prevent any new follicles from maturing, and (2) they maintain the lining of the uterus—now the corpus luteum is not needed. There is usually no menstruation during pregnancy.

The placenta has a fetal side contributed by the chorion and a maternal side consisting of uterine tissues. Notice in Figure 22.17 how the chorionic villi are surrounded by maternal blood; yet maternal and fetal blood never mix, since exchange always takes place across plasma membranes. Carbon dioxide and other wastes move from the fetal side to the maternal side of the placenta, and nutrients and oxygen move from the maternal side to the fetal side. The umbilical cord stretches between the placenta and the fetus. Although the umbilical cord

may seem to travel from the placenta to the intestine, actually the umbilical cord is simply taking fetal blood to and from the placenta. The umbilical cord is the lifeline of the fetus because it contains the umbilical arteries and vein, which transport waste molecules (carbon dioxide and urea) to the placenta for disposal and take oxygen and nutrient molecules from the placenta to the rest of the fetal circulatory system.

Harmful chemicals can also cross the placenta. This is of particular concern during the embryonic period, when various structures are first forming. Each organ or part seems to have a sensitive period during which a substance can alter its normal development. For example, if a woman takes the drug thalidomide, a tranquilizer, between days 27 and 40 of her pregnancy, the infant is likely to be born with deformed limbs. If thalidomide is taken after day 40, however, the infant is born with normal limbs.

During mammalian development, the embryo and later the fetus depend on the placenta for gas exchange, for acquiring nutrients, and for ridding the body of wastes.

Birth

The uterus has contractions throughout pregnancy. At first, these are light, lasting about 20–30 seconds and occurring every 15–20 minutes. Near the end of pregnancy, the contractions may become stronger and more frequent so that a woman thinks she is in labor. However, the onset of true labor is marked by uterine contractions that occur regularly every 15–20 minutes and last for 40 seconds or longer.

A positive feedback mechanism can explain the onset and continuation of labor. Uterine contractions are induced by a stretching of the cervix, which also brings about the release of oxytocin from the posterior pituitary. Oxytocin stimulates the uterine muscles, both directly and through the action of prostaglandins. Uterine contractions push the fetus downward, and the cervix stretches even more. This cycle keeps repeating itself until birth occurs.

Stage 1

Prior to or at the first stage of **parturition,** which is the process of giving birth to an offspring, there can be a "bloody show" caused by expulsion of a mucous plug from the cervical canal. This plug prevents bacteria and sperm from entering the uterus during pregnancy.

At first, the uterine contractions of labor occur in such a way that the cervical canal slowly disappears as the lower part of the uterus is pulled upward toward the baby's head (Fig. 22.18b). This process is called effacement, or "taking up the cervix." With further contractions, the baby's head acts as a wedge to assist cervical dilation. If the amniotic membrane has not already ruptured, it is apt to do so during this stage, releasing the amniotic fluid, which leaks out the vagina (an event sometimes referred to as "breaking water"). The first stage of parturition ends once the cervix is dilated completely.

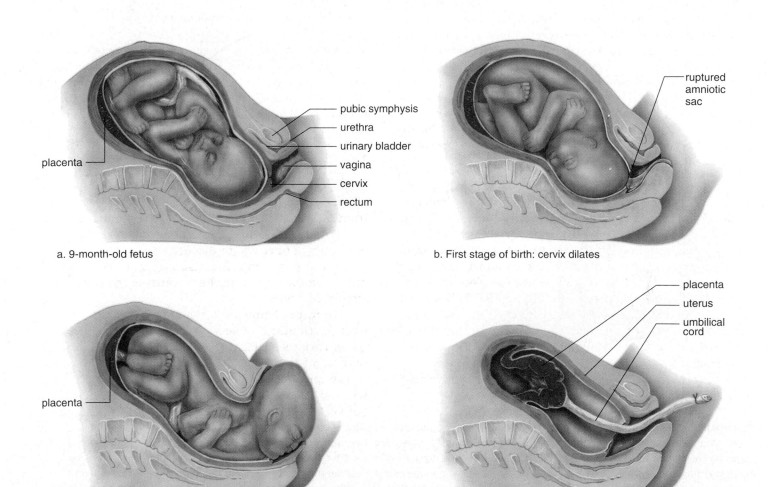

a. 9-month-old fetus

— pubic symphysis
— urethra
— urinary bladder
placenta —
— vagina
— cervix
— rectum

b. First stage of birth: cervix dilates

— ruptured amniotic sac

c. Second stage of birth: baby emerges

placenta —

d. Third stage of birth: expelling afterbirth

— placenta
— uterus
— umbilical cord

Figure 22.18   Three stages of parturition (birth).
a. Position of fetus just before birth begins. **b.** Dilation of cervix. **c.** Birth of baby. **d.** Expulsion of afterbirth.

Preventing Birth Defects

Not on test

It is believed that at least 1 in 16 newborns has a birth defect, either minor or serious, and the actual percentage may be even higher. Most likely, only 20% of all birth defects are due to heredity. Those that are hereditary can sometimes be detected before birth. Amniocentesis allows the fetus to be tested for abnormalities of development; chorionic villi sampling allows the embryo to be tested; and a new method has been developed for screening eggs to be used for in vitro fertilization (Fig. 22A).

It is recommended that all females take everyday precautions to protect any future and/or presently developing embryos and fetuses from defects. Proper nutrition is a must (deficiency in folic acid causes neural tube defects). X-ray diagnostic therapy should be avoided during pregnancy because X rays cause mutations in the developing embryo or fetus. Children born to women who received X-ray treatment are apt to have birth defects and/or to develop leukemia later. Toxic chemicals such as pesticides and many organic industrial chemicals, which are also mutagenic, can cross the placenta. Cigarette smoke not only contains carbon monoxide but also other fetotoxic chemicals. Babies born to smokers are often underweight and subject to convulsions.

Pregnant Rh⁻ women should receive an Rh immunoglobulin injection to prevent the production of Rh antibodies. These antibodies can cause nervous system and heart defects.

Sometimes, birth defects are caused by microbes. Females can be immunized before the childbearing years for rubella (German measles), which in particular causes birth defects such as deafness. Unfortunately, immunization for sexually transmitted diseases is not possible. The AIDS virus can cross the placenta, and over 1,500 babies who contracted AIDS while in their mother's womb are now mentally retarded. When a mother has herpes, gonorrhea, or chlamydia, newborns can become infected as they pass through the birth canal. Blindness and other physical and mental defects may develop. Birth by cesarean section could prevent these occurrences.

Pregnant women should not take any type of drug without a doctor's permission. Certainly illegal drugs, such as marijuana, cocaine, and heroin, should be completely avoided. "Cocaine babies" now make up 60% of drug-affected babies. Severe fluctuations in blood pressure which are produced by the use of cocaine temporarily deprive the developing brain of oxygen. Cocaine babies have visual problems, lack coordina-

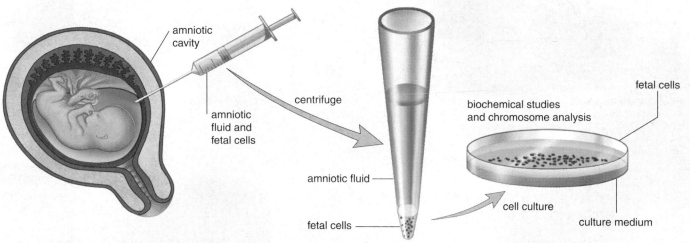

a. Amniocentesis

Figure 22A Three methods for genetic defect testing before birth.
a. Amniocentesis is usually performed from the 15th to the 17th week of pregnancy. A long needle is passed through the abdominal wall to withdraw a small amount of amniotic fluid, along with fetal cells. Since there are only a few cells in the amniotic fluid, testing may be delayed as long as four weeks until cell culture produces enough cells for testing purposes. About 40 tests are available for different defects. **b.** Chorionic villi sampling is usually performed from the 8th to the 12th week of pregnancy. The doctor inserts a long, thin tube through the vagina into the uterus. With the help of ultrasound, which gives a picture of the uterine contents, the tube is placed between the lining of the uterus and the chorion. Then a sampling of the chorionic villi cells is obtained by suction. Chromosome analysis and biochemical tests for genetic defects can be done immediately on these cells. **c.** Screening eggs for genetic defects is a new technique. Preovulatory eggs are removed by aspiration after a laparoscope (optical telescope) is inserted into the abdominal cavity through a small incision in the region of the navel. The first polar body is tested. If the woman is heterozygous (*Aa*) and the defective gene (*a*) is found in the polar body, then the egg must have received the normal gene (*A*). Normal eggs then undergo in vitro fertilization and are placed in the prepared uterus.

tion, and are mentally retarded. The drugs aspirin, caffeine (present in coffee, tea, and cola), and alcohol should be severely limited. It is not unusual for babies of drug addicts and alcoholics to display withdrawal symptoms and to have various abnormalities. Babies born to women who have about 45 drinks a month and as many as 5 drinks on one occasion are apt to have fetal alcohol syndrome (FAS). These babies have decreased weight, height, and head size, with malformation of the head and face. Mental retardation is common in FAS infants.

Medications can also cause problems. When the synthetic hormone DES was given to pregnant women to prevent mis-

carriage, their daughters showed various abnormalities of the reproductive organs and an increased tendency toward cervical cancer. Other sex hormones, including birth-control pills, can possibly cause abnormal fetal development, including abnormalities of the sex organs. The tranquilizer thalidomide is well known for having caused deformities of the arms and legs in children born to women who took the drug. Therefore, a woman has to be very careful about taking medications while pregnant.

Now that physicians and laypeople are aware of the various ways in which birth defects can be prevented, it is hoped that the incidence of birth defects will decrease in the future.

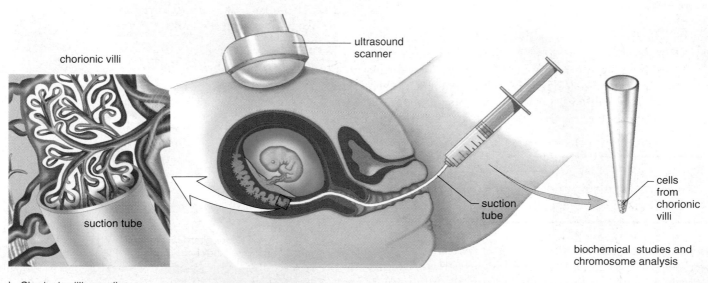

b. Chorionic villi sampling

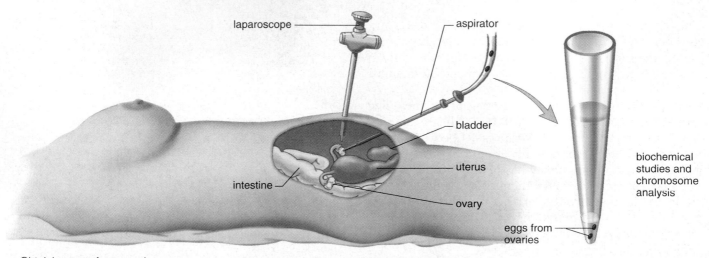

c. Obtaining eggs for screening

Stage 2

During the second stage of parturition, the uterine contractions occur every 1–2 minutes and last about one minute each. They are accompanied by a desire to push, or bear down. As the baby's head gradually descends into the vagina, the desire to push becomes greater. When the baby's head reaches the exterior, it turns so that the back of the head is uppermost (Fig. 22.18c). Since the vaginal orifice may not expand enough to allow passage of the head, an **episiotomy** is often performed. This incision, which enlarges the opening, is sewn together later. As soon as the head is delivered, the baby's shoulders rotate so that the baby faces either to the right or the left. At this time, the physician may hold the head and guide it downward, while one shoulder and then the other emerges. The rest of the baby follows easily.

Once the baby is breathing normally, the umbilical cord is cut and tied, severing the child from the placenta. The stump of the cord shrivels and leaves a scar, which is the umbilicus.

Stage 3

The placenta, or **afterbirth,** is delivered during the third stage of parturition (Fig. 22.18d). About 15 minutes after delivery of the baby, uterine muscular contractions shrink the uterus and dislodge the placenta. The placenta is then expelled into the vagina. As soon as the placenta and its membranes are delivered, the third stage of parturition is complete.

During the first stage of parturition, the cervix dilates; during the second stage, the child is born; and during the third stage, the afterbirth is expelled.

Female Breast and Lactation

A female breast contains 15 to 25 lobules, each with a milk duct, which begins at the nipple and divides into numerous other ducts that end in blind sacs called alveoli (Fig. 22.19).

During pregnancy, the breasts enlarge as the ducts and alveoli increase in number and size. The same hormones that affect the mother's breasts can also affect the child's. Some newborns, including males, even secrete a small amount of milk for a few days.

Usually, no milk is produced during pregnancy. The hormone prolactin is needed for lactation to begin, and the production of this hormone is suppressed because of the feedback control that the increased amount of estrogen and progesterone during pregnancy has on the pituitary. Once the baby is delivered, however, the pituitary begins secreting prolactin. It takes a couple of days for milk production to begin, and in the meantime, the breasts produce **colostrum,** a thin, yellow, milky fluid rich in protein, including antibodies.

The continued production of milk requires a suckling child. When a breast is suckled, the nerve endings in the areola are stimulated, and a nerve impulse travels along neural pathways from the nipples to the hypothalamus, which directs the pituitary gland to release the hormone oxytocin. When this hormone arrives at the breast, it causes contraction of the lobules so that milk flows into the ducts (called milk letdown), where it may be drawn out of the nipple by the suckling child.

Whether to breast-feed or not is a private decision based in part on a woman's particular circumstances. However, it is well known that breast milk contains antibodies produced by the mother that can help a baby survive. Babies have immature immune systems, less stomach acid to destroy foreign antigens, and also unsanitary habits. Breast-fed babies are less likely to develop stomach and intestinal illnesses, including diarrhea, during the first thirteen weeks of life. Breast-feeding also has physiological benefits for the mother. Suckling causes uterine contractions that can help the uterus return to its normal size, and breast-feeding uses up calories and can help a woman return to her normal weight.

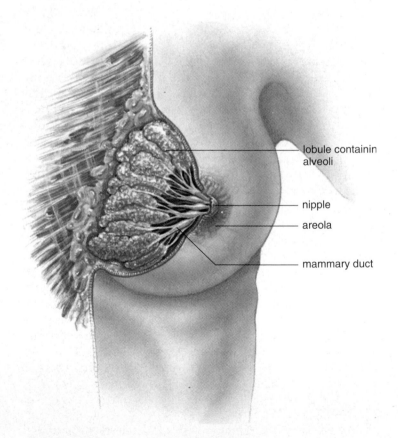

lobule containin alveoli

nipple

areola

mammary duct

Figure 22.19   Female breast anatomy.
The female breast contains lobules consisting of ducts and alveoli. The alveoli are lined by milk-producing cells in the lactating (milk-producing) breast.

22.4 Human Development After Birth

Development does not cease once birth has occurred but continues throughout the stages of life: infancy, childhood, adolescence, and adulthood. **Aging** encompasses these progressive changes that contribute to an increased risk of infirmity, disease, and death (Fig. 22.20).

Today, there is great interest in **gerontology,** the study of aging, because our society includes more older individuals than ever before, and the number is expected to rise dramatically. In the next half-century, the number of people over age 75 will rise from the present 8 million to 14.5 million, and the number over age 80 will rise from 5 million to 12 million. The human life span is judged to be a maximum of 120–125 years. The present goal of gerontology is not necessarily to increase the life span, but to increase the health span, the number of years an individual enjoys the full functions of all body parts and processes.

Hypotheses About Aging

There are many hypotheses about what causes aging. Three of these are considered here.

Genetic in Origin

Several lines of evidence indicate that aging has a genetic basis: (1) The number of times a cell divides is species-specific. For example, the maximum number of times human cells divide is around 50. Perhaps as we grow older, more and more cells are unable to divide, and instead, they undergo degenerative changes and die. (2) Some cell lines may become nonfunctional long before the maximum number of divisions has occurred. Whenever DNA replicates, mutations can occur, and this can lead to the production of nonfunctional proteins. Eventually, the number of inadequately functioning cells can build up, which contributes to the aging process. (3) The children of long-lived parents tend to live longer than those of short-lived parents. Recent work suggests that when an animal produces fewer free radicals, it lives longer. Free radicals are unstable molecules that carry an extra electron. In order to stabilize themselves, free radicals donate an electron to another molecule such as DNA or proteins (e.g., enzymes) or lipids found in plasma membranes. Eventually, these molecules are unable to function, and the cell is destroyed. Certain genes code for antioxidant enzymes that detoxify free radicals. Research suggests that animals with particular forms of these genes—and therefore more efficient antioxidant enzymes—live longer.

Whole-Body Process

A decline in the hormonal system can affect many different organs of the body. For example, type II diabetes is common in older individuals. The pancreas makes insulin, but the cells lack the receptors that enable them to respond. Menopause in women occurs for a similar reason. There is plenty of folli-

Figure 22.20 Aging.
Aging is a slow process during which the body undergoes changes that eventually bring about death, even if no marked disease or disorder is present. Medical science is trying to extend the human life span and the health span, the length of time the body functions normally.

cle-stimulating hormone in the bloodstream, but the ovaries do not respond. Perhaps aging results from the loss of hormonal activities and a decline in the functions they control.

The immune system, too, no longer performs as it once did, and this can affect the body as a whole. The thymus gland gradually decreases in size, and eventually most of it is replaced by fat and connective tissue. The incidence of cancer increases among the elderly, which may signify that the immune system is no longer functioning as it should. This idea is also substantiated by the increased incidence of autoimmune diseases in older individuals.

It is possible, though, that aging is not due to the failure of a particular system that can affect the body as a whole, but to a specific type of tissue change that affects all organs and even the genes. It has been noticed for some time that proteins—such as the collagen fibers present in many support tissues—become increasingly cross-linked as people age. Undoubtedly, this cross-linking contributes to the stiffening and loss of elasticity characteristic of aging tendons and ligaments. It may also account for the inability of such organs as the blood vessels, the heart, and the lungs to function as they once did. Some researchers have now found that glucose has the tendency to attach to any type of protein, which is the first step in a cross-linking process. They are presently experimenting with drugs that can prevent cross-linking.

Extrinsic Factors

The current data about the effects of aging are often based on comparisons of the elderly to younger age groups. But perhaps today's elderly were not as aware when they were younger of the importance of, for example, diet and exercise to general health. It is possible, then, that much of what we attribute to aging is instead due to years of poor health habits.

Consider, for example, osteoporosis. This condition is associated with a progressive decline in bone density in both males and females so that fractures are more likely to occur after only minimal trauma. Osteoporosis is common in the elderly—by age 65, one-third of women will have vertebral fractures, and by age 81, one-third of women and one-sixth of men will have suffered a hip fracture. While there is no denying that a decline in bone mass occurs as a result of aging, certain extrinsic factors are also important. The occurrence of osteoporosis itself is associated with cigarette smoking, heavy alcohol intake, and inadequate calcium intake. Not only is it possible to eliminate these negative factors by personal choice, but it is also possible to add a positive factor. A moderate exercise program has been found to slow down the progressive loss of bone mass.

Even more important, a sensible exercise program and a proper diet that includes at least five servings of fruits and vegetables a day will most likely help eliminate cardiovascular disease. Experts no longer believe that the cardiovascular system necessarily suffers a large decrease in functioning ability with age. Persons 65 years of age and older can have well-functioning hearts and open coronary arteries if their health habits are good and they continue to exercise regularly.

Effect of Age on Body Systems

Data about how aging affects body systems are necessarily based on past events. It is possible that, in the future, age will not have these effects or at least not to same degree as those described here.

Skin

As aging occurs, skin becomes thinner and less elastic because the number of elastic fibers decreases and the collagen fibers undergo cross-linking, as discussed previously. Also, there is less adipose tissue in the subcutaneous layer; therefore, older people are more likely to feel cold. The loss of thickness partially accounts for sagging and wrinkling of the skin.

Homeostatic adjustment to heat is also limited because there are fewer sweat glands for sweating to occur. There are fewer hair follicles, so the hair on the scalp and the extremities thins out. The number of oil (sebaceous) glands is reduced, and the skin tends to crack. Older people also experience a decrease in the number of melanocytes, making their hair gray and skin pale. In contrast, some of the remaining pigment cells are larger, and pigmented blotches appear on the skin.

Processing and Transporting

Cardiovascular disorders are the leading cause of death today. The heart shrinks because of a reduction in cardiac muscle cell size. This leads to loss of cardiac muscle strength and reduced cardiac output. Still, it is observed that the heart, in the absence of disease, is able to meet the demands of increased activity. It can double its rate or triple the amount of blood pumped each minute even though the maximum possible output declines.

Because the middle layer of arteries contains elastic fibers, which are most likely subject to cross-linking, the arteries become more rigid with time, and their size is further reduced by plaque, a buildup of fatty material. Therefore, blood pressure readings gradually rise. Such changes are common in individuals living in Western industrialized countries but not in agricultural societies. A diet low in cholesterol and saturated fatty acid has been suggested as a way to control degenerative changes in the cardiovascular system.

Blood flow to the liver is reduced, and this organ does not metabolize drugs as efficiently as before. This means that, as a person gets older, less medication is needed to maintain the same level in the bloodstream.

Cardiovascular problems are often accompanied by respiratory disorders, and vice versa. Growing inelasticity of lung tissue means that ventilation is reduced. Because we rarely use the entire vital capacity, these effects are not noticed unless the demand for oxygen increases.

Blood supply to the kidneys is also reduced. The kidneys become smaller and less efficient at filtering wastes. Salt and water balance are difficult to maintain, and the elderly dehydrate faster than young people. Difficulties involving urination include incontinence (lack of bladder control) and the inability to urinate. In men, the prostate gland may enlarge and reduce the diameter of the urethra, making urination so difficult that surgery is often needed.

The loss of teeth, which is frequently seen in elderly people, is more apt to be the result of long-term neglect than aging. The digestive tract loses tone, and secretion of saliva and gastric juice is reduced, but there is no indication of reduced absorption. Therefore, an adequate diet, rather than vitamin and mineral supplements, is recommended. Elderly people commonly complain of constipation, increased gas, and heartburn; gastritis, ulcers, and cancer can also occur.

Integration and Coordination

While most tissues of the body regularly replace their cells, some at a faster rate than others, the brain and the muscles ordinarily do not. However, contrary to previous opinion, recent studies show that few neural cells of the cerebral cortex are lost during the normal aging process. This means that cognitive skills remain unchanged even though a loss in short-term memory characteristically occurs. Although the elderly learn more slowly than the young, they can acquire and remember new material. It has been noted that when more time is given for the subject to respond, age differences in learning decrease.

Neurons are extremely sensitive to oxygen deficiency, and if neuron death does occur, it may be due not to aging itself but to reduced blood flow in narrowed blood vessels. Specific disorders, such as depression, Parkinson disease, and Alzheimer disease, are sometimes seen, but they are not common. Reaction time, however, does slow, and more stimulation is needed for hearing, taste, and smell receptors to function as before. After age 50, the ability to hear tones at higher frequencies decreases gradually, and this can make it difficult to identify individual voices and to understand conversation in a group. The lens of the eye does not accommodate as well and also may develop a cataract. Glaucoma, the buildup of pressure due to increased fluid, is more likely to develop because of a reduction in the size of the anterior cavity of the eye.

Loss of skeletal muscle mass is not uncommon, but it can be controlled by a regular exercise program. The capacity to do heavy labor decreases, but routine physical work should be no problem. A decrease in the strength of the respiratory muscles and inflexibility of the rib cage contribute to the inability of the lungs to expand as before, and reduced muscularity of the urinary bladder contributes to an inability to empty the bladder completely, and therefore to the occurrence of urinary infections.

As noted before, aging is accompanied by a decline in bone density. Osteoporosis, characterized by a loss of calcium and mineral from bone, is not uncommon, but evidence indicates that proper health habits can prevent its occurrence. Arthritis, which causes pain upon movement of a joint, is also seen.

Weight gain occurs because the basal metabolism decreases and inactivity increases. Muscle mass is replaced by stored fat and retained water.

The Reproductive System

Females undergo menopause, and thereafter the level of female sex hormones in the blood falls markedly. The uterus and the cervix are reduced in size, and the walls of the oviducts and the vagina become thinner. The external genitals become less pronounced. In males, the level of androgens falls gradually over the age span of 50–90, but sperm production continues until death.

It is of interest that, as a group, females live longer than males. Although their health habits may be better, it is also

Figure 22.21   Remaining active.
The aim of gerontology is to allow the elderly to enjoy living.

possible that the female sex hormone estrogen offers women some protection against cardiovascular disorders when they are younger. Males suffer a marked increase in heart disease in their forties, but an increase is not noted in females until after menopause, when women lead men in the incidence of stroke. Men are still more likely than women to have a heart attack, however.

Conclusion

We have listed many adverse effects of aging, but it is important to emphasize that while such effects are seen, they are not an inevitable occurrence (Fig. 22.21). We must discover any extrinsic factors that precipitate these adverse effects and guard against them. Just as it is wise to make the proper preparations to remain financially independent when older, it is also wise to realize that biologically successful old age begins with the health habits developed when we are younger.

The deterioration of organ systems associated with aging can possibly be prevented in part by utilizing good health habits.

Bioethical Focus Maternal Health Habits

Because maternal health habits can affect a child before it is born, there has been a growing acceptance of prosecuting women when a newborn has a condition, such as fetal alcohol syndrome, that could only have been caused by the drinking habits of the mother. Employers have also become aware that they might be subject to prosecution if the workplace exposes pregnant employees to toxins. To protect themselves, Johnson Controls, a U.S. battery manufacturer, developed a fetal protection policy. No woman who could bear a child was offered a job that might expose her to toxins that could negatively affect the development of her baby. To get such a job, a woman had to show that she had been sterilized or was otherwise incapable of having children. In 1991, the U.S. Supreme Court declared this policy unconstitutional on the basis of sexual discrimination. The decision was hailed as a victory for women, but was it? The decision was written in such a way that women alone, and not an employer, are responsible for any harm done to the fetus by workplace toxins.

Decide Your Opinion

1. Do you believe a woman should be prosecuted if her child is born with a preventable condition? Why or why not?
2. Should the employer or the woman be held responsible when a workplace toxin does harm to an unborn child?
3. Should sexually active women who can bear a child be expected to avoid substances or situations that could possibly harm an unborn child, even if they are using birth control? Why or why not?

Summarizing the Concepts

22.1 Early Developmental Stages

Development occurs after fertilization. Only one sperm actually enters the egg, and this sperm's nucleus fuses with the egg nucleus. The early developmental stages in animals include the following events. During cleavage, division occurs, but there is no overall growth. The result is a morula, which becomes the blastula when an internal cavity (the blastocoel) appears. During the gastrula stage, invagination of cells into the blastocoel results in formation of the germ layers: ectoderm, mesoderm, and endoderm. Later development of organs can be related to these layers.

The development of three types of animals (lancelet, frog, and chick) can be compared. The first three stages (cleavage, blastulation, and gastrulation) differ according to the amount of yolk in the egg.

During neurulation, the nervous system develops from midline ectoderm, just above the notochord. At this point, it is possible to draw a typical cross section of a chordate embryo (see Fig. 22.5).

22.2 Developmental Processes

Two important mechanisms—cytoplasmic segregation and induction—bring about cellular differentiation and morphogenesis as development occurs. The egg contains chemical signals called maternal determinants that are parceled out during cell division. After the first cleavage of a frog embryo, only a daughter cell that receives a portion of the gray crescent is able to develop into a complete embryo. This illustrates the importance of cytoplasmic segregation to the early development of a frog.

Induction is the ability of one embryonic tissue to influence the development of another tissue. For example, the notochord induces the formation of the neural tube in frog embryos. The reciprocal induction that occurs between the lens and the optic vesicle is another good example of induction. Induction occurs because the inducing cells give off chemical signals that influence their neighbors.

C. elegans and *Drosophila* are two model organisms for the study of developmental genetics. Fate maps and the development of the vulva in *C. elegans* have shown that induction is an ongoing process in which one tissue after the other regulates the development of another through chemical signals coded for by particular genes. Apoptosis is also necessary to normal development.

Work with *Drosophila* has allowed researchers to identify morphogen genes that determine the shape and form of the body. An important concept has emerged: during development, sequential sets of master genes code for morphogen gradients that activate the next set of master genes in turn.

Homeotic genes control pattern formation, such as the presence of antennae, wings, and limbs on the segments of *Drosophila*. Homeotic genes code for proteins that contain a homeodomain, a particular sequence of 60 amino acids. The homeodomain is the portion of the protein that binds to DNA. Homologous homeotic genes have been found in a wide variety of organisms, and therefore they must have arisen early in the history of life and been conserved.

22.3 Human Embryonic and Fetal Development

Human development before birth can be divided into embryonic development (months 1 and 2) and fetal development (months 3–9). The first stages of human development resemble those of the chick. The similarities are probably due to their evolutionary relationship, not to the amount of yolk the eggs contain, because the human egg has little yolk.

The extraembryonic membranes appear early in human development. The trophoblast of the blastocyst is the first sign of the chorion, which goes on to become the fetal part of the placenta. Exchange of gases, nutrients, and wastes occurs between fetal and maternal blood at the placenta. The amnion contains amniotic fluid, which cushions and protects the embryo. The yolk sac and allantois are also present.

Fertilization occurs in the oviduct, and cleavage occurs as the embryo moves toward the uterus. The morula becomes the blastocyst before implanting in the endometrium of the uterus. Organ development begins with neural tube and heart formation. There follows a steady progression of organ formation during embryonic development. During fetal development, refinement of features occurs, and the fetus adds weight. Birth occurs about 280 days after the start of the mother's last menstruation.

22.4 Human Development After Birth

Development after birth consists of infancy, childhood, adolescence, and adulthood. Young adults are at their prime, and then the aging process begins. Aging may be due to cellular repair changes, which are genetic in origin. Other factors that may affect aging are changes in body processes and certain extrinsic factors.

Testing Yourself

Choose the best answer for each question.

1. When all three germ layers are present (ectoderm, endoderm, and mesoderm) the embryo is termed a
   a. blastula.   d. morula.
   b. archenteron.   e. blastopore.
   c. gastrula.

2. In the following sequence (a–e), which stage is out of order?
   a. cleavage   d. gastrula
   b. blastula   e. neurula
   c. morula

3. Which of these pairs is mismatched?
   a. cleavage—cell division
   b. blastula—gut formation
   c. gastrula—three germ layers
   d. neurula—nervous system
   e. Both b and c are mismatched.

4. Which of the germ layers is best associated with development of the heart?
   a. ectoderm   d. neurula
   b. mesoderm   e. All of these are correct.
   c. endoderm

5. The nervous system develops from the
   a. ectoderm.   d. blastopore.
   b. mesoderm.   e. neurula.
   c. endoderm.

6. The ability of one embryonic tissue to influence the growth and development of another tissue is termed
   a. morphogenesis.   d. cellular differentiation.
   b. pattern formation.   e. induction.
   c. apoptosis.

7. In many embryos, differentiation begins at what stage?
   a. cleavage   d. neurula
   b. blastula   e. after the completion of
   c. gastrula   stages a–d

8. Morphogenesis is associated with
   a. protein gradients.   c. homeotic genes.
   b. induction.   d. All of these are correct.

9. Genes that control the structural pattern of an animal are
   a. homeotic genes.   d. All of these are correct.
   b. morphogen genes.   e. Both a and b are correct.
   c. induction genes.

10. In human development, which part of the blastocyst will develop into a fetus?
    a. morula   d. chorion
    b. trophoblast   e. yolk sac
    c. inner cell mass

11. In humans, the placenta develops from the chorion. This indicates that human development
    a. resembles that of the chick.
    b. is associated with extraembryonic membranes.
    c. cannot be compared to the development of lower animals.
    d. begins only upon implantation.
    e. Both a and b are correct.

12. In humans, the fetus
    a. has four extraembryonic membranes.

b. has developed organs and is recognizably human.
c. is dependent upon the placenta for excretion of wastes and acquisition of nutrients.
d. All of these are correct.

13. The term totipotent refers to
    a. cells becoming specialized in structure and function.
    b. the partitioning of cellular materials for mitosis.
    c. the influence of one tissue on the development of other tissues.
    d. a cell containing the full complement of genes.

14. The placenta does not
    a. produce estrogen and progesterone.
    b. exchange dissolved gases.
    c. supply nutrients.
    d. cause the embryo to implant itself.

15. Developmental changes
    a. require growth, differentiation, and morphogenesis.
    b. stop occurring when a person is grown.
    c. are dependent upon a parceling out of genes into daughter cells.
    d. are dependent upon activation of master genes in an orderly sequence.
    e. Both a and d are correct.

16. Which of these pairs is mismatched?
    a. brain—ectoderm   d. lens—endoderm
    b. gut—endoderm   e. heart—mesoderm
    c. bone—mesoderm

17. Which hormone can be administered to begin the process of childbirth?
    a. estrogen   d. testosterone
    b. oxytocin   e. Both b and d are correct.
    c. prolactin

18. Which hormone(s) play(s) a direct role in lactation?
    a. estrogen   d. prostaglandins
    b. oxytocin   e. Both b and c are correct.
    c. prolactin

19. Label this diagram illustrating the placement of the extraembryonic membranes, and give a function for each membrane in humans:

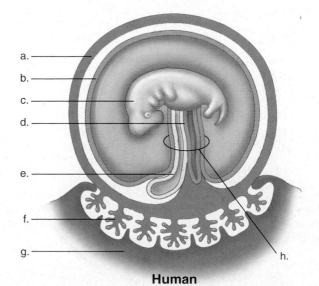

Human

e-Learning Connection   www.mhhe.com/maderinquiry10

Concepts	Questions	Media Resources*
22.1 Early Developmental Stages		

• Development begins when a sperm fertilizes an egg.
• The first stages of embryonic development in animals lead to the establishment of the embryonic germ layers.
• The presence of yolk affects the manner in which animal embryos go through the early developmental stages.
• In vertebrates, the nervous system develops above the notochord after formation of a neural tube.
	1. What is the role of the acrosome during fertilization?
	2. In what way does the amount of yolk in an egg make a difference during development?
		Essential Study Partner
		    Fertilization
		    Cell Differentiation
		
		Art Quiz
		    Gastrulation—Mammal
		
		General Biology Weblinks
		    Developmental Biology
		
		Case Study
		    Development

22.2 Developmental Processes		

• Cytoplasmic segregation and induction help bring out cellular differentiation and morphogenesis.
• Developmental genetics has benefited from research into the development of *Caenorhabditis elegans*, a roundworm, and *Drosophila melanogaster*, a fruit fly.
• Homeotic genes are involved in shaping the outward appearance of animals.
	1. Explain what is meant by "cellular differentiation" and "morphogenesis."
	2. What is the function of homeotic genes?
		Art Quizzes
		    Embryos and Evolutionary
		        History
		    Allometric Growth
		
		BioCourse Study Guide
		    Mammalian Development I

22.3 Human Embryonic and Fetal Development		

• Humans, like chicks, are dependent upon extraembryonic membranes that perform various services that contribute to development.
• Human development consists of embryonic and fetal development.
• Humans are placental mammals; the placenta is a unique organ where exchange between fetal blood and mother's blood takes place.
• Birth is a multistage process that includes delivery of the child and the extraembryonic membranes.
	1. In what way are extraembryonic membranes evolutionarily significant?
	2. During what stage of development are all the main organ systems present if not fully developed?
		Essential Study Partner
		    Early Development
		    Human Development
		    Hormones and Pregnancy
		    Preembryonic Development
		    Embryonic and Fetal
		        Development
		    Parturition
		
		Art Quizzes
		    Placenta
		
		Labeling Exercises
		    Human Extraembryonic
		        Membranes
		    Anatomy of the Breast
		
		Case Studies
		    Frozen Embryos

22.4 Human Development After Birth		

• Investigation into aging shows hope of identifying underlying causes of degeneration and prolonging the health span of individuals.	1. What habits can be adopted to ensure health as an individual ages?	Essential Study Partner
		    Postnatal Period
		    Aging Skin
		
		Case Study
		    Treatment of Critically Ill
		        Newborns

*For additional Media Resources, see the Online Learning Center.

CHAPTER 23

Patterns of Gene Inheritance

Chapter Concepts

The genes we inherit from our parents determine our characteristics, including our physical traits and even our behavior.

*M*ichael Douglas strikingly resembles his father Kirk Douglas. A look-alike relationship between one generation and the next has been noted for some time, but now we know the reason for it. Genes! Through the process of sexual reproduction, Kirk Douglas passed on a copy of half of his genes to his son. Why doesn't Michael Douglas look exactly like this father? For precisely the same reason: Kirk Douglas gave Michael only half his genes; his mother gave him the other half.

We know today that many hundreds of genes are located on each of the chromosomes that come together when the sperm fertilizes the egg. Each gene controls some particular characteristic of the cell or individual. Why do the Douglases have a cleft chin? The genes involved in development of the chin brought this about. We now recognize the power of the gene, and have learned how to manipulate genes for our own purposes. What would Kirk Douglas do if he wanted a son who looked exactly like him? He would think about having just his genes used to start a new life. That's called cloning.

23.1 Mendel's Laws

Today, most people know that DNA is the genetic material, and they may have heard that scientists have just sequenced all the bases in the DNA of human cells. In contrast, they may never have heard of Gregor Mendel, an Austrian monk who in 1860 developed certain laws of heredity after doing crosses between garden pea plants (Fig. 23.1). But Gregor Mendel investigated genetics at the organismal level, and this is still the level that intrigues most of us on a daily basis. We observe, for example, that facial and other features run in families, and we would like some convenient way of explaining this observation. And so it is appropriate to begin our study of genetics at the organismal level and learn to use Mendel's laws of heredity.

Gregor Mendel

Mendel's parents were farmers, so he no doubt acquired the practical experience he needed to grow pea plants during childhood. Mendel was also a mathematician; he kept careful and complete records even though he crossed and catalogued some 24,034 plants through several generations. He concluded that the plants transmitted distinct factors (now called genes) to their offspring. The particulate model of heredity based on his studies assumes that genes are sections of chromosomes. In Figure 23.2, the letters on the homologous chromosomes stand for genes that control a trait, such as color of hair, type of fingers, or length of nose. The genes are in definite sequence and remain in their spots, or **loci,** on the chromosomes. Alternative forms of a gene having the same position on a pair of homologous chromosomes and affecting the same trait are called **alleles.** In Figure 23.2, *G* is an allele of *g*, and vice versa; *R* is an allele of *r*, and vice versa. *G* could never be an allele of *R* because *G* and *R* are at different loci.

Mendel's work is described in the Science Focus on the next page. He said that pea plants have two factors for every trait, such as stem length. He observed that one of the factors controlling the same trait can be dominant over the other, which is recessive. For example, he found that a pea plant could be tall even if one factor was for shortness. In Mendel's experiments, a tall pea plant was sometimes the parent of a short plant. Therefore, he reasoned that, while the individual plant has two factors for each trait, the gametes (i.e., sperm and egg) contain only one factor for each trait. This is now known as Mendel's **law of segregation.**

The law of segregation states the following:

- Each individual has two factors for each trait.
- The factors segregate (separate) during the formation of the gametes.
- Each gamete contains only one factor from each pair of factors.
- Fertilization gives each new individual two factors for each trait.

Figure 23.1   Mendel working in his garden.
Mendel grew and tended the pea plants he used for his experiments. For each experiment, he observed as many offspring as possible. For a cross that required him to count the number of round seeds to wrinkled seeds, he observed and counted a total of 7,324 peas!

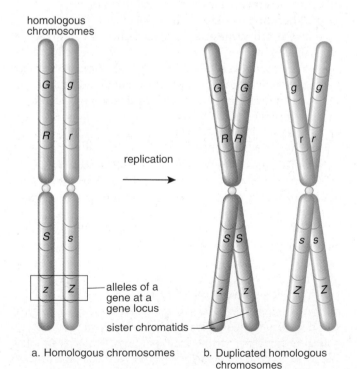

a. Homologous chromosomes   b. Duplicated homologous chromosomes

Figure 23.2   Gene locus.
a. Each allelic pair, such as *Gg* or *Zz*, is located on homologous chromosomes at a particular gene locus. **b.** Following replication, each sister chromatid carries the same alleles in the same order.

23.3 Beyond Patterns

Certain traits, such a
of simple dominant c
other, more complica

Polygenic Inheri

Polygenic inheritanc
two or more sets of al
allelic pairs, possibly
chromosomes. Each do
on the phenotype, and
is a continuous variat
tribution of these phe

a.

b.

Figure 23.13 Polyge
When you record the heigh
values follow a bell-shaped
is due to control of a trait b
effects are also involved.

Science Focus

The Investigations of Gregor Mendel

Mendel's use of pea plants as his experimental material was a good choice because pea plants are easy to cultivate, have a short generation time, and can be self-pollinated or cross-pollinated at will. Mendel selected certain traits for study and, before beginning his experiments, made sure his parental (P generation) plants bred true—in other words, he observed that when these plants self-pollinated, the offspring were like one another and like the parent plant. For example, a parent with yellow seeds always had offspring with yellow seeds; a plant with green seeds always had offspring with green seeds. Then Mendel cross-pollinated the plants by dusting the pollen of plants with yellow seeds on the stigma of plants with green seeds whose own anthers had been removed, and vice versa (Fig. 23A). Either way, the offspring (called F_1, or the first filial generation) resembled the parents with yellow seeds. Mendel then allowed the F_1 plants to self-pollinate. Once he had obtained an F_2 generation, he observed the color of the peas produced. He counted over 8,000 plants and found an approximate 3:1 ratio (about three plants with yellow seeds for every plant with green seeds) in the F_2 generation. Mendel realized that these results were explainable, assuming (a) there are two factors for every trait; (b) one of the factors can be dominant over the other, which is recessive; and (c) the factors separate when the gametes are formed.

He believed that the F_2 plants with yellow seeds carried a dominant factor because his results could be related to the binomial expression $a^2 + 2ab + b^2$. He said if $a = Y$ and $b = y$, then the four F_2 plants were $YY + 2Yy + yy$. And three plants with yellow seeds are expected for every plant with green seeds.

As a test to determine if the F_1 generation was indeed Yy, Mendel backcrossed it with the recessive parent, yy. His results of 1:1 indicated that he had reasoned correctly. Today, when a one-trait testcross is done, a suspected heterozygote is crossed with the recessive phenotype because it has a known genotype.

Mendel performed a second series of experiments in which he crossed true-breeding plants that differed in two traits. For example, he crossed plants with yellow, round peas with plants with green, wrinkled peas. The F_1 generation always had both dominant characteristics; therefore, he allowed the F_1 plants to self-pollinate. Among the F_2 generation, he achieved an almost perfect ratio of 9:3:3:1. For example, for every plant that had green, wrinkled peas, approximately nine had yellow, round peas, and so forth. Mendel saw that these results were explainable if pairs of factors separate independently from one another when the gametes form, allowing all possible combinations of factors to occur in the gametes. This would mean that the probability of achieving any two factors together in the F_2 offspring is the product of their chance of occurring separately. Therefore, since the chance of yellow peas was ¾ (in a one-trait cross) and the chance of round peas was ¾ (in a one-trait cross), the chance of their occurring together was 9⁄16, and so forth.

Mendel achieved his success in genetics by studying large numbers of offspring, keeping careful records, and treating his data quantitatively. He showed that the application of mathematics to biology is extremely helpful in producing testable hypotheses.

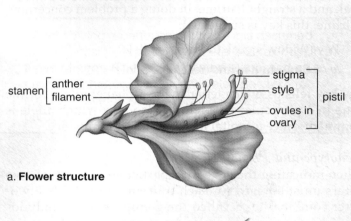

a. **Flower structure**

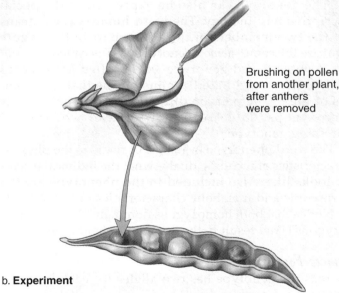

Brushing on pollen from another plant, after anthers were removed

b. **Experiment**

Figure 23A Garden pea anatomy and traits.
a. In the garden pea, pollen grains produced in the anther contain sperm, and ovules in the ovary contain eggs. **b.** When Mendel performed crosses, he brushed pollen from one plant onto the stigma of another plant. After sperm fertilized eggs, the ovules developed into seeds (peas). The open pod shows all the results of a cross between plants with round, yellow seeds and plants with wrinkled, green seeds. (Although the artist depicted all possible seed types in one pod, each plant produces just one of these seed types.)

Autosomal Don

Of the many autosom
only two.

Neurofibromatosis

Neurofibromatosis, s
disease, is one of th
affects roughly one i
every racial and ethn

At birth or later, t
more large, tan spots
in size and number a
(lumps) called neurofi
erings of nerves may
variable expressivity.
patients live a normal
are severe. Some patie:
a large head, and eye
blindness and hearing
matosis have learning

In 1990, research
matosis, which was k
gene controls the pro
bromin that normally
division. Any numbe
bromin that fails to bl
formation of tumor
inserted genes that do
can move from one lo
first discovered in ot
humans.

Huntington Disease

Huntington disease is
progressive degener
causes severe muscle
(Fig. 23.12). The diseas
of the gene for a prot
appear normal until
already had children
Occasionally, the first
eration is seen in teena
is no effective treatm
years after the onset o

Several years ago,
Huntington disease w
was developed for the
want to know if they h
is no cure. At least no
from a mutation that c
copies of the amino aci
the huntingtin protein
glutamines. If hunting
changes shape and form
worse, it attracts and ca

---

e-Learning Connection

www.mhhe.com/maderinquiry10

Concepts	Questions	Media Resources*
23.1 Mendel's Law		

• Today it is known that alleles (alternate forms of a gene), located on chromosomes, control the traits of individuals.
 • Mendel discovered certain laws of heredity after doing experiments with garden peas during the mid-1800s.
 • The law of segregation states that each organism contains two factors for each trait, and the factors segregate during formation of gametes.
 • A testcross can be used to determine the genotype of an individual with the dominant phenotype.
 • The law of independent assortment states that every possible combination of parental factors is present in the gametes.
	1. Explain the difference between "genotype" and "phenotype."
	 2. When considering two alleles for each of two traits that are located on different chromosomes, why will gametes contain every possible combination of the traits?
		Essential Study Partner
		  Introduction
		  Genetics
		  Monohybrid Cross
		  Dihybrid Cross
		  Recombination
		
		 Art Quizzes
		  Mendel's Experiment Results
		  Testcross
		  Dihybrid Cross
		
		 General Biology Weblinks
		  Mendelian Genetics
		
		 Explorations
		  Constructing a Genetic Map
		  Heredity in Families
		  Gene Segregation within
		   Families

23.2 Genetic Disorders		
• Many genetic disorders are inherited according to Mendel's laws. • The pattern of inheritance indicates whether the disorder is a recessive or a dominant disorder. • Dominant genetic disorders appear if a single dominant allele is inherited; recessive genetic disorders require the inheritance of two recessive alleles.	1. When considering genetic diseases, what is a "carrier?"	Exploration Cystic Fibrosis
23.3 Beyond Simple Inheritance Patterns		

• Polygenic traits include skin color, behavior, and various syndromes.
 • Blood type is controlled by multiple alleles and exhibits codominance.
 • Some traits, like sickle-cell disease, are incompletely dominant.
	1. Are there any human polygenic disorders? Explain.
	 2. If a woman has type O blood, could a man with type AB conceive children with type O blood? Explain.
		Essential Study Partner
		  Beyond Mendel
		
		 Art Quizzes
		  Continuous Variation
		  Incomplete Dominance
		  Multiple Alleles—ABO
		  Blood Groups

*For additional Media Resources, see the Online Learning Center.

Patterns of Chromosome Inheritance

Chapter Concepts

Chris Burke is a professional actor and musician. Chris, shown here with his girlfriend, has Down syndrome.

Chris Burke was born with three copies of chromosome 21. He has Down syndrome. His parents were advised to put him in an institution.

The abnormalities associated with Down syndrome are numerous and diverse: all body parts tend to be short, the face is usually broad and flat, the nose is typically small, the eyes slant, and the eyelids have a fold. The tongue is large and furrowed. Malformations of the heart, digestive tract, kidneys, thyroid gland, and adrenal glands commonly occur. And intelligence is usually below normal.

But Chris' parents persevered and they gave Chris the same loving care and attention they gave their other children, and it paid off. Chris is remarkably talented. He is a playwright, actor, and musician. He starred in Life Goes On, a TV series written just for him, and he is frequently asked to be a guest star in a number of TV shows. His love of music and collaboration with other musicians has led to the release of several albums—like Chris, the songs are uplifting and inspirational. You can read more about this remarkable young man in his autobiography, A Special Kind of Hero. This book is sold in bookstores everywhere.

24.1 Viewing the Chromosomes

As you know from previous discussions, the genes are on the chromosomes. So it is proper to speak of inherited chromosomes as much as inherited genes. It is possible to view an individual's chromosomes by constructing a **karyotype,** a visual display of the chromosomes arranged by size, shape, and banding pattern. Normally, both males and females have 23 pairs of chromosomes; 22 pairs are autosomes and one pair is the sex chromosomes. These are called the **sex chromosomes** because they differ between the sexes. In humans, males have the sex chromosomes X and Y, and females have two X chromosomes.

Various human disorders result from abnormal chromosome number or structure. Such disorders often result in a **syndrome,** which is a group of symptoms that always occur together. Table 24.1 lists several syndromes that are due to an abnormal chromosome number. Doing a karyotype will reveal such abnormalities.

Any cell in the body except red blood cells, which lack a nucleus, can be a source of chromosomes for karyotyping. In adults, it is easiest to use white blood cells separated from a blood sample for this purpose. In fetuses, whose chromosomes are often examined in order to be forewarned of a syndrome, cells can be obtained by either amniocentesis or chorionic villi sampling.

Amniocentesis

Amniocentesis is a procedure for obtaining a sample of amniotic fluid from the uterus of a pregnant woman. Blood tests and age of the mother are used to determine when the procedure should be done. Amniocentesis is not usually performed until about the fourteenth to the seventeenth week of pregnancy. A long needle is passed through the abdominal and uterine walls to withdraw a small amount of fluid, which also contains fetal cells (Fig. 24.1a). Testing of the cells and karyotyping the chromosomes may be delayed as long as four weeks so that the cells can be cultured to increase their number. As many as 400 chromosome and biochemical problems can be detected by testing the cells and the amniotic fluid.

The risk of spontaneous abortion increases by about 0.3% due to amniocentesis, and doctors only use the procedure if it is medically warranted.

Chorionic Villi Sampling

Chorionic villi sampling (CVS) is a procedure for obtaining chorionic cells in the region where the placenta will develop. This procedure can be done as early as the fifth week of pregnancy. A long, thin suction tube is inserted through the vagina into the uterus (Fig. 21.4b). Ultrasound, which gives a picture of the uterine contents, is used to place the tube between the uterine lining and the chorionic villi. Then a sampling of chorionic cells is obtained by suction. The cells do not have to be cultured, and karyotyping can be done immediately. This sampling procedure does not include any amniotic fluid so the biochemical tests done on the amniotic fluid following amniocentesis are not possible. Also, CVS carries a greater risk of spontaneous abortion than amniocentesis—0.8% compared to 0.3%. The advantage of CVS is getting the results of karyotyping at an earlier date.

Karyotyping

After a cell sample has been obtained, the cells are stimulated to divide in a culture medium. A chemical is used to stop mitosis during metaphase when chromosomes are the most highly compacted and condensed. The cells are then killed, spread on a microscope slide, and dried. Stains are applied to the slides and the cells are photographed. Staining produces dark and light cross-bands of varying widths and these can be used in addition to size and shape to help pair up the chromosomes. Today a computer is used to arrange the chromosomes in pairs (Fig. 21.4c). Comparing a normal karyotype with that of a person who has Down syndrome reveals the most common chromosome abnormality in humans (Fig. 21.4d and e).

Amniocentesis and chorionic villi sampling are used to obtain cells for karyotyping if it is suspected that the fetus may have a chromosome abnormality.

Table 24.1	Syndromes from Abnormal Chromosome Numbers				
Syndrome	Sex	Disorder	Chromosome Number	Frequency	
				Spontaneous Abortions	Live Births
Down	M or F	Trisomy 21	47	1/40	1/800
Poly-X	F	XXX (or XXXX)	47 or 48	0	1/1,500
Klinefelter	M	XXY (or XXXY)	47 or 48	1/300	1/800
Jacobs	M	XYY	47	?	1/1,000
Turner	F	XO	45	1/18	1/2,500

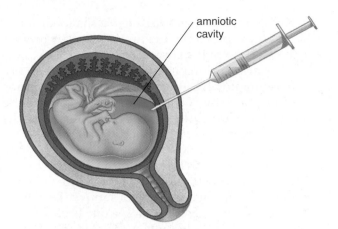

a. During amniocentesis, a long needle is used to withdraw amniotic fluid containing fetal cells.

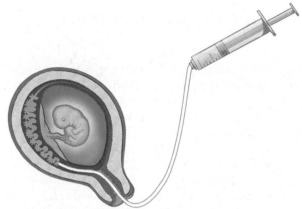

b. During chorionic villi sampling, a suction tube is used to remove cells from the chorion, where the placenta will develop.

c. Cells are microscopically examined and photographed. Computer arranges the chromosomes into pairs.

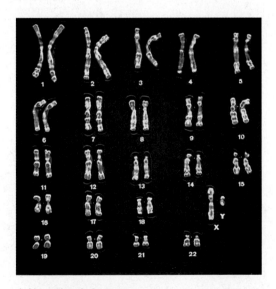

d. Normal male karyotype with 46 chromosomes

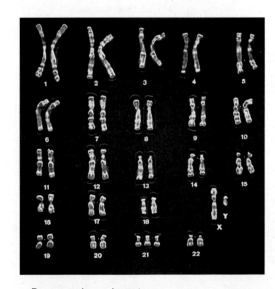

e. Down syndrome karyotype with an extra chromosome 21

Figure 24.1   Human karyotype preparation.

A karyotype is an arrangement of an individual's chromosomes into numbered pairs according to their size, shape, and banding pattern. **a.** Amniocentesis and (**b**) chorionic villi sampling provide cells for karyotyping to determine if the unborn child has a chromosome abnormality. **c.** After cells are treated as described in the text, a computer constructs the karyotype. **d.** Karyotype of a normal male. **e.** Karyotype of a male with Down syndrome. A Down syndrome karyotype has three number 21 chromosomes.

24.2 Changes in Chromosome Number

Normally, an individual receives 22 pairs of autosomes and two sex chromosomes. Sometimes individuals are born with either too many or too few autosomes or sex chromosomes, most likely due to nondisjunction during meiosis. **Nondisjunction** occurs during meiosis I when both members of a homologous pair go into the same daughter cell, or during meiosis II when the sister chromatids fail to separate and both daughter chromosomes go into the same gamete. Figure 24.2 assumes that nondisjunction has occurred during oogenesis; some abnormal eggs have 24 chromosomes, while others have only 22 chromosomes. If an egg with 24 chromosomes is fertilized with a normal sperm, the result is a **trisomy,** so called because one type of chromosome is present in three copies. If an egg with 23 chromosomes is fertilized with a normal sperm, the result is a **monosomy,** so called because one type of chromosome is present in a single copy.

Normal development depends on the presence of exactly two of each kind of chromosome. Too many chromosomes is tolerated better than a deficiency of chromosomes, and several trisomies are known to occur in humans. Among autosomal trisomies, only trisomy 21 (Down syndrome) has a reasonable chance of survival after birth. This is probably due to the fact that chromosome 21 is the smallest of the chromosomes.

The chances of survival are greater when trisomy or monosomy involves the sex chromosomes. In normal XX females, one of the X chromosomes becomes a darkly staining mass of chromatin called a **Barr body** (after the person who discovered it). A Barr body is an inactive X chromosome; therefore, we now know that the cells of females function with a single X chromosome just as those of males do. This is most likely the reason why a zygote with one X chromosome (Turner syndrome) can survive. Then too, all extra X chromosomes beyond a single one become Barr bodies, and this explains why poly-X females and XXY males are seen fairly frequently. An extra Y chromosome, called Jacobs syndrome, is tolerated in humans; most likely because the Y chromosome carries few genes. Jacobs syndrome (XYY) is due to nondisjunction during meiosis II of spermatogenesis. We know this because two Ys are only present during meiosis II in males.

Nondisjunction changes the chromosome number in gametes. Offspring sometimes inherit an extra chromosome (trisomy) or are missing a chromosome (monosomy).

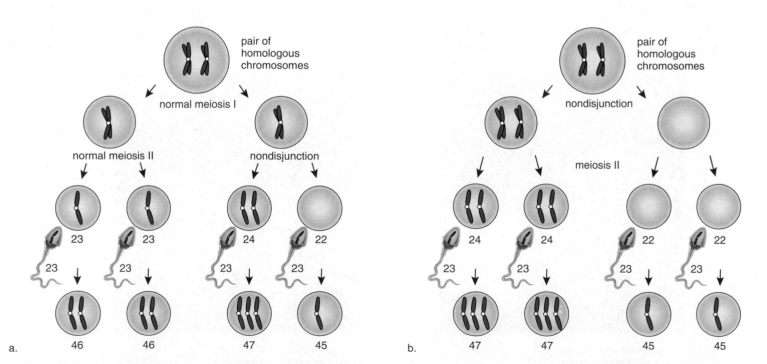

Figure 24.2   Nondisjunction of chromosomes during oogenesis, followed by fertilization with normal sperm.
a. Nondisjunction can occur during meiosis II if the sister chromatids separate, but the resulting chromosomes go into the same daughter cell. Then the egg will have one more (24) or one less (23) than the usual number of chromosomes. Fertilization of these abnormal eggs with normal sperm produces an abnormal zygote with 47 or 45 chromosomes. **b.** Nondisjunction can also occur during meiosis I and result in abnormal eggs that also have one more or one less than the normal number of chromosomes. Fertilization of these abnormal eggs with normal sperm results in a zygote with an abnormal chromosome number and the syndromes listed in Table 24.1.

Down Syndrome

The most common autosomal trisomy seen among humans is trisomy 21, also called Down syndrome (Fig. 24.3). This syndrome is easily recognized by these characteristics: short stature; an eyelid fold; a flat face; stubby fingers; a wide gap between the first and second toes; a large, fissured tongue; a round head; a palm crease, the so-called simian line; and, unfortunately, mental retardation, which can sometimes be severe.

Persons with Down syndrome usually have three copies of chromosome 21 because the egg had two copies instead of one. (In 23% of the cases studied, however, the sperm had the extra chromosome 21.) The chances of a woman having a Down syndrome child increase rapidly with age, starting at about age 40, and the reasons for this are still being determined.

Although an older woman is more likely to have a Down syndrome child, most babies with Down syndrome are born to women younger than age 40 because this is the age group having the most babies. Karyotyping can detect a Down syndrome child. However, young women are not routinely encouraged to undergo the procedures necessary to get a sample of fetal cells (i.e., amniocentesis or chorionic villi sampling) because the risk of complications is greater than the risk of having a Down syndrome child. Fortunately, a test based on substances in maternal blood can help identify fetuses who may need to be karyotyped.

The genes that cause Down syndrome are located on the bottom third of chromosome 21 (Fig. 24.3b), and extensive investigative work has been directed toward discovering the specific genes responsible for the characteristics of the syndrome. Thus far, investigators have discovered several genes that may account for various conditions seen in persons with Down syndrome. For example, they have located genes most likely responsible for the increased tendency toward leukemia, cataracts, accelerated rate of aging, and mental retardation. The gene for mental retardation, dubbed the *Gart* gene, causes an increased level of purines in the blood, a finding associated with mental retardation. One day it may be possible to control the expression of the *Gart* gene even before birth so that at least this symptom of Down syndrome does not appear.

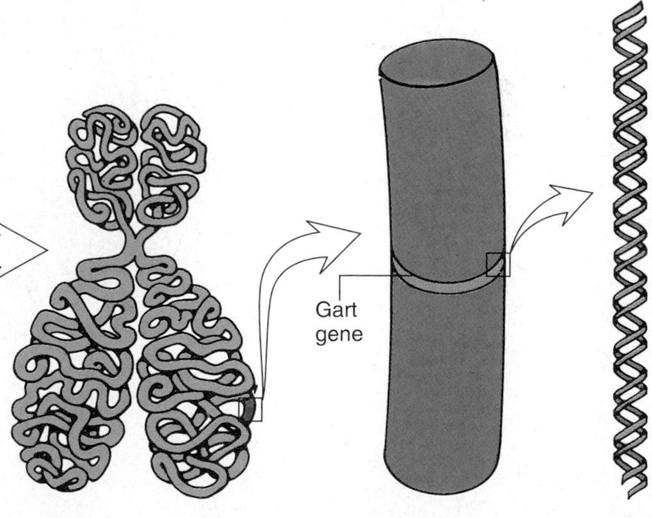

Figure 24.3   Abnormal autosomal chromosome number.
Persons with Down syndrome have an extra chromosome 21. **a.** Common characteristics of the syndrome include a wide, rounded face and a fold on the upper eyelids. Mental retardation, along with an enlarged tongue, makes it difficult for a person with Down syndrome to speak distinctly. **b.** Karyotype of an individual with Down syndrome shows an extra chromosome 21. More sophisticated technologies allow investigators to pinpoint the location of specific genes associated with the syndrome. An extra copy of the *Gart* gene, which leads to a high level of purines in the blood, may account for the mental retardation seen in persons with Down syndrome.

Changes in Sex Chromosome Number

An abnormal sex chromosome number is the result of inheriting too many or too few X or Y chromosomes. Figure 24.2 can be used to illustrate nondisjunction of the sex chromosomes during oogenesis if you assume that the chromosomes shown represent X chromosomes. Nondisjunction during oogenesis or spermatogenesis can result in gametes that have too few or too many X or Y chromosomes. After fertilization, the syndromes listed in Table 24.1 (other than Down syndrome which is autosomal) are possibilities.

A person with Turner syndrome (XO) is a female and a person with Klinefelter syndrome (XXY) is a male. This shows that in humans the presence of a Y chromosome, not the number of X chromosomes, determines maleness. The *SRY* gene on the short arm of the Y chromosome produces a hormone called testis-determining factor, which plays a critical role in the development of male genitals.

Turner Syndrome

From birth, an XO individual with Turner syndrome has only one sex chromosome, an X; the O signifies the absence of a second sex chromosome. Turner females are short, with a broad chest and folds of skin on the back of the neck. The ovaries, oviducts, and uterus are very small and underdeveloped. Turner females do not undergo puberty or menstruate, and their breasts do not develop (Fig. 24.4a). However, some have given birth following in vitro fertilization using donor eggs. They usually are of normal intelligence and can lead fairly normal lives if they receive hormone supplements.

Klinefelter Syndrome

A male with Klinefelter syndrome has two or more X chromosomes in addition to a Y chromosome. Counting the number of Barr bodies can tell the number of extra X chromosomes. The person has underdeveloped testes and prostate gland and no facial hair. But there may be some breast development (Fig. 24.4b). Affected individuals have large hands and feet and very long arms and legs. They are usually slow to learn but not mentally retarded unless they inherit more than two X chromosomes. No matter how many X chromosomes there are, an individual with a Y chromosome is a male.

The Health Focus on the next page tells of the experiences of a person with Klinefelter syndrome who suggests that it is best for parents to know right away that they have a child with this disorder because much can be done to help them lead a normal life.

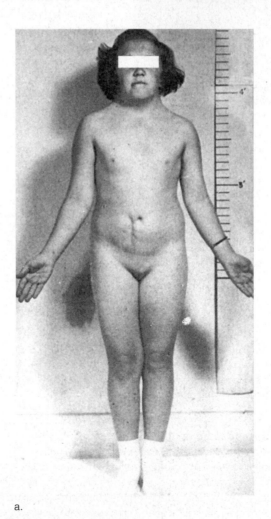
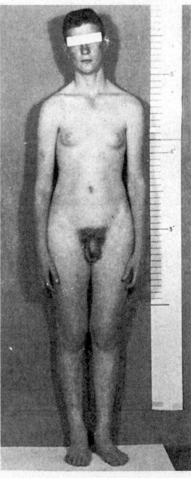

Figure 24.4   Abnormal sex chromosome number.
a. A female with Turner (XO) syndrome has a short thick neck, short stature, and lack of breast development. **b.** A male with Klinefelter (XXY) syndrome has immature sex organs and some development of the breasts.

a.

b.

Poly-X Females

A poly-X female has more than two X chromosomes and extra Barr bodies in the nucleus. Females with three X chromosomes have no distinctive phenotype aside from a tendency to be tall and thin. Although some have delayed motor and language development, most poly-X females are not mentally retarded. Some may have menstrual difficulties, but many menstruate regularly and are fertile. Children usually have a normal karyotype.

Females with more than three X chromosomes occur rarely. Unlike XXX females, XXXX females are usually tall and severely retarded. Various physical abnormalities are seen but they may menstruate normally.

Jacobs Syndrome

XYY males with Jacobs syndrome can only result from nondisjunction during spermatogenesis. Affected males are usually taller than average, suffer from persistent acne, and tend to have speech and reading problems. At one time, it was suggested that these men were likely to be criminally aggressive, but it has since been shown that the incidence of such behavior among them may be no greater than among XY males.

Several syndromes are due to an abnormal number of sex chromosomes (see also Table 24.1).

Health Focus

Living with Klinefelter Syndrome

In 1996, at the age of 25, I was diagnosed with Klinefelter syndrome (KS). Being diagnosed has changed my life for the better.

I was a happy baby, but when I was still very young, my parents began to believe that there was something wrong with me. I knew something was different about me, too, as early on as five years old. I was very shy and had trouble making friends. One minute I'd be well behaved, and the next I'd be picking fights and flying into a rage. Many psychologists, therapists, and doctors tested me because of school and social problems and severe mood changes. Their only diagnosis was "learning disabilities" in such areas as reading comprehension, abstract thinking, word retrieval, and auditory processing. No one could figure out what the real problem was, and I hated the tutoring sessions I had. In the seventh grade, a psychologist told me that I was stupid and lazy, I would probably live at home for the rest of my life, and I would never amount to anything. For the next five years, he was basically right, and I barely graduated from high school.

I believe, though, that I have succeeded because I was told that I would fail. I quit the tutoring sessions when I enrolled at a community college; I decided I could figure things out on my own. I received an associate degree there, then transferred to a small liberal arts college. I never told anyone about my learning disabilities and never sought special help. However, I never had a semester below a 3.0, and I graduated with two B.S. degrees. I was accepted into a graduate program but decided instead to accept a job as a software engineer even though I did not have an educational background in this field. As I later learned, many KS'ers excel in computer skills. I had been using a computer for many years and had learned everything I needed to know on my own, through trial and error.

Around the time I started the computer job, I went to my physician for a physical. He sent me for blood tests because he noticed that my testes were smaller than usual. The results were conclusive: Klinefelter syndrome with sex chromosomes XXY. I initially felt denial, depression, and anger, even though I now had an explanation for many of the problems I had experienced all my life. But then I decided to learn as much as I could about the condition and treatments available. I now give myself a testosterone injection once every two weeks, and it has made me a different person, with improved learning abilities and stronger thought processes in addition to a more outgoing personality.

I found, though, that the best possible path I could take was to help others live with the condition. I attended my first support group meeting four months after I was diagnosed. By spring 1997, I had developed an interest in KS that was more than just a part-time hobby. I wanted to be able to work with this condition and help people forever. I have been very involved in KS conferences and have helped to start support groups in the U.S., Spain, and Australia.

Since my diagnosis, it has been my dream to have a son with KS, although when I was diagnosed, I found out it was unlikely that I could have biological children. Through my work with KS, I had the opportunity to meet my fiancee Chris. She has two wonderful children: a daughter, and a son who has the same condition that I do. There are a lot of similarities between my stepson and me, and I am happy I will be able to help him get the head start in coping with KS that I never had. I also look forward to many more years of helping other people seek diagnosis and live a good life with Klinefelter syndrome.

Stefan Schwarz

stefan13@mail.ptd.net

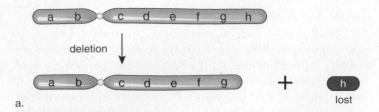

a.

b.

Figure 24.5   Deletion.

a. When chromosome 7 loses an end piece, the result is Williams syndrome. **b.** These children, although unrelated, have the same appearance, health, and behavioral problems.

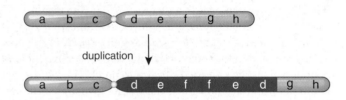

a.

b.

Figure 24.6   Duplication.

a. When a piece of chromosome 15 is duplicated and inverted, **(b)** a syndrome results in which the child has poor muscle tone and autistic characteristics.

24.3 Changes in Chromosome Structure

A mutation is a permanent genetic change. A change in chromosome structure that can be detected microscopically is a **chromosome mutation.** Chromosome mutations occur when chromosomes suffer breaks. Various environmental agents—radiation, certain organic chemicals, or even viruses—can cause chromosomes to break apart. Ordinarily, when breaks occur in chromosomes, the segments reunite to give the same sequence of genes. But their failure to reunite correctly can result in one of several types of mutations: deletion, duplication, translocation, or inversion. Chromosome mutations can occur during meiosis, and if the offspring inherits the abnormal chromosome, a syndrome may very well develop.

Deletions and Duplications

A **deletion** occurs when a single break causes a chromosome to lose an end piece or when two simultaneous breaks lead to the loss of an internal chromosome segment. An individual who inherits a normal chromosome from one parent and a chromosome with a deletion from the other parent no longer has a pair of alleles for each trait, and a syndrome can result.

Williams syndrome occurs when chromosome 7 loses a tiny end piece (Fig. 24.5). Children who have this syndrome look like pixies because they have turned-up noses, wide mouths, a small chin, and large ears. Although their academic skills are poor, they exhibit excellent verbal and musical abilities. The gene that governs the production of the protein elastin is missing, and this affects the health of the cardiovascular system and causes their skin to age prematurely. Such individuals are very friendly but need an ordered life, perhaps because of the loss of a gene for a protein that is normally active in the brain.

Cri du chat (cat's cry) syndrome is seen when chromosome 5 is missing an end piece. The affected individual has a small head, is mentally retarded, and has facial abnormalities. Abnormal development of the glottis and larynx results in the most characteristic symptom—the infant's cry resembles that of a cat.

In a **duplication,** a chromosome segment is repeated in the same chromosome or in a nonhomologous chromosome. In any case, the individual has more than two alleles for certain traits. An inverted duplication is known to occur in chromosome 15. Inverted means that the duplicated segment joins in the direction opposite from normal. Children with this syndrome, called inv dup 15 syndrome, have poor muscle tone, mental retardation, seizures, a curved spine, and autistic characteristics, including poor speech, hand flapping, and lack of eye contact (Fig. 24.6).

Translocation

A **translocation** is the exchange of chromosome segments between two, nonhomologous chromosomes. A person who has both of the involved chromosomes has the normal amount of genetic material and is healthy, unless the chromosome exchange breaks an allele into two pieces. The person who inherits only one of the translocated chromosomes will no doubt have only one copy of certain alleles and three copies of certain other alleles. A genetic counselor begins to suspect a translocation has occurred when spontaneous abortions are commonplace and family members suffer from various syndromes. A special microscopic technique allows a technician to determine that a translocation has occurred.

In 5% of cases, a translocation that occurred in a previous generation between chromosomes 21 and 14 is the cause of Down syndrome. The affected person inherits two normal chromosomes 21 and an abnormal chromosome 14 that contains a segment of chromosome 21. In these cases, Down syndrome is not related to the age of the mother, but instead tends to run in the family of either the father or the mother.

Figure 24.7 shows a father and son who have a translocation between chromosomes 2 and 20. Although they have the normal amount of genetic material, they have the distinctive face, abnormalities of the eyes and internal organs, and severe itching characteristic of Alagille syndrome. People with this syndrome ordinarily have a deletion on chromosome 20; therefore, it can be deduced that the translocation disrupted an allele on chromosome 20 in the father. The symptoms of Alagille syndrome range from mild to severe, so some people may not be aware they have the syndrome. This father did not realize it until he had a child with the syndrome.

Inversion

An **inversion** occurs when a segment of a chromosome is turned 180 degrees. You might think this is not a problem because the same genes are present, but the reverse sequence of alleles can lead to altered gene activity.

Crossing-over between an inverted chromosome and the noninverted homologue can lead to recombinant chromosomes that have both duplicated and deleted segments. This happens because alignment between the two homologues is only possible when the inverted chromosome forms a loop (Fig. 24.8).

Chromosome mutations can lead to various syndromes among offspring when the mutation produces chromosomes that have deleted, duplicated, translocated, and inverted segments.

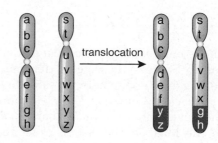

a.

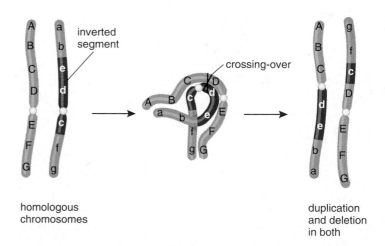

b.

Figure 24.7   Translocation.
a. When chromosomes 2 and 20 exchange segments, **(b)** Alagille syndrome, with distinctive facial features, sometimes results because the translocation disrupts an allele on chromosome 20.

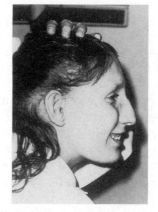 

homologous
chromosomes

duplication
and deletion
in both

Figure 24.8   Inversion.
(Left) A segment of one homologue is inverted. Notice that in the shaded segment *edc* occurs instead of *cde*. *(Middle)* The two homologues can pair only when the inverted sequence forms an internal loop. After crossing over, a duplication and a deletion can occur. *(Right)* The homologue on the left has *AB* and *ab* sequences and neither *fg* nor *FG* genes. The homologue on the right has *gf* and *FG* sequences and neither *AB* nor *ab* genes.

24.4 Sex-Linked Traits

The sex chromosomes contain genes just as the autosomal chromosomes do. Some of these genes determine whether the individual is a male or a female. Investigators have now discovered a series of genes on the Y chromosome that determine the development of male genitals, and at least one on the X chromosome that seems to be necessary for the development of female genitals.

Traits controlled by genes on the sex chromosomes are said to be **sex-linked;** an allele on an X chromosome is **X-linked,** and an allele on the Y chromosome is Y-linked. Most sex-linked genes are only on the X chromosomes, and the Y chromosome is blank for these. Very few alleles have been found on the Y chromosome, as you might predict, since it is much smaller than the X chromosome.

Many of the genes on the X chromosomes are unrelated to the gender of the individual, and we will look at a few of these in depth. It would be logical to suppose that a sex-linked trait is passed from father to son or from mother to daughter, but this is not the case. A male always receives a sex-linked allele from his mother, from whom he inherited an X chromosome. *The Y chromosome from the father does not carry an allele for the trait.* Usually a sex-linked genetic disorder is recessive; therefore, a female must receive two alleles, one from each parent, before she has the condition.

X-Linked Alleles

When considering X-linked traits, the allele on the X chromosome is shown as a letter attached to the X chromosome. For example, this is the key for red-green color blindness, a well-known X-linked recessive disorder:

X^B = normal vision

X^b = color blindness

The possible genotypes and phenotypes in both males and females are:

Genotypes	Phenotypes
$X^B X^B$	Female who has normal color vision
$X^B X^b$	Carrier female who has normal color vision
$X^b X^b$	Female who is color blind
$X^B Y$	Male who has normal vision
$X^b Y$	Male who is color blind

Recall that carriers are individuals who appear normal but can pass on an allele for a genetic disorder. Note here that the second genotype is a carrier female because, although a female with this genotype appears normal, she is capable of passing on an allele for color blindness. Color-blind females are rare because they must receive the allele from both parents; color-blind males are more common because they need only one recessive allele to be color blind. The allele for

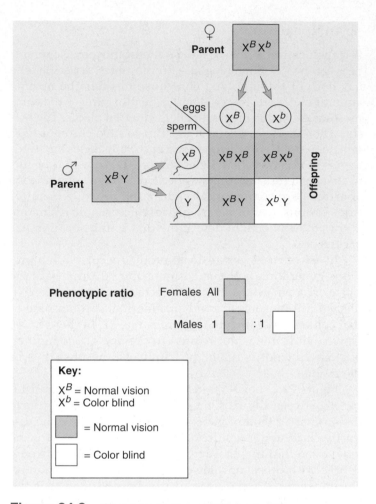

Figure 24.9 **Cross involving an X-linked allele.**
The male parent is normal, but the female parent is a carrier; an allele for color blindness is located on one of her X chromosomes. Therefore, each son has a 50% chance of being color blind. The daughters will appear normal, but each one has a 50% chance of being a carrier.

color blindness has to be inherited from their mother because it is on the X chromosome; males only inherit the Y chromosome from their father.

Now let us consider a mating between a man with normal vision and a heterozygous woman (Fig. 24.9). What is the chance that this couple will have a color-blind daughter? A color-blind son? All daughters will have normal color vision because they all receive an X^B from their father. The sons, however, have a 50% chance of being color blind, depending on whether they receive an X^B or an X^b from their mother. The inheritance of a Y chromosome from their father cannot offset the inheritance of an X^b from their mother. Notice in Figure 20.19 that the phenotypic results for sex-linked traits are given separately for males and females.

The X chromosome carries alleles that are not on the Y chromosome. Therefore, a recessive allele on the X chromosome is expressed in males.

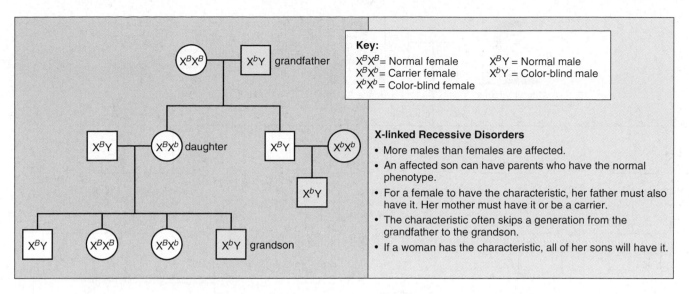

Figure 24.10   X-linked recessive pedigree chart.
The list gives ways of recognizing an X-linked recessive disorder.

X-Linked Disorders

The pedigree chart in Figure 24.10 shows the usual pattern of inheritance for an X-linked recessive genetic disorder. More males than females have the trait because recessive alleles on the X chromosome are expressed in males. The disorder often passes from grandfather to grandson through a carrier daughter.

As previously mentioned, color blindness is a common X-linked recessive disorder; two others are muscular dystrophy and hemophilia.

Color Blindness

In humans, the receptors for color vision in the retina of the eyes are three different classes of cone cells. Only one type of pigment protein is present in each class of cone cell; there are blue-sensitive, red-sensitive, and green-sensitive cone cells. The allele for the blue-sensitive protein is autosomal, but the alleles for the red- and green-sensitive proteins are on the X chromosome. About 8% of Caucasian men have red-green color blindness. Most of these see brighter greens as tans, olive greens as browns, and reds as reddish-browns. A few cannot tell reds from greens at all. They see only yellows, blues, blacks, whites, and grays.

Muscular Dystrophy

Muscular dystrophy, as the name implies, is characterized by a wasting away of the muscles. The most common form, Duchenne muscular dystrophy, is X-linked and occurs in about one out of every 3,600 male births. Symptoms, such as waddling gait, toe walking, frequent falls, and difficulty in rising, may appear as soon as the child starts to walk. Muscle weakness intensifies until the individual is confined to a wheelchair. Death usually occurs by age 20; therefore,

affected males are rarely fathers. The recessive allele remains in the population through passage from carrier mother to carrier daughter.

Recently, the allele for Duchenne muscular dystrophy was isolated, and it was discovered that the absence of a protein now called dystrophin is the cause of the disorder. Much investigative work determined that dystrophin is involved in the release of calcium from the sarcoplasmic reticulum in muscle fibers. The lack of dystrophin causes calcium to leak into the cell, which promotes the action of an enzyme that dissolves muscle fibers. When the body attempts to repair the tissue, fibrous tissue forms, and this cuts off the blood supply so that more and more cells die.

A test is now available to detect carriers of Duchenne muscular dystrophy. Also, various treatments are being attempted. Immature muscle cells can be injected into muscles, and for every 100,000 cells injected, dystrophin production occurs in 30–40% of muscle fibers. The allele for dystrophin has been inserted into the thigh muscle cells of mice, and about 1% of these cells then produced dystrophin.

Hemophilia

About one in 10,000 males is a hemophiliac. There are two common types of hemophilia: hemophilia A is due to the absence or minimal presence of a clotting factor known as factor IX, and hemophilia B is due to the absence of clotting factor VIII. Hemophilia is called the bleeder's disease because the affected person's blood either does not clot or clots very slowly. Although hemophiliacs bleed externally after an injury, they also bleed internally, particularly around joints. Hemorrhages can be stopped with transfusions of fresh blood (or plasma) or concentrates of the clotting protein. Also, factor VIII is now available as a biotechnology product.

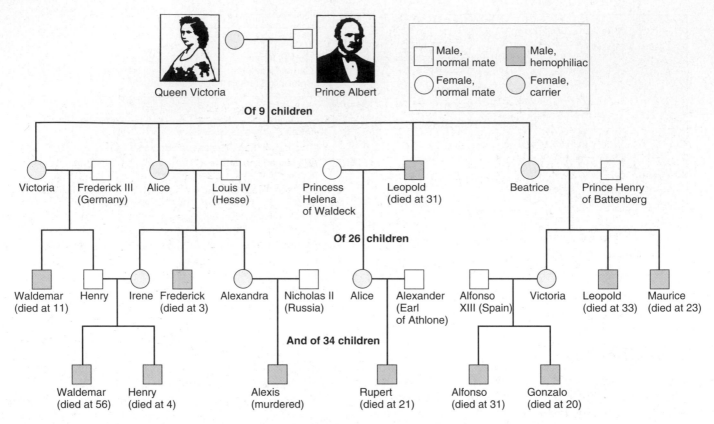

Figure 24.11 **A simplified pedigree showing the X-linked inheritance of hemophilia in European royal families.**
Because Queen Victoria was a carrier, each of her sons had a 50% chance of having the disease, and each of her daughters had a 50% chance of being a carrier. This pedigree shows only the affected descendants. Many others are unaffected, including the members of the present British royal family.

At the turn of the century, hemophilia was prevalent among the royal families of Europe, and all of the affected males could trace their ancestry to Queen Victoria of England. Figure 24.11 shows that, of Queen Victoria's 26 grandchildren, four grandsons had hemophilia and four granddaughters were carriers. Because none of Queen Victoria's or relatives were affected, it seems that the faulty allele she carried arose by mutation either in Victoria or in one of her parents. Her carrier daughters Alice and Beatrice introduced the allele into the ruling houses of Russia and Spain, respectively. Alexis, the last heir to the Russian throne before the Russian Revolution, was a hemophiliac. There are no hemophiliacs in the present British royal family because Victoria's eldest son, King Edward VII, did not receive the allele.

Fragile X Syndrome

Fragile X syndrome is an X-linked genetic disorder with an unusual pattern of inheritance. As discussed in the Health Focus on the next page, fragile X syndrome is due to base triplet repeats in a gene on the X chromosome. We now know that other disorders, such as Huntington disease, are also due to base triplet repeats.

Practice Problems*

1. Both the mother and the father of a male hemophiliac appear to be normal. From whom did the son inherit the allele for hemophilia? What are the genotypes of the mother, the father, and the son?

2. A woman is color blind. What is the chances that her sons will be color blind? If she is married to a man with normal vision, what are the chances that her daughters will be color blind? Will be carriers?

3. Both the husband and wife have normal vision. The wife gives birth to a color-blind daughter. What can you deduce about the girl's parentage?

*Answers to Practice Problems appear in Appendix A.

Certain traits that have nothing to do with the gender of the individual are controlled by genes on the X chromosomes. Males have only one X chromosome, and therefore X-linked recessive disorders are more likely in males.

Fragile X Syndrome

Fragile X syndrome is one of the most common genetic causes of mental retardation, second only to Down syndrome. It affects about one in 1,500 males and one in 2,500 females and is seen in all ethnic groups. As children, fragile X syndrome individuals may be hyperactive or autistic; their speech is delayed in development and often repetitive in nature. As adults, males have large testes and big, usually protruding ears (Fig. 24A*a*). They are short in stature, but the jaw is prominent and the face is long and narrow. Stubby hands, lax joints, and a heart defect may also occur.

In 1991, the DNA sequence at the fragile site was isolated and found to have base triplet repeats: CGG was repeated over and over again. An unaffected person has only 6 to 50 repeats, while a person with fragile X syndrome has 230 to 2,000 repeats. This mutation affects a gene located at the site, and the result is mental retardation. It is called fragile X syndrome because its diagnosis used to be dependent upon observing an X chromosome whose tip is attached to the rest of the chromosome by only a thin thread (Fig. 24A*b*).

Fragile X syndrome is an X-linked condition, and therefore you would expect all males having a fragile X to show the condition. However, one-fifth of these males with a fragile X do not have symptoms, and investigators wanted to discover why not. Analysis of DNA in these males shows that they have only an intermediate number of repeats—from 50 to 230 copies. A female who inherits this number of repeats from her father may have mild symptoms of retardation and may have sons with fragile X syndrome. Therefore, anyone with an intermediate number of repeats is said to have a premutation. And premutations can lead to full-blown mutations in future generations (Fig. 24A*c*).

How does the pattern of inheritance differ from that of the usual pattern for an X-linked genetic condition? First, a fragile X can have from 50 to 2,000 repeats. If the number is below 230, the individual has no symptoms, and if the number is above 230, the individual has symptoms. Second, a male with no symptoms can transmit the condition. For some unknown reason, a premutation leads to a full-blown mutation.

This type of mutation—called by some a dynamic mutation because it changes, and by others an expanded trinucleotide repeat because the number of triplet copies increases—is now known to characterize other conditions. Huntington disease is caused by a base triplet repeat of CAG, for example. What might cause repeats to occur in the first place, and why does this cause a syndrome? Repeats might arise during DNA replication prior to cell division, and their presence undoubtedly leads to nonfunctioning or malfunctioning proteins.

Scientists have developed a new technique that can identify repeats in DNA, and they hope that this technique will help them find the genes for other human disorders. They expect expanded trinucleotide repeats to be a very common mutation indeed.

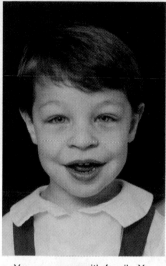

a. Young person with fragile X syndrome

Same individual when mature

b. Fragile X chromosome

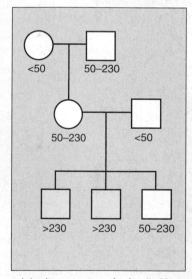

c. Inheritance pattern for fragile X syndrome

Figure 24A   Fragile X syndrome.
a. A young person with fragile X syndrome appears normal, but with age develops an elongated face with a prominent jaw and ears that noticeably protrude. **b.** An arrow points out the fragile site of this fragile X chromosome. **c.** The numbers indicate the number of base triplet repeats at the fragile site. A grandfather who has a premutation with 50–230 repeats has no symptoms but transmits the condition to his grandsons through his daughter. Two grandsons have full-blown mutations with more than 230 base repeats.

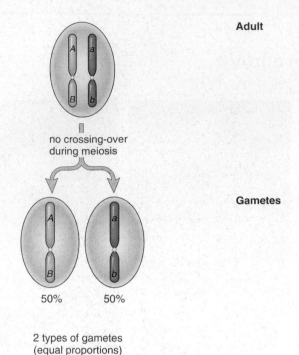

no crossing-over
during meiosis

Adult

Gametes

50%   50%

2 types of gametes
(equal proportions)

a. Complete linkage

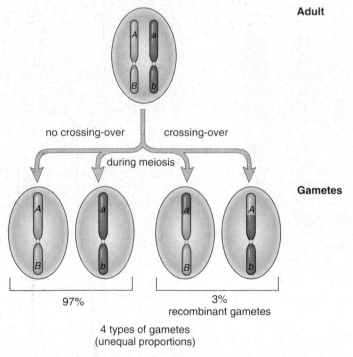

Adult

no crossing-over   crossing-over
during meiosis

Gametes

97%   3%
recombinant gametes

4 types of gametes
(unequal proportions)

b. Incomplete linkage

Figure 24.12   Linkage group.
In this individual, alleles *A* and *B* are on one member of a
homologous pair, and alleles *a* and *b* are on the other member.
a. When linkage is complete, this dihybrid produces only two types of
gametes in equal proportion. **b.** When linkage is incomplete, this
dihybrid produces four types of gametes because crossing-over has
occurred. The recombinant gametes occur in reduced proportion
because crossing-over occurs infrequently.

24.5 Linked Genes

The chromosome theory of inheritance predicts that each
chromosome contains a long series of alleles in a definite
sequence. All the alleles on one chromosome form a **linkage
group** because they tend to be inherited together. Figure
24.12*a* shows the results of a cross when linkage is complete: a dihybrid produces only two types of gametes in
equal proportion.

During crossing-over, you will recall the nonsister chromatids exchange genetic material and therefore genes. If
crossing-over occurs between the two alleles of interest, a
dihybrid produces four types of gametes instead of two
(Fig. 24.12*b*). Recombinant gametes (with recombined alleles) occur in reduced number because crossing-over is infre-
quent. Still, all possible phenotypes will occur among the
offspring.

To take an actual example, the genes for ABO blood
types and the gene for a condition called nail-patella syndrome (NPS) are on the same chromosome. A person with
NPS has fingernails and toenails that are reduced or absent
and a kneecap (patella) that is small. NPS (*N*) is dominant
while the normal condition (*n*) is recessive. Figure 24.13
shows the predicted results of a cross involving these genes.

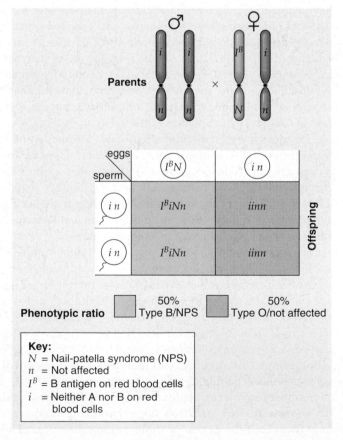

Figure 24.13   Cross involving linked genes.
Linked genes reduce the number of expected phenotypes among the
offspring.

Linkage was not complete; that is, crossing-over occurred, and 10% of the children had recombinant phenotypes: 5% had type B blood and no NPS, and 5% had type O blood and NPS:

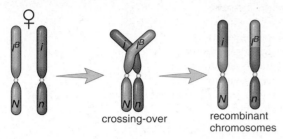

crossing-over   recombinant
                           chromosomes

The occurrence of crossing-over helps tell the sequence of genes on a chromosome because crossing-over occurs more often between distant genes than between genes that are close together on a chromosome. For example, consider these homologous chromosomes:

pair of homologous chromosomes

We expect recombinant gametes to include G and z more often than R and s. In keeping with this observation, investigators began to use recombination frequencies to map the chromosomes. Each 1% of crossing-over is equivalent to one map unit between genes. Linkage data have been used to map the chromosomes of the fruit fly *Drosophila*. For example, if we know that black body and purple eye are 6 map units apart, purple eye and vestigial wings are 12.5 units apart, and black body and vestigial wings are 18.5 units apart, then the sequence must be: black body, purple eye, and vestigial wings. In other words, black body is 6 map units from purple eyes, and purple eyes is 12.5 map units from vestigial wings.

The possibility of using linkage data to map human chromosomes is limited because we can only work with matings that have occurred by chance. This, coupled with the fact that humans tend not to have numerous offspring, means that additional methods must be used to sequence the genes on human chromosomes. Today, it is customary to also rely on biochemical methods to map the human chromosomes, as we shall discuss in Chapter 26.

The presence of linkage groups changes the expected results of genetic crosses. The frequency of recombinant gametes that occurs due to the process of crossing-over has been used to map chromosomes.

Bioethical Focus   Choosing Gender

Do you approve of choosing a baby's gender even before it is conceived? As you know, the sex of a child is dependent upon whether an X-bearing sperm or a Y-bearing sperm enters the egg. A new technique has been developed that can separate X-bearing sperm from Y-bearing sperm. First, the sperm are dosed with a DNA-staining chemical. Because the X chromosome has slightly more DNA than the Y chromosome, it takes up more dye. When a laser beam shines on the sperm, the X-bearing sperm shine a little more brightly than the Y-bearing sperm. A machine sorts the sperm into two groups on this basis. The results are not perfect. Following artificial insemination, there's about an 85% success rate for a girl and about a 65% success rate for a boy.

Some might argue that while it is acceptable to use vaccines to prevent illnesses, or to give someone a heart transplant, it goes against nature to choose gender. But what if the mother is a carrier of an X-linked genetic disorder such as hemophilia or Duchenne muscular dystrophy? Is it more acceptable to bring a child into the world with a genetic disorder that may cause an early death? Would it be better to select sperm for a girl, who at worst would be a carrier like her mother? Previously a pregnant woman with these concerns had to wait for the results of an amniocentesis test, and then abort the pregnancy if it were a boy. Is it better to increase the chances of a girl to begin with?

Some authorities do not find gender selection acceptable for any reason. Even if it doesn't lead to a society with far more members of one sex than another, there could be a problem. Once you separate reproduction from the sex act, they say, it opens the door to children that have been genetically designed in every way.

Decide Your Opinion

1. Do you think it is acceptable to choose the gender of a baby? Even if it requires artificial insemination at a clinic? Why or why not?
2. Do you see any difference between choosing gender and choosing eggs or embryos free of a genetic disease for reproduction purposes? Explain.
3. Do you think it is acceptable one day to genetically design children before they are born?

Summarizing the Concepts

24.1 Viewing the Chromosomes

It is possible to karyotype the chromosomes of a cell arrested in metaphase of mitosis. A karyotype shows that humans usually have 22 pairs of autosomes and two sex chromosome from each parent. Females have two X chromosomes, and males have one X and one Y chromosome.

Amniocentesis and chorionic villi sampling can provide fetal cells for the karyotyping of chromosomes when a chromosome abnormality may have been inherited.

24.2 Changes in Chromosome Number

The genes are on the chromosomes, and therefore the inheritance of an abnormal chromosome number can dramatically affect the phenotype. Nondisjunction during meiosis can result in an abnormal number of autosomes or sex chromosomes in the gametes.

Down syndrome results when an individual inherits three copies of chromosome 21. Females who are XO have Turner syndrome, and those who are XXX are poly-X females. Males with Klinefelter syndrome are XXY. Males who are XYY have Jacobs syndrome.

24.3 Changes in Chromosome Structure

Changes in chromosome structure also affect the phenotype. Chromosome mutations include deletions, duplications, translocations, and inversions. In Williams syndrome, one copy of chromosome 7 has a deletion. In cri du chat syndrome, one copy of chromosome 5 has a deletion. In inv dup 15 syndrome, chromosome 15 has an inverted duplication. Translocations do not necessarily cause any difficulties if the person has inherited both translocated chromosomes. However, the translocation can disrupt a particular allele, and then a syndrome will follow.

An inversion can lead to chromosomes that have a deletion and a duplication when the inverted piece loops back to align with the non-inverted homologue and crossing-over follows.

24.4 Sex-Linked Traits

Some traits are sex linked, meaning that although they do not determine gender, they are carried on the sex chromosomes. Most of the alleles for these traits are carried on the X chromosome and the Y is blank. It is customary to show an X-linked allele as a superscript on the X chromosome; for example X^b. The phenotypic results of crosses are given for females and males separately.

Because males normally receive only one X chromosome, they are subject to X-linked recessive genetic disorders caused by the inheritance of recessive alleles on the X chromosome. For example, in a cross between a normal male and a carrier female, only the male children could have the X-linked disorder color blindness. Other well-known X-linked disorders are hemophilia and Duchenne muscular dystrophy.

24.5 Linked Genes

All the genes on one chromosome form a linkage group, which is broken only when crossing-over occurs. Genes that are linked tend to go together into the same gamete. If crossing-over occurs, a dihybrid cross gives all possible phenotypes among the offspring, but the expected ratio is greatly changed because recombinant phenotypes are in reduced number.

Crossing-over data are used to map the chromosomes of animals, such as fruit flies, but it is not possible in humans to prearrange matings in order to map the chromosomes.

Testing Yourself

Choose the best answer for each question.
For questions 1–3, match the conditions in the key with the descriptions below.

Key:
a. Down syndrome
b. Turner syndrome
c. Klinefelter syndrome
d. Jacobs syndrome

1. Male with underdeveloped testes and some breast development

2. Trisomy 21

3. XO female

4. In a karyotype, the chromosomes are arranged
   a. in no particular order.
   b. according to numbered pairs.
   c. according to those inherited from the mother and those inherited from the father.
   d. from the smallest to the largest.
   e. Both b and d are correct.

5. A karyotype could be helpful to identify
   a. Klinefelter syndrome.
   b. hemophilia.
   c. identical twins.
   d. which children belong to which parents.

6. An individual with a Y chromosome cannot
   a. have a chromosome abnormality.
   b. be a male if there are two X chromosomes present.
   c. have a second Y chromosome.
   d. be a female.
   e. Both b and d are true.

7. An abnormal number of chromosomes could result during meiosis because of
   a. recombination.
   b. a carrier.
   c. nondisjunction.
   d. inversion.
   e. translocation.

8. Which of the following is the chromosome condition of a person with Jacobs syndrome?
   a. XO   d. XYY
   b. XXY   e. OY
   c. XYO

9. Which of the following conditions is due to a change in chromosome structure?
   a. cri du chat syndrome
   b. Down syndrome
   c. cystic fibrosis
   d. Klinefelter syndrome
   e. hemophilia

10. Which of the following conditions is not X-linked?
    a. color blindness
    b. hemophilia
    c. Down syndrome
    d. fragile X syndrome
    e. muscular dystrophy

11. Which type fertilization could result in Turner syndrome?
    a. normal sperm and egg that has two X chromosomes
    b. sperm that has no sex chromosome and normal egg
    c. sperm that has only one sex chromosome and normal egg
    d. All of these are correct.
    e. None of these are correct.

12. Which of the following statements is NOT true regarding sex-linked traits?
    a. Women can be carriers because they can be heterozygous for the trait.
    b. X-linked traits are more common in men.
    c. Males inherit X-linked traits from their fathers.
    d. Males are never carriers since they only receive one X chromosome.
    e. Both c and d are not correct.

13. Down syndrome
    a. can be caused by nondisjunction of chromosome 21.
    b. shows no overt abnormalities.
    c. does not occur in mothers younger than 40.
    d. is a sex-linked disorder resulting in retardation.
    e. All of these are correct.

14. If linkage is complete, how many phenotypes are seen among the offspring of two dihybrids with *AB* on one chromosome and *ab* on the other chromosome
    a. four   c. two
    b. three.   d. one

15. Which chromosome mutation can be a cause of Down syndrome?
    a. inversion
    b. translocation
    c. duplication
    d. deletion
    e. Both b and c are correct.

16. If a man has an X-linked recessive disorder, which of the following statements is not likely?
    a. Both parents are unaffected.
    b. Only the males in a pedigree chart have the disorder.
    c. Only females in previous generations have the disorder.
    d. Both a and c are not likely.

17. John has hemophilia, but his parents do not. Using *H* for normal and *h* for hemophilia, give the genotype of his father, mother, and himself in that order.
    a. *Hh, Hh, hh*
    b. X^HY, hh, X^HY
    c. X^HY, X^HX^h, X^hY
    d. X^hY, X^HX^H, X^hY
    e. X^HY, X^hY, X^hy

18. This pedigree chart pertains to color blindness. The genotype of the starred individual is _____.

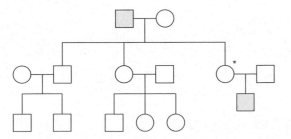

Additional Genetics Problems

X-linked Problems

1. A boy has an X-linked condition known as severe combined immune deficiency syndrome. What key should you use? What are the genotypes of the parents, who have the normal phenotype?

2. A woman is color blind, and her spouse has normal vision. If they produce a son and a daughter, which child will be color blind?

3. If a female who carries an X-linked allele for hemophilia reproduces with a normal man, what are the chances that male children will have hemophilia? That female children will have the condition?

4. A girl has hemophilia. What is the genotype of her father? What is the genotype of her mother, who has a normal phenotype?

5. John is the only member of his family with hemophilia. What is the chance that a newborn brother will also be a hemophiliac?

6. A normal woman whose father had hemophilia marries a man who has hemophilia. What is the chance their sons will be hemophiliacs? What is the chance their daughters will be hemophiliacs?

7. In fruit flies, X^R = red eye and X^r = white eye.
   a. If a white-eyed male reproduces with a homozygous red-eyed female, what phenotypic ratio is expected for males? For females?
   b. If a white-eyed female reproduces with a red-eyed male, what phenotypic ratio is expected for males? For females?

Mixed Problems

8. What is the genotype of a man homozygous for brown eyes (autosomal dominant) and who is color blind? What is the genotype of a woman with blue eyes who is not color blind but whose father was color blind?

9. What is the chance that a daughter born to the couple described in question 8 will have brown eyes and be colorblind? What is the chance a son will have blue eyes and not be color blind?

10. A man who is color blind and has a straight hairline reproduces with a woman who is homozygous dominant for normal color vision and widow's peak. What are the possible genotypes and phenotypes of the children?

11. A man who is homozygous for curling the tongue (dominant) is color blind. He reproduces with a woman with homozygous normal color vision who is heterozygous for tongue-curling. What is the chance this couple will have a color-blind son?

12. In fruit flies, gray (*G*) versus black body (*g*) is autosomal; red eye (*R*) versus white eye (*r*) is X-linked. A female fly heterozygous for both gray body and red eyes reproduces with a red-eyed male heterozygous for gray body. What phenotypic ratio is expected for males? For females?

Linkage Problem

13. It is known that *A* and *B* are 10 map units apart; *A* and *C* are 6 units apart; *A* and *D* are 18 units apart; and *C* and *D* are 8 units apart. What is the order of the genes on the chromosome?

e-Learning Connection

Concepts	Questions	Media Resources*
24.1 Viewing the Chromosomes		

• The human karyotype contains 22 pairs of autosomes and one pair of sex chromosomes for a total of 46 chromosomes.
• Amniocentesis and chorionic villi sampling provide fetal cells for karyotyping purposes.
	1. What are autosomes and sex chromosomes and what are the genotypes of human males and females?
	2. What is a syndrome?
		Essential Study Partner
		    Introduction
		    Chromosomes
		    Karyotype
		    Sex Chromosomes

24.2 Changes in Chromosome Number		

• Inheritance of an abnormal chromosome number is most likely due to nondisjunction.
• Down syndrome is due to the inheritance of three copies of chromosome 21.
• Several syndromes are the result of inheriting an incorrect number of sex chromosomes.
	1. What is the most common cause of an abnormal number of chromosomes?
	2. What is a Barr body?
		Essential Study Partner
		    Abnormal Chromosomes
		
		Art Quiz
		    Nondisjunction and Sex
		        Chromosomes
		
		Exploration
		    Exploring Meiosis: Down
		        Syndrome

24.3 Changes in Chromosome Structure		
• Chromosome mutations include deletions, duplications, translocations, and inversions. • Chromosome mutations are associated with various syndromes.	1. What possible causes are associated with changes in chromosome structure? 2. Why do chromosomal inversions cause problems when the same genetic material is still present?	Essential Study Partner Abnormal Chromosomes
24.4 Sex-Linked Traits		
• Certain traits, unrelated to the gender of the individual, are controlled by genes located on the sex chromosomes. • Males who inherit an allele for an X-linked recessive disorder exhibit the disorder because they inherit only one X chromosome.	1. Why are sex-linked disorders, such as red-green colorblindness, much more common in males than in females?	Art Quiz Barr Bodies
24.5 Linked Genes		
• Alleles that occur on the same chromosome form a linkage group and tend to be inherited together.	1. How is it possible to easily determine whether two genes are linked on the same chromosome or whether they occur on different chromosomes?	

*For additional Media Resources, see the Online Learning Center.

Molecular Basis of Inheritance

Chapter Concepts

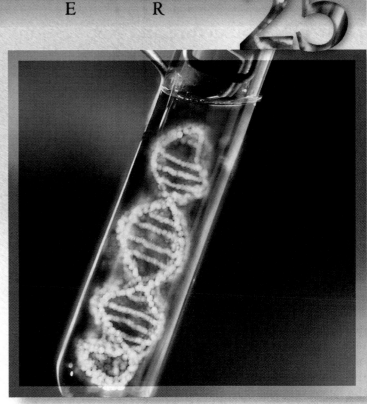

Test tube containing DNA whose structure resembles a spiral staircase (digital composite).

One of the most exciting periods of scientific activity in history occurred during the thirty short years between the 1930s and 1960s. Geneticists knew that chromosomes contain protein and DNA (deoxyribonucleic acid). Of these two organic molecules, proteins are seemingly more complicated; they consist of countless sequences of 20 amino acids, which can coil and fold into complex shapes. DNA, on the other hand, contains only four different nucleotides. Surely, the diversity of life forms on earth must be the result of the endless varieties of proteins.

Due to several elegantly executed experiments, by the mid-1950s researchers realized that DNA, not protein, is the genetic material. But this finding only led to another fundamental question—what exactly is the structure of DNA? The biological community at the time knew that whoever determined the structure of DNA would get a Nobel Prize, and would go down in history. Consequently, researchers were racing against time and each other. The story of the discovery of DNA resembles a mystery, with each clue adding to the total picture until the breathtaking design of DNA—a double helix—was finally unraveled.

25.2 Gene Expression

As mentioned in Chapter 24, an individual with a genetic disorder has a missing or malfunctioning enzymatic or structural protein in the cell. Persons with Duchenne muscular dystrophy, for example, are missing the protein called dystrophin, which is normally present in muscle cells. In the genetic disorder Huntington disease, the protein huntingtin is altered and unable to carry out its usual functions. People with cystic fibrosis have a malfunctioning Cl^- channel protein in the plasma membrane. By studying various metabolic disorders, geneticists have confirmed many times over that proteins are the link between genotype and phenotype.

Table 25.1	DNA Structure Compared to RNA Structure	
	DNA	**RNA**
Sugar	Deoxyribose	Ribose
Bases	Adenine, guanine, thymine, cytosine	Adenine, guanine, uracil, cytosine
Strands	Double stranded with base pairing	Single stranded
Helix	Yes	No

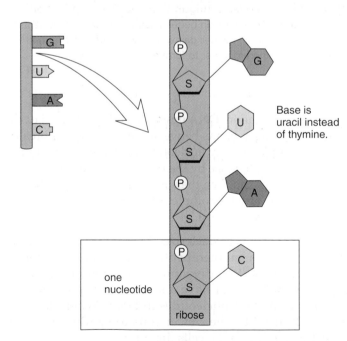

Figure 25.5   Structure of RNA.
Like DNA, RNA is a polymer of nucleotides. In an RNA nucleotide, the sugar ribose is attached to a phosphate molecule and to a base, either G, U, A, or C. Notice that in RNA, the base uracil replaces thymine as one of the pyrimidine bases. RNA is single stranded, whereas DNA is double stranded.

Granted that there is a connection between genes and proteins, what exactly is it that genes do? A **gene** is a segment of DNA that specifies the amino acid sequence of a protein. This is the information that DNA stores and the reason DNA activity brings about the development of the unique structures that make up a particular organism. In this chapter, you will learn how a difference in base sequence causes a difference in protein structure and determines, for example, whether you have blue, brown, or hazel eye pigments.

A gene does not directly control protein synthesis; instead, it passes its genetic information on to RNA, which is more directly involved in protein synthesis.

RNA

Like DNA, **RNA (ribonucleic acid)** is a polynucleotide (Fig. 25.5). However, the nucleotides in RNA contain the sugar ribose, not deoxyribose. Also, the bases in RNA are adenine (A), cytosine (C), guanine (G), and uracil (U). In other words, the base uracil replaces the thymine found in DNA. Finally, RNA is single stranded and does not form a double helix in the same manner as DNA (Fig. 25.5). Table 25.1 summarizes the similarities and differences between DNA and RNA.

There are three major classes of RNA, each with specific functions in protein synthesis:

Messenger RNA (mRNA) takes a message from DNA to the ribosomes.

Ribosomal RNA (rRNA), along with proteins, makes up the ribosomes, where proteins are synthesized.

Transfer RNA (tRNA) transfers amino acids to the ribosomes.

Gene expression requires two steps, known as transcription and translation. First, information is transferred from DNA to RNA during **transcription.** Consider that transcription means making a close copy of a document. During transcription, one type of polynucleotide (DNA) becomes another type of polynucleotide (RNA). Second, during **translation,** an RNA transcript directs the sequence of amino acids in a polypeptide. Consider that a translator needs to understand two languages. Similarly, the cell understands two different languages: nucleotide sequences and amino acid sequences. During translation, a nucleotide sequence directs an amino acid sequence (see Fig. 25.7).

With the help of RNA, a gene (a segment of DNA) specifies the sequence of amino acids in a polypeptide. In this way, genes control the structure and the metabolism of cells.

The Genetic Code

DNA has a particular sequence of bases, and a polypeptide has a particular sequence of amino acids. This suggests that DNA contains coded information. Can four bases provide enough combinations to code for 20 amino acids? If the code were a doublet (any two bases stand for one amino acid), it would not be possible to code for 20 amino acids, but if the code were a triplet, then the four bases could supply 64 different triplets, far more than needed to code for 20 different amino acids. It should come as no surprise, then, to learn that the code is a **triplet code.**

To crack the code, a cell-free experiment was done: artificial RNA was added to a medium containing bacterial ribosomes and a mixture of amino acids. Comparison of the bases in the RNA with the resulting polypeptide allowed investigators to decipher the code. Each three-letter unit of an mRNA molecule is called a **codon.** All 64 mRNA codons have been determined (Fig. 25.6). Sixty-one triplets correspond to a particular amino acid; the remaining three are stop codons, which signal polypeptide termination. The one codon that stands for the amino acid methionine is also a start codon signaling polypeptide initiation. Notice too that most amino acids have more than one codon; leucine, serine, and arginine have six different codons, for example. This offers some protection against possibly harmful mutations that change the sequence of the bases.

The genetic code is just about universal in living things. This suggests that the code dates back to the very first organisms on earth and that all living things are related.

Central Concept

The central concept of genetics can be summarized as in Figure 25.7. DNA has a sequence of bases that is transcribed into a sequence of bases in mRNA. Every three bases is a codon that stands for a particular amino acid. In this way, DNA specifies the sequence of amino acids in a protein when translation occurs at the ribosomes.

First Base	Second Base				Third Base
	U	C	A	G	
U	UUU phenylalanine	UCU serine	UAU tyrosine	UGU cysteine	U
	UUC phenylalanine	UCC serine	UAC tyrosine	UGC cysteine	C
	UUA leucine	UCA serine	UAA stop	UGA stop	A
	UUG leucine	UCG serine	UAG stop	UGG tryptophan	G
C	CUU leucine	CCU proline	CAU histidine	CGU arginine	U
	CUC leucine	CCC proline	CAC histidine	CGC arginine	C
	CUA leucine	CCA proline	CAA glutamine	CGA arginine	A
	CUG leucine	CCG proline	CAG glutamine	CGG arginine	G
A	AUU isoleucine	ACU threonine	AAU asparagine	AGU serine	U
	AUC isoleucine	ACC threonine	AAC asparagine	AGC serine	C
	AUA isoleucine	ACA threonine	AAA lysine	AGA arginine	A
	AUG (start) methionine	ACG threonine	AAG lysine	AGG arginine	G
G	GUU valine	GCU alanine	GAU aspartate	GGU glycine	U
	GUC valine	GCC alanine	GAC aspartate	GGC glycine	C
	GUA valine	GCA alanine	GAA glutamate	GGA glycine	A
	GUG valine	GCG alanine	GAG glutamate	GGG glycine	G

Figure 25.6 Messenger RNA codons.
Notice that in this chart, each of the codons (white rectangles) is composed of three letters representing the first base, second base, and third base. For example, find the rectangle where C for the first base and A for the second base intersect. You will see that U, C, A, or G can be the third base. CAU and CAC are codons for histidine; CAA and CAG are codons for glutamine.

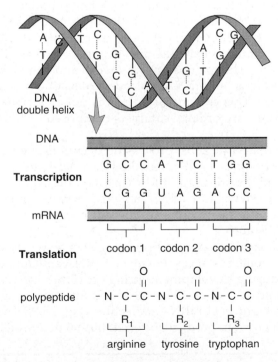

Figure 25.7 Overview of gene expression.
Transcription occurs when DNA acts as a template for mRNA synthesis. Translation occurs when the sequence of the mRNA codons determines the sequence of amino acids in a protein.

Transcription

During transcription, a segment of the DNA helix unwinds and unzips, and complementary RNA nucleotides from an RNA nucleotide pool in the nucleus pair with the DNA nucleotides of one strand. The RNA nucleotides are joined by an enzyme called **RNA polymerase,** and an RNA molecule results. Although all three classes of RNA are formed by transcription, we will focus on transcription to form mRNA (Fig. 25.8). When mRNA forms, it has a sequence of bases complementary to DNA; wherever A, T, G, or C is present in the DNA template, U, A, C, or G is incorporated into the mRNA molecule. In this way, the code is transcribed, or copied. Now mRNA has a sequence of codons that are complementary to the DNA triplet code.

> Following transcription, mRNA has a sequence of bases complementary to one of the DNA strands. Now mRNA contains codons that are complementary to the DNA triplet code.

Processing of mRNA

Most genes in humans are interrupted by segments of DNA that are not part of the gene. These portions are called *introns* because they are *intra*gene segments. The other portions of the gene are called *exons* because they are ultimately *ex*pressed. Only exons result in a protein product.

When DNA is transcribed, the mRNA contains bases that are complementary to both exons and introns, but before the mRNA exits the nucleus, it is *processed.* During processing, the introns are removed, and the exons are joined to form an mRNA molecule consisting of continuous exons. This splicing of mRNA is done by ribozymes. **Ribozymes** are organic catalysts composed of RNA, not protein.

In eukaryotic cells, processing occurs in the nucleus. The newly formed mRNA is called the *primary* mRNA molecule, and the processed mRNA is called the *mature* mRNA molecule. The mature mRNA molecule passes from the cell nucleus into the cytoplasm. There it becomes associated with ribosomes.

Ordinarily, processing brings together all the exons of a gene. In some instances, however, mRNA splicing, particularly during development, brings together just some of exons and not others. This so-called alternative splicing produces different mRNA molecules (Fig. 25.9). Then the cell will go on to produce different but related proteins.

> In eukaryotic cells, the primary mRNA molecule is processed; introns are removed, so that the mature mRNA molecule contains only exons. Mature RNA leaves the nucleus and becomes associated with ribosomes.

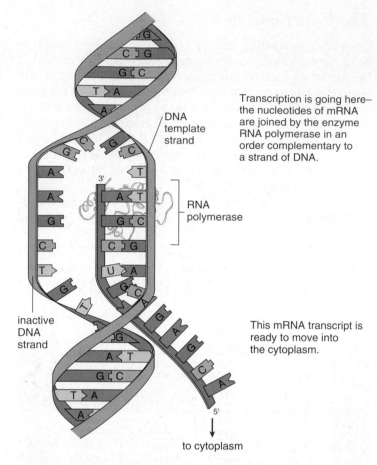

Transcription is going here—the nucleotides of mRNA are joined by the enzyme RNA polymerase in an order complementary to a strand of DNA.

DNA template strand

RNA polymerase

inactive DNA strand

This mRNA transcript is ready to move into the cytoplasm.

to cytoplasm

Figure 25.8   Transcription to form mRNA.
During transcription, complementary RNA is made from a DNA template. A portion of DNA unwinds and unzips at the point of attachment of RNA polymerase. A strand of mRNA is produced when complementary bases join in the order dictated by the sequence of bases in DNA. Transcription occurs in the nucleus, and the mRNA passes out of the nucleus to enter the cytoplasm.

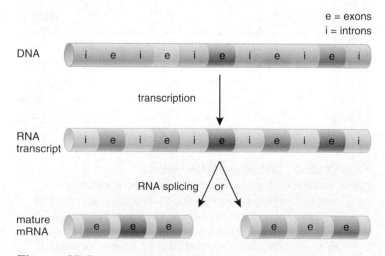

e = exons
i = introns

DNA

transcription

RNA transcript

RNA splicing / or

mature mRNA

Figure 25.9   Function of introns.
Introns allow alternative splicing and therefore the production of different versions of mature mRNA from the same gene.

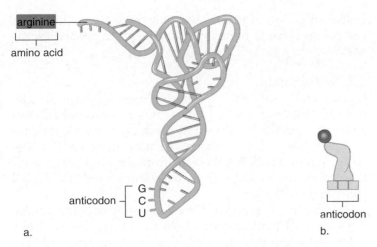

Figure 25.10 Transfer RNA: amino acid carrier.
a. A tRNA is a polynucleotide that folds into a bootlike shape because of complementary base pairing. At one end of the molecule is its specific anticodon—in this case GCU; at the other end an amino acid attaches that corresponds to this anticodon—in this case arginine. **b.** tRNA will be represented like this in the illustrations that follow.

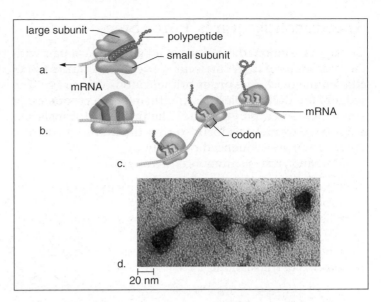

Figure 25.11 Polyribosome structure and function.
a. Side view of a ribosome shows positioning of mRNA and growing protein. **b.** Frontal view of a ribosome. **c.** Several ribosomes, collectively called a polyribosome, move along an mRNA at one time. Therefore, several proteins can be made at the same time. **d.** Electron micrograph of a polyribosome.

Translation

Translation is the second step by which gene expression leads to protein synthesis. During translation, the sequence of codons in mRNA specifies the order of amino acids in a protein. Translation requires several enzymes and two other types of RNA: transfer RNA and ribosomal RNA.

Transfer RNA

Transfer RNA (tRNA) molecules bring amino acids to the ribosomes. Each is a single-stranded nucleic acid that doubles back on itself to create regions where complementary base pairing results in a bootlike shape (Fig. 25.10). There is at least one tRNA molecule for each of the twenty amino acids found in proteins. The amino acid binds to one end of the molecule. The opposite end of the molecule contains an **anticodon,** a group of three bases that is complementary to a specific codon of mRNA. The entire complex is designated as tRNA–amino acid. One area of active research is to determine how the correct amino acid becomes attached to the correct tRNA molecule.

When a tRNA–amino acid complex comes to the ribosome, its anticodon pairs with an mRNA codon. Let us consider an example: if the codon is ACC, what is the anticodon, and what amino acid will be attached to the tRNA molecule? From Figure 25.6, we can determine this:

Codon	Anticodon	Amino Acid
ACC	UGG	Threonine

The order of the codons of the mRNA determines the order that tRNA–amino acids come to a ribosome, and therefore the final sequence of amino acids in a protein.

Ribosomal RNA

Ribosomal RNA (rRNA) is called structural RNA because it is found in the **ribosomes,** small structural bodies. In eukaryotic cells, ribosomal RNA is produced in a nucleolus within the nucleus. There it joins with proteins manufactured in the cytoplasm to form two ribosomal subunits, one large and one small. Each subunit contains an rRNA molecule and many different types of proteins. The subunits leave the nucleus and join together in the cytoplasm to form a ribosome just as protein synthesis begins.

A ribosome has a binding site for mRNA as well as binding sites for two tRNA molecules at a time. These binding sites facilitate complementary base pairing between tRNA anticodons and mRNA codons. As the ribosome moves down the mRNA molecule, new tRNAs arrive, and a polypeptide forms and grows longer. Translation terminates once the polypeptide is fully formed; the ribosome dissociates into its two subunits and falls off the mRNA molecule.

As soon as the initial portion of mRNA has been translated by one ribosome, and the ribosome has begun to move down the mRNA, another ribosome attaches to the same mRNA. Therefore, several ribosomes are often attached to and translating the same mRNA. The entire complex is called a **polyribosome** (Fig. 25.11.)

During translation, the sequence of bases in mRNA determines the order that tRNA amino acids come to a ribosome and therefore the order of amino acids in a particular polypeptide.

Translation Requires Three Steps

During translation, the codons of an mRNA base-pair with the anticodons of tRNA molecules. Each tRNA carries a specific amino acid. The order of the codons determines the order of the tRNA molecules and therefore the sequence of amino acids in a polypeptide. The process of translation must be extremely orderly so that the amino acids of a polypeptide are sequenced correctly.

Protein synthesis involves three steps:

1. Chain initiation—The steps necessary to begin the process of translation.
2. Chain elongation—The steps necessary to bring about polypeptide synthesis.
3. Chain termination—The steps necessary to end the process of translation.

It should be kept in mind that enzymes are required for each of the steps to occur, and that energy is needed for the first two steps also.

Chain Initiation

During chain initiation, a small ribosomal subunit, the mRNA, an initiator tRNA, and a large ribosomal subunit all come together. First, a small ribosomal subunit attaches to the mRNA in the vicinity of the start codon (AUG). The anticodon of a tRNA, called the initiator RNA, pairs with this codon. Then a large ribosomal subunit joins to the small subunit (Fig. 25.12a).

Notice that a ribosome has two binding sites for tRNAs. They are called binding sites because this is where a tRNA is located when its anticodon binds to a codon of mRNA. The initiator tRNA is always at the first binding site because its anticodon binds to the codon AUG.

When chain initiation occurs, the first tRNA has come to a ribosome.

Chain Elongation

During chain elongation, the tRNA at the first binding site usually bears an attached peptide. Why? Because the initiator tRNA passes its amino acid to a tRNA–amino acid complex that has come to the second binding site. Then the ribosome moves forward—the tRNA at the second binding site is now at the first binding site (Fig. 25.12b). This sequence of events is called translocation.

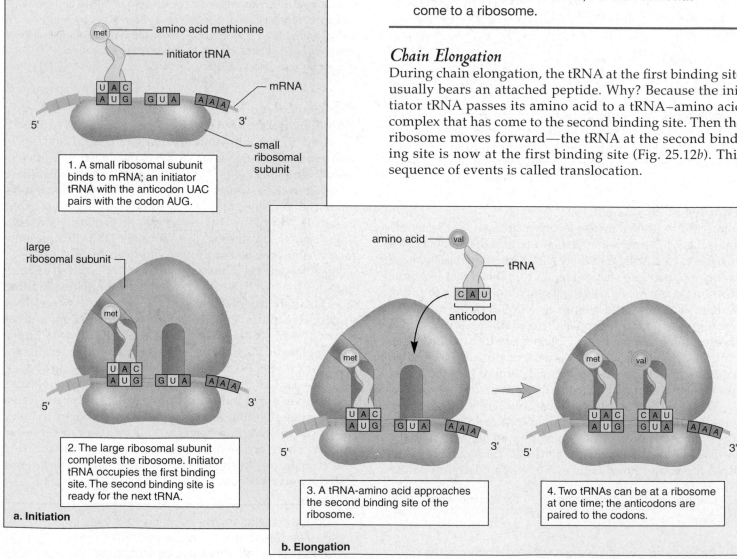

1. A small ribosomal subunit binds to mRNA; an initiator tRNA with the anticodon UAC pairs with the codon AUG.

2. The large ribosomal subunit completes the ribosome. Initiator tRNA occupies the first binding site. The second binding site is ready for the next tRNA.

a. Initiation

3. A tRNA-amino acid approaches the second binding site of the ribosome.

4. Two tRNAs can be at a ribosome at one time; the anticodons are paired to the codons.

b. Elongation

Figure 25.12   Protein synthesis.

Translocation occurs again and again during chain elongation. Each time, the growing polypeptide is transferred and attached by peptide bond formation to the newly arrived amino acid. Bringing about this transfer requires energy and a ribozyme, which is a part of the larger ribosomal subunit. After translocation occurs, the outgoing tRNA molecule will pick up another amino acid before returning to the ribosome.

The complete cycle—complementary base pairing of new tRNA, transfer of peptide chain and translocation—is repeated at a rapid rate (about 15 times each second in *Escherichia coli*).

During chain elongation, amino acids are added one at a time to the growing polypeptide.

Chain Termination

Chain termination occurs when protein synthesis comes to an end. Termination occurs at a stop codon—that is, a codon which does not code for an amino acid (Fig. 25.12c). The polypeptide is enzymatically cleaved from the last tRNA by a release factor.

Now the tRNA and polypeptide leave the ribosome, which dissociates into its two subunits (see also Fig. 25.11).

During chain termination, the ribosome separates into its two subunits and the polypeptide is released.

A newly synthesized polypeptide may function alone, or it may become a part of a protein that has more than one polypeptide. Proteins play a role in the anatomy and physiology of cells, as discussed earlier in this text. The plasma membrane of all cells contains proteins that carry out various functions, and many proteins are enzymes that participate in cellular metabolism. The tissues of multicellular animals are distinguished by the uniqueness of their proteins. Properly functioning proteins are of paramount importance to the cell and to the organism. Organisms inherit genes that code for their own particular mix of proteins in their cells. If an organism inherits a mutated gene, the result can be a genetic disorder or a propensity toward cancer, in which proteins are not fulfilling their usual functions. This topic is explored later in this chapter.

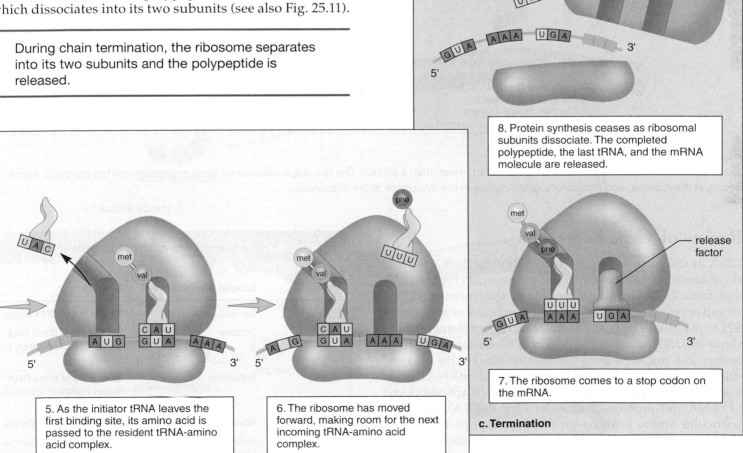

8. Protein synthesis ceases as ribosomal subunits dissociate. The completed polypeptide, the last tRNA, and the mRNA molecule are released.

release factor

7. The ribosome comes to a stop codon on the mRNA.

c. Termination

5. As the initiator tRNA leaves the first binding site, its amino acid is passed to the resident tRNA-amino acid complex.

6. The ribosome has moved forward, making room for the next incoming tRNA-amino acid complex.

Figure 25.12   Protein synthesis—*continued*.

26.1 Cloning of a Gene

In biology, **cloning** is the production of identical copies through some asexual means. When an underground stem or root sends up new shoots the resulting plants are clones of one another. The members of a bacterial colony on a petri dish are clones because they all came from the division of the same original cell. And human identical twins are clones. The first two cells of the embryo separated and each became a complete individual.

Gene cloning is the production of many identical copies of the same gene. Scientists clone genes for a number of reasons. They might want to determine the differ-ence in base sequence between a normal gene and a mutated gene; for example. Or they might use the genes to alter the phenotype of other organisms in a beneficial way. When the organism is a human, we call it gene therapy. If the organisms are not human, we call them transgenic organisms.

Scientists might instead be interested in the protein coded for by the cloned gene. If so, there are several unique ways they can acquire products of cloned genes, as we shall see. First, we want to consider two ways to clone a gene, that is, recombinant DNA technology and the polymerase chain reaction.

Recombinant DNA Technology

Recombinant DNA (rDNA) contains DNA from two or more different sources (Fig. 26.1). To make rDNA, a tech-nician needs a **vector,** by which rDNA will be introduced into a host cell. One common vector is a plasmid. **Plasmids** are small accessory rings of DNA from bacteria. The ring is not part of the bacterial chromosome and replicates on its own. Plasmids were discovered by investigators studying the bacterium *Escherichia coli* (*E. coli*).

Two enzymes are needed to introduce foreign DNA into vector DNA: (1) a **restriction enzyme,** which cleaves DNA and (2) an enzyme called **DNA ligase,** which seals DNA into an opening created by the restriction enzyme. Hundreds of restriction enzymes occur naturally in bacte-ria, where they cut up an viral DNA that enters the cell. They are called restriction enzymes because they *restrict* the growth of viruses, but they also act as molecular scissors to cleave any piece of DNA at a specific site. For example, the restriction enzyme called *Eco*RI always cuts double-stranded DNA in this manner when it has this sequence of bases:

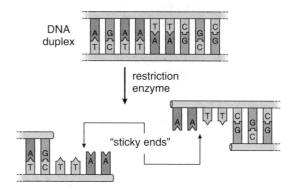

Figure 26.1 Cloning of a human gene.
Human DNA and plasmid DNA are cleaved by a specific type of restriction enzyme. Human DNA (e.g., containing the insulin gene) is spliced into a plasmid by the enzyme DNA ligase. Gene cloning is achieved after a bacterium takes up the plasmid. If the gene functions normally as expected, the product (e.g., insulin) may also be retrieved.

Notice that there is now a gap into which a piece of foreign DNA can be placed if it ends in bases complementary to those exposed by the restriction enzyme. To assure this, it is only necessary to cleave the foreign DNA with the same type of restriction enzyme. The single-stranded, but com-plementary, ends of the two DNA molecules are called "sticky ends" because they can bind a piece of foreign DNA by complementary base pairing. Sticky ends facilitate the insertion of foreign DNA into vector DNA.

Now genetic engineers use DNA ligase to seal the foreign piece of DNA into the vector. DNA splicing is now complete; an rDNA molecule has been prepared (Fig. 26.1). Bacterial cells take up recombinant plasmids, especially if they are treated to make them more permeable. Thereafter, as the plasmid replicates, the gene is cloned.

In order for human gene to express itself in a bacterium, the gene has to be accompanied by regulatory regions unique to bacteria. Also, the gene should not contain introns because bacteria don't have introns. However, it's possible to make a human gene that lacks introns. The enzyme called reverse transcriptase can be used to make a DNA copy of mRNA. This DNA molecule, called **complementary DNA** (cDNA), does not contain introns. Alternatively, it is possible to manufacture small pieces of DNA in the laboratory. A machine called a DNA synthesizer joins together the correct sequence of nucleotides, and this resulting DNA also lacks introns.

The Polymerase Chain Reaction

The **polymerase chain reaction (PCR)** can create copies of a piece of DNA very quickly in a test tube. The original sample of DNA is usually just a portion of the entire genome which is all of an individual's genetic material. And PCR is very specific—it amplifies (makes copies) of a targeted DNA sequence which can be less than one part in a million of the total DNA sample!

PCR requires the use of DNA polymerase, the enzyme that carries out DNA replication, and a supply of nucleotides for the new DNA strands. PCR is a chain reaction because the targeted DNA is repeatedly replicated as long as the process continues. The colors in Figure 26.2 distinguish old from new DNA. Notice that the amount of DNA doubles with each replication cycle.

PCR has been in use for several years, and now almost every laboratory has automated PCR machines to carry out the procedure. Automation became possible after a temperature-insensitive (thermostable) DNA polymerase was extracted from the bacterium *Thermus aquaticus*, which lives in hot springs. The enzyme can withstand the high temperature used to separate double-stranded DNA; therefore, replication does not have to be interrupted by the need to add more enzyme.

Analyzing DNA Segments

DNA amplified by PCR is often analyzed for various purposes. Mitochondrial DNA sequences in modern living populations were used to decipher the evolutionary history of human populations. Since so little DNA is required for PCR to be effective, it has even been possible to sequence DNA taken from a 76,000-year-old mummified human brain.

Also, following PCR, DNA can be subjected to   When a piece of DNA is treated with restriction enzymes, the result is a unique collection of different-sized fragments. During a process called gel electrophoresis, the fragments can be sep-

arated according to their charge/size ratios, and the result is a pattern of distinctive bands. If two DNA patterns match, there is a high probability that the DNA came from the same person.

DNA fingerprinting has many uses. When the DNA matches that of a virus or mutated gene, we know that a viral infection, genetic disorder, or cancer is present. Fingerprinting DNA from a single sperm is enough to identify a suspected rapist. Human remains can be identified by comparing a sample of DNA to that of a cell left on a personal item (i.e., toothbrush, cigarette butt) or to the DNA of the parents. A DNA fingerprint resembles that of one's parents because it is inherited.

Recombinant DNA technology and the polymerase chain reaction are two ways to clone a gene.

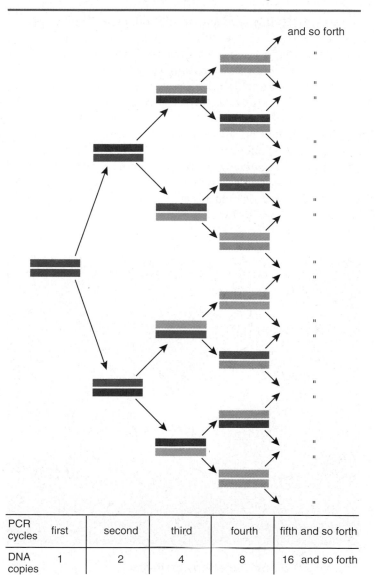

PCR cycles	first	second	third	fourth	fifth and so forth
DNA copies	1	2	4	8	16 and so forth

Figure 26.2   Polymerase chain reaction (PCR).
PCR allows the production of many identical copies of DNA in a laboratory test tube.

26.2 Biotechnology Products

Today, bacteria, plants, and animals are genetically engineered to make biotechnology products (Fig. 26.3). Organisms that have had a foreign gene inserted into them are called **transgenic organisms.**

Transgenic Bacteria

Recombinant DNA technology is used to produce transgenic bacteria, which are grown in huge vats called bioreactors. The gene product is collected from the medium. Biotechnology products that are produced by bacteria include insulin, human growth hormone, t-PA (tissue plasminogen activator), and hepatitis B vaccine are now on the market. Transgenic bacteria have many other uses as well. Some have been produced to promote the health of plants. For example, bacteria that normally live on plants and encourage the formation of ice crystals have been changed from frost-plus to frost-minus bacteria. Also, a bacterium that normally colonizes the roots of corn plants has now been endowed with genes (from another bacterium) that code for an insect toxin. The toxin protects the roots from insects.

Bacteria can be selected for their ability to degrade a particular substance, and this ability can then be enhanced by genetic engineering. For instance, naturally occurring bacteria that eat oil can be genetically engineered to do an even better job of cleaning up beaches after oil spills. Industry has found that bacteria can be used as biofilters

Figure 26.3   Biotechnology products.
Products such as clotting factor VIII, which is administered to hemophiliacs, can be made by transgenic bacteria, plants, or animals. After being processed and packaged, it is sold as a commercial product.

to prevent airborne chemical pollutants from being vented into the air. Bacteria can also remove sulfur from coal before it is burned and help clean up toxic waste dumps. One such strain was given genes that allowed it to clean up levels of toxins that would have killed other strains. Further, these bacteria were given "suicide" genes that caused them to self-destruct when the job had been accomplished.

Organic chemicals are often synthesized by having catalysts act on precursor molecules or by using bacteria to carry out the synthesis. Today, it is possible to go one step further and to manipulate the genes that code for these enzymes. For instance, biochemists discovered a strain of bacteria that is especially good at producing phenylalanine, an organic chemical needed to make aspartame, the dipeptide sweetener better known as NutraSweet. They isolated, altered, and formed a vector for the appropriate genes so that various bacteria could be genetically engineered to produce phenylalanine.

Many major mining companies already use bacteria to obtain various metals. Genetic engineering can enhance the ability of bacteria to extract copper, uranium, and gold from low-grade sources. Some mining companies are testing genetically engineered organisms that have improved bioleaching capabilities.

Transgenic Plants

Techniques have been developed to introduce foreign genes into immature plant embryos or into plant cells called protoplasts that have had the cell wall removed. It is possible to treat protoplasts with an electric current while they are suspended in a liquid containing foreign DNA. The electric current makes tiny, self-sealing holes in the plasma membrane through which genetic material can enter. Protoplasts go on to develop into mature plants.

Foreign genes transferred to cotton, corn, and potato strains have made these plants resistant to pests because their cells now produce an insect toxin. Similarly, soybeans have been made resistant to a common herbicide. Some corn and cotton plants are both pest- and herbicide-resistant. These and other genetically engineered crops are now sold commercially.

Plants are also being engineered to produce human proteins, such as hormones, clotting factors, and antibodies, in their seeds. One type of antibody made by corn can deliver radioisotopes to tumor cells, and another made by soybeans can be used to treat genital herpes.

A weed called mouse-eared cress has been engineered to produce a biodegradable plastic (polyhydroxybutyrate, or PHB) in cell granules.

Transgenic Animals

Techniques have been developed to insert genes into the eggs of animals. It is possible to microinject foreign genes into eggs by hand, but another method uses vortex mixing. The eggs are placed in an agitator with DNA and silicon-carbide needles, and the needles make tiny holes through which the DNA can enter. When these eggs are fertilized, the resulting offspring are transgenic animals. Using this technique, many types of animal eggs have taken up the gene for bovine growth hormone (bGH). The procedure has been used to produce larger fishes, cows, pigs, rabbits, and sheep.

Gene pharming, the use of transgenic farm animals to produce pharmaceuticals, is being pursued by a number of firms. Genes that code for therapeutic and diagnostic proteins are incorporated into the animal's DNA, and the proteins appear in the animal's milk. Plans are underway to produce drugs for the treatment of cystic fibrosis, cancer, blood diseases, and other disorders by this method. Figure 26.4 outlines the procedure for producing transgenic mammals: DNA containing the gene of interest is injected into donor eggs. Following in vitro fertilization, the zygotes are placed in host females where they develop. After female offspring mature, the product is secreted in their milk.

Cloning Transgenic Animals

For many years, it was believed that adult vertebrate animals could not be cloned. Although each cell contains a copy of all the genes, certain genes are turned off in mature, specialized cells. Cloning of an adult vertebrate requires that all the genes of an adult cell be turned on again if development is to proceed normally. This had long been thought impossible.

In 1997, however, Scottish scientists announced that they had produced a cloned sheep called Dolly. Since then, calves and goats have also been cloned, as described in Figure 26.4. After enucleated eggs from a donor are microinjected with 2n nuclei from the same transgenic animal, they are coaxed to begin development in vitro. Development continues in host females until the clones are born. The offspring are clones because all have the genotype and phenotype of the adult that donated the 2n nuclei. In a procedure that produced cloned mice, the 2n nuclei were taken from corona radiata cells. Corona radiata cells are those that cling to an egg after ovulation occurs. Now that scientists have a way to clone animals, this procedure will undoubtedly be used routinely to procure biotechnology products.

Animal Organs as Biotechnology Products

The Health Focus on the next page discusses how it may be possible to use genetically engineered pigs to serve as a source of organs for human transplant operation. Alternatively, scientists are learning how to stimulate human cells to construct organs in the laboratory.

Genetically engineered bacteria, plants, and animals are used to make biotechnology products. Procedures have also been developed to clone these animals. Organs made in the laboratory are also biotechnology products.

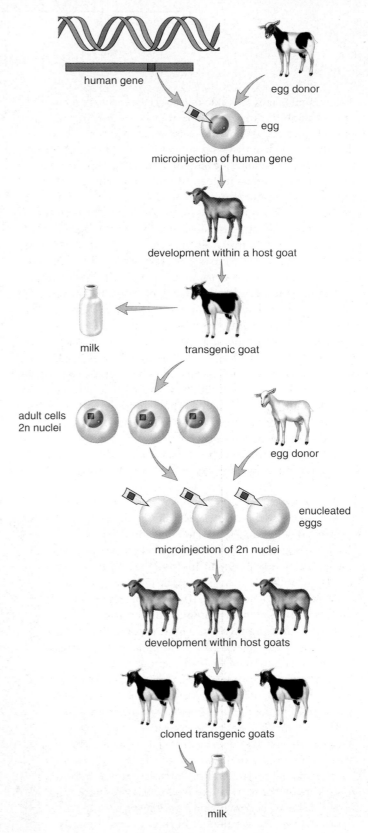

Figure 26.4   Transgenic animals.
A genetically engineered egg develops in a host to give a transgenic goat that produces a biotechnology product in its milk. Nuclei from the transgenic goat are transferred into donor eggs, which develop into cloned transgenic goats.

Health Focus

Organs for Transplant

Although it is now possible to transplant various organs, there are not enough human donors to go around. Thousands of patients die each year while waiting for an organ. It's no wonder, then, that scientists are suggesting we get organs from a source other than another human. **Xenotransplantation** is the use of animal organs instead of human organs in transplant patients. You might think that apes, such as the chimpanzee or the baboon, would be a scientifically suitable species for this purpose. But apes are slow breeders and probably cannot be counted on to supply all the organs needed. Also, many people might object to using apes for this purpose. In contrast, animal husbandry has long included the raising of pigs as a meat source, and pigs are prolific. A female pig can become pregnant at six months of age and can have two litters a year, each averaging about ten offspring.

Ordinarily, the human body would violently reject transplanted pig organs. Genetic engineering, however, can make these organs less antigenic. Scientists have produced a strain of pigs whose organs would most likely, even today, survive for a few months in humans. They could be used to keep a patient alive until a human organ was available. The ultimate goal is to make pig organs as widely accepted by humans as type O blood. (A person with type O blood is called a universal donor because the red blood cells carry no A or B antigens.)

As xenotransplantation draws near, other concerns have been raised. Some experts fear that animals might be infected with viruses, akin to Ebola virus or the "mad cow" disease virus. After infecting a transplant patient, these viruses might spread into the general populace and begin an epidemic. As an indication of this possibility, scientists believe that HIV was spread to humans from monkeys when humans ate monkey meat. Advocates of using pigs for xenotransplantation point out that pigs have been around humans for centuries without infecting them with any serious diseases.

An alternative to xenotransplantation also exists. Just a few years ago, scientists believed that transplant organs had to come from humans or other animals. Now, however, tissue engineering is demonstrating that it is possible to make some bioartificial organs—hybrids created from a combination of living cells and biodegradable polymers. Presently, lab-grown hybrid tissues are on the market. For example, a product composed of skin cells growing on a polymer is used to temporarily cover the wounds of burn patients. Similarly, damaged cartilage can be replaced with a hybrid tissue produced after chondrocytes are harvested from a patient. Another connective tissue product made from fibroblasts and collagen is available to help heal deep wounds without scarring. Soon to come are a host of other products, including replacement corneas, heart valves, bladder valves, and breast tissue.

Tissue engineers have also created implants—cells producing a useful product encapsulated within a narrow plastic tube or a capsule the size of a dime or quarter (Fig. 26A). The pores of the container are large enough to allow the product to diffuse out but too small for immune cells to enter and destroy the cells. An implant whose cells secrete natural pain killers will survive for months in the spinal cord and can be easily withdrawn when desired. A "bridge to a liver transplant" is a bedside vascular apparatus. The patient's blood passes through porous tubes surrounded by pig liver cells. These cells convert toxins in the blood to nonpoisonous substances.

The goal of tissue engineering is to produce fully functioning organs for transplant. After nine years, a Harvard Medical School team headed by Anthony Atala has produced a working urinary bladder in the laboratory. After testing the bladder in laboratory animals, the Harvard group is ready to test it in humans whose own bladders have been damaged by accident or disease, or will not function properly due to a congenital birth defect. Another group of scientists has been able to grow arterial blood vessels in the laboratory. Tissue engineers are hopeful that they will one day produce larger internal organs such as the liver or kidney.

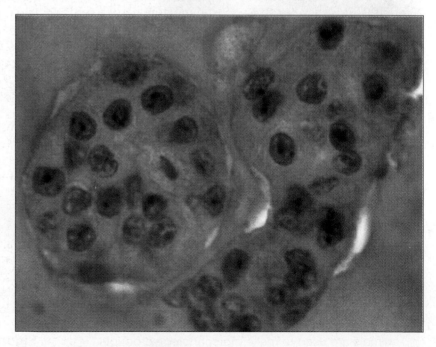

Figure 26A Microreactors.
These pancreatic cells from pigs are in a plastic container whose pores are too small for immune cells to enter and destroy them.

26.3 The Human Genome Project

A genome is all the genetic information of an individual or a species. The goals of the Human Genome Project are to map the human chromosomes in two ways: (1) to construct a map that shows the sequence of base pairs along our chromosomes and (2) to construct a map that shows the sequence of genes along the human chromosomes.

The Base Sequence Map

Researchers have now completed the first goal. They know the sequence of the three billion base pairs, one after the other, along the length of our chromosomes. It took some fifteen years for researchers to complete this monumental task. Two rival groups have been at work on the project. The International Human Genome Sequencing Consortium, which consists of laboratories in many different countries, depends on the support of public funds, including substantial contributions from the United States Government. On the other hand, Celera Genomics, a private company which is supported by a pharmaceutical firm, has been sequencing the genome for only a few years. These competing groups used slightly different techniques, but their data matches.

Even though we now know the sequence of bases in the human genome, much work still needs to be done to make sense out of what we have discovered. We do know that there is little difference between the sequence of our bases and other organisms whose DNA sequences are also known. From this we can conclude that we share a large number of genes with much simpler organisms such as bacteria! It's possible that eventually we will discover that our uniqueness is due to the regulation of these genes.

The Genetic Map

The genetic map will tell the location of genes along each chromosome. Figure 26.5 shows the loci of significant mutant genes on chromosome 17. Many genes have had

their loci determined. Still, we do not know the sequence of all the genes on any particular chromosome.

Completing the chromosomal genetic map should accelerate now that the base sequence map is complete. Researchers need only know a short sequence of bases in a gene of interest in order for the computer to search the genome for a match. Then, the computer will tell the researcher where this gene is located.

A question still being hotly debated is the total number of human genes. Much of our DNA consists of nucleotide repeats that do not code for a protein. So far, researchers have found only 30,000 genes that do code for proteins. This number seems terribly low; that is, a roundworm has 20,000 genes, so a human, which is certainly more complex than a roundworm, should have many more genes. Some researchers think more genes are yet to be identified. Others, believing they have found most of our genes, speculate that each of these genes could code for about three proteins, simply by using different combinations of exons.

As discussed in the Health Focus on page 532, researchers are hopeful that mapping the human chromosomes will help them not only discover mutant genes for many more human disorders but also develop medicines to treat these disorders. It may also be possible to locate genes that can enhance the lives of people. Such genes could possibly lead to germline therapy before a child is born.

There are many ethical questions regarding how our knowledge of the human genome should be used. Therefore, it is imperative that everyone be educated about the human genome because in the end it is the public which must decide these issues.

The Human Genome Project has two goals: to construct a base-pair sequence map and to construct a genetic map of the chromosomes. We now know the sequence of base pairs along the chromosomes. This knowledge may help speed up the construction of the genetic map.

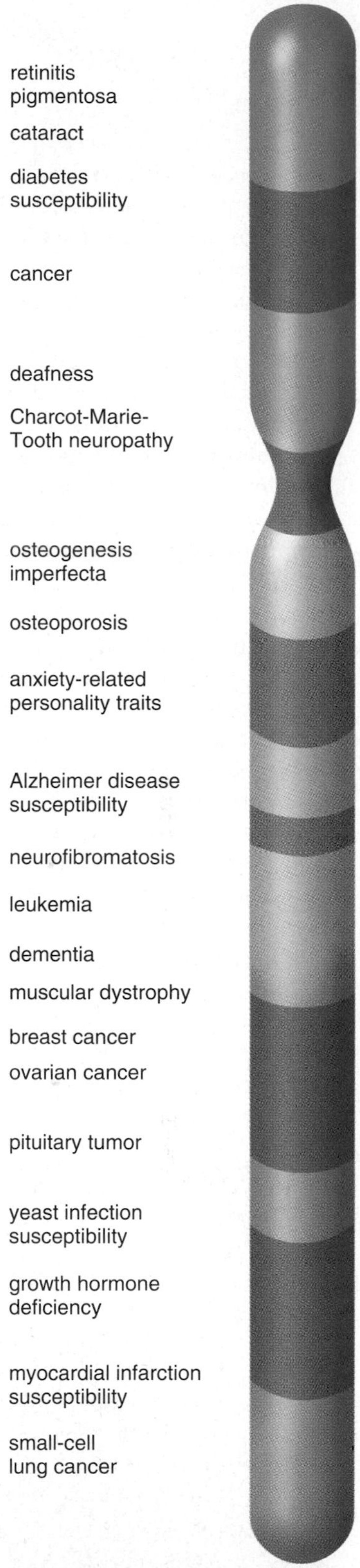

retinitis pigmentosa

cataract

diabetes susceptibility

cancer

deafness

Charcot-Marie-Tooth neuropathy

osteogenesis imperfecta

osteoporosis

anxiety-related personality traits

Alzheimer disease susceptibility

neurofibromatosis

leukemia

dementia

muscular dystrophy

breast cancer

ovarian cancer

pituitary tumor

yeast infection susceptibility

growth hormone deficiency

myocardial infarction susceptibility

small-cell lung cancer

Figure 26.5 Genetic map of chromosome 17.
This map shows the sequence of mutant genes that cause the diseases noted.

New Cures on the Horizon

Back in the 1980s, Leroy Hood couldn't get funding for the DNA sequencer he was developing. Biologists didn't like the idea of just "collecting facts," and it took several years before an entrepreneur decided to fund the project. Without ever better DNA sequencers, the Human Genome Project would never have completed its monumental task of determining the sequence of bases in our DNA.

Now that we know the sequence of all the bases in the DNA of all our chromosomes, biologists all over the world believe that this knowledge will result in rapid medical advances for ourselves and our children. At least four categories of improvement are expected: many more medicines will be available to keep us healthy; physicians will be able to prescribe medicines more effectively; a longer life span, even to over 100 years, may become common place; and we will be able to shape the genotype of our offspring.

First prediction: Many new medicines will be coming.

Genome sequence data will allow scientists to determine all the proteins that are active only during development plus all those that are still active in adults.

Most drugs are either proteins or small chemicals that are able to interact with proteins. Many of these small chemicals target proteins that act as signals between cells or within the cytoplasm of cells. Today's drugs were usually discovered in a hit-or-miss fashion, but now we can take a more systematic approach to finding effective drugs. For example, it is known that all receptor and signaling proteins start with the same ten-amino-acid sequence. Now, it is possible to scan the human genome for all genes that code for this sequence of ten amino acids, and thereby find all the signaling proteins. Thereafter, they must be tested.

In a recent search for a protein that makes wounds heal, researchers cultured skin cells with fourteen proteins (found by chance) that can cause skin cells to grow. Only one of these proteins made skin cells grow and did nothing else. They expect this protein to become an effective drug for conditions such as venous ulcers, skin lesions that affect many thousands of people in the United States. Such tests, leading to effective results, can be carried out with all the signaling proteins scientists will discover from scanning the human genome.

People's genotypes differ. We all have mutations that account for our various illnesses. Knowing each patient's mutations will allow physicians to match the right drugs to the particular patient.

Second prediction: Genotype testing will be standard.

Genome sequence data will allow us to discover genetically different subgroups of the population. Physicians will be looking for two types of mutations in particular. One type is called single nucleotide polymorphisms (SNPs, pronounced "snips"), in which individuals differ by only one nucleotide. The other type of mutation is nucleotide repeats—the same three bases, repeated over and over again, interrupt a gene and affect its expression. It is not yet clear how many SNPs will be medically significant, but the present estimate is on the order of 300,000.

How will a physician be able to determine which of the 300,000 SNPs and other types of mutations are in your genome? The use of a gene chip will quickly and efficiently give them knowledge of your genotype. A gene chip is an array of thousands of genes on one or several glass slides packaged together. After the gene chip is exposed to an individual's DNA, a technician can note any mutant sequences present in the individual's genes. Soon a chip will be able to hold all the genes carried within the human genome.

Some disorders such as sickle-cell disease are caused by a single SNP, but many disorders seem to require more than one, and possibly in different combinations. In one study, researchers found that a series of SNPs, numbered 1–12, were associated with the development of asthma. A particular drug, called albuterol, was effective for patients with certain combinations and not others. This example and others show that many diseases are polygenic, and that only a genome scan is able to detect which

First prediction
Many new medicines will be coming.

Second prediction
Genotype testing will be standard.

mutations are causing an individual to have the disease, and how it should be properly treated.

Genome scans are also expected to make drugs safer to take. As you know, many drugs potentially have unwanted side effects. Why do some people and not others have one or more of the side effects? Most likely because people have different genotypes. It is expected that a physician will be able to match patients to drugs that are safe for them on the basis of their genotypes.

Biologists also hope that gene chips will single out the specific oncogenes and mutated tumor-suppressor genes that cause the various types of cancer. The protein products of these genes will become targets against which chemists can try to develop drugs. If so, the current methods of treating cancer—surgery, radiation, and chemotherapy to kill all dividing cells—will no longer be necessary.

Third prediction: A longer and healthier life will be yours.

Genome sequence data may allow scientists to determine which genes enable people to live longer. Investigators have already found evidence for genes that extend the life span of animals such as roundworms and fruit flies. The sequencing of the human genome makes it possible for scientists to find such genes in humans also.

For example, we know that the presence of free radicals causes cellular molecules to become unstable and cells to die.

Certain genes are believed to code for antioxidant enzymes that detoxify free radicals. It could be that human beings with particular forms of these genes have more efficient antioxidant enzymes, and therefore live longer. If so, researchers will no doubt be able to locate these genes and also others that promote a longer life.

Consider, too, that natural selection favors phenotypes that result in the greatest number of fertile offspring in the next generation.

**Third prediction
A longer and healthier
life will be yours.**

Since children are usually born to younger individuals, natural selection is indifferent to genes that protect the body from the deleterious affects of aging. Researchers can possibly find such genes, however, in individuals who have a long life span. Use of these genes would possibly oppose a destiny, determined so far only by evolution.

Possible stem cell therapy has generated much interest of late. Stem cells are embryonic cells and also some adult cells, such as those in red bone marrow, that are nondifferentiated. These cells have the potential to become any type of tissue, depending on which signaling molecules are used. Genome sequence data will eventually give scientists knowledge of all the signaling molecules humans possess. Stem cells could also be subjected to gene therapy in order to correct any defective genes before scientists use them to create the tissues or organs of the body. Use of these tissues and organs to repair and/or replace worn-out structures could no doubt expand the human life span.

Fourth prediction: You will be able to design your children.

Genome sequence data will be used to identify many more mutant genes that cause genetic disorders than are presently known. In the future, it may be possible to cure genetic disorders before the child is born by adding a normal gene to any egg that carries a mutant gene. Or an artificial chromosome, constructed to carry a large number of corrective genes, could automatically be placed in eggs. In vitro fertilization would have to be utilized in order to take advantage of such measures for curing genetic disorders before birth.

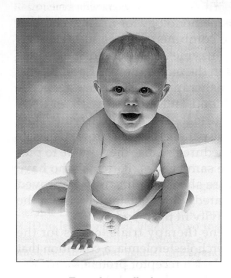

Genome sequence data can also be used to identify polygenic genes for traits such as height, intelligence, or behavioral characteristics. A couple could decide on their own which genes they wish to use to enhance a child's phenotype. In other words, the sequencing of the human genome may bring about a genetically just society, in which all types of genes would be accessible to all parents.

**Fourth prediction
You will be able
to design your children.**

Evolution and Diversity

See the Online Learning Center for careers in Botany and Zoology.

Evolution of Life

Chapter Concepts

Mammals are adapted to reproduce on land. The penis protects sperm from drying out when the male passes them to the female.

*A*daptations to specific environments are wonderful examples of the evolutionary process. Although most animals are adapted to reproduce in the water, some, such as ourselves, are adapted to reproduce on land.

When frogs reproduce in water, no need exists to protect the gametes or the embryo from drying out. The male simply clasps the female during amplexus, causing her to release her eggs in the water. Thereafter, the male releases sperm. Fertilization occurs in the water, and the zygote develops externally into a swimming larva known as a tadpole. Metamorphosis produces the adult form that can move up onto land.

When mammals, which are adapted to living on land, reproduce, the gametes and the embryo are protected from drying out. The male has a penis and he passes sperm directly to the female during sexual intercourse. Eggs never leave the body of the female and the fertilized zygote develops internally within the uterus. The developing offspring is protected within the watery amniotic sac and does not breathe air until birth occurs.

27.1 Evidence of Evolution

Evolution is all the changes that have occurred in living things since the beginning of life. Table 27.1 indicates that earth is about 4.5 billion years old and that prokaryotes evolved about 3.5 billion years ago. The eukaryotic cell arose about 2.1 billion years ago, but multicellularity didn't begin until 700 million years ago. This means that for 80% of the time since life arose, only unicellular organisms were present. Animal evolution may not have begun in earnest until the Cambrian explosion, a possible burst of animal evolution at the start of the Cambrian period. Most of the evolutionary events we will be discussing in future chapters occurred in less than 20% of the history of life!

Evolution encompasses common descent and adaptation to the environment. Because of common descent, all living things share the same fundamental characteristics: they are made of cells, take chemicals and energy from the environment, respond to external stimuli, reproduce, and evolve. Life is diverse because the various types of living things are adapted to different ways of life.

Let us look at the varied evidence that evolution has occurred. Many fields of biology give evidence that evolution has occurred.

Fossil Evidence

Fossils are the remains and traces of past life or any other direct evidence of past life. Most fossils consist only of hard parts, such as shells, bones, or teeth, because these are usually not destroyed after an organism dies. The soft parts of a dead organism are often consumed by scavengers or decomposed by bacteria. Occasionally, however the organism is buried quickly and in such a way that decomposition is never completed or is completed so slowly that the soft parts leave an imprint of their structure. Traces include trails, footprints, burrows, worm casts, or even preserved droppings.

The great majority of fossils are found embedded in sedimentary rock. Sedimentation, a process that has been going on since the earth was formed, can take place on land or in bodies of water. Weathering and erosion of rocks produces an accumulation of particles that vary in size and nature and are called sediment. Sediment becomes a stratum (pl., strata), a recognizable layer in a stratigraphic sequence. Any given stratum is older than the one above it and younger than the one immediately below it.

Paleontologists are biologists who discover and study the fossil record and from it make decisions about the history of life. Particularly interesting are the fossils that serve as **transitional links** between groups. For example, the famous fossils of *Archaeopteryx* are intermediate between reptiles and birds (Fig. 27.1). The dinosaur-like skeleton of this fossil has reptilian features, including jaws with teeth and a long, jointed tail, but *Archaeopteryx* also had feathers and wings. Other transitional links among fossil vertebrates include the amphibious fish *Eustheopteron*, the reptile-like amphibian *Seymouria*, and the mammal-like reptiles, or therapsids. These fossils allow us to deduce that fishes evolved before amphibians, which evolved before reptiles, which evolved before both birds and mammals in the history of life.

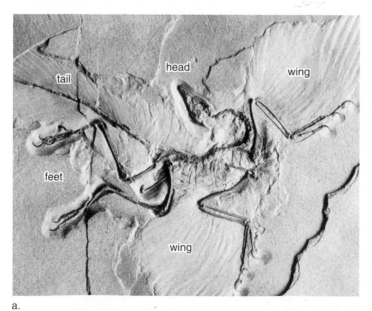

a.

b.

Figure 27.1 Transitional fossils.
a. *Archaeopteryx* was a transitional link between reptiles and birds. Fossils indicate it had feathers and wing claws. Most likely, it was a poor flier. Perhaps it ran over the ground on strong legs and climbed up into trees with the assistance of these claws. **b.** *Archaeopteryx* also had a feather-covered, reptilian-type tail that shows up well in this artist's representation.

| Table 27.1 | The Geological Time Scale: Major Divisions of Geological Time with Some of the Major Evolutionary Events of Each Geological Period | | | | |

Era	Period	Epoch	Millions of Years Ago	Plant Life	Animal Life
Cenozoic* (from the present to 66.4 million years ago)	Neogene	Holocene	0–0.01	Destruction of tropical rain forests by humans accelerates extinctions	AGE OF HUMAN CIVILIZATION
				Significant Mammalian Extinction	
		Pleistocene	0.01–2	Herbaceous plants spread and diversify	Modern humans appear
		Pliocene	2–6	Herbaceous angiosperms flourish	First hominids appear
		Miocene	6–24	Grasslands spread as forests contract	Apelike mammals and grazing mammals flourish; insects flourish
	Paleogene	Oligocene	24–37	Many modern families of flowering plants evolve	Browsing mammals and monkeylike primates appear
		Eocene	37–58	Subtropical forests thrive with heavy rainfall	All modern orders of mammals are represented
		Paleocene	58–66	Angiosperms diversify	Primitive primates, herbivores, carnivores, and insectivores appear
				Mass Extinction of Dinosaurs and Most Reptiles	
Mesozoic (from 66.4 to 245 million years ago)	Cretaceous		66–144	Flowering plants spread; coniferous trees decline	Placental mammals appear; modern insect groups appear
	Jurassic		144–208	Cycads and other gymnosperms flourish	Dinosaurs flourish; birds appear
				Mass Extinction of ~~All~~ Many Life Forms	
	Triassic		208–245	Cycads and ginkgoes appear; forests of gymnosperms and ferns dominate	First mammals appear; first dinosaurs appear; corals and molluscs dominate seas
				Mass Extinction of ~~All~~ Many Life Forms	
Paleozoic (from 245 to 570 million years ago)	Permian		245–286	Conifers appear	Reptiles diversify; amphibians decline
	Carboniferous		286–360	Age of great coal-forming forests: club mosses, horsetails, and ferns flourish	Amphibians diversify; first reptiles appear; first great radiation of insects
				Mass Extinction of ~~All~~ Many Life Forms	
	Devonian		360–408	First seed ferns appear	Jawed fishes diversify and dominate the seas; first insects and first amphibians appear
	Silurian		408–438	Low-lying vascular plants appear on land	First jawed fishes appear
				Mass Extinction of ~~All~~ Many Life Forms	
	Ordovician		438–505	Marine algae flourish	Invertebrates spread and diversity; jawless fishes, first vertebrates appear
	Cambrian		505–570	Marine algae flourish	Invertebrates with skeletons are dominant
Precambrian time (from 570 to 4,600 million years ago)			700	Multicellular organisms appear	
			2,100	First complex (eukaryotic) cells appear	
			3,100–3,500	First prokaryotic cells in stromatolites appear	
			4,500	Earth forms	

*Many authorities divide the Cenozoic era into the Tertiary period (contains Paleocene, Eocene, Oligocene, Miocene, and Pliocene) and the Quaternary period (contains Pleistocene and Holocene).

Charles Darwin's Theory of Natural Selection

At the age of 22, Charles Darwin signed on as a naturalist with the HMS *Beagle*, a ship that took a five-year trip around the world in the first half of the nineteenth century. Because the ship sailed in the tropics of the Southern Hemisphere, where living things are more abundant and varied, Darwin encountered forms of life very different from those in his native England.

Even though it was not his original intent, Darwin began to realize and to gather evidence that living things change over time and from place to place. He read a book by Charles Lyell, a geologist who suggested the world is very old and has been undergoing gradual changes for many, many years. Darwin found the remains of a giant ground sloth and an armadillo on the east coast of South America and wondered if this extinct species was related to the living forms of these animals. When he compared the animals of Africa to those of South America, he noted that the African ostrich and the South American rhea, although similar in appearance, were actually different animals. He reasoned that they had a different line of descent because they were on different continents. When Darwin arrived at the Galápagos Islands, he began to study the diversity of finches (see Fig. 27.19), whose adaptations could best be explained by assuming they had diverged from a common ancestor. With this type of evidence, Darwin concluded that species evolve (change) with time.

When Darwin returned home, he spent the next 20 years gathering data to support the principle of organic evolution. His most significant contribution to this principle was his theory of natural selection, which explains how a species becomes adapted to its environment. Before formulating the theory, he read an essay on human population growth written by Thomas Malthus. Malthus observed that although the reproductive potential of humans is great, many environmental factors, such as availability of food and living space, tend to keep the human population within bounds. Darwin applied these ideas to all populations of organisms. For example, he calculated that a single pair of elephants could have 19 million descendants in 750 years. He realized that other organisms have even greater reproductive potential than this pair of elephants; yet, usually the number of each type of organism remains about the same. Darwin decided there is a constant struggle for existence, and only a few members of a population survive to reproduce. The ones that survive and contribute to the evolutionary future of the species are by and large the better-adapted individuals. This so-called survival of the fittest causes the next generation to be better adapted than the previous generation.

Darwin's theory of natural selection was nonteleological. That is, rather than believing that organisms strive to adapt themselves to the environment, Darwin concluded that the environment acts on organisms to select those individuals that are best adapted. These are the ones that have been "naturally selected" to pass on their characteristics to the next generation. In order to emphasize the nonteleological nature of Darwin's theory, it is often contrasted with the ideas of Jean-Baptiste Lamarck, another nineteenth-century naturalist (Fig. 27A). The Lamarckian explanation for the long neck of the giraffe was based on the assumption that the ancestors of the modern giraffe were trying to reach into the trees to browse on high-growing vegetation. Continual stretching of the neck caused it to become longer, and this acquired characteristic was passed on to the next generation. Lamarck's proposal is teleological because, according to him, the outcome is known ahead of time. This type of explanation has not stood the test of time, but Darwin's theory of evolution by natural selection has been fully substantiated by later investigations.

These are the critical elements of Darwin's theory:

- **Variations.** Individual members of a species vary in physical characteristics. Physical variations can be passed from generation to generation. (Darwin was never aware of genes, but we know today that the inheritance of the genotype determines the phenotype.)
- **Struggle for existence.** The members of all species compete with each other for limited resources. Certain members are able to capture these resources better than others.
- **Survival of the fittest.** Just as humans carry on artificial breeding programs to select which plants and animals will reproduce, natural selection by the environment determines which organisms survive and reproduce. While Darwin emphasized the importance of survival, modern evolutionists emphasize the importance of unequal reproduction. In any case, however, the selection process is nonteleological. Certain members of the population are selected to produce more offspring

simply because they happen to have a variation that makes them better suited to the environment.

- **Adaptation.** Natural selection causes a population of organisms, and ultimately a species, to become adapted to the environment. The process is slow, but each subsequent generation includes more individuals that are better adapted to the environment.

Can natural selection account for the origin of new species and for the great diversity of life? Yes, if we are aware that life has been evolving for a very long time and that variously adapted populations can arise from a common ancestor.

Darwin was prompted to publish his findings only after he received a letter from another naturalist, Alfred Russel Wallace, who had come to the exact same conclusions about evolution. Although both scientists subsequently presented their ideas at the same meeting of the famed Royal Society in London in 1858, only Darwin later gathered detailed evidence in support of his theory. He described his experiments and reasoning at great length in *The Origin of Species by Means of Natural Selection,* a book still studied by many biologists today.

Figure 27A   Mechanism of evolution.
This diagram contrasts (**a**) Jean-Baptiste Lamarck's proposal of acquired characteristics with (**b**) Charles Darwin's theory of natural selection. Only Darwin's theory is supported by data.

Early giraffes probably had short necks that they stretched to reach food.

Early giraffes probably had necks of various lengths.

Their offspring had longer necks that they stretched to reach food.

Natural selection due to competition led to survival of the longer-necked giraffes and their offspring.

Eventually, the continued stretching of the neck resulted in today's giraffe.

Eventually, only long-necked giraffes survived the competition.

a. Lamarck's proposal

b. Darwin's theory

Gene Flow

Gene flow is the movement of alleles between populations, such as occurs when breeding individuals migrate from one population to the other. Adult plants are not able to migrate, but their gametes are often either blown by the wind or carried by insects. The wind, in particular, can carry pollen for long distances and can therefore be a factor in gene flow between plant populations.

Gene flow between two populations keeps their gene pools similar. It also prevents close adaptation to a local environment.

Nonrandom Mating

Nonrandom mating occurs when individuals pair up, not by chance, but according to their genotypes or phenotypes. Inbreeding, or mating between relatives to a greater extent than by chance, is an example of nonrandom mating. Inbreeding decreases the proportion of heterozygotes and increases the proportions of both homozygotes at all gene loci. In a human population, inbreeding increases the frequency of recessive abnormalities (see Fig. 27.14).

Natural Selection

Natural selection is the process by which populations become adapted to their environment. The Science Focus on pages 552–53 outlines how Charles Darwin, the father of evolution, explained evolution by natural selection. Here, we restate these steps in the context of modern evolutionary theory. Evolution by natural selection requires:

1. **Variation.** The members of a population differ from one another.
2. **Inheritance.** Many of these differences are heritable genetic differences.
3. **Differential adaptedness.** Some of these differences affect how well an organism is adapted to its environment.
4. **Differential reproduction.** Individuals that are better adapted to their environment are more likely to reproduce, and their fertile offspring will make up a greater proportion of the next generation.

In evolution by natural selection, the **fitness** of an individual is measured by the number of fertile offspring. Where do mutations come from that can lead to increased fitness? Gene mutations are the ultimate source of variation because they provide new alleles. However, in sexually reproducing organisms, variations can come about due to crossing-over and independent assortment of chromosomes during meiosis and also fertilization when gametes are combined. A different combination of alleles can lead to a new and different phenotype.

In this context, consider that most of the traits on which natural selection acts are polygenic and controlled by more than one pair of alleles. Such traits have a range of phenotypes that follow a bell-shaped curve.

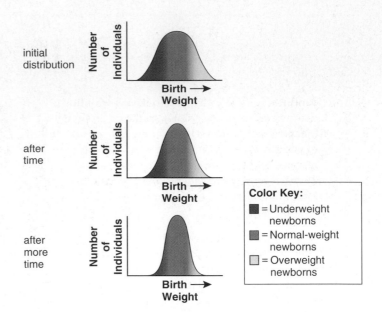

Figure 27.15 Stabilizing selection.
This occurs when natural selection favors the intermediate phenotype over the extremes. For example, most human newborns are of intermediate weight (about 3.2 kg, or 7 lb), and very few babies are either very small or very large.

Three types of natural selection are known: stabilizing selection, directional selection, and disruptive selection.

Stabilizing Selection   Stabilizing selection occurs when an intermediate phenotype is favored (Fig. 27.15). It can improve adaptation of the population to those aspects of the environment that remain constant. With stabilizing selection, extreme phenotypes are selected against, and individuals near the average are favored. As an example, consider the birth weight of human infants, which ranges from 0.89 to 4.9 kilograms (2–10.8 lb). The death rate is higher for infants whose birth weights are at the extremes of this range and lowest for babies who have a birth weight between 3.1 kilograms and 3.5 kilograms. Most babies have an intermediate birth weight, which gives the best chance of survival. Similar results have been found in other animals, also.

Directional Selection   Directional selection occurs when an extreme phenotype is favored and the distribution curve shifts in that direction (Fig. 27.16). Such a shift can occur when a population is adapting to a changing environment. For example, the gradual increase in the size of the modern horse, *Equus*, can be correlated with a change in the environment from forest conditions to grassland conditions. *Hyracotherium*, the ancestor of the modern horse, was about the size of a dog and was adapted to the forestlike environment of the Eocene, an epoch of the Paleogene period. This animal could have hidden among the trees for protection, and its low-crowned teeth were appropriate for browsing on leaves. In the Miocene and Pliocene epochs, however, grasslands began to replace the forests. Then the ancestors of *Equus* were subject to selective pressure for the development of strength, intelligence, speed, and durable grinding teeth. A larger size provided the strength needed for combat, a larger skull made room for a larger brain, elongated legs ending in hooves gave speed for escaping from enemies, and the durable grinding teeth enabled the animals to feed efficiently on grasses. Nevertheless, the evolution of the horse should not be viewed as a straight line of descent; there were many side branches that became extinct.

The evolution of peppered moths discussed previously (see section 27.2) is another good example of directional selection in which the selective agent is known. Moths rest on the trunks of trees during the day; if they are seen by predatory

birds, they are eaten. As long as the tree trunks in the environment are light in color, the light-colored moths live to reproduce. But if the tree trunks turn black from industrial pollution, the dark-colored moths survive and reproduce to a greater extent than the light-colored moths. The dark-colored phenotype then becomes the more frequent one in the population (see Fig. 27.12). However, if pollution is reduced and the trunks of the trees regain their normal color, the light-colored moths again increase in number.

Pesticides and antibiotics are selective agents for insects and bacteria, respectively. The forms that survive exposure to these agents give rise to future generations that are resistant to these toxic substances.

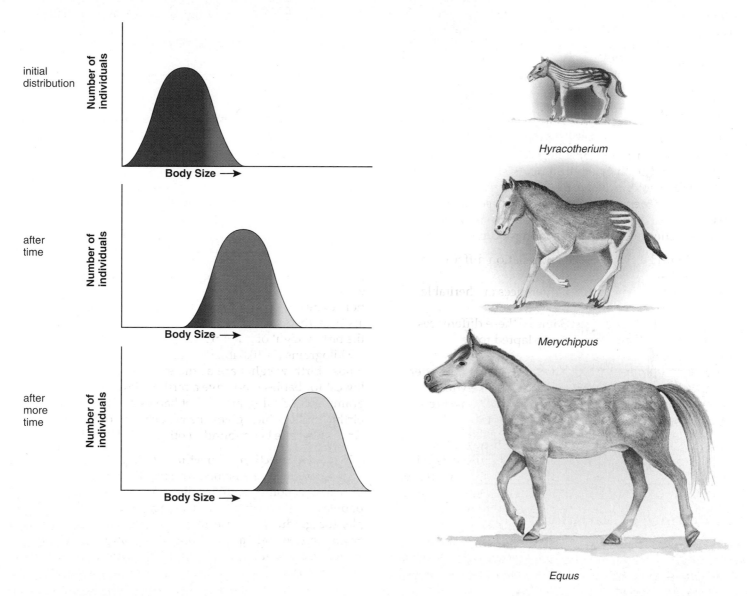

Figure 27.16   Directional selection.
Directional selection occurs when natural selection favors one extreme phenotype, resulting in a shift in the distribution curve. For example, *Equus*, the modern-day horse, which is adapted to a grassland habitat, is much larger than its ancestor *Hyracotherium*, which was adapted to a forest habitat.

Figure 27.17   Disruptive selection.
This occurs when natural selection favors two extreme phenotypes. For example, it is observed today that British land snails mainly comprise two different phenotypes, each adapted to a particular habitat.

Disruptive Selection   In **disruptive selection,** two or more extreme phenotypes are favored over any intermediate phenotype (Fig. 27.17). For example, British land snails (*Cepaea nemoralis*) have a wide habitat range that includes low-vegetation areas (grass fields and hedgerows) and forest areas. In low-vegetation areas, thrushes feed mainly on snails with dark shells that lack light bands, and in forest areas, they feed mainly on snails with light-banded shells. Therefore, the two different habitats have resulted in two different phenotypes in the population.

The agents of evolutionary change are mutations, genetic drift, gene flow, nonrandom mating, and natural selection. These processes cause changes in the gene pool frequencies of a population. Of these, only natural selection results in adaptation to the environment.

Maintenance of Variation

Sickle-cell disease exemplifies how genetic variation is sometimes maintained in a population. Persons with sickle-cell disease have sickle-shaped red blood cells, leading to hemorrhaging and organ destruction. In parts of Africa, there is a high incidence of malaria caused by a parasite that lives in and destroys red blood cells. Sickle-cell disease

tends to be more common in such areas. A study of the three genotypes and phenotypes involved explains why:

Genotype	Phenotype	Result
$Hb^A Hb^A$	Normal	Dies due to malarial infection
$Hb^A Hb^S$	Sickle-cell trait	Lives due to protection from both
$Hb^S Hb^S$	Sickle-cell disease	Dies due to sickle-cell disease

Persons with sickle-cell trait are more likely to survive to reproduce for two reasons. Most of the time, they do not have circulatory problems because their red blood cells have a normal shape. Even so, the malarial parasite cannot survive in their red blood cells. When the cells are sickled, they lose potassium and the parasite dies.

The frequency of the sickle-cell allele in some parts of Africa is 0.40, while among African Americans, it is only 0.05 due to lowered incidence of malaria in the United States. The ability of the heterozygote to survive accounts for the greater frequency of the sickle-cell allele in Africa. The favored heterozygote keeps the two homozygotes equally present in a population. When the ratio of two or more phenotypes remains the same in each generation, it is called balanced polymorphism.

27.4 Speciation

Usually, a species occupies a certain geographical range, within which there are several subpopulations. For our present discussion, **species** is defined as a group of inter-breeding subpopulations that share a gene pool and are isolated reproductively from other species. The subpopulations of the same species exchange genes, but different species do not exchange genes. Reproductive isolation of the gene pools of similar species is accomplished by such mechanisms as those listed in Table 27.2. If **premating isolating mechanisms** are in place, reproduction is never attempted. If **postmating isolating mechanisms** are in place, reproduction may take place, but it does not produce fertile offspring.

Process of Speciation

Whenever reproductive isolation develops, speciation has occurred. Figure 27.18 outlines how reproductive isolation is believed to come about. In the first frame, a species is represented by two populations that are experiencing gene flow. However, when the populations become separated by a geographic barrier, gene flow is no longer possible. A geographic barrier could be a new canal recently built by humans, an upheaval caused by an earthquake, or some other physical factor. Now different variations arise in the

two populations due to independent mutations, drift, and selection so that first postmating, and then, if enough time passes, premating reproductive isolation occurs. Even if the geographic barrier is subsequently removed, the two populations will not be able to reproduce with one another, and therefore, what was once one species has become two species. This model of speciation is called **allopatric speciation.**

It is also possible that a single population could suddenly divide into two reproductively isolated groups without the need for geographic isolation. The best evidence for this type of speciation, called **sympatric speciation,** is found among plants, where multiplication of the chromosome number in one plant prevents it from successfully reproducing with others of its kind. But self-reproduction could lead to a number of offspring with the new chromosome number.

Speciation is the origin of species. This usually requires geographic isolation followed by reproductive isolation.

Table 27.2	Reproductive Isolating Mechanisms
Isolating Mechanism	**Example**
Premating	
Habitat isolation	Species at same locale occupy different habitats
Temporal isolation	Species reproduce at different seasons or different times of day
Behavioral isolation	In animals, courtship behavior differs, or they respond to different songs, calls, pheromones, or other signals
Mechanical isolation	Genitalia unsuitable for one another
Postmating	
Gamete isolation	Sperm cannot reach or fertilize egg
Zygote mortality	Fertilization occurs, but zygote does not survive
Hybrid sterility	Hybrid survives but is sterile and cannot reproduce
F_2 fitness	Hybrid is fertile, but F_2 hybrid has reduced fitness

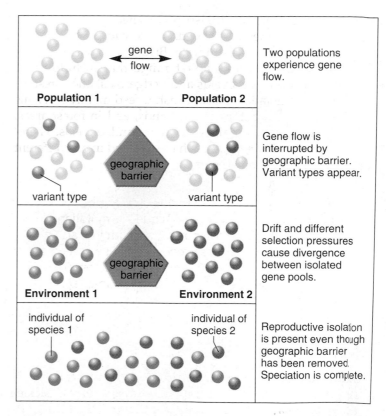

Figure 27.18 Allopatric speciation.
Allopatric speciation occurs after a geographic barrier prevents gene flow between populations that originally belonged to a single species.

Visual Focus

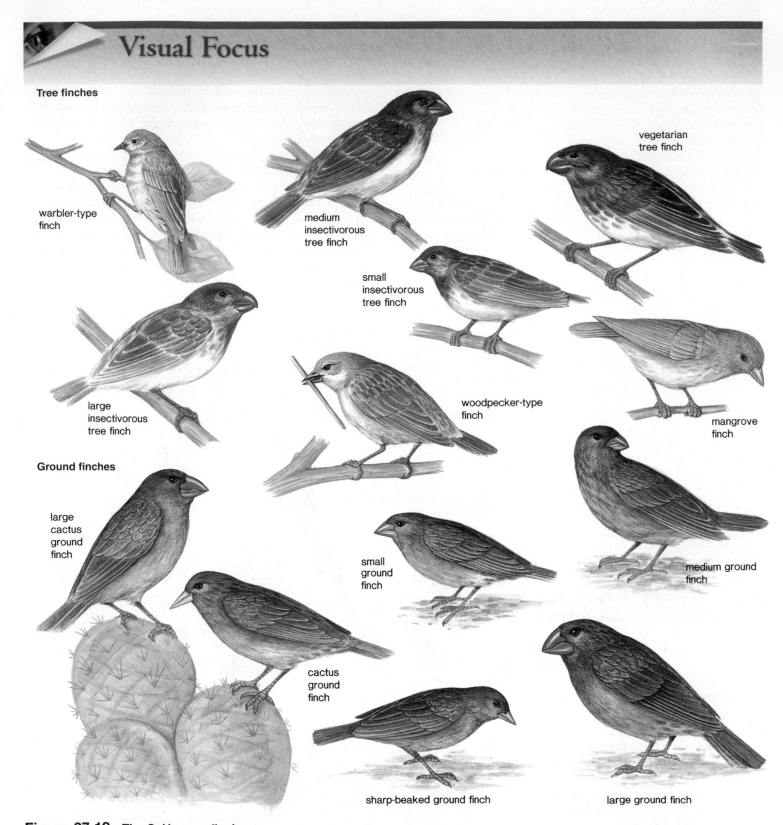

Tree finches

warbler-type finch

medium insectivorous tree finch

vegetarian tree finch

large insectivorous tree finch

small insectivorous tree finch

woodpecker-type finch

mangrove finch

Ground finches

large cactus ground finch

cactus ground finch

small ground finch

medium ground finch

sharp-beaked ground finch

large ground finch

Figure 27.19   The Galápagos finches.
Each of these finches is adapted to gathering and eating a different type of food. Tree finches have beaks largely adapted to eating insects and, at times, plants. The woodpecker-type finch, a tool-user, uses a cactus spine or twig to probe in the bark of a tree for insects. Ground finches have beaks adapted to eating prickly-pear cactus or different-sized seeds.

Adaptive Radiation

One of the best examples of speciation is provided by the finches on the Galápagos Islands, which are often called Darwin's finches because Darwin first realized their significance as an example of how evolution works. The Galápagos Islands, located 600 miles west of Ecuador, South America, are volcanic, but they do have forest regions at higher elevations. The 13 species of finches (Fig. 27.19), placed in three genera, are believed to be descended from mainland finches that migrated to one of the islands. Therefore, Darwin's finches are an example of **adaptive radiation,** or the proliferation of a species by adaptation to different ways of life. We can imagine that after the original population of a single island increased, some individuals dispersed to other islands. The islands are ecologically different enough to have promoted divergent feeding habits. This is apparent because, although the birds physically resemble each other in many respects, they have different beaks, each adapted to gathering and eating a different type of food. There are seed-eating ground finches, with beaks appropriate to cracking small-, medium-, or large-sized seeds; cactus-eating ground finches, with beaks appropriate for eating prickly-pear cacti; insect-eating tree finches, also with different-sized beaks; and a warbler-type tree finch, with a beak adapted to eating insects and gathering nectar. Among the tree finches, there is a woodpecker type, which lacks the long tongue of a true woodpecker but makes up for this by using a cactus spine or a twig to ferret out insects.

The Pace of Speciation

Currently there are two hypotheses about the pace of speciation, and therefore evolution. One hypothesis is called the phyletic gradualism model, and the other is called the punctuated equilibrium model. Each model gives a different answer to the question of why there are so few transitional links in the fossil record.

Traditionally, evolutionists have supported a model called **phyletic gradualism,** which says that change is very slow but steady within a lineage before and after a divergence (splitting of the line of descent) (Fig. 27.20a). Therefore, it is not surprising that few transitional links such as *Archaeopteryx* (see Fig. 27.1) have been found. Indeed, the fossil record, even if it were complete, may be unable to show when speciation has occurred. Why? Because a new species comes about after reproductive isolation, and reproductive isolation cannot be detected in the fossil record!

In recent years, a new model of evolution called **punctuated equilibrium** has been proposed. It says that stasis, a period of no visible change, is punctuated by speciation. With reference to the length of the fossil record (about 3.5 billion years), speciation occurs relatively rapidly, and this can explain why few transitional links are found. Speciation occurs so rapidly that few ever became fossils.

Then, too, speciation most likely involves only an isolated population at one locale. Only when a new species evolves and displaces the existing species is the new species likely to show up in the fossil record.

Adaptive radiation is an example of allopatric speciation that is easily observable. Whether speciation occurs slowly or rapidly is being debated.

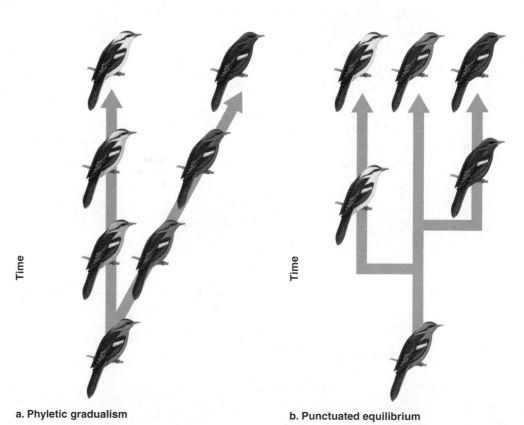

a. Phyletic gradualism   **b. Punctuated equilibrium**

Figure 27.20   Phyletic gradualism versus punctuated equilibrium.
The differences between (**a**) phyletic gradualism and (**b**) punctuated equilibrium are reflected in these patterns of time versus speciation.

27.5 Classification

Recall that each type organism is given a scientific name and the scientific name for modern humans is *Homo sapiens*. The first word, *Homo*, is the genus, a classification category that contains many species. The second word is the specific epithet, which may tell something descriptive about the organism. The specific epithet *sapiens* refers to a large brain. When species are classified, they are placed in a hierarchy of categories: **species, genus, family, order, class, phylum**, and **kingdom**. This text uses one further classification category called a domain.

Five-Kingdom System

For many years, most biologists favored a five-kingdom classification system consisting of the kingdoms Plantae, Animalia, Fungi, Protista, and Monera. Organisms were placed into these kingdoms on the basis of type of cell (prokaryotic or eukaryotic), level of organization (unicellular or multicellular), and mode of nutrition. In this system, organisms in the kingdom Monera are distinguished by their structure—they are prokaryotic (lack a membrane-bounded nucleus)—whereas the organisms in the other kingdoms are eukaryotic (have a membrane-bounded nucleus). As you can see in Figure 27.21, which depicts the five-kingdom system, monerans are at the base of the tree of life. It is suggested that protists evolved from the monerans, and that the fungi, plants, and animals evolved from the protists via three separate lines of evolution.

The five-kingdom system of classification is based on structural differences and also on modes of nutrition among the eukaryotes.

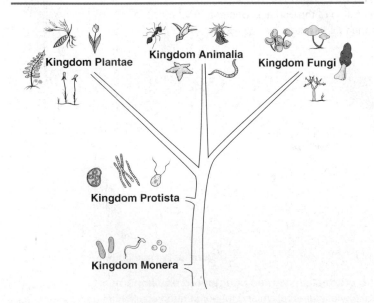

Figure 27.21   Five-kingdom system of classification.
All prokaryotes are in the kingdom Monera. The eukaryotes are in kingdoms Protista, Fungi, Plantae, and Animalia. The diagram shows the lines of descent.

Three-Domain System

Within the past ten years, new information has called into question the five-kingdom system of classification. The molecule rRNA probably changes only slowly during evolution, and indeed, many change only when there is a major evolutionary event. Molecular data based on the sequencing of rRNA suggest that there are three groups of organisms: the bacteria, the archaea, and the eukarya. Therefore, the three-domain system recognizes three domains: domains **Bacteria, Archaea,** and **Eukarya.**

The bacteria diverged first in the tree of life, followed by the archaea and then the eukarya. The archaea and eukarya are more closely related to each other than either is to the bacteria (Fig. 27.22).

Table 27.3 gives detailed information about the differences between the three domains. Bacteria and archaea are both unicellular prokaryotes that lack a membrane-bounded nucleus. Bacteria and archaea are distinguishable from each other on the basis of lipid and cell wall biochemistry. The biochemical attributes of archaea allow them to live in very hostile environments such as anaerobic swamps, salty bodies of water or even hot, acidic environments like hot springs and geysers.

Eukarya are unicellular to multicellular organisms, whose cells have a membrane-bounded nucleus. This domain contains the kingdoms **Protista, Fungi, Plantae,** and **Animalia** as illustrated in Figure 27.23.

The three-domain system of classification is based on biochemical differences that show there are three vastly different groups of organisms.

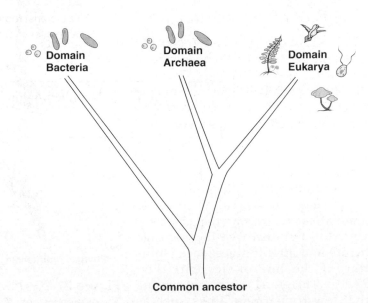

Figure 27.22   Three-domain system of classification.
Bacteria and archaea (both prokaryotes) are in separate domains. All eukaryotes (protists, fungi, plants, and animals) are in the domain Eukarya.

Figure 27.23 **The three domains of life.**
This pictorial representation of the domains Bacteria, Archaea, and Eukarya includes an example for each of the four kingdoms in the domain Eukarya.

Table 27.3	Classification Criteria for Three Domains				
Domains Bacteria and Archaea		**Domain Eukarya**			
		Kingdom Protista	*Kingdom Fungi*	*Kingdom Plantae*	*Kingdom Animalia*
Type of cell	Prokaryotic	Eukaryotic	Eukaryotic	Eukaryotic	Eukaryotic
Complexity	Unicellular	Unicellular (usually)	Multicellular (usually)	Multicellular	Multicellular
Type of nutrition	Autotrophic or heterotrophic	Photosynthetic or heterotrophic by various	Heterotrophic saprotrophs	Photosynthetic	Heterotrophic by ingestion
Motility	Sometimes by flagella	Sometimes by flagella (or cilia)	Nonmotile	Nonmotile	Motile by contractile fibers
Life cycle	Asexual (usually)	Various	Haplontic	Alternation of generations	Diplontic
Internal protection of zygote	No	No	No	Yes	Yes
Nervous system	None	Conduction of stimuli	None	None	Present

Bioethical Focus   The Theory of Evolution

The term "theory" in science is reserved for those ideas that scientists have found to be all-encompassing because they are based on data collected in a number of different fields. Evolution is a scientific theory. So is the cell theory, which says that all organisms are composed of cells, and so is the atomic theory, which says that all matter is composed of atoms. No one argues that schools should teach alternatives to the cell theory or the atomic theory. Yet confusion reigns over the use of the expression "the theory of evolution."

No wonder most scientists in our country are dismayed when state legislatures or school boards rule that teachers must put forward a variety of "theories" on the origin of life, including one that runs contrary to the mass of data that supports the theory of evolution. An organization in California called the Institute for Creation Research advocates that students be taught an "intelligent-design theory," which says that DNA could never have arisen without the involvement of an "intelligent agent," and that gaps in the fossil record mean that species arose fully developed with no antecedents.

Since our country forbids the mingling of church and state—no purely religious ideas can be taught in the schools—the advocates for an intelligent-design theory are careful not to mention the Bible or any strictly religious ideas (i.e., God created the world in seven days). Still, teachers who have a solid scientific background do not feel comfortable teaching an intelligent-design theory because it does not meet the test of a scientific theory. Science is based on hypotheses that have been tested by observation and/or experimentation. A scientific theory has stood the test of time—that is, no hypotheses have been supported by observation and/or experimentation that run contrary to the theory. On the contrary, the theory of evolution is supported by data collected in such wide-ranging fields as development, anatomy, geology, and biochemistry.

The polls consistently show that nearly half of all Americans prefer to believe the Old Testament account of creation. That, of course, is their right, but should schools be required to teach an intelligent-design theory that traces its roots back to the Old Testament and is not supported by observation and experimentation?

Decide Your Opinion

1. Should teachers be required to teach an intelligent-design theory of the origin of life in schools? Why or why not?
2. Should schools rightly teach that science is based on data collected by the testing of hypotheses by observation and experimentation? Why or why not?
3. Should schools be required to show that the intelligent-design theory does not meet the test of being scientific? Why or why not?

Summarizing the Concepts

27.1 Evidence of Evolution
The fossil record and biogeography, as well as comparative anatomy, development, and biochemistry, all give evidence of evolution. The fossil record gives us the history of life in general and allows us to trace the descent of a particular group. Biogeography shows that the distribution of organisms on earth is explainable by assuming organisms evolved in one locale. Comparing the anatomy and the development of organisms reveals homologous structures among those that are closely related. All organisms have certain biochemical molecules in common, and chemical similarities indicate the degree of relatedness.

27.2 Origin of Life
A chemical evolution is believed to have resulted in the first cell(s). Inorganic chemicals, probably derived from the primitive atmosphere, reacted to form small organic molecules. These reactions occurred in the ocean, either on the surface or in the region of hydrothermal vents, deep within the ocean.

After small organic molecules such as glucose, amino acids, and nucleotides arose, they polymerized to form the macromolecules. Amino acids joined to form proteins, and nucleotides joined to form nucleic acids. Perhaps RNA was the first nucleic acid. The RNA-first hypothesis is supported by the discovery of ribozymes, RNA enzymes. The protein-first hypothesis is supported by the observation that amino acids polymerize abiotically when exposed to dry heat.

Once there was a plasma membrane, the protocell came into being. Eventually, the DNA → RNA → protein self-replicating system evolved, and a true cell came into being.

27.3 Process of Evolution
Evolution is described as a process that involves a change in gene frequencies within the gene pool of a sexually reproducing population. The Hardy-Weinberg law states that the gene pool frequencies arrive at an equilibrium that is maintained generation after generation unless disrupted by mutations, genetic drift, gene flow, nonrandom mating, or natural selection. Any change from the initial allele frequencies in the gene pool of a population signifies that evolution has occurred.

27.4 Speciation
Speciation is the origin of species. This usually requires geographic isolation, followed by reproductive isolation. The evolution of several species of finches on the Galápagos Islands is an example of adaptive radiation because each one has a different way of life.

Currently, there are two hypotheses about the pace of speciation. Traditionalists support phyletic gradualism—slow, steady change leading to speciation. A new model, punctuated equilibrium, says that a long period of stasis is interrupted by speciation.

27.5 Classification
Classification involves the assignment of species to a hierarchy of categories: species, genus, family, order, class, phylum, kingdom, and in this text, domain. The five-kingdom system of classification recognizes these kingdoms: Monera (the bacteria), Protista (e.g., algae, protozoa), Fungi, Plantae, and Animalia. The three-domain system used by this text is based on molecular data and recognizes three domains: Bacteria, Archaea, and Eukarya. Both bacteria and archaea are prokaryotes. Members of the kingdoms Protista, Fungi, Plantae, and Animalia are eukaryotes.

Testing Yourself

Choose the best answer for each question.

1. The fossil record offers direct evidence for common descent because you can
   a. see that the types of fossils change over time.
   b. sometimes find common ancestors.
   c. trace the ancestry of a particular group.
   d. Only b and c are correct.
   e. All of these are correct.

2. Organisms adapted to the same way of life
   a. always have homologous structures.
   b. do not need to have homologous structures.
   c. always live in the same biogeographical region.
   d. Both a and c are correct.
   e. Both b and c are correct.

3. If evolution occurs, we would expect different biogeographical regions with similar environments to
   a. all contain the same mix of plants and animals.
   b. have all land masses connected.
   c. each have its own specific mix of plants and animals.
   d. have plants and animals that have similar adaptations.
   e. Both c and d are correct.

4. Continental drift helps explain the
   a. occurrence of mass extinctions.
   b. distribution of fossils on earth.
   c. geological upheavals such as earthquakes.
   d. Only a and b are correct.
   e. All of these are correct.

5. Which of these did Stanley Miller place in his experimental system to show that organic molecules could have arisen from inorganic molecules on the primitive earth?
   a. microspheres
   b. clay and water
   c. purines and pyrimidines
   d. the primitive gases
   e. All of these are correct.

6. Which of these is the chief reason the protocell was probably a fermenter?
   a. It didn't have any enzymes.
   b. It didn't have a nucleus.
   c. The atmosphere didn't have any oxygen.
   d. Fermentation provides the greatest amount of energy.
   e. All of these are correct.

7. Evolution of the DNA → RNA → protein system was a milestone because the protocell
   a. was a heterotrophic fermenter.
   b. could now reproduce.
   c. lived in the ocean.
   d. needed energy to grow.
   e. All of these are correct.

8. Assuming a Hardy-Weinberg equilibrium, 21% of a population is homozygous dominant, 50% is heterozygous, and 29% is homozygous recessive. What percentage of the next generation is predicted to be homozygous recessive?
   a. 21%   d. 25%
   b. 50%   e. 42%
   c. 29%

9. Which of the following cannot occur if a population is to maintain an equilibrium of allele frequencies?
   a. People leave one country and relocate in another.
   b. A disease wipes out the majority of a herd of deer.
   c. Members of an Indian tribe only allow the two tallest people in the tribe to marry each spring.
   d. Large black rats are the preferred males in a population of rats.
   e. All of these are correct.

10. A human population has a higher-than-usual percentage of individuals with a genetic disease. The most likely explanation is
    a. gene flow.   d. genetic drift.
    b. stabilizing selection.   e. All of these are correct.
    c. directional selection.

11. Which of these is/are necessary to natural selection?
    a. variations   d. Only b and c are correct.
    b. differential reproduction   e. All of these are correct.
    c. inheritance of differences

12. Which of these is a premating isolating mechanism?
    a. habitat isolation   d. hybrid sterility
    b. temporal isolation   e. Both a and b are correct.
    c. gamete isolation

13. Allopatric but not sympatric speciation requires
    a. reproductive isolation.   d. prior hybridization.
    b. geographic isolation.   e. rapid rate of mutation.
    c. spontaneous differences in males and females.

14. The many species of Galápagos finches were each adapted to eating different foods. This is an example of
    a. gene flow.   d. Only b and c are correct.
    b. adaptive radiation.   e. All of these are correct.
    c. sympatric speciation.

15. The classification category below the level of family is
    a. class.   d. phylum.
    b. order.   e. genus.
    c. species.

16. Which kingdom is mismatched?
    a. Protista—domain Bacteria
    b. Protista—multicellular algae
    c. Plantae—flowers and mosses
    d. Animalia—arthropods and humans
    e. Fungi—molds and mushrooms

17. The following diagrams represent an original distribution of phenotypes in a population. Superimpose on each of these diagrams another one to show that in (a) disruptive selection has occurred; in (b) stabilizing selection has occurred; and in (c) directional selection has occurred.

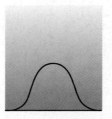 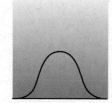 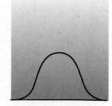

a. Disruptive selection   **b.** Stabilizing selection   **c.** Directional selection

e-Learning Connection   www.mhhe.com/maderinquiry10

Concepts	Questions	Media Resources*
27.1 Evidence of Evolution		
• Many fields of biology provide evidence that common descent has occurred.	1. List the categories of evidence in support of evolution. 2. How long have unicellular prokaryotes existed compared to multicellular eukaryotes?	Essential Study Partner 　Evidence for Evolution 　Geologic Time 　Fossils 　Continental Drift Art Quizzes 　Cytochrome *c* Evolution Animation Quizzes 　Plate Tectonics 　Molecular Clock
27.2 Origin of Life		
• A chemical evolution proceeded from small organic molecules to macromolecules to protocell(s). • The first cell was bounded by a membrane and contained a replicating system, consisting of DNA, RNA, and proteins.	1. How was the primitive atmosphere, that allowed simple organic molecules to evolve, different from the atmosphere of earth today? 2. Considering the origin of cells, which likely arose first—autotroph or heterotroph?	Essential Study Partner 　Origin of Life 　Key Events Art Quizzes 　Miller-Urey Experiment 　Miller-Urey Experiment 　　Results 　Current Bubble Hypothesis
27.3 Process of Evolution		
• Populations evolve and not individuals; evolution is defined in terms of population genetic changes. • There are several agents of evolutionary change, one of which is natural selection. • Natural selection is the mechanism resulting in adaptation to the environment.	1. List the five agents of evolutionary change. 2. How do pesticides and antibiotics lead to resistant insects and bacteria, respectively?	Essential Study Partner 　Natural Selection 　Types of Selection 　Variation 　Adaptation 　Before Darwin
27.4 Speciation		
• New species come about when populations are reproductively isolated from other similar populations. • Species are classified into seven hierarchical categories according to shared characteristics. At each level above species, there are more and more species in the category.	1. How do subpopulations become separate species? 2. What are the two hypotheses about the pace of evolution and how do they differ?	Essential Study Partner 　Introduction 　Allopatric Speciation 　Sympatric Speciation 　Hierarchies Art Quizzes 　Evolutionary Relationships 　　Among Kingdoms
27.5 Classification		
• Species are classified into seven hierarchical categories according to shared characteristics. At each level above species, there are more and more types of species in the category.		Essential Study Partner 　Hierarchies 　Kingdoms 　Three Domains Art Quizzes 　Taxonomic Hierarchy 　Evolutionary Relationships 　　Among Kingdoms

*For additional Media Resources, see the Online Learning Center.

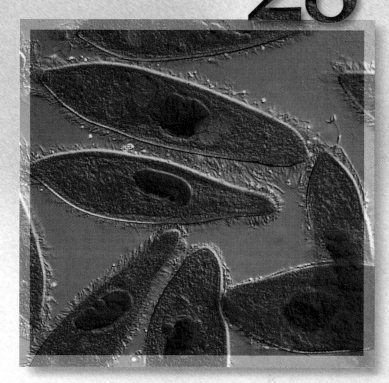

CHAPTER 28

Microbiology

Chapter Concepts

A skilled microscopist has captured the cilia and internal anatomy of these unicellular paramecia. Notice the obvious nucleus; paramecia are eukaryotes like ourselves.

Place a single drop of pond water under a microscope and you can see an amazing menagerie of unicellular organisms, such as many types of protists. Among protists are the motile protozoans, such as paramecia, and the colorful algae. Turn the magnification very high and you may be able to make out the even smaller prokaryotes, which abound in nature. Prokaryotes are found in all sort of environments, even water boiling out of the ocean's volcanic vents and within rocks miles below the surface of the earth. Closer to home, microbes can live in the crevices of your kitchen cutting board or within the food you're eating. Set a loaf of bread, free of preservatives, on the counter and you'll observe colonies of bacteria and fungi in no time.

Some viruses, bacteria, and fungi make a home inside the bodies of people. To combat these pathogens, researchers desperately hunt for drugs, such as antiviral agents which kill viruses and antibiotics, which kill bacteria and fungi. Some antibiotics are extracted from microbes themselves, and scientists are now hopeful of reviving the decades-old idea of using bacteria-destroying viruses against bacteria.

This chapter will introduce you to the viruses and the smallest of the organisms that are members of the domains discussed in the previous chapter: domains Bacteria, Archaea, and Eukarya.

28.1 Viruses

Viruses are not included in the classification table found in Appendix B because they are noncellular and should not be classified with organisms which are always cellular. Viruses are generally smaller than 200 nm in diameter and therefore are comparable in size to a large protein macromolecule. Many can be purified and crystallized, and the crystals can be stored just as chemicals are stored.

Structure of Viruses

A virus always has at least two parts: an outer capsid composed of protein units, and an inner core of nucleic acid—either DNA or RNA (Fig. 28.1). The viral genome has at most several hundred genes; a human cell, as you know, contains thousands of genes. A virus may also contain various enzymes for nucleic acid replication. The capsid is often surrounded by an outer membranous envelope, which is actually partially composed of their host's plasma membrane. The classification of viruses is based on (1) type of nucleic acid, including whether it is single stranded or double stranded, (2) viral size and shape, and (3) the presence or absence of an outer envelope.

Parasitic Nature

Viruses are *obligate intracellular parasites*. Viruses infect all sorts of cells—from bacterial cells to human cells—but each type is very specific. For example, bacteriophages infect only bacteria, the tobacco mosaic virus infects only plants, and the rabies virus infects only mammals. Some human viruses even specialize in a particular tissue. Human immunodeficiency viruses (HIV) enter specific types of blood cells, the polio virus reproduces in spinal nerve cells, and the hepatitis viruses infect only liver cells. In order to have a ready supply of animal viruses in the laboratory, they are sometimes injected into live chick embryos. Today, however, it is more customary to infect cells that are maintained in tissue culture.

The remarkable parasite–host-cell relationship exemplified by viruses has led to the hypothesis that the nucleic acids of viruses are derived from host-cell genomes! If so, viruses must have evolved after cells came into existence, and new viruses may be evolving even now.

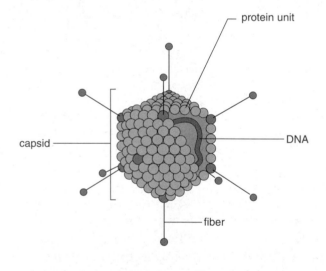
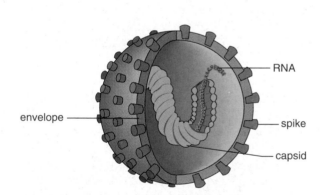
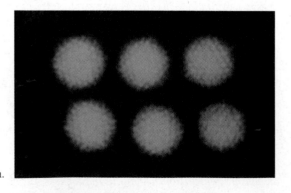
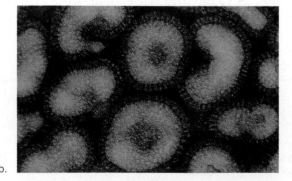

Figure 28.1 Adenovirus.
a. An adenovirus is a DNA virus with a polyhedral capsid and a fiber at each corner. **b.** An influenza virus is an RNA virus with a helical capsid surrounded by an envelope with glycoprotein spikes.

Viruses, like organisms, can mutate, and this habit can be quite troublesome because a vaccine that is effective today may not be effective tomorrow. Flu viruses are well known for mutating, and this is why you have to have a flu shot every year—antibodies generated from last year's shot are not expected to be effective this year.

Viruses are nonliving particles that reproduce and mutate only inside specific host cells.

Replication of Viruses

Viruses are specific to a particular host cell because portions of the capsid (or the spikes of the envelope) bind in a lock-and-key manner with a receptor on the host-cell plasma membrane. After viral nucleic acid enters the cell, it *takes over the metabolic machinery of the host cell* so that more viruses are produced.

Replication of Bacteriophages

Bacteriophages, or simply phages, are viruses that parasitize bacteria; the bacterium in Figure 28.2 could be *Escherichia coli,* which lives in our intestines. The bacteriophage, termed lambda, is capable of carrying out two cycles. In the lytic cycle, the host cell undergoes *lysis,* a breaking open of the cell to release new viruses. In the lysogenic cycle, viral replication does not immediately occur, but replication may take place sometime in the future.

Lytic Cycle The **lytic cycle** may be divided into five stages: attachment, penetration, biosynthesis, maturation, and release. During attachment, portions of the capsid combine with a receptor on the rigid bacterial cell wall in a lock-and-key manner. During penetration, a viral enzyme digests away part of the cell wall, and viral DNA is injected into the bacterial cell. Biosynthesis of viral components begins after the virus brings about inactivation of host genes not necessary to viral replication. The virus takes over the machinery of the cell in order to carry out viral DNA replication and production of multiple copies of the capsid protein subunits. During maturation, viral DNA and capsids are assembled to produce several hundred viral particles. Lysozyme, an enzyme coded for by a viral gene, is produced; this disrupts the cell wall, and new viruses are released. The bacterial cell dies as a result.

During the lytic cycle, a bacteriophage takes over the machinery of the cell so that viral replication and release occur.

Lysogenic Cycle In the **lysogenic cycle,** the infected bacterium does not immediately produce viruses but may do so

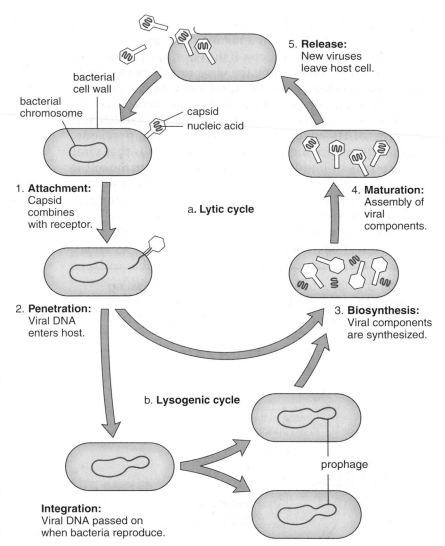

Figure 28.2 Lytic and lysogenic cycles.
a. In the lytic cycle, viral particles escape when the cell is lysed (broken open).
b. In the lysogenic cycle, viral DNA is integrated into host DNA. At some time in the future, the lysogenic cycle can be followed by the lytic cycle.

sometime in the future. In the meantime, the phage is latent—not actively replicating. Following attachment and penetration, viral DNA becomes integrated into bacterial DNA with no destruction of host DNA. While latent, the viral DNA is called a *prophage.* The prophage is replicated along with the host DNA, and all subsequent cells, called lysogenic cells, carry a copy of the prophage. Certain environmental factors, such as ultraviolet radiation, can induce the prophage to enter the lytic stage of biosynthesis, followed by maturation and release.

During the lysogenic cycle, the phage becomes a prophage that is integrated into the host genome. At a later time, the phage may reenter the lytic cycle and replicate itself.

Figure 28.15   Diversification among the brown algae.
Laminaria and *Fucus* are seaweeds known as kelps. They live along rocky coasts of the north temperate zone. *Sargassum*, the other brown alga shown, lives at sea where floating masses form a home for many organisms.

harvested for human food and for fertilizer. They are also a source of algin, a pectinlike material that is added to ice cream, sherbet, cream cheese, and other products to give them a stable, smooth consistency.

Diatoms are a type of unicellular golden brown algae (phylum Chrysophyta, 11,000 species). The structure of a diatom is often compared to a box because the cell wall has two halves, or valves, with the larger valve acting as a "lid" for the smaller valve (Fig. 28.16*a*). When diatoms reproduce, each receives only one old valve. The new valve fits inside the old one.

The cell wall of a diatom has an outer layer of silica, a common ingredient of glass. The valves are covered with a great variety of striations and markings, which form beautiful patterns when observed under the microscope. Diatoms are among the most numerous of all unicellular algae in the oceans. As such, they serve as an important source of food for other organisms. In addition, they produce a major portion of the earth's oxygen supply. In ancient times, diatoms were also present in astronomical numbers. Their remains, raised above sea level by geological upheavals, are now mined as diatomaceous earth for use as filtering agents, soundproofing materials, and scouring powders.

Dinoflagellates

Many **dinoflagellates** (phylum Pyrrophyta, 1,000 species) are bounded by protective cellulose plates (Fig. 28.16*b*). Most have two flagella; one is free, but the other is located in a transverse groove. The beating of the flagella causes the organism to spin like a top. Occasionally, when surface waters are warm and nutrients are high, there are so many of these unicellular organisms in the ocean that they cause a condition called "red tide." Toxins in red tides cause widespread fish kills and can paralyze humans who eat shellfish that have fed on the dinoflagellates.

Dinoflagellates are an important source of food for small animals in the ocean. They also live as symbiotes within the bodies of some invertebrates. For example, because corals usually contain large numbers of dinoflagellates, corals grow much faster than they would otherwise.

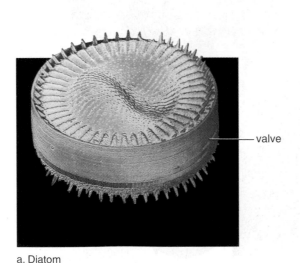

a. Diatom

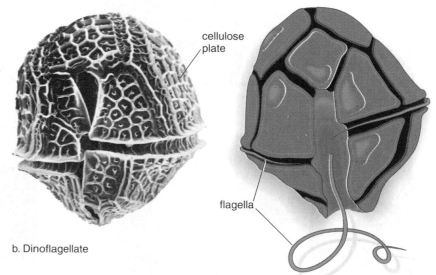

b. Dinoflagellate

Figure 28.16   Diatoms and dinoflagellates.
a. Diatoms may be variously colored, but their chloroplasts contain a unique golden brown pigment in addition to chlorophylls *a* and *c*. The beautiful pattern results from markings on the silica-embedded wall. **b.** Dinoflagellates have an internal skeleton of cellulose plates *(left)* and two flagella *(right)*. Gonyaulax shown here contains a red pigment and is responsible for occasional "red tides."

Euglenoids

Euglenoids (phylum Euglenophyta, 1,000 species) are small (10–500 μm) freshwater unicellular organisms that typify the problem of classifying protists. One-third of all genera have chloroplasts; the rest do not. Those that lack chloroplasts ingest or absorb their food. Euglenoids grown in the absence of light have been known to lose their chloroplasts and become heterotrophic. The chloroplasts are surrounded by three rather than two membranes. A structure called the pyrenoid produces an unusual type of carbohydrate polymer (paramylon) not seen in green algae.

Euglenoids have two flagella, one of which is typically much longer than the other and projects out of an anterior vase-shaped invagination (Fig. 28.17). It is called a tinsel flagellum because it has hairs on it. Near the base of this flagellum is an eyespot, which shades a photoreceptor for detecting light. Because euglenoids are bounded by a flexible *pellicle* composed of protein strips lying side by side, they can assume different shapes as the underlying cytoplasm undulates and contracts. As in certain protozoans, there is a contractile vacuole for ridding the body of excess water. Euglenoids reproduce by longitudinal cell division, and sexual reproduction is not known to occur.

Euglenoids have both plant- and animal-like characteristics. They have chloroplasts but lack a cell wall and swim by means of flagella.

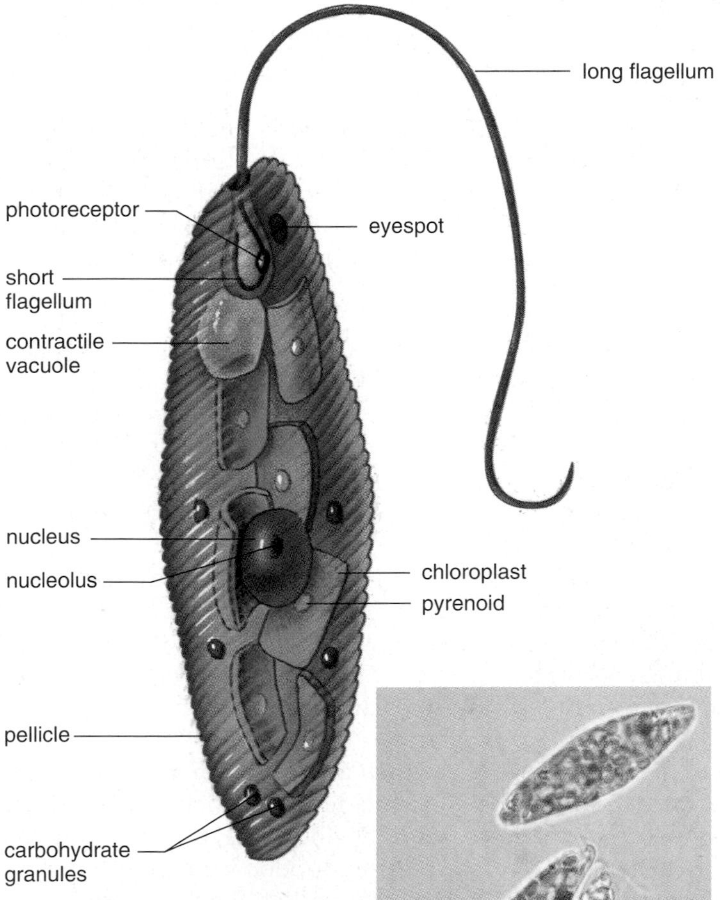

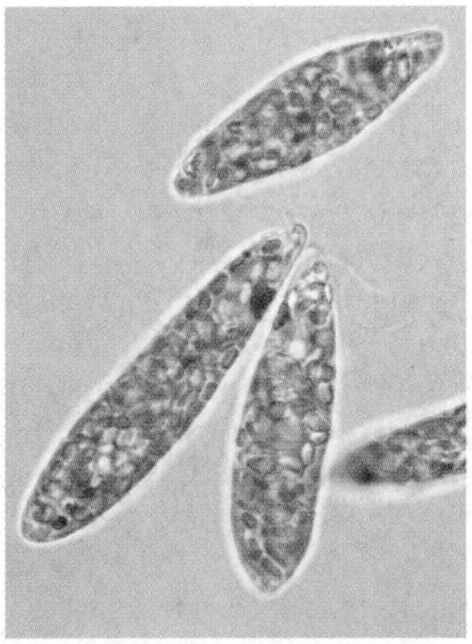

Red Algae

Like the brown algae, **red algae** (phylum Rhodophyta, 4,000 species) are multicellular, but they live chiefly in warmer seawater, growing in both shallow and deep waters. Red algae are usually much smaller and more delicate than brown algae, although they can be up to a meter long. Some forms of red algae are simple filaments, but more often they are complexly branched, with the branches having a feathery, flat, or expanded, ribbonlike appearance (Fig. 28.18). Coralline algae are red algae that have cell walls impregnated with calcium carbonate. In some instances, they contribute as much to the growth of coral reefs as do coral animals.

Like brown algae, red algae are seaweeds of economic importance. The mucilaginous material in the cell walls of certain genera of red algae is a source of agar used commercially to make capsules for vitamins and drugs, as a material for making dental impressions, and as a base for cosmetics. In the laboratory, agar is a culture medium for bacteria. When purified, it becomes the gel for electrophoresis, a procedure that separates proteins and nucleotides. Agar is also used in food preparation, both as an antidrying agent for baked goods and to make jellies and desserts set rapidly.

Many red algae have filamentous branches or are multicellular.

Figure 28.17 *Euglena*.
a. In *Euglena*, a very long flagellum propels the body, which is enveloped by a flexible pellicle. A photoreceptor shaded by an eyespot allows *Euglena* to find light, after which photosynthesis can occur in the numerous chloroplasts. Pyrenoids synthesize a reserve carbohydrate, which is stored in the chloroplasts and also in the cytoplasm. **b.** Micrograph of several specimens.

Figure 28.18 Red alga.
Red algae are smaller and more delicate than brown algae.

Protozoans

Protozoans are typically heterotrophic, motile, unicellular organisms of small size (2–1,000 µm). They are not animals because animals in the classification used by this text are multicellular and undergo embryonic development.

Protozoans usually live in water, but they can also be found in moist soil or inside other organisms. Some protozoans engulf whole food and are termed holozoic; others are saprotrophic and absorb nutrient molecules across the plasma membrane. Still others are parasitic and are responsible for several significant human infections.

Most protozoans are unicellular, but they should not be considered simple organisms. Each cell alone must carry out all the functions performed by specialized tissues and organs in more complex organisms. They have organelles for purposes we have not seen before. For example, their food is digested inside food vacuoles, and freshwater protozoans have "contractile" vacuoles for the elimination of water. Although asexual reproduction involving binary fission and mitosis is the rule, many protozoans also reproduce sexually during some part of their life cycle. The protozoans we will study can be placed in four groups according to their type of locomotor organelle:

Name	Locomotion	Example
Amoeboids	Pseudopods	*Amoeba*
Ciliates	Cilia	*Paramecium*
Zooflagellates	Flagella	*Trypanosoma*
Sporozoans	No locomotion	*Plasmodium*

Amoeboids

The amoeboids (phylum Rhizopoda, 40,000 species) are protists that move and engulf their prey with **pseudopods.** *Amoeba proteus* is a commonly studied freshwater member of this group (Fig. 28.19a). When amoeboids feed, they **phagocytize;** the pseudopods surround and engulf the prey, which may be algae, bacteria, or other protozoans. Digestion then occurs within a food vacuole. Some white blood cells in humans are amoeboid, and they phagocytize debris, parasites, and worn-out cells. Freshwater amoeboids, including *Amoeba proteus,* also have contractile vacuoles where excess water from the cytoplasm collects before the vacuole appears to "contract," releasing the water through a temporary opening in the plasma membrane.

Entamoeba histolytica is a parasite that can infect the human intestine and cause amoebic dysentery. Complications arise when this parasite invades the intestinal lining and reproduces there. If the parasites enter the body proper, the resulting liver and brain impairment can be fatal.

The foraminifera, which are largely marine, have an external calcareous shell (made up of calcium carbonate) with foramina, holes through which long, thin pseudopods extend (Fig. 28.19b). The pseudopods branch and join to form a net where the prey is digested. Foraminifera live in the sediment of the ocean floor in incredible numbers—there may be as many as 50,000 shells in a single gram of sediment. Deposits for millions of years, followed by a geological upheaval, formed the White Cliffs of Dover along the southern coast of England. Also, the great Egyptian pyramids are built of foraminiferan limestone. Today, oil geologists look for foraminifera in sedimentary rock as an indicator of organic deposits, which are necessary for the formation of oil.

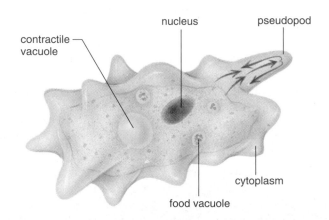

a.

b.

Figure 28.19   Amoeboid protozoans.
a. *Amoeba proteus* is common in freshwater ponds. Bacteria and other microorganisms are digested in food vacuoles, and contractile vacuoles rid the body of excess water. The arrows show the streaming of cytoplasm into and out of a projection, called a pseudopod.
b. Pseudopods of a live foraminiferan project through holes in the calcium carbonate shell. These shells were so numerous that they became a large part of the White Cliffs of Dover when a geological upheaval occurred.

Ciliates

The **ciliates** (phylum Ciliophora, 8,000 species), including those in the genus *Paramecium*, are the most complex of the protozoans (Fig. 28.20). Hundreds of cilia, which beat in a coordinated rhythmic manner, project through tiny holes in a semirigid outer covering, or pellicle. Numerous oval capsules lying in the cytoplasm just beneath the pellicle contain **trichocysts.** Upon mechanical or chemical stimulation, trichocysts discharge long, barbed threads, useful for defense and for capturing prey. When a paramecium feeds, food is swept down a gullet, below which food vacuoles form. Following digestion, the soluble nutrients are absorbed by the cytoplasm, and the indigestible residue is eliminated at the anal pore.

During asexual reproduction, ciliates divide by transverse binary fission. Ciliates have two types of nuclei: a large macronucleus and one or more small micronuclei. The macronucleus controls the normal metabolism of the cell; during sexual reproduction, two ciliates exchange a micronucleus.

The diversity of ciliates is quite remarkable. Barrelshaped didinia expand to consume paramecia much larger than themselves. Suctoria have tentacles, which they use like straws to suck their prey dry. *Stentor* looks like a blue vase decorated with stripes.

Zooflagellates

Protozoans that move by means of flagella are called **zooflagellates** (phylum Zoomastigophora, 1,500 species) to distinguish them from flagellated unicellular algae. Many zooflagellates enter into symbiotic relationships. *Trichonympha collaris* lives in the gut of termites; it contains a bacterium that enzymatically converts the cellulose of wood to soluble carbohydrates that are easily digested by the insect. *Giardia lamblia*, whose cysts are transmitted through contaminated water, causes severe diarrhea. *Trichomonas vaginalis*, a sexually transmitted organism, infects the vagina and urethra of women and the prostate, seminal vesicles, and urethra of men. A **trypanosome**, *Trypanosoma brucei*, transmitted by the bite of the tsetse fly, is the cause of African sleeping sickness (Fig. 28.21). The white blood cells in an infected animal accumulate around the blood vessels leading to the brain and cut off circulation. The lethargy characteristic of the disease is caused by an inadequate supply of oxygen to the brain.

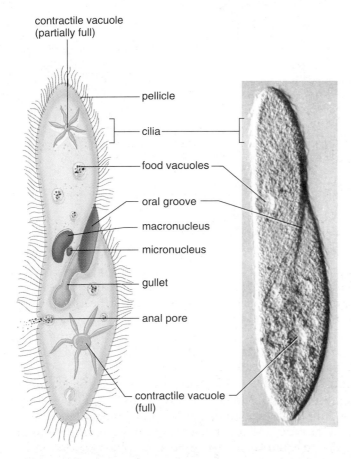

Figure 28.20   Ciliated protozoans.
Structure of *Paramecium*, adjacent to an electron micrograph. Ciliates are the most complex of the protozoans. Note the oral groove, the gullet, and the anal pore.

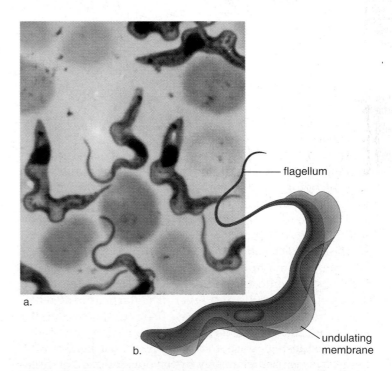

Figure 28.21   Zooflagellates.
a. Photograph of *Trypanosoma brucei*, the cause of African sleeping sickness, among red blood cells. **b.** The drawing shows its general structure.

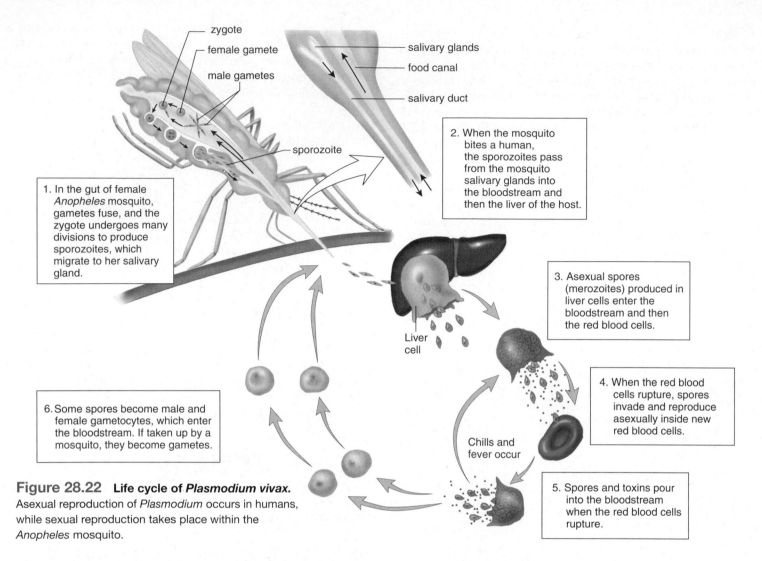

Figure 28.22   Life cycle of *Plasmodium vivax.*
Asexual reproduction of *Plasmodium* occurs in humans, while sexual reproduction takes place within the *Anopheles* mosquito.

Labels and step boxes within the figure:

zygote
female gamete
male gametes
salivary glands
food canal
salivary duct
sporozoite
Liver cell
Chills and fever occur

1. In the gut of female *Anopheles* mosquito, gametes fuse, and the zygote undergoes many divisions to produce sporozoites, which migrate to her salivary gland.

2. When the mosquito bites a human, the sporozoites pass from the mosquito salivary glands into the bloodstream and then the liver of the host.

3. Asexual spores (merozoites) produced in liver cells enter the bloodstream and then the red blood cells.

4. When the red blood cells rupture, spores invade and reproduce asexually inside new red blood cells.

5. Spores and toxins pour into the bloodstream when the red blood cells rupture.

6. Some spores become male and female gametocytes, which enter the bloodstream. If taken up by a mosquito, they become gametes.

Sporozoans

Sporozoans (phylum Apicomplexa, 3,600 species) are non-motile, but they cause serious diseases in humans. Their name recognizes that these organisms form spores at some point in their life cycle.

Pneumocystis carinii causes the type of pneumonia seen primarily in AIDS patients. The most widespread human parasite is *Plasmodium vivax*, the cause of one type of malaria. When a human is bitten by an infected female *Anopheles* mosquito, the parasite eventually invades the red blood cells. The chills and fever of malaria appear when the infected cells burst and release toxic substances into the blood (Fig. 28.22). Malaria is still a major killer of humans, despite extensive efforts to control it. A resurgence of the disease was caused primarily by the development of insecticide-resistant strains of mosquitoes and by parasites resistant to current antimalarial drugs.

The protozoans ingest their food and are motile. Protozoans are classified according to the type of locomotor organelle employed.

Water Molds and Slime Molds

Water molds (phylum Oomycota, 580 species) live in the water, where they parasitize fish, forming furry growths on their gills. Others live on land and parasitize insects and plants; a water mold was responsible for the 1840s potato famine in Ireland. Most water molds, like fungi, are saprotrophic and have a filamentous body, but they have the diplontic life cycle (see Fig. 28.10*c*), whereas fungi have the haplontic cycle (see Fig. 28.10*a*).

Slime molds (phylum Gymnomycota, 560 species) might look like molds of the kingdom Fungi (see section 28.4), but their vegetative state is amoeboid, whereas fungi are filamentous. Also, fungi are saprotrophic, whereas slime molds are heterotrophic by ingestion. When conditions are unfavorable to growth, however, slime molds produce and release spores that are resistant to environmental extremes. Fungi also produce such spores.

Usually plasmodial slime molds exist as a **plasmodium,** a diploid multinucleated cytoplasmic mass enveloped by a slime sheath that creeps along, phagocytizing decaying plant material in a forest or agricultural field (Fig. 28.23). Under unfavorable conditions, the plasmodium forms sporangia, structures that produce spores which are dispersed by the wind. The spores germinate to produce gametes that join to form a zygote. This zygote begins the cycle again. Cellular slime molds, as you might expect, exist as individual amoeboid cells. Each lives by phagocytizing bacteria and yeasts. As the food supply runs out, the cells release a chemical that causes them to aggregate into a pseudoplasmodium that produces spores within sporangia.

Water molds and slime molds have a unique combination of traits that distinguish them from fungi.

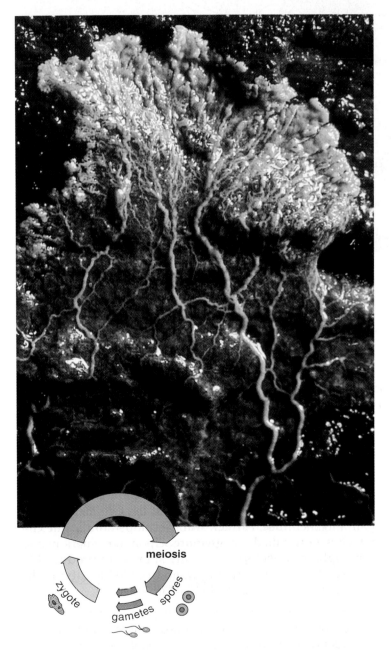

Figure 28.23 Plasmodium and life cycle of a yellow slime mold, *Hemitrichia stipitata*.

28.4 The Fungi

Fungi are members of the domain Eukarya and the kingdom Fungi. They are multicellular eukaryotes that are heterotrophic by absorption. They send out digestive enzymes into the immediate environment, and then, when organic matter is broken down, they absorb nutrient molecules. Like bacteria, most fungi are saprotrophic decomposers that break down the waste products and dead remains of plants and animals.

Although yeasts are unicellular fungi, the body of a fungus is usually a multicellular structure known as a mycelium. A **mycelium** is a network of filaments called hyphae (sing., **hypha**):

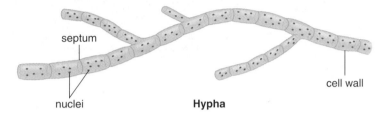

Fungal cells are quite different from plant cells, not only by lacking chloroplasts but also by having a cell wall that contains **chitin,** not cellulose. Chitin is a polymer of glucose, but each glucose molecule has an amino group attached to it. (Chitin is also found in the external skeletons of insects and all arthropods.) How can a nonmotile terrestrial organism such as a fungus ensure that the species will be dispersed to new locations? Fungi produce nonflagellated spores during both sexual and asexual reproduction, which are dispersed by the wind.

Classification

Domain Eukarya, Kingdom Fungi

Multicellular eukaryotes; heterotrophic by absorption; lack flagella; nonmotile spores form during both asexual and sexual reproduction

Division Zygomycota: zygospore fungi

Soil and dung molds, black bread molds (*Rhizopus*)

Division Ascomycota: sac fungi

Many small, wood-decaying fungi, yeasts (*Saccharomyces*), molds (*Neurospora*), morels, cup fungi, truffles; plant parasites: powdery mildews, ergots

Division Basidiomycota: club fungi

Mushrooms, stinkhorns, puffballs, bracket and shelf fungi, coral fungi; plant parasites: rusts, smuts

Division Deuteromycota: imperfect fungi (i.e., no known means of sexual reproduction)

Athlete's foot, ringworm, candidiasis

29.3 Seedless Vascular Plants

All the other plants we will study are **vascular plants.** Vascular tissue in these plants consists of **xylem,** which conducts water and minerals up from the soil, and **phloem,** which transports organic nutrients from one part of the plant to another. The vascular plants usually have true roots, stems, and leaves. The roots absorb water from the soil, and the stem conducts water to the leaves. Xylem, with its strong-walled cells, supports the body of the plant against the pull of gravity. The leaves are fully covered by a waxy cuticle except where it is interrupted by stomata, little pores whose size can be regulated to control water loss.

Figure 29.5   **The Carboniferous period.**
Growing in the swamp forests of the Carboniferous period were treelike club mosses *(left)*, treelike horsetails *(right)*, and lower, fernlike foliage *(left)*. When the trees fell, they were covered by water and did not decompose well. Sediment built up and turned to rock, whose pressure caused the organic material to become coal, a fossil fuel which still helps run our industrialized society today.

The sporophyte is the dominant generation in vascular plants. This is advantageous because the sporophyte is the generation with vascular tissue. Another advantage of having a dominant sporophyte relates to its being diploid. If a faulty gene is present, it can be masked by a functional gene. Then, too, the greater the amount of genetic material, the greater the possibility of mutations that will lead to increased variety and complexity. Indeed, vascular plants are complex, extremely varied, and widely distributed.

Ferns and Their Allies

Some vascular plants do not produce seeds. Seedless vascular plants include whisk ferns, club mosses, horsetails, and ferns, which disperse the species by producing wind-blown spores. When the spores germinate, they produce a relatively large gametophyte that is independent of the sporophyte for its nutrition. In these plants, flagellated sperm are released by antheridia and swim in a film of external water to the archegonia, where fertilization occurs. Because spores disperse the species and the nonvascular gametophyte is independent of the sporophyte, these plants cannot wholly benefit from the adaptations of the sporophyte to a terrestrial environment.

The seedless vascular plants formed the great swamp forests of the Carboniferous period (Fig. 29.5). A large number of these plants died but did not decompose completely. Instead, they were compressed to form the coal that we still

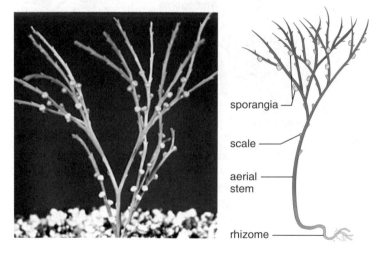

Figure 29.6   **Whisk fern, *Psilotum*.**
Whisk ferns have no roots or leaves—the branches carry on photosynthesis. The sporangia are yellow.

mine and burn today. (Oil has a similar origin, but it most likely formed in marine sedimentary rocks and included animal remains.)

Whisk Ferns

The **psilotophytes** (division Psilotophyta, several species) are represented by *Psilotum*, a whisk fern (Fig. 29.6). Whisk ferns, named for their resemblance to whisk brooms, are found in Arizona, Texas, Louisiana, and Florida, as well as Hawaii and Puerto Rico. *Psilotum* looks like a rhyniophyte,

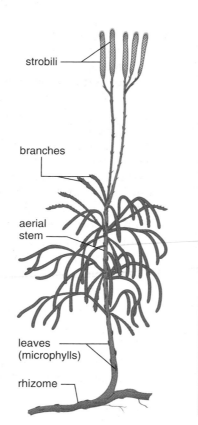

Figure 29.7   Club moss, *Lycopodium*.
Green photosynthetic stems are covered by scalelike leaves, and sporangia are found on leaves arranged as strobili.

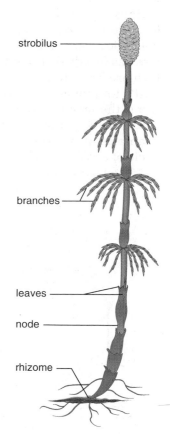

Figure 29.8   Horsetail, *Equisetum*.
Whorls of branches and tiny leaves occur at the joints of the stem. The sporangia are borne in strobili.

a vascular plant that is known only from the fossil record. An erect stem forks repeatedly and is attached to a *rhizome*, a fleshy horizontal stem that lies underground. There are no leaves, and the branches carry on photosynthesis. Sporangia located at the ends of short branches produce spores that disperse the species. The independent gametophyte, which is found underground and is penetrated by a mycorrhizal fungus, produces flagellated sperm.

Club Mosses

The **club mosses** (division Lycopodophyta, 1,000 species) are common in moist woodlands of the temperate zone where they are known as ground pines. Typically, a branching rhizome sends up aerial stems less than 30 cm tall. Tightly packed, scalelike leaves cover stems and branches, giving the plant a mossy look (Fig. 29.7). In club mosses, the sporangia are borne on terminal clusters of leaves, called *strobili* (sing., strobilus), which are club-shaped.

The majority of club mosses live in the tropics and subtropics where many of them are epiphytes—plants that live on, but do not parasitize, trees. The closely related spike mosses *(Selaginella)* are extremely varied and include the

resurrection plant, which curls up into a tight ball when dry but unfurls as if by magic when moistened.

Horsetails

Horsetails (division Equisetophyta, 15 species), which thrive in moist habitats around the globe, are represented by *Equisetum*, the only genus in existence today (Fig. 29.8). A rhizome produces aerial stems that stand about 1.3 meters. In some species, the whorls of slender green side branches at the joints (nodes) of the stem make the plant bear a fanciful resemblance to a horse's tail. The leaves are small and scalelike. Many horsetails have strobili at the tips of branch-bearing stems; others send up special buff-colored, naked stems that bear the strobili.

The stems are tough and rigid because of silica deposited in cell walls. Early Americans, in particular, used horsetails to scour pots and called them "scouring rushes." Today, they are still used as ingredients in a few abrasive powders.

Ferns

Ferns (division Pteridophyta, 12,000 species) are a widespread group of plants. They are most abundant in warm,

30.3 Molluscs

Molluscs, along with annelids and arthropods, are protostomes (Fig. 30.10). In protostomes, the first embryonic opening becomes the mouth. Because the true coelom (see Fig. 30.8c) forms by the splitting of the mesoderm, protostomes are also schizocoelomates. Certain (but not all) protostomes, have a trochophore larva (top-shaped with a band of cilia at the midsection).

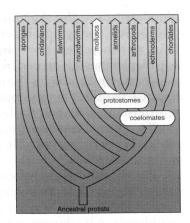

Characteristics of Molluscs

Molluscs (phylum Mollusca, about 110,000 species) are a very large and diversified group; however, they all have a body composed of at least three distinct parts:

1. *Visceral mass,* the soft-bodied portion that contains internal organs.
2. *Foot,* the strong, muscular portion used for locomotion.
3. *Mantle,* the membranous or sometimes muscular covering that envelops but does not completely enclose the visceral mass. The *mantle cavity* is the space between the two folds of the mantle. The mantle may secrete a shell.

In addition to these three parts, many molluscs have a head region with eyes and other sense organs.

The division of the body into distinct areas may have contributed to diversification of animals in this phylum. There are many different types of molluscs adapted to various ways of life (Fig. 30.11). Molluscan groups can be distinguished by a modification of the foot. In the **gastropods** (meaning stomach-footed), including nudibranchs, conchs, and snails, the foot is ventrally flattened, and the animal moves by muscle contractions that pass along the foot. While nudibranchs, also called sea slugs, lack a shell, conchs and snails have a coiled shell in which the visceral mass spirals. Some types of snails are adapted to life on land. For example, their mantle is richly supplied with blood vessels and functions as a lung when air is moved in and out through respiratory pores.

Cephalopods

In **cephalopods** (meaning head-footed), including octopuses and squids, the foot has evolved into tentacles about the head. Aside from the tentacles, which seize prey, cephalopods have a powerful beak and a radula (toothy tongue) to tear prey apart. Cephalization aids these animals in recognizing prey and in escaping enemies. The eyes are superficially similar to those of vertebrates—they have a lens and a retina with photoreceptors. However, the eye is con-

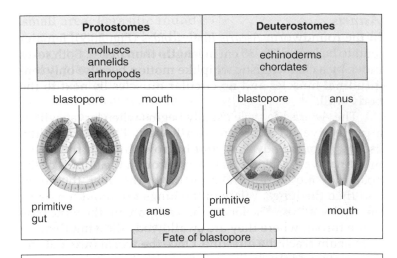

Fate of blastopore

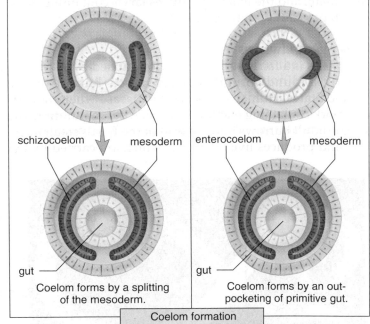

Coelom formation

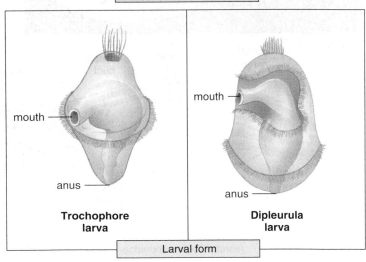

Larval form

Figure 30.10   Protostomes versus deuterostomes.
In protostomes, the first embryonic opening, called the blastopore, becomes the mouth. The deuterostomes will be discussed in Chapter 31.

a. Chiton, *Tonicella*

b. Chambered nautilus, *Nautilus*

c. Scallop, *Pecten*

d. Spanish shawl nudibranch, *Flabellina*

Figure 30.11   Molluscan diversity.
a. A chiton has a flattened foot and a shell that consists of eight articulating valves. **b.** A chambered nautilus achieves buoyancy by regulating the amount of air in the chambers of its shell. **c.** A scallop has sensory tentacles extended between the valves. **d.** A nudibranch (sea slug) lacks a shell, gills, and a mantle cavity. Dorsal projections function in gas exchange.

structed so differently from the vertebrate eye that we believe the so-called camera-type eye actually evolved twice—once in the molluscs and once in the vertebrates. In cephalopods, the brain is formed from a fusion of ganglia, and nerves leaving the brain supply various parts of the body. An especially large pair of nerves controls the rapid contraction of the mantle, allowing these animals to move quickly by a jet propulsion of water. Rapid movement and the secretion of a brown or black pigment from an ink gland help cephalopods escape their enemies. Octopuses have no shell, and squid have only a remnant of one concealed beneath the skin.

Bivalves

Clams, oysters, and scallops are called **bivalves** because there are two parts to their shells. In a clam, such as the freshwater clam *Anodonta*, the shell, secreted by the mantle, is composed of protein and calcium carbonate, with an inner layer of mother-of-pearl. If a foreign body is placed between the mantle and the shell, pearls form as concentric layers of shell are deposited about the particle.

The adductor muscles hold the valves of the shell together (Fig. 30.12). Within the mantle cavity, the gills, an organ for gas exchange in aquatic forms, hang down on either side of the visceral mass, which lies above the foot.

The clam is a filter feeder. Food particles and water enter the mantle cavity by way of the incurrent siphon, a posterior opening between the two valves. Mucous secretions cause smaller particles to adhere to the gills, and ciliary action sweeps them toward the mouth. This method of feeding does not require rapid movement.

The Visceral Mass

The heart of a clam lies just below the hump of the shell within the pericardial cavity, the only remains of the

a. Flat-backed millipede, *Sigmoria*

b. Tarantula, *Aphonopelma*

c. Dungeness crab, *Cancer*

d. Paper wasp, *Polistes*

e. Stone centipede, *Lithobius*

Figure 30.15   Arthropod diversity.
a. A millipede has only one pair of antennae, and the head is followed by a series of segments, each with two pairs of appendages. **b.** The hairy tarantulas of the genus *Aphonopelma* are dark in color and sluggish in movement. Their bite is harmless to people. **c.** A crab is a crustacean with a calcified exoskeleton, one pair of claws and four other pairs of walking legs. **d.** A wasp is an insect with two pairs of wings, both used for flying, and three pairs of walking legs. **e.** A centipede has only one pair of antennae, and the head is followed by a series of segments, each with a single pair of appendages.

30.5 Arthropods

Arthropods (phylum Arthropoda, over one million species) are extremely diverse (Fig. 30.15). Over one million species have been discovered and described, but some experts suggest that as many as 30 million arthropods could exist—most of them insects.

What characteristics account for the success of arthropods? Arthropod literally means "jointed foot," but actually they have freely movable jointed appendages. The exoskeleton of arthropods is com-

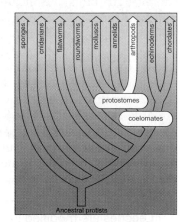

posed primarily of **chitin,** a strong, flexible, nitrogenous polysaccharide. The exoskeleton serves many functions, including protection, attachment for muscles, locomotion, and prevention of desiccation. However, because this exoskeleton is hard and nonexpandable, arthropods must undergo **molting,** or shedding of the exoskeleton, as they grow larger. Before molting, the body secretes a new, larger exoskeleton, which is soft and wrinkled, underneath the old one. After enzymes partially dissolve and weaken the old exoskeleton, the animal breaks it open and wriggles out. The new exoskeleton then quickly expands and hardens.

Arthropods are segmented, but some segments are fused into regions, such as a head, a thorax, and an abdomen. Trilobites, an early and now extinct arthropod, had a pair of appendages on each body segment. In modern arthropods, appendages are specialized for such functions as walking, swimming, reproducing, eating, and sensory reception. These modifications account for much of the diversity of arthropods.

Arthropods have a well-developed nervous system that includes a brain and a ventral solid nerve cord. The head bears various types of sense organs, including antennae (or feelers) and two types of eyes—compound and simple. The compound eye is composed of many complete visual units, each of which operates independently. The lens of each visual unit focuses an image on the light-sensitive membranes of a small number of photoreceptors within that unit. Vision is not acute, but it is much better for details of movement than with our eyes.

Crustaceans

Crustaceans are a group of largely marine arthropods that include barnacles, shrimps, lobsters, and crabs. There are also some freshwater crustaceans, including the crayfish, and some terrestrial ones, including the sowbug, or "roly-poly bug."

Crustaceans are named for their hard shells; the exoskeleton is calcified to a greater degree in some forms than in others. Although crustacean anatomy is extremely diverse, the head usually bears a pair of compound eyes and five pairs of appendages. The first two pairs, called antennae, lie in front of the mouth and have sensory functions. The other three pairs are mouthparts used in feeding.

In crayfish such as *Cambarus*, the thorax bears five pairs of walking legs. The first walking leg is a pinching claw (Fig. 30.16). The *gills* are situated above the walking legs. The head and thorax are fused into a cephalothorax, which is covered on the top and sides by a nonsegmented carapace. The abdominal segments, containing much musculature, are equipped with swimmerets, small paddlelike structures. The last two segments bear the uropods and the telson, which make up a fan-shaped tail to propel the crayfish backward.

Internal Organs

The digestive system includes a stomach, which is divided into two main regions: an anterior portion called the gastric mill, equipped with chitinous teeth to grind coarse food, and a posterior region, which acts as a filter to prevent coarse particles from entering the digestive glands where absorption takes place. Green glands lying in the head region, anterior to the esophagus, excrete metabolic wastes through a duct that opens externally at the base of the antennae. The coelom is reduced to a space around the reproductive system. A heart within a pericardial cavity pumps blood containing the respiratory pigment hemocyanin into a **hemocoel** consisting of

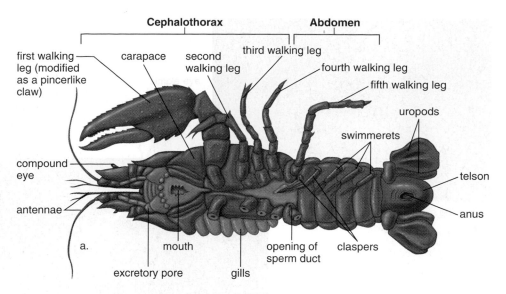

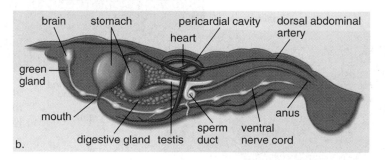

Figure 30.16   Male crayfish, *Cambarus*.
a. Externally, it is possible to observe the jointed appendages, including the swimmerets, the walking legs, and the claws. These appendages, plus a portion of the carapace, have been removed from the right side so that the gills are visible. **b.** Internally, the parts of the digestive system are particularly visible. The circulatory system can also be clearly seen. Note the ventral solid nerve cord.

sinuses (open spaces), where the hemolymph flows about the organs. (Whereas hemoglobin is a red, iron-containing pigment, hemocyanin is a blue, copper-containing pigment.) This is an open circulatory system because blood is not contained within blood vessels.

The nervous system of the crayfish is very similar to that of the earthworm. It has a brain and a ventral nerve cord that passes posteriorly. Along the length of the nerve cord, ganglia in segments give off 8 to 19 paired lateral nerves.

The sexes are separate in the crayfish, and the gonads are located just ventral to the pericardial cavity. In the male, a coiled sperm duct opens to the outside at the base of the fifth walking leg. Sperm transfer is accomplished by the first two pairs of swimmerets, which are enlarged and quite strong. In the female, the ovaries open at the bases of the third walking legs. A stiff fold between the bases of the fourth and fifth pairs of walking legs serves as a seminal receptacle. Following fertilization, the eggs are attached to the swimmerets of the female.

Concepts	Questions	Media Resources*
30.1 Evolution and Classification of Animals		

• Animals are multicellular heterotrophs exhibiting at least some mobility.
• Animals are grouped according to level of organization, symmetry, body plan, pattern of embryonic development, and presence or absence of segmentation.
	1. What is cephalization?
	2. What ancestor gave rise to all animals?
		Essential Study Partner
		   Characteristics
		
		Art Quiz
		   Animal Kingdom Phylogeny
		
		Animation Quiz
		   Symmetry in Nature

30.2 Introducing the Invertebrates		

• Sponges are multicellular, with limited mobility and no symmetry.
• Cnidarians are radially symmetrical, with two tissue layers.
• Planarians are bilaterally symmetrical, with a definite head region.
• Roundworms have a pseudocoelom and the tube-within-a-tube body plan.
	1. What level of organization is exhibited in the sponges? In the flatworms?
	2. What two anatomical features are present in the roundworms that are not present in sponges, cnidarians, or flatworms?
		Essential Study Partner
		   Sponges
		   Radial Phyla
		   Bilateral/No Coelom
		   Pseudocoelomates
		
		Art Quizzes
		   Cnidarian Body Plan
		   Body Plans for Bilaterally
		      Symmetrical Animals

30.3 Molluscs		

• Molluscs have a muscular foot (variously modified), and a visceral mass enveloped by a mantle.	1. What two significant evolutionary advances are seen in the molluscs?
	2. Bivalves exhibit an open circulatory system. How does an open circulatory system differ from one that is closed?
		Essential Study Partner
		   Molluscs
		
		Art Quiz
		   Mollusc Body Plan

30.4 Annelids		
• Annelids are segmented, with a well-developed, true coelom.	1. What evolutionary advancement is evident for the first time in the annelids? 2. What specialization that resulted from the tube-within-a-tube body plan is seen among the annelids?	Essential Study Partner Annelids
30.5 Arthropods		

• Arthropods have jointed appendages and an exoskeleton that must be periodically shed.	1. Why are arthropods, especially insects, so successful and abundant?	Essential Study Partner
		   Arthropods
		
		Art Quizzes
		   Insect Structure—Internal
		   Decapod Crustacean—
		      Lobster
		   Metamorphosis
		
		Case Studies
		   Cicada Woodstock
		   Mosquito Coast

*For additional Media Resources, see the Online Learning Center.

31

Animals: Part II

Chapter Concepts

Three jawless lampreys are using their suction-cup type mouths to take nourishment from a carp.

As Wendy waited in a camouflaged blind for her next target, the wildlife photojournalist thought about her recent assignments. Last month, she had donned scuba gear to photograph parasitic lamprey attached to a carp, and also worked underwater in a cage to capture pictures of the fierce hammerhead shark. Her cameras had documented alligators chasing down prey and turtles returning to the ocean after burying their eggs on the seashore. Using a hang glider, Wendy had even photographed a bald eagle flying around its nest on the side of a steep mountain. Other jobs had her tracking kangaroos in Australia and apes in Africa. Today, she was waiting for a lion to attack the herd of antelopes at the watering hole near her blind.

From fish to birds, reptiles to mammals, the vertebrate world offers Wendy a seemingly endless number of photographic subjects. Many of those creatures will make an appearance in this chapter. We'll also detail how, through evolution, one line of descent gave rise to the uniquely intelligent species called Homo sapiens, *otherwise known as modern humans.*

31.1 Echinoderms

In Chapter 30, we introduced the invertebrates, the large group of animals character- ized by their lack of an endoskeleton of bone or car- tilage. This chapter will focus mainly on the echinoderms and the chordates which include the vertebrates.

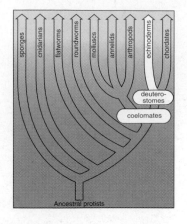

Among animals, chor- dates are most closely related to the **echinoderms** (phylum Echinodermata, about 6,000 species). In both echinoderms and chordates, the second embryonic opening becomes the mouth. Therefore, echinoderms and chordates are called **deuterostomes.** In both as well, the coelom forms by out- pocketing of the primitive gut, and this is the reason both echinoderms and chordates are called enterocoelomates. Then, too, echinoderms and close relatives of chordates have a dipleurula larva (bands of cilia placed as shown in Figure 30.10).

Characteristics of Echinoderms

Echinoderms are a diverse group of marine animals; there are no terrestrial echinoderms. They have an endoskeleton (internal skeleton) consisting of spine-bearing, calcium-rich plates. The spines, which stick out through their delicate skin, account for their name (Fig. 31.1).

It may seem surprising that echinoderms, although related to chordates, lack those features we associate with vertebrates (e.g., human beings). For example, the echino- derms are often radially, not bilaterally, symmetrical. Their larva is a free-swimming filter feeder with bilateral sym- metry, but it metamorphoses into a radially symmetrical adult. Recall that with radial symmetry, it is possible to obtain two identical halves, no matter how the animal is sliced longitudinally, whereas in bilaterally symmetrical ani- mals, only one longitudinal cut gives two identical parts.

Echinoderm Diversity

Echinoderms are quite diverse. The stalked sea lilies and the motile feather stars are classified in the class Crinoidea. The sea cucumbers, in the class Holothuroidea, have a long leathery body that resembles a cucumber, except for the feeding tentacles about the mouth. Brittle stars in the class Ophiuroidea, have a central disk from which long, flexible arms radiate. You may be more familiar with sea urchins and sand dollars in the class Echinoidea, which have spines for locomotion, defense, and burrowing. Sea urchins are food for sea otters off the coast of California (see page 101). Finally, the sea stars in the class Asteroidea are studied in detail in the next section.

Figure 31.1   Echinoderm diversity.
a. Sea urchins have large, colored, external spines for protection.
b. Sea cucumbers look like a cucumber; they lack arms but have tentacle-like tube feet with suckers around the mouth. **c.** A feather star extends its arms to strain suspended food particles from the sea.

b. Sea cucumber, *Parastichopus*

a. Purple sea urchin, *Strongylocentrotus*

c. Feather star, *Comanthus*

Sea Stars

Sea stars are commonly found along rocky coasts where they feed on clams, oysters, and other bivalve molluscs. The five-rayed body has an oral, or mouth, side (the underside) and an aboral, or anus, side (the upper side) (Fig. 31.2). Various structures project through the body wall: (1) spines from the endoskeletal plates offer some protection; (2) pincerlike structures called pedicellariae keep the surface free of small particles; and (3) skin gills, tiny fingerlike extensions of the skin, are used for gas exchange. On the oral surface, each arm has a groove lined by small tube feet.

To feed, a sea star positions itself over a bivalve and attaches some of its tube feet to each side of the shell. By working its tube feet in alternation, it pulls the shell open. A very small crack is enough for the sea star to evert its cardiac stomach and push it through the crack, so that it contacts the soft parts of the bivalve. The stomach secretes enzymes, and digestion begins even while the bivalve is attempting to close its shell. Later, partly digested food is taken into the sea star's body, where digestion continues in the pyloric stomach using enzymes from the digestive glands found in each arm. A short intestine opens at the anus on the aboral side.

In each arm, the well-developed coelomic cavity contains not only a pair of digestive glands, but also gonads (either male or female), which open on the aboral surface by very small pores. The nervous system consists of a central nerve ring that gives off radial nerves in each arm. A light-sensitive eyespot is at the tip of each arm. Sea stars are capable of coordinated but slow responses and body movements.

Locomotion depends on the **water vascular system.** Water enters this system through a structure on the aboral side called the sieve plate, or madreporite. From there it passes down a stone canal into a ring canal, which surrounds the mouth. The ring canal gives off a radial canal in each arm. From the radial canals, water enters the ampullae. Contraction of the ampulla forces water into the tube foot, expanding it. When the foot touches a surface, the center is withdrawn, giving it suction so that it can adhere to the surface. By alternating the expansion and contraction of the tube feet, a sea star moves slowly along.

Echinoderms don't have a respiratory, excretory, or circulatory system. Fluids within the coelomic cavity and the water vascular system carry out many of these functions. For example, gas exchange occurs across the skin gills and the tube feet. Nitrogenous wastes diffuse through the coelomic fluid and the body wall. Cilia on the peritoneum lining the coelom keep the coelomic fluid moving.

Sea stars reproduce both asexually and sexually. If the body is fragmented, each fragment can regenerate a whole animal. Fishermen who try to get rid of sea stars by cutting them up and tossing them overboard are merely propagating more sea stars! Sea stars also spawn, releasing either eggs or sperm. The dipleurula larva is bilateral and metamorphoses to become the radially symmetrical adult.

Echinoderms have a well-developed coelom and internal organs despite being radially symmetrical. Spines project from their endoskeleton, and they have a unique water vascular system.

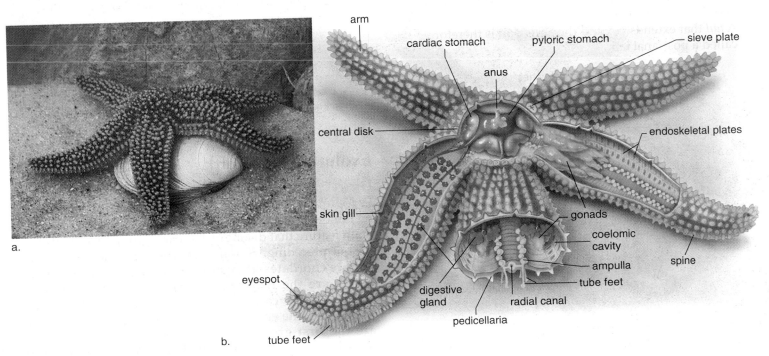

Figure 31.2   Sea star anatomy and behavior.
a. A sea star uses the suction of its tube feet to open a clam, its primary source of food. **b.** Each arm of a sea star contains digestive glands, gonads, and portions of the water vascular system. This system (colored yellow) terminates in tube feet.

Behavior and Ecology

See the Online Learning Center for careers in Ecology.

Animal Behavior

Chapter Concepts

A female satin bowerbird has chosen to mate with this male. Most likely, she was attracted by his physique and the blue decorations of his bower.

*A*t the start of the breeding season, male bowerbirds use small sticks and twigs to build elaborate display areas called bowers. They clear the space around the bower and decorate the area with fresh flowers, fruits, pebbles, shells, bits of glass, tinfoil, and any bright baubles they can find. Each species has its own preference in decorations. The satin bowerbird of eastern Australia prefers blue objects, a color that harmonizes with the male's glossy blue-black plumage.

After the bower is complete, a male spends most of his time near his bower, calling to females, renewing his decorations, and guarding his work against possible raids by other males. After inspecting many bowers and their owners, a female approaches one and the male begins a display. He faces her, fluffs up his feathers, and flaps his wings to the beat of a call. The female enters the bower, and if she crouches, the two mate.

Behavior is studied in the same objective manner as any other field of science. A behaviorist would determine how a male satin bowerbird is organized to perform this behavior and how the behavior helps him secure a mate.

Figure 32.6   A male olive baboon displaying full threat.
In olive baboons, males are larger than females and have enlarged canines. Competition between males establishes a dominance hierarchy for the distribution of resources.

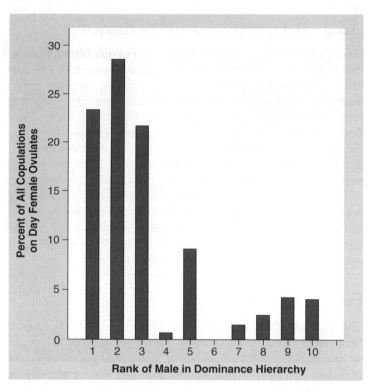

Figure 32.7   Female choice and male dominance among baboons.
Although it may appear that females mate indiscriminately, they mate more often with a dominant male when they are most fertile.

Male Competition

Males can father many offspring because they continuously produce sperm in great quantity. We expect males to compete in order to inseminate as many females as possible. Studies have been done to determine if the benefit of access to mating is worth the cost of competition among males. Only if the positive effects outweigh the negative effects will a behavior continue.

Dominance Hierarchy

Baboons, a type of Old World monkey, live together in a troop. Males and females have separate **dominance hierarchies** in which a higher-ranking animal has greater access to resources than a lower-ranking animal. Dominance is decided by confrontations, resulting in one animal giving way to the other.

Baboons are dimorphic; the males are larger than the females, and they can threaten other members of the troop with their long, sharp canines (Fig. 32.6). One or more males become dominant by frightening the other males. The male baboon pays a cost for his dominant position. Being larger means that he needs more food, and being

willing and able to fight predators means that he may get hurt, and so forth. Is there a reproductive benefit to his behavior? Yes, in that dominant males do indeed monopolize females when they are most fertile (Figure 32.7). Females undergo a period known as estrus, during which they ovulate and are willing to mate. At this time, a female approaches a dominant male, and they form a mating pair for several hours or days.

Nevertheless, there are possibly other avenues to fathering offspring. A male may act as a helper to a female and her offspring; the next time she is in estrus, she may mate preferentially with him instead of a dominant male. Or subordinate males may form a friendship group that opposes a dominant male, making him give up a receptive female.

Territoriality

A **territory** is an area that is defended against competitors. As discussed in the Ecology Reading on page 670, scientists are able to track an animal in the wild in order to determine its home range or territory. **Territoriality** includes the type of defensive behavior needed to defend a territory. Baboons travel within a home range, foraging for food each day and sleeping in trees at night. Dominant males decide where and when the troop will move. If the troop is threatened, dominant males protect the troop as it retreats and attack intruders when necessary. Vocalization and displays, rather than outright fighting, may be sufficient to defend a territory. In songbirds, for example, males use singing to announce their willingness to defend a territory. Other males of the species become reluctant to make use of the same area.

Red deer stags (males) on the Scottish island of Rhum compete to be the harem master of a group of hinds (females) that mate only with them. The reproductive group occupies a territory that the harem master defends against other stags. Harem masters first attempt to repel challengers by roaring. If the challenger remains, the two lock antlers and push against one another (Fig. 32.8). If the challenger then withdraws, the master pursues him for a short distance, roaring the whole time. If the challenger wins, he becomes the harem master.

A harem master can father two dozen offspring at most, because he is at the peak of his fighting ability for only a short time. And there is a cost to being a harem master. Stags must be large and powerful in order to fight; therefore, they grow faster and have less body fat. During bad times, they are more likely to die of starvation, and in general, they have shorter lives. Harem master behavior will only persist in the population if its cost (reduction in the potential number of offspring because of a shorter life) is less than its benefit (increased number of offspring due to harem access).

Evolution by sexual selection occurs when females have the opportunity to select among potential mates and/or when males compete among themselves for access to reproductive females.

Figure 32.8   Competition between male red deer.
Male red deer compete for a harem within a particular territory. **a.** Roaring alone may frighten off a challenger, but (**b**) outright fighting may be necessary, and the victor is most likely the stronger of the two animals.

Tracking an Animal in the Wild

Miniature radio transmitters that emit a radio signal allow researchers to track an animal in the wild. The transmitter is encapsulated along with a battery in a protective epoxy resin covering, and the package is either attached to an animal with a collar or clip, or implanted surgically into the animal's main body cavity. In either case, it is important to capture the animal carefully and to use sedation or anesthesia to calm the animal during tagging or surgical implantation. The animal is permitted to recover fully prior to its release. A radio transmitter device is generally no more than 10% of the animal's body weight, and therefore it should not interfere with normal activities.

To track an animal, a researcher needs a radio receiver equipped with an antenna and earphones. The strongest signal comes from the direction of the animal as the antenna is rotated above the head. The resulting data allow the researcher to obtain a series of fixes (to determine where the animal is) and calculate its rate of movement. Plotting the sequence of fixes on a map of the appropriate scale gives information on the area that is used by the animal in the course of a night or several nights (Fig. 32B).

The distance over which the signal travels varies with the size of the transmitter and the strength of the battery. For small rodents, it may be necessary to be within 5–10 meters of the animal to hear the signal; for larger collars like some of those used on elk or caribou, the signal can actually be received by a satel-lite orbiting above the earth. Later, the signals are sent in a radio beam by antenna to the ground. Satellite transmission allows researchers to work with animals that are far away, and prevents any possibility of interfering with the animal's activities.

A number of important findings have been made utilizing radiotelemetry, and the following examples illustrate the kinds of results obtained. We now know that elk in Yellowstone National Park in Wyoming have summer home range areas that are 5–10 times larger than in winter. Areas occupied by many mice and voles get larger as the animal ages, from the time the animals are weaned until they are adults, but these increases are greater for males than for females. Winter dens of rattlesnakes (and other snakes) are often some distance from the areas where they spend their summer months. Some crows migrate each spring and fall, whereas others remain as residents in the same locale throughout the year.

Radio transmitters can also be used to obtain data on physiological functions (heart rate, body temperature) in conjunction with observations of behavior while animals continue to engage in normal activities. When physiological measurements are taken, a device is included in the package that is sensitive to temperature or heart rate and can translate that information to the transmitter. Altogether, the use of radio transmitters has added a new dimension to the ability to explore the lives of animals.

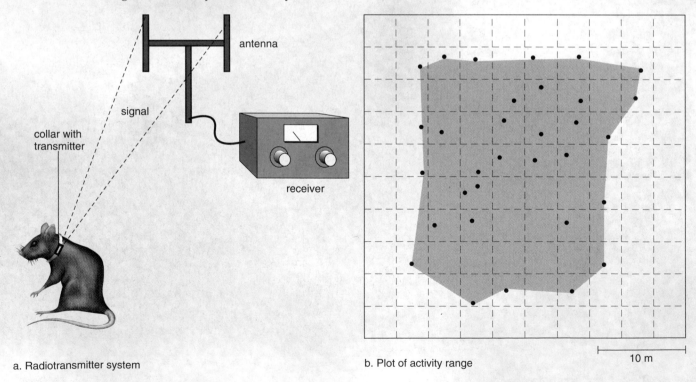

antenna

signal

collar with transmitter

receiver

a. Radiotransmitter system

b. Plot of activity range

10 m

Figure 32B **Electronic animal tracking.**
a. By using a miniature radio transmitter in a neck collar, an antenna, and a receiver, it is possible to record an animal's location approximately every 15 minutes throughout a 24-hour period. **b.** These data are then used to plot the animal's activity range for that day. (Some data in the plot represent multiple fixes at the same location.)

32.4 Animal Societies

Animals exhibit a wide diversity of social behaviors. Some animals are largely solitary and join with a member of the opposite sex only for the purpose of reproduction. Others pair, bond, and cooperate in raising offspring. Still others form a **society** in which members of species are organized in a cooperative manner, extending beyond sexual and parental behavior. We have already mentioned the social groups of baboons and red deer. Social behavior in these and other animals requires that they communicate with one another.

Communicative Behavior

Communication is an action by a sender that influences the behavior of a receiver. The communication can be purposeful, but does not have to be. Bats send out a series of sound pulses and listen for the corresponding echoes in order to find their way through dark caves and locate food at night. Some moths have an ability to hear these sound pulses, and they begin evasive tactics when they sense that a bat is near. Are the bats purposefully communicating with the moths? No, bat sounds are simply a cue to the moths that danger is near.

Communication is an action by a sender that affects the behavior of a receiver.

Chemical Communication

Chemical signals have the advantage of being effective both night and day. The term **pheromone** designates chemical signals in low concentration that are passed between members of the same species. For example, female moths secrete chemicals from special abdominal glands, which are detected downwind by receptors on male antennae. The antennae are especially sensitive, and this assures that only male moths of the correct species (not predators) will be able to detect them.

Cheetahs and other cats mark their territories by depositing urine, feces, and anal gland secretions at the boundaries (Fig. 32.9). Klipspringers (small antelope) use secretions from a gland below the eye to mark twigs and grasses of their territory.

Auditory Communication

Auditory (sound) communication has some advantages over other kinds of communication (Fig. 32.10). It is faster than chemical communication, and it too is effective both night and day. Further, auditory communication can be modified not only by loudness but also by pattern, duration, and repetition. In an experiment with rats, a researcher discovered that an intruder can avoid attack by increasing the frequency with which it makes an appeasement sound.

Male crickets have calls, and male birds have songs for a number of different occasions. For example, birds may have one song for distress, another for courting, and still another for marking territories. Sailors have long heard the songs of humpback whales transmitted through the hull of a ship. But only recently has it been shown that the song has six basic themes, each with its own phrases, that can vary in length and be interspersed with sundry cries and chirps. The purpose of the song is probably sexual, serving to advertise the availability of the singer. Language is the ultimate auditory communication, but only humans have the biological ability to produce a large number of different

Figure 32.9   Use of a pheromone.
This male cheetah is spraying a pheromone onto a tree in order to mark his territory.

Figure 32.10   A chimpanzee with a researcher.
Chimpanzees are unable to speak but can learn to use a visual language consisting of symbols. Some researchers believe chimps only mimic their teachers and never understand the cognitive use of language. Here the experimenter shows Nim the sign for "drink." Nim copies.

e-Learning Connection   www.mhhe.com/maderinquiry10

Concepts	Questions	Media Resources*
32.1 Genetic Basis of Behavior		
• An animal is organized to carry out behaviors that help it survive and reproduce. • Behaviors have a genetic basis but can also be influenced by environmental factors. • The nervous and endocrine systems have immediate control over behaviors.	1. If behavior for obtaining and carrying different types of nesting material is inherited, what behavior occurs in the hybrid offspring of two different species of birds? 2. Is the endocrine system, in addition to the nervous system, responsible for behavior?	Essential Study Partner Introduction Nature/Nurture Innate Behavior
32.2 Development of Behavior		
• Behaviors sometimes undergo development after birth, as when learning affects behavior.	1. What is operant conditioning? 2. What is/are the possible purpose(s) of imprinting?	Essential Study Partner Learning
32.3 Adaptiveness of Behavior		
• Natural selection influences such behaviors as methods of feeding, selecting a home, and reproducing.	1. Define "sexual selection." 2. What is a dominance hierarchy?	Essential Study Partner Adaptive Value Mating Aggression
32.4 Animal Societies		
• Animals living in societies have various means of communicating with one another.	1. List and give examples of the various categories of communication that have been observed in different species of animals.	Essential Study Partner Communication
32.5 Sociobiology and Animal Behavior		
• Apparently, altruistic behavior only occurs if it actually benefits the animal.	1. What is altruism and what does an individual gain through altruistic behavior? 2. What is inclusive fitness?	Essential Study Partner Altruism and Sociality

*For additional Media Resources, see the Online Learning Center.

Population Growth and Regulation

Chapter Concepts

A group of elephants on the plain in Africa.

*E*lephants have a large size, are social, live a long time, and produce few offspring. Females live in social family units, and the much larger males visit them only during a breeding season. Females give birth about every five years to a single calf that is well cared for and has a good chance of meeting the challenges of its lifestyle. Normally, an elephant population exists at the carrying capacity of the environment.

A population ecologist studies the distribution and abundance of organisms and relates the hard, cold statistics to the species life history in order to determine what causes population growth or decline. Elephants are threatened because of human population growth and also because males, in preference to females, are killed for their tusks. Males don't tend to breed until they have reached their largest size—the size that makes them prized by humans. By now, even a moratorium on killing males would not help much. So few breeding males are left that elephant populations are expected to continue to decline for quite some time. The study of population ecology is necessary to preservation of species.

33.1 Scope of Ecology

Ecology is the study of the interactions of organisms with each other and with the physical environment. Ecology, like so many biological disciplines, is wide-ranging and involves several levels of study (Fig. 33.1).

At the simplest level, ecologists study how the individual organism is adapted to its environment. For example, they might study why a particular species of fish in a coral reef lives only in warm tropical waters and how it feeds. Most organisms do not exist singly; rather, they are part of a **population,** which is defined as all the organisms of the same species interacting with the environment at a particular locale. At the population level of study, ecologists are interested in describing the size of populations over time. For example, in a coral reef an ecologist might study the relative sizes of parrotfishes over time. A **community** consists of all the various populations at a particular locale. A coral reef contains numerous populations of fishes, crustaceans, corals, and so forth. At the community level, ecologists want to study how various extrinsic and intrinsic factors affect the size of these populations. An **ecosystem** encompasses a community of populations as well as the nonliving environment. For example, energy flow and chemical cycling in a coral reef can affect the success of the organisms that inhabit it. Finally, the **biosphere** is that portion of the entire earth's surface—air, water, land—where living things exist. Knowing the composition and diversity of the coral reef ecosystem is important to the dynamics of the biosphere. Table 33.1 summarizes the terms commonly used in the study of ecology.

Modern ecology is not just descriptive, it also develops hypotheses that can be tested. A central goal of modern ecology is to develop models which explain and predict the distribution and abundance of populations. Ultimately, ecology considers not one particular area, but the distribution and abundance of populations in the biosphere.

Community Composition and Diversity

The *composition* of a community is simply a listing of the various populations in the community. The *diversity* includes both the number of different populations in a community

Table 33.1	Ecological Terms
Term	**Definition**
Ecology	Study of the interactions of organisms with each other and with the physical environment
Population	All the members of the same species that inhabit a particular area
Community	All the populations found in a particular area
Ecosystem	A community and its physical environment, including both nonliving (abiotic) and living (biotic) components
Biosphere	All the communities on earth whose members exist in air and water and on land

Organism   Population   Community   Ecosystem

Figure 33.1   Levels of organization in a coral reef.
The study of ecology encompasses various levels, from the individual organism to the population, community, and ecosystem. The biosphere includes all the different ecosystems of planet Earth.

and the relative abundance of individuals in each population. These two attributes allow us to compare communities.

Just glancing at Figure 33.2 makes it easy to see that two forests have different compositions. Pictorially, we can see that narrow-leaved evergreen trees are present in a coniferous forest, and broad-leaved evergreen trees are numerous in a tropical rain forest. Mammals also differ between the two communities, as those listed demonstrate.

Ecologists can determine diversity by counting the number and abundance of each type of population in the community. To take an extreme example: a deciduous forest in West Virginia has, among other species, 76 yellow poplar trees but only one American elm. If we walked through this forest, we might miss seeing the American elm. If, instead, the forest had 36 poplar trees and 41 American elms, the forest would seem more diverse to us and indeed would be more diverse. The greater the diversity, the greater the number and the more even the populations.

Models of Community Composition

Why do populations assemble together in the same place at the same time? For many years, most ecologists supported the *interactive model* to explain community composition. According to this model, a community is the highest level of organization—that is, from cell to tissue, to organism, to population, and finally, to a community. Just as the parts of an organism are dependent on one another, so populations are dependent on biotic interactions, such as those of a food chain. Further, like an organism, a community remains stable because of homeostatic mechanisms. This theory predicts that the community composition of, say, a coniferous forest in a particular locale will always be the same.

The *individualistic model* of community composition instead hypothesizes that populations assemble according to species' tolerance for abiotic factors. The range of a species is based on its tolerance for such abiotic factors as temperature, light, water availability, salinity, and so forth. For example, the number of different populations in terrestrial communities increases as we move from the northern latitudes to the equator, most likely because the weather at the equator is warmer and there is more precipitation. It's possible that populations assemble because their species tolerance ranges simply overlap. The individualistic model predicts that community compositions are not constant, and that the boundaries between communities will not be distinct from one another. In other words, it predicts that no two coniferous forests will have the same composition and that it would be difficult to tell where a coniferous forest begins and ends.

Ecology is the study of the interactions of populations between each other and with the physical environment. Similarly, community composition is dependent on both biotic interactions and abiotic factors.

a.

moose
lynx
snowshoe hare
bear
red fox
wolf

monkey
sloth
anteater
kinkajou
jaguar
tapir
bat

b.

Figure 33.2   Two terrestrial communities.
Diversity of communities is described by the number of populations and their relative abundance, as witnessed by the plants and animals in these terrestrial communities. **a.** A coniferous forest. Some mammals found here are listed to the *left*. **b.** A tropical rain forest. Some mammals found here are listed to the *right*.

Ecological Succession

Communities are subject to disturbances that can range in severity from a beaver damming a pond to a volcanic eruption. We know from observation that following these disturbances, we'll see changes in the area over time. **Ecological succession** is a change in community composition over time. On land, *primary succession* occurs in areas where there is no soil formation, such as following a volcanic eruption or a glacial retreat. *Secondary succession* begins in areas where soil is present, as when a cultivated cornfield returns to a natural state.

The first species to begin secondary succession are called *pioneer species*—that is, plants that are invaders of disturbed areas—and then the succession progresses through a series of stages as described in Figure 33.3. Notice that such a series begins with grasses and proceeds from shrub stages to a mixture of shrubs and trees, until finally there are only trees.

Succession also occurs in aquatic communities, as when lakes and ponds undergo a series of stages by which they disappear and become filled in.

Models of Succession

Ecologists have developed various hypotheses to explain succession and predict future events. The *climax-pattern model* of succession says that particular areas will always lead to the same type of community, called a **climax community.** This model is based on the observation that climate, in particular, determines whether a desert, a grassland, or a particular type of forest results. Therefore, for example, coniferous forests occur in northern latitudes, deciduous forests in temperate zones, and tropical rain forests in the tropics. The climax-pattern model of succession is now being modified to accommodate the idea that the exact composition of a community need not always be the same. That is, while we might expect to see a coniferous forest as opposed to a tropical rain forest in northern latitudes, the exact mix of plants and animals after each event of succession can vary.

Does each stage in succession facilitate or inhibit the next stage? To support a *facilitation model* it can be observed that shrubs can't grow on dunes until dune grass has caused soil to develop. Similarly, in the example given in Figure 33.3, shrubs can't arrive until grasses have made the soil suitable for them. Thus, it's possible that each successive community prepares the way for the next, so that grass-shrub-forest development occurs in a sequential way.

On the other hand, the *inhibition model* says that colonists hold on to their space and inhibit the growth of other plants until the colonists die or are damaged. Still another possible model, the *tolerance model,* predicts that different types of plants can colonize an area at the same time. Sheer chance determines which seeds arrive first, and successional stages may simply reflect the length of time it takes species to mature. This alone could account for the grass-shrub-forest development seen in Figure 33.3. The length of time it takes for trees to develop might simply give the impression that there is a recognizable series of plant communities, from the simple to the complex. But in reality, the models we have mentioned are not mutually exclusive, and succession is probably a complex process.

Ecological succession, which occurs after a disturbance, probably involves complex processes, and the result cannot always be foretold.

| grass | low shrub | high shrub | shrub-tree | low tree | high tree |

Figure 33.3 Secondary succession in a forest.
In secondary succession in a large conifer plantation in central New York state, certain species are common to particular stages. However, the process of regrowth shows approximately the same stages as secondary succession in a former cornfield.

33.2 Population Characteristics and Growth

The populations within a community are characterized by a particular density and distribution. *Population density* is the number of individuals per unit area or volume. For example, if we calculated the density of the human population, we would know how many individuals there are per square mile. From this we might get the impression that members of the human population are uniformly distributed, but we know full well that most people live in cities. Even within a city, more people live in particular neighborhoods than others.

Population distribution is the pattern of dispersal of individuals within a particular locale. There are three patterns of distribution: uniform, random, and clumped (Fig. 33.4). Most organisms, including humans, have a clumped pattern of distribution. Today, ecologists have classified the reasons for the spatial distribution of organisms into *abiotic* (nonliving) and *biotic* (living) factors. An abiotic factor might be the presence of a particular inorganic nutrient. For example, a study of the distribution of hard clams in a bay on the south shore of Long Island, New York, revealed that clams are apt to occur where the sediment contains oyster shells because oyster shells provide calcium carbonate for the formation of clam shells. Hopefully, this information can be used to transform areas of low abundance to areas of high abundance of clams. With reference to humans, we know that many cities have sprung up at the junction of rivers or near inlets that make good harbors for ships.

Other abiotic factors, such as precipitation and temperature, can be limiting factors for the distribution of an organism. *Limiting factors* are those factors that particularly determine whether an organism lives in an area. For example, within a particular river, trout tend to congregate where the water is cool and the oxygen content is high, while carp and catfish are found near the coast because they can tolerate waters that are warm and have a low concentration of oxygen. On a mountain, trees do not exist above the timberline because they cannot grow where a low temperature keeps the water frozen most of the year.

Biotic factors, such as the availability of food, affect the distribution of populations. The red kangaroo tends to live where it can feed on arid grasses, and also more humans live where food is abundant.

Each population has a particular density (number of individuals per unit area) and distribution, whether uniform, random, or clumped.

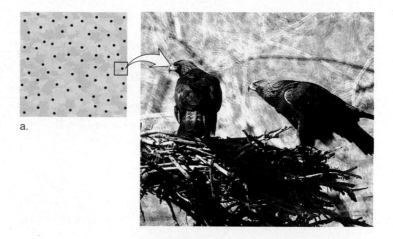
a.

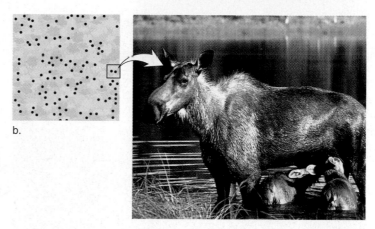
b.

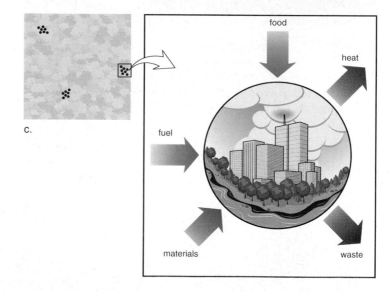
c.

Figure 33.4   Patterns of distribution within a population.
Members of a population may be distributed uniformly, randomly, or in clumps. **a.** Golden eagle pair distribution is uniform over a suitable habitat area due to the territoriality of the birds. **b.** The distribution of female moose with calves is random over a suitable habitat. **c.** Human beings tend to be clumped in cities where many people take up residence. Cities take resources from and send their waste to surrounding regions. The clumped pattern is the most common.

Patterns of Population Growth

Each population has a particular pattern of growth. Populations have a certain size, and the size can stay the same from year to year, increase, or decrease, according to a *per capita rate of increase*. Suppose, for example, a human population presently has a size of 1,000 individuals, the birthrate is 30 per year, and the death rate is 10 per year. The per capita rate of increase per year will be:

$$\frac{30-10}{1,000} = 0.02 = 2.0\% \text{ per year}$$

(Notice that this per capita rate of increase disregards both immigration and emigration, which for the purpose of our discussion can be assumed to be equivalent.) The highest possible per capita rate of increase for a population is called its **biotic potential** (Fig. 33.5). Whether the biotic potential is high or low depends on such factors as the following: └ how much children u can have

  animals ↑ humans ↓

1. Usual number of offspring per reproduction
2. Chances of survival until age of reproduction
3. How often each individual reproduces
4. Age at which reproduction begins

Suppose we are studying the growth of a population of insects that are capable of infesting and taking over an area. Under these circumstances, **exponential growth** is expected. An exponential pattern of population growth results in a J-shaped curve (Fig. 33.6*a*). This pattern of population growth can be likened to compound interest at the bank: the more your money increases, the more interest you will get. If the insect population has 2,000 individuals and the per capita rate of increase is 20% per month, there will be 2,400 insects after one month, 2,880 after two months, 3,456 after three months, and so forth.

Notice that a J-shaped curve has these phases:

desert ↑ farmable land ↑

lag phase: During this phase, growth is slow because the population is small.

exponential growth phase: During this phase, growth is accelerating, and the population is exhibiting its biotic potential.

Usually, exponential growth cannot continue for long because of environmental resistance. **Environmental resistance** is all those environmental conditions, such as a limited supply of food, an accumulation of waste products, increased competition, or predation, that prevent populations from achieving their biotic potential. Due to environmental resistance, growth levels off, and a pattern of population growth called logistic growth is expected.

Figure 33.5   Biotic potential.
Animal husbandry relies on biotic potential. If a single female pig had her first litter at nine months, and produced two litters a year, each of which contained an average of four females (which in turn reproduced at the same rate), there would be 2,220 pigs by the end of three years.

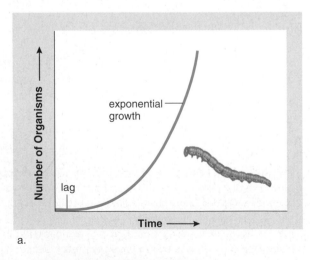

a.

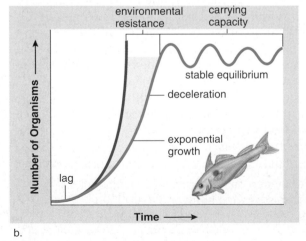

b.

Figure 33.6   Patterns of population growth.
a. Exponential growth results in a J-shaped growth curve because the growth rate is positive. **b.** Logistic growth results in an S-shaped growth curve because environmental resistance causes the population size to level off and be in a steady state.

Logistic growth results in an S-shaped growth curve (Fig. 33.6*b*).

Notice that an S-shaped curve has these phases:

lag phase: During this phase, growth is slow because the population is small.

exponential growth phase: During this phase, growth is accelerating due to biotic potential.

deceleration phase: During this phase, the rate of population growth slows down.

stable equilibrium phase: During this phase, little if any growth takes place because births and deaths are about equal.

The stable equilibrium phase is said to occur at the **carrying capacity** of the environment. The carrying capacity is the number of individuals the environment can normally support.

Our knowledge of logistic growth has practical implications. The model predicts that exponential growth will occur only when population size is much lower than the carrying capacity. So, as a practical matter, if humans are using a fish population as a continuous food source, it would be best to maintain that population size in the exponential phase of growth when biotic potential is having its full effect and the birth rate is the highest. If we overfish, the fish population will sink into the lag phase, and it will be years before exponential growth recurs. On the other hand, if we are trying to limit the growth of a pest, it is best to reduce the carrying capacity rather than reduce the population size. Reducing the population size only encourages exponential growth to begin once again. Farmers can reduce the carrying capacity for a pest by alternating rows of different crops rather than growing one type of crop throughout the entire field.

Exponential growth produces a J-shaped curve because population growth accelerates over time. Logistic growth produces an S-shaped curve because the population size stabilizes when the carrying capacity of the environment has been reached.

Survivorship ✗

Population growth patterns assume that populations are made up of identical individuals. Actually, the individuals are in different stages of their life spans. Let us consider how many members of an original group of individuals born at the same time, called a **cohort,** are still alive after certain intervals of time. If we plot the number surviving, a survivorship curve is produced.

For the sake of discussion, three types of idealized survivorship curves are recognized (Fig. 33.7*a*). The type I curve is characteristic of a population of humans, in which most individuals survive well past the midpoint, and death does not come until near the end of the life span. On the

other hand, the type III curve is typical for a population of oysters, in which most individuals die very young. In the type II curve, survivorship decreases at a constant rate throughout the life span. This has been found typical of a population of songbirds.

Sometimes populations do not fit any of these curves exactly. For example, in a cohort of *Poa annua* plants, most individuals survive till six to nine months, and then the chances of survivorship diminish at an increasing rate.

Much can be learned about the life history of a species by studying its survivorship curve. Would you predict that most or only a few members of a population with a type III survivorship curve are contributing offspring to the next generation? Obviously, since death comes early for most members, only a few are living long enough to reproduce. In the other two types of survivorship curves, which members contribute the most offspring to the next generation?

Populations have a pattern of survivorship that becomes apparent from studying the survivorship curve of a cohort.

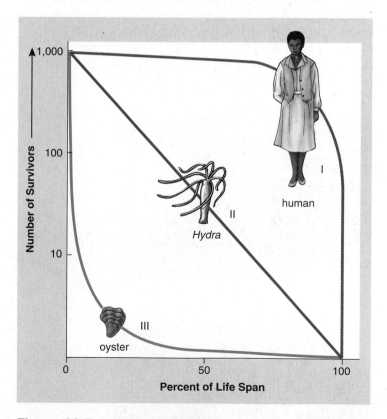

Figure 33.7   Survivorship curves.
Human beings have a type I survivorship curve: the individual usually lives a normal life span, and then death is increasingly expected. Hydras have a type II curve in which the chances of surviving are the same for any particular age. Oysters have a type III curve: most deaths occur during the free-swimming larva stage, but oysters that survive to adulthood usually live a normal life span.

Human Population Growth

The human population has an exponential pattern of growth and a J-shaped growth curve (Fig. 33.8). It is apparent from the position of 2001 on the growth curve in Figure 33.8c that growth is still quite rapid. The equivalent of a medium-sized city (225,000) is added to the world's population every day, and 82 million (the equivalent of the combined populations of the United Kingdom, Norway, Ireland, Iceland, Finland, Netherlands, and Denmark) are added every year.

The present situation can be appreciated by considering the doubling time. The **doubling time**—the length of time it takes for the population size to double—is now estimated to be 53 years. Such an increase in population size will put extreme demands on our ability to produce and distribute resources. In 53 years, the world will need double the amount of food, jobs, water, energy, and so on just to maintain the present standard of living.

Many people are gravely concerned that the amount of time needed to add each additional billion persons to the world population has become shorter. The first billion didn't occur until 1800; the second billion arrived in 1930; the third billion in 1960; and today there are over 6 billion. Only if the per capita rate of increase declines can there be zero population growth, when the birthrate equals the death rate, and population size remains steady. The world's population may level off at 8, 10.5, or 14.2 billion, depending on the speed with which the per capita rate of increase declines.

More-Developed Versus Less-Developed Countries

The countries of the world can be divided into two groups. The **more-developed countries (MDCs),** typified by countries in North America and Europe, are those in which population growth is low and the people enjoy a good standard of living (Fig. 33.8a). The **less-developed countries (LDCs),** such as countries in Latin America, Africa, and Asia, are those in which population growth is expanding rapidly and the majority of people live in poverty (Fig. 33.8b). (Sometimes the term *third-world countries* is used to mean the less-developed countries. This term was introduced by those who thought of the United States and Europe as the first world and the former USSR as the second world.)

The MDCs doubled their populations between 1850 and 1950. This was largely due to a decline in the death rate, the development of modern medicine, and improved socioeconomic conditions. The decline in the death rate was followed shortly thereafter by a decline in the birthrate, so that populations in the MDCs experienced only modest growth between 1950 and 1975. This sequence of events (i.e., decreased death rate followed by decreased birthrate) is termed a **demographic transition.**

Yearly growth of the MDCs as a whole has now stabilized at about 0.1%. The populations of several of the MDCs, including Germany, Greece, Italy, Hungary, and Sweden, are not growing or are actually decreasing in size. In contrast, there is no leveling off and no end in sight to

a.

b.

c.

Source: Population Reference Bureau.

Figure 33.8   World population growth.
People in the **(a)** more-developed countries have a high standard of living and will contribute the least to world population growth, while people in the **(b)** less-developed countries have a low standard of living and will contribute the most to world population growth. **c.** The graph shows world population in the past with estimates to 2150.

U.S. population growth. Although yearly growth of the United States is only 0.6%, many people immigrate to the United States each year. In addition, an unusually large number of babies were born between 1947 and 1964 (called a baby boom). Therefore, a large number of women are still of reproductive age.

Although the death rate began to decline steeply in the LDCs following World War II with the importation of modern medicine from the MDCs, the birthrate remained high. The yearly growth of the LDCs peaked at 2.5% between 1960 and 1965. Since that time, a demographic transition has occurred: the decline in the death rate slowed, and the birthrate fell. The yearly growth is now 1.9%. Still, because of exponential growth, the population of the LDCs may explode from 4.9 billion today to 11 billion in 2100. Most of this growth will occur in Africa, Asia, and Latin America. Ways to greatly reduce the expected increase have been suggested:

1. Establish and/or strengthen family planning programs. A decline in growth is seen in countries with good family planning programs supported by community leaders. Currently, 25% of women in sub-Saharan Africa say they would like to delay or stop childbearing, yet they are not practicing birth control; likewise, 15% of women in Asia and Latin America have an unmet need for birth control.
2. Use social progress to reduce the desire for large families. Many couples in the LDCs presently desire as many as four to six children. But providing education, raising the status of women, and reducing child mortality are desirable social improvements that could cause them to think differently.
3. Delay the onset of childbearing. A delay in the onset of childbearing and wider spacing of births could cause a temporary decline in the birthrate and reduce the present reproductive rate.

Age Distributions ✗

The **age-structure diagrams** of MDCs and LDCs in Figure 33.9 divide the population into three age groups: dependency, reproductive, and postreproductive. The LDCs are experiencing a population momentum because they have more women entering the reproductive years than older women leaving them.

Laypeople are sometimes under the impression that if each couple has two children, **zero population growth** (no increase in population size) will take place immediately. However, **replacement reproduction,** as it is called, will still cause most countries today to continue growing due to the age structure of the population. If there are more young women entering the reproductive years than there are older women leaving them, replacement reproduction will still result in population growth.

Many MDCs have a stable age structure, but most LDCs have a youthful profile—a large proportion of the

population younger than 15. This means that the LDC populations will still expand greatly, even after replacement reproduction is attained. The more quickly replacement reproduction is achieved, however, the sooner zero population growth will result.

Currently, the less-developed countries are expanding dramatically because of exponential growth.

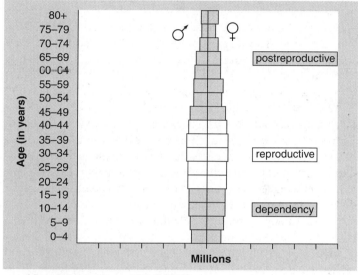

a. More developed countries (MDCs)

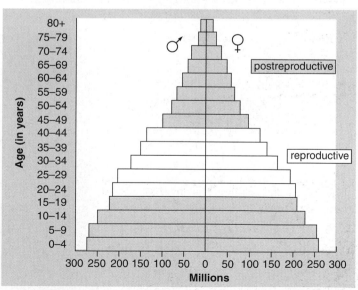

b. Less developed countries (LDCs)

Figure 33.9   Age-structure diagrams (1998).
The diagrams illustrate that (**a**) the MDCs are approaching stabilization, whereas (**b**) the LCDs will expand rapidly due to their age distributions.

Science Focus

The United States Population ✗

The United States Constitution requires that a national census be taken every ten years. The latest census concluded that the U.S. population numbered 281,421,906 persons in April 2000. This is a gain of 32.7 million since the last census in 1990 and the largest increase ever between censuses. It is clear that the United States, unlike other more developed countries, is still undergoing exponential growth. Most industrialized countries now have a growth rate at or lower than 0.1 percent but the growth rate for the United States is 0.6 percent. This positive difference results in the addition of many persons within a short period of time because the population is very large. Also, many persons immigrate to the United States each year. By 2001, the United States population was believed to be around 285 million.

Minorities

Most people in the United States are Caucasian but their percent of total population is decreasing because the major minorities—the Hispanic, Black, and Asian populations—are increasing at a faster rate than the Caucasian population. The Hispanic population grew the most over the past ten years, followed by the Asian and then Black population.

The American Indian population increased only slightly from 0.6 to 0.7 percent of the population. For the first time, Americans were allowed to identify with more than one ethnic group and about 2 percent did so. A significant number of these chose American Indian as one of the ethnic groups and then they could not be counted as belonging to this single minority.

California is now a "minority majority" state meaning that those Californians who considered themselves Caucasian make up less than one-half the population. Other statistics regarding minorities are also of interest.

Hispanics

Hispanics who indicate they are from Mexico account for nearly 60 percent of this particular segment of the total population. Whereas ten years ago Hispanics were 9 percent of the total population, now they are 12.5 percent. This is the largest percent increase of all the minorities.

Altogether, the percent of persons identified as Hispanic is now slightly larger than those who consider themselves Black. The Hispanic population overtook the Black population as the largest minority because this population has a higher birth rate and higher immigration rate than do Blacks.

Hispanics are now more dispersed than formerly. Hispanics no longer tend to live in the Southwest and the West and cities such as Miami, New York, New Jersey, and Chicago. They are dispersing to smaller cities and even rural areas in the Midwest, South, and Northeast.

Asians

The Asian population includes people from countries such as China, Philippines, Japan, Korea, India, Pakistan, and so forth. The Asian population increased by two-thirds, but even

Figure 33A Percent changes in U.S. population.
Over the past ten years the Caucasian population dropped in percent of total population while the Hispanic, Asian, and Black populations increased in percent of total population. Not shown is the American Indian population, which rose modestly from 0.6 to 0.7 percent. About 2 percent of the total population identified with more than one ethnicity, according to the 2000 census.

Ethnicity: Caucasian
2000 69.1%
1990 75.6%
−6.5%

Ethnicity: Hispanic
2000 12.5%
1990 9.0%
+3.5%

so they account for only 3.7 percent of the year 2000 population. About 60% of Asian Americans are foreign born; therefore, immigration accounts for most of the increase in the Asian population.

The Asian population is also more widely dispersed than before. As many as 36 percent live in California, but there are also large Asian American populations now living in Georgia, Pennsylvania, Minnesota, and several other states. Many have dispersed to where they can find jobs rather than where they can find others of the same ethnic background.

Blacks

Most Blacks in the United States identify themselves as African Americans. The Black population increased at a faster rate than the Caucasian population, but had a much slower rate of increase than the Hispanics and also a slower rate than the Asians. Unlike these other minority populations, there is little immigration from other countries, and this accounts for why the Black population did not increase as fast as other minorities.

African Americans remain the predominant minority group in the South, however. While they are about 12 percent of the U.S. population as a whole, they are now about 19 percent of the population in the South. During the 1990s, many blacks left the North and decided to return to the South because of the region's booming economy, attractive life style, improved racial climate, and because they already have relatives there.

What's Ahead?

The 2000 Census found that nearly 40 percent of the population under age 18 (i.e., children) belong to a minority group. This means that minorities are sure to increase their percent of the total population during the next decade. More than one-half of the children in California, Arizona, Hawaii, New Mexico, and Texas are members of a minority. By the next census, all these states will likely be minority majorities as California is now.

As mentioned, this was the first time the census allowed individuals to identify with more than one ethnicity. While 2 percent of the total population said they were multiracial, 4 percent of the children indicated they were multiracial. Most were a mixture of any two of these three ethnicities: Caucasian, Black, and Asian. It would appear then that we can safely predict that the percent of people identifying themselves as multiracial will increase over the next ten years.

The number of people living in the Western and Southern states continued to increase faster than those living in the Northeastern states. At present, the South at 36 percent has the largest share of the population. But Nevada, Arizona, Colorado, Utah, and Idaho—all in the West—were the five fastest growing states over the decade. The other states mentioned grew at a more modest rate. The West (now 22 percent) will most likely overtake the Midwest (now 23 percent) before the next census, just as it overtook the Northeast (now 19 percent) after the 1990 Census.

Ethnicity: Asian
2000 3.7%
1990 2.8%
 +0.9%

Ethnicity: Black
2000 12.1%
1990 11.7%
 +0.4%

33.3 Regulation of Population Growth

Ecologists want to determine the factors that regulate population growth. They have observed two types of life history patterns: an *opportunistic pattern* and an *equilibrium pattern.* Members of opportunistic populations are small in size, mature early, and have a short life span. They tend to produce many relatively small offspring and to forgo parental care in favor of a greater number of offspring (Fig. 33.10*a*). The more offspring, the more likely it is that some of them will survive a population crash. Classic examples of such opportunistic species are many insects and weeds.

In contrast, the size of equilibrium populations remains pretty much at the carrying capacity (Fig. 33.10*b*). Resources such as food and shelter are relatively scarce for these individuals, and those who are best able to compete will have the largest number of offspring. These organisms allocate energy to their own growth and survival and to the growth and survival of their offspring. Therefore, they are fairly large, are slow to mature, and have a fairly long life span. They are specialists rather than colonizers and tend to become extinct when their normal way of life is destroyed. The best possible examples of equilibrium species are found among birds and mammals. The Florida panther is the largest animal in the Florida Everglades, requires a very large range, and produces few offspring, which must be cared for. Currently, the Florida panther is unable to compensate for a reduction in its range, and is therefore on the verge of extinction.

For some time, ecologists have recognized that the environment contains both abiotic and biotic components. They suggested that abiotic factors such as weather and natural disasters were **density-independent.** By this they meant that the number of organisms present did not influence the effect of the factor. For example, fires don't necessarily kill a larger percentage of individuals as the population increases in size. On the other hand, biotic factors such as competition, predation, and parasitism were designated as **density-dependent.** Populations that have the opportunistic life history pattern tend to be regulated by density-independent factors, and those that have the equilibrium life history pattern tend to be regulated by density-dependent factors.

Density-independent and density-dependent factors can often explain the population dynamics of natural populations.

Figure 33.10   Life history patterns.
Dandelions are an opportunistic species with the characteristics noted, and bears are an equilibrium species with the characteristics noted. Often the distinctions between these two possible life history patterns are not as clear-cut as they may seem.

Opportunistic Pattern

Small individuals
Short life span
Fast to mature
Many offspring
Little or no care of offspring

Equilibrium Pattern

Large individuals
Long life span
Slow to mature
Few offspring
Much care of offspring

Competition

Competition occurs when members of different species try to utilize a resource (such as light, space, or nutrients) that is in limited supply. According to the **competitive exclusion principle,** no two species can occupy the same ecological niche at the same time if resources are limiting. The **ecological niche** is the role the organism plays in the community, including its **habitat (where the organism lives)** and its interactions with other organisms. The niche includes the resources an organism uses to meet its energy, nutrient, and survival demands.

While it may seem as if several populations living in the same area are occupying the same niche, it is usually possible to find slight differences. When two species of paramecia are grown separately, each survives; but when they are grown in one test tube, resources are limited and only one species survives (Fig. 33.11). What does it take to have different ecological niches so that extinction of one species is avoided? In another laboratory experiment, two species of paramecia continue to occupy the same tube when one species feeds on bacteria at the bottom of the tube and the other feeds on bacteria suspended in solution. Under these circumstances, **resource partitioning** decreases competition between the two species. As another example, consider that swallows, swifts, and martins all eat flying insects and parachuting spiders. These birds even frequently fly in mixed flocks. But each type of bird has different nesting sites and migrates at a slightly different time of year. Therefore, they are not competing for the same food source when they are feeding their young.

On the Scottish coast, a small barnacle (*Chthamalus stellatus*) lives on the high part of the intertidal zone, and a large barnacle (*Balanus balanoides*) lives on the lower part (Fig. 33.12). Free-swimming larvae of both species attach themselves to rocks at any point in the intertidal zone, where they develop into the sessile adult forms. In the lower zone, the large *Balanus* barnacles seem to either force the smaller *Chthamalus* individuals off the rocks or grow over them. Competition is therefore restricting the range of *Chthamalus* on the rocks. *Chthamalus* is more resistant to drying out than is *Balanus;* therefore, it has an advantage that permits it to grow in the upper intertidal zone.

Competition may result in resource partitioning. When similar species seem to be occupying the same ecological niche, it is usually possible to find differences that indicate resource partitioning has occurred.

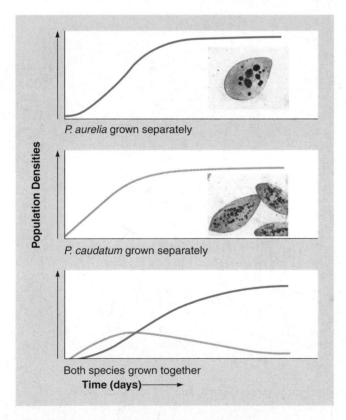

Figure 33.11 **Competition between two laboratory populations of *Paramecium*.**
When grown alone in pure culture (top two graphs), *Paramecium caudatum* and *Paramecium aurelia* exhibit logistic growth. When the two species are grown together in mixed culture (bottom graph), *P. aurelia* is the better competitor, and *P. caudatum* dies out.

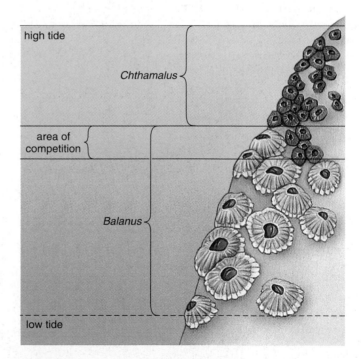

Figure 33.12 **Competition between two species of barnacles.**
Competition prevents two species of barnacles from occupying as much of the intertidal zone as possible. Both exist in the area of competition between *Chthamalus* and *Balanus*. Above this area, only *Chthamalus* survives, and below it only *Balanus* survives.

Predation

Predation occurs when one living organism, called the **predator,** feeds on another, called the **prey.** In the broadest sense, predaceous consumers include not only animals like lions that kill zebras, but also filter-feeding blue whales that strain krill from ocean waters, parasitic ticks that suck blood from victims, and even herbivorous deer that browse on trees and bushes.

Predator-Prey Population Dynamics

Do predators reduce the population density of prey? On the face of it, you would probably hypothesize that they do, and your hypothesis would certainly be supported by a laboratory study in which the protozoans *Paramecium caudatum* (prey) and *Didinium nasutum* (predator) were grown together in a culture medium. *Didinium* ate all the *Paramecium* and then died of starvation. In nature, we can find a similar example. When a gardener brought prickly-pear cactus to Australia from South America, the cactus spread out of control until millions of acres were covered with nothing but cacti. The cacti were brought under control when a moth from South America, whose caterpillar feeds only on the cactus, was introduced. Now both cactus and moth are found at greatly reduced densities in Australia.

Mathematical formulas can predict a cycling of predator and prey populations instead of a steady state. Cycling could occur if (1) the predator population overkills the prey and then the predator population also declines in number, and (2) the prey population overshoots the carrying capacity and suffers a crash, and then the predator population follows suit because of a lack of food. In either case, the result would be a series of peaks and valleys, with the predator population lagging slightly behind the prey.

A famous case of predator/prey cycles occurs between the snowshoe hare and the Canadian lynx, a type of small cat (Fig. 33.13). The snowshoe hare is a common herbivore in the coniferous forests of North America, where it feeds on terminal twigs of various shrubs and small trees. The Canadian lynx feeds on snowshoe hares but also on ruffed grouse and spruce grouse, two types of birds. Investigators at first assumed that the lynx had brought about the decline of the hare population, as would be explained per formula (1). But others noted that the decline in snowshoe hare abundance was accompanied by low growth and reproductive rates that could be signs of a food shortage per formula (2). It appears that both explanations apply to the data. In other words, both a predator-hare cycle and a hare-food cycle have combined to produce an overall effect, which is observed in Figure 33.13. It's interesting to note that the population densities of the grouse populations also cycle, perhaps because the lynx switches to this food source when the hare population declines. Predators and prey do not normally exist as simple, two-species systems, and therefore abundance patterns should be viewed with the complete community in mind.

Interactions between predator and prey and between prey and its food source can produce complex cycles.

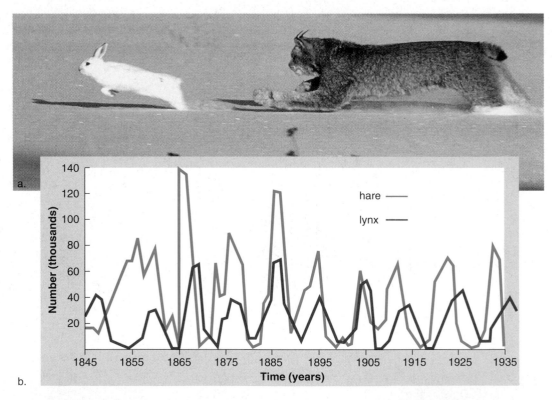

Figure 33.13   Predator-prey interaction between a lynx and a snowshoe hare.

a. A Canadian lynx *(Lynx canadensis)* is a solitary predator. A long, strong forelimb with sharp claws grabs its main prey, the snowshoe hare *(Lepus americanus)*. **b.** The number of pelts received yearly by the Hudson Bay Company for almost 100 years shows a pattern of ten-year cycles in population densities. The snowshoe hare population reaches a peak abundance before that of the lynx by a year or more.

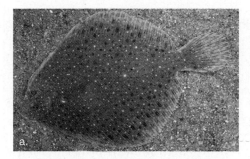

Figure 33.14   **Antipredator defenses.**
a. Concealment. Flounders can take on the same coloration as their background. **b.** Fright. The South American lantern fly has a large false head that resembles that of an alligator. This may frighten a predator into thinking it is facing a dangerous animal. **c.** Warning coloration. The skin secretions of dart-poison frogs are so poisonous that they were used by natives to make their arrows instant lethal weapons. The coloration of these frogs warns others, such as birds, to beware.

Antipredator Defenses

While predators have evolved strategies to secure the maximum amount of food with minimal expenditure of energy, prey organisms have evolved strategies to escape predation. **Coevolution** is present when two species adapt in response to selective pressure imposed by the other.

In plants, the sharp spines of the cactus, the pointed leaves of holly, and the tough, leathery leaves of the oak tree all discourage predation by insects. Plants even produce poisonous chemicals, some of which are hormone analogues that interfere with the development of insect larvae. Animals have varied antipredator defenses. Some effective defenses are concealment, fright, and warning coloration (Fig. 33.14), as well as mimicry.

Mimicry   **Mimicry** occurs when one species resembles another that possesses an antipredator defense. Mimicry can help a predator capture food or a prey avoid capture. For example, snapping turtles have tongues, and angler fishes have lures that resemble worms for the purpose of bringing fish within reach. To avoid capture, there are inchworms that resemble twigs and caterpillars that can transform themselves into shapes resembling snakes.

Batesian mimicry (named for Henry Bates, who discovered it) occurs when a prey mimics another species that has a successful antipredator defense. Many examples of Batesian mimicry involve warning coloration. Among flies of the family Syrphidae, which feed on the nectar and pollen of flowers, one species resembles the wasp *Vespula arenaria* so closely that it is difficult to tell them apart (Fig. 33.15). Once a predator experiences the defense of the wasp, it remembers the coloration and avoids all animals that look similar. There are also examples of species that have the same defense and resemble each other. For example, many coral snake species have brilliant red, black, and yellow body rings. And the stinging insects—bees, wasps, and hornets—all have the famil-

Figure 33.15   **Mimicry.**
Flies of the family Syrphidae are called flower flies because they are likely to be found on flowers, where they drink nectar and eat pollen. Some species mimic a wasp, which is protected from predation by its sting.

iar black and yellow color bands. Mimics that share the same protective defense are called Müllerian mimics after Fritz Müller, who discovered this form of mimicry.

Just as with other antipredator defenses, behavior plays a role in mimicry. Mimicry works better if the mimic acts like the model. For example, beetles that resemble a wasp actively fly from place to place and spend most of their time in the same habitat as the wasp model. Their behavior makes them resemble a wasp to an even greater degree.

Prey escape predation by utilizing concealment, fright, flocking together, warning coloration, and mimicry.

Symbiosis

Symbiosis refers to close interactions between members of two populations. Three types of symbiotic relationships have traditionally been defined—parasitism, commensalism, and mutualism. Table 33.2 lists these three categories in terms of their benefits to one species or the other; of the three types, only mutualism may increase the population size of both species. However, some ecologists now consider it wasted effort to try to classify symbiotic relationships into these categories, since the amount of harm or good two species do one another is dependent on what the investigator chooses to measure. Although the following discussion describes the traditional classification system, bear in mind that symbiotic relationships do not always fall neatly into these three categories.

Parasitism

Parasitism is a symbiotic relationship in which the *parasite* derives nourishment from another organism called the *host.* Therefore, the parasite benefits and the host is harmed. Parasites occur in all kingdoms of life. Bacteria (e.g., strep infection), protists (e.g., malaria), fungi (e.g., rusts and smuts), plants (e.g., mistletoe), and animals (e.g., leeches) all contain parasitic species. The effects of parasites on the health of the host can range from slight weakening to actually killing them over time.

In addition to providing nourishment, host organisms also provide their parasites with a place to live and reproduce, as well as a mechanism for dispersing offspring to new hosts. Many parasites have both a primary and secondary host. The secondary host may be a vector that transmits the parasite to the next primary host. As an example, consider the deer ticks *Ixodes dammini* and *I. ricinus* in the eastern and western United States, respectively. Deer ticks are arthropods that go through a number of stages (egg, larva, nymph, adult). They are so named because adults feed and mate on white-tailed deer in the fall. The female lays her eggs on the ground, and when the eggs hatch in the spring, they become larvae that feed primarily on white-footed mice. If a mouse is infected with the bacterium *Borrelia burgdorferi*, the larvae become infected also. The fed larvae overwinter and molt the next spring to become nymphs that can, by chance, take a blood meal from a human. At this time, the tick may pass the bacterium on to a human who subsequently comes down with Lyme disease, characterized by arthritic-like symptoms. The fed nymphs develop into adults, and the cycle begins again.

Commensalism

Commensalism is a symbiotic relationship between two species in which one species is benefited and the other is neither benefited nor harmed. Often one species provides a

Table 33.2	Symbiosis	
	Species 1	**Species 2**
Parasitism*	Benefited	Harmed
Commensalism	Benefited	No effect
Mutualism	Benefited	Benefited

*Can be considered a type of predation.

Figure 33.16   Egret symbiosis.
Cattle egrets eat insects off and around various animals, such as an African cape buffalo.

Figure 33.17   Cleaning symbiosis.
Cleaners remove parasites from their clients. Here a cleaner wrasse is entering the mouth of a spotted sweetlip.

home and/or transportation for the other species, such as when barnacles attach themselves to the backs of whales. Remoras are fishes that attach themselves to the bellies of sharks by means of a modified dorsal fin acting as a suction cup. The remoras obtain a free ride and also feed on the remains of the shark's meals. Clownfishes live within the waving mass of tentacles of sea anemones. Because most fishes avoid the poisonous tentacles of the anemones, clownfishes are protected from predators.

If clownfishes attract other fishes on which the anemone can feed, this relationship borders on mutualism. Other examples of commensalism may also be mutualistic. For example, cattle egrets benefit from grazing near cattle because the cattle flush insects and other animals from the vegetation as they graze (Fig. 33.16).

Mutualism

Mutualism is a symbiotic relationship in which both members of the association benefit. Mutualistic relations need not be equally beneficial to both species. We can imagine that the relationship between plants and their animal pollinators began when herbivores, such as insects, feasted on pollen. The provision of nectar may have spared the pollen and at the same time allowed the animal to become an instrument of pollination. Lichens can grow on rocks because their fungal member conserves water and leaches minerals that are provided to the algal partner, which photosynthesizes and provides organic food for both populations. When each is grown separately in the laboratory, the algae seem to do fine, but the fungus does poorly. For that reason, it's been suggested that the fungus is parasitic at least to a degree on the algae.

Ants form mutualistic relationships with both plants and insects. In tropical America, the bullhorn acacia tree is adapted to provide a home for ants of the species *Pseudomyrmex ferruginea*. Unlike other acacias, this species has swollen thorns with a hollow interior, where ant larvae can grow and develop. In addition to housing the ants, the acacias provide them with food. The ants feed from nectaries at the base of the leaves and eat fat and protein–containing nodules called Beltian bodies, which are found at the tops of some of the leaves. The ants constantly protect the plant from caterpillars of moths and butterflies and from other plants that might shade it because, unlike other ants, they are active twenty-four hours a day. Indeed, when the ants on experimental trees were poisoned, the trees died.

Cleaning symbiosis is a symbiotic relationship in which the individual being cleaned is often a vertebrate. Crustaceans, fish, and birds act as cleaners and are associated with a variety of vertebrate clients. Large fish in coral reefs line up at cleaning stations and wait their turn to be cleaned by small fish that even enter the mouths of the large fish (Fig. 33.17). It's been suggested that cleaners may be exploiting the relationship by feeding on host tissues as well as on ectoparasites. On the other hand, cleaning could ultimately lead to net gains in client fitness.

Symbiotic relationships occur between species, but it may be too simplistic to divide them into parasitic, commensalistic, and mutualistic relationships.

Bioethical Focus   Population Growth in Less-Developed Countries

In less-developed countries, population experts have discovered that women who have equal rights with men tend to have fewer children. Also, family planning leads to healthier women, and healthier women have healthier children, and the cycle continues. Women no longer have to have many babies for a few to survive. More education is also helpful because better educated people are more interested in postponing childbearing and promoting women's rights.

"There isn't any place where women who have the choice haven't chosen to have fewer children," says Beverly Winikoff at the Population Council in New York City. Bangladesh is a case in point.

Bangladesh is one of the densest and poorest countries in the world. In 1990, the birthrate was 4.9 children per woman, and now it is 3.3. This achievement was due in part to the Dhaka-based Grameen Bank, which loans small amounts of money, mostly to destitute women to start a business. The bank discovered that when women start making decisions about their lives, they also start making decisions about the size of their families. Family planning within Grameen families is twice as common as the national average; in fact, those women who get a loan promise to keep their families small! Also helpful has been the network of village clinics that counsel women who want to use contra-

ceptives. The expression "contraceptives are the best contraceptives" refers to the fact that you don't have to wait for social changes to get people to use contraceptives—the two feed back on each other.

Decide Your Opinion

1. Do you think less-developed countries should simply make contraception available, or should more persuasive methods be employed? Explain.
2. Do you think that more-developed countries should be concerned about population growth in the less-developed countries? Why or why not?
3. Are you in favor of foreign aid to help countries develop family planning programs? Why or why not?

Summarizing the Concepts

33.1 Scope of Ecology

Ecology is the study of the interactions of organisms with other organisms and with the physical environment. Ecology encompasses several levels of study: organism, population, community, ecosystem, and biosphere.

Communities differ in their composition (populations found there) and their diversity (species richness and relative abundance). Abiotic factors such as latitude and environmental gradients seem to largely control community composition. A change in community composition over time is called ecological succession. A climax community is associated with a particular geographic area.

33.2 Population Characteristics and Growth

Population density is simply the number of individuals per unit area or volume. Distribution of these individuals can be uniform, random, or clumped. The members of most populations, including the human population, are clumped. Limiting factors such as water, temperature, and availability of organic nutrients often determine a population's distribution.

Future population size is dependent upon the per capita rate of increase. The per capita rate of increase is calculated by subtracting the number of deaths from the number of births and dividing by the number of individuals in the population. (Immigration and emigration are usually considered to be equal.) Every population has a biotic potential, the greatest possible per capita rate of increase under ideal circumstances.

Two possible patterns of population growth are considered. Exponential growth results in a J-shaped curve because, as the population increases in size, so does the number of new members. Most environments restrict growth, and exponential growth cannot continue indefinitely. Under these circumstances, logistic growth occurs, and an S-shaped growth curve results. When the population reaches carrying capacity, the population stops growing because environmental resistance opposes biotic potential.

Populations tend to have one of three types of survivorship curves, depending on whether most individuals live out the normal life span, die at a constant rate regardless of age, or die early.

The human population is expanding exponentially, and it is unknown when the population size will level off. Most of the expected increase will occur in certain LDCs (less-developed countries) of Africa, Asia, and Latin America. Support for family planning, human development, and delayed childbearing could help lessen the expected increase.

33.3 Regulation of Population Growth

The patterns of population growth patterns have been used to suggest that life history patterns vary from those species that are opportunists to those that are in equilibrium with the carrying capacity of the environment. Opportunistic species produce many young within a short period of time and rely on rapid dispersal to new, unoccupied environments. Their population size is regulated by density-independent factors. Equilibrium species produce a limited number of young, which they nurture for a long time, and their population size is regulated by density-dependent factors such as competition and predation.

According to the competitive exclusion principle, no two species can occupy the same niche at the same time when resources are limiting. When such resources as food and living space are divided between two or more species, resource partitioning has occurred. Barnacles competing on the Scottish coast are an example of ongoing competition in which living space is partitioned.

Predator-prey interactions between two species are influenced by environmental factors. Sometimes predation can cause prey populations to decline and remain at relatively low densities, or a cycling of population densities may occur. Prey defenses take many forms: concealment, use of fright, and warning coloration are three possible mechanisms. Batesian mimicry usually occurs when one species has the warning coloration but lacks the defense. Müllerian mimicry occurs when two species with the same warning coloration have the same defenses.

In a parasitic relationship, the parasite benefits and the host is harmed. In a commensalistic relationship, neither party is harmed. And in a mutualistic relationship, both partners benefit. Parasites often utilize more than one host, as is the case with deer ticks. In a commensalistic relationship, one species often provides a home and/or transportation for another species. Mutualistic relationships are quite varied. Flowers and their pollinators, algae and fungi in a lichen, ants who have a plant partner, and cleaning symbiosis are all examples of mutualistic relationships.

Testing Yourself

Choose the best answer for each question.

1. Which of these levels of ecological study involves an interaction between abiotic and biotic components?
   a. organisms
   b. populations
   c. communities
   d. ecosystem
   e. All of these are correct.

2. The more diverse community would have
   a. the greater number of different populations.
   b. unequal numbers of individuals in each population.
   c. fairly equal numbers of individuals in each population.
   d. rare plant and animal populations.
   e. Both a and c are important factors.

3. The individualistic model of community composition states that
   a. the individuals in the community are present due to individual preferences, which are not dependent on any biotic or abiotic factor present.
   b. populations interact heavily with one another, and therefore become dependent upon each other within the community.
   c. populations inhabit a community because of the abiotic factors present, not necessarily because of other populations that are present.
   d. All of these are true.

4. A J-shaped growth curve should be associated with
   a. exponential growth.
   b. biotic potential.
   c. no environmental resistance.
   d. high per capita rate of increase.
   e. All of these are correct.

5. An S-shaped growth curve
   a. occurs when there is no environmental resistance.
   b. includes an exponential growth phase.
   c. occurs if survivorship is short-lived.
   d. occurs in natural populations but not in laboratory ones.
   e. All of these are correct.

6. If a population has a type I survivorship curve (most of its members live the entire life span), which of these would you also expect?
   a. a single reproductive event per adult
   b. most individuals reproduce
   c. sporadic reproductive events
   d. reproduction occurring near the end of the life span
   e. None of these are correct.

7. A pyramid-shaped age distribution means that
   a. the prereproductive group is the largest group.
   b. the population will grow for some time in the future.
   c. more young women are entering the reproductive years than older women leaving them.
   d. the country is more likely an LDC than an MDC.
   e. All of these are correct.

8. Which of these is a population-independent regulating factor?
   a. competition   d. weather
   b. predation   e. resource availability
   c. size of population

9. An equilibrium life history pattern does NOT include
   a. large individuals.
   b. long life span.
   c. individuals slow to mature.
   d. few offspring.
   e. little or no care of offspring.

10. The human population
    a. is undergoing exponential growth.
    b. is not subject to environmental resistance.
    c. fluctuates from year to year.
    d. only grows if emigration occurs.
    e. All of these are correct.

11. Six species of monkeys are found in a tropical forest. Most likely, they
    a. occupy the same ecological niche.
    b. eat different foods and occupy different ranges.
    c. spend much time fighting each other.
    d. are from different stages of succession.
    e. All of these are correct.

12. Leaf cutter ants keep fungal gardens. The ants provide food for the fungus but also feed on the fungus. This is an example of
    a. competition.   d. parasitism.
    b. predation.   e. mutualism.
    c. commensalism.

13. Clownfishes live among sea anemone tentacles, where they are protected. If the clownfish provides no service to the anemone, this is an example of
    a. competition.   d. commensalism.
    b. predation.   e. mutualism.
    c. parasitism.

14. Two species of barnacles vie for space in the intertidal zone. The one that remains is
    a. the better competitor.
    b. better adapted to the area.
    c. the better predator on the other.
    d. the better parasite on the other.
    e. Both a and b are correct.

15. A bullhorn acacia provides a home and nutrients for ants. If the relationship is mutualistic,
    a. the plant is under the control of pheromones produced by the ants.
    b. the ants protect the plant.
    c. the plants and the ants compete with one another.
    d. the plants and the ants have coevolved to occupy different ecological niches.
    e. All of these are correct.

16. An opportunistic life history pattern does NOT include
    a. small individuals.
    b. long life span.
    c. fast to mature.
    d. many offspring.
    e. little or no care of offspring.

17. The frilled lizard of Australia suddenly opened its mouth wide and unfurled folds of skin around its neck. Most likely, this was a way to
    a. conceal itself.
    b. warn that it was noxious to eat.
    c. scare a predator.
    d. scare its prey.
    e. All of these are correct.

18. When one species mimics another species, the mimic sometimes
    a. lacks the defense of the model.
    b. possesses the defense of the model.
    c. is brightly colored.
    d. competes with another mimic.
    e. All of these are correct.

19. Which of these models of succession are mismatched?
    a. climax-pattern model—last stage is the one typical of that community.
    b. facilitation model—each stage helps the next stage take place.
    c. inhibition model—each stage hinders the next stage from occurring.
    d. tolerance model—chance determines which plants are present, and life cycles determine how long each stage is present.
    e. None of these are mismatched.

20. The ecological niche of a population
    a. includes its adaptations.
    b. includes its way of life.
    c. includes its habitat.
    d. avoids direct competition.
    e. All of these are correct.

21. Label this S-shaped growth curve:

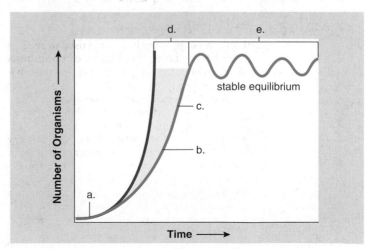

e-Learning Connection   www.mhhe.com/maderinquiry10

Concepts	Questions	Media Resources*
33.1 Scope of Ecology		

• Ecology is the study of the interactions of organisms with each other and the physical environment.
• The study of ecology encompasses the individual, the population, the community, the ecosystem, and the biosphere.
• Abiotic and biotic factors influence community composition and diversity.
• Ecological succession is a change in species composition and community structure and organization over time.
	1. Describe the individualistic model of community composition.
	2. How does primary succession differ from secondary succession?
		Essential Study Partner
		  Introduction
		  Organization
		  Succession
		
		General Biology Weblinks
		  Ecology

33.2 Population Characteristics and Growth		

• Population size depends upon births, deaths, immigration, and emigration.
• Two patterns of population growth (exponential and logistic) have been developed.
• Mortality within a population is often illustrated by a survivorship curve.
• The human population is still growing exponentially, and how long this can continue is not known.
	1. How does environmental resistance influence the rate of growth of a population?
	2. What is a demographic transition?
		Essential Study Partner
		  Introduction
		  Characteristics
		  Growth
		  Human Population
		  Control of Human
		    Populations
		
		Art Quizzes
		  Two Models of Population
		    Growth
		  Survivorship Curves
		  History of Human
		    Population Size
		  Population Pyramids from
		    1990
		  Geometric and Arithmetic
		    Progressions
		
		Animation Quizzes
		  Exponential Population
		    Growth
		  Stages of Population Growth

33.3 Regulation of Population Growth		

• Life history patterns range from one in which many young receive little care to one in which few young receive much care.
• Factors that affect population size are classified as density-independent and density-dependent.
• Competition often leads to resource partitioning, which reduces competition between species.
• Predation often reduces prey population density, which in turn can lead to a reduction in predator population density.
• Symbiotic relationships include parasitism, commensalism, and mutualism.
	1. Does competition tend to lead toward competitive exclusion or resource partitioning?
	2. What is coevolution?
		Essential Study Partner
		  Size Regulation
		  Life History
		
		Art Quizzes
		  Competition and Niches
		  Predator-Prey Cycle
		
		Case Study
		  The Wolf in Yellowstone
		    National Park

*For additional Media Resources, see the Online Learning Center.

34

Ecosystems and Human Interferences

Chapter Concepts

34.1 The Nature of Ecosystems
- What are the two types of components in an ecosystem? 698
- Why are autotrophs called producers and heterotrophs called consumers? What are the different types of consumers in ecosystems? 698
- What two characteristics determine the organization of ecosystems? Explain. 699

34.2 Energy Flow
- What type of diagram represents the various paths of energy flow in a community? 700–1
- What type of diagram represents a single path of energy flow in a community? 700–1
- What type of diagram illustrates that energy is lost in ecosystems? 701

34.3 Global Biogeochemical Cycles
- Chemical cycling in the biosphere may involve what three components? 702
- What are two examples of gaseous biogeochemical cycles? What is an example of a sedimentary biogeochemical cycle? 702
- How do water, carbon, nitrogen, and phosphorus cycle in the biosphere? 703–9
- What ecological problems are associated with each of the cycles? Explain. 703–9

Piñon trees are a population of organisms in an ecosystem that also includes deer mice, which feed on piñon nuts.

When people in the Southwest first began dying of an unknown cause in 1993, no one knew to blame the warm, wet weather or the fat little deer mice scampering about. The unusually heavy spring rains nourished trees carrying piñon nuts, a favorite food of the deer mouse. The result was a tenfold increase in the deer mouse population. Suddenly, mice were everywhere—inside garages, in backyards, even on Indian reservations in the region. With the mice came exposure to hantaviruses carried in their feces and urine. A new strain of hantavirus led to hantavirus pulmonary syndrome, and many deaths.

Natural ecosystems are composed of populations of organisms—deer mice, piñon trees, and yes, humans—that interact among themselves and the physical environment. The saying goes that in an ecosystem everything is connected to everything else. Thus, in this example, warm, wet weather led to plentiful piñon nuts, which led to a deer mouse explosion, and finally, to illness in humans. This chapter examines the interworkings of ecosystems and how they have been impacted by human beings.

34.1 The Nature of Ecosystems

An **ecosystem** possesses both nonliving (abiotic) and living (biotic) components. The abiotic components include resources, such as sunlight and inorganic nutrients, and conditions, such as type of soil, water availability, prevailing temperature, and amount of wind. The biotic components of an ecosystem are the various populations of organisms.

Biotic Components of an Ecosystem

The populations of an ecosystem are categorized according to their food source (Fig. 34.1). Some populations are autotrophs and some are heterotrophs.

Autotrophs require only inorganic nutrients and an outside energy source to produce organic nutrients for their own use and for all the other members of a community. Therefore, they are called **producers**—they produce food. Photosynthetic organisms produce most of the organic nutrients for the biosphere. Algae of all types possess chlorophyll and carry on photosynthesis in freshwater and marine habitats. Algae make up the phytoplankton, which are photosynthesizing organisms suspended in water. Green plants are the dominant photosynthesizers on land.

Some autotrophic bacteria are chemosynthetic. They obtain energy by oxidizing inorganic compounds such as ammonia, nitrites, and sulfides, and they use this energy to synthesize organic compounds. Chemosynthesizers have been found to support communities in some caves and also at hypothermal vents along deep-sea oceanic ridges. The chemosynthesizers that function in the nitrogen cycle will be discussed in this chapter on page 706.

Heterotrophs need a preformed source of organic nutrients. They are the **consumers**—they consume food. **Herbivores** are animals that graze directly on plants or algae. In terrestrial habitats, insects are small herbivores, while in aquatic habitats, zooplankton, such as protozoans, play that role. **Carnivores** feed on other animals; birds that feed on insects are carnivores, and so are hawks that feed on birds. This example allows us to mention that there are primary consumers (e.g., insects), secondary consumers (e.g., birds), and tertiary consumers (e.g., hawks). Sometimes tertiary consumers are called top predators. **Omnivores** are animals that feed both on plants and animals. As you most likely know, humans are omnivores.

The **decomposers** are nonphotosynthetic bacteria and fungi, such as molds and mushrooms, that break down dead organic matter, including animal wastes. They perform a very valuable service because they release inorganic nutrients that are then taken up by plants once more. Otherwise, plants would have to wait for minerals to be released from rocks. **Detritus** is partially decomposed matter in the water or soil. Fan worms feed on detritus floating in marine waters, while clams take it from the substratum. Earthworms and some beetles, termites, and maggots are soil detritus feeders.

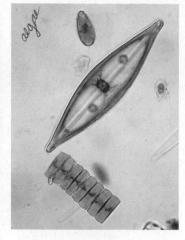

a. Producers

b. Herbivores

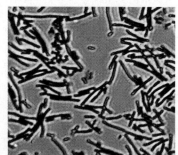

c. Carnivores

d. Decomposers

Figure 34.1   Biotic components.
a. Green plants and diatoms are producers. **b.** Giraffes and caterpillars are herbivores. **c.** An osprey and a praying mantis are carnivores. **d.** Mushrooms and bacteria are decomposers.

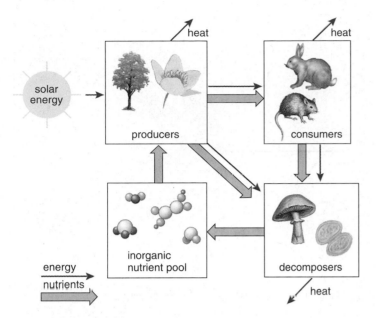

Figure 34.2   Nature of an ecosystem.
Chemicals cycle, but energy flows through an ecosystem. As energy transformations repeatedly occur, all the energy derived from the sun eventually dissipates as heat.

Energy Flow and Chemical Cycling

When we diagram all the biotic components of an ecosystem, it is possible to illustrate that every ecosystem is characterized by two fundamental phenomena: energy flow and chemical cycling. Energy flow begins when producers absorb solar energy, and chemical cycling begins when producers take in inorganic nutrients from the physical environment. Thereafter, producers make organic nutrients (food) directly for themselves and indirectly for the other populations of the ecosystem. Energy flow occurs because as nutrients pass from one population to another all the energy content is eventually converted to heat, which dissipates in the environment. Therefore, most ecosystems cannot exist without a continual supply of solar energy. Chemicals cycle when inorganic nutrients are returned to the producers from the atmosphere or soil, as appropriate (Fig. 34.2).

Only a portion of the organic nutrients made by autotrophs is passed on to heterotrophs because plants use organic molecules to fuel their own cellular respiration. Similarly, only a small percentage of nutrients taken in by heterotrophs is available to higher-level consumers. Figure 34.3 shows why. A certain amount of the food eaten by a herbivore is never digested and is eliminated as feces. Metabolic wastes are excreted as urine. Of the assimilated energy, a large portion is utilized during cellular respiration and thereafter becomes heat. Only the remaining food, which is converted into increased body weight (or additional offspring), becomes available to carnivores.

The elimination of feces and urine by a heterotroph, and indeed the death of all organisms, does not mean that substances are lost to an ecosystem. They represent the

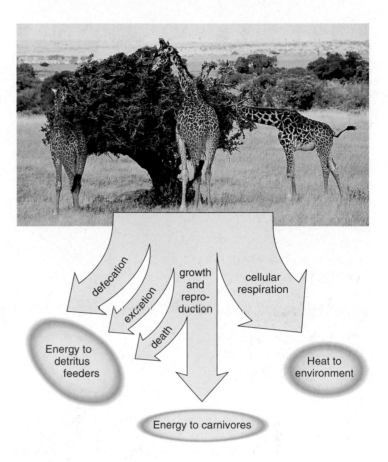

Figure 34.3   Energy balances.
Only about 10% of the food energy taken in by a herbivore is passed on to carnivores. A large portion goes to detritus feeders in the ways indicated, and another large portion is used for cellular respiration.

nutrients made available to decomposers. Since decomposers can be food for other heterotrophs of an ecosystem, the situation can get a bit complicated. Still, we can conceive that all the solar energy that enters an ecosystem eventually becomes heat. And this is consistent with the observation that ecosystems are dependent on a continual supply of solar energy.

The laws of thermodynamics support the concept that energy flows through an ecosystem. The first law states that energy cannot be created (or destroyed). This explains why ecosystems are dependent on a continual outside source of energy, usually solar energy, which is used by photosynthesizers to produce organic nutrients. The second law states that, with every transformation, some energy is degraded into a less available form such as heat. Because plants carry on cellular respiration, for example, only about 55% of the original energy absorbed by plants is available to an ecosystem.

Energy flows through the populations of an ecosystem, while chemicals cycle within and between ecosystems.

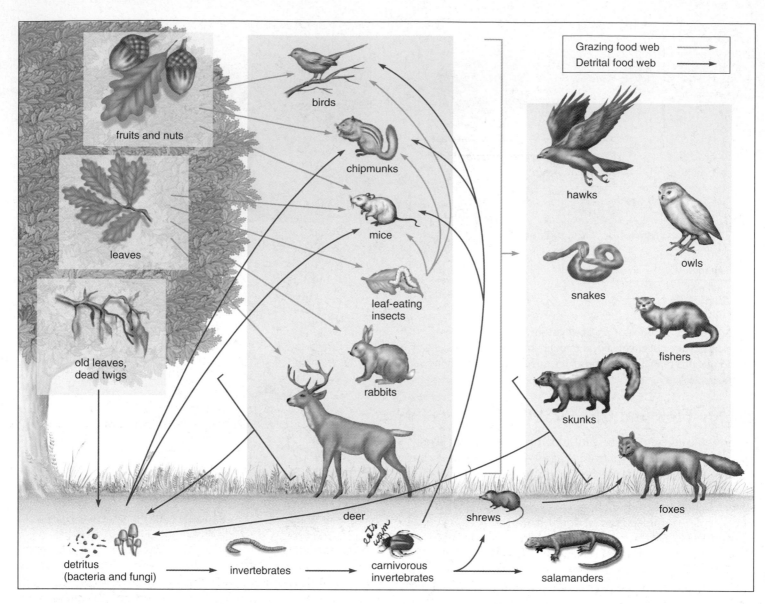

Figure 34.4   Forest food webs.
Two linked food webs are shown for a forest ecosystem: a grazing food web and a detrital food web.

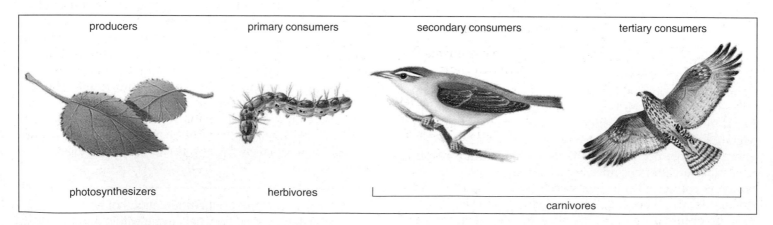

Figure 34.5   Food chain.
Trace a similar grazing food chain in the grazing food web depicted in Figure 34.4.

34.2 Energy Flow

The principles we have been discussing can now be applied to an actual ecosystem—a forest. The various interconnecting paths of energy flow are represented by diagramming a **food web,** as shown in Figure 34.4. The upper part of Figure 34.4 (green arrows) is a **grazing food web** because it begins with trees, such as the oak trees depicted. A **detrital food web** (brown arrows) begins with detritus, partially decayed matter in the soil. Insects in the form of caterpillars feed on leaves, while mice, rabbits, and deer feed on leaf tissue at or near the ground. Birds, chipmunks, and mice feed on fruits and nuts, but they are in fact omnivores because they also feed on caterpillars. These herbivores and omnivores all provide food for a number of different carnivores.

In the detrital food web, detritus, along with the bacteria and fungi of decay, is food for larger decomposers. Because some of these, such as shrews and salamanders, become food for aboveground carnivores, the detrital and the grazing food webs are joined.

We naturally tend to think that aboveground plants like trees are the largest storage form of organic matter and energy, but this is not necessarily the case. In this particular forest, the organic matter lying on the forest floor and mixed into the soil contains over twice as much energy as the leaf matter of living trees. Therefore, more energy in a forest may be funneling through the detrital food web than through the grazing food web.

Trophic Levels

You can see that Figure 34.4 would allow us to link organisms one to another in a straight line, according to who eats whom. Such diagrams that show a single path of energy flow are called **food chains** (Fig. 34.5). For example, in the grazing food web, we could find this **grazing food chain:**

> leaves → caterpillars → tree birds → hawks

And in the detrital food web, we could find this **detrital food chain:**

> dead organic matter → soil microbes → earthworms

A **trophic level** is composed of all the organisms that feed at a particular link in a food chain. In the grazing food web in Figure 34.4, going from left to right, the trees are primary producers (first trophic level), the first series of animals are primary consumers (second trophic level), and the next group of animals are secondary consumers (third trophic level).

Ecological Pyramids

The shortness of food chains can be attributed to the loss of energy between trophic levels. In general, only about 10% of the energy of one trophic level is available to the

next trophic level. Therefore, if a herbivore population consumes 1,000 kg of plant material, only about 100 kg is converted to herbivore tissue, 10 kg to first-level carnivores, and 1 kg to second-level carnivores. The so-called 10% rule of thumb explains why few carnivores can be supported in a food web. The flow of energy with large losses between successive trophic levels is sometimes depicted as a **ecological pyramid** (Fig. 34.6).

Energy losses between trophic levels also results in pyramids based on the number of organisms or the amount of biomass at each trophic level. When constructing such pyramids, problems arise, however. For example, in Figure 34.4, each tree would contain numerous caterpillars; therefore, there would be more herbivores than autotrophs! The explanation, of course, has to do with size. Autotrophs can be tiny, like microscopic algae, or they can be big, like beech trees; similarly, herbivores can be as small as caterpillars or as large as elephants.

Pyramids of biomass eliminate size as a factor because **biomass** is the number of organisms multiplied by their weight. You would certainly expect the biomass of the producers to be greater than the biomass of the herbivores, and that of the herbivores to be greater than that of the carnivores. In aquatic ecosystems, such as lakes and open seas where algae are the only producers, the herbivores may have a greater biomass than the producers when you take

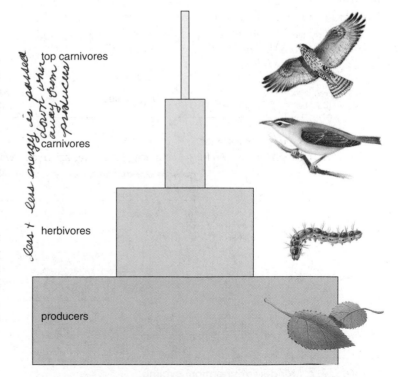

Figure 34.6   Ecological pyramid.
An ecological pyramid reflects the loss of energy from one trophic level to the next. A pyramid of energy, number of organisms, or biomass are all due to this loss of energy.

their measurements. Why? The reason is that, over time, the algae reproduce rapidly, but they are also consumed at a high rate. Any pyramids like this one, which have more herbivores than producers, are called inverted pyramids:

Such problems as these are making ecologists hesitant about using pyramids to describe ecological relationships. One more problem is what to do with the decomposers, which are rarely included in pyramids, even though a large portion of energy becomes detritus in many ecosystems.

The flow of energy through the populations explains in large part the organization of an ecosystem depicted in food webs, food chains, and ecological pyramids.

34.3 Global Biogeochemical Cycles

All organisms require a variety of organic and inorganic nutrients. Carbon dioxide and water are necessary for photosynthesis. Nitrogen is a component of all the structural and functional proteins and nucleic acids that sustain living tissues. Phosphorus is essential for ATP and nucleotide production. In contrast to energy, inorganic nutrients are used over and over again by autotrophs.

Since the pathways by which chemicals circulate through ecosystems involve both living (biotic) and nonliving (geological) components, they are known as **biogeochemical cycles.** For each element, chemical cycling may involve (1) a reservoir—a source normally unavailable to producers, such as fossilized remains, rocks, and deep-sea sediments; (2) an exchange pool—a source from which organisms do generally take chemicals, such as the atmosphere or soil; and (3) the biotic community—through which chemicals move along food chains, perhaps never entering a pool (Fig. 34.7).

With the exception of water, which exists as a gas, liquid, and solid, there are two types of biogeochemical cycles. In a gaseous cycle, exemplified by the carbon and nitrogen cycles, the element returns to and is withdrawn from the atmosphere as a gas. In a sedimentary cycle, exemplified by the phosphorus cycle, the element is absorbed from the sediment by plant roots, passed to heterotrophs, and eventually returned to the soil by decomposers, usually in the same general area.

The diagrams on the next few pages make it clear that nutrients can flow between terrestrial and aquatic ecosystems. In the nitrogen and phosphorus cycles, these nutrients run off from a terrestrial to an aquatic ecosystem and in that way enrich the aquatic ecosystem. Decaying organic matter in aquatic ecosystems can be a source of nutrients for intertidal inhabitants such as fiddler crabs. Sea birds feed on fish but deposit guano (droppings) on land, and in that way phosphorus from the water is deposited on land. It would seem that anything put into the environment in one ecosystem could find its way to another ecosystem. As proof, scientists find the soot from urban areas and pesticides from agricultural fields in the snow and animals of the Arctic.

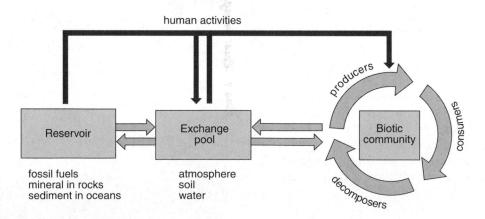

Figure 34.7 Model for chemical cycling.
Chemical nutrients cycle between these components of ecosystems. Reservoirs, such as fossil fuels, minerals in rocks, and sediments in oceans, are normally relatively unavailable sources, but exchange pools such as those in the atmosphere, soil, and water are available sources of chemicals for the biotic community. When human activities remove chemicals from reservoirs and pools and make them available to the biotic community, pollution can result.

The Water Cycle

The **water (hydrologic) cycle** is described in Figure 34.8. Fresh water is distilled from salt water. The sun's rays cause fresh water to evaporate from seawater, and the salts are left behind. Vaporized fresh water rises into the atmosphere, cools, and falls as rain over the oceans and the land.

Water evaporates from land and from plants (evaporation from plants is called transpiration). It also evaporates from bodies of fresh water, but since land lies above sea level, gravity eventually returns all fresh water to the sea. In the meantime, water is contained within standing waters (lakes and ponds), flowing water (streams and rivers), and groundwater.

Some of the water from **precipitation** (e.g., rain, snow, sleet, hail, and fog) sinks, or percolates, into the ground and saturates the earth to a certain level. The top of the saturation zone is called the groundwater table, or simply, the water table. Sometimes groundwater is also located in **aquifers,** rock layers that contain water and release it in

appreciable quantities to wells or springs. Aquifers are recharged when rainfall and melted snow percolate into the soil. In some parts of the United States, especially the arid West and southern Florida, withdrawals from aquifers exceed any possibility of recharge. This is called "groundwater mining." In these locations, the groundwater is dropping, and residents may run out of groundwater, at least for irrigation purposes, within a few short years. Fresh water, which makes up only about 3% of the world's supply of water, is called a renewable resource because a new supply is always being produced. But it is possible to run out of fresh water when the available supply is not adequate or is polluted so that it is not usable.

In the water cycle, fresh water evaporates from the bodies of water. Precipitation on land enters the ground, surface waters, or aquifers. Water ultimately returns to the ocean—even the quantity that remains in aquifers for some time.

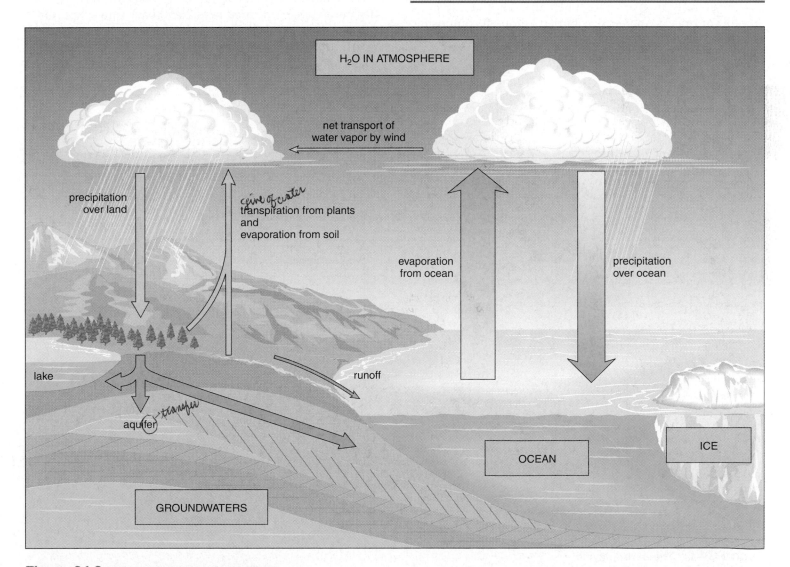

Figure 34.8   **The water (hydrologic) cycle.**

The Carbon Cycle

In the **carbon cycle**, organisms exchange carbon dioxide with the atmosphere (Fig. 34.9). On land, plants take up carbon dioxide from the air, and through photosynthesis they incorporate carbon into organic nutrients that are used for other living things. When organisms (e.g., plants, animals, and decomposers) respire, a portion of this carbon is returned to the atmosphere as carbon dioxide. In aquatic ecosystems, the exchange of carbon dioxide with the atmosphere is indirect. Carbon dioxide from the air combines with water to produce bicarbonate ion (HCO_3^-), a source of carbon for protists, which also produce organic nutrients through photosynthesis. And when aquatic organisms respire, the carbon dioxide they give off becomes bicarbonate ion.

Living and dead organisms are reservoirs for carbon. If dead remains fail to decompose, they are subject to physical processes that transform them into coal, oil, and natural gas. We call these reservoirs for carbon the **fossil fuels.** Most of the fossil fuels were formed during the Carboniferous period, 286 to 360 million years ago, when an exceptionally large amount of organic matter was buried before decomposing. Another reservoir for carbon is calcium carbonate shells from marine organisms that accumulate in ocean bottom sediments.

Carbon Dioxide and Global Warming

A **transfer rate** is defined as the amount of a nutrient that moves from one component of the environment to another within a specified period of time. The width of the arrows in Figure 34.9 indicates the transfer rate of carbon dioxide (CO_2). The transfer rates due to photosynthesis and respiration, which includes decay, are just about even. However, more carbon dioxide is now being deposited in the atmosphere than is being removed. In 1850, atmospheric carbon dioxide was about 280 parts per million (ppm), and today it is about 350 ppm. This increase is largely due to the burning of wood and fossil fuels and the destruction of forests to make way for farmland and pasture.

Other gases are also being emitted due to human activities, including nitrous oxide (N_2O, from fertilizers and animal wastes) and methane (CH_4, from bacterial decomposition, particularly in the guts of animals, in sediments, and in flooded rice paddies). These gases are known as **greenhouse gases,** because just like the panes of a greenhouse, they allow solar radiation to pass through but hinder the escape of infrared rays (heat) back into space. Thus, these emissions are contributing significantly to an overall rise in the earth's ambient temperature, a phenomenon

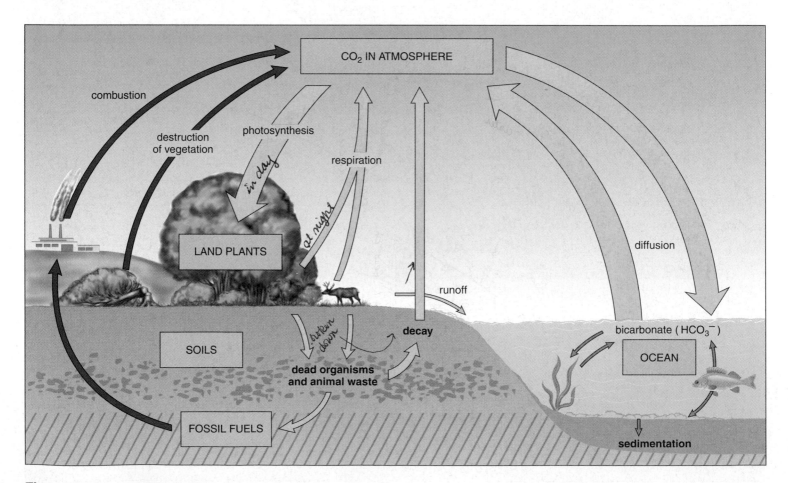

Figure 34.9   The carbon cycle.
Purple arrows represent human activities; gray arrows represent natural events.

called **global warming,** and their effect is known as the **greenhouse effect.**

Figure 34.10 shows the earth's radiation balances. One thing to be learned from this diagram is that water vapor is a greenhouse gas, so clouds (which are composed of water vapor) also reradiate heat back to earth. If the earth's temperature rises, more water will evaporate, forming more clouds and setting up a positive feedback effect that could increase global warming still more.

Today, data collected around the world show a steady rise in the concentration of greenhouse gases. For example, methane is increasing by about 1% a year. Such data have been used to generate computer models that predict the environmental temperature may become warmer than ever before. The global climate has already warmed about 0.6°C since the industrial revolution. Computer models are unable to consider all possible variables, but the earth's temperature may rise 1.5–4.5°C by 2060 if greenhouse emissions continue at the current rates.

Global warming will bring about other effects, which computer models attempt to forecast. It is predicted that, as the oceans warm, temperatures in the polar regions will rise to a greater degree than in other regions. If so,

glaciers will melt, and sea levels will rise, not only due to this melting but also because water expands as it warms. Water evaporation will increase, and most likely there will be increased rainfall along the coasts and dryer conditions inland. The occurrence of droughts will reduce agricultural yields and also cause trees to die off. Expansion of forests into arctic areas might not offset the loss of forests in the temperate zones. Coastal agricultural lands, such as the deltas of Bangladesh and China, will be inundated, and billions of dollars will have to be spent to keep coastal cities such as New York, Boston, Miami, and Galveston from disappearing into the sea. Species extinction is also likely, as will be discussed in Chapter 36.

The atmosphere is an exchange pool for carbon dioxide. Fossil fuel combustion in particular has increased the amount of carbon dioxide in the atmosphere. Global warming is predicted because carbon dioxide and other gases impede the escape of infrared radiation from the surface of the earth.

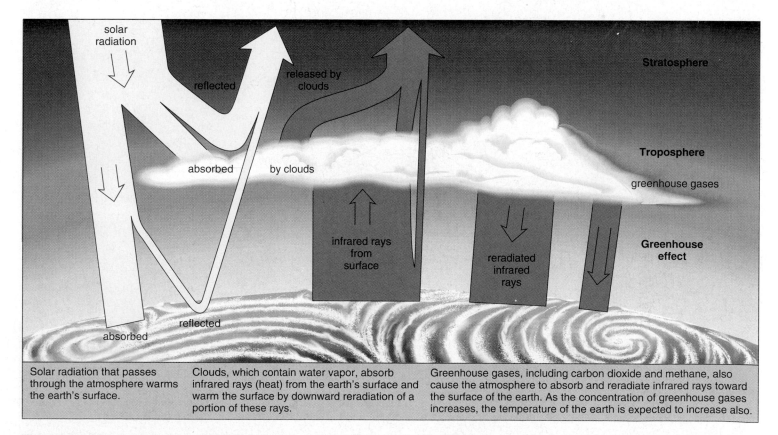

| Solar radiation that passes through the atmosphere warms the earth's surface. | Clouds, which contain water vapor, absorb infrared rays (heat) from the earth's surface and warm the surface by downward reradiation of a portion of these rays. | Greenhouse gases, including carbon dioxide and methane, also cause the atmosphere to absorb and reradiate infrared rays toward the surface of the earth. As the concentration of greenhouse gases increases, the temperature of the earth is expected to increase also. |

Figure 34.10   Earth's radiation balances.
The contribution of greenhouse gases *(far right)* to the earth's surface causes global warming.

The Nitrogen Cycle

Nitrogen is an abundant element in the atmosphere. Nitrogen makes up about 78% of the atmosphere by volume, but even so, nitrogen deficiency sometimes limits plant growth. Plants cannot incorporate nitrogen gas into organic compounds, and therefore they depend on various types of bacteria to make nitrogen available to them through the **nitrogen cycle** (Fig. 34.11).

Nitrogen fixation occurs when nitrogen (N_2) is converted to a form that plants can use. Some nitrogen-fixing bacteria live in nodules on the roots of legumes (plants of the pea family, such as peas, beans, and alfalfa). They make nitrogen-containing organic compounds available to a host plant. Free-living bacteria in bodies of water and in the soil are able to fix nitrogen gas as ammonium (NH_4^+). Plants can use NH_4^+ and nitrate (NO_3^-) from the soil. After NO_3^- is taken up, it is enzymatically reduced to NH_4^+, which is used to produce amino acids and nucleic acids.

Nitrification is the production of nitrates. Nitrogen gas (N_2) is converted to nitrate (NO_3^-) in the atmosphere when cosmic radiation, meteor trails, and lightning provide the high energy needed for nitrogen to react with oxygen. Ammonium (NH_4^+) in the soil is converted to nitrate by chemoautotrophic soil bacteria in a two-step process. First, nitrite-producing bacteria convert ammonium to nitrite (NO_2^-), and then nitrate-producing bacteria convert nitrite to nitrate. Notice the subcycle in the nitrogen cycle that involves dead organisms and animal wastes, ammonium, nitrites, nitrates, and plants. This subcycle does not necessarily depend on nitrogen gas at all (Fig. 34.11).

Denitrification is the conversion of nitrate to nitrous oxide and nitrogen gas. Denitrifying bacteria exist in both aquatic and terrestrial ecosystems. Denitrification balances nitrogen fixation, but not completely.

Nitrogen and Air Pollution

Human activities significantly alter transfer rates in the nitrogen cycle. Because we produce fertilizers, thereby converting N_2 to NO_3^-, and burn fossil fuels, the atmosphere contains three times more nitrogen oxides (NO_x) than it would otherwise. Fossil fuel combustion also pumps much sulfur dioxide (SO_2) into the atmosphere (Fig. 34.12b). Both nitrogen oxides and sulfur dioxide are converted to acids

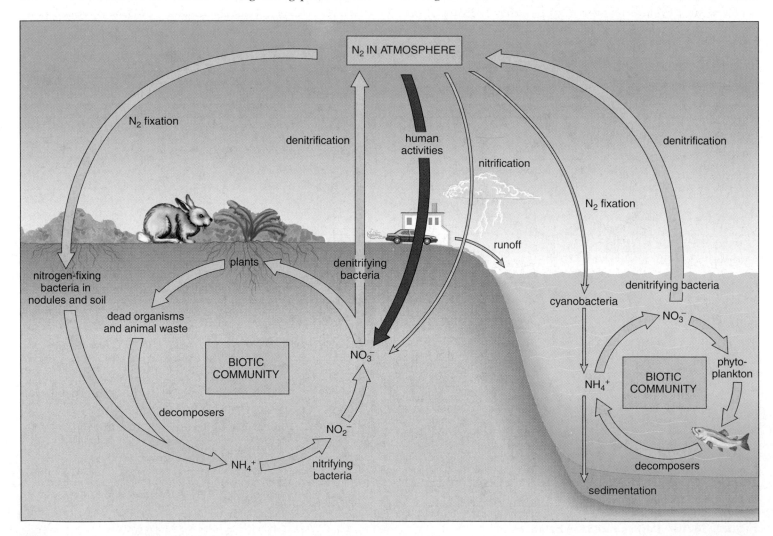

Figure 34.11   The nitrogen cycle.

when they combine with water vapor in the atmosphere. These acids return to earth as either wet deposition (acid rain or snow) or dry deposition (sulfate and nitrate salts).

Increased deposition of acids has drastically affected forests and lakes in northern Europe, Canada, and the northeastern United States because their soils are naturally acidic and their surface waters are only mildly alkaline (basic) to begin with. The forests in these areas are dying (Fig. 34.12*a*), and their waters cannot support normal fish populations. **Acid deposition** reduces agricultural yields and corrodes marble, metal, and stonework, an effect that is noticeable in cities.

Nitrogen oxides (NO_x) and hydrocarbons (HC) react with one another in the presence of sunlight to produce **photochemical smog,** which contains ozone (O_3) and **PAN (peroxyacetylnitrate).** Hydrocarbons come from fossil fuel combustion, but additional amounts come from various other sources, including paint solvents and pesticides. Breathing ozone affects the respiratory and nervous systems, resulting in respiratory distress, headache, and exhaustion. These symptoms are particularly apt to appear in young people. Ozone is especially damaging to plants, resulting in leaf mottling and reduced growth.

Normally, warm air near the earth is able to escape into the atmosphere, taking pollutants with it. However, during a **thermal inversion,** pollutants are trapped near the earth beneath a layer of warm, stagnant air. Because the air does not circulate, pollutants can build up to dangerous levels. Areas surrounded by hills are particularly susceptible to the effects of a temperature inversion because the air tends to stagnate, and little turbulent mixing can occur (Fig. 34.13).

Fertilizer use also results in the release of nitrous oxide (N_2O), a greenhouse gas and a contributor to ozone shield depletion. The ozone shield is a layer of ozone high in the atmosphere that protects the earth from dangerous levels of solar radiation (see the Health Focus on page 710).

Atmospheric N_2 is a reservoir and exchange pool for nitrogen, but atmospheric nitrogen must be fixed by bacteria in order to make this mineral available to plants. Environmental problems are associated with the release of nitrous oxide (N_2O) due to the action of bacteria on fertilizers, and nitrogen oxides (NO_x) due to fossil fuel combustion.

Figure 34.12   Acid deposition.
a. Many forests in higher elevations of northeastern North America and Europe are dying due to acid deposition. **b.** Air pollution due to emissions from factories and fossil fuel burning is the major cause of acid deposition, which contains nitric acid (H_2NO_3) and sulfuric acid (H_2SO_4).

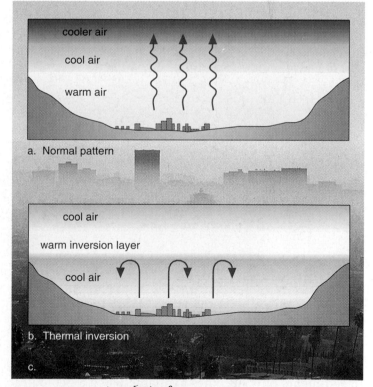

Figure 34.13 ** Know this **Thermal inversion.
a. Normally, pollutants escape into the atmosphere when warm air rises. **b.** During a thermal inversion, a layer of warm air (warm inversion layer) overlies and traps pollutants in cool air below. **c.** Los Angeles is particularly susceptible to thermal inversions, and this accounts for why this city is the "air pollution capital" of the United States.

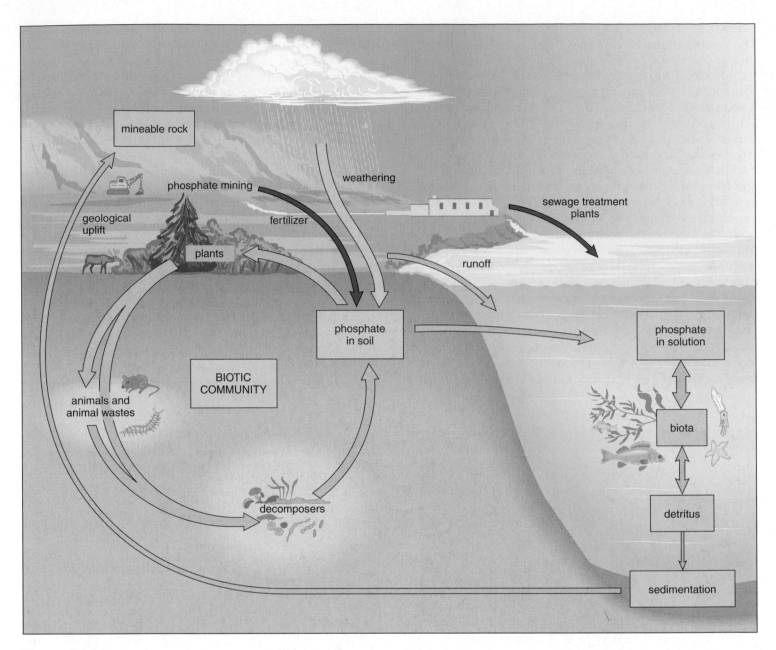

Figure 34.14   The phosphorus cycle.

The Phosphorus Cycle

On land, the weathering of rocks makes phosphate ions (PO_4^{3-} and HPO_4^{2-}) available to plants, which take up phosphate from the soil (Fig. 34.14). Some of this phosphate runs off into aquatic ecosystems where algae take it up from the water before it becomes trapped in sediments. Phosphate in sediments only becomes available when a geological upheaval exposes sedimentary rocks to weathering once more. Phosphorus does not enter the atmosphere; therefore, the **phosphorus cycle** is called a sedimentary cycle.

The phosphate taken up by producers is incorporated into a variety of organic molecules, including phospholipids and ATP or the nucleotides that become a part of DNA and RNA. Animals eat producers and incorporate some of the phosphate into teeth, bones, and shells that do not decompose for very long periods. However, death and decay of all organisms and also decomposition of animal wastes do make phosphate ions available to producers once again. Because available phosphate is generally taken up very quickly, it is a limiting inorganic nutrient in ecosystems. A limiting nutrient is one that regulates the growth of organisms because it is in shorter supply than other nutrients in the environment.

Phosphorus and Water Pollution

Human beings boost the supply of phosphate by mining phosphate ores for fertilizer and detergent production.

Runoff of phosphate and nitrogen due to fertilizer use, animal wastes from livestock feedlots, and discharge from sewage treatment plants results in **eutrophication** (overenrichment) of waterways. Eutrophication can lead to an algal bloom, apparent when green scum floats on the water. When the algae die off, decomposers use up all the available oxygen during cellular respiration. The result is a massive fish kill.

Figure 34.15 lists the various sources of water pollution. Point sources of pollution are specific, and nonpoint sources are those caused by runoff from the land. Industrial wastes can include heavy metals and organochlorides, such as those in some pesticides. These materials are not degraded readily under natural conditions or in conventional sewage treatment plants. They enter bodies of water and are subject to **biological magnification** because they remain in the body and are not excreted. Therefore, they become more concentrated as they pass along a food chain. Biological magnification occurs more readily in aquatic food chains, which have more links than terrestrial food chains. Humans are the final consumers in food chains, and in some areas, human milk contains detectable amounts of DDT and PCBs, which are organochlorides.

Coastal regions are the immediate receptors for local pollutants and the final receptors for pollutants carried by rivers that empty at a coast. Waste dumping occurs at sea, but ocean currents sometimes transport both trash and pollutants back to shore. Offshore mining and shipping add pollutants to the oceans. Some 5 million metric tons of oil a year—or more than one gram per 100 square meters of the oceans' surfaces—end up in the oceans. Large oil spills kill plankton, fish fry, and shellfishes, as well as birds and marine mammals. The largest tanker spill in U.S. territorial waters occurred on March 24, 1989, when the tanker *Exxon Valdez* struck a reef in Alaska's Prince William Sound and leaked 44 million liters of crude oil.

In the last 50 years, we have polluted the seas and exploited their resources to the point that many species are at the brink of extinction. Fisheries once rich and diverse, such as George's Bank off the coast of New England, are in severe decline. Haddock was once the most abundant species in this fishery, but now it accounts for less than 2% of the total catch. Cod and bluefin tuna have suffered a 90% reduction in population size. In warm, tropical regions, many areas of coral reefs are now overgrown with algae because the fish that normally keep the algae under control have been killed off.

Sedimentary rock is a reservoir for phosphorus; for the most part, producers are dependent on decomposers to make phosphate available to them. Fertilizer production and other human activities add phosphate to aquatic ecosystems, contributing to water pollution.

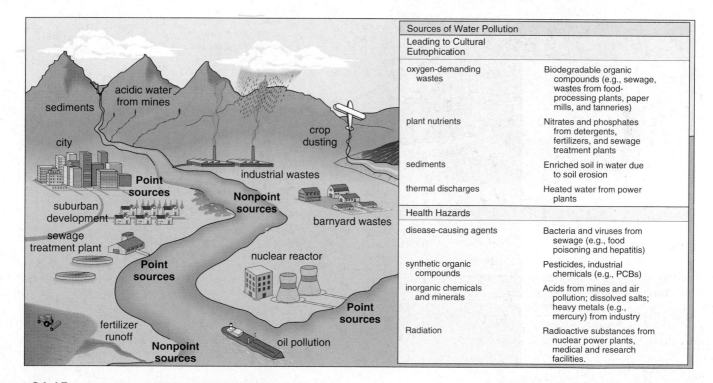

Figure 34.15   Sources of surface water pollution.
Many bodies of water are dying due to the introduction of pollutants from point sources, which are easily identifiable, and nonpoint sources, which cannot be specifically identified.

Ecology Focus

Ozone Shield Depletion

The earth's atmosphere is divided into layers. The troposphere is the layer that envelops us as we go about our day-to-day lives. When ozone is present in the troposphere (called ground-level ozone), it is considered a pollutant because it adversely affects a plant's ability to grow and our ability to breathe oxygen (O_2). In the stratosphere, some 50 kilometers above the earth, ozone forms the **ozone shield**, a layer of ozone that absorbs much of the ultraviolet (UV) rays of the sun so that fewer rays strike the earth. Ozone forms when ultraviolet radiation from the sun splits oxygen molecules (O_2), and then the oxygen atoms (O) combine with other oxygen molecules to give ozone (O_3).

The absorption of UV radiation by the ozone shield is critical for living things. In humans, UV radiation causes mutations that can lead to skin cancer and can make the lens of the eye develop cataracts. In addition, it adversely affects the immune system and our ability to resist infectious diseases. UV radiation also impairs crop and tree growth and kills off algae (phytoplankton) and tiny shrimplike animals (krill) that sustain oceanic life. Without an adequate ozone shield, therefore, our health and food sources are threatened.

It became apparent in the 1980s that depletion of ozone had occurred worldwide and that the depletion was most severe above the Antarctic every spring. Here ozone depletion became so great that it covered an area two and a half times the size of Europe, and exposed not only Antarctica but also the southern tip of South America and vast areas of the Pacific and Atlantic oceans to harmful ultraviolet rays. In the popular press, severe depletions of the ozone layer are called **"ozone holes"** (Fig. 34A*a*). Of even greater concern, an ozone hole has now appeared above the Arctic as well, and ozone holes were also detected within northern and southern latitudes, where many people live. Whether or not these holes develop in the spring depends on prevailing winds, weather conditions, and the type of particles in the atmosphere. A United Nations Environmental

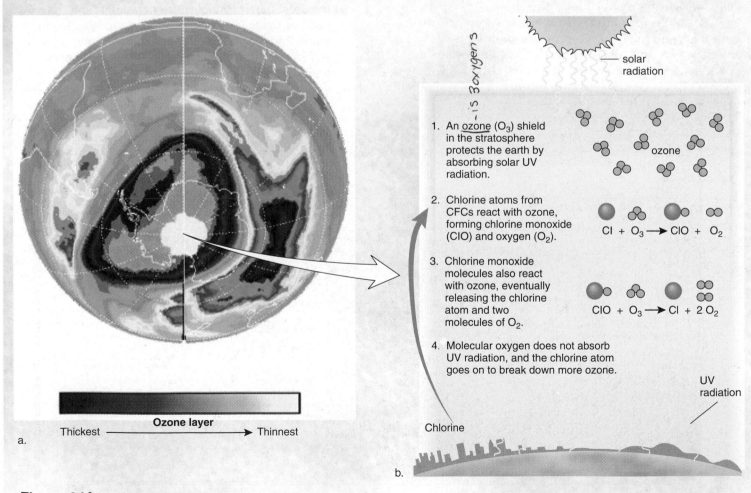

— is 3 oxygens

1. An ozone (O_3) shield in the stratosphere protects the earth by absorbing solar UV radiation.

2. Chlorine atoms from CFCs react with ozone, forming chlorine monoxide (ClO) and oxygen (O_2).

$$Cl + O_3 \rightarrow ClO + O_2$$

3. Chlorine monoxide molecules also react with ozone, eventually releasing the chlorine atom and two molecules of O_2.

$$ClO + O_3 \rightarrow Cl + 2\,O_2$$

4. Molecular oxygen does not absorb UV radiation, and the chlorine atom goes on to break down more ozone.

Ozone layer
Thickest ———→ Thinnest

a.

b.

Figure 34A   Ozone shield depletion.
a. Map of ozone levels in the atmosphere of the Southern Hemisphere, September 2000. The ozone depletion, often called an ozone hole (pale lavender to pale blue), is larger than the size of Europe. **b.** The release of chlorine atoms from chlorofluorocarbons (CFCs) contributes to the occurrence of the ozone hole.

Program report predicts a 26% rise in cataracts and non-melanoma skin cancers for every 10% drop in the ozone level. A 26% increase translates into 1.75 million additional cases of cataracts and 300,000 more skin cancers every year, worldwide.

The seriousness of the situation caused scientists around the globe to begin studying the cause of ozone depletion. A vortex of cold wind (a whirlpool in the atmosphere) circles the poles during the winter months, creating polar clouds. These clouds contain ice crystals in which chemical reactions occur that break down ozone. The cause of ozone depletion can be traced to the release of chlorine atoms (Cl) into the stratosphere. Chlorine atoms combine with ozone and form chlorine monoxide (ClO), which also breaks down ozone before releasing the same chlorine atom once again (Fig. 34A*b*). One atom of chlorine can destroy up to 100,000 molecules of ozone before settling to the earth's surface as chloride years later.

The chlorine atoms that enter the troposphere and eventually reach the stratosphere come primarily from the breakdown of **chlorofluorocarbons** (CFCs), chemicals much in use by humans. The best-known CFC is Freon, a coolant found in refrigerators and air conditioners. CFCs are also used as cleaning agents and as foaming agents during the production of styrofoam coffee cups, egg cartons, insulation, and paddings. Formerly, CFCs were used as propellants in spray cans, but this application is now banned in the United States and several European countries. Other molecules, such as the cleaning solvent methyl chloroform, are also sources of harmful chlorine atoms.

Most of the countries of the world have stopped using CFCs, and the United States halted production in 1995. Since that time, satellite measurements indicate that the amount of harmful chlorine pollution in the stratosphere has started to decline. Scientists determined that chlorine concentrations peaked between 1992 and 1994. But because it takes several years for air from the troposphere to leak up into the stratosphere, chlorine concentrations didn't peak in the stratosphere until 1997.

The decline in chlorine pollution should lead to an increase in the ozone layer, but other factors are involved. Researchers report that during the winter of 2000, there were more and longer-lasting polar clouds than previously. Why might that be? As the earth's surface warms due to global warming (see Fig. 34.10), less heat reradiates into the stratosphere and the stratosphere, is becoming cooler than usual. Mathematical modeling suggests that stratospheric clouds could last twice as long over the Arctic by the year 2010, when the coldest winter ever is expected.

Lingering clouds may inhibit the possible recovery of the ozone shield, despite lower amounts of chlorine pollution (Fig. 34B). Chlorine monoxide (ClO) ordinarily leads to ozone depletion, but it can also react with nitric oxide (NO), forming nitrogen dioxide (NO_2), which then breaks down, releasing an oxygen atom. The oxygen atom reacts with molecular oxygen to form more ozone, doing away with at least one part of the damaging ozone breakdown cycle caused by chlorine. Unfortunately, polar clouds drip nitrogen, and thereby lower the nitrogen concentration in the stratosphere. It is speculated that once polar stratospheric clouds become twice as persistent, there could be an ozone loss of 30% due to reduction in nitric oxide levels. It is clear, then, that recovery of the ozone shield may take several more years and involve other pollution-fighting approaches, aside from lowering chlorine pollution.

Figure 34B Polar clouds.
Global warming is expected to make the stratosphere (as opposed to the troposphere) cool. As the stratosphere cools, polar clouds will increase and last longer. Cloud cover contributes to the breakdown of the ozone shield by chlorine pollution.

Bioethical Focus   Curtailing Greenhouse Gases

Global warming will upset normal weather cycles, which will most likely lead to outbreaks of hantavirus as well as malaria, dengue and yellow fevers, filariasis, encephalitis, schistosomiasis, and cholera. Clearly, greenhouse gases should be curtailed.

In December 1997, 159 countries met in Kyoto, Japan, to work out a protocol that would reduce greenhouse gases worldwide. Greenhouse gases are those like carbon dioxide and methane, which allow the sun's rays to pass through but then trap the heat from escaping. It is believed that the emission of greenhouse gases, especially from power plants, will cause earth's temperature to rise 1.5°–4.5°

by 2060. The U.S. Senate does not want to ratify the agreement because it does not include a binding emissions commitment from the less-developed countries. While the U.S. presently emits a large proportion of the greenhouse gases, China is expected to pass the United States in about 2020 to become the biggest source of greenhouse emissions.

Negotiations with the developing countries is still going on and some creative ideas have been put forward. Why not have a trading program that allows companies to buy and sell emission credits across international boundaries? Greenhouse reduction techniques would also be for sale. Presumably, if it became

monetarily worth their while, companies in developing countries would have an incentive to use such techniques in order to reduce greenhouse emissions.

Decide Your Opinion

1. Should all countries be expected to reduce their greenhouse emissions? Why or why not?
2. Do you approve of giving companies monetary incentives to reduce greenhouse emissions? Why or why not?
3. If you were a CEO, would you be willing to reduce greenhouse emissions simply because they cause a deterioration of the environment and probably cause human illness? Why or why not?

Summarizing the Concepts

34.1 The Nature of Ecosystems

An ecosystem is composed of populations of organisms plus the physical environment. Some populations are producers and some are consumers. Producers are autotrophs that produce their own organic food. Consumers are heterotrophs that take in organic food. Consumers may be herbivores, carnivores, omnivores, or decomposers.

Ecosystems are characterized by energy flow and chemical cycling. Inorganic nutrients are not lost from the biosphere as is energy. They recycle within and between ecosystems. Decomposers return some proportion of inorganic nutrients to autotrophs, and other portions are imported or exported between ecosystems in global cycles.

Producers transform solar energy into food for themselves and all consumers. As herbivores feed on plants (or algae), and carnivores feed on herbivores, energy is converted to heat. Feces, urine, and dead bodies become food for decomposers. Eventually, all the solar energy that enters an ecosystem is converted to heat, and thus ecosystems require a continual supply of solar energy.

34.2 Energy Flow

Ecosystems contain food webs in which the various organisms are connected by eating relationships. In a grazing food web, food chains begin with a producer. In a detrital food web, food chains begin with detritus. The two food webs are joined when the same consumer is a link in both a grazing and a detrital food chain. A trophic level is all the organisms that feed at a particular link in a food chain. Ecological pyramids show trophic levels stacked one on top of the other like building blocks. Generally they show that energy content, and therefore numbers of organisms and biomass, decreases from one trophic level to the next.

34.3 Global Biogeochemical Cycles

Biogeochemical cycles contain reservoirs, which are components of ecosystems such as fossil fuels, sediments, and rocks, that contain

elements available on a limited basis to living things. Exchange pools are components of ecosystems, such as the atmosphere, soil, and water, which are ready sources of nutrients for living things.

In the water cycle, evaporation over the ocean is not compensated for by rainfall. Evaporation from terrestrial ecosystems includes transpiration from plants. Rainfall over land results in bodies of fresh water plus groundwater, including aquifers. Eventually, all water returns to the oceans.

In the carbon cycle, organisms add as much carbon dioxide to the atmosphere as they remove. Shells in ocean sediments, organic compounds in living and dead organisms, and fossil fuels are reservoirs for carbon. Human activities such as burning fossil fuels and trees are adding carbon dioxide to the atmosphere. Like the panes of a greenhouse, carbon dioxide and other gases allow the sun's rays to pass through but impede the release of infrared wavelengths. It is predicted that a buildup of these "greenhouse gases" will lead to global warming; a rise in sea level and a change in climate patterns could follow.

In the nitrogen cycle certain bacteria in water, soil, and root nodules can fix atmospheric nitrogen. Other bacteria return nitrogen to the atmosphere. Human activities convert atmospheric nitrogen to fertilizer, which is broken down by soil bacteria; humans also burn fossil fuels. In this way, a large quantity of nitrogen oxide (NO_x) and sulfur dioxide (SO_2) is added to the atmosphere where it reacts with water vapor to form acids that contribute to acid deposition. Acid deposition can kill lakes and forests and corrode marble, metal, and stonework. Nitrogen oxides and hydrocarbons (HC) react to form smog, which contains ozone and PAN (peroxyacetylnitrate). These oxidants are harmful to animal and plant life.

In the phosphorus cycle, the biotic community recycles phosphorus back to the producers, and only limited quantities are made available by the weathering of rocks. Phosphates are mined for fertilizer production; when phosphates and nitrates enter lakes and ponds, overenrichment occurs. Many kinds of wastes enter the rivers and then flow to the oceans, which have now become degraded from added pollutants.

Testing Yourself

Choose the best answer for each question.

1. Of the total amount of energy that passes from one trophic level to another, about 10% is
   a. respired and becomes heat.
   b. passed out as feces or urine.
   c. stored as body tissue.
   d. recycled to autotrophs.
   e. All of these are correct.

2. Compare this food chain:
   algae → water fleas → fish → green herons
   to this food chain:
   trees → tent caterpillars → red-eyed vireos → hawks.

   Both water fleas and tent caterpillars are
   a. carnivores.
   b. primary consumers.
   c. detritus feeders.
   d. present in grazing and detrital food webs.
   e. Both a and b are correct.

3. Which of the following contribute(s) to the carbon cycle?
   a. respiration
   b. photosynthesis
   c. fossil fuel combustion
   d. decomposition of dead organisms
   e. All of these are correct.

4. How do plants contribute to the carbon cycle?
   a. When plants respire, they release CO_2 into the atmosphere.
   b. When plants photosynthesize, they consume CO_2 from the atmosphere.
   c. When plants photosynthesize, they provide oxygen to heterotrophs.
   d. When plants emigrate, they transport carbon molecules between ecosystems.
   e. Both a and b are correct.

5. How do nitrogen-fixing bacteria contribute to the nitrogen cycle?
   a. They return nitrogen (N_2) to the atmosphere.
   b. They change ammonium to nitrate.
   c. They change N_2 to ammonium.
   d. They withdraw nitrate from the soil.
   e. They decompose and return nitrogen to autotrophs.

6. In what way are decomposers like producers?
   a. Either may be the first member of a grazing or a detrital food chain.
   b. Both produce oxygen for other forms of life.
   c. Both require a source of nutrient molecules and energy.
   d. Both are present only in the lithosphere.
   e. Both produce organic nutrients for other members of ecosystems.

7. Choose the statement that is true concerning this food chain:
   grass → rabbits → snakes → hawks
   a. Each predator population has a greater biomass than its prey population.
   b. Each prey population has a greater biomass than its predator population.
   c. Each population is omnivorous.

   d. Each population returns inorganic nutrients and energy to the producer.
   e. Both a and c are correct.

For questions 8–10, match the terms with those in the key.
Key:
   a. sulfate salts   c. carbon dioxide
   b. ozone   d. pesticides

8. acid deposition

9. greenhouse effect

10. photochemical smog

11. Which of these pairs is mismatched?
    a. fossil fuel burned—carbon dioxide given off
    b. nuclear power—radioactive wastes
    c. fertilizer use—phosphate in atmosphere
    d. aerosol sprays—chlorine in atmosphere
    e. pesticides—pollutants in water

12. Acid deposition causes
    a. lakes and forests to die.
    b. acid indigestion in humans.
    c. the greenhouse effect to lessen.
    d. pests to increase decomposition.
    e. All of these are correct.

13. Water is a renewable resource, and
    a. there will always be a plentiful supply.
    b. the oceans can never become polluted.
    c. it is still subject to pollution.
    d. primary sewage treatment plants assure clean drinking water.
    e. Both a and c are correct.

14. What is biological magnification?
    a. overenrichment of water systems
    b. pollutants trapped below a stagnant layer of warm air
    c. materials that are not degraded and become more concentrated in subsequent organisms along food chains
    d. the amount of nutrients moved from one environment to another in a specific time frame.

15. Label the following diagram of an ecosystem:

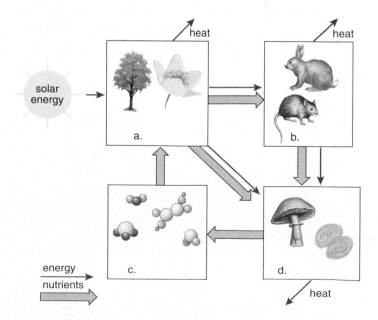

e-Learning Connection

www.mhhe.com/maderinquiry10

Concepts	Questions	Media Resources*
34.1 The Nature of Ecosystems		
• An ecosystem is a community of organisms along with their physical and chemical environment. • Living components contribute to an ecosystem in their own way. Some are autotrophic and produce organic nutrients. Others are heterotrophic and consume organic nutrients. • An ecosystem is characterized by energy flow and chemical cycling.	1. What are omnivores? 2. Why is it not possible for energy to cycle within an ecosystem?	Essential Study Partner Introduction
34.2 Energy Flow		

• Energy flow begins when autotrophs use solar energy to produce organic nutrients for themselves and all living things. Eventually these nutrients are broken down and solar energy returns to the atmosphere as heat.	1. What is a detrital food web?
	2. Is there an advantage of representing trophic levels using a pyramid of biomass instead of a pyramid of numbers?
		Essential Study Partner
		    Energy Flow
		
		Art Quizzes
		    Path of Energy Through
		        Food Webs
		    Trophic Levels Within a Food
		        Chain
		    Food Web in Cayuga Lake

34.3 Global Biogeochemical Cycles		

• Nutrients cycle within and between ecosystems in global biogeochemical cycles.
• Biogeochemical cycles are gaseous (carbon cycle and nitrogen cycle) or sedimentary (phosphorus cycle).
• The addition of carbon dioxide (and other gases) to the atmosphere is associated with global warming.
• The production of fertilizers from nitrogen gas is associated with acid deposition, photochemical smog, and temperature inversions.
• Fertilizer also contains mined phosphate; fertilizer runoff is associated with water pollution.
	1. How have humans interfered with the global carbon cycle?
	2. What global biogeochemical cycle is particularly dependent on the activity of various groups of bacteria?
		Essential Study Partner
		    Nutrient Cycles
		    Biomagnification
		    Acid Rain
		    Pesticides
		    Air Pollution Sources
		    Hazardous Waste
		
		Art Quizzes
		    The Water Cycle
		    The Carbon Cycle
		    The Nitrogen Cycle
		    Biological Magnification
		        of DDT
		    Greenhouse Effect
		
		Animation Quizzes
		    Global Warming
		    Ozone Layer Depletion
		    Carbon Cycle
		    Nitrogen Cycle
		    Acid Rain
		    Bioaccumulation
		
		Case Studies
		    Nitrogen Controversy

*For additional Media Resources, see the Online Learning Center.

The Biosphere

Chapter Concepts

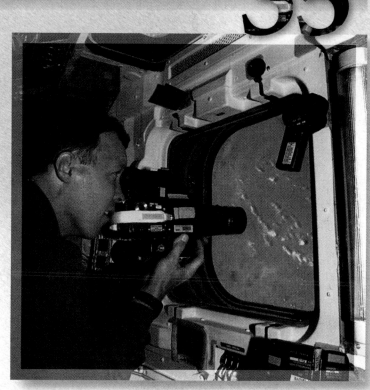

An astronaut takes pictures of planet Earth from his space shuttle.

*T*he astronauts inside the space shuttle trained a variety of cameras and other surveillance instruments on the planet below them. The first target is the frigid land near the North Pole. Having little rainfall and devoid of light most of the year, this Arctic tundra only teems with life in summer, when animals such as caribou migrate to the region. The shuttle scientists next concentrate their attention on a tropical rain forest in South America, home to a tremendous diversity of animal and plant species—many still undocumented. The astronauts then direct their devices toward the Sahara, the massive desert that spans northern Africa. As dry as the Arctic tundra, but considerably warmer, the desert is nevertheless home to an impressive variety of hardy plants and animals. Finally, the shuttle passes above the waters of the South Pacific, where the scientists try to determine if the weather phenomenon called El Niño will strike again. In its brief time in orbit, the shuttle has studied strikingly different parts of the earth's biosphere, the thin layer of water, land, and air inhabited by living organisms.

35.1 Climate and the Biosphere

Climate refers to the prevailing weather conditions in a particular region. Climate is dictated by temperature and rainfall, which are influenced by the following factors: (1) variations in solar radiation distribution due to the tilt of a spherical earth as it orbits about the sun; (2) other effects such as topography and whether a body of water is nearby.

Effect of Solar Radiation

Because the earth is a sphere, the sun's rays are more direct at the equator and more spread out at polar regions (Fig. 35.1a). Therefore, the tropics are warmer than temperate regions. The tilt of the earth as it orbits around the sun causes one pole or the other to be closer to the sun (except at the spring and fall equinoxes), and this accounts for the seasons that occur in all parts of the earth except at the equator (Fig. 35.1b). When the Northern Hemisphere is having winter, the Southern Hemisphere is having summer, and vice versa.

If the earth were standing still and were a solid, uniform ball, all air movements—which we call winds—would be in two directions. Warm equatorial air would rise and move directly to the poles, creating a zone of lower pressure that would be filled by cold polar air moving equatorward.

However, because the earth rotates on its axis daily and its surface consists of continents and oceans, the flows of warm and cold air are modified into three large circulation cells in each hemisphere (Fig. 35.2). At the equator, the sun heats the air and evaporates water. The warm, moist air rises, cools, and loses most of its moisture as rain. The greatest amounts of rainfall on earth are near the equator. The rising air flows toward the poles, but at about 30° north and south latitude, it sinks toward the earth's surface and reheats. As the air descends and warms, it becomes very dry, creating zones of low rainfall. The great deserts of Africa, Australia, and the Americas occur at these latitudes. At the earth's surface, the air flows both poleward and equatorward. At about 60° north and south latitude, the air rises and cools, producing additional zones of high rainfall. This moisture supports the great forests of the temperate zone. Part of this rising air flows equatorward, and part continues poleward, descending near the poles, which are zones of low precipitation.

Besides affecting precipitation, the spinning of the earth also affects the winds, so that the major global circulation systems flow toward the east or west rather than directly north or south (Fig. 35.2). Between about 30° north latitude and 30° south latitude, the winds blow from the east-southeast in the Southern Hemisphere and from the east-northeast in the Northern Hemisphere (the east coasts

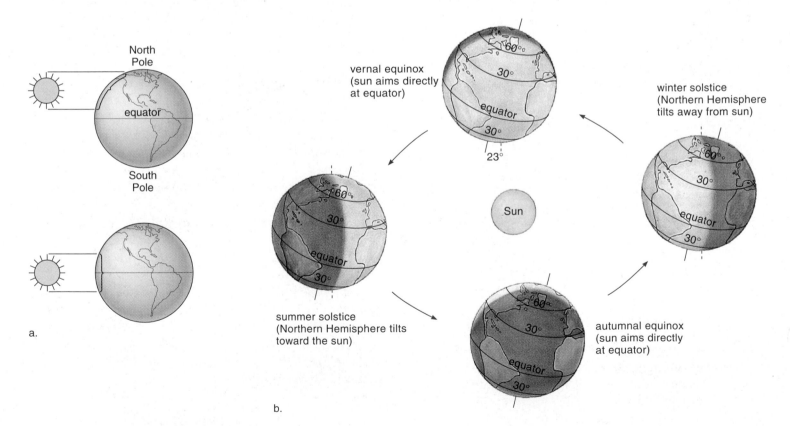

Figure 35.1   Distribution of solar energy.
a. Since the earth is a sphere, beams of solar energy striking the earth near one of the poles are spread over a wider area than similar beams striking the earth at the equator. **b.** The seasons of the Northern and Southern Hemispheres are due to the tilt of the earth on its axis as it rotates about the sun.

of continents at these latitudes are wet). These are called trade winds because sailors depended upon them to fill the sails of their trading ships. Between 30° and 60° north and south latitude, strong winds, called the prevailing westerlies, blow from west to east. The west coasts of the continents at these latitudes are wet, as is the Pacific Northwest where a massive evergreen forest is located. Weaker winds, called the polar easterlies, blow from east to west at still higher latitudes of their respective hemispheres.

> The distribution of solar energy caused by a spherical earth, and the rotation and path of the earth around the sun, affect how the winds blow and the amount of rainfall various regions receive.

Other Effects

Topography means the physical features, or "the lay," of the land. One physical feature that affects climate is the presence of mountains. As air blows up and over a mountain range, it rises and cools. One side of the mountain, called the windward side, receives more rainfall than the other side, called the leeward side. On the leeward side, the air descends, picks up moisture, and produces clear weather (Fig. 35.3). The difference between the windward side and the leeward side can be quite dramatic. In the Hawaiian Islands, for example, the windward side of the mountains receives more than 750 cm of rain a year, while the leeward side, which is in a **rain shadow,** gets on the average only 50 cm of rain and is generally sunny. In the United States, the western side of the Sierra Nevada Mountains is lush, while the eastern side is a semidesert.

The oceans are slower to change temperature—that is, to gain or lose their heat—than landmasses. This causes coasts to have a unique weather pattern that is not seen inland. During the day, the land warms more quickly than the ocean, and the air above the land rises. Then a cool sea breeze blows in from the ocean. At night, the reverse happens; the breeze blows from the land to the sea.

India and some other countries in southern Asia have a **monsoon** climate, in which wet ocean winds blow onshore for almost half the year. The land heats more rapidly than the waters of the Indian Ocean during spring. The difference in temperature between the land and the ocean causes a gigantic circulation of air: warm air rises over the land, and cooler air comes in off the ocean to replace it. As the warm air rises, it loses its moisture and the monsoon season begins. As just discussed, rainfall is particularly heavy on the windward side of hills. Cherrapunji in northern India receives an annual average of 1,090 cm of rain a year because of its high altitude. The weather pattern has reversed by November. The land is now cooler than the ocean; therefore, dry winds blow from the Asian continent across the Indian Ocean. In the winter, the air over the land is dry, the skies cloudless, and temperatures pleasant. The chief crop of India is rice, which starts to grow when the monsoon rains begin.

In the United States, people often speak of the "lake effect," meaning that in the winter, arctic winds blowing over the Great Lakes become warm and moisture-laden. When these winds rise and lose their moisture, snow begins to fall. Places such as Buffalo, New York, get heavy snowfalls due to the lake effect, and snow is on the ground there for an average of 90 to 140 days every year.

> Atmospheric circulations between the ocean and the landmasses influence regional climate conditions.

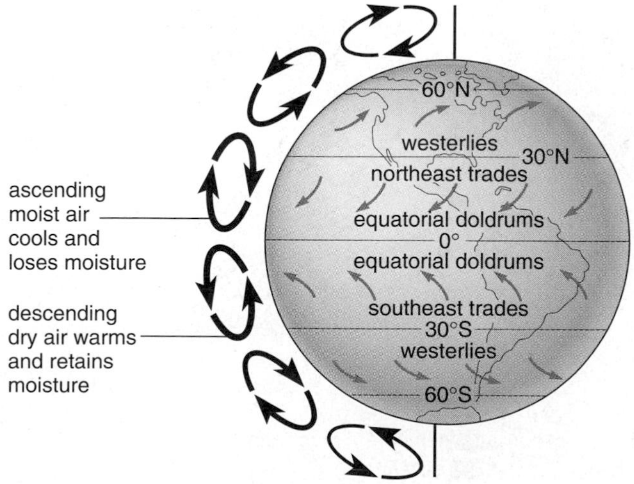

Figure 35.2   Global wind circulation.
Air ascends and descends as shown because the earth rotates on its axis. Also, the trade winds move from the northeast to the west in the Northern Hemisphere, and from the southeast to the west in the Southern Hemisphere. The westerlies move toward the east.

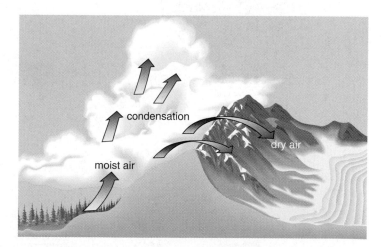

Figure 35.3   Formation of a rain shadow.
When winds from the sea cross a coastal mountain range, they rise and release their moisture as they cool this side of a mountain, which is called the windward side. The leeward side of a mountain receives relatively little rain and is therefore said to lie in a "rain shadow."

Figure 35.4   Pattern of biome distribution.

a. Pattern of world biomes in relation to temperature and moisture. The dashed line encloses a wide range of environments in which either grasses or woody plants can dominate the area, depending on the soil type. **b.** The same type of biome can occur in different regions of the world, as shown on this global map.

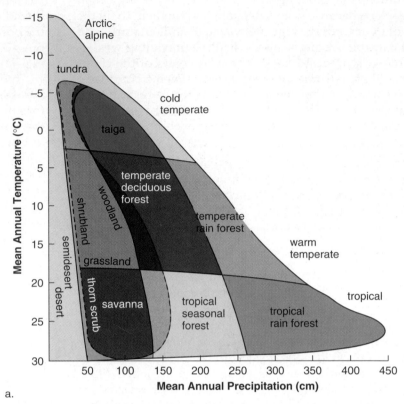

a.

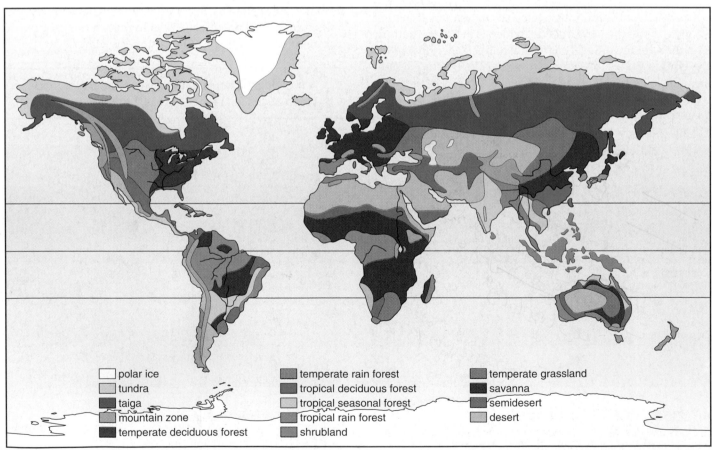

b.

35.2 Terrestrial Communities

A major type of terrestrial community is called a **biome.** A biome has a particular mix of plants and animals that are adapted to living under certain environmental conditions, of which climate has an overriding influence. For example, when terrestrial biomes are plotted according to their mean annual temperature and mean annual rainfall, a particular pattern results (Fig. 35.4*a*). The distribution of biomes is shown in Figure 35.4*b*. Even though Figure 35.4 shows definite demarcations, the biomes gradually change from one type to the other. Also, although we will be discussing each type of biome separately, we should remember that each biome has inputs from and outputs to all the other terrestrial and aquatic communities of the biosphere.

The distribution of the biomes—and hence, the pattern of life on earth—is determined principally by differences in climate due to the distribution of solar radiation and the topographical features just discussed. The effect of a temperature gradient can be seen not only when we consider latitude but also when we consider altitude. If you travel from the equator to the North Pole, it is possible to observe first a tropical rain forest, followed by a temperate deciduous forest, a coniferous forest, and tundra, in that order, and this sequence is also seen when ascending a mountain (Fig. 35.5). The coniferous forest of a mountain is called a **montane coniferous forest,** and the tundra near the peak of a mountain is called an **alpine tundra.** When going from the equator to the South Pole, you would not reach a region corresponding to a coniferous forest and tundra of the Northern Hemisphere. Why not? Look at the distribution of the landmasses—they are shifted toward the north.

The distribution of biomes is determined by physical factors such as climate (principally temperature and rainfall), which varies according to latitude and altitude.

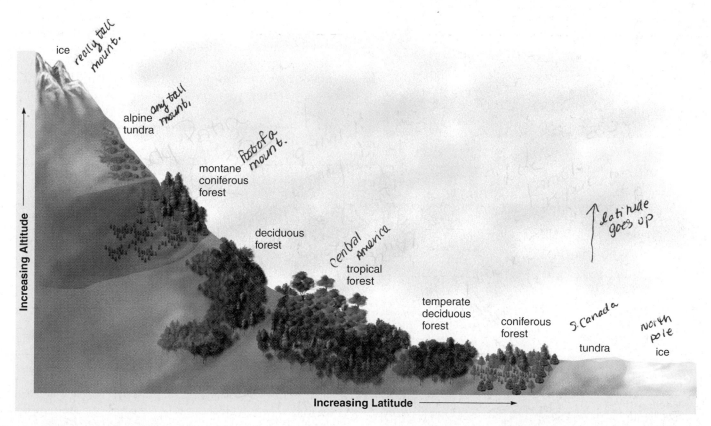

Figure 35.5   Climate and biomes.
Biomes change with altitude just as they do with latitude because vegetation is partly determined by temperature. Rainfall also plays a significant role, which is one reason grasslands, instead of tropical or deciduous forests, are sometimes found at the base of mountains.

a.

b.

c.

Tundra

The **Arctic tundra** biome, which encircles the earth just south of ice-covered polar seas in the Northern Hemisphere, covers about 20% of the earth's land surface (Fig. 35.6). (A similar community, called the alpine tundra, occurs above the timberline on mountain ranges.) The Arctic tundra is cold and dark much of the year. Because rainfall amounts to only about 20 cm a year, the tundra could possibly be considered a desert, but melting snow creates a landscape of pools and mires in the summer, especially because so little evaporates. Only the topmost layer of earth thaws; the **permafrost** beneath this layer is always frozen, and therefore, drainage is minimal.

Trees are not found in the tundra because the growing season is too short, their roots cannot penetrate the permafrost, and they cannot become anchored in the boggy soil of summer. In the summer, the ground is covered with short grasses and sedges, as well as numerous patches of lichens and mosses. Dwarf woody shrubs, such as dwarf birch, flower and seed quickly while there is plentiful sun for photosynthesis.

A few animals live in the tundra year-round. For example, the mouselike lemming stays beneath the snow; the ptarmigan, a grouse, burrows in the snow during storms; and the musk ox conserves heat because of its thick coat and short, squat body. In the summer, the tundra is alive with numerous insects and birds, particularly shorebirds and waterfowl that migrate inland. Caribou and reindeer also migrate to and from the tundra, as do the wolves that prey upon them. Polar bears are common near the coast.

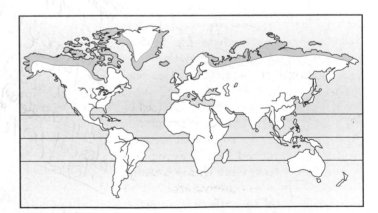

Figure 35.6   The tundra.
a. In this biome, which is nearest the polar regions, the vegetation consists principally of lichens, mosses, grasses, and low-growing shrubs. **b.** Pools of water that do not evaporate or drain into the permanently frozen ground attract many birds that feed on the plentiful insects in the summer. **c.** Caribou, more plentiful in the summer than in the winter, feed on lichens, grasses, and shrubs.

Coniferous Forests

Coniferous forests are found in three locations: in the **taiga,** which extends around the world in the northern part of North America and Eurasia; near mountain tops (where it is called a montane coniferous forest); and along the Pacific coast of North America, as far south as northern California.

The taiga (Fig. 35.7) typifies the coniferous forest with its cone-bearing trees, such as spruce, fir, and pine. These trees are well adapted to the cold because both the leaves and bark have thick coverings. Also, the needlelike leaves can withstand the weight of heavy snow. There is a limited understory of plants, but the floor is covered by low-lying mosses and lichens beneath a layer of needles. Birds harvest the seeds of the conifers, and bears, deer, moose, beaver, and muskrat live around the cool lakes and along the streams. Wolves prey on these larger mammals. A montane coniferous forest also harbors the wolverine and the mountain lion.

The coniferous forest that runs along the west coast of Canada and the United States is sometimes called a **temperate rain forest.** The prevailing winds moving in off the Pacific Ocean lose their moisture when they meet the coastal mountain range. The plentiful rainfall and rich soil have produced some of the tallest conifer trees ever in existence, including the coastal redwoods. This forest is also called an old-growth forest because some trees are as old as 800 years. It truly is an evergreen forest because mosses, ferns, and other plants grow on all the tree trunks. Whether the limited portion of the forest that remains should be preserved from logging has been quite controversial. Unfortunately, the controversy has centered around the northern spotted owl, which is endemic to this area, rather than around the larger issue, the conservation of this particular ecosystem.

bull moose,
Alces americanus

Figure 35.7   The taiga.
The taiga, which means swampland, spans northern Europe, Asia, and North America. The appellation "spruce-moose" refers to the dominant presence of spruce trees and moose, which frequent the ponds.

Temperate Deciduous Forests

Temperate deciduous forests are found south of the taiga in eastern North America (Fig. 35.8), eastern Asia, and much of Europe. The climate in these areas is moderate, with relatively high rainfall (75–150 cm per year). The seasons are well defined, and the growing season ranges between 140 and 300 days. The trees, such as oak, beech, and maple, have broad leaves and are termed deciduous trees; they lose their leaves in the fall and grow them in the spring.

The tallest trees form a canopy, an upper layer of leaves that are the first to receive sunlight. Even so, enough sunlight penetrates to provide energy for another layer of trees called understory trees. Beneath these trees are shrubs that may flower in the spring before the trees have put forth their leaves. Still another layer of plant growth—mosses, lichens, and ferns—resides beneath the shrub layer. This stratification provides a variety of habitats for insects and birds. Ground life is also plentiful. Squirrels, cottontail rabbits, shrews, skunks, woodchucks, and chipmunks are small herbivores. These and ground birds such as turkeys, pheasants, and grouse are preyed on by red foxes. White-tail deer and black bears have increased in number of late. In contrast to the taiga, amphibians and reptiles occur in this biome because the winters are not as cold. Frogs and turtles prefer an aquatic existence, as do the beaver and muskrat, which are mammals.

Autumn fruits, nuts, and berries provide a supply of food for the winter, and the leaves, after turning brilliant colors and falling to the ground, contribute to the rich layer of humus. The minerals within the rich soil are washed far into the ground by spring rains, but the deep tree roots capture these and bring them back up into the forest system again.

Figure 35.8 **Temperate deciduous forest.**
A temperate deciduous forest is home to many varied plants and animals. Millipedes can be found among leaf litter, chipmunks feed on acorns, and bobcats prey on these and other small mammals.

Ecology Focus

Wildlife Conservation and DNA

After DNA analysis, scientists were amazed to find that some 60% of loggerhead turtles drowning in the nets and hooks of fisheries in the Mediterranean Sea were from beaches in the U.S. Southeast. Since the unlucky creatures were a good representative sample of the turtles in the area, that meant more than half the young turtles living in the Mediterranean Sea had hatched from nests on beaches in Florida, Georgia, and South Carolina. Some 20,000 to 50,000 loggerheads die each year due to the Mediterranean fisheries, which may partly explain the decline in loggerheads nesting on U.S. Southeast beaches for the last 25 years.

At the University of Alaska's Institute of Arctic Biology in Fairbanks, graduate student Sandra Talbot recently finished sequencing DNA by hand from Alaskan brown bears. Wildlife geneticist Gerald Shields, who heads the program, and Talbot have identified two types of brown bears in Alaska. One type resides only on southeastern Alaska's Admiralty, Baranof, and Chichagof islands, known as the ABC Islands (Fig. 35A).

The other brown bear in Alaska is found throughout the rest of the state, as well as in Siberia and western Asia. A third distinct type of brown bear, known as the Montana grizzly, resides in other parts of North America. These three types comprise all of the known brown bears in the New World.

The ABC bears' uniqueness may be bad news for the timber industry, which has expressed interest in logging parts of the ABC Islands. Says Shields, "Studies show that when roads are built and the habitat is fragmented, the population of brown bears declines. Our genetic observations suggest they are truly unique, and we should consider their heritage. They could never be replaced by transplants. . . ."

In what will become a classic example of how DNA analysis might be used to protect endangered species from future ruin, scientists from the United States and New Zealand recently carried out discreet experiments in a Japanese hotel room on whale sushi bought in local markets. Sushi, a staple of the Japanese diet, is a rice and meat concoction wrapped in seaweed. Armed with a miniature DNA sampling machine, the scientists found that of the 16 pieces of whale sushi they examined, many were from whales that are endangered or protected under an international moratorium on whaling. "Their findings demonstrated the true power of DNA studies," says David Woodruff, a conservation biologist at the University of California, San Diego.

One sample was from an endangered humpback, four were from fin whales, one was from a northern minke, and another from a beaked whale. Stephen Palumbi of the University of Hawaii says the technique could be used for monitoring and verifying catches. Until then, he says, "no species of whale can be considered safe."

Meanwhile, Ken Goddard, director of the unique U.S. Fish and Wildlife Service Forensics Laboratory in Ashland, Oregon, is already on the watch for wildlife crimes in the United States and 122 other countries that send samples to him for analysis. "DNA is one of the most powerful tools we've got," says Goddard, a former California police crime-lab director.

The lab has blood samples, for example, for all of the wolves being released into Yellowstone Park—"for the obvious reason that we can match those samples to a crime scene," says Goddard. The lab has many cases currently pending in court that he cannot discuss. But he likes to tell the story of the lab's first DNA-matching case. Shortly after the lab opened in 1989, California wildlife authorities contacted Goddard. They had seized the carcass of a trophy-size deer from a hunter. They believed the deer had been shot illegally on a 3,000-acre preserve owned by actor Clint Eastwood. The agents found a gut pile on the property but had no way to match it to the carcass. The hunter had two witnesses to deny the deer had been shot on the preserve.

Goddard's lab analysis made a perfect match between tissue from the gut pile and tissue from the carcass. Says Goddard: "We now have a cardboard cutout of Clint Eastwood at the lab saying 'Go ahead: Make my DNA.'"

Figure 35A Brown bear diversity.
These two brown bears appear similar, but DNA studies recently revealed that one type, known as an ABC bear, resides only on southeastern Alaska's Admiralty, Baranof, and Chichagof islands.

Tropical Forests

In the **tropical rain forests** of South America, Africa, and the Indo-Malayan region near the equator, the weather is always warm (between 20° and 25°C), and rainfall is plentiful (with a minimum of 190 cm per year). This may be the richest biome, both in terms of number of different kinds of species and their abundance.

A tropical rain forest has a complex structure, with many levels of life (Fig. 35.9). Some of the broadleaf evergreen trees grow from 15 to 50 meters or more. These tall trees often have trunks buttressed at ground level to prevent their toppling over. Lianas, or woody vines, which encircle the tree as it grows, also help strengthen the trunk. The diversity of species is enormous—a 10-km^2 area of tropical rain forest may contain 750 species of trees and 1,500 species of flowering plants.

Although some animals live on the ground (e.g., pacas, agoutis, peccaries, and armadillos), most live in the trees (Fig. 35.10). Insect life is so abundant that the majority of species have not been identified yet. Termites play a vital role in the decomposition of woody plant material, and ants are found everywhere, particularly in the trees. The various birds, such as hummingbirds, parakeets, parrots, and toucans, are often beautifully colored. Amphibians and reptiles are well represented by many types of frogs, snakes, and lizards. Lemurs, sloths, and monkeys are well-known primates that feed on the fruits of the trees. The largest carnivores are the big cats—the jaguars in South America and the leopards in Africa and Asia.

Many animals spend their entire life in the canopy, as do some plants. **Epiphytes** are plants that grow on other plants but usually have roots of their own that absorb moisture and minerals leached from the canopy; others catch rain and debris in hollows produced by overlapping leaf bases. The most common epiphytes are related to pineapples, orchids, and ferns.

Figure 35.9 Levels of life in a tropical rain forest.
The primary levels within a tropical rain forest are the canopy, the understory, and the forest floor. But the canopy (solid layer of leaves) contains levels as well, and some organisms spend their entire life in one particular level. Long lianas (hanging vines) climb into the canopy, where they produce leaves. Epiphytes are air plants that grow on the trees but do not parasitize them.

While we usually think of tropical forests as being nonseasonal rain forests, tropical forests that have wet and dry seasons are found in India, Southeast Asia, West Africa, South and Central America, the West Indies, and northern Australia. Here, there are deciduous trees, with many layers of growth beneath them. In addition to the animals just mentioned, some of these forests also contain elephants, tigers, and hippopotamuses.

Whereas the soil of a temperate deciduous forest biome is rich enough for agricultural purposes, the soil of a tropical rain forest biome is not. Nutrients are cycled directly from the litter to the plants again. Productivity is high because of high temperatures, a yearlong growing season, and the rapid recycling of nutrients from the litter. (In humid tropical forests, iron and aluminum oxides occur at the surface, causing a reddish residue known as laterite. When the trees are cleared, laterite bakes in the hot sun to a bricklike consistency that will not support crops.) Swidden agriculture, often called slash-and-burn agriculture, has been successful, but also destructive, in the tropics. Trees are felled and burned, and the ashes provide enough nutrients for several harvests. Thereafter, the forest must be allowed to regrow, and a new section must be cut and burned.

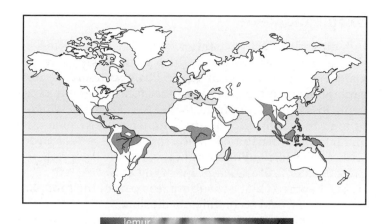

Figure 35.10   Animals of the tropical rain forest.

Shrublands

It is difficult to define a shrub, but in general, shrubs are shorter than trees (4.5–6 m) with a woody persistent stem and no central trunk. Shrubs have small but thick evergreen leaves that are often coated with a waxy material that prevents loss of moisture from the leaves. Their thick underground roots can survive dry summers and frequent fires and take deep moisture from the soil. Shrubs are adapted to withstand arid conditions and can also quickly sprout new growth after a fire. As a point of interest, you will recall from Chapter 33 that a shrub stage is part of the process of both primary and secondary succession.

Shrublands tend to occur along coasts that have dry summers and receive most of their rainfall in the winter. Shrubland are found along the cape of South Africa, the western coast of North America, and the southwestern and southern shores of Australia, as well as around the Mediterranean Sea, and in central Chile. The dense shrubland that occurs in California is known as **chaparral** (Fig. 35.11). This type of shrubland, called Mediterranean, lacks an understory and ground litter, and is highly flammable. The seeds of many species require the heat and scarring action of fire to induce germination. Other shrubs sprout from the roots after a fire.

There is also a northern shrub area that lies west of the Rocky Mountains. This area is sometimes classified as a cold desert, but the region is dominated by sagebrush and other hardy plants. Some of the birds found here are dependent upon sagebrush for their existence.

Figure 35.11   Shrubland.
Shrublands, such as chaparral in California, are subject to raging fires, but the shrubs are adapted to quickly regrow.

Grasslands

Grasslands occur where rainfall is greater than 25 cm but generally insufficient to support trees. For example, in temperate areas, where rainfall is between 10 and 30 inches it is too dry for forests and too wet for deserts to form. Natural grasslands once covered more than 40% of the earth's land surface, but many areas that once were grasslands are now used to grow crops such as wheat and corn.

Grasses are well adapted to a changing environment and can tolerate a high degree of grazing, flooding, drought, and sometimes fire. Where rainfall is high, large tall grasses that reach more than 2 meters in height (e.g., pampas grass) can flourish. In drier areas, shorter grasses between 5 and 10 cm are dominant. Low-growing bunch grasses (e.g., grama grass) grow in the United States near deserts. Grasses also generally grow in different seasons; some grassland animals migrate, and ground squirrels hibernate, when there is little grass for them to eat.

The temperate grasslands include the Russian steppes, the South American pampas, and the North American prairies (Fig. 35.12). When traveling across the United States from east to west, the line between the temperate deciduous forest and a tall-grass prairie is roughly along the border between Illinois and Indiana. The tall-grass prairie requires more rainfall than does the short-grass prairie that occurs near deserts. Large herds of bison—estimated at hundreds of thousands—once roamed the prairies, as did herds of pronghorn antelope. Now, small mammals, such as mice, prairie dogs, and rabbits, typically live below ground, but usually feed above ground. Hawks, snakes, badgers, coyotes, and foxes feed on these mammals. Virtually all of these grasslands, however, have been converted to agricultural lands.

Savannas

Savannas, which are grasslands that contain some trees, occur in regions where a relatively cool dry season is followed by a hot, rainy one (Fig. 35.13). One tree that can survive the severe dry season is the flat-topped acacia, which sheds its leaves during a drought. The African savanna supports the greatest variety and number of large herbivores of all the biomes. Elephants and giraffes are browsers that feed on tree vegetation. Antelopes, zebras, wildebeests, water buffalo, and rhinoceroses are grazers that feed on grasses. Any plant litter that is not consumed by grazers is attacked by a variety of small organisms, among them termites. Termites build towering nests in which they tend fungal gardens, their source of food. The herbivores support a large population of carnivores. Lions and hyenas sometimes hunt in packs, cheetahs hunt singly by day, and leopards hunt singly by night.

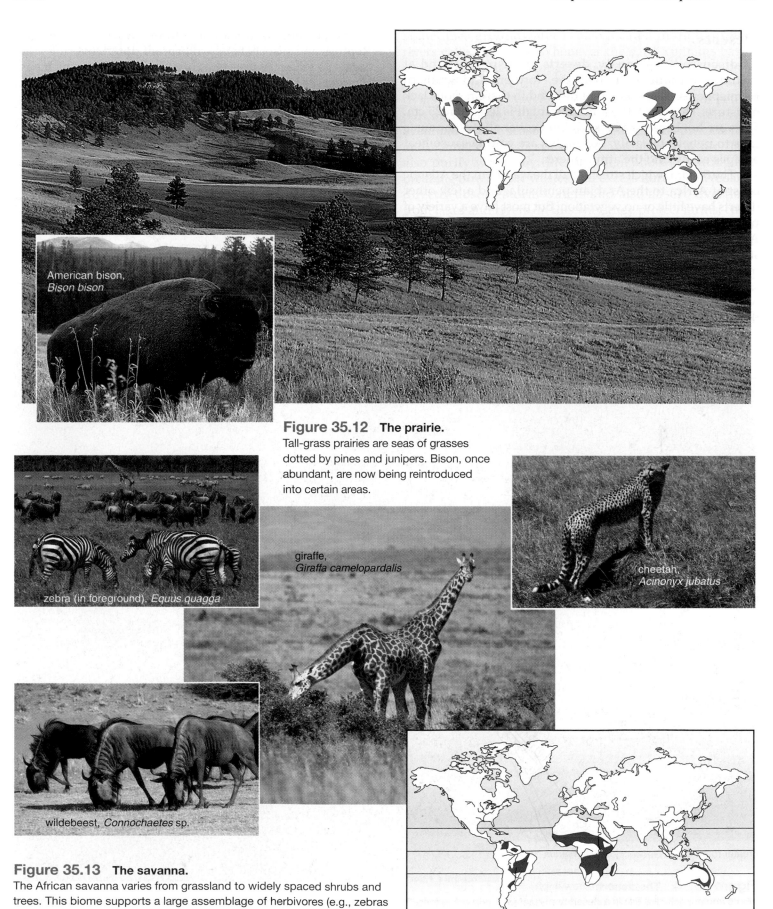

Figure 35.12   **The prairie.**
Tall-grass prairies are seas of grasses dotted by pines and junipers. Bison, once abundant, are now being reintroduced into certain areas.

Figure 35.13   **The savanna.**
The African savanna varies from grassland to widely spaced shrubs and trees. This biome supports a large assemblage of herbivores (e.g., zebras wildebeests, and giraffes). Carnivores (e.g., cheetahs) prey on these.

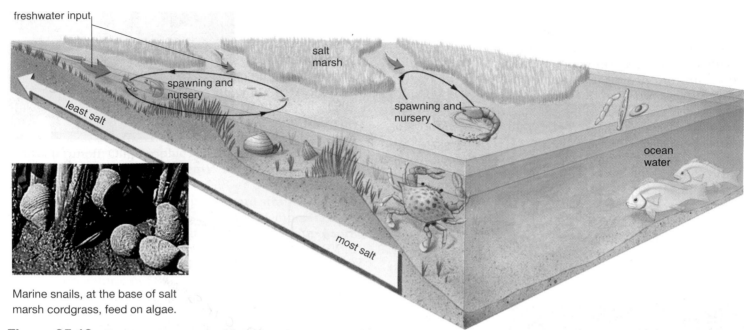

Marine snails, at the base of salt marsh cordgrass, feed on algae.

Figure 35.19   Estuary structure and function.

Since an estuary is located where a river flows into the ocean, it receives nutrients from the land. Estuaries serve as a nursery for the spawning and rearing of the young for many species of fishes, shrimp and other crustaceans, and molluscs.

Coastal Communities

Near the mouth of a river, a salt marsh in the temperate zone or a mangrove swamp in the subtropical and tropical zones is likely to develop. Also, the silt carried by a river may form mudflats. It is proper to think of seacoasts and mudflats, salt marshes, and mangrove swamps as belonging to one ecological system.

Estuaries

An **estuary** is a partially enclosed body of water where fresh water and seawater meet and mix (Fig. 35.19). A river brings fresh water into the estuary, and the sea, because of the tides, brings salt water. Coastal bays, tidal marshes, fjords (an inlet of water between high cliffs), some deltas (a triangular-shaped area of land at the mouth of a river), and lagoons (a body of water separated from the sea by a narrow strip of land) are all examples of estuaries.

Organisms living in an estuary must be able to withstand constant mixing of waters and rapid changes in salinity. Not many organisms are suited to this environment, but those that are suited find an abundance of nutrients. An estuary acts as a nutrient trap because the sea prevents the rapid escape of nutrients brought by a river.

Although only a few small fish permanently reside in an estuary, many develop there, creating a constant abundance of larval and immature fish. It has been estimated that well over half of all marine fishes develop in the protective environment of an estuary, which explains why estuaries are called the nurseries of the sea. Estuaries are also feeding grounds for many birds, fish, and shellfish because they offer a ready supply of food.

a.

b.

Figure 35.20   Types of estuaries.

Some of the many types of regions that qualify as estuaries include the salt marsh depicted in Figure 35.19, and mudflats (**a**), which are frequented by migrant birds, and mangrove swamps (**b**) skirting the coastlines of many tropical and subtropical lands. The tangled roots of mangrove trees trap sediments and nutrients that sustain many immature forms of sea life.

— upper littoral zone

— mid littoral zone

— lower littoral zone

a.

b.

c.

Figure 35.21   Seacoasts.
a. The littoral zone of a rocky coast, where the tide comes in and out, has different types of shelled and algal organisms at its upper, middle, and lower portions. **b.** Some organisms of a rocky coast live in tidal pools. **c.** A sandy shore looks devoid of life except for the birds that feed there.

Salt marshes dominated by salt marsh cordgrass are often associated with estuaries. So are mudflats and mangrove swamps, where sediment and nutrients from the land collect (Fig. 35.20).

Seashores
Both rocky and sandy shores are constantly bombarded by the sea as the tides roll in and out (Fig. 35.21). The **littoral zone** lies between the high and low water marks. The littoral zone of a rocky beach is divided into subzones. In the upper portion of the littoral zone, barnacles are glued so tightly to the stone by their own secretions that their calcareous outer plates remain in place even after the enclosed shrimplike animal dies. In the midportion of the littoral zone, brown algae known as rockweed may overlie the barnacles. In the lower portions of the littoral zone, oysters and mussels attach themselves to the rocks by filaments called *byssal threads*. Also

present are snails called limpets and periwinkles. But periwinkles have a coiled shell and secure themselves by hiding in crevices or under seaweeds, while limpets press their single flattened cone tightly to a rock. Below the littoral zone, macroscopic seaweeds, which are the main photosynthesizers, anchor themselves to the rocks by holdfasts.

Organisms cannot attach themselves to shifting, unstable sands on a sandy beach; therefore, nearly all the permanent residents dwell underground. They either burrow during the day and surface to feed at night, or they remain permanently within their burrows and tubes. Ghost crabs and sandhoppers (amphipods) burrow themselves above the high tide mark and feed at night when the tide is out. Sandworms and sand (ghost) shrimp remain within their burrows in the littoral zone and feed on detritus whenever possible. Still lower in the sand, clams, cockles, and sand dollars are found.

36.2 Value of Biodiversity

Conservation biology strives to reverse the trend toward the possible extinction of thousands of plants and animals. To bring this about, it is necessary to make all people aware that biodiversity is a resource of immense value.

Direct Value

Various individual species perform services for human beings and contribute greatly to the value we should place on biodiversity. Only some of the most obvious values are discussed here and illustrated in Figure 36.3.

Medicinal Value

Most of the prescription drugs used in the United States were originally derived from living organisms. The rosy periwinkle from Madagascar is an excellent example of a tropical plant that has provided us with useful medicines. Potent chemicals from this plant are now used to treat two forms of cancer: leukemia and Hodgkin disease. Because of these drugs, the survival rate for childhood leukemia has gone from 10% to 90%, and Hodgkin disease is usually curable. Although the value of saving a life cannot be calculated, it is still sometimes easier for us to appreciate the worth of a resource if it is explained in monetary terms. Thus, researchers tell us that, judging from the success rate in the past, an additional 328 types of drugs are yet to be found in tropical rain forests, and the value of this resource to society is probably $147 billion.

You may already know that the antibiotic penicillin is derived from a fungus and that certain species of bacteria produce the antibiotics tetracycline and streptomycin. These drugs have proven to be indispensable in the treatment of diseases, including certain sexually transmitted diseases.

Leprosy is among those diseases for which there is as yet no cure. The bacterium that causes leprosy will not grow in the laboratory, but scientists discovered that it grows naturally in the nine-banded armadillo. Having a source for the bacterium may make it possible to find a cure for leprosy. The blood of horseshoe crabs contains a substance called limulus amoebocyte lysate, which is used to ensure that medical devices such as pacemakers, surgical implants, and prosthetic devices are free of bacteria. Blood is taken from 250,000 crabs a year, and then they are returned to the sea unharmed.

Agricultural Value

Crops such as wheat, corn, and rice are derived from wild plants that have been modified to be high producers. The same high-yield, genetically similar strains tend to be grown worldwide. When rice crops in Africa were being devastated by a virus, researchers grew wild rice plants from thousands of seed samples until they found one that contained a gene for resistance to the virus. These wild plants were then used in a breeding program to transfer the gene into high-yield rice plants. If this variety of wild rice had become extinct before it could be discovered, rice cultivation in Africa might have collapsed.

Biological pest controls—natural predators and parasites—are often preferable to using chemical pesticides. When a rice pest called the brown planthopper became resistant to pesticides, farmers began to use natural brown planthopper enemies instead. The economic savings were calculated at well over $1 billion. Similarly, cotton growers in Cañete Valley, Peru, found that pesticides were no longer working against the cotton aphid because of resistance. Research identified natural predators that are now being used to an ever greater degree by cotton farmers. Again, savings have been enormous.

Most flowering plants are pollinated by animals, such as bees, wasps, butterflies, beetles, birds, and bats. The honeybee, *Apis mellifera*, has been domesticated, and it pollinates almost $10 billion worth of food crops annually in the United States. The danger of this dependency on a single species is exemplified by mites that have now wiped out more than 20% of the commercial honeybee population in the United States. Where can we get resistant bees? From the wild, of course. The value of wild pollinators to the U.S. agricultural economy has been calculated at $4.1 to $6.7 billion a year.

Consumptive Use Value

We have had much success cultivating crops, keeping domesticated animals, growing trees in plantations, and so forth. But so far, aquaculture, the growing of fish and shellfish for human consumption, has contributed only minimally to human welfare—instead, most freshwater and marine harvests depend on the catching of wild animals, such as fishes (e.g., trout, cod, tuna, and flounder), crustaceans (e.g., lobsters, shrimps, and crabs), and mammals (e.g., whales). Obviously, these aquatic organisms are an invaluable biodiversity resource.

The environment provides all sorts of other products that are sold in the marketplace worldwide, including wild fruits and vegetables, skins, fibers, beeswax, and seaweed. Also, some people obtain their meat directly from the environment. In one study, researchers calculated that the economic value of wild pig in the diet of native hunters in Sarawak, East Malaysia, was approximately $40 million per year.

Similarly, many trees are still felled in the natural environment for their wood. Researchers have calculated that a species-rich forest in the Peruvian Amazon is worth far more if the forest is used for fruit and rubber production than for timber production. Fruit and the latex needed to produce rubber can be brought to market for an unlimited number of years, whereas once the trees are gone, no more timber can be harvested.

Wild species directly provide us with all sorts of goods and services whose monetary value can sometimes be calculated.

Wild species, like the lesser long-nosed bat, *Leptonycteris curasoae*, are pollinators of agricultural and other plants.

Wild species, like the rosy periwinkle, *Catharanthus roseus*, are sources of many medicines.

Wild species, like rubber trees, *Hevea*, can provide a product indefinitely if the forest is not destroyed.

Wild species, like many marine species, provide us with food.

Wild species, like the nine-banded armadillo, *Dasypus novemcinctus*, play a role in medical research.

Wild species, like lady bugs, *Coccinella*, play a role in biological control of agricultural pests.

Figure 36.3 Direct value of wildlife.
The direct services of wild species benefit human beings immensely, and it is sometimes possible to calculate the monetary value, which is always surprisingly large.

Indirect Value

The wild species we have been discussing live in ecosystems. If we want to preserve them, it is more economical to save the ecosystems than the individual species. Ecosystems perform many services for modern humans, who increasingly live in cities. These services are said to be indirect because they are pervasive and not easily discernible (Fig. 36.4). Even so, our very survival depends on the functions that ecosystems perform for us.

Biogeochemical Cycles

You'll recall from Chapter 34 that ecosystems are characterized by energy flow and chemical cycling. The biodiversity within ecosystems contributes to the workings of the water, carbon, nitrogen, phosphorus, and other biogeochemical cycles. We are dependent on these cycles for fresh water, removal of carbon dioxide from the atmosphere, uptake of excess soil nitrogen, and provision of phosphate. When human activities upset the usual workings of biogeochemical cycles, the dire environmental consequences include excess pollutants that are harmful to us. Technology is unable to artificially contribute to or create any of the biogeochemical cycles.

Waste Disposal

Decomposers break down dead organic matter and other types of wastes to inorganic nutrients that are used by the producers within ecosystems. This function aids humans immensely because we dump millions of tons of waste material into natural ecosystems each year. If it were not for decomposition, waste would soon cover the entire surface of our planet. We can build sewage treatment plants, but they are expensive, and few of them break down solid wastes completely to inorganic nutrients. It is less expensive and more efficient to water plants and trees with partially treated wastewater and let soil bacteria cleanse it completely.

Biological communities are also capable of breaking down and immobilizing pollutants, such as heavy metals and pesticides, that humans release into the environment. A review of wetland functions in Canada assigned a value of $50,000 per hectare (100 acres or 10,000 square meters) per year to the ability of natural areas to purify water and take up pollutants.

Provision of Fresh Water

Few terrestrial organisms are adapted to living in a salty environment—they need fresh water. The water cycle continually supplies fresh water to terrestrial ecosystems. Humans use fresh water in innumerable ways, including drinking it and irrigating their crops. Freshwater ecosystems such as rivers and lakes also provide us with fish and other types of organisms for food.

Unlike other commodities, there is no substitute for fresh water. We can remove salt from seawater to obtain fresh water, but the cost of desalination is about four to eight times the average cost of fresh water acquired via the water cycle.

Forests and other natural ecosystems exert a "sponge effect." They soak up water and then release it at a regular rate. When rain falls in a natural area, plant foliage and dead leaves lessen its impact, and the soil slowly absorbs it, especially if the soil has been aerated by organisms. The water-holding capacity of forests reduces the possibility of flooding. The value of a marshland outside Boston, Massachusetts, has been estimated at $72,000 per hectare per year solely on its ability to reduce floods. Forests release water slowly for days or weeks after the rains have ceased. Rivers flowing through forests in West Africa release twice as much water halfway through the dry season, and between three and five times as much at the end of the dry season, as do rivers from coffee plantations.

Prevention of Soil Erosion

Intact ecosystems naturally retain soil and prevent soil erosion. The importance of this ecosystem attribute is especially observed following deforestation. In Pakistan, the world's largest dam, the Tarbela Dam, is losing its storage capacity of 12 billion cubic meters many years sooner than expected because silt is building up behind the dam due to deforestation. At one time, the Philippines were exporting $100 million worth of oysters, mussels, clams, and cockles each year. Now silt carried down rivers following deforestation is smothering the mangrove ecosystem that serves as a nursery for the sea. Most coastal ecosystems are not as bountiful as they once were because of deforestation and a myriad of other assaults.

Regulation of Climate

At the local level, trees provide shade and reduce the need for fans and air conditioners during the summer.

Globally, forests ameliorate the climate because they take up carbon dioxide. The leaves of trees use carbon dioxide when they photosynthesize, and the bodies of the trees store carbon. When trees are cut and burned, carbon dioxide is released into the atmosphere. Carbon dioxide makes a significant contribution to global warming, which is expected to be stressful for many plants and animals. Only a small percentage of wildlife will be able to move northward, where the weather will be suitable for them.

Ecotourism

Almost everyone prefers to vacation in the natural beauty of an ecosystem. In the United States, nearly 100 million people enjoy vacationing in a natural setting. To do so, they spend $4 billion each year on fees, travel, lodging, and food. Many tourists want to go sport fishing, whale watching, boat riding, hiking, birdwatching, and the like. Some want to merely immerse themselves in the beauty of a natural environment.

a.

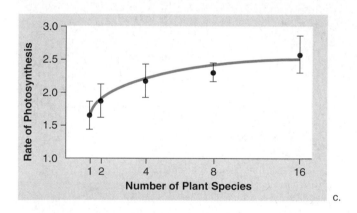

c.

b.

Figure 36.4 **Indirect value of ecosystems.**
a. Natural ecosystems provide (**b**) human-impacted ecosystems with many ecological services. **c.** Research results show that the higher the biodiversity (measured by number of plant species), the greater the rate of photosynthesis of an experimental community.

Biodiversity and Natural Ecosystems

Massive changes in biodiversity, such as deforestation, have a significant impact on ecosystem services like those we have been discussing. Researchers are interested in determining whether a high degree of biodiversity also helps ecosystems function more efficiently. To test the benefits of biodiversity in a Minnesota grassland habitat, researchers sowed plots with seven levels of plant diversity. Their study found that ecosystem performance improves with increasing species richness. The more diverse the plots, the lower the concentrations of inorganic soil nitrogen, indicating a higher level of nitrate uptake. A similar study in California also showed greater overall resource use in more diverse plots.

Another group of experimenters tested the effects of an increase in diversity at four levels: producers, herbivores, parasites, and decomposers. They found that the rate of photosynthesis increased as diversity increased (Fig. 36.4c).

A computer simulation has shown that the response of a deciduous forest to elevated carbon dioxide is a function of species diversity. The more complex community, composed of nine tree species, exhibited a 30% greater amount of photosynthesis than a community composed of a single species.

More studies are needed to test whether biodiversity maximizes resource acquisition and retention within an ecosystem. Also, are more diverse ecosystems better able to withstand environmental changes and invasions by other species, including pathogens? Then, too, how does fragmentation affect the distribution of organisms within an ecosystem and the functioning of an ecosystem?

Ecosystems perform services that most likely depend on a high degree of biodiversity.

36.3 Causes of Extinction

In order to stem the tide of extinction, it is first necessary to identify its causes. Researchers examined the records of 1,880 threatened and endangered wild species in the United States and found that habitat loss was involved in 85% of the cases (Fig. 36.5*a*). Alien species had a hand in nearly 50%, pollution was a factor in 24%, overexploitation in 17%, and disease in 3%. The percentages add up to more than 100% because most of these species are imperiled for more than one reason. Macaws are a good example that a combination of factors can lead to a species decline (Fig. 36.5*b*). Not only has their habitat been reduced by encroaching timber and mining companies, but macaws are also hunted for food and collected for the pet trade.

Habitat Loss

Habitat loss has occurred in all ecosystems, but concern has now centered on tropical rain forests and coral reefs because they are particularly rich in species. A sequence of events in Brazil offers a fairly typical example of the manner in which rain forest is converted to land noninhabitable for wildlife. The construction of a major highway into the forest first provided a way to reach the interior of the forest (Fig. 36.5*c*). Small towns and industries sprang up along the highway, and roads branching off the main highway gave rise

to even more roads. The result was fragmentation of the once immense forest. The government offered subsidies to anyone willing to take up residence in the forest, and the people who came cut and burned trees in patches (Fig. 36.5*d*). Tropical soils contain limited nutrients, but when the trees are burned, nutrients are released that support a lush growth for the grazing of cattle for about three years. However, once the land was degraded (Fig. 36.5*e*), the farmer and his family moved on to another portion of the forest.

Loss of habitat also affects freshwater and marine biodiversity. Coastal degradation is largely due to the large concentration of people living there. Already, 60% of coral reefs have been destroyed or are on the verge of destruction; it's possible that all coral reefs may disappear during the next 40 years. Mangrove forest destruction is also a problem; Indonesia, with the most mangrove acreage, has lost 45% of its mangroves, and the percentage is even higher for other tropical countries. Wetland areas, estuaries, and seagrass beds are also being rapidly destroyed.

Figure 36.5   Habitat loss.
a. In a study examining records of imperiled U.S. plants and animals, habitat loss emerged as the greatest threat to wildlife. **b.** Macaws that reside in South American tropical rain forests are endangered for the reasons listed in the graph in (**a**). **c.** The construction of roads in an area in Brazil opened up the rain forest and subjected it to fragmentation. **d.** The result was patches of forest and degraded land. **e.** Wildlife could not live in destroyed portions of the forest.

c.

d.

e.

Ecology Focus

Alien Species Wreak Havoc

While some foreign species live peaceably amid their new neighbors, many threaten entire economies and ecosystems. The brown tree snake slipped onto Guam from southwestern Pacific islands in the late 1940s. Since then, it has wiped out nine of eleven native bird species, leaving the forests eerily quiet. Indeed, some conservationists now rank invasive species among the top menaces to endangered species (Fig. 36A). "We're losing more habitat here to pests than to bull-dozers," says Alan Holt of the Nature Conservancy of Hawaii. As a remote archipelago, Hawaii was particularly vulnerable to ecological disruption when Polynesian voyagers arrived 1,600 years ago. Having evolved in isolation for millions of years, many native species had discarded evolutionary adaptations that deter predators. The islands abounded with snails with no shells, plants with no thorns, and birds that nested on the ground. Polynesian hunters promptly wiped out several species of large, flightless birds, but a stowaway in their canoes also did serious damage. The Polynesian rat flourished, decimating dozens of species of ground-nesting birds. The first Europeans and their many plant and animal companions unleashed an even larger wave of extinctions. In 1778, for instance, Captain James Cook brought ashore goats, which soon went feral, devastating native plants.

Biologists Art Medeiros and Lloyd Loope of Haleakala National Park on Maui often feel they are fighting an endless ground war. The park is home to a legion of endangered plants and animals found nowhere else in the world. Six years ago, officials completed a 50-mile, $2.4 million fence to keep out feral goats and pigs. But just as the forest understory was beginning to recover, rabbits released into the park by a bored pet owner launched their own assault. Staffers got to work with rifle and snare, but soon after they'd bagged the last of the rabbits, axis deer—miniature elk from India—began hopping the fence.

Those are just the warmblooded invaders. Medeiros and Loope also are developing chemical weapons for their, thus far, losing battle against the Argentine ant. This tiny terminator threatens to wipe out the park's native insects and the rare native plants that depend on the insects for pollination. "It's an eraser," says Loope. "Shake down a flowering bush inside ant territory and you'll get five species [of native insects]; outside their range, you'll get 10 times that."

Another potential "eraser" at the park gates is the Jackson's chameleon, a colorful Kenyan native that dines on insects and snails. Then there's the dreaded miconia. Since it was introduced to Tahiti as an ornamental in 1937, the tree that locals call the "green cancer" has overrun more than half the island. Its dense foliage shades out other plants—from competing trees to the mosses that anchor soils and hold rainwater. As a result, many plants have been pushed to the brink of extinction, and the mountainsides are eroding, silting over coral reefs that help sustain fisheries.

Siccing pests on pests poses its own problems. Some recruits have run amok, doing as much ecological damage as the pests they were meant to control. In the 1880s, for instance, Hawaiian sugar-cane growers brought in mongooses to prune mice and rat populations. Prune they did, but they also preyed heavily on native birds. Happily, biocontrol efforts have been more successful in recent years. Nearly 90% of the agents released in the past two decades have been known to attack only the target pest, according to a study conducted jointly by researchers at the Hawaii Department of Agriculture and the University of Hawaii.

By all accounts, preventing invaders from gaining a foothold in the first place is an even better strategy. It's been recommended that, among other measures, importers be made liable for damages caused by the alien species they introduce and that emergency-response teams be established to jump on new infestations.

a.

b.

Figure 36A   Alien species.
a. The brown tree snake has devastated endemic bird populations after being introduced into many Pacific islands. **b.** Purple loosestrife arrived in U. S. Atlantic ports 200 years ago and has since spread steadily westward, crowding out native species.

Figure 36.6   Alien Species.
Mongooses were introduced into Hawaii to control rats, but they also prey on native birds.

Alien Species

Alien species, sometimes called exotics, are nonnative species that migrate into a new ecosystem or are deliberately or accidentally introduced by humans. Ecosystems around the globe are characterized by unique assemblages of organisms that have evolved together in one location. Migrating to a new location is not usually possible because of barriers such as oceans, deserts, mountains, and rivers. Humans, however, have introduced alien species into new ecosystems chiefly due to:

Colonization. Europeans, in particular, brought various familiar species with them when they colonized new places. For example, the pilgrims brought the dandelion to the United States as a familiar salad green.

Horticulture and agriculture. Some exotics now taking over vast tracts of land have escaped from cultivated areas. Kudzu is a vine from Japan that the U. S. Department of Agriculture thought would help prevent soil erosion. The plant now covers much landscape in the South, including even walnut, magnolia, and sweet gum trees.

Accidental transport. Global trade and travel accidentally bring many new species from one country to another. Researchers found that the ballast water released from ships into Coos Bay, Oregon, contained 367 marine species from Japan. The zebra mussel from the Caspian Sea was accidentally introduced into the Great Lakes in 1988. It now forms dense beds that squeeze out native mussels.

Alien species can disrupt food webs. As mentioned earlier in this chapter, opossum shrimp introduced into a lake in Montana led to an additional trophic level that in the end meant less food for bald eagles and grizzly bears (see Fig. 36.2).

Exotics on Islands

Islands are particularly susceptible to environmental discord caused by the introduction of alien species. Islands have unique assemblages of native species that are closely adapted to one another and cannot compete well against exotics. Myrtle trees, *Myrica faya,* introduced into the Hawaiian Islands from the Canary Islands, are symbiotic with a type of bacterium that is capable of nitrogen fixation. This feature allows the species to establish itself on nutrient-poor volcanic soil, a distinct advantage in Hawaii. Once established, myrtle trees call a halt to the normal succession of native plants on volcanic soil.

The brown tree snake has been introduced onto a number of islands in the Pacific Ocean (see Fig. 36A*a*). The snake eats eggs, nestlings, and adult birds. On Guam, it has reduced 10 native bird species to the point of extinction. On the Galápagos Islands, black rats have reduced populations of giant tortoise, while goats and feral pigs have changed the vegetation from highland forest to pampaslike grasslands and destroyed stands of cactus. Mongooses introduced into the Hawaiian Islands to control rats also prey on native birds (Fig. 36.6). The Ecology Focus on the previous page offers more examples of disruption by alien species.

Pollution

In the present context, **pollution** can be defined as any environmental change that adversely affects the lives and health of living things. Pollution has been identified as the third main cause of extinction. Pollution can also weaken organisms and lead to disease, the fifth main cause of extinction. Biodiversity is particularly threatened by the following environmental pollution:

Acid Deposition. Both sulfur dioxide from power plants and nitrogen oxides in automobile exhaust are converted to acids when they combine with water vapor in the atmosphere. These acids return to earth as either wet deposition (acid rain or snow) or dry deposition (sulfate and nitrate salts). Sulfur dioxide and nitrogen oxides are emitted in one locale, but deposition occurs across state and national boundaries. Acid deposition causes trees to weaken and increases their susceptibility to disease and insects. It also kills small invertebrates and decomposers so that the entire ecosystem is threatened. Many lakes in the northern United States are now lifeless because of the effects of acid deposition.

Eutrophication. Lakes are also under stress due to overenrichment. When lakes receive excess nutrients due to runoff from agricultural fields and wastewater from sewage treatment, algae begin to grow in abundance. An algal bloom is apparent as a green scum or excessive mats of filamentous algae. Upon death, the decomposers break down the algae, but in so doing they use up oxygen. A decreased amount of oxygen is available to fish, leading sometimes to a massive fish kill.

Ozone Depletion. The ozone shield is a layer of ozone (O_3) in the stratosphere, some 50 kilometers above the earth. The ozone shield absorbs most of the wavelengths of harmful ultraviolet (UV) radiation so that they do not strike the earth. The cause of ozone depletion can be traced to chlorine atoms (Cl^-) that come from the breakdown of chlorofluorocarbons

(CFCs). The best-known CFC is Freon, a heat transfer agent still found in refrigerators and air conditioners today. Severe ozone shield depletion can impair crop and tree growth and also kill plankton (microscopic plant and animal life) that sustain oceanic life. The immune system and the ability of all organisms to resist infectious diseases will most likely be weakened.

Organic Chemicals. Our modern society makes use of organic chemicals in all sorts of ways. Organic chemicals called nonylphenols are used in products ranging from pesticides to dishwashing detergents, cosmetics, plastics, and spermicides. These chemicals mimic the effects of hormones, and in that way most likely harm wildlife. Salmon are born in fresh water but mature in salt water. After investigators exposed young fish to nonylphenol, they found that 20–30% were unable to make the transition between fresh and salt water. Nonylphenols cause the pituitary to produce prolactin, a hormone that most likely prevents saltwater adaptation.

Global Warming. The expression **global warming** refers to an expected increase in average temperature during the twenty-first century. You may recall from Chapter 34 that carbon dioxide is a gas that comes from the burning of fossil fuels, and methane is a gas that comes from oil and gas wells,

rice paddies, and animals. These gases are known as greenhouse gases because, just like the panes of a greenhouse, they allow solar radiation to pass through but hinder the escape of its heat back into space. Data collected around the world show a steady rise in the concentration of the various greenhouse gases. These data are used to generate computer models that predict the earth may warm to temperatures never before experienced by living things (Fig. 36.7a).

As the oceans warm, temperatures in the polar regions will rise to a greater degree than in other regions. The sea level will then rise because glaciers will melt and water expands as it warms. A one-meter rise in sea level in the next century could inundate 25–50% of U.S. coastal wetlands. This loss of habitat could be higher if wetlands cannot move inward because of coastal development and levees.

The tropics will also feel the effects of global warming. The growth of corals is very dependent upon mutualistic algae living in their walls. When the temperature rises by 4 degrees, corals expel their algae and are said to be "bleached" (Fig. 36.7b). Almost no growth or reproduction occurs until the algae return. Also, coral reefs prefer shallow waters, and if sea levels rise, they may "drown." Multiple assaults on coral are even now causing them to be stricken with various diseases.

Global warming could very well cause many extinctions on land also. As temperatures rise, regions of suitable climate for various species will shift toward the poles and higher elevations. The present assemblages of species in ecosystems will be disrupted as some species migrate northward, leaving others behind. Plants migrate when seeds disperse and growth occurs in a new locale. For example, to remain in a favorable habitat, it's been calculated that the rate of beech tree migration would have to be 40 times faster than has ever been observed. It seems unlikely that beech or any other type of tree would be able to meet the pace required. Then, too, many species of organisms are confined to relatively small habitat patches that are surrounded by agricultural or urban areas they would not be able to cross. And even if they have the capacity to disperse to new sites, suitable habitats may not be available.

Figure 36.7   Global warming.
a. Mean global temperature change is expected to rise due to the introduction of greenhouse gases into the atmosphere. Global warming has the potential to significantly affect the world's biodiversity. **b.** A temperature rise of only a few degrees causes coral reefs to "bleach" and become lifeless.

Overexploitation

Overexploitation occurs when the number of individuals taken from a wild population is so great that the population becomes severely reduced in numbers. A positive feedback cycle explains overexploitation: the smaller the population, the more valuable its members, and the greater the incentive to capture the few remaining organisms.

A decorative plant and exotic pet market supports both legal and illegal trade in wild species. Rustlers dig up rare cacti such as the single-crested saguaro and sell them to gardeners. Parakeets and macaws are among the birds taken from the wild for sale to pet owners. For every bird delivered alive, many more have died in the process. The same holds true for tropical fish, which often come from the coral reefs of Indonesia and the Philippines. Divers dynamite reefs or use plastic squeeze-bottles of cyanide to stun them; in the process, many fish die.

Declining species of mammals, such as the Siberian tiger, are still hunted for their hides, tusks, horns, or bones. Because of its rarity, a single Siberian tiger is now worth more than $500,000—its bones are pulverized and used as a medicinal powder. The horns of rhinoceroses become ornate carved daggers, and their bones are ground up to sell as a medicine. The ivory of an elephant's tusk is used to make art objects, jewelry, or piano keys. The fur of a Bengal tiger sells for as much as $100,000 in Tokyo.

The U.N. Food and Agricultural organization tells us that we have now overexploited 11 of 15 major oceanic fishing areas. Fish are a renewable resource if harvesting does not exceed the ability of the fish to reproduce. But our society uses larger and more efficient fishing fleets to decimate fishing stocks. Pelagic species such as tuna are captured by purse-seine fishing, in which a very large net surrounds a school of fish, and then the net is closed in the same manner as a drawstring purse. Dolphins that swim above schools of tuna are often captured and then killed in this type of net. Other fishing boats drag huge trawling nets, large enough to accommodate 12 jumbo jets, along the seafloor to capture bottom-dwelling fish (Fig. 36.8a). Only large fish are kept; undesirable small fish and sea turtles are discarded, dying, back into the ocean. Trawling has been called the marine equivalent of clear-cutting trees because after the net goes by, the sea bottom is devastated (Fig. 36.8b). Today's fishing practices don't allow fisheries to recover. Cod and haddock, once the most abundant bottom-dwelling fish along the northeast coast, are now often outnumbered by dogfish and skate.

A marine ecosystem can be disrupted by overfishing, as exemplified on the U.S. West Coast. When sea otters began to decline in numbers, investigators found that they were being eaten by orcas (killer whales). Usually orcas prefer seals and sea lions to sea otters, but they began eating sea otters when few seals and sea lions could be found. What caused a decline in seals and sea lions? Their preferred foodsources—perch and herring—were no longer plentiful due to overfishing. Ordinarily, sea otters keep the population of sea urchins, which feed on kelp, under control. But with fewer sea otters around, the sea urchin population exploded and decimated the kelp beds. Thus, overfishing set in motion a chain of events that detrimentally altered the food web of an ecosystem.

The five main causes of extinction are: habitat loss, introduction of alien species, pollution, overexploitation, and disease.

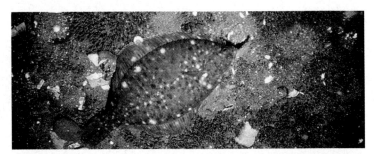

a.

b.

Figure 36.8 Trawling.
a. These Alaskan pollock were caught by dragging a net along the seafloor. **b.** Appearance of the seabed before *(top)* and after *(bottom)* the net went by.

— ice
— alpine tundra
— montane coniferous forest
— decidous forest
Tropical R. forest — Temperate forest
Temperate rainfor — Coniferous forest
coniferous — Tremperat
tundra — ice
ice

N pole ⎰ polar ice
↓ ⎱ Tundra
Taiga
Temperate F.
Tropical F.

est

orbnur

X

est

est

1 ice
2 alpine tundra
3 montane coniferous fo
4 deciduous forest
5 tropical forest
6 temperate deciduou
7 coniferous fores t
8   tundra
9   ice

ice
alpine tundra
montaneous coniferous fo
deciduous forest
tropical forest
temperate deciduous
coniferous forest
tundra
ice

seeds of trees. When bats are killed off and their roosts destroyed, the trees fail to reproduce. The grizzly bear is a keystone species in the northwestern United States and Canada (Fig. 36.9a). Bears disperse the seeds of berries; as many as 7,000 seeds may be in one dung pile. Grizzlies kill the young of many hoofed animals and thereby keep their populations under control. Grizzlies are also a principal mover of soil when they dig up roots and prey upon hibernating ground squirrels and marmots.

Metapopulations

The grizzly bear population is actually a **metapopulation**—that is, a population subdivided into several small, isolated populations due to habitat fragmentation. Originally there were probably 50,000 to 100,000 grizzlies south of Canada, but this number has been reduced because communities have encroached on their home range and bears have been killed by frightened homeowners. Now there are six virtually isolated subpopulations totaling about 1,000 individuals. The Yellowstone National Park population numbers 200, but the others are even smaller.

Saving metapopulations sometimes requires determining which of the populations is a source and which are sinks. A **source population** is one that most likely lives in a favorable

c. Northern spotted owl, *Strix occidentalis caurina*

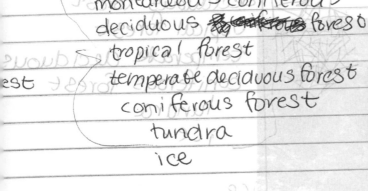

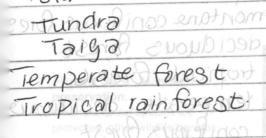

area, and its birthrate is most likely higher than its death rate. Individuals from source populations move into **sink populations** where the environment is not as favorable and where the birthrate equals the death rate at best. When trying to save the northern spotted owl, conservationists determined that it was best to avoid having owls move into sink habitats. The northern spotted owl reproduces successfully in old-growth rain forests of the Pacific Northwest (Fig. 36.9b,c) but not in nearby immature forests that are in the process of recovering from logging. Distinct boundaries that hindered the movement of owls into these sink habitats proved to be beneficial in maintaining source populations.

Landscape Dynamics

Grizzly bears inhabit a number of different types of ecosystems, including plains, mountains, and rivers. Saving any one of these particular types of ecosystems alone would not be sufficient to preserve grizzly bears. Instead, it is necessary to save diverse ecosystems that are at least connected by corridors. You will recall that a landscape encompasses different types of ecosystems. An area called the Greater Yellowstone Ecosystem, where bears are free to roam, has now been defined. It contains millions of acres in Yellowstone National Park; state lands in Montana, Idaho, and Wyoming; five different national forests; various wildlife refuges; and even private lands.

Landscape protection for one species is often beneficial for other wildlife that share the same space. The last of the contiguous 48 states' harlequin ducks, bull trout, westslope cutthroat trout, lynx, pine martens, wolverines, mountain caribou, and great gray owls are found in areas occupied by grizzlies. The recent return of gray wolves has occurred in this territory also. Then, too, grizzly range overlaps with 40% of Montana's vascular plants of special conservation concern.

The Edge Effect When preserving landscapes, it is necessary to consider the **edge effect.** An edge reduces the amount of habitat typical of an ecosystem because the edges around a patch have a habitat slightly different from the interior of a patch. For example, forest edges are brighter, warmer, drier, and windier, with more vines, shrubs, and weeds than the forest interior. Also, Figure 36.10a shows that a small and a large patch of habitat have the same amount of edge; therefore, the effective habitat shrinks as a patch gets smaller.

The edge effect can have a serious impact on population size. Songbird populations west of the Mississippi have been declining of late, and ornithologists have noticed that the nesting success of songbirds is quite low at the edge of a forest. The cause turns out to be the brown-headed cowbird, a social parasite of songbirds. Adult cowbirds prefer to feed in open agricultural areas, and they only briefly enter the forest when searching for a host nest in which to lay their eggs (Fig. 36.10b). Cowbirds are therefore benefited, while songbirds are disadvantaged, by the edge effect.

Computer Analyses

Two types of computer analyses, in particular, are now available to help conservationists plan how best to protect a species.

Gap analysis is the use of the computer to find gaps in preservation—places where biodiversity is high outside of preserved areas. First, computerized maps are drawn up showing the topography, vegetation, hydrology, and land ownership of a region. Then computer maps are done showing the geographic distribution of a region's animal and plant species. Once the distribution maps are superimposed onto the land-use maps, it is obvious where preserved habitats still need to be and/or could be located.

A **population viability analysis** can help researchers determine how much habitat a species requires to maintain itself. First it is necessary to calculate the minimum population size needed to prevent extinction. This size should protect the species from unforeseen events such as natural catastrophes or chance swings in the birth and death rates. Another

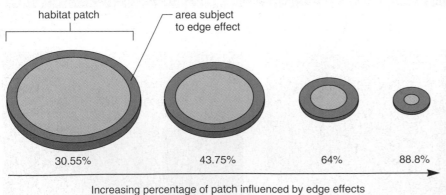

a.   b.

Figure 36.10   Edge effect.
a. The smaller the patch, the greater the proportion that is subject to the edge effect. **b.** Cowbirds lay their eggs in the nest of songbirds (yellow warblers). A cowbird is bigger than a warbler nestling and will be able to acquire most of the food brought by the warbler parent.

component to consider is the size needed to protect genetic diversity. This number varies according to the species. For example, analysis of red-cockaded woodpecker populations showed that an adult population of about 1,323 is needed to result in a genetically effective population of 500 because of the breeding system of the species. After you know the minimum population size, you can determine how much total acreage is needed for that population.

All life history characteristics of organisms must be taken into account when a population viability analysis is done. For example, female grizzlies don't give birth until they are 5 or 6, and then they typically wait three years before reproducing again. After doing one of the first population viability analyses, Mark Shaffer of the Wilderness Society predicted that a total grizzly bear population of 70 to 90 individuals, each with a suitable home range of 6,000 miles, will have about a 95% chance of surviving for 100 years. But Fred Allendorf pointed out that because only a few dominant males breed, the population needs to be larger than this to protect genetic diversity. Also, to prevent inbreeding, he recommended the introduction of one or two unrelated bears each decade into populations of 100 individuals. The bottom line is that dispersal among subpopulations is needed to prevent inbreeding and extinction.

Habitat Restoration

Restoration ecology is a new subdiscipline of conservation biology that seeks scientific ways to return ecosystems to their former state. Three principles have so far emerged. First, it's best to begin as soon as possible before remaining fragments of the original habitat are lost. These fragments are sources of wildlife and seeds from which to restock the restored habitat. Second, once the natural history of the habitat is understood, it is best to use biological techniques that mimic natural processes to bring about restoration. This might take the form of controlled burns to bring back grassland habitats, biological pest controls to rid the area of aliens, or bioremediation techniques to clean up pollutants. Third, the goal is **sustainable development,** the ability of an ecosystem to maintain itself while providing services to human beings. We will use the Everglades ecosystem in Florida to illustrate these principles.

The Everglades

Originally, the Everglades encompassed the whole of southern Florida, from Lake Okeechobee down to Florida Bay. This ecosystem is a vast sawgrass prairie, interrupted occasionally by a cypress dome or hardwood tree island. Within these islands, both temperate and tropical evergreen trees grow amongst dense and tangled vegetation. Mangroves are found along sloughs (creeks) and at the shoreline. The prop roots of red mangroves protect over 40 different types of juvenile fishes as they grow to maturity. During the wet season, from May to November, animals disperse throughout the region, but in the dry season, from December to April, they congregate wherever pools of water are found. Alligators are famous for making "gator holes" where water collects, and fish, shrimp, crabs, birds, and a host of other living things survive until the rains come again. The Everglades once supported millions of large and beautiful birds (Fig. 36.11).

Restoration Plan   A restoration plan has been developed that will sustain the Everglades ecosystem while maintaining the services society requires. The U.S. Army Corps of Engineers is to redesign local flood control measures so that the Everglades receive a more natural flow of water from Lake Okeechobee. This will require the flooding of the Everglades Agricultural Area and the growth of only crops that can tolerate these conditions, such as sugarcane and rice. These measures have the benefit of stopping the loss of topsoil and preventing possible residential development in the area. There will also be an extended buffer zone between an expanded Everglades and the urban areas on the east coast of Florida. The buffer zone will contain a contiguous system of interconnected marsh areas, detention reservoirs, seepage barriers, and water treatment areas. This plan is expected to stop the decline of the Everglades, while still allowing agriculture to continue and providing water and flood control to the eastern coast. Sustainable development will maintain the ecosystem indefinitely and still meet human needs.

Today, landscape preservation is commonly needed to protect metapopulations. Often the preserved area must be restored before sustainable development is possible.

Alligator beside his "gator hole"

White ibis in breeding coloration

Roseate spoonbill wading

Figure 36.11   Restoration of the Everglades.
If the Everglades are restored, the chances of survival for these wild animals will improve.

Bioethical Focus   The Endangered Species Act

The Endangered Species Act requires the federal government to identify endangered and threatened species and to protect their habitats, even to the extent of purchasing the land in question. Some feel that the act protects wildlife at the expense of jobs for U.S. citizens. In an effort to allow development to proceed, it is now possible to get approval for a Habitat Conservation Plan (HCP). An HCP permits the use of one part of an environmentally sensitive area if another part is conserved for wildlife. Or developers can help the government buy habitat someplace else.

The nation's first HCP was approved in 1980. It permitted housing construction on San Bruno Mountain near San Francisco, because 97% of the habitat for the endangered mission blue butterfly was preserved. By this time, hundreds of HCPs have been approved, and the conservation requirement is not as stringent as it used to be. Tim Cullinan, a director of the National Audubon Society, recently found that logging companies in the Pacific Northwest are proposing the exchange of conserved public land for the right to log privately owned old forests. In other words, nothing has been given up because public land is already owned by the public. By now, there are so many HCPs in the works, they are being rubber-stamped by government officials with no public review at all.

Some wonder if the federal government should have the authority to approve HCPs, or if it isn't the public's responsibility to do so instead.

Decide Your Opinion

1. What are the concerns of developers versus environmentalists with regard to natural areas?
2. Is it short-sighted to stress the importance of jobs over the rights of wildlife? Why or why not?
3. In what ethical ways can each side make their concerns known to the general public?

Summarizing the Concepts

36.1 Conservation Biology and Biodiversity

Conservation biology is the scientific study of biodiversity and its management for sustainable human welfare. The present unequaled rate of extinctions has drawn together scientists and environmentalists in basic and applied fields to address the problem.

Biodiversity, the variety of life on earth, must also be preserved at the genetic, community (ecosystem), and landscape levels of organization. Conservationists have discovered that biodiversity is not evenly distributed in the biosphere, and therefore saving certain areas may protect more species than saving other areas.

36.2 Value of Biodiversity

The direct value of biodiversity is evidenced by the observable services of individual wild species. Wild species are our best source of new medicines to treat human ills, and they meet other medical needs as well: for example, the bacterium that causes leprosy grows naturally in armadillos, and horseshoe crab blood contains a bacteria-fighting substance.

Wild species have agricultural value. Domesticated plants and animals are derived from wild species, which also serve as a source of genes for the improvement of their phenotypes. Instead of pesticides, wild species can be used as biological controls, and most flowering plants make use of animal pollinators. Much of our food, particularly fish and shellfish, is still caught in the wild. Hardwood trees from natural forests supply us with lumber for various purposes, such as making furniture.

The indirect services provided by ecosystems are largely unseen but absolutely necessary to our well-being. These services include the workings of biogeochemical cycles, waste disposal, provision of fresh water, prevention of soil erosion, and regulation of climate. Many people enjoy vacationing in natural settings. Various studies show that more diverse ecosystems function better than less diverse systems.

36.3 Causes of Extinction

Researchers have identified the major causes of extinction. Habitat loss is the most frequent cause, followed by introduction of alien species, pollution, overexploitation, and disease. (Pollution often leads to disease, so these were discussed at the same time.) Habitat loss has occurred in all parts of the biosphere, but concern has now centered on tropical rain forests and coral reefs where biodiversity is especially high. Alien species have been introduced into foreign ecosystems because of colonization, horticulture or agriculture, and accidental transport. Among the various causes of pollution (acid rain, eutrophication, and ozone depletion), global warming is expected to cause the most instances of extinction. Overexploitation is exemplified by commercial fishing, which is so efficient that fisheries of the world are collapsing.

36.4 Conservation Techniques

To preserve species, it is necessary to preserve their habitat. Some emphasize the need to preserve biodiversity hotspots because of their richness. Often today it is necessary to save metapopulations because of past habitat fragmentation. If so, it is best to determine the source populations and save those instead of the sink populations. A keystone species like the grizzly bear requires the preservation of a landscape consisting of several types of ecosystems over millions of acres of territory. Obviously, in the process, many other species will also be preserved.

Conservation today is assisted by two types of computer analysis in particular. A gap analysis tries for a fit between biodiversity concentrations and land still available to be preserved. A population viability analysis indicates the minimum size of a population needed to prevent extinction from happening.

Since many ecosystems have been degraded, habitat restoration may be necessary before sustainable development is possible. Three principles of restoration are: (1) start before sources of wildlife and seeds are lost; (2) use simple biological techniques that mimic natural processes; and (3) aim for sustainable development so that the ecosystem fulfills the needs of humans while remaining intact.

Testing Yourself

Choose the best answer for each question.

1. Which of these would NOT be within the realm of conservation biology?
   a. helping to manage a national park
   b. a government board charged with restoring an ecosystem
   c. writing textbooks and/or popular books on the value of biodiversity
   d. introducing endangered species back into the wild
   e. All of these are concerns of conservation biology.

2. Which of these pairs is NOT a valid contrast with regard to the number of species?
   a. temperate zone—tropical zone
   b. hotspots—cold spots
   c. rain forest canopy—rain forest floor
   d. pelagic zone—deep-sea benthos

3. The value of wild pollinators to the U.S. agricultural economy has been calculated as $4.1 to $6.7 billion a year. What is the implication?
   a. Society could easily replace the need for wild pollinators by domesticating various types of pollinators.
   b. Pollinators may be valuable, but that doesn't mean that all species provide us with a valuable service.
   c. If we did away with all natural ecosystems, we wouldn't be dependent on wild pollinators.
   d. Society doesn't always appreciate the services that wild species provide us naturally and without any fanfare.
   e. All of these statements are correct.

4. The services provided to us by ecosystems are unseen. This means
   a. they are not valuable.
   b. they are noticed particularly when the service is disrupted.
   c. biodiversity is not needed for ecosystems to keep on functioning as before.
   d. we should be knowledgeable about them and protect them.
   e. Both b and d are correct.

5. Which of these is a true statement?
   a. Habitat loss is the most frequent cause of extinctions today.
   b. Alien species are often introduced into ecosystems by accidental transport.
   c. Global warming is expected to cause many extinctions in the twenty-first century.
   d. Overexploitation of fisheries could very well lead to a complete collapse of the fishing industry.
   e. All of these statements are true.

6. Which of these results is NOT expected because of global warming?
   a. the inability of species to migrate to cooler climates as environmental temperatures rise
   b. the bleaching and drowning of coral reefs
   c. rise in sea levels and loss of wetlands
   d. preservation of species because cold weather causes hardships
   e. All of these are expected.

7. What chemical is most associated with ozone depletion?
   a. chlorofluorocarbons
   b. carbon dioxide
   c. sulfur dioxide
   d. freon
   e. Both a and d are correct.

8. Which statement indicates that the grizzly bear is a keystone species existing as a metapopulation?
   a. Grizzly bears require many thousands of miles of preserved land because they are large animals.
   b. Grizzly bears have functions that increase biodiversity, but presently the population is subdivided into isolated subpopulations.
   c. When grizzly bears are preserved, so are many other types of species within a diverse landscape.
   d. Grizzly bears are a source population for many other types of organisms across several population types.
   e. All of these statements are correct.

9. A population in an unfavorable area with a high infant mortality rate would be a
   a. metapopulation.
   b. source population.
   c. sink population.
   d. new population.

10. What is the edge effect?
    a. More species live near the edge of an ecosystem where more resources are available to them.
    b. New species originate at the edge of ecosystems due to interactions with other species.
    c. Edge of an ecosystem is not a typical habitat and may be an area where survival is more difficult.
    d. More species are found at the edge of a rain forest due to deforestation of the forest interior.

11. Sustainable development of the Everglades will mean that
    a. the various populations that make up the Everglades will continue to exist indefinitely.
    b. human needs will also be met while successfully managing the ecosystem.
    c. the means used to maintain the Everglades will mimic the processes that naturally maintain the Everglades.
    d. the restoration plan is a workable plan.
    e. All of these statements are correct.

12. Which statement accepted by conservation biologists best shows that they support ethical principles?
    a. Biodiversity is the variety of life observed at various levels of biological organization.
    b. Wild species directly provide us with all sorts of goods and services.
    c. Reduction in the burning of fossil fuels would help reduce the effects of global warming.
    d. Three principles of restoration biology need to be adhered to in order to restore ecosystems.
    e. Biodiversity is desirable and has value in and of itself, regardless of any practical benefit.

13. Complete the following graph by labeling each bar with a cause of extinction, from the most influential to the least influential.

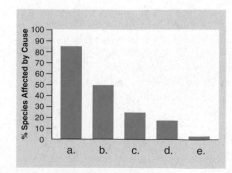

e-Learning Connection   www.mhhe.com/maderinquiry10

Concepts	Questions	Media Resources*
36.1 Conservation Biology and Biodiversity		

• Conservation biology addresses a crisis—the loss of biodiversity.
• Conservation biology is an applied, goal-oriented, multidisciplinary field.
• Extinction rates have risen to many times their natural levels, and many types of ecosystems are disappearing.
• Biodiversity includes species diversity, genetic diversity, community diversity, and landscape diversity in marine, freshwater, and terrestrial habitats.
	1. Define "conservation biology."
	2. List the various aspects of biodiversity.
		Essential Study Partner
		    Sustainability
		
		BioCourse Study Guide
		    Biodiversity and
		    Conservation

36.2 Value of Biodiversity		

• Biodiversity has both direct and indirect value.	1. What are some of the direct values associated with biodiversity?
	2. What are some of the indirect values of preserving ecosystems?
		Essential Study Partner
		    Species
		    Biodiversity—Communities

36.3 Causes of Extinction		

• Habitat loss, introduction of alien species, pollution, overexploitation, and disease are now largely responsible for the loss of biodiversity.
• Global warming will shift the optimal range of many species northward and disrupt many coastal ecosystems.
	1. How does the introduction of alien species into a new habitat cause extinction of some of the native organisms?
	2. How will global warming lead to widespread loss of species?
		Essential Study Partner
		    Extinction
		    Biodiversity—Human Impact
		
		Art Quiz
		    Greenhouse Effect
		
		Animation Quizzes
		    Global Warming
		    Acid Rain
		
		Case Studies
		    Killer Seaweed Begins
		        U.S. Invasion
		    Radioactive Wastes From
		        Medical Facilities: Where
		        Do They Go?

36.4 Conservation Techniques		

• Because of fragmented habitats, it is often necessary to conserve subdivided populations today.
• Identifying and conserving biodiversity hotspots and/or keystone species can save many other species as well.
• Computer analyses can be done to select areas for preservation and to determine the minimal population size needed for survival.
• Ecological preservation often involves restoration of habitats today.
	1. What are "biodiversity hotspots" and "keystone species?"
	2. How can computer analyses assist in the field of conservation?
		Essential Study Partner
		    Pest Control Alternatives
		    Air Pollution Control
		    Water Pollution Control
		    Solar Energy
		    Other Renewable Energy
		        Sources
		
		Case Studies
		    The Wolf in Yellowstone
		        National Park
		    Averting Disaster in
		        Biosphere 2

*For additional Media Resources, see the Online Learning Center.

Appendix A

Answer Key

This appendix contains the answers to the Testing Yourself questions, which appear at the end of each chapter, and the Practice Problems and Additional Genetics Problems, which appear in Chapters 23–25.

Chapter 1
Testing Yourself
1. d; 2. a; 3. c; 4. e; 5. a; 6. b; 7. a; 8. c; 9. b; 10. d; 11. d; 12. c; 13. e; 14. a; 15. d; 16. c; 17. b; 18. b; 19. **a.** scientific theory; **b.** energy; **c.** adaptation; **d.** homeostasis

Chapter 2
Testing Yourself
1. b; 2. c; 3. a; 4. b; 5. c; 6. e; 7. a; 8. b; 9. b; 10. c; 11. b; 12. d; 13. c; 14. d; 15. c; 16. a; 17. d; 18. a; 19. **a.** monomers; **b.** condensation synthesis; **c.** polymer; **d.** hydrolysis

Chapter 3
Testing Yourself
1. c; 2. c; 3. c; 4. c; 5. a; 6. c; 7. a; 8. b; 9. e; 10. d; 11. e; 12. d; 13. e; 14. c; 15. a; 16. **a.** nucleus—DNA specifies protein synthesis; **b.** nucleolus—RNA helps form ribosomes; **c.** smooth ER transports; **d.** rough ER produces protein; **e.** Golgi apparatus packages and secretes.

Chapter 4
Testing Yourself
1. **a.** glycolipid; **b.** glycoprotein; **c.** carbohydrate chain; **d.** hydrophilic head; **e.** hydrophobic tails; **f.** phospholipid bilayer; **g.** filaments of the cytoskeleton; **h.** peripheral protein; **i.** cholesterol; **j.** integral protein; 2. b; 3. b; 4. a; 5. e; 6. d; 7. c; 8. c; 9. c; 10. b; 11. a; 12. a; 13. e; 14. e; 15. e; 16. c; 17. a; 18. b; 19. **a.** hypertonic solution—net movement of water toward outside, cell shrivels. **b.** hypotonic solution—net movement of water toward inside; vacuoles fill with water and turgor pressure develops, causing chloroplasts to push against the cell wall.

Chapter 5
Testing Yourself
1. e; 2. e; 3. d; 4. a; 5. c 6. c; 7. b; 8. c; 9. c; 10. b; 11. c; 12. c; 13. a; 14. e; 15. c; 16. **a.** chromatid; **b.** centrosome (or centriole); **c.** spindle fiber (or aster); **d.** nuclear envelope. 17. The right cell represents metaphase I because homologous pairs are at the metaphase plate.

Chapter 6
Testing Yourself
1. b; 2. e; 3. e; 4. d; 5. e; 6. d; 7. d; 8. e; 9. c; 10. e; 11. c; 12. d; 13. b; 14. b; 15. e; 16. c; 17. d; 18. a; 19. b; 20. e; 21. **a.** active site; **b.** substrates; **c.** product; **d.** enzyme; **e.** enzyme-substrate complex; **f.** enzyme. The shape of an enzyme is important to its activity because it allows an enzyme-substrate complex to form. 22 **a.** metabolism; **b.** cofactor; **c.** kinetic energy; **d.** vitamin; **e.** denatured

Chapter 7
Testing Yourself
1. e; 2. b; 3. b; 4. c; 5. d; 6. d; 7. c; 8. b; 9. b; 10. a; 11. c; 12. c; 13. a; 14. b; 15. c; 16. b; 17. d; 18. c; 19. **a.** citric acid cycle; **b.** anaerobic; **c.** pyruvate; **d.** fermentation; 20. **a.** cristae; **b.** matrix; **c.** outer membrane; **d.** intermembrane space; **e.** inner membrane

Chapter 8
Testing Yourself
1. e; 2. e; 3. b; 4. e; 5. a; 6. e; 7. a; 8. e; 9. c; 10. e; 11. e; 12. e; 13. d; 14. e; 15. d; 16. **a.** thylakoid membrane; **b.** O_2; **c.** stroma; **d.** Calvin cycle; **e.** granum; **f.** thylakoid; **g.** stroma; 17. **a.** water; **b.** oxygen; **c.** carbon dioxide; **d.** carbohydrate; **e.** ADP + $\textcircled{P}$, ATP; **f.** NADP$^+$, NADPH

Chapter 9
Testing Yourself
1. c; 2. b; 3. c; 4. c; 5. b; 6. b; 7. b; 8. c; 9. c; 10. a; 11. a; 12. b; 13. d; 14. e; 15. b; 16. **a.** epidermis; **b.** cortex; **c.** endodermis; **d.** phloem; **e.** xylem; 17. **a.** upper epidermis; **b.** palisade mesophyll; **c.** leaf vein; **d.** spongy mesophyll; **e.** lower epidermis

Chapter 10
Testing Yourself
1. c; 2. c; 3. b; 4. e; 5. e; 6. a; 7. b; 8. d; 9. b; 10. d; 11. a; 12. c; 13. d; 14. c; 15. d; 16. **a.** H_2O; **b.** H_2O, **c.** K$^+$, **d.** stoma, **e.** guard cell; **f.** H_2O, **g.** H_2O, **h.** K$^+$. (*Left*) after K$^+$ (dots) enters guard cells, water follows by osmosis, causing guard cells to become turgid, opening the stoma; (*Right*) After K$^+$ (dots) exit guard cells, water follows by osmosis, causing guard cells to become flaccid and close. 17. **a.** sporophyte; **b.** meiosis; **c.** megaspore; **d.** microspore; **e.** megagametophyte (embryo sac); **f.** microgametophyte (pollen grain); **g.** gametes; **h.** fertilization; **i.** seed

Chapter 11
Testing Yourself
1. b; 2. b; 3. a; 4. e; 5. a; 6. d; 7. e; 8. e; 9. e; 10. b; 11. e; 12. c; 13. d; 14. e; 15. d; 16. c; 17. d; 18. c; 19. **a.** columnar epithelium, lining of intestine (digestive tract), protection and absorption; **b.** cardiac muscle, wall of heart, pumps blood; **c.** compact bone, skeleton, support and protection

Chapter 12
Testing Yourself
1. b; 2. d; 3. b; 4. e; 5. c; 6. a; 7. b; 8. e; 9. d; 10. d; 11. d; 12. e; 13. c; 14. c; 15. a; 16. b; 17. d; 18. **a.** salivary glands; **b.** esophagus; **c.** stomach; **d.** liver; **e.** gallbladder; **f.** pancreas; **g.** small intestine; **h.** large intestine; **i.** glucose and amino acids; **j.** lipids; **k.** water

Chapter 13
Testing Yourself
1. b; 2. c; 3. e; 4. b; 5. d; 6. a; 7. e; 8. a; 9. d; 10. c; 11. e; 12. b; 13. e; 14. e; 15. b; 16. e; 17. a. aorta; b. left pulmonary arteries; c. pulmonary trunk; d. left pulmonary veins; e. left atrium; f. semilunar valves; g. atrioventricular (mitral) valve; h. left ventricle; i. septum; j. inferior vena cava; k. right ventricle; l. chordae tendineae; m. atrioventricular (tricuspid) valve; n. right atrium; o. right pulmonary veins; p. right pulmonary arteries; q. superior vena cava. See also Figure 13.4, page 243.

Chapter 14
Testing Yourself
1. a; 2. b; 3. b; 4. e; 5. e; 6. e; 7. a; 8. b; 9. c; 10. a; 11. a; 12. a; 13. e; 14. b; 15. e; 16. b; 17. c; 18. a. antigen-binding sites; b. light chain; c. heavy chain; d. V stands for variable region; C stands for constant region.

Chapter 15
Testing Yourself
1. d; 2. d; 3. c; 4. d; 5. c; 6. b; 7. e; 8. a; 9. b; 10. b; 11. e; 12. b; 13. d; 14. a; 15. a. nasal cavity; b. nose; c. pharynx; d. epiglottis; e. glottis; f. larynx; g. trachea; h. bronchus; i. bronchiole. See also Figure 15.1, p. 282.

Chapter 16
Testing Yourself
1. c; 2. c; 3. a; 4. a; 5. c; 6. b; 7. d; 8. a; 9. d; 10. d; 11. a; 12. c; 13. d; 14. d; 15. c; 16. a; 17. a. glomerulus; b. efferent arteriole; c. afferent arteriole; d. proximal convoluted tubule; e. loop of the nephron; f. descending limb; g. ascending limb; h. peritubular capillary network; i. distal convoluted tubule; j. renal vein; k. renal artery; l. collecting duct

Chapter 17
Testing Yourself
1. b; 2. c; 3. a; 4. b; 5. c; 6. d; 7. b; 8. c; 9. b; 10. d; 11. d; 12. c; 13. e; 14. a. sensory neuron (or fiber); b. interneuron; c. motor neuron (or fiber); d. sensory receptor; e. cell body; f. dendrites; g. axon; h. nucleus of Schwann cell; i. node of Ranvier (neurofibril node); j. effector

Chapter 18
Testing Yourself
1. d; 2. e; 3. e; 4. c; 5. d; 6. d; 7. c; 8. b; 9. e; 10. d; 11. b; 12. c; 13. a; 14. e; 15. a. retina—contains receptors; b. choroid—absorbs stray light; c. sclera—protects and supports eyeball; d. optic nerve—transmits impulses to brain; e. fovea centralis—makes acute vision possible; f. ciliary body—holds lens in place, accommodation; g. lens—refracts and focuses light rays; h. iris—regulates light entrance; i. pupil—admits light; j. cornea—refracts light rays

Chapter 19
Testing Yourself
1. b; 2. f; 3. c; 4. e; 5. b; 6. b; 7. a; 8. b; 9. c; 10. c; 11. e; 12. c; 13. c; 14. b; 15. d; 16. d; 17. a; 18. e; 19. a. T tubule; b. sarcoplasmic reticulum; c. myofibril; d. Z line; e. sarcomere; f. sarcolemma

Chapter 20
Testing Yourself
1. f; 2. b; 3. c; 4. a; 5. e; 6. d; 7. d; 8. b; 9. e; 10. d; 11. e; 12. a; 13. e; 14. e; 15. b; 16. e; 17. e; 18. a; 19. a; 20. a. inhibits; b. inhibits; c. releasing hormone; d. stimulating hormone; e. target gland hormone

Chapter 21
Testing Yourself
1. a. seminal vesicle; b. ejaculatory duct; c. prostate gland; d. bulbourethral gland; e. anus; f. vas deferens; g. epididymis; h. testis; i. scrotum; j. foreskin; k. glans penis; l. penis; m. urethra; n. vas deferens; o. urinary bladder. Path of sperm: h, g, f, n, m. See also Figure 21.1, page 414. 2. c; 3. d; 4. c; 5. c; 6. b; 7. d; 8. c; 9. d; 10. d; 11. e; 12. c; 13. d; 14. c; 15. d; 16. a; 17. c; 18. e; 19. c; 20. b; 21. e

Chapter 22
Testing Yourself
1. c; 2. b; 3. b; 4. b; 5. a; 6. e; 7. a; 8. d; 9. e; 10. c; 11. e; 12. d; 13. d; 14. d; 15. e; 16. d; 17. b; 18. e; 19. a. chorion (contributes to forming placenta where wastes are exchanged for nutrients with the mother); b. amnion (protects and prevents desiccation); c. embryo; d. allantois (blood vessels become umbilical blood vessels); e. yolk sac (first site of blood cell formation); f. fetal portion of placenta; g. maternal portion of placenta; h. umbilical cord (connects developing embryo to the placenta). See also Figure 22.11, page 448.

Chapter 23
Practice Problems 1
1. a. W; b. WS, Ws; c. T, t; d. Tg, tg; e. AB, Ab, aB, ab
2. a. gamete; b. genotype; c. gamete; d. genotype

Practice Problems 2
1. 75% or 3/4; 2. mother and father: Ee and Ee; 3. father: DD; mother: dd; children: Dd

Practice Problems 3
1. Dihybrid; 2. 1/16; 3. child: ddff; father: DdFf; mother: ddff

Practice Problems 4
1. cc; Cc and Cc; 2. 25% or 1/4; 3. heterozygous

Practice Problems 5
1. Woman: Hh; husband: hh; 2. 50%, 50%; 3. Children with the genotype Hh, because the gene (H) that causes the condition is dominant. 4. homozygous dominant or heterozygous

Practice Problems 6
1. AABBCC and aabbcc; 2. very light; 3. child: ii, mother: $I^A i$, father: $I^A i$, $I^B i$; 4. baby 1 = Doe; baby 2 = Jones

Testing Yourself
1. b; 2. e; 3. b; 4. b; 5. a; 6. c; 7. d; 8. c; 9. b; 10. c; 11. b; 12. e; 13. d; 14. c; 15. a; 16. autosomal recessive

Additional Genetics Problems
1. 50%; 2. John: cc; parents: Cc. 3. father: Wwcc, mother: wwCc, child: wwcc; 4. 75%; 5. 9/16; 6. Darkest skin possible is light; lightest skin possible is very light; 7. A, B, AB, O; 8. man and woman: HH', child: H'H'

Chapter 24

Testing Yourself

1. c; **2.** a; **3.** b; **4.** b; **5.** a; **6.** d; **7.** c; **8.** d; **9.** a; **10.** c; **11.** b; **12.** c; **13.** a; **14.** c; **15.** b; **16.** c; **17.** c; **18.** $X^B X^b$

Additional Genetics Problems

1. Key: X^S = unaffected, X^s = syndrome; father: $X^S Y$; mother: $X^S X^s$
2. son; **3.** 50%, none; **4.** $X^h Y$, $X^H X^h$; **5.** 50%; **6.** 50% for both sons and daughters; **7. a.** Males: all red eyes; females: all red eyes; **b.** Males: all white eyes; females: all red eyes; **8.** man: $BBX^b Y$, woman: $bbX^B X^b$; **9.** 50% for daughter with brown eyes and color blind: 0% for son with blue eyes and not color blind; **10.** boys = $X^B YWw$, girls = $X^B X^b Ww$; **11.** 0%; **12.** Males: 3 gray body with red eyes: 1 black body with red eyes: 3 gray body with white eyes: 1 black body with white eyes; females: 3 gray body with red eyes: 1 black body with red eyes; **13.** A, C, B, D

Practice Problems

1. His mother; $X^H X^h$, $X^H Y$, $X^h Y$
2. 100%; none; 100%
3. The husband is not the father.

Chapter 25

Testing Yourself

1. e; **2.** e; **3.** c; **4.** c; **5.** d; **6.** c; **7.** d ; **8.** b; **9.** b; **10.** d; **11.** c; **12.** b; **13.** d; **14.** b; **15.** e; **16.** d; **17.** b; **18.** c; **19.** d; **20. a.** DNA, **b.** regulator gene; **c.** promoter; **d.** operator; **e.** mRNA; **f.** active protein repressor

Chapter 26

Testing Yourself

1. c; **2.** d; **3.** c; **4.** d; **5.** e; **6.** e; **7.** c; **8.** e; **9.** a; **10.** e; **11.** b; **12.** d; **13.** a; **14.** d; **15.** d; **16.** e; **17.** AATT

Chapter 27

Testing Yourself

1. e; **2.** b; **3.** e; **4.** e; **5.** d; **6.** c; **7.** b; **8.** c; **9.** e; **10.** d; **11.** e; **12.** e; **13.** b; **14.** b; **15.** e; **16.** a;
17.

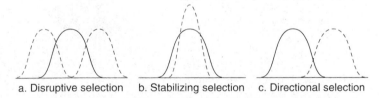

a. Disruptive selection   b. Stabilizing selection   c. Directional selection

Chapter 28

Testing Yourself

1. d; **2.** b; **3.** c; **4.** c; **5.** b; **6.** d; **7.** e; **8.** d; **9.** b; **10.** e; **11.** a; **12.** e; **13.** e; **14.** e; **15. a.** sexual reproduction; **b.** isogametes pairing; **c.** zygote (2n); **d.** zygospore (2n); **e.** asexual reproduction; **f.** zoospores (n); **g.** nucleus; **h.** chloroplast; **i.** pyrenoid; **j.** starch granule; **k.** flagellum; **l.** eyespot; **m.** gamete formation. See also Figure 28.11, page 576. **n.** Asexual: one parent, no gametes produced, no genetic recombination. Sexual: two parents, gametes produced, genetic recombination

Chapter 29

Testing Yourself

1. e; **2.** a; **3.** c; **4.** e; **5.** d; **6.** c; **7.** c; **8.** e; **9.** b; **10.** b; **11.** b; **12.** e; **13.** a; **14.** e; **15.** e; **16.** a; **17.** d; **18. a.** sporophyte (2n); **b.** meiosis; **c.** gametophyte (n); **d.** fertilization. See also Figure 29.2a, page 593.

Chapter 30

Testing Yourself

1. a. tentacle; **b.** gastrovascular cavity; **c.** mouth; **d.** mesoglea; **2.** e; **3.** c; **4.** e; **5.** e; **6.** b; **7.** c; **8.** a; **9.** a; **10.** c; **11.** d; **12.** a; **13.** a; **14.** d; **15.** a; **16. a.** head; **b.** antenna; **c.** simple eye; **d.** compound eye; **e.** thorax; **f.** tympanum; **g.** abdomen; **h.** forewing; **i.** hindwing; **j.** ovipositor; **k.** spiracles; **l.** air sac; **m.** spiracle; **n.** tracheae. See also Figure 30.18, page 629.

Chapter 31

Testing Yourself

1. b; **2.** e; **3.** e; **4.** b; **5.** e; **6.** c; **7.** e; **8.** a; **9.** a; **10.** d; **11.** d; **12.** a; **13.** b; **14.** a, b, d; **15.** b; **16.** e; **17.** c; **18.** b; **19. a.** pharyngeal pouches; **b.** dorsal tubular nerve cord; **c.** notochord; **d.** post-anal tail

Chapter 32

Testing Yourself

1. b; **2.** b; **3.** c; **4.** a; **5.** e; **6.** c; **7.** c; **8.** e; **9.** c; **10.** b; **11.** b; **12.** a; **13.** e; **14.** b; **15.** e; **16.** a

Chapter 33

Testing Yourself

1. d; **2.** e; **3.** c; **4.** e; **5.** b; **6.** b; **7.** e; **8.** d; **9.** e; **10.** a; **11.** b; **12.** e; **13.** d; **14.** e; **15.** b; **16.** b; **17.** c; **18.** e; **19.** e; **20.** e; **21. a.** lag; **b.** exponential growth; **c.** deceleration; **d.** environmental resistance; **e.** carrying capacity

Chapter 34

Testing Yourself

1. c; **2.** b; **3.** e; **4.** e; **5.** c; **6.** c; **7.** b; **8.** a; **9.** c; **10.** b; **11.** c; **12.** a; **13.** c; **14.** c; **15. a.** producers; **b.** consumers; **c.** inorganic nutrient pool; **d.** decomposers

Chapter 35

Testing Yourself

1. b; **2.** d; **3.** a; **4.** c; **5.** d; **6.** d; **7.** a; **8.** e; **9.** d; **10.** e; **11.** e; **12.** b; **13.** a; **14.** a; **15.** a; **16.** c; **17.** d; **18.** e; **19.** b; **20.** c

Chapter 36

Testing Yourself

1. e; **2.** b; **3.** d; **4.** e; **5.** e; **6.** d; **7.** e; **8.** b; **9.** c; **10.** c; **11.** e; **12.** e; **13. a.** habitat loss; **b.** alien species; **c.** pollution; **d.** overexploitation; **e.** disease

Appendix B

Classification of Organisms

Domain Bacteria

Prokaryotic, unicellular organisms which lack a membrane-bounded nucleus and reproduce asexually. Metabolically diverse being heterotrophic by absorption; autotrophic by chemosynthesis or by photosynthesis. Motile forms move by flagella consisting of a single filament. (571)

Domain Archaea

Prokaryotic, unicellular organisms which lack a membrane-bounded nucleus and reproduce asexually. Many are autotrophic by chemosynthesis; some are heterotrophic by absorption. Most live in extreme or anaerobic environments. Archaea are distinguishable from bacteria by their unique rRNA base sequence and their distinctive plasma membrane and cell wall chemistry. (572)

Domain Eukarya

Eukaryotic, unicellular to multicellular organisms which have a membrane-bounded nucleus containing several chromosomes. Sexual reproduction is common. Phenotypes and nutrition are diverse; each kingdom has specializations that distinguish it from the other kingdoms. Flagella, if present, have a 9 + 2 organization. (575)

Kingdom Protista

Eukaryotic, unicellular organisms and their immediate multicellular descendants. Asexual reproduction is common, but sexual reproduction as a part of various life cycles does occur. Phenotypically and nutritionally diverse, being either photosynthetic or heterotrophic by various means. Locomotion, if present, utilizes flagella, cilia, or pseudopods.

Algae*

Phylum Chlorophyta: green algae (576)

Phylum Phaeophyta: brown algae (577)

Phylum Rhodophyta: red algae (579)

Phylum Chrysophyta: golden-brown algae (578)

Phylum Pyrrophyta: dinoflagellates (578)

Phylum Euglenophyta: euglenoids (579)

Protozoans*

Phylum Rhizopoda: amoeboids (580)

Phylum Zoomastigophora: zooflagellates (581)

Phylum Ciliophora: ciliates (581)

Phylum Apicomplexa: sporozoans (582)

Slime Molds*

Phylum Gymnomycota: slime molds (583)

Water Molds*

Phylum Oomycota: water molds (582)

Kingdom Fungi

Multicellular eukaryotes which form nonmotile spores during both asexual and sexual reproduction as a part of the haplontic life cycle. The only multicellular forms of life to be heterotrophic by absorption. They lack flagella in all life cycle stages.

Division Zygomycota: zygospore fungi (584)

Division Ascomycota: sac fungi (585)

Division Basidiomycota: club fungi (586)

Division Deuteromycota: imperfect fungi (means of sexual reproduction is not known) (587)

*Not a classification category, but added for clarity

Kingdom Plantae

Multicellular, primarily terrestrial, eukaryotes with well-developed tissues. Plants have an alternation of generations life cycle and are usually photosynthetic. Like green algae, they contain chlorophylls *a* and *b*, carotenoids; store starch in chloroplasts; and have a cell wall which contains cellulose.

Nonvascular Plants*

Division Hepatophyta: liverworts (594)

Division Bryophyta: mosses (594)

Seedless Vascular Plants*

Division Psilotophyta: whisk ferns (596)

Division Lycopodophyta: club mosses, spike mosses, quillworts (597)

Division Equisetophyta: horsetails (597)

Division Pteridophyta: ferns (597)

Gymnosperms*

Division Cycadophyta: cycads (600)

Division Ginkgophyta: maidenhair tree (600)

Division Pinophyta: conifers, such as pines, firs, yews, redwoods, spruces (600)

Angiosperms*

Division Magnoliophyta: flowering plants (604)

Class Liliopsida: monocots (604)

Class Magnoliopsida: dicots (604)

Kingdom Animalia

Multicellular organisms with well-developed tissues that have the diplontic life cycle. Animals tend to be mobile and are heterotrophic by ingestion, generally in a digestive cavity. Complexity varies; the more complex forms have well-developed organ systems. More than a million species have been described.

Invertebrates*

Phylum Porifera: sponges (613)

Phylum Cnidaria: *cnidarians* (614)

Class Anthozoa: sea anemones, corals (614)

Class Hydrozoa: *Hydra, Obelia* (615)

Class Scyphozoa: *jellyfishes* (614)

Phylum Platyhelminthes: flatworms (616)

Class Turbellaria: planarians (616)

Class Trematoda: flukes (617)

Class Cestoda: tapeworms (617)

Phylum Nematoda: roundworms: *Ascaris* (619)

Phylum Rotifera: rotifers

Phylum Mollusca: molluscs (620)

Class Polyplacophora: chitons (621)

Class Bivalvia: clams, scallops, oysters, mussels (621)

Class Cephalopoda: squids, chambered nautilus, octopus (620)

Class Gastropoda: snails, slugs, nudibranchs (620)

Phylum Annelida: annelids (623)

Class Polychaeta: clam worms, tube worms (623)

Class Oligochaeta: earthworms (624)

Class Hirudinea: leeches (625)

Phylum Arthropoda: arthropods (626)

Subphylum Crustacea: crustaceans *(shrimps, crabs, lobsters, barnacles)* (627)

Subphylum Uniramia: millipedes, centipedes, insects (626, 628)

Subphylum Chelicerata: spiders, scorpions, horseshoe crabs (631)

Phylum Echinodermata: echinoderms (636)

Class Crinoidea: sea lillies, feather stars (636)

Class Asteroidea: sea stars (636)

Class Ophiuroidea: brittle stars (636)

Class Echinoidea: sea urchins, sand dollars (636)

Class Holothuroidea: sea cucumbers (636)

Phylum Chordata: chordates (638)

Subphylum Cephalochordata: lancelets (640)

Subphylum Urochordata: tunicates (640)

Vertebrates*

Subphylum Vertebrata: Vertebrates (641)

Class Agnatha: jawless fishes (hagfishes, lampreys) (642)

Class Chondrichthyes: cartilaginous fishes (sharks, skates) (642)

Class Osteichthyes: bony fishes (bony fishes—lobe-skinned fishes and ray-finned fishes) (642)

Class Amphibia: amphibians (frogs, toads) (643)

Class Reptilia: reptiles (turtles, snakes, lizards, crocodiles) (645)

Class Aves: birds (owls, woodpeckers, kingfishers, cuckoos) (647)

Class Mammalia: mammals (monotremes—duckbill platypus, spiny anteater; marsupials—opposums, kangaroos, koalas; placental mammals—shrews, whales, rats, rabbits, dogs, cats) (648)

Order Primates: primates (prosiminians, monkeys, apes) (650)

Family hominids (650)

Genus *Homo:* humans (655)

*Not a classification category, but added for clarity

Appendix C

Metric System

Unit and Abbreviation	Metric Equivalent	Approximate English-to-Metric Equivalents	Units of Temperature
Length			
nanometer (nm)	$= 10^{-9}\,m\ (10^{-3}\,\mu m)$		
micrometer (µm)	$= 10^{-6}\,m\ (10^{-3}\,mm)$		
millimeter (mm)	$= 0.001\ (10^{-3})\,m$		
centimeter (cm)	$= 0.01\ (10^{-2})\,m$	1 inch = 2.54 cm 1 foot = 30.5 cm	
meter (m)	$= 100\ (10^2)\,cm$ $= 1{,}000\,mm$	1 foot = 0.30 m 1 yard = 0.91 m	
kilometer (km)	$= 1{,}000\ (10^3)\,m$	1 mi = 1.6 km	
Weight (mass)			
nanogram (ng)	$= 10^{-9}\,g$		
microgram (µg)	$= 10^{-6}\,g$		
milligram (mg)	$= 10^{-3}\,g$		
gram (g)	$= 1{,}000\,mg$	1 ounce = 28.3 g 1 pound = 454 g	
kilogram (kg)	$= 1{,}000\ (10^3)\,g$	= 0.45 kg	
metric ton (t)	$= 1{,}000\,kg$	1 ton = 0.91 t	
Volume			
microliter (µl)	$= 10^{-6}\,l\ (10^{-3}\,ml)$		
milliliter (ml)	$= 10^{-3}\,liter$ $= 1\,cm^3\ (cc)$ $= 1{,}000\,mm^3$	1 tsp = 5 ml 1 fl oz = 30 ml	
liter (l)	$= 1{,}000\,ml$	1 pint = 0.47 liter 1 quart = 0.95 liter 1 gallon = 3.79 liter	
kiloliter (kl)	$= 1{,}000\,liter$		

Temperature scale (°F / °C):

°F 230, 220, 212° —210, 200, 190, 180, 170, 160°—160, 150, 140, 134° 131°—130, 120, 110, 105.8°, 98.6°—100, 90, 80, 70, 60, 56.66°—, 50, 40, 32°—30, 20, 10, 0, −10, −20, −30, −40

°C 110, 100—100°, 90, 80, 70—71°, 60, 57°, 50, 41°, 40 37°, 30, 20, 13.7°, 10, 0—0°, −10, −30, −40

°C	°F	
100	212	Water boils at standard temperature and pressure.
71	160	Flash pasteurization of milk
57	134	Highest recorded temperature in the United States, Death Valley, July 10, 1913
41	105.8	Average body temperature of a marathon runner in hot weather
37	98.6	Human body temperature
13.7	56.66	Human survival is still possible at this temperature.
0	32.0	Water freezes at standard temperature and pressure.

To convert temperature scales:

$$°C = \frac{5(°F - 32)}{9}$$

$$°F = \frac{9°C}{5} + 32$$

Appendix D

Periodic Table of the Elements

Atomic number

Atomic mass

Chemical symbol

1	1
H	
hydrogen	

group Ia

																	VIIIa
1 1																	**2** 4
H																	**He**
hydrogen	**IIa**											**IIIa**	**IVa**	**Va**	**VIa**	**VIIa**	helium
3 7	**4** 9											**5** 11	**6** 12	**7** 14	**8** 16	**9** 19	**10** 20
Li	**Be**											**B**	**C**	**N**	**O**	**F**	**Ne**
lithium	beryllium											boron	carbon	nitrogen	oxygen	fluorine	neon
11 23	**12** 24											**13** 27	**14** 28	**15** 31	**16** 32	**17** 35	**18** 40
Na	**Mg**	**IIIb**	**IVb**	**Vb**	**VIb**	**VIIb**		**VIIIb**		**Ib**	**IIb**	**Al**	**Si**	**P**	**S**	**Cl**	**Ar**
sodium	magnesium											aluminum	silicon	phosphorus	sulfur	chlorine	argon
19 39	**20** 40	**21** 45	**22** 48	**23** 51	**24** 52	**25** 55	**26** 56	**27** 59	**28** 59	**29** 64	**30** 65	**31** 70	**32** 73	**33** 75	**34** 79	**35** 80	**36** 84
K	**Ca**	**Sc**	**Ti**	**V**	**Cr**	**Mn**	**Fe**	**Co**	**Ni**	**Cu**	**Zn**	**Ga**	**Ge**	**As**	**Se**	**Br**	**Kr**
potassium	calcium	scandium	titanium	vanadium	chromium	manganese	iron	cobalt	nickel	copper	zinc	gallium	germanium	arsenic	selenium	bromine	krypton
37 85	**38** 88	**39** 89	**40** 91	**41** 93	**42** 96	**43** 98	**44** 101	**45** 103	**46** 106	**47** 108	**48** 112	**49** 115	**50** 119	**51** 122	**52** 128	**53** 127	**54** 131
Rb	**Sr**	**Y**	**Zr**	**Nb**	**Mo**	**Tc**	**Ru**	**Rh**	**Pd**	**Ag**	**Cd**	**In**	**Sn**	**Sb**	**Te**	**I**	**Xe**
rubidium	strontium	yttrium	zirconium	niobium	molybdenum	technetium	ruthenium	rhodium	palladium	silver	cadmium	indium	tin	antimony	tellurium	iodine	xenon
55 133	**56** 137	**57** 139	**72** 178	**73** 181	**74** 184	**75** 186	**76** 190	**77** 192	**78** 195	**79** 197	**80** 201	**81** 204	**82** 207	**83** 209	**84** 210	**85** 210	**86** 222
Cs	**Ba**	**La**	**Hf**	**Ta**	**W**	**Re**	**Os**	**Ir**	**Pt**	**Au**	**Hg**	**Tl**	**Pb**	**Bi**	**Po**	**At**	**Rn**
cesium	barium	lanthanum	hafnium	tantalum	tungsten	rhenium	osmium	iridium	platinum	gold	mercury	thallium	lead	bismuth	polonium	astatine	radon
87 223	**88** 226	**89** 227	**104** 261	**105** 260	**106** 263	**107** 261	**108** 265	**109** 266	**110** 269	**111** 272	**112** 277						
Fr	**Ra**	**Ac**	**Rf**	**Db**	**Sg**	**Bh**	**Hs**	**Mt**	★★★	★★★	★★★						
francium	radium	actinium	rutherfordium	dubnium	seaborgium	bohrium	hassium	meitnerium									

58 140	**59** 141	**60** 144	**61** 147	**62** 150	**63** 152	**64** 157	**65** 159	**66** 163	**67** 165	**68** 167	**69** 169	**70** 173	**71** 175
Ce	**Pr**	**Nd**	**Pm**	**Sm**	**Eu**	**Gd**	**Tb**	**Dy**	**Ho**	**Er**	**Tm**	**Yb**	**Lu**
cerium	praseodymium	neodymium	promethium	samarium	europium	gadolinium	terbium	dysprosium	holmium	erbium	thulium	ytterbium	lutetium
90 232	**91** 231	**92** 238	**93** 237	**94** 242	**95** 243	**96** 247	**97** 247	**98** 249	**99** 254	**100** 253	**101** 256	**102** 254	**103** 257
Th	**Pa**	**U**	**Np**	**Pu**	**Am**	**Cm**	**Bk**	**Cf**	**Es**	**Fm**	**Md**	**No**	**Lr**
thorium	protactinium	uranium	neptunium	plutonium	americium	curium	berkelium	californium	einsteinium	fermium	mendelevium	nobelium	lawrencium

★★★ These elements have not yet been named

Glossary

A

abscisic acid (ABA) Plant hormone that causes stomata to close and that initiates and maintains dormancy.   179

abscission Dropping of leaves, fruits, or flowers from a plant.   179

acetyl-CoA Molecule made up of a two-carbon acetyl group attached to coenzyme A. During cellular respiration, the acetyl group enters the citric acid cycle for further breakdown.   120

acetylcholine (ACh) Neurotransmitter active in both the peripheral and central nervous systems.   323

acetylcholinesterase (AChE) Enzyme that breaks down acetylcholine within a synapse.   323

acid Molecules tending to raise the hydrogen ion concentration in a solution and to lower its pH numerically.   28

acid deposition Return to earth as rain or snow of the sulfate or nitrate salts of acids produced by commercial and industrial activities.   707

acquired immunodeficiency syndrome (AIDS) Disease caused by a retrovirus and transmitted via body fluids; characterized by failure of the immune system.   429

acrosome Cap at the anterior end of a sperm that partially covers the nucleus and contains enzymes that help the sperm penetrate the egg.   417

actin Muscle protein making up the thin filaments in a sarcomere; its movement shortens the sarcomere, yielding muscle contraction.   381

action potential Electrochemical changes that take place across the axomembrane; the nerve impulse.   320

active site Region on the surface of an enzyme where the substrate binds and where the reaction occurs.   107

active transport Use of a plasma membrane carrier protein and energy to move a substance into or out of a cell from lower to higher concentration.   75

adaptation Organism's modification in structure, function, or behavior suitable to the environment.   5

adaptive radiation Evolution of several species from a common ancestor into new ecological or geographical zones.   559

adenine (A) One of four nitrogen-containing bases in nucleotides composing the structure of DNA and RNA.   41

adhesion junction Junction between cells in which the adjacent plasma membranes do not touch but are held together by intercellular filaments attached to buttonlike thickenings.   196

adipose tissue Connective tissue in which fat is stored.   196

ADP (adenosine diphosphate) Nucleotide with two phosphate groups that can accept another phosphate group and become ATP.   41, 104

adrenal gland Gland that lies atop a kidney; the adrenal medulla produces the hormones epinephrine and norepinephrine, and the adrenal cortex produces the glucocorticoid and mineralocorticoid hormones.   399

adrenocorticotropic hormone (ACTH) Hormone secreted by the anterior lobe of the pituitary gland that stimulates activity in the adrenal cortex.   394

adventitious root Fibrous roots which develop from stems or leaves, such as prop roots of corn or holdfast roots of ivy.   156

afterbirth The placenta and the extraembryonic membranes, which are delivered (expelled) during the third stage of parturition.   458

age-structure diagram In demographics, a display of the age groups of a population; a growing population has a pyramid-shaped diagram.   685

agglutination Clumping of red blood cells due to a reaction between antigens on red blood cell plasma membranes and antibodies in the plasma.   276

agranular leukocyte White blood cell that does not contain distinctive granules.   251

albumin Plasma protein of the blood having transport and osmotic functions.   250

aldosterone Hormone secreted by the adrenal cortex that decreases sodium and increases potassium excretion; raises blood volume and pressure.   311, 400

algae (sing., alga) Type of protist that carries on photosynthesis; unicellular forms are a part of phytoplankton and multicellular forms are called seaweed.   575

allantois Extraembryonic membrane that accumulates nitrogenous wastes in reptiles and birds and contributes to the formation of umbilical blood vessels in mammals, including humans.   448

allele Alternative form of a gene—alleles occur at the same locus on homologous chromosomes.   466

allergen Foreign substance capable of stimulating an allergic response.   274

allergy Immune response to substances that usually are not recognized as foreign.   274

allopatric speciation Origin of new species in populations that are separated geographically.   557

alternation of generations Life cycle, typical of plants, in which a diploid sporophyte alternates with a haploid gametophyte.   170, 577, 593

altruism Social interaction that has the potential to decrease the lifetime reproductive success of the member exhibiting the behavior.   673

alveolus (pl., alveoli) Air sac of a lung.   285

amino acid Monomer of a protein; takes its name from the fact that it contains an amino group ($—NH_2$) and an acid group ($—COOH$).   37

amniocentesis Procedure for removing amniotic fluid surrounding the developing fetus for the testing of the fluid or cells within the fluid.   484

amnion Extraembryonic membrane of reptiles, birds, and mammals that forms an enclosing, fluid-filled sac.   448

amniote egg Egg that has an amnion, as seen during the development of reptiles, birds, and mammals.   641

amphibian Member of a class of vertebrates that includes frogs, toads, and salamanders; they are still tied to a watery environment for reproduction.   643

ampulla Base of a semicircular canal in the inner ear.   361

amygdala Portion of the limbic system which functions to add emotional overtones to memories.   330

anabolic steroid Synthetic steroid that mimics the effect of testosterone.   404

anaerobic Growing or metabolizing in the absence of oxygen.   118

analogous structure Structure that has a similar function in separate lineages but differs in anatomy and ancestry.   544

androgen Male sex hormone (e.g., testosterone).   404

aneurysm Ballooning of a blood vessel.   255

angiogenesis Formation of new blood vessels; one mechanism by which cancer spreads.   518

angiosperm Flowering plant that produces seeds within an ovary that develops into a fruit; therefore, the seeds are covered.   148, 604

annelid Member of a phylum of invertebrates that contains segmented worms, such as the earthworm and the clam worm.   623

annual ring Layer of wood (secondary xylem) usually produced during one growing season.   161

anterior pituitary Portion of the pituitary gland that is controlled by the hypothalamus and produces six types of hormones, some of which control other endocrine glands.   394

anther In flowering plants, pollen-bearing portion of stamen.   170, 606

antheridium Sperm-producing structure, as in the moss life cycle.   594

antibody Protein produced in response to the presence of an antigen; each antibody combines with a specific antigen.   266

antibody-mediated immunity Specific mechanism of defense in which plasma cells derived from B cells produce antibodies that combine with antigens.   267

anticodon Three-base sequence in a transfer RNA molecule that pairs with a complementary codon in mRNA.   509

antidiuretic hormone (ADH) Hormone secreted by the posterior pituitary that increases the permeability of the collecting ducts in a kidney.   310, 394

antigen Foreign substance, usually a protein or a polysaccharide, that stimulates the immune system to react, such as to produce antibodies.   266

antigen-presenting cell (APC) Cell that displays the antigen to certain cells of the immune system so they can defend the body against that particular antigen.   270

antigen receptor Receptor proteins in the plasma membrane of immune system cells whose shape allows them to combine with a specific antigen.   266

aorta Major systemic artery that receives blood from the left ventricle of the aorta.   246

aortic bodies Sensory receptors in the aortic arch sensitive to oxygen content, carbon dioxide content, and blood pH.   288

apoptosis Programmed cell death involving a cascade of specific cellular events leading to death and destruction of the cell.   82, 267, 446, 518

appendicular skeleton Portion of the skeleton forming the pectoral girdle and upper extremities, and the pelvic girdle and lower extremities.   374

appendix Small, tubular appendage that extends outward from the cecum of the large intestine.   220

aqueous humor Clear, watery fluid between the cornea and lens of the eye.   350

aquifer Rock layers that contain water and will release it in appreciable quantities to wells or springs.   703

arachnid Group of arthropods that contains spiders and scorpions.   631

Archaea One of the three domains of life containing prokaryotic cells that often live in extreme habitats and which have unique genetic, biochemical, and physiological characteristics.   7, 573

archegonium Egg-producing structure as in the moss life cycle.   594

arterial duct Fetal connection between the pulmonary artery and the aorta; ductus arteriosus.   452

arteriole Vessel that takes blood from an artery to capillaries.   240

artery Vessel that takes blood away from the heart to arterioles; characteristically possessing thick elastic and muscular walls.   240

arthropod Member of a phylum of invertebrates that contains among other groups crustaceans and insects that have an exoskeleton and jointed appendages.   626

articular cartilage Hyaline cartilaginous covering over the articulating surface of the bones of synovial joints.   366

association area Region of the cerebral cortex related to memory, reasoning, judgment, and emotional feelings.   328

aster Short, radiating fibers produced by the centrosomes in animal cells.   86

astigmatism Blurred vision due to an irregular curvature of the cornea or the lens.   355

atom Smallest particle of an element that displays the properties of the element.   20

atomic mass Mass of an atom equal to the number of protons plus the number of neutrons within the nucleus.   21

atomic number Number of protons within the nucleus of an atom.   20

ATP (adenosine triphosphate) Nucleotide with three phosphate groups. The breakdown of ATP into ADP + ℗ makes energy available for energy-requiring processes in cells.   41, 104

atrial natriuretic hormone (ANH) Hormone secreted by the heart that increases sodium excretion and therefore lowers blood volume and pressure.   311, 400

atrioventricular bundle Group of specialized fibers that conduct impulses from the atrioventricular node to the ventricles of the heart; AV bundle.   244

atrioventricular valve Valve located between the atrium and the ventricle.   242

atrium One of the upper chambers of either the left atrium or the right atrium of the heart which receive blood.   242

atrophy Wasting away or decrease in size of an organ or tissue.   387

auditory tube Extension from the middle ear to the nasopharynx which equalizes air pressure on the eardrum.   293, 358

autoimmune disease Disease that results when the immune system mistakenly attacks the body's own tissues.   277

autonomic system Branch of the peripheral nervous system that has control over the internal organs; consists of the sympathetic and parasympathetic systems.   337

autosome Any chromosome other than the sex chromosomes.   474

autotroph Organism that can capture energy and synthesize organic molecules from inorganic nutrients.   548, 698

auxin Plant hormone regulating growth, particularly cell elongation; most often indoleacetic acid (IAA).   178

AV (atrioventricular) node Small region of neuromuscular tissue that transmits impulses received from the SA node to the ventricular walls.   244

axial skeleton Portion of the skeleton that supports and protects the organs of the head, the neck, and the trunk.   370

axillary bud Bud located in the axil of a leaf.   149

axon Fiber of a neuron that conducts nerve impulses away from the cell body.   318

axon bulb Small swelling at the tip of one of many endings of the axon.   323

B

Bacteria One of three domains of life; prokaryotic cells other than archaea with unique genetic, biochemical, and physiological characteristics.   7, 571

bacteriophage Virus that infects bacteria.   567

ball and socket joint Most freely movable type of joint (e.g., the shoulder or hip joint).   376

basal nuclei Subcortical nuclei deep within the white matter that serve as relay stations for motor impulses and produce dopamine to help control skeletal muscle activities.   328

base Molecules tending to lower the hydrogen ion concentration in a solution and raise the pH numerically.   28

basement membrane Layer of nonliving material that anchors epithelial tissue to underlying connective tissue.   194

behavior Observable, coordinated responses to environmental stimuli.   4, 662

benthic division Ocean or lake floor from the high-tide mark to the deepest depths; which supports a unique set of organisms.   737

bicarbonate ion Ion that participates in buffering the blood, and the form in which carbon dioxide is transported in the bloodstream.   290

bilateral symmetry Body plan having two corresponding or complementary halves.   612

bile Secretion of the liver that is temporarily stored and concentrated in the gallbladder before being released into the small intestine, where it emulsifies fat.   219

binary fission Bacterial reproduction into two daughter cells without the utilization of a mitotic spindle.   570

binomial name Scientific name of an organism, the first part of which designates the genus and second part of which designates the specific epithet.   466

biodiversity Total number of species, the variability of their genes, and the ecosystems in which they live.   9, 742

biodiversity hotspot Region of the world that contains unusually large concentrations of species.   9, 743

biogeochemical cycle Circulating pathway of elements such as carbon and nitrogen involving exchange pools, storage areas, and biotic communities.   702

biogeography Study of the geographical distribution of organisms.   543

biological magnification Process by which nonexcreted substances like DDT become more concentrated in organisms in the higher trophic levels of the food chain.   709

biomass The number of organisms multiplied by their weight.   701

biome One of the biosphere's major communities characterized in particular by certain climatic conditions and particular types of plants.   719

biosphere That portion of the surface of the earth (air, water, and land) where living things exist.   8, 678

biotic potential Maximum population growth rate under ideal conditions.   682

bipedalism Walking erect on two feet.   650

birth-control pill Oral contraception containing estrogen and progesterone.   425

bivalve Type of mollusc with a shell composed of two valves; includes clams, oysters, and scallops.   621

blade Broad expanded portion of a plant leaf that may be single or compound.   149

blastocoel Fluid-filled cavity of a blastula.   440

blastocyst Early stage of human embryonic development that consists of a hollow fluid-filled ball of cells.   449

blastula Hollow ball of cells occurring during animal development prior to gastrula formation.   440

blind spot Region of the retina lacking rods or cones and where the optic nerve leaves the eye.   353

blood Type of connective tissue in which cells are separated by a liquid called plasma.   198

blood pressure Force of blood pushing against the inside wall of an artery.   248

B lymphocyte Lymphocyte that matures in the bone marrow and, when stimulated by the presence of a specific antigen, gives rise to antibody-producing plasma cells.   266

bone Connective tissue having protein fibers and a hard matrix of inorganic salts, notably calcium salts.   197

bony fish Member of a class of vertebrates (class Osteichthyes) containing numerous, diverse fishes, with a bony rather than cartilaginous skeleton.   642

bottleneck effect Type of genetic drift, in which a majority of genotypes is prevented from participating in the production of the next generation as a result of a natural disaster or human interference.   551

brain Enlarged superior portion of the central nervous system located in the cranial cavity of the skull.   326

brain stem Portion of the brain consisting of the medulla oblongata, pons, and midbrain.   328

bronchiole Smaller air passages in the lungs that begin at the bronchi and terminate in alveoli.   285

bronchus (pl., bronchi) One of two major divisions of the trachea leading to the lungs.   285

buffer Substance or group of substances that tend to resist pH changes of a solution, thus stabilizing its relative acidity and basicity.   29

bulbourethral gland Either of two small structures located below the prostate gland in males; adds secretions to semen.   414

bursa Saclike, fluid-filled structure, lined with synovial membrane, that occurs near a joint.   376

C

C$_3$ plant Plant that fixes carbon dioxide via the Calvin cycle; the first stable product of C$_3$ photosynthesis is a three-carbon compound.   142

C$_4$ plant Plant that fixes carbon dioxide to produce a C$_4$ molecule that releases carbon dioxide to the Calvin cycle.   142

calcitonin Hormone secreted by the thyroid gland that increases the blood calcium level.   398

calorie Amount of heat energy required to raise the temperature of water 1°C.   27

Calvin cycle Primary pathway of the light-independent reactions of photosynthesis; converts carbon dioxide to carbohydrate.   135, 139

cancer Malignant tumor whose nondifferentiated cells exhibit loss of contact inhibition, uncontrolled growth, and the ability to invade tissues and metastasize.   518

capillary Microscopic vessel connecting arterioles to venules; exchange of substances between blood and tissue fluid occur across their thin walls.   194, 240

capsule Gelatinous layer surrounding the cells of blue-green algae and certain bacteria.   62

carbaminohemoglobin Hemoglobin carrying carbon dioxide.   290

carbohydrate Class of organic compounds characterized by the presence of CH$_2$O groups; includes monosaccharides, disaccharides, and polysaccharides.   32

carbon cycle Continuous process by which carbon circulates in the air, water, and organisms of the biosphere.   704

carbon dioxide fixation Photosynthetic reaction in which carbon dioxide is attached to an organic compound.   140

carbonic anhydrase Enzyme in red blood cells that speeds the formation of carbonic acid from water and carbon dioxide.   290

carcinogenesis Development of cancer.   518

carcinoma Cancer arising in epithelial tissue.   194

cardiac cycle One complete cycle of systole and diastole for all heart chambers.   244

cardiac muscle Striated, involuntary muscle found only in the heart.   199

carnivore Secondary or higher consumer in a food chain, that therefore eats other animals.   698

carotenoid Yellow or orange pigment that serves as an accessory to chlorophyll in photosynthesis. 133

carotid body Structure located at the branching of the carotid arteries and that contain chemoreceptors sensitive to the hydrogen ion concentration but also the level of carbon dioxide and oxygen in blood. 288

carrier Heterozygous individual who has no apparent abnormality but can pass on an allele for a recessively inherited genetic disorder. 474

carrier protein Protein molecule that combines with a substance and transports it through the plasma membrane. 69, 74

carrying capacity Largest number of organisms of a particular species that can be maintained indefinitely in an ecosystem. 683

cartilage Connective tissue in which the cells lie within lacunae separated by a flexible proteinaceous matrix. 196

cartilaginous fish Member of a class of vertebrates (class Chondrichthyes) with a cartilaginous rather than bony skeleton, including sharks, rays, and skates. 642

Casparian strip Layer of impermeable lignin and suberin bordering four sides of root endodermal cells; prevents water and solute transport between adjacent cells. 155

cataract Opaqueness of the lens of the eye, making the lens incapable of transmitting light. 351

cecum Small pouch. In humans, a cecum lies below the entrance of the small intestine, and is the blind end of the large intestine. 220

cell Structural and functional unit of an organism; the smallest structure capable of performing all the functions necessary for life. 3, 46

cell body Portion of a neuron that contains a nucleus and from which dendrites and an axon extend. 318

cell cycle Repeating sequence of events in eukaryotes that involves cell growth and nuclear division; consists of the stages G_1, S, G_2, and M. 82

cell-mediated immunity Specific mechanism of defense in which T cells destroy antigen-bearing cells. 271

cell plate Structure that precedes the formation of the cell wall as a part of cytokinesis in plant cells. 88

cell theory One of the major theories of biology which states that all organisms are made up of cells; cells are capable of self-reproduction and cells come only from pre-existing cells. 46

cellular differentiation Process and the developmental stages by which a cell becomes specialized for a particular function. 172, 444

cellular respiration Metabolic reactions that use the energy primarily from carbohydrate but also fatty acid or amino acid breakdown to produce ATP molecules. 116

cellulose Polysaccharide composed of glucose molecules; the chief constituent of a plant's cell wall. 33

cell wall Structure that surrounds a plant, protistan, fungal, or bacterial cell and maintains the cell's shape and rigidity. 49, 62

central nervous system (CNS) Portion of the nervous system consisting of the brain and spinal cord. 318

centriole Cell organelle, existing in pairs, that occurs in the centrosome and may help organize a mitotic spindle for chromosome movement during animal cell division. 60

centromere Constricted region of a chromosome where sister chromatids are attached to one another and where the chromosome attaches to a spindle fiber. 85, 87

centrosome Major microtubule organizing center of cells, consisting of granular material. In animal cells, it contains two centrioles. 58

cephalization Having a well-recognized anterior head with a brain and sensory receptors. 612

cerebellum Part of the brain located posterior to the medulla oblongata and pons that coordinates skeletal muscles to produce smooth, graceful motions. 328

cerebral cortex Outer layer of cerebral hemispheres; receives sensory information and controls motor activities. 327

cerebral hemisphere One of the large, paired structures that together constitute the cerebrum of the brain. 327

cerebrospinal fluid Fluid found in the ventricles of the brain, in the central canal of the spinal cord, and in association with the meninges. 324

cerebrum Main part of the brain consisting of two large masses, or cerebral hemispheres; the largest part of the brain in mammals. 327

cervix Narrow end of the uterus, which leads into the vagina. 418

channel protein Forms a channel to allow a particular molecule or ion to cross the plasma membrane. 69

chemical energy Energy associated with the interaction of atoms in a molecule. 102

chemiosmosis Ability of certain membranes to use a hydrogen ion gradient to drive ATP formation. 123

chemoreceptor Sensory receptor that is sensitive to chemical stimulation— for example, sensory receptors for taste and smell. 344

chitin Strong but flexible nitrogenous polysaccharide found in fungal cells and in the exoskeleton of arthropods. 583, 626

chlorophyll Green pigment that captures solar energy during photosynthesis. 133

chloroplast Membranous organelle that contains chlorophyll and is the site of photosynthesis. 56, 134

chordae tendineae Tough bands of connective tissue that attach the papillary muscles to the atrioventricular valves within the heart. 242

chordate Member of the phylum Chordata, which includes lancelets, tunicates, fishes, amphibians, reptiles, birds, and mammals; characterized by a notochord, dorsal tubular nerve cord, pharyngeal gill pouches and post-anal tail at some point in the life cycle. 638

chorion Extraembryonic membrane functioning for respiratory exchange in reptiles and birds; contributes to placenta formation in mammals. 448

chorionic villi Treelike extensions of the chorion, an extraembryonic membrane, projecting into the maternal tissues at the placenta. 451

chorionic villi sampling Prenatal test in which a sample of chorionic villi cells are removed for diagnostic purposes. 484

choroid Vascular, pigmented middle layer of the eyeball. 350

chromatin Network of fibrils consisting of DNA and associated proteins observed within a nucleus that is not dividing. 52, 85

chromosome Rodlike structure in the nucleus seen during cell division; contains the hereditary units, or genes. 52, 85

chromosome mutation Variation in regard to the normal number of chromosomes inherited or in regard to the normal sequence of alleles on a chromosome; the sequence can be inverted, translocated from a nonhomologous chromosome, deleted, or duplicated. 490

chyme Thick, semi-liquid food material that passes from the stomach to the small intestine. 218

ciliary body In the eye, the structure that contains the ciliary muscle, which controls the shape of the lens. 350

ciliate Complex unicellular protist that moves by means of cilia and digests food in food vacuoles. 581

cilium (pl., cilia) Motile, short, hairlike extensions on the exposed surfaces of cells. 60, 194

circadian rhythm Regular physiological or behavioral event that occurs on an approximately 24-hour cycle.   405

citric acid cycle Cyclical metabolic pathway found in the matrix of mitochondria that participates in cellular respiration; breaks down acetyl groups to carbon dioxide and hydrogen.   117, 121

cleavage Cell division without cytoplasmic addition or enlargement; occurs during first stage of animal development.   440

cleavage furrow Indentation that begins the process of cleavage, by which animal cells undergo cytokinesis.   89

climate Weather condition of an area including especially prevailing temperature and average daily/yearly rainfall.   716

climax community In ecology, community that results when succession has come to an end.   680

clonal selection theory The concept that an antigen selects which lymphocyte will undergo clonal expansion and produce more lymphocytes bearing the same type of antigen receptor.   267

cloning Production of identical copies; can be either the production of identical individuals or in genetic engineering, the production of identical copies of a gene.   526

clotting Process of blood coagulation, usually when injury occurs.   252

club moss Type of seedless vascular plant that are also called ground pine because they appear to be miniature pine trees.   597

cnidarian Invertebrate in the phylum Cnidaria existing as either a polyp or medusa with two tissue layers and radial symmetry.   614

cochlea Portion of the inner ear that resembles a snail's shell and contains the spiral organ, the sense organ for hearing.   358

cochlear canal Canal within the cochlea that bears the spiral organ.   359

cochlear nerve Either of two cranial nerves that carry nerve impulses from the spiral organ to the brain; auditory nerve.   359

codominance Pattern of inheritance in which both alleles of a gene are equally expressed.   479

codon Three-base sequence in messenger RNA that causes the insertion of a particular amino acid into a protein or termination of translation.   507

coelom Embryonic body cavity lying between the digestive tract and body wall that is completely lined by mesoderm; in humans, the embryonic coelom becomes the thoracic and abdominal cavities.   201, 612

coenzyme Nonprotein organic molecule that aids the action of the enzyme to which it is loosely bound.   109

coevolution Interaction of two species such that each influences the evolution of the other species.   691

cofactor Nonprotein adjunct required by an enzyme in order to function; many cofactors are metal ions, while others are coenzymes.   109

cohesion-tension theory Explanation for upward transportation of water in xylem based upon transpiration-created tension and the cohesive properties of water molecules.   182

cohort Group of individuals having a statistical factor in common, such as year of birth, in a population study.   683

collagen fiber White fiber in the matrix of connective tissue, giving flexibility and strength.   196

collecting duct Duct within the kidney that receives fluid from several nephrons; the reabsorption of water occurs here.   307

collenchyma Plant tissue composed of cells with unevenly thickened walls; supports growth of stems and petioles.   152

color vision Ability to detect the color of an object, dependent on three kinds of cone cells.   352

colostrum Thin, milky fluid rich in proteins, including antibodies, that is secreted by the mammary glands a few days prior to or after delivery before true milk is secreted.   458

columnar epithelium Type of epithelial tissue with cylindrical cells.   194

commensalism Symbiotic relationship in which one species is benefited and the other is neither harmed nor benefited.   692

community Assemblage of populations interacting with one another within the same environment.   8, 678

compact bone Type of bone that contains osteons consisting of concentric layers of matrix and osteocytes in lacunae.   197, 366

competitive exclusion principle The conclusion that no two species can occupy the same niche at the same time.   689

complement system Group of plasma proteins that form a nonspecific defense mechanism, often by puncturing microbes; it complements the antigen-antibody reaction.   266

complementary base pairing Hydrogen bonding between particular bases; in DNA, thymine (T) pairs with adenine (A) and guanine (G) pairs with cytosine (C); in RNA uracil (U) pairs with A and G pairs with C.   504

complementary DNA (cDNA) DNA that has been synthesized from mRNA by the action of reverse transcriptase.   527

compound Substance having two or more different elements united chemically in fixed ratio.   23

concentration gradient Gradual change in chemical concentration from one point to another.   70

conclusion Statement made following an experiment as to whether the results support or falsify the hypothesis.   11

condensation synthesis Chemical reaction resulting in a covalent bond with the accompanying loss of a water molecule.   32

condom Sheath used to cover the penis during sexual intercourse; used as a contraceptive and, if latex, to minimize the risk of transmitting infection.   425

cone cell Photoreceptor in retina of eye that responds to bright light; detects color and provides visual acuity.   352

conidiospore Spore produced by sac, club, and imperfect fungi during asexual reproduction.   587

conifer Cone-bearing gymnosperm plants that include pine, cedar, and spruce trees.   600

conjugation Transfer of genetic material from one cell to another.   577

connective tissue Type of tissue characterized by cells separated by a matrix that often contains fibers.   196

conservation biology Scientific discipline that seeks to understand the effects of human activities on species, communities, and ecosystems and to develop practical approaches to preventing the extinction of species and the destruction of ecosystems.   742

consumer Organism that feeds on another organism in a food chain; primary consumers eat plants, and secondary (or higher) consumers eat animals.   132, 698

continental drift Movement of continents with respect to one another over the earth's surface.   543

contraceptive Medication or device used to reduce the chance of pregnancy.   425

control Sample that goes through all the steps of an experiment but lacks the factor or is not exposed to the factor being tested; a standard against which results of an experiment are checked.   11

coral Group of cnidarians having a calcium carbonate skeleton that participate in the formation of coral reefs.   614

coral reef Structure found in tropical waters that is formed by the buildup of coral skeletons and where many and various types of organisms reside.   736

cork Outer covering of bark of trees; made up of dead cells that may be sloughed off.   151

cork cambium Lateral meristem that produces cork.   161

cornea Transparent, anterior portion of the outer layer of the eyeball.   350

coronary artery Artery that supplies blood to the wall of the heart.   247

corpus luteum Yellow body that forms in the ovary from a follicle that has discharged its secondary oocyte; it secretes progesterone and some estrogen.   421

cortex In plants, ground tissue bounded by the epidermis and vascular tissue in stems and roots; in animals, outer layer of an organ such as the cortex of the kidney or adrenal gland.   155

cortisol Glucocorticoid secreted by the adrenal cortex that responds to stress on a long-term basis; reduces inflammation and promotes protein and fat metabolism.   400

cotyledon Seed leaf for embryo of a flowering plant; provides nutrient molecules for the developing plant before photosynthesis begins. 150, 172

coupled reactions Reactions that occur simultaneously; one is an exergonic reaction that releases energy and the other is an endergonic reaction that requires an input of energy in order to occur.   105

covalent bond Chemical bond in which atoms share one pair of electrons.   24

cranial nerve Nerve that arises from the brain.   334

creatinine Nitrogenous waste, the end product of creatine phosphate metabolism.   303

creatine phosphate Compound unique to muscles that contains a high-energy phosphate bond.   385

cristae (sing., crista) Short, fingerlike projections formed by the folding of the inner membrane of mitochondria.   57

crossing-over Exchange of corresponding segments of genetic material between nonsister chromatids of homologous chromosomes during synapsis of meiosis I.   91

crustacean Member of a group of marine arthropods that contains among others shrimps, crabs, crayfish, and lobsters.   627

cuboidal epithelium Type of epithelial tissue with cube-shaped cells.   194

culture Total pattern of human behavior; includes technology and the arts and is dependent upon the capacity to speak and transmit knowledge.   653

cuticle Waxy layer covering the epidermis of plants that protects the plant against water loss and disease-causing organisms.   151, 592

cyanobacterium Photosynthetic bacterium that contains chlorophyll and releases O_2; formerly called a blue-green alga.   572

cycad Type of gymnosperm with palmate leaves and massive cones; cycads are most often found in the tropics and subtropics.   600

cyclic electron pathway Portion of the light-dependent reaction that involves only photosystem I and generates ATP.   136

cytokinesis Division of the cytoplasm following mitosis and meiosis.   82

cytokinin Plant hormone that promotes cell division; often works in combination with auxin during organ development in plant embryos.   178

cytoplasm Contents of a cell between the nucleus and the plasma membrane that contains the organelles.   49

cytosine (C) One of four nitrogen-containing bases in nucleotides composing the structure of DNA and RNA; pairs with guanine.   41

cytoskeleton Internal framework of the cell, consisting of microtubules, actin filaments, and intermediate filaments.   58

cytotoxic T cell T lymphocyte that attacks and kills antigen-bearing cells.   271

D

data Facts or pieces of information collected through observation and/or experimentation.   11

daughter chromosome Separated chromatids become daughter chromosomes during anaphase of mitosis and anaphase II of meiosis.   85

day-neutral plant Plant whose flowering is not dependent on day length, i.e., tomato and cucumber.   180

decomposer Organism, usually a bacterium or fungus, that breaks down organic matter into inorganic nutrients that can be recycled in the environment.   571, 698

defecation Discharge of feces from the rectum through the anus.   220

delayed allergic response Allergic response initiated at the site of the allergen by sensitized T cells, involving macrophages and regulated by cytokines.   274

demographic transition Decline in the birthrate following a reduction in the death rate so that the population growth rate is lowered.   684

denaturation Loss of an enzyme's normal shape so that it no longer functions; caused by a less than optimal pH and temperature.   38, 108

dendrite Part of a neuron that sends signals toward the cell body.   318

denitrification Conversion of nitrate or nitrite to nitrogen gas by bacteria in soil.   706

dense fibrous connective tissue Type of connective tissue containing many collagen fibers packed together, and found in tendons and ligaments, for example.   196

density-dependent factor Biotic factor, such as disease and competition, that affects population size according to the population's density.   688

density-independent factor Abiotic factors, such as fire and flood, that affects population size independent of the population's density.   688

dermis Region of skin that lies beneath the epidermis.   205

desert Ecological biome characterized by a limited amount of rainfall; deserts have hot days and cool nights.   728

detrital food chain Straight-line linking of organisms according to who eats whom, beginning detritus.   701

detrital food web Complex pattern of interlocking and crisscrossing food chains beginning with detritus.   701

detritus Partially decomposed organic matter derived from tissues and animal wastes.   698

deuterostome Group of coelomate animals in which the second embryonic opening is associated with the mouth; the first embryonic opening, the blastopore, is associated with the anus.   612, 636

development Series of stages by which a zygote becomes an organism or by which an organism changes during its life span; includes puberty and aging for example.   5, 172

diaphragm Dome-shaped horizontal sheet of muscle and connective tissue that divides the thoracic cavity from the abdominal cavity. Also, a birth-control device consisting of a soft rubber or latex cup that fits over the cervix.   288, 425

diastole Relaxation period of a heart chamber during the cardiac cycle.   244

diastolic pressure Arterial blood pressure during the diastolic phase of the cardiac cycle.   248

diatom Golden brown alga with a cell wall in two parts, or valves; significant part of phytoplankton.   578

dicot Abbreviation of dicotyledon. Flowering plant group; members have two embryonic leaves (cotyledons), net-veined leaves, cylindrical arrangement of vascular bundles, flower parts in fours or fives, and other characteristics.   150

diencephalon Portion of the brain in the region of the third ventricle that includes the thalamus and hypothalamus.   328

differentially permeable Ability of plasma membranes to regulate the passage of substances into and out of the cell, allowing some to pass through and preventing the passage of others.   70

diffusion Movement of molecules or ions from a region of higher to lower concentration; it requires no energy and stops when the distribution is equal.   71

dihybrid Individual that is heterozygous for two traits; shows the phenotype governed by the dominant alleles but carries the recessive alleles.   472

dinoflagellate Photosynthetic unicellular protist, with two flagella, one whiplash and the other located within a groove between protective cellulose plates; significant part of phytoplankton.   578

diploid 2n number of chromosomes; twice the number of chromosomes found in gametes.   85

diplontic life cycle Life cycle typical of animals in which the adult is always diploid and meiosis produces the gametes.   577

directional selection Natural selection in which an extreme phenotype is favored, usually in a changing environment.   554

disaccharide Sugar that contains two units of a monosaccharide; e.g., maltose.   32

disruptive selection Natural selection in which extreme phenotypes are favored over the average phenotype, leading to more than one distinct form.   556

distal convoluted tubule Final portion of a nephron that joins with a collecting duct; associated with tubular secretion.   307

diuretic Drug used to counteract hypertension by causing the excretion of water.   311

DNA (deoxyribonucleic acid) Nucleic acid found in cells; the genetic material that specifies protein synthesis in cells.   502

DNA ligase Enzyme that links DNA fragments; used during production of recombinant DNA to join foreign DNA to vector DNA.   526

domain Largest of the categories, or taxa, used by taxonomists to group species; the three domains are Archaea, Bacteria, and Eukarya.   7, 560

dominance hierarchy Organization of animals in a group that determines the order in which the animals have access to resources.   668

dominant allele Allele that exerts its phenotypic effect in the heterozygote; it masks the expression of the recessive allele.   468

dormancy In plants, a cessation of growth under conditions that seem appropriate for growth.   174

dorsal-root ganglion Mass of sensory neuron cell bodies located in the dorsal root of a spinal nerve.   334

double fertilization In flowering plants, one sperm joins with polar nuclei within the embryo sac to produce a 3n endosperm nucleus, and another sperm joins with an egg to produce a zygote.   171, 604

double helix Double spiral; describes the three-dimensional shape of DNA.   504

doubling time Number of years it takes for a population to double in size.   684

E

echinoderm Phylum of marine animals that includes sea stars, sea urchins, and sand dollars; characterized by radial symmetry and a water vascular system.   636

ecological niche Role an organism plays in community, including its habitat and its interactions with other organisms.   689

ecological pyramid Pictorial graph based on energy content (and also biomass or number of organisms) of various trophic levels in a food web—from the producer to the final consumer populations.   701

ecological succession Gradual replacement of communities in an area following a disturbance (secondary succession) or the creation of new soil (primary succession).   680

ecology Study of the interactions of organisms with other organisms and the physical environment.   678

ecosystem Biological community together with the associated abiotic environment; characterized by energy flow and chemical cycling.   8, 678, 698

ectoderm Outer germ layer of the embryonic gastrula; it gives rise to the nervous system and skin.   440

edema Swelling due to tissue fluid accumulation in the intercellular spaces.   262

edge effect Edges around a patch that have a slightly different habitat than the favorable habitat in the interior of the patch.   754

egg Nonflagellate female gamete; also referred to as ovum.   171, 418

elastic cartilage Type of cartilage composed of elastic fibers, allowing greater flexibility.   197

elastic fiber Yellow fiber in the matrix of connective tissue, providing flexibility.   196

electrocardiogram Recording of the electrical activity associated with the heartbeat.   245

electron Subatomic particle that has almost no weight and carries a negative charge; orbits in a shell about the nucleus of an atom.   20

electron transport system Chain of electron carriers in the cristae of mitochondria and thylakoid membrane of chloroplasts. As the electrons pass from one carrier to the next, energy is released and used to establish a hydrogen ion gradient. This gradient is associated with the production of ATP molecules.   117, 122, 136

element Substance that cannot be broken down into substances with different properties; composed of only one type atom.   20

El Niño Warming of water in the Eastern Pacific equatorial region such that the Humboldt Current is displaced with possible negative results such as reduction in marine life.   734

embolus Moving blood clot that is carried through the bloodstream.   255

embryo Stage of a multicellular organism that develops from a zygote and before it becomes free living; in seed plants the embryo is part of the seed.   171, 440

embryonic disk During human development, flattened area during gastrulation from which the embryo arises.   450

embryonic period From approximately the second to the eighth week of human development, during which the major organ systems are organized.   448

embryo sac Megagametophyte of flowering plants that produces an egg cell.   170

emulsification Breaking up of fat globules into smaller droplets by the action of bile salts or any other emulsifier.   34

endergonic reaction Chemical reaction that requires an input of energy; opposite of exergonic reaction.   104

endocrine gland Ductless organ that secretes (a) hormone(s) into the bloodstream.   392

endocytosis Process by which substances are moved into the cell from the environment by phagocytosis (cellular eating) or pinocytosis (cellular drinking; includes receptor-mediated endocytosis).   76

endoderm Inner germ layer of the embryonic gastrula that becomes the lining of the digestive and respiratory tract and associated organs.   440

endodermis Plant root tissue that forms a boundary between the cortex and the vascular cylinder.   155

endometrium Lining of the uterus, which becomes thickened and vascular during the uterine cycle.   419

endoplasmic reticulum (ER) Membranous system of tubules, vesicles, and sacs in cells, sometimes having attached ribosomes. Rough ER has ribosomes; smooth ER does not.   53

endosperm In angiosperms, the 3n tissue that nourishes the embryo and seedling and is formed as a result of a sperm joining with two polar nuclei.   171, 604

endosymbiotic hypothesis Possible explanation of the evolution of eukaryotic organelles by phagocytosis of prokaryotes.   63

endothermic Maintenance of a constant body temperature independent of the environmental temperature.   645

energy Capacity to do work and bring about change; occurs in a variety of forms.   4, 102

energy of activation Energy that must be added to cause molecules to react with one another.   106

entropy Measure of disorder or randomness.   103

environmental resistance Sum total of factors in the environment that limit the numerical increase of a population in a particular region.   682

enzymatic protein Protein that catalyzes a specific reaction.   69

enzyme Organic catalyst, usually a protein that speeds up a reaction in cells due to its particular shape.   37, 106

enzyme inhibition Means by which cells regulate enzyme activity; there is competitive and noncompetitive inhibition.   109

epidermis In plants, tissue that covers roots and leaves and stems of nonwoody organism; in mammals, the outer protective region of the skin.   151, 155, 204

epididymis Coiled tubule next to the testes where sperm mature and may be stored for a short time.   414

epiglottis Structure that covers the glottis and closes off the air tract during the process of swallowing.   216, 284

epinephrine Hormone secreted by the adrenal medulla in times of stress; also called adrenaline.   399

epiphyte Plant that takes its nourishment from the air because its placement in other plants gives it an aerial position.   184, 724

episodic memory Capacity of brain to store and retrieve information with regard to persons and events.   330

epithelial tissue Type of tissue that lines hollow organs and covers surfaces; epithelium.   194

erythrocyte See red blood cell.

erythropoietin Hormone, produced by the kidneys, that speeds red blood cell formation.   251, 303

esophagus Muscular tube for moving swallowed food from the pharynx to the stomach.   216

essential amino acids Amino acids required in the human diet because the body cannot make them.   228

estrogen Female sex hormone, which, along with progesterone, maintains sexual organs and secondary sex characteristics.   404, 422

estuary Portion of ocean located where a river enters and fresh water mixes with salt water.   732

ethylene Plant hormone that causes ripening of fruit and is also involved in abscission.   179

Eukarya One of the three domains of life, consisting of organisms in the kingdoms Protista, Fungi, Plantae, and Animalia.   7

eukaryotic cell Type of cell that has a membrane-bounded nucleus and membranous organelles; found in organisms within the domain Eukarya.   49

eutrophication Enrichment of water by inorganic nutrients used by phytoplankton; often overenrichment caused by human activities leading to excessive bacterial growth and oxygen depletion.   709, 730

evolution Changes that occur in the members of a species with the passage of time, often resulting in increased adaptation of organisms to the environment.   5, 540

excretion Removal of metabolic wastes from the body.   302

exergonic reaction Chemical reaction that releases energy; opposite of endergonic reaction.   104

exocytosis Process in which an intracellular vesicle fuses with the plasma membrane so that the vesicle's contents are released outside the cell.   76

expiration Act of expelling air from the lungs; exhalation.   282

expiratory reserve volume Volume of air that can be forcibly exhaled after normal exhalation.   286

exponential growth Growth, particularly of a population, in which the increase occurs in the same manner as compound interest.   682

external respiration Exchange of oxygen and carbon dioxide between alveoli and blood.   290

exteroceptor Sensory receptor that detects stimuli from outside the body (e.g., taste, smell, vision, hearing, and equilibrium).   344

extraembryonic Membrane that is not a part of the embryo but is necessary to the continued existence and health of the embryo.   448

F

facilitated transport Use of a plasma membrane carrier to move a substance into or out of a cell from higher to lower concentration: no energy required.   74

FAD Flavin adenine dinucleotide; a coenzyme of oxidation-reduction that becomes $FADH_2$ as oxidation of substrates occurs and then delivers electrons to the electron transport system in mitochondria during cellular respiration.   116

fall overturn Mixing process that occurs in fall in stratified lakes whereby the oxygen-rich top waters mix with nutrient-rich bottom waters.   730

fat Organic molecule that contains glycerol and fatty acids and is found in adipose tissue.   34

fate map Diagram that traces the differentiation of cells during development from their origin to their final structure and function.   446

fatty acid Molecule that contains a hydrocarbon chain and ends with an acid group.   34

fermentation Anaerobic breakdown of glucose that results in a gain of two ATP and end products such as alcohol and lactate.   117, 125

fern Member of a group of plants that have large fronds; in the sexual life cycle the independent gametophyte produces flagellated sperm and the vascular sporophyte produces windblown spores.   597

fertilization Union of a sperm nucleus and an egg nucleus, which creates the zygote with the diploid number of chromosomes.   90, 440

fiber Structure resembling a thread; also plant material that is nondigestible.   227

fibrin Insoluble protein threads formed from fibrinogen during blood clotting.   252

fibrinogen Plasma protein that is converted into fibrin threads during blood clotting.   252

fibroblast Cell in connective tissues which produces fibers and other substances.   196

fibrocartilage Cartilage with a matrix of strong collagenous fibers.   197

fibrous root system In most monocots, a mass of similarly sized roots that cling to the soil.   156

filament End to end chains of cells that form as cell division occurs in only one plane; in plants, the elongated stalk of a stamen.   170, 577

filter feeder Method of obtaining nourishment by certain animals which strain minute organic particles from the water in a way that deposits them in the digestive tract.   613

fimbria Fingerlike extension from the oviduct near the ovary.   418

fin In fish and other aquatic animals, membranous, winglike or paddlelike process used to propel, balance, or guide the body.   641

fitness Ability of an organism to produce fertile offspring; measured against the ability of other organisms to reproduce in the same environment.   554

flagellum (pl., flagella) Slender, long extension used for locomotion by some bacteria, protozoans, and sperm.   60, 62

flower Reproductive organ of plants that contains the structures for the production of pollen grains and covered seeds.   170, 606

fluid-mosaic model Model for the plasma membrane based on the changing location and pattern of protein molecules in a fluid phospholipid bilayer.   68

fluke Group of parasitic flatworms having suckers that live in internal organ where they receive nourishment; blood flukes and liver flukes are examples.   617

focused Bending of light rays by the cornea, lens, and humors so that they converge and create an image on the retina.   351

follicle Structure in the ovary that produces the egg and, in particular, the female sex hormones, estrogen and progesterone.   421

follicle-stimulating hormone (FSH) Hormone secreted by the anterior pituitary gland that stimulates the development of an ovarian follicle in a female or the production of sperm in a male.   417

fontanel Membranous region located between certain cranial bones in the skull of a fetus or infant.   370, 452

food chain The order in which one population feeds on another in an ecosystem, from detritus (detrital food chain) or producer (grazing food chain) to final consumer.   701

food web Complex pattern of energy flow in an ecosystem represented by interlocking and crisscrossing food chains.   701

foramen magnum Opening in the occipital bone of the vertebrate skull through which the spinal cord passes.   370

formed element Constituent of blood that is either cellular (red blood cells and white blood cells) or at least cellular in origin (platelets).   249

fossil Any past evidence of an organism that has been preserved in the earth's crust.   540

fossil fuel Remains of once living organisms that are burned to release energy, such as coal, oil, and natural gas.   704

founder effect Type of genetic drift, in which only a fraction of the total genetic diversity of the original gene pool is represented as a result of a few individuals founding a colony.   551

fovea centralis Region of the retina consisting of densely packed cones that is responsible for the greatest visual acuity.   351

free energy Useful energy in a system that is capable of performing work.   104

fruit In flowering plants, the structure that forms from an ovary and associated tissues and encloses seeds.   171, 173, 604

fruiting body Spore-producing and spore-disseminating structure found in sac and club fungi.   585

functional group Specific cluster of atoms attached to the carbon skeleton of organic molecules that enters into reactions and behaves in a predictable way.   31

fungus (pl., fungi) Saprotrophic decomposer; the body is made up of filaments called hyphae that form a mass called a mycelium; for example mushrooms and molds.   583

G

gamete Haploid sex cell; the egg or a sperm which join during fertilization to form a zygote.   90, 428

gametophyte Haploid generation of the alternation of generations life cycle of a plant; produces gametes that unite to form a diploid zygote.   170, 593

ganglion Collection or bundle of neuron cell bodies usually outside the central nervous system.   334

gap analysis Use of computers to discover places where biodiversity is high outside preserved areas.   754

gap junction Region between cells formed by the joining of two adjacent plasma membranes; it lends strength and allows ions, sugars, and small molecules to pass between cells.   196

gastric gland Gland within the stomach wall that secretes gastric juice.   218

gastrovascular cavity Blind digestive cavity that also serves a circulatory (transport) function in animals that lack a circulatory system.   615

gastrula In animal development, the embryonic stage following formation of the germ layers.   440

gastrulation Process during animal development during which the germ layers form.   441

gene Unit of heredity existing as alleles on the chromosomes; in diploid organisms typically two alleles are inherited—one from each parent.   4

gene flow Sharing of genes between two populations through interbreeding.   554

gene pool Total of all the genes of all the individuals in a population.   549

genetic drift Mechanism of evolution due to random changes in the allelic frequencies of a population; more likely to occur in small populations or when only a few individuals of a large population reproduce.   551

genome Full set of genetic information of a species or a virus.   531

genotype Genes of an individual for a particular trait or traits; often designated by letters, for example, *BB* or *Aa*.   468

germinate Beginning of growth of a seed, spore, or zygote, especially after a period of dormancy.   174

germ layer Developmental layer of the body—that is, ectoderm, mesoderm, or endoderm.   440

gerontology Study of aging.   459

gibberellin Plant hormone promoting increased stem growth; also involved in flowering and seed germination.   178

gill Respiratory organ in most aquatic animals; in fish, an outward extension of the pharynx.   638

gland Epithelial cell or group of epithelial cells that are specialized to secrete a substance.   194

global warming Predicted increase in the earth's temperature, due to the greenhouse effect, which will lead to the melting of polar ice and a rise in sea levels.   704, 751

glomerular capsule Double-walled cup that surrounds the glomerulus at the beginning of the nephron.   307

glomerular filtrate Filtered portion of blood contained within the glomerular capsule.   309

glomerular filtration Movement of small molecules from the glomerulus into the glomerular capsule due to the action of blood pressure.   309

glomerulus Capillary network within the glomerular capsule in a nephron, where glomerular filtration takes place. 306

glottis Opening for airflow in the larynx. 216, 284

glucagon Hormone secreted by the pancreas which causes the liver to break down glycogen and raises the blood glucose level. 402

glucocorticoid Type of hormone secreted by the adrenal cortex that influences carbohydrate, fat, and protein metabolism; see cortisol. 399

glucose Six-carbon sugar that organisms degrade as a source of energy during cellular respiration. 32

glycogen Storage polysaccharide, found in animals, that is composed of glucose molecules joined in a linear fashion but having numerous branches. 32

glycolipid Lipid in plasma membranes that bears a carbohydrate chain attached to a hydrophobic tail. 68

glycolysis Anaerobic metabolic pathway found in the cytoplasm that participates in cellular respiration and fermentation; it converts glucose to two molecules of pyruvate. 118

glycoprotein Protein in plasma membranes that bears a carbohydrate chain. 53, 68

Golgi apparatus Organelle, consisting of flattened saccules and also vesicles, that processes, packages, and distributes molecules about or from the cell. 54

gonad Organ that produces gametes; the ovary produces eggs, and the testis produces sperm. 404

gonadotropic hormone Substance secreted by anterior pituitary that regulates the activity of the ovaries and testes; principally, follicle-stimulating hormone (FSH) and luteinizing hormone (LH). 394

gonadotropin-releasing hormone (GnRH) Hormone secreted by the hypothalamus that stimulates the anterior pituitary to secrete follicle-stimulating hormone and luteinizing hormone. 417

granular leukocyte White blood cell with prominent granules in the cytoplasm. 251

granum Stack of chlorophyll-containing thylakoids in a chloroplast. 57

gravitational equilibrium Maintenance of balance when the head and body are motionless. 361

gray crescent Gray area that appears in an amphibian egg after being fertilized by the sperm, thought to contain chemical signals that turn on the genes that control development. 445

gray matter Nonmyelinated nerve fibers and cell bodies in the central nervous system. 324

grazing food chain A flow of energy to a straight-line linking of organisms according to who eats whom. 701

grazing food web Complex pattern of interlocking and crisscrossing food chains that begins with populations of autotrophs serving as producers. 701

greenhouse effect Reradiation of solar heat toward the earth, caused by gases in the atmosphere. 705

greenhouse gases Gases such as carbon dioxide, methane, nitrous oxide, water vapor, ozone, and nitrous oxide in the atmosphere which are involved in the greenhouse effect. 704

ground tissue Tissue that constitutes most of the body of a plant; consists of parenchyma, collenchyma, and sclerenchyma cells which function in storage, basic metabolism, and support. 151

growth factor Chemical signal that regulates mitosis and differentiation of cells that have receptors for it; important in such processes as fetal development, tissue maintenance and repair, and hematopoiesis; sometimes a contributing factor in cancer. 405

growth hormone (GH) Substance secreted by the anterior pituitary; controls size of individual by promoting cell division, protein synthesis, and bone growth. 394

growth plate Cartilaginous layer within an epiphysis of a long bone that permits growth of bone to occur. 368

guanine (G) One of four nitrogen-containing bases in nucleotides composing the structure of DNA and RNA; pairs with cytosine. 41

guard cell One of two cells that surround a leaf stoma; changes in the turgor pressure of these cells cause the stoma to open or close. 185

gymnosperm Type of woody seed plant in which the seeds are not enclosed by fruit and are usually borne in cones, such as those of conifers. 600

H

habitat Place where an organism lives and is able to survive and reproduce. 689

hair cell Cell with stereocilia (long microvilli) that is sensitive to mechanical stimulation; mechanoreceptor for hearing and equilibrium in the inner ear. 358

hair follicle Tubelike depression in the skin in which a hair develops. 205

halophile Type of archaea that lives in extremely salty habitats. 573

haploid n number of chromosomes; half the diploid number; the number characteristic of gametes which contain only one set of chromosomes. 85

haplontic life cycle Life cycle typical of protists in which the adult is always haploid because meiosis occurs after zygote formation. 576

hard palate Bony, anterior portion of the roof of the mouth. 214

helper T cell T lymphocyte that releases cytokines and stimulates certain other immune cells to perform their respective functions. 271

hemoglobin Iron-containing pigment in red blood cells that combines with and transports oxygen. 250, 290

hepatic portal system Portal system that begins at the villi of the small intestine and ends at the liver. 247

hepatic portal vein Vein leading to the liver and formed by the merging blood vessels leaving the small intestine. 247

hepatic vein Vein that runs between the liver and the inferior vena cava. 247

herbaceous stem Nonwoody stem. 158

herbivore Primary consumer in a food chain; a plant eater. 698

hermaphrodite Animal having both male and female sex organs. 617

heterospore Spore that is dissimilar from another produced by the same plant; microspores and megaspores are produced by seed plants. 593

heterotroph Organism that cannot synthesize organic molecules from inorganic nutrients and therefore must take in organic nutrients (food). 548, 698

heterozygous Having two different alleles (as *Aa*) for a given trait. 468

hexose Six-carbon sugar. 32

hinge joint Type of joint that allows movement as a hinge does, such as the movement of the knee. 376

hippocampus Part of the cerebral cortex where memories form. 330

histamine Substance, produced by basophils and mast cells in connective tissue, that causes capillaries to dilate. 264

HLA (human leukocyte associated) antigen Plasma membrane protein that identifies the cell as belonging to a particular individual and acts as an antigen in other individuals. 270

homeobox Nucleotide sequence located in all homeotic genes and serving to identify portions of the genome, in many different types of organisms, that are active in pattern formation. 447

homeostasis Maintenance of normal internal conditions in a cell or an organism by means of self-regulating mechanisms. 4

homeotic genes Genes that control the overall body plan by controlling the fate of groups of cells during development. 446

homologous chromosome Similarly constructed chromosomes with the same shape and that contain genes for the same traits; also called homologues.   90

homologous structure Structure that is similar in two or more species because of common ancestry.   544

homologue Member of a homologous pair of chromosomes.   90

Homo sapiens Modern humans.   655

homozygous Having identical alleles (as *AA* or *aa*) for a given trait; pure breeding.   468

hormone Chemical signal produced in one part of the body that controls the activity of other parts.   220, 392

host Organism that provides nourishment and/or shelter for a parasite.   692

human chorionic gonadotropin (HCG) Hormone produced by the chorion that functions to maintain the uterine lining.   424, 449

hyaline cartilage Cartilage whose cells lie in lacunae separated by a white translucent matrix containing very fine collagen fibers.   196

hydrogen bond Weak bond that arises between a slightly positive hydrogen atom of one molecule and a slightly negative atom of another molecule or between parts of the same molecule.   26

hydrolysis Splitting of a covalent bond by the addition of water.   32

hydrolytic enzyme Enzyme that catalyzes a reaction in which the substrate is broken down by the addition of water.   224

hydrophilic Type of molecule that interacts with water by dissolving in water and/or forming hydrogen bonds with water molecules.   26

hydrophobic Type of molecule that does not interact with water because it is nonpolar.   26

hydrostatic skeleton Fluid-filled body compartment which provides support for muscle contraction resulting in movement; seen in cnidarians, flatworms, roundworms, and segmented worms.   618

hydrothermal vent Hot springs in the sea floor along ocean ridges where heated seawater and sulfate react to produce hydrogen sulfide; here chemosynthetic bacteria support a community of varied organisms.   737

hypertension Elevated blood pressure, particularly the diastolic pressure.   256

hypertonic solution Higher solute concentration (less water) than the cell; causes cell to lose water by osmosis.   73

hypertrophy Increase in muscle size following long-term exercise.   387

hypha (pl., hyphae) Filament of the vegetative body of a fungus.   583

hypothalamic-inhibiting hormone One of many hormones produced by the hypothalamus that inhibits the secretion of an anterior pituitary hormone.   394

hypothalamic-releasing hormone One of many hormones produced by the hypothalamus that stimulates the secretion of an anterior pituitary hormone.   394

hypothalamus Part of the brain located below the thalamus that helps regulate the internal environment of the body and produces releasing factors that control the anterior pituitary.   328, 394

hypothesis Supposition that is formulated after making an observation; it can be tested by obtaining more data, often by experimentation.   10

hypotonic solution Lower solute (more water) concentration than the cytosol of a cell; causes cell to gain water by osmosis.   72

I

immediate allergic response Allergic response that occurs within seconds of contact with an allergen, caused by the attachment of the allergen to IgE antibodies.   274

immune system All the cells in the body which protect the body against foreign organisms and substances and also cancerous cells.   202, 264

immunity Ability of the body to protect itself from foreign substances and cells, including disease-causing agents.   264

immunization Use of a vaccine to protect the body against specific disease-causing agents.   272

immunoglobulin (Ig) Globular plasma protein that functions as an antibody.   267

implantation Attachment and penetration of the embryo into the lining of the uterus (endometrium).   448

imprinting Learning to make a particular response to only one type of animal or object.   665

inclusive fitness Fitness that results from personal reproduction and from helping nondescendant relatives reproduce.   673

incomplete dominance Inheritance pattern in which the offspring has an intermediate phenotype compared to its parents; for example a normal individual and an individual with sickle-cell disease can produce a child with sickle-cell trait.   479

incus Middle of three ossicles of the ear that serve to conduct vibrations from the tympanic membrane to the oval window of the inner ear.   358

independent assortment Alleles of unlinked genes assort independently of each other during meiosis so that the gametes contain all possible combinations of alleles.   91, 471

induced-fit model Change in the shape of an enzyme's active site which enhances the fit between the enzyme's active site and its substrate(s).   107

induction Ability of a chemical or a tissue to influence the development of another tissue.   445

inferior vena cava Large vein that enters the right atrium from below and carries blood from the trunk and lower extremities.   246

inflammatory reaction Tissue response to injury that is characterized by redness, swelling, pain, and heat.   264

inner ear Portion of the ear consisting of a vestibule, semicircular canals, and the cochlea where equilibrium is maintained and sound is transmitted.   358

inorganic molecule Type of molecule that is not an organic molecule; not derived from a living organism.   31

insect Member of a group of arthropods in which the head has antennae, compound eyes, and simple eyes; a thorax has three pairs of legs and often wings; and an abdomen has internal organs.   628

insertion End of a muscle that is attached to a movable bone.   377

inspiration Act of taking air into the lungs; inhalation.   282

inspiratory reserve volume Volume of air that can be forcibly inhaled after normal inhalation.   286

insulin Hormone secreted by the pancreas that lowers the blood glucose level by promoting the uptake of glucose by cells and the conversion of glucose to glycogen by the liver and skeletal muscles.   402

integration Summing up of excitatory and inhibitory signals by a neuron or by some part of the brain.   323, 345

integumentary system Organ system consisting of skin various organs, such as hair, which are found in skin.   202

intercalated disks Region that holds adjacent cardiac muscle cells together and that appear as dense bands at right angles to the muscle striations.   199

interferon Protein formed by a cell infected with a virus that blocks the infection of another cell.   266

interkinesis Period of time between meiosis I and meiosis II during which no DNA replication takes place.   92

interleukin Cytokine produced by macrophages and T lymphocytes that functions as a regulator of the immune response.   273

internal respiration Exchange of oxygen and carbon dioxide between blood and tissue fluid.   290

interneuron Neuron, located within the central nervous system, conveying messages between parts of the central nervous system.   318

internode In vascular plants, the region of a stem between two successive nodes.   149

interoceptor Sensory receptor that detects stimuli from inside the body (e.g., pressoreceptors, osmoreceptors, chemoreceptors).   344

interphase Stages of the cell cycle (G_1, S, G_2) during which growth and DNA synthesis occur when the nucleus is not actively dividing.   82

interstitial cell-stimulating hormone (ICSH) Name sometimes given to luteinizing hormone in males; controls the production of testosterone by interstitial cells.   417

intervertebral disk Layer of cartilage located between adjacent vertebrae.   372

intrauterine device (IUD) Birth-control device consisting of a small piece of molded plastic inserted into the uterus, and believed to alter the uterine environment so that fertilization does not occur.   425

invertebrate Referring to an animal without a serial arrangement of vertebrae.   610

ion Charged particle that carries a negative or positive charge.   23

ionic bond Chemical bond in which ions are attracted to one another by opposite charges.   23

iris Muscular ring that surrounds the pupil and regulates the passage of light through this opening.   350

isotonic solution Solution that is equal in solute concentration to that of the cell; causes cell to neither lose nor gain water by osmosis.   72

isotope Atom having the same atomic number but a different atomic mass due to the number of neutrons.   21

J

jawless fish Type of fish that has no jaws, includes today's hagfishes and lampreys.   642

joint Union of two or more bones; an articulation.   366

juxtaglomerular apparatus Structure located in the walls of arterioles near the glomerulus that regulates renal blood flow.   311

K

karyotype Chromosomes arranged by pairs according to their size, shape, and general appearance in mitotic metaphase.   484

keystone species Species whose activities have a significant role in determining community structure.   753

kinetic energy Energy associated with motion.   102

kinin Chemical mediator, released by damaged tissue cells and mast cells, which causes the capillaries to dilate and become more permeable.   264

L

lacteal Lymphatic vessel in an intestinal villus; it aids in the absorption of lipids.   219

lactose intolerance Inability to digest lactose because of an enzyme deficiency.   224

lacuna Small pit or hollow cavity, as in bone or cartilage, where a cell or cells are located.   196

lancelet Invertebrate chordate with a body that resembles a lancet and has the four chordate characteristics as an adult.   640

landscape A number of interacting ecosystem fragments.   743

lanugo Short, fine hair that is present during the later portion of fetal development.   452

larynx Cartilaginous organ located between pharynx and trachea that contains the vocal cords; voice box.   284

law Theory that is generally accepted by an overwhelming number of scientists.   11

leaf Lateral appendage of a stem, highly variable in structure, often containing cells that carry out photosynthesis.   149

leaf vein Vascular tissue within a leaf.   152

learning Relatively permanent change in an animal's behavior that results from practice and experience.   330, 664

leech Blood-sucking annelid, usually found in fresh water, with a sucker at each end of a segmented body.   625

lens Clear membranelike structure found in the eye behind the iris; brings objects into focus.   350

leptin Hormone produced by adipose tissue that acts on the hypothalamus to signal satiety.   405

less-developed country (LDC) Country in which population growth is expanding rapidly and the majority of people live in poverty.   684

leukocyte See white blood cell.

lichen Fungi and algae coexisting in a symbiotic relationship.   587

ligament Tough cord or band of dense fibrous connective tissue that joins bone to bone at a joint.   196, 366

light-dependent reactions Portion of photosynthesis that captures solar energy and takes place in thylakoid membranes.   135

light-independent reactions Portion of photosynthesis that takes place in the stroma of chloroplasts and can occur in the dark; it uses the products of the light-dependent reactions to reduce carbon dioxide to a carbohydrate.   135

limbic system Association of various brain centers including the amygdala and hippocampus; governs learning and memory and various emotions such as pleasure, fear, and happiness.   329

lineage Evolutionary line of descent.   650

linkage group Alleles of different genes that are located on the same chromosome and tend to be inherited together.   496

lipid Organic compound that is insoluble in water; notably fats, oils, and steroids.   34

littoral zone Shore zone between high-tide mark and low-tide mark; also shallow water of a lake where light penetrates to the bottom.   733

logistic growth Population increase that results in an S-shaped curve; growth is slow at first, steepens, and then levels off due to environmental resistance.   683

long-day plant Plant which flowers when day length is longer than a critical length, i.e., wheat, barley, clover, spinach.   180

long-term potentiation (LTP) Enhanced response at synapses within the hippocampus, likely essential to memory storage.   332

loop of the nephron Portion of the nephron lying between the proximal convoluted tubule and the distal convoluted tubule that functions in water reabsorption.   307

loose fibrous connective tissue Tissue composed mainly of fibroblasts widely separated by a matrix containing collagen and elastic fibers.   196

luteinizing hormone (LH) Hormone produced by the anterior pituitary gland that stimulates the development of the corpus luteum in females and the production of testosterone in males.   417

lymph Fluid, derived from tissue fluid, that is carried in lymphatic vessels.   254, 262

lymphatic vessel Vessel that carries lymph.   262

lymph nodes Mass of lymphoid tissue located along the course of a lymphatic vessel.   262

lymphocyte Specialized white blood cell that functions in specific defense; occurs in two forms—T lymphocyte and B lymphocyte.   251

lymphoid organ Organ other than a lymphatic vessel that is part of the lymphatic system; includes lymph nodes, tonsils, spleen, thymus gland, and bone marrow.   262

lysogenic cycle Bacteriophage life cycle in which the viral DNA is incorporated into host cell DNA without viral reproduction and lysis.   567

lysosome Membrane-bounded vesicle that contains hydrolytic enzymes for digesting macromolecules.   55

lytic cycle Bacteriophage life cycle in which viral reproduction takes place and host cell lysis does occur.   567

M

macrophage Large phagocytic cell derived from a monocyte that ingests microbes and debris.   251, 264

malleus First of three ossicles of the ear that serve to conduct vibrations from the tympanic membrane to the oval window of the inner ear.   358

Malpighian tubule Blind, threadlike excretory tubule attached to the gut of an insect.   629

maltase Enzyme produced in small intestine that breaks down maltose to two glucose molecules.   224

mammal Homeothermic vertebrate characterized especially by the presence of hair and mammary glands.   648

marsupial Member of a group of mammals bearing immature young nursed in a marsupium, or pouch; for example, kangaroo and opossum.   648

mast cell Cell to which antibodies, formed in response to allergens, attach, bursting the cell and releasing allergy mediators, which cause symptoms.   264

matrix Unstructured semifluid substance that fills the space between cells in connective tissues or inside organelles.   57, 196

matter Anything that takes up space and has mass.   20

mechanoreceptor Sensory receptor that is sensitive to mechanical stimulation, such as that from pressure, sound waves, and gravity.   344

medulla oblongata Part of the brain stem that is continuous with the spinal cord; controls heartbeat, blood pressure, breathing, and other vital functions.   329

megagametophyte In seed plants, the gametophyte that produces an egg; in flowering plants, an embryo sac.   600

megakaryocyte Large cell that gives rise to blood platelets.   252

megaspore One of the two types of spores produced by seed plants; develops into a megagametophyte (embryo sac).   170, 593

meiosis Type of nuclear division that occurs as part of sexual reproduction in which the daughter cells receive the haploid number of chromosomes in varied combinations.   90

melanocyte Specialized cell in the epidermis that produces melanin, the pigment responsible for skin color.   205

melanocyte-stimulating hormone (MSH) Substance that causes melanocytes to secrete melanin in lower vertebrates.   394

melatonin Hormone, secreted by the pineal gland, that is involved in biorhythms.   405

memory Capacity of the brain to store and retrieve information about past sensations and perceptions; essential to learning.   330

meninges (sing., meninx) Protective membranous coverings about the central nervous system.   201, 324

menisci (sing., meniscus) Cartilaginous wedges that separate the surfaces of bones in synovial joints.   376

menopause Termination of the ovarian and uterine cycles in older women.   424

menstruation Loss of blood and tissue from the uterus at the end of a uterine cycle.   422

meristem Undifferentiated, embryonic tissue in the active growth regions of plants.   151

mesoderm Middle germ layer of embryonic gastrula; gives rise to the muscles, the connective tissue, and the circulatory system.   440

mesophyll Inner, thickest layer of a leaf consisting of palisade and spongy mesophyll; the site of most of photosynthesis.   164

messenger RNA (mRNA) Ribonucleic acid whose sequence of codons specifies the sequence of amino acids during protein synthesis.   506

metabolic pathway Series of linked reactions, beginning with a particular reactant and terminating with an end product.   106

metabolic pool Metabolites that result from catabolism and subsequently can be used for anabolism.   126

metabolism All of the chemical changes that occur within a cell.   4, 104

metamorphosis Change in shape and form that some animals, such as amphibians and insects, undergo during development.   629

metaphase plate A disc formed during metaphase in which all of a cell's chromosomes lie in a single plane at right angles to the spindle fibers.   87

metapopulation Population subdivided into several small and isolated populations due to habitat fragmentation.   753

metastasis Spread of cancer from the place of origin throughout the body; caused by the ability of cancer cells to migrate and invade tissues.   518

methanogen Type of archaea that lives in oxygen-free habitats, such as swamps, and releases methane gas.   572

microevolution Change in gene frequencies within a population over time.   549

microgametophyte In seed plants, the gametophyte that produces sperm; a pollen grain.   600

microsphere Body formed from proteinoids exposed to water; has properties similar to today's cells.   547

microspore In seed plants, one of the two types of spores; develops into a microgametophyte (pollen grain).   170, 593

microtubule Small cylindrical structure that contains thirteen rows of the protein tubulin about an empty central core; present in the cytoplasm, centrioles, cilia, and flagella.   58

microvillus Cylindrical process that extends from an epithelial cell of a villus of the intestinal wall and serves to increase the surface area of the cell.   194

midbrain Part of the brain located below the thalamus and above the pons; contains reflex centers and tracts.   328

middle ear Portion of the ear consisting of the tympanic membrane, the oval and round windows, and the ossicles; where sound is amplified.   358

mimicry Superficial resemblance of two or more species; a mechanism that avoids predation by appearing to be noxious.   691

mineral Naturally occurring inorganic substance containing two or more elements; certain minerals are needed in the diet.   183, 232

mineralocorticoid Hormones secreted by the adrenal cortex that regulate salt and water balance, leading to increases in blood volume and blood pressure.   399

mitochondrion Membrane-bounded organelle in which ATP molecules are produced during the process of cellular respiration.   56, 120

mitosis Type of cell division in which daughter cells receive the exact chromosome and genetic makeup of the parent cell; occurs during growth and repair.   82, 85

model Simulation of a process that aids conceptual understanding until the process can be studied firsthand; a hypothesis that describes how a particular process could possibly be carried out.   11

molecular clock Idea that the rate at which mutational changes accumulate in certain genes is constant over time and is not involved in adaptation to the environment.   650

molecule Union of two or more atoms of the same element; also the smallest part of a compound that retains the properties of the compound.   23

mollusc Member of the phylum Mollusca that includes squids, clams, snails, and chitons; characterized by a visceral mass, a mantle, and a foot.   620

molting Periodic shedding of the exoskeleton in arthropods.   626

monoclonal antibody One of many antibodies produced by a clone of hybridoma cells which all bind to the same antigen.   274

monocot Abbreviation of monocotyledon. Flowering plant group; among other characteristics, members have one embryonic leaf, parallel-veined leaves, and scattered vascular bundles.   150

monocyte Type of a granular leukocyte that functions as a phagocyte particularly after it becomes a macrophage which is also an antigen-presenting cell.   251

monohybrid Individual that is heterozygous for one trait; shows the phenotype of the dominant allele but carries the recessive allele.   469

monosaccharide Simple sugar; a carbohydrate that cannot be decomposed by hydrolysis.   32

monosomy One less chromosome than usual.   486

monotreme Egg-laying mammal—for example, duckbill platypus and spiny anteater.   648

monsoon Climate in India and southern Asia caused by wet ocean winds that blow onshore for almost half the year.   717

more-developed country (MDC) Country in which population growth is low and the people enjoy a good standard of living.   684

morphogenesis Emergence of shape in tissues, organs, or entire embryo during development.   444

morphogen gene Unit of inheritance that controls a gradient which influences morphogenesis.   446

morula Spherical mass of cells resulting from cleavage during animal development prior to the blastula stage.   440

mosaic evolution Concept that human characteristics did not evolve at the same rate; e.g., some body parts are more humanlike than others in early hominids.   652

motor molecule Protein that moves along either actin filaments or microtubules and translocates organelles.   58

motor neuron Nerve cell that conducts nerve impulses away from the central nervous system and innervates effectors (muscle and glands).   318

motor unit Motor neuron and all the muscle fibers it innervates.   384

mucous membrane Membrane that lines a cavity or tube that opens to the outside of the body; mucosa.   201

multiple allele Inheritance pattern in which there are more than two alleles for a particular trait; each individual has only two of all possible alleles.   478

multiregional continuity hypothesis Proposal that modern humans evolved independently in at least three different places: Asia, Africa, and Europe.   655

muscle fiber Cell with myofibrils containing actin and myosin filaments arranged within sarcomeres; a group of muscle fibers is a muscle.   381

muscle twitch Contraction of a whole muscle in response to a single stimulus.   384

muscular (contractile) tissue Type of tissue composed of fibers that can shorten and thicken.   199

mutagen Agent, such as radiation or a chemical, that brings about a mutation in DNA.   516

mutation Alteration in chromosome structure or number and also an alteration in a gene due to a change in DNA composition.   551

mutualism Symbiotic relationship in which both species benefit.   693

mycelium Mass of hyphae that makes up the body of a fungus.   583

mycorrhiza Mutually beneficial symbiotic relationship between a fungus and the roots of vascular plants.   184, 587

myelin sheath White, fatty material—derived from the membrane of Schwann cells that forms a covering for nerve fibers.   318

myocardium Cardiac muscle in the wall of the heart.   242

myofibril Specific muscle cell organelle containing a linear arrangement of sarcomeres, which shorten to produce muscle contraction.   381

myoglobin Pigmented compound in muscle tissue that stores oxygen.   385

myogram Recording of a muscular contraction.   384

myosin Muscle protein making up the thick filaments in a sarcomere; it pulls actin to shorten the sarcomere, yielding muscle contraction.   381

N

NAD⁺ (nicotinamide adenine dinucleotide) Coenzyme that functions as a carrier of electrons and hydrogen ions, especially in cellular respiration.   116

NADP⁺ (nicotinamide adenine dinucleotide phosphate) Coenzyme that functions as a carrier of electrons and hydrogen ions during photosynthesis.   135

nasopharynx Region of the pharynx associated with the nasal cavity.   216

natural killer (NK) cell Lymphocyte that causes an infected or cancerous cell to burst.   266

natural selection Mechanism of evolution caused by environmental selection of organisms most fit to reproduce; results in adaptation to the environment.   554

negative feedback Mechanism of homeostatic response by which the output of a system suppresses or inhibits activity of the system.   206

nematocyst In cnidarians, a capsule that contains a threadlike fiber whose release aids in the capture of prey.   614

nephridium Segmentally arranged, paired excretory tubules of many invertebrates, as in the earthworm.   623

nephron Anatomical and functional unit of the kidney; kidney tubule.   305

nerve Bundle of nerve fibers outside the central nervous system.   334

nerve impulse Action potential (electrochemical change) traveling along a neuron.   320

nervous tissue Tissue that contains nerve cells (neurons), which conduct impulses, and neuroglial cells, which support, protect, and provide nutrients to neurons.   200

neural plate Region of the dorsal surface of the chordate embryo that marks the future location of the neural tube.   443

neural tube Tube formed by closure of the neural groove during development. In vertebrates, the neural tube develops into the spinal cord and brain.   443

neuroglia Nonconducting nerve cells that are intimately associated with neurons and function in a supportive capacity.   200

neuromuscular junction Region where an axon bulb approaches a muscle fiber; contains a presynaptic membrane, a synaptic cleft, and a postsynaptic membrane.   382

neuron Nerve cell that characteristically has three parts: dendrites, cell body, and axon.   200

neurotransmitter Chemical stored at the ends of axons that is responsible for transmission across a synapse.   323

neutron Subatomic particle that has a weight of one atomic mass unit, carries no charge, and is found in the nucleus of an atom.   20

neutrophil Granular leukocyte that is the most abundant of the white blood cells; first to respond to infection.   251

nitrification Process by which nitrogen in ammonia and organic compounds is oxidized to nitrites and nitrates by soil bacteria.   706

nitrogen cycle Continuous process by which nitrogen circulates in the air, soil, water, and organisms of the biosphere.   706

nitrogen fixation Process whereby free atmospheric nitrogen is converted into compounds, such as ammonium and nitrates, usually by bacteria.   706

node In plants, the place where one or more leaves attach to a stem.   149

node of Ranvier Gap in the myelin sheath around a nerve fiber.   318

noncyclic electron pathway Portion of the light-dependent reaction of photosynthesis that involves both photosystem I and photosystem II. It generates both ATP and NADPH.   136

nondisjunction Failure of homologous chromosomes or daughter chromosomes to separate during meiosis I and meiosis II respectively.   486

nonvascular plant Bryophytes such as mosses and liverworts that have no vascular tissue and either occur in moist locations or have special adaptations for living in dry locations.   594

norepinephrine (NE) Neurotransmitter of the postganglionic fibers in the sympathetic division of the autonomic system; also a hormone produced by the adrenal medulla.   323, 399

notochord Dorsal supporting rod that exists in all chordates sometime in their life history; replaced by the vertebral column in vertebrates.   443

nuclear envelope Double membrane that surrounds the nucleus and is continuous with the endoplasmic reticulum.   52

nuclear pore Opening in the nuclear envelope which permits the passage of proteins into the nucleus and ribosomal subunits out of the nucleus.   52

nuclei Masses of cell bodies in the CNS.   329

nucleoid Region of a bacterium where the bacterial chromosome is found; it is not bounded by a nuclear envelope.   62, 570

nucleolus Dark-staining, spherical body in the nucleus that produces ribosomal subunits.   52

nucleoplasm Semifluid medium of the nucleus, containing chromatin.   52

nucleotide Monomer of DNA and RNA consisting of a five-carbon sugar bonded to a nitrogen-containing base and a phosphate group.   40

nucleus Membrane-bounded organelle within a eukaryotic cell that contains chromosomes and controls the structure and function of the cell.   52

O

observation Step in the scientific method by which data are collected before a conclusion is drawn.   10

oil Triglyceride usually of plant origin, composed of glycerol and three fatty acids that is liquid in consistency because there are many unsaturated bonds in the hydrocarbon chains of the fatty acids.   34

oil gland Gland of the skin, associated with hair follicle, that secretes sebum; sebaceous gland.   205

olfactory cell Modified neuron that is a sensory receptor for the sense of smell.   349

omnivore Organism in a food chain that feeds on both plants and animals.   698

oncogene Cancer-causing gene.   520

oogenesis Production of an egg in females by the process of meiosis and maturation.   96, 418

operant conditioning Learning that results from rewarding or reinforcing a particular behavior.   664

operator In an operon the sequence of DNA that binds tightly to a repressor and thereby regulates the expression of structural genes.   513

operon Group of structural and regulating genes that function as a single unit.   513

optic nerve Cranial nerve that carries nerve impulses from the retina of the eye to the brain, thereby contributing to the sense of sight.   351

organ Combination of two or more different tissues performing a common function.   3, 148

organelle Small, often membranous structure in the cytoplasm, having a specific structure and function.   46, 49

organic molecule Molecule that always contains carbon and hydrogen and often oxygen; organic molecules are associated with living things.   31

organ system Group of related organs working together.   3

origin End of a muscle that is attached to a relatively immovable bone.   377

oscilloscope Apparatus that records changes in voltage by graphing them on a screen; used to study the nerve impulse.   320

osmosis Diffusion of water through a selectively permeable membrane.   72

osmotic pressure Measure of the tendency of water to move across a differentially permeable membrane; visible as an increase in liquid on the side of the membrane with higher solute concentration.   72

ossicle One of the small bones of the middle ear—malleus, incus, stapes.   358

osteocyte Mature bone cell located within the lacunae of bone.   366

osteon Cylindrical-shaped unit containing bone cells that surround an osteonic canal; Haversian system.   366

otolith Calcium carbonate granule associated with ciliated cells in the utricle and the saccule.   361

outer ear Portion of ear consisting of the pinna and auditory canal.   358

out-of-Africa hypothesis Proposal that modern humans originated only in Africa; then they migrated out of Africa and supplanted populations of early *Homo* in Asia and Europe about 100,000 years ago.   655

oval opening In the fetus, a shunt for the flow of blood from the right atrium to the left atrium and in that way by-passing the lungs.   452

oval window Membrane-covered opening between the stapes and the inner ear.   358

ovarian cycle Monthly follicle changes occurring in the ovary that control the level of sex hormones in the blood and the uterine cycle.   421

ovary In animals, the female gonad, the organ that produces eggs, estrogen, and progesterone; in flowering plants, the base of the pistil that protects ovules and along with associated tissues becomes a fruit.   170, 404, 418, 606

oviduct Tube that transports eggs to the uterus; also called uterine tube.   418

ovulation Release of a secondary oocyte from the ovary; if fertilization occurs the secondary oocyte becomes an egg.   418

ovule In seed plants, a structure where the megaspore becomes an egg-producing megagametophyte and which develops into a seed following fertilization.   170, 600

oxidation Loss of one or more electrons from an atom or molecule; in biological systems, generally the loss of hydrogen atoms.   110

oxidative phosphorylation Process by which ATP production is tied to an electron transport system that uses oxygen as the final acceptor; occurs in mitochondria.   122

oxygen debt Amount of oxygen needed to metabolize lactate, a compound that accumulates during vigorous exercise.   125, 385

oxyhemoglobin Compound formed when oxygen combines with hemoglobin.   290

oxytocin Hormone released by the posterior pituitary that causes contraction of uterus and milk letdown.   394

ozone shield Accumulation of O_3, formed from oxygen in the upper atmosphere; a filtering layer that protects the earth from ultraviolet radiation.   710

P

pacemaker See SA (sinoatrial) node.

palisade mesophyll In a plant leaf, the layer of mesophyll containing elongated cells with many chloroplasts.   164

PAN (peroxyacetylnitrate) Type of noxious chemical found in photochemical smog.   707

pancreatic amylase Enzyme in the pancreas that digests starch to maltose.   224

pancreatic islets (of Langerhans) Distinctive group of cells within the pancreas that secretes insulin and glucagon.   402

parasitism Symbiotic relationship in which one species (parasite) benefits in terms of growth and reproduction to the harm of the other species (host).   692

parasympathetic division That part of the autonomic system that is active under normal conditions; uses acetylcholine as a neurotransmitter.   337

parathyroid gland One of four glands embedded in the posterior surface of the thyroid gland; produces parathyroid hormone.   398

parathyroid hormone (PTH) Hormone secreted by the four parathyroid glands that increases the blood calcium level and decreases the blood phosphate level.   398

parenchyma cell Thin-walled, minimally differentiated cell that photosynthesizes or stores the products of photosynthesis.   152

parturition Processes that lead to and include birth and the expulsion of the afterbirth.   455

pathogen Disease-causing agent such as viruses, parasitic bacteria, fungi, and animals.   194, 250

pattern formation Positioning of cells during development that determines the final shape of an organism.   444

pelagic division Open portion of the sea.   736

penis External organ in males through which the urethra passes and that serves as the organ of sexual intercourse.   415

pentose Five-carbon sugar; deoxyribose is the pentose sugar found in DNA; ribose is a pentose sugar found in RNA.   32

peptide bond Covalent bond that joins two amino acids.   38

peptide hormone Type of hormone that is a protein, a peptide, or is derived from an amino acid.   408

perception Mental awareness of sensory stimulation.   345

perennial Flowering plant that lives more than one growing season because the underground parts regrow each season.   148

perforin Molecule secreted by a cytotoxic T cell that perforates the plasma membrane of a target cell so that water and salts enter, causing the cell to swell and burst.   271

pericardium Protective serous membrane that surrounds the heart.   242

pericycle Layer of cells surrounding the vascular tissue of roots; produces branch roots.   155

periosteum Fibrous connective tissue covering the surface of bone.   366

peripheral nervous system (PNS) Nerves and ganglia that lie outside the central nervous system.   318

peristalsis Wavelike contractions that propel substances along a tubular structure like the esophagus.   201, 217

peritubular capillary network Capillary network that surrounds a nephron and functions in reabsorption during urine formation.   306

permafrost Permanently frozen ground usually occurring in the tundra, a biome of Arctic regions.   720

peroxisome Enzyme-filled vesicle in which fatty acids and amino acids are metabolized to hydrogen peroxide that is broken down to harmless products.   55

petal A flower part that occurs just inside the sepals; often conspicuously colored to attract pollinators.   170

petiole Part of a plant leaf that connects the blade to the stem.   149

PGAL (glyceraldehyde-3-phosphate) Significant metabolite in both the glycolytic pathway and the Calvin cycle; PGAL is oxidized during glycolysis and reduced during the Calvin cycle.   139

phagocytize To ingest extracellular particles by engulfing; cell eating.   580

phagocytosis Process by which amoeboid-type cells engulf large substances, forming an intracellular vacuole.   76

pharynx Portion of the digestive tract between the mouth and the esophagus which serves as a passageway for food and also air on its way to the trachea.   216, 283

phenotype Outward appearance of an organism caused by the genotype and environmental influences.   468

pheromone Chemical signal that works at a distance and alters the behavior of another member of the same species.   407, 671

phloem Vascular tissue that conducts organic solutes in plants; contains sieve-tube elements and companion cells.   152, 186, 596

phospholipid Molecule that forms the bilayer of the cell's membranes; has a polar, hydrophilic head bonded to two nonpolar, hydrophobic tails.   35

photochemical smog Air pollution that contains nitrogen oxides and hydrocarbons which react to produce ozone and PAN (peroxyacetylnitrate).   707

photoreceptor Sensory receptor in retina that responds to light stimuli.   344

photosynthesis Process by which plants and algae make their own food using the energy of the sun.   132

photosystem Cluster of light-absorbing pigment molecules within thylakoid membranes.   136

phototropism Growth response of plant stems to light; stems demonstrate positive phototropism.   180

pH scale Measure of the hydrogen ion concentration $[H^+]$; any pH below 7 is acidic and any pH above 7 is basic.   29

phyletic gradualism Evolutionary model that proposes evolutionary change resulting in a new species can occur gradually in an unbranched lineage.   559

phytochrome Photoreversible plant pigment that is involved in photoperiodism and other responses of plants such as etiolation.   181

pineal gland Endocrine gland located in the third ventricle of the brain that produces melatonin.   405

pinocytosis Process by which vesicle formation brings macromolecules into the cell.   76

pistil Structure in a flower that consists of a stigma, a style, and an ovule-containing ovary; the ovule becomes a seed and the ovary becomes fruit.   170, 606

pith Parenchyma tissue in the center of some stems and roots.   155

pituitary gland Endocrine gland that lies just inferior to the hypothalamus;

consists of the anterior pituitary and posterior pituitary. 394

placenta Structure that forms from the chorion and the uterine wall and allows the embryo, and then the fetus, to acquire nutrients and rid itself of wastes. 424, 454

placental mammal Member of mammalian subclass characterized by the presence of a placenta during the development of an offspring. 649

plankton Freshwater and marine organisms that are suspended on or near the surface of the water; includes phytoplankton and zooplankton. 731

plant hormone Chemical signal that is produced by various plant tissues and coordinates the activities of plant cells. 177

plaque Accumulation of soft masses of fatty material, particularly cholesterol, beneath the inner linings of the arteries. 229, 255

plasma Liquid portion of blood. 198, 249

plasma cell Cell derived from a B-cell lymphocyte that is specialized to mass-produce antibodies. 267

plasma membrane Membrane surrounding the cytoplasm that consists of a phospholipid bilayer with embedded proteins; functions to regulate the entrance and exit of molecules from cell. 49, 62

plasmid Self-duplicating ring of accessory DNA in the cytoplasm of bacteria. 62, 526

plasmodesma (pl., plasmodesmata) Cytoplasmic strand that extends through a pore in the cell wall and connects the cytoplasms of adjacent cells. 186

plasmodium In slime molds, free-living mass of cytoplasm that moves by pseudopods on a forest floor or in a field feeding on decaying plant material by phagocytosis; reproduces by spore formation. 582

plasmolysis Contraction of the cell contents due to the loss of water. 73

platelet Cell fragment that is necessary to blood clotting; also called a thrombocyte. 198, 252

pleura Serous membrane that encloses the lungs. 201, 288

plumule In flowering plants, the embryonic plant shoot that bears young leaves. 175

polar body Nonfunctioning daughter cell, formed during oogenesis, that has little cytoplasm. 96

polar nuclei Two nuclei in the embryo sac (megagametophyte) of flowering plants that combines with a sperm nucleus to form the triploid

endosperm which nourishes the developing plant embryo. 170

pollen grain In seed plants, the sperm-producing microgametophyte. 170, 600

pollen sac In flowering plants, that part of the anther where microspores are produced and become pollen grains. 170

pollination In seed plants, the delivery of pollen to the vicinity of the egg-producing megagametophyte. 170, 600

pollution Any environmental change that adversely affects the lives and health of living things. 750

polygenic inheritance Inheritance pattern in which a trait is controlled by several allelic pairs; each dominant allele contributes to the phenotype in an additive and like manner. 477

polymerase chain reaction (PCR) Technique that uses the enzyme DNA polymerase to produce copies of a particular piece of DNA within a test tube. 527

polyp Small, abnormal growth that arises from the epithelial lining. 221

polypeptide Polymer of many amino acids linked by peptide bonds. 38

polyribosome String of ribosomes simultaneously translating regions of the same mRNA strand during protein synthesis. 53

polysaccharide Polymer made from sugar monomers; the polysaccharides starch and glycogen are polymers of glucose monomers. 32

pons Portion of the brain stem above the medulla oblongata and below the midbrain; assists the medulla oblongata in regulating the breathing rate. 328

population Group of organisms of the same species occupying a certain area and sharing a common gene pool. 8, 549

population viability analysis Calculation of the minimum population size needed to prevent extinction. 678, 754

positive feedback Mechanism of homeostatic response in which the output intensifies and increases the likelihood of response, instead of countering it and canceling it. 208, 394

posterior pituitary Portion of the pituitary gland that stores and secretes oxytocin and antidiuretic hormone which are produced by the hypothalamus. 394

potential energy Stored energy as a result of location or spatial arrangement. 102

precipitation Water deposited on the earth in the form of rain, snow, sleet, hail, or fog. 703

predation Interaction in which one organism uses another, called the prey, as a food source. 690

predator Organism that practices predation. 690

prediction Step of the scientific method which follows the formulation of a hypothesis and assists in creating the experimental design. 10

prefrontal area Association area in the frontal lobe that receives information from other association areas and uses it to reason and plan actions. 328

pressure-flow theory Explanation for phloem transport; osmotic pressure following active transport of sugar into phloem brings about a flow of sap from a source to a sink. 186

prey Organism that provides nourishment for a predator. 690

primary motor area Area in the frontal lobe where voluntary commands begin; each section controls a part of the body. 327

primary root Original root that grows straight down and remains the dominant root of the plant; contrast fibrous root system. 156

primary somatosensory area Area dorsal to the central sulcus where sensory information arrives from skin and skeletal muscles. 327

primate Member of the order Primate; includes prosimians, monkeys, apes, and hominids all of whom have adaptations for living in trees. 650

principle Theory that is generally accepted by an overwhelming number of scientists. Also called law. 11

prion Infectious particle consisting of protein only and no nucleic acid which is believed to be linked to several diseases of the central nervous system; (contraction for a proteinlike infectious agent). 569

producer Photosynthetic organism at the start of a grazing food chain that makes its own food (e.g., green plants on land and algae in water). 132, 698

product Substance that forms as a result of a reaction. 104

progesterone Female sex hormone that helps maintain sexual organs and secondary sex characteristics. 404, 422

prokaryotic cell Lacking a membrane-bounded nucleus and organelles; the cell type within the domain Bacteria and Archaea. 62

prolactin (PRL) Hormone secreted by the anterior pituitary that stimulates the production of milk from the mammary glands. 394

promoter In an operon, a sequence of DNA where RNA polymerase binds prior to transcription. 513

prostate gland Gland located around the male urethra below the urinary bladder; adds secretions to semen. 414

protein Organic macromolecule that is composed of either one or several polypeptides. 37

proteinoid Abiotically polymerized amino acids that when exposed to water become microspheres which have cellular characteristics. 547

prothrombin Plasma protein that is converted to thrombin during the steps of blood clotting. 252

prothrombin activator Enzyme that catalyzes the transformation of the precursor prothrombin to the active enzyme thrombin. 252

protocell In biological evolution, a possible cell forerunner that became a cell once it could reproduce. 546

proton Positive subatomic particle, located in the nucleus and having a weight of approximately one atomic mass unit. 20

proto-oncogene Normal gene that can become an oncogene through mutation. 520

protostome Group of coelomate animals in which the first embryonic opening (the blastopore) is associated with the mouth. 612

protozoan Heterotrophic unicellular protist that moves by flagella, cilia, pseudopodia, or are immobile. 580

proximal convoluted tubule Highly coiled region of a nephron near the glomerular capsule, where tubular reabsorption takes place. 307

pseudocoelom Body cavity lying between the digestive tract and body wall that is incompletely lined by mesoderm. 612

pseudopod Cytoplasmic extension of amoeboid protists; used for locomotion and engulfing food. 580

pulmonary artery Blood vessel that takes blood away from the heart to the lungs. 243

pulmonary circuit Circulatory pathway that consists of the pulmonary trunk, the pulmonary arteries, and the pulmonary veins; takes O_2-poor blood from the heart to the lungs and O_2-rich blood from the lungs to the heart. 246

pulmonary fibrosis Accumulation of fibrous connective tissue in the lungs; caused by inhaling irritating particles, such as silica, coal dust, or asbestos. 295

pulmonary vein Blood vessel that takes blood to the heart from the lungs. 243

punctuated equilibrium Evolutionary model that proposes there are periods of rapid change dependent on speciation followed by long periods of stasis. 559

Punnett square Gridlike device used to calculate the expected results of simple genetic crosses. 469

pupil Opening in the center of the iris of the eye. 350

purine Type of nitrogen-containing base, such as adenine and guanine, having a double-ring structure. 504

Purkinje fibers Specialized muscle fibers that conduct the cardiac impulse from AV bundle into the ventricles. 244

pyrimidine Type of nitrogen-containing base, such as cytosine, thymine, and uracil, having a single-ring structure. 504

pyruvate End product of glycolysis; its further fate, involving fermentation or entry into a mitochondrion, depends on oxygen availability. 117

R

radial symmetry Body plan in which similar parts are arranged around a central axis, like spokes of a wheel. 610

radicle In seed plants, a portion of the plant embryo that becomes the root. 172, 184

radioactive isotope Unstable form of an atom that spontaneously emits radiation in the form of radioactive particles or radiant energy. 21

rain shadow Lee side (side sheltered from the wind) of a mountainous barrier, which receives much less precipitation than the windward side. 717

reactant Substance that participates in a reaction. 104

receptor-mediated endocytosis Selective uptake of molecules into a cell by vacuole formation after they bind to specific receptor proteins in the plasma membrane. 76

receptor protein Protein located in the plasma membrane or within the cell that binds to a substance that alters some metabolic aspect of the cell. 69

recessive allele Hereditary factor that expresses itself in the phenotype only when the genotype is homozygous. 468

recombinant New combination of alleles as a result of crossing-over or biotechnology. 496

recombinant DNA (rDNA) DNA that contains genes from more than one source. 526

red blood cell Formed element that contains hemoglobin and carries oxygen from the lungs to the tissues; erythrocyte. 198, 250

red bone marrow Vascularized modified connective tissue that is sometimes found in the cavities of spongy bone; site of blood cell formation. 263, 366

redox reaction Oxidation-reduction reaction; one molecule loses electrons (oxidation) while another molecule simultaneously gains electrons (reduction). 110

reduced hemoglobin Hemoglobin that is carrying hydrogen ions. 290

reduction Chemical reaction that results in addition of one or more electrons to an atom, ion, or compound. Reduction of one substance occurs simultaneously with oxidation of another. 110

referred pain Pain perceived as having come from a site other than that of its actual origin. 347

reflex Automatic, involuntary response of an organism to a stimulus. 216, 329, 335

refractory period Time following an action potential when a neuron is unable to conduct another nerve impulse. 320

regulator gene In an operon, a gene that codes for a protein that regulates the expression of other genes. 513

renal artery Vessel that originates from the aorta and delivers blood to the kidney. 302

renal cortex Outer portion of the kidney that appears granular. 305

renal medulla Inner portion of the kidney that consists of renal pyramids. 305

renal pelvis Hollow chamber in the kidney that receives freshly prepared urine from the collecting ducts. 305

renal vein Vessel that takes blood from the kidney to the inferior vena cava. 302

renin Enzyme released by kidneys that leads to the secretion of aldosterone and a rise in blood pressure. 311, 400

replacement reproduction Population in which each person is replaced by only one child. 685

replication Making an exact copy, as when a DNA molecule is duplicated by a complementary base pairing process. 505

reproduce To produce a new individual of the same kind. 4

reptile Member of a class of terrestrial vertebrates with internal fertilization, scaly skin, and an egg with a leathery shell; includes snakes, lizards, turtles, and crocodiles. 645

residual volume Amount of air remaining in the lungs after a forceful expiration. 286

resource partitioning Mechanism that increases the number of niches by apportioning the supply of a resource. 689

respiratory center Group of nerve cells in the medulla oblongata that send out nerve impulses on a rhythmic basis, resulting in involuntary inspiration on an ongoing basis. 288

resting potential Polarity across the plasma membrane of a resting neuron due to an unequal distribution of ions. 320

restoration ecology Subdiscipline of conservation biology that seeks ways to return ecosystems to their former state. 755

restriction enzyme Bacterial enzyme that stops viral reproduction by cleaving viral DNA; used to cut DNA at specific points during production of recombinant DNA. 526

reticular fiber Very thin collagen fibers in the matrix of connective tissue, highly branched and forming delicate supporting networks. 196

reticular formation Complex network of nerve fibers within the brain stem that arouses the cerebrum. 329

retina Innermost layer of the eyeball, which contains the rod cells and the cone cells. 351

retinal Light-absorbing molecule, which is a derivative of vitamin A and a component of rhodopsin. 352

retrovirus RNA virus containing the enzyme reverse transcriptase that carries out RNA/DNA transcription. 568

rhizome Rootlike, underground stem. 162

rhodopsin Visual pigment found in the rods whose activation by light energy leads to vision. 352

rib cage Contains ribs and intercostal muscles. 288

ribosomal RNA (rRNA) Type of RNA found in ribosomes where protein synthesis occurs. 506

ribosome RNA and protein in two subunits; site of protein synthesis in the cytoplasm. 53, 62, 508

ribozyme Enzyme that carries out mRNA processing. 508

RNA (ribonucleic acid) Nucleic acid produced from covalent bonding of nucleotide monomers that contain the sugar ribose; occurs in three forms: messenger RNA, ribosomal RNA, and transfer RNA. 40

RNA polymerase During transcription, an enzyme that joins nucleotides complementary to a portion of DNA. 508

rod cell Photoreceptor in retina of eyes that responds to dim light. 352

root apical meristem Undifferentiated, embryonic tissue located at the apex of the stem. 154

root hair Extension of a root epidermal cell that increases the surface area for the absorption of water and minerals. 151

root nodule Structure on plant root that contains nitrogen-fixing bacteria. 156, 184

root system Includes the main root and any and all of its lateral (side) branches. 148

rotational equilibrium Maintenance of balance when the head and body are suddenly moved or rotated. 361

round window Membrane-covered opening between the inner ear and the middle ear. 358

RuBP (ribulose bisphosphate) Five-carbon compound that combines with and fixes carbon dioxide during the Calvin cycle and is later regenerated by the same cycle. 139

S

saccule Saclike cavity in the vestibule of the inner ear; contains sensory receptors for gravitational equilibrium. 361

salivary amylase Secreted from the salivary glands; the first enzyme to act on starch. 215, 224

salivary gland Gland associated with the oral cavity that secretes saliva. 214

SA (sinoatrial) node Small region of neuromuscular tissue that initiates the heartbeat; also called the pacemaker. 244

saprotroph Heterotroph such as a bacterium or a fungus that externally digests dead organic matter before absorbing the products. 571

sarcolemma Plasma membrane of a muscle fiber; also forms the tubules of the T system involved in muscular contraction. 381

sarcomere One of many units, arranged linearly within a myofibril, whose contraction produces muscle contraction. 381

sarcoplasmic reticulum Smooth endoplasmic reticulum of skeletal muscle cells; surrounds the myofibrils and stores calcium ions. 381

saturated fatty acid Molecule that lacks double bonds between the carbons of its hydrocarbon chain. The chain bears the maximum number of hydrogens. 34

savanna Terrestrial biome that is a grassland in Africa, characterized by few trees and a severe dry season. 726

Schwann cell Cell that surrounds a fiber of a peripheral nerve and forms the myelin sheath. 318

scientific theory Concept supported by a broad range of observations, experiments, and conclusions. 11

sclera White, fibrous, outer layer of the eyeball. 350

sclerenchyma Plant tissue composed of cells with heavily lignified cell walls and functions in support. 152

scrotum Pouch of skin that encloses the testes. 414

secondary oocyte In oogenesis the functional product of meiosis I; becomes the egg. 96

secretion Releasing of a substance by exocytosis from a cell that may be gland or part of gland. 54

seed Mature ovule that contains a sporophyte embryo with stored food enclosed by a protective coat. 171, 600

seed coat Protective covering for the seed, formed by the hardening of the ovule wall. 171

segmentation Repetition of body parts as segments along the length of the body; seen in annelids, arthropods, and chordates. 612

segregation Separation of alleles from each other during meiosis so that the gametes contain one from each pair. Each resulting gamete has an equal chance of receiving either allele. 466

semantic memory Capacity of the brain to store and retrieve information with regard to words or numbers and such. 330

semen Thick, whitish fluid consisting of sperm and secretions from several glands of the male reproductive tract. 414

semicircular canal One of three tubular structures within the inner ear that contain sensory receptors responsible for the sense of rotational equilibrium. 358

semilunar valve Valve resembling a half moon located between the ventricles and their attached vessels. 242

seminal vesicle Convoluted, saclike structure attached to the vas deferens near the base of the urinary bladder in males; adds secretions to semen. 414

seminiferous tubule Highly coiled duct within the male testes that produces and transports sperm. 417

sensation Conscious awareness of a stimulus due to nerve impulse sent to the brain from a sensory receptor by way of sensory neurons. 345

sensory neuron Nerve cell that transmits nerve impulses to the central nervous system after a sensory receptor has been stimulated. 318

sensory receptor Structure that receives either external or internal environmental stimuli and is a part of a sensory neuron or transmits signals to a sensory neuron. 344

sepal Outermost, sterile, leaflike covering of the flower; usually green in color. 170

serous membrane Membrane that covers internal organs and lines cavities without an opening to the outside of the body; serosa.   201

serum Light yellow liquid left after clotting of blood.   252

sessile Animal that tends to stay in one place.   612

sex chromosome Chromosome that determines the sex of an individual; in humans, females have two X chromosomes and males have an X and Y chromosome.   484, 492

sex-linked Allele that occurs on the sex chromosomes but may control a trait that has nothing to do with sexual characteristics of an individual.   492

sexual selection Changes in males and females, often due to male competition and female selectivity leading to increased fitness.   667

shoot apical meristem Group of actively dividing embryonic cells at the tips of plant shoots.   158

shoot system Aboveground portion of a plant consisting of the stem, leaves, and flowers.   148

short-day plant Plant which flowers when day length is shorter than a critical length, i.e., cocklebur, poinsettia, and chrysanthemum.   180

sieve plate Perforated end wall of a sieve-tube element.   152

sieve-tube element Member that joins with others in the phloem tissue of plants as a means of transport for nutrient sap.   152

simple goiter Condition in which an enlarged thyroid produces low levels of thyroxine.   397

sink population Population found in an unfavorable area where at best the birthrate equals the death rate; sink populations receive new members from source populations.   186, 754

sinus Cavity or hollow space in an organ such as the skull.   370

sister chromatid One of two genetically identical chromosomal units that are the result of DNA replication and are attached to each other at the centromere.   85

skeletal muscle Striated, voluntary muscle tissue that comprises skeletal muscles; also called striated muscle.   199

skill memory Capacity of the brain to store and retrieve information necessary to perform motor activities, such as riding a bike.   330

sliding filament model An explanation for muscle contraction based on the movement of actin filaments in relation to myosin filaments.   381

slime layer Gelatinous sheath surrounding the cell wall of certain bacteria.   62

smooth (visceral) muscle Nonstriated, involuntary muscle tissue found in the walls of internal organs.   199

society Group in which members of species are organized in a cooperative manner, extending beyond sexual and parental behavior.   671

sociobiology Application of evolutionary principles to the study of social behavior of animals, including humans.   673

sodium-potassium pump Carrier protein in the plasma membrane that moves sodium ions out of and potassium into cells; important in nerve and muscle cells.   75, 320

soft palate Entirely muscular posterior portion of the roof of the mouth.   214

solute Substance that is dissolved in a solvent, forming a solution.   71

solvent Fluid, such as water, that dissolves solutes.   71

somatic cell A body cell; excludes cells that undergo meiosis and become a sperm or egg.   82

somatic system That portion of the peripheral nervous system containing motor neurons that control skeletal muscles.   335

source population Population that can provide members to other populations of the species because it lives in a favorable area, and the birthrate is most likely higher than the death rate.   753

species Group of similarly constructed organisms capable of interbreeding and producing fertile offspring; organisms that share a common gene pool.   5, 557

sperm Male sex cell with three distinct parts at maturity: head, middle piece, and tail.   417

spermatogenesis Production of sperm in males by the process of meiosis and maturation.   96, 417

sphincter Muscle that surrounds a tube and closes or opens the tube by contracting and relaxing.   217

spinal cord Part of the central nervous system; the nerve cord that is continuous with the base of the brain and housed within the vertebral column.   324

spindle Microtubule structure that brings about chromosomal movement during nuclear division.   86

spiral organ Organ in the cochlear duct of the inner ear which is responsible for hearing; organ of Corti.   359

spongy bone Type of bone that has an irregular meshlike arrangement of thin plates of bone; often contains red bone marrow.   197, 366

spongy mesophyll Layer of tissue in a plant leaf containing loosely packed cells increasing the amount of surface area for gas exchange.   164

sporangium (pl., sporangia) Structure that produces spores.   584, 593

spore Haploid reproductive cell, sometimes resistant to unfavorable environmental conditions, that is capable of producing a new individual which is also haploid.   170, 576, 593

sporophyte Diploid generation of the alternation of generations life cycle of a plant; produces haploid spores that develop into the haploid generation.   170, 593

spring overturn Mixing process that occurs in spring in stratified lakes whereby the oxygen-rich top waters mix with nutrient-rich bottom waters.   731

squamous epithelium Type of epithelial tissue that contains flat cells.   194

stabilizing selection Outcome of natural selection in which extreme phenotypes are eliminated and the average phenotype is conserved.   554

stamen In flowering plants, the portion of the flower that consists of a filament and an anther containing pollen sacs where pollen is produced.   170, 606

stapes Last of three ossicles of the ear that serve to conduct vibrations from the tympanic membrane to the oval window membrane of the inner ear.   358

starch Storage polysaccharide found in plants that is composed of glucose molecules joined in a linear fashion with few side chains.   32

stem Usually the upright, vertical portion of a plant, which transports substances to and from the leaves.   149

stem cell Any cell that can divide and differentiate into more functionally specific cell types; embryonic cells or adult cells, such as those in red bone marrow that give rise to blood cells.   253

steroid Type of lipid molecule having a complex of four carbon rings; examples are cholesterol, progesterone, and testosterone.   35

steroid hormone Type of hormone that has a complex of four carbon rings but different side chains from other steroid hormones.   408

stigma In flowering plants, portion of the pistil where pollen grains adhere and germinate before fertilization can occur.   170

stimulus Change in the internal or external environment that a sensory receptor can detect leading to nerve impulses in sensory neurons.   344

stolon Stem that grows horizontally along the ground and may give rise to new plants where it contacts the soil (e.g., the runners of a strawberry plant).   162

stoma Microscopic opening bordered by guard cells in the leaves of plants through which gas exchange takes place.   57, 151, 185

striated Having bands; in cardiac and skeletal muscle, alternating light and dark crossbands produced by the distribution of contractile proteins.   199

stroma Fluid within a chloroplast that contains enzymes involved in the synthesis of carbohydrates during photosynthesis.   57, 135, 151

structural gene Gene that directs the synthesis of an enzyme or a structural protein in the cell.   513

style Elongated, central portion of the pistil between the ovary and stigma.   170

subcutaneous layer A sheet that lies just beneath the skin and consists of loose connective and adipose tissue.   205

substrate-level phosphorylation Process in which ATP is formed by transferring a phosphate from a metabolic substrate to ADP.   118

superior vena cava Large vein that enters the right atrium from above and carries blood from the head, thorax, and upper limbs to the heart.   246

sustainable development Management of an ecosystem so that it maintains itself while providing services to human beings.   755

suture Type of immovable joint articulation found between bones of the skull.   376

sweat gland Skin gland that secretes a fluid substance for evaporate cooling; sudoriferous gland.   205

symbiosis Relationship that occurs when two different species live together in a unique way; it may be beneficial, neutral, or detrimental to one and/or the other species.   692

sympathetic division That part of the autonomic system that usually promotes activities associated with emergency (fight or flight) situations; uses norepinephrine as a neurotransmitter.   337

sympatric speciation Origin of new species in populations that overlap geographically.   557

synapse Junction between neurons consisting of the presynaptic (axon) membrane, the synaptic cleft, and the postsynaptic (usually dendrite) membrane.   323

synapsis Pairing of homologous chromosomes during prophase I of meiosis.   90

synaptic cleft Small gap between presynaptic and postsynaptic membranes of a synapse.   323

syndrome Group of symptoms that appear together and tend to indicate the presence of a particular disorder.   484

synovial joint Freely moving joint in which two bones are separated by a cavity.   376

synovial membrane Membrane that forms the inner lining of the capsule of a freely movable joint.   201, 376

systemic circuit That part of the circulatory system that serves body parts other than the gas-exchanging surfaces in the lungs.   246

systole Contraction period of a heart during the cardiac cycle.   244

systolic pressure Arterial blood pressure during the systolic phase of the cardiac cycle.   248

T

taiga Terrestrial biome that is a coniferous forest extending in a broad belt across northern Eurasia and North America.   721

tapeworm Type of parasitic flatworm whose body consists of a scolex and proglottids and who attach themselves to the intestinal wall, where they receive nourishment from the host.   617

taproot Main axis of a root that penetrates deeply and is used by certain plants such as carrots for food storage.   156

taste bud Sense organ containing the receptors associated with the sense of taste.   348

taxonomy Branch of biology concerned with identifying, describing, and naming organisms.   6

tectorial membrane Membrane that lies above and makes contact with the hair cells in the spiral organ.   359

telomere Tip of the end of a chromosome that shortens with each cell division and may thereby regulate the number of times a cell can divide.   520

template Pattern or guide used to make copies; parental strand of DNA serves as a guide for the production of daughter DNA strands and DNA also serves as guide for the production of messenger RNA.   505

tendon Strap of fibrous connective tissue that connects skeletal muscle to bone.   196, 366

terminal bud Bud that develops at the apex of a shoot.   158

territoriality Behavior related to the act of marking or defending a particular area against invasion by another species member; area often used for the purpose of feeding, mating, and caring for young.   668

testis Male gonad which produces sperm and the male sex hormones.   404, 414

testosterone Male sex hormone that helps maintain sexual organs and secondary sex characteristics.   404, 417

tetanus Sustained muscle contraction without relaxation.   384

tetany Severe twitching caused by involuntary contraction of the skeletal muscles due to a calcium imbalance.   398

thalamus Part of the brain located in the lateral walls of the third ventricle that serves as the integrating center for sensory input; it plays a role in arousing the cerebral cortex.   328

thermal inversion Temperature inversion in which warm air traps cold air and pollutants near the earth.   707

thermoacidophile Type of archaea that lives in hot and acidic aquatic habitats, such as hot springs or near hydrothermal vents.   573

thermoreceptor Sensory receptor that is sensitive to changes in temperature.   344

threshold Electrical potential level (voltage) at which an action potential or nerve impulse is produced.   320

thrombin Enzyme that converts fibrinogen to fibrin threads during blood clotting.   252

thrombus Blood clot that remains in the blood vessel where it formed.   255

thylakoid Flattened sac within a granum whose membrane contains chlorophyll and where the light-dependent reactions of photosynthesis occur.   57, 62, 135

thymine (T) One of four nitrogen-containing bases in nucleotides composing the structure of DNA; pairs with adenine.   41

thymus gland Lymphoid organ, located along the trachea behind the sternum, involved in the maturation of T lymphocytes mature in the thymus gland. Secretes hormones called thymosins, which aid the maturation of T cells and perhaps stimulate immune cells in general.   263, 405

thyroid gland Endocrine gland in the neck that produces several important hormones, including thyroxine, triiodothyronine, and calcitonin.   397

thyroid-stimulating hormone (TSH) Substance produced by the anterior pituitary that causes the thyroid to secrete thyroxine and triiodothyronine.   394

thyroxine (T_4) Hormone secreted from the thyroid gland that promotes growth and development; in general, it increases the metabolic rate in cells.   397

tidal volume Amount of air normally moved in the human body during an inspiration or expiration.   286

tight junction Region between cells where adjacent plasma membrane proteins join to form an impermeable barrier.   194

tissue Group of similar cells which perform a common function.   3, 194

tissue culture Process of growing tissue artificially in usually a liquid medium in laboratory glassware.   176, 194

tissue fluid Fluid that surrounds the body's cells; consists of dissolved substances that leave the blood capillaries by filtration and diffusion. 254

T lymphocyte Lymphocyte that matures in the thymus; cytotoxic T cells kill antigen-bearing cells outright; helper T cells release cytokines that stimulate other immune system cells.   266

tone Continuous, partial contraction of muscle.   384

tonsils Partially encapsulated lymph nodules located in the pharynx. 263, 293

totipotent Cell that has the full genetic potential of the organism and has the potential to develop into a complete organism.   444

trachea Windpipe from the larynx to the bronchi; also, tubes in insects that transport air to the tissues.   284, 629

tracheid In flowering plants, type of cell in xylem that has tapered ends and pits through which water and minerals flow.   152

tract Bundle of myelinated axons in the central nervous system.   324

transcription Process resulting in the production of a strand of mRNA that is complementary to a segment of DNA.   506

transcription factor Protein that initiates transcription by RNA polymerase and thereby starts the process that results in gene expression.   515

transfer rate Amount of a substance that moves from one component of the environment to another within a specified period of time.   704

transfer RNA (tRNA) Type of RNA that transfers a particular amino acid to a ribosome during protein synthesis; at one end it binds to the amino acid and at the other end it has an anticodon that binds to an mRNA codon.   506

transgenic organisms Free-living organism in the environment that has a foreign gene in its cells.   528

transition reaction Reaction that oxidizes pyruvate with the release of carbon dioxide; results in the acetyl-CoA and connects glycolysis to the citric acid cycle.   117, 120

translation Process by which the sequence of codons in mRNA dictates the sequence of amino acids in a polypeptide.   506

transpiration Plant's loss of water to the atmosphere, mainly through evaporation at leaf stomata.   182

triglyceride Neutral fat composed of glycerol and three fatty acids.   34

triplet code During gene expression, each sequence of three nucleotide bases stands for a particular amino acid. 507

trisomy One more chromosome than usual.   486

trophic level Feeding level of one or more populations in a food web.   701

trophoblast Outer membrane surrounding the embryo in mammals; when thickened by a layer of mesoderm, it becomes the chorion, an extraembryonic membrane.   449

tropism In plants, a growth response toward or away from a directional stimulus.   176, 180

tropomyosin Protein that blocks muscle contraction until calcium ions are present.   383

trypanosome Parasitic zooflagellate that causes severe disease in human beings and domestic animals, including a condition called sleeping sickness.   581

T (transverse) tubule Membranous channel which extends inward from a muscle fiber membrane and passes through the fiber.   381

tubular reabsorption Movement of primarily nutrient molecules and water from the contents of the nephron into blood at the proximal convoluted tubule.   309

tubular secretion Movement of certain molecules from blood into the distal convoluted tubule of a nephron so that they are added to urine.   309

tumor Cells derived from a single mutated cell that has repeatedly undergone cell division; benign tumors remain at the site of origin and malignant tumors metastasize. 518

tumor-suppressor gene Gene that codes for a protein that ordinarily suppresses cell division; inactivity can lead to a tumor.   520

turgor pressure In plant cells, pressure of the cell contents against the cell wall when the central vacuole is full.   72

tympanic membrane Located between the outer and middle ear and receives sound waves; the eardrum.   358

U

ulcer Open sore in the lining of the stomach; frequently caused by bacterial infection.   218

umbilical arteries and vein Fetal blood vessels that travel to and from the placenta.   452

umbilical cord Cord connecting the fetus to the placenta through which blood vessels pass.   451

unsaturated fatty acid Fatty acid molecule that has one or more double bonds between the atoms of its carbon chain.   34

upwelling Upward movement of deep, nutrient-rich water along coasts; it replaces surface waters that move away from shore when the direction of prevailing wind shifts.   734

uracil (U) One of four nitrogen-containing bases in nucleotides composing the structure of RNA; pairs with adenine.   41

urea Primary nitrogenous waste of humans derived from amino acid breakdown.   303

ureter One of two tubes that take urine from the kidneys to the urinary bladder.   302

urethra Tubular structure that receives urine from the bladder and carries it to the outside of the body.   302, 414

uric acid Waste product of nucleotide metabolism.   303

urinary bladder Organ where urine is stored before being discharged by way of the urethra.   302

uterine cycle Monthly occurring changes in the characteristics of the uterine lining (endometrium).   422

uterus Organ located in the female pelvis where the fetus develops; the womb. 418

utricle Saclike cavity in the vestibule of the inner ear that contains sensory receptors for gravitational equilibrium.   361

V

vaccine Antigens prepared in such a way that they can promote active immunity without causing disease.   272, 431

vacuole Membrane-bounded sac that holds fluid and a variety of other substances.   55

vagina Organ that leads from the uterus to the vestibule and serves as the birth canal and organ of sexual intercourse in females.   419

valve Membranous extension of a vessel or the heart wall that opens and closes, ensuring one-way flow.   241

vas deferens Tube that leads from the epididymis to the urethra in males. 414

vascular bundle In plants, primary phloem and primary xylem enclosed by a bundle sheath.   152

vascular cambium In plants, lateral meristem that produces secondary phloem and secondary xylem.   158

vascular cylinder In dicot roots, a core of tissues bounded by the endodermis, consisting of vascular tissues and pericycle.   152

vascular plant Plant that has vascular tissue (xylem and phloem); includes seedless vascular plants (e.g., ferns) and seed plants (gymnosperms and angiosperms). 596

vascular tissue Transport tissue in plants consisting of xylem and phloem. 151, 155

vector In genetic engineering, a means to transfer foreign genetic material into a cell—for example, a plasmid. 526

vein Vessel that takes blood to the heart from venules; characteristically having nonelastic walls; in plants, veins consist of vascular bundles that branch into the leaves. 152, 240

venous duct Fetal connection between the umbilical vein and the inferior vena cava; ductus venosus. 452

ventilation Process of moving air into and out of the lungs; breathing. 288

ventricle Cavity in an organ, such as a lower chamber of the heart; or the ventricles of the brain, which are interconnecting cavities that produce and serve as a reservoir for cerebrospinal fluid. 242, 324

venule Vessel that takes blood from capillaries to a vein. 241

vernix caseosa Cheeselike substance covering the skin of the fetus. 452

vertebral column Portion of the vertebrate endoskeleton that houses the spinal cord; consists of many vertebrae separated by intervertebral disks. 372

vertebrate Chordate in which the notochord is replaced by a vertebral column. 641

vesicle Small, membrane-bounded sac that stores substances within a cell. 53

vessel element Cell which joins with others to form a major conducting tube found in xylem. 152

vestigial structure Underdeveloped structure that was functional in some ancestor but is no longer functional in a particular organism. 544

villus (pl., villi) Fingerlike projection from the wall of the small intestine that functions in absorption. 219

viroid Infectious strand of RNA devoid of a capsid and much smaller than a virus. 569

virus Noncellular obligate parasite of living cells consisting of an outer capsid and an inner core of nucleic acid. 566

visual accommodation Ability of the eye to focus at different distances by changing the curvature of the lens. 351

visual field Area of vision for each eye. 354

vital capacity Maximum amount of air moved in or out of the human body with each breathing cycle. 286

vitamin Essential organic requirement in the diet, needed in small amounts. They are often part of coenzymes. 109, 230

vitreous humor Clear gelatinous material between the lens of the eye and the retina. 351

vocal cord Fold of tissue within the larynx; creates vocal sounds when it vibrates. 284

vulva External genitals of the female that surround the opening of the vagina. 419

W

water (hydrologic) cycle Interdependent and continuous circulation of water from the ocean, to the atmosphere, to the land, and back to the ocean. 703

water vascular system Series of canals that takes water to the tube feet of an echinoderm, allowing them to expand. 637

white blood cell Leukocyte, of which there are several types, each having a specific function in protecting the body from invasion by foreign substances and organisms. 198, 251

white matter Myelinated axons in the central nervous system. 324

wood Secondary xylem which builds up year after year in woody plants and becomes the annual rings. 161

X

X chromosome Female sex chromosome that carries genes involved in sex determination; see Y chromosome. 484, 488

xenotransplantation Use of animal organs, instead of human organs, in human transplant patients. 536

X-linked Allele located on an X chromosome but may control a trait that has nothing to do with the sex characteristics of an individual. 492

xylem Vascular tissue that transports water and mineral solutes upward through the plant body; it contains vessel elements and tracheids. 152, 182, 596

Y

Y chromosome Male sex chromosome that carries genes involved in sex determination; see X chromosome. 484, 488

yolk Dense nutrient material that is present in the egg of a bird or reptile. 442

yolk sac Extraembryonic membrane that encloses the yolk of birds; in humans it is the first site of blood cell formation. 448

Z

zero population growth No increase in population size. 685

zooflagellate Nonphotosynthetic protist that moves by flagella; typically zooflagellates enter into symbiotic relationships and some are parasitic. 581

zygospore Thick-walled, resting cell formed during sexual reproduction of zygospore fungi. 584

zygote Diploid cell formed by the union of sperm and egg; the product of fertilization. 96, 171, 418

Credits

Line Art and Readings

Chapter 6

Figure 6.7: From Peter H. Raven & George B. Johnson, *Biology*, Fifth Edition. Copyright 1996 The McGraw-Hill Companies. All Rights Reserved.

Chapter 9

Figure 9A: Data from St. Regis Paper Company, New York, NY, 1966. Ecology Focus, page 163: Courtesy of Charles Horn. Figure 9.18b: From Burton S. Guttman, *Biology*. Copyright 1999 The McGraw-Hill Companies. All Rights Reserved.

Chapter 10

Figures 10.5 and 10.6: From Kinsgley R. Stern, *Introductory Plant Biology*, 6th edition. Copyright 1994 The McGraw-Hill Companies. All Rights Reserved.

Chapter 11

Figure 11.4: From John J.W. Hole, Jr., *Human Anatomy & Physiology*, 6th edition. Copyright 1994 The McGraw-Hill Companies. All Rights Reserved. Figure 11.13: From Shier et al., *Hole's Human Anatomy & Physiology*, 8th edition. Copyright 1996 The McGraw-Hill Companies. All Rights Reserved.

Chapter 12

Figure 12.15: Data from T.T. Shintani, *Eat More, Weigh Less ™ Diet*, 1983. Table 12.8: Source: David C. Nieman et al., *Nutrition*, Copyright 1990 The McGraw-Hill Companies.

Chapter 13

Figure 13.2: From Kent M. Van De Graaff and Stuart Ira Fox, *Concepts of Human Anatomy and Physiology*, Third Edition, Copyright 1992 The McGraw-Hill Companies. All Rights Reserved.

Chapter 14

Figure 14.9a: Source: Approved by the Advisory Committee on Immunization Practices (ACIP), the American Academy of Pediatrics (AAP), and the American Academy of Family Physicians.

Chapter 15

Health Focus, page 297: From *The Most Often Asked Questions About Smoking Tobacco and Health and the Answers*, REVISED July 1993. Copyright American Cancer Society, Inc., Atlanta, GA. Used with permission.

Chapter 16

Table 16.1: From A.J. Vander et al. *Human Physiology*, 4th edition. Copyright 1985 The McGraw-Hill Companies. All Rights Reserved.

Chapter 19

Figures 19.2 and 19.5a,c: From Shier et al., *Hole's Human Anatomy & Physiology*, 4th edition. The McGraw-Hill Companies.

Chapter 20

Health Focus, page 406: Adapted from Robert L. Sack, Alfred J. Lewy, and Mary L. Blood, *Journal of Visual Impairment and Blindness 92:* 145-161, March 1998.

Chapter 21

Ecology Focus, page 427: Courtesy of Dr. John M. Matter, Juniata College. Figure 21.1: From John W. Hole, Jr., *Human Anatomy & Physiology*, 6th Edition, Figure 22.1a, page 807. The McGraw-Hill Companies.
Figure 21.2: From Kent Van De Graaff & Stuart Ira Fox, *Concepts in Anatomy & Physiology*, 4th Edition, Figure 28.22, page 857. The McGraw-Hill Companies.
Figures 21.14, 21.15, 21.16 and 21.17: Data from Division of STD Prevention, Sexually Transmitted Disease Surveillance, 1998. U.S. Department of Health and Human Services, Public Health Service, Atlanta: Centers for Disease Control and Prevention, September 2000.

Chapter 22

Figure 22.8: From Ricki Lewis, *Life*, 3rd edition. Copyright 1997 The McGraw-Hill Companies. All Rights Reserved. Figures 22.18 and 22.20: From Kent Van De Graaff and Stuart Ira Fox, *Concepts of Human Anatomy and Physiology*, 4th edition. Copyright 1995 The McGraw-Hill Companies. All Rights Reserved. Figure 22.19: From Kent M. Van De Graaff and Stuart Ira Fox, *Concepts of Human Anatomy & Physiology*, 4th edition. Copyright 1995 The McGraw-Hill Companies. All Rights Reserved.

Chapter 24

Health Focus, page 489: Courtesy of Stefan Schwarz.

Chapter 25

Science Focus, page 503: Courtesy of Joyce Haines.

Chapter 27

Figure 27.20: From Peter H. Raven & George B. Johnson, *Biology*, 5th edition. Copyright 1996 The McGraw-Hill Companies. All Rights Reserved.

Chapter 29

Ecology Focus, pages 602-603: Charles N. Horn. Figure 29.14: From Moore, Clark, & Vodopich, *Botany*, 2nd edition. Copyright 1995 The McGraw-Hill Companies. All Rights Reserved.

Chapter 31

Figure 31.11a: From Burton S. Guttman, *Biology*. Copyright 1999 The McGraw-Hill Companies. All Rights Reserved.
Figure 31.4: From Stephen Miller & John Harley, *Zoology*, 9th edition. Copyright 1996 The McGraw-Hill Companies. All Rights Reserved.
Figure 31.17: From C. Hickman, *Integrated Principles of Zoology*, 9th edition. Copyright 1997 The McGraw-Hill Companies. All Rights Reserved.

Chapter 32

Figure 32.2(graph): Data from S.J. Arnold, "The Microevolution of Feeding Behavior" in *Foraging Behavior: Ecology, Ethological, and Psychological Approaches*, edited by A. Kamil and T. Sargent, 1980, Garland Publishing Company, New York, NY. Figure 32.7: Data from G. Hausfater, "Dominance and Reproduction in Baboons (Papio cynocephalus): A Quantilive Analysis," *Contributions in Primatology*, 7:1-150, 1975, page 367; Ecology Focus, page 670: Courtesy of Lee Drikamer.

Chapter 33

Figure 33.3: After R.L. Smith, 1960; Figure 33.9: United Nations Population Division, 1998.

Chapter 35

Ecology Focus, page 723: Copyright 1995 by the National Wildlife Federation. Reprinted with permission from *National Wildlife*, magazine's October/November, 1995 issue.
Figures 35.1b, 35.2, 35.4b, 35.6(map), 35.7(map), 35.8(map), 35.10(map), 35.12(map), 35.13(map), and 35.14(map): From Peter H. Raven and George B. Johnson, *Biology*, 5th edition. Copyright 1996 The McGraw-Hill Companies. All Rights Reserved.

Chapter 36

Figure 36.2: Redrawn from *Shrimp Stocking, Salmon Collapse, and Eagle Displacement: Cascading Interactions in the Food Web of a Large Aquatic Ecosystem*, by C.N. Spencer, B.R. McCleeland, and J.A. Stanford, *Bioscience*, 41(1): 14-21. Ecology Focus, page 749: Copyright 1995 U.S. News & World Report, L.P. Reprinted with permission. Figure 36.7: Data from David M. Gates, *Climate Change and Its Biological Consequences*, 1993, Sinauer Associates, Inc. Sunderland, MA.

Photographs

History of Biology

Leeuwenhoek: © Corbis-Bettmann; Darwin, Pasteur, Koch, Pavlov, Lorenz: © Corbis-Bettmann; McClintock: © AP/Wide World Photos; Franklin: © Cold Springs Harbor Laboratory.

Part Openers

Two: © Mark Mattlock/Masterfile; Three: © Biophoto Associates/Photo Researchers, Inc.; Four: © Hank Morgan/Photo Researchers, Inc.; Five: Courtesy Fred Ross; Six: © Jeff Foott/Bruce Coleman, Inc.; Seven: © Jon Feingersh/The Stock Market; Eight: © Connie Coleman/Stone; Nine: © Richard Thom/Visuals Unlimited; Ten: © C.C. Lockwood/Animals Animals/Earth Scenes; Eleven: © AP Photo/Obed Zilwa; Twelve: © Michael Newman/PhotoEdit; Thirteen: © Michelle Del Guercio/Photo Researchers, Inc.; Fourteen: © Bob Daemmrich/Stock Boston; Fifteen: © Bruce Ayres/Stone; Sixteen: © Richard Clintsman/Stone; Seventeen: © John Terence Turner/FPG International; Eighteen: © Jeff Greenberg/Peter Arnold, Inc.; Nineteen: © Stephen Simpson/FPG International; Twenty: © Bob Daemmrich/Stock Boston; Twenty-One: © Bob Daemmrich/Stock Boston; Twenty-Two a: Lennart Nilsson, "A Child is Born" 1990 Delacorte Press, Pg. 63; Twenty-Two b: Lennart Nilsson, "A Child is Born" 1990 Delacorte Press, Pg. 81; Twenty-Two c: Lennart Nilsson, "A

Index

W

X

Y

Z

History of Biology

Year	Name	Country	Contribution
1628	William Harvey	Britain	Demonstrates that the blood circulates and the heart is a pump.
1665	Robert Hooke	Britain	Uses the word *cell* to describe compartments he sees in cork under the microscope.
1668	Francesco Redi	Italy	Shows that decaying meat protected from flies does not spontaneously produce maggots.
1673	Antonie van Leeuwenhoek	Holland	Uses microscope to view living microorganisms.
1735	Carolus Linnaeus	Sweden	Initiates the binomial system of naming organisms.
1809	Jean B. Lamarck	France	Supports the idea of evolution but thinks there is inheritance of acquired characteristics.
1825	Georges Cuvier	France	Founds the science of paleontology and shows that fossils are related to living forms.
1828	Karl E. von Baer	Germany	Establishes the germ layer theory of development.
1838	Matthias Schleiden	Germany	States that plants are multicellular organisms.
1839	Theodor Schwann	Germany	States that animals are multicellular organisms.
1851	Claude Bernard	France	Concludes that a relatively constant internal environment allows organisms to survive under varying conditions.
1858	Rudolf Virchow	Germany	States that cells come only from preexisting cells.
1858	Charles Darwin	Britain	Presents evidence that natural selection guides the evolutionary process.
1858	Alfred R. Wallace	Britain	Independently comes to same conclusions as Darwin.
1865	Louis Pasteur	France	Disproves the theory of spontaneous generation for bacteria; shows that infections are caused by bacteria, and develops vaccines against rabies and anthrax.
1866	Gregor Mendel	Austria	Proposes basic laws of genetics based on his experiments with garden peas.
1882	Robert Koch	Germany	Establishes the germ theory of disease and develops many techniques used in bacteriology.
1900	Walter Reed	United States	Discovers that the yellow fever virus is transmitted by a mosquito.
1902	Walter S. Sutton Theodor Boveri	United States Germany	Suggest that genes are on the chromosomes, after noting the similar behavior of genes and chromosomes.
1903	Karl Landsteiner	Austria	Discovers ABO blood types.
1904	Ivan Pavlov	Russia	Shows that conditioned reflexes affect behavior, based on experiments with dogs.
1910	Thomas H. Morgan	United States	States that each gene has a locus on a particular chromosome, based on experiments with *Drosophila*.
1922	Sir Frederick Banting Charles Best	Canada	Isolate insulin from the pancreas.
1924	Hans Spemann Hilde Mangold	Germany	Show that induction occurs during development, based on experiments with frog embryos.
1927	Hermann J. Muller	United States	Proves that X rays cause mutations.
1929	Sir Alexander Fleming	Britain	Discovers the toxic effect of a mold product he called penicillin on certain bacteria.

Antonie van Leeuwenhoek

Charles Darwin

Louis Pasteur

Robert Koch

Ivan Pavlov